S0-BWR-330

AR

ANNUAL REVIEW OF CELL BIOLOGY

ANNUAL REVIEW OF CELL BIOLOGY

VOLUME 9, 1993

GEORGE E. PALADE, *Editor*
University of California, San Diego

BRUCE M. ALBERTS, *Associate Editor*
University of California, San Francisco

JAMES A. SPUDICH, *Associate Editor*
Stanford University School of Medicine

ANNUAL REVIEWS INC 4139 EL CAMINO WAY P.O. BOX 10139 PALO ALTO, CALIFORNIA 94303-0897

ANNUAL REVIEWS INC.
Palo Alto, California, USA

International Standard Serial Number: 0743–4634
International Standard Book Number: 0–8243–3109-5

∞ The paper used in this publication meets the minimum requirements of American National Standard for Information Sciences—Permanence of Paper for Printed Library Materials, ANSI Z39.48-1984.

Typesetting by Kachina Typesetting Inc., Tempe, Arizona; John Olson, President; Marty Mullins, Typesetting Coordinator; and by the Annual Reviews Inc. Editorial Staff

PRINTED AND BOUND IN THE UNITED STATES OF AMERICA

PREFACE

The ninth volume of the *Annual Reviews of Cell Biology* reflects, like the preceding volumes of this series, what is new, exciting and challenging in the vast and relentlessly growing field of cellular and molecular biology. Again, as in the preceding volumes, all the sectors of this amazingly active field are represented in the belief that whatever happens at the specialized periphery affects the central core and illuminates the inner working of the quintessential cell.

Caught as we are in the relentless struggle to secure the means needed to do our work under highly competitive conditions, we rarely—if ever—take the time to consider our position in the history of the field. The second half of this century has been the golden era of sciences, especially of life sciences, which in the last four to five decades have advanced in breadth and depth faster than over centuries before. We can recognize two quantum jumps—the first, cellular, and the second, molecular—that explain these spectacular advances. To find a comparable example of human creativity focused on a special field and leaving behind a priceless heritage, we have to go back to the quatrocento and cinquecento of the Italian renaissance. The golden flower that bloomed at that time was in the plastic and visual arts. The new golden flower has been in the sciences. It is still blooming with amazing vigor. The efforts of countless scientists have made its blooming possible. Therefore, we can take pride in what we have accomplished and be grateful for the privilege of living and working through such an unforgettable era. The *Annual Review of Cell Biology* has been perforce the privileged chronicle of these momentous accomplishments.

I know it is not easy, but perhaps we can take solace in dreaming about golden flowers whenever our grant applications do not receive fundable priority. In any case, we can say, "*Et in Arcadia ego,*" I, too, was there when it happened.

George E. Palade
Editor

Annual Review of Cell Biology
Volume 9, 1993

CONTENTS

INDEXES

OTHER REVIEWS OF INTEREST TO CELL BIOLOGISTS

From the *Annual Review of Biochemistry*, Volume 62 (1993):

Eukaryotic DNA Replication: Anatomy of an Origin, Melvin L. DePamphilis
Analysis of Glycoprotein-Associated Oligosaccharides, Raymond A. Dwek, Christopher J. Edge, David J. Harvey, Mark R. Wormald, and Raj B. Parekh
Protein Tyrosine Phosphatases, Kevin M. Walton and Jack E. Dixon
The Structure and Biosynthesis of Glycosyl Phosphatidylinositol Protein Anchors, Paul T. Englund
Human Gene Therapy, Richard A. Morgan and W. French Anderson
Nucleocytoplasmic Transport in the Yeast Saccharomyces Cerevisiae, Mark A. Osborne and Pamela A. Silver
hnRNP Proteins and the Biogenesis of mRNA, Gideon Dreyfuss, Michael J. Matunis, Serafin Piñol-Roma, and Christopher G. Burd
Membrane Partitioning During Cell Division, Graham Warren
Molecular Chaperone Functions of Heat-Shock Proteins, Joseph P. Hendrick and Franz-Ulrich Hartl
Cytoplasmic Microtubule-Associated Motors, R. A. Walker and M. P. Sheetz
Signalling by Receptor Tyrosine Kinases, Wendy J. Fantl, Daniel E. Johnson, and Lewis T. Williams
Membrane-Anchored Growth Factors, Joan Massagué and Atanasio Pandiella
The Tumor Suppressor Genes, Arnold J. Levine
Pathways of Protein Folding, C. Robert Matthews
The Receptors for Nerve Growth Factor and Other Neurotrophins, Simona Raffioni, Ralph A. Bradshaw, and Stephen E. Buxser
Function and Regulation of RAS, Douglas R. Lowy and Berthe M. Willumsen

From the *Annual Review of Biophysics and Biomolecular Structure*, Volume 22 (1993):

Models of Lipid-Protein Interactions in Membranes, Ole G. Mouritsen and Myer Bloom
Mechanisms of Membrane Fusion, Joshua Zimmerberg, Steven S. Vogel, and Leonid V. Chernomordik
Prolyl Isomerase: Enzymatic Catalysis of Slow Protein-Folding Reactions, Franz X. Schmid
The Effects of Phosphorylation on the Structure and Function of Proteins, L. N. Johnson and D. Barford
Structure and Function of Human Growth Hormone: Implications for the Hematopoietins, James A. Wells and Abraham M. de Vos
Interactions of Proteins with Lipid Headgroups: Lessons from Protein Kinase C, Alexandra C. Newton

Glycoprotein Motility and Dynamic Domains in Fluid Plasma Membranes, Michael P. Sheetz
What Does Electron Cryomicroscopy Provide that Crystallography and NMR Spectroscopy Cannot? W. Chiu

From the *Annual Review of Immunology,* Volume 11 (1993):

Specialization, Tolerance, Memory, Competition, Latency, and Strife Among T Cells, N. A. Mitchison
Sequences and Factors: A Guide to MHC Class-II Transcription, Laurie H. Glimcher and Catherine J. Kara
The Human T Cell Receptor in Health and Disease, Paul A. H. Moss, William M. C. Rosenberg, and John I. Bell
Antigen Receptors on B Lymphocytes, Michael Reth
In Vitro Antibodies: Strategies for Production and Application, Sherie L. Morrison
Apoptosis and Programmed Cell Death in Immunity, J. John Cohen, Richard C. Duke, Valerie A. Fadok, and Karen S. Sellins
Cytokine Receptors and Signal Transduction, Atsushi Miyajima, Toshio Kitamura, Nobuyuki Harada, Takashi Yokota, and Ken-ichi Arai
Cyclosporin A, FK-506, and Rapamycin: Pharmacologic Probes of Lymphocyte Signal Transduction, Nolan S. Sigal and Francis J. Dumont
Physiological and Molecular Mechanisms of Lymphocyte Homing, Louis J. Picker and Eugene C. Butcher
The T Cell Receptor as a Multicomponent Signalling Machine: CD4/CD8 Coreceptors and CD45 in T Cell Activation, Charles A. Janeway, Jr.
Lymphocyte Development from Stem Cells, Koichi Ikuta, Nobuko Uchida, Jeff Friedman, and Irving L. Weissman
Complement Deficiencies, Harvey R. Colten and Fred S. Rosen

From the *Annual Review of Microbiology,* Volume 47 (1993):

Transport of Nucleic Acids Through Membrane Channels: Snaking Through Small Holes, Vitaly Citovsky and Patricia Zambryski
Chromosome Segregation in Yeast, Barbara D. Page and Michael Snyder
ATP-Dependent Transport Systems in Bacteria and Humans: Relevance to Cystic Fibrosis and Multidrug Resistance, Carl A. Doige and Giovanna Ferro-Luzzi Ames
Regulation of the Heat-Shock Response in Bacteria, T. Yura, H. Nagai, and H. Mori
Enzymes and Proteins From Organisms that Grow Near and Above 100°C, Michael W. W. Adams

From the *Annual Review of Neuroscience,* Volume 16 (1993):

Transsynaptic Control of Gene Expression, Robert C. Armstrong and Marc R. Montminy
Molecular Mechanisms of Developmental Neuronal Death, E. M. Johnson, Jr. and T. L. Deckwerth
Neurotransmitter Transporters: Recent Progress, Susan G. Amara and Michael J. Kuhar
Protein Targeting in the Neuron, Regis B. Kelly and Eric Grote
Recent Advances in the Molecular Biology of Dopamine Receptors, Jay A. Gingrich and Marc G. Caron
Molecular Basis of Neural-Specific Gene Expression, Gail Mandel and David McKinnon
Regulation of Ion Channel Distribution at Synapses, Stanley C. Froehner
Inhibitors of Neurite Growth, M. E. Schwab, J. P. Kapfhammer, and C. E. Bandtlow

From the *Annual Review of Pharmacology and Toxicology,* Volume 33 (1993):

Calmodulin in Neurotransmitter and Hormone Action, Margaret E. Gnegy
Antiinflammatory Peptide Agonists, Edward T. Wei and Holly A. Thomas
Covalent Modifications of G-Proteins, Bernard Kwok-Keung Fung and Harvey K. Yamane

From the *Annual Review of Physiology,* Volume 55 (1993):

Transcriptional Regulation During Cardiac Growth and Development, Kenneth R. Chien, Hong Zhu, Kirk U. Knowlton, Wanda Miller-Hance, Marc van-Bilsen, Terrence X. O'Brien, and Sylvia M. Evans
The Insulin-like Growth Factors, W. S. Cohick and D. R. Clemmons
Neutrophil-Endothelial Cell Interactions in the Lung, Marlys H. Gee and Kurt H. Albertine
Regulation of the Renal Brush Border Membrane Na^+-H^+ Exchanger, E. J. Weinman and S. Shenolikar
The Renal H-K-ATPase: Physiological Significance and Role in Potassium Homeostasis, Charles S. Wingo and Brian D. Cain
The Intestinal Na^+/Glucose Cotransporter, Ernest M. Wright
The Cystic Fibrosis Transmembrane Conductance Regulator, J. R. Riorden
Controlling Cell Chemistry with Caged Compounds, Stephen R. Adams and Roger Y. Tsien

From the *Annual Review of Plant Physiology and Plant Molecular Biology*, Volume 44 (1993):

Organelle Movements, Richard E. Williamson
Biochemistry of Phosphoinositides, Gary G. Coté and Richard C. Crain
Vacuolar H^+-Translocating Pyrophosphatase, Philip A. Rea and Ronald J. Poole
Protein Degradation in Plants, Richard D. Vierstra
Proton-Coupled Sugar and Amino Acid Transporters in Plants, Daniel R. Bush

ANNUAL REVIEWS INC. is a nonprofit scientific publisher established to promote the advancement of the sciences. Beginning in 1932 with the *Annual Review of Biochemistry,* the Company has pursued as its principal function the publication of high quality, reasonably priced *Annual Review* volumes. The volumes are organized by Editors and Editorial Committees who invite qualified authors to contribute critical articles reviewing significant developments within each major discipline. The Editor-in-Chief invites those interested in serving as future Editorial Committee members to communicate directly with him. Annual Reviews Inc. is administered by a Board of Directors, whose members serve without compensation.

Annu. Rev. Cell Biol. 1993. 9:1–26

CELL MEDIATED EVENTS THAT CONTROL BLOOD COAGULATION AND VASCULAR INJURY

Charles T. Esmon

Cardiovascular Biology Research Program, Oklahoma Medical Research Foundation; Departments of Pathology and Biochemistry, University of Oklahoma Health Sciences Center, and Howard Hughes Medical Institute, Oklahoma City, Oklahoma 73104

KEY WORDS: thrombosis, endothelial cell, membranes, thrombin, hemostasis

---

CONTENTS

INTRODUCTION

One of the major challenges confronting investigators interested in thrombosis and hemostasis is to understand the mechanisms that control clot formation. This process involves at least 13 coagulation proteins and nearly as many protein inhibitors. While the coagulation cascade can be assembled and made to function on model membranes in the absence of intact cells

0743–4634/93/1115–0001$05.00

(Mann et al 1990), it is clear that cellular contributions provide mechanisms for control of this process not available in the cell free model system. Cells participate in several key events in blood coagulation; initiation, propagation of the initial stimulus, and interaction with fibrinogen and fibrin to modify the structure of the blood clot (Kieffer & Phillips 1990). These cellular interactions not only provide a focal point for controlling the extent to which coagulation is amplified, but they also provide mechanisms for localizing the hemostatic response. All of these responses require cell activation. Furthermore, the adhesive nature of the activated cells undoubtedly contributes to localization of the hemostatic plug.

Opposing the clotting process are a variety of extremely potent anticoagulants and fibrinolytic agents, primarily localized to the endothelium. This review examines the nature and mechanism of the cell activation events, the assembly of coagulation complexes on cells, and some of the cross talk and interactions of the inflammatory pathways with coagulation. Space limitations prevent a detailed discussion of adhesive reactions and some of the growth-promoting aspects of coagulation.

REVIEW OF THE COAGULATION FACTORS

Blood coagulation requires the sequential activation of a series of protease zymogens that must combine with a protein cofactor on membrane surfaces to accelerate the subsequent activation event that ultimately terminates in the formation of thrombin (Davie et al 1991) (Figure 1). Recent biochemical and molecular biological advances have resulted in the isolation and characterization of all of the known coagulation factors and their inhibitors (Table 1) (reviewed in Furie & Furie 1988). Studies in isolated systems have allowed a relatively detailed description of the interactions and functions of the proteins both in terms of protein-protein interactions and phospholipid-protein interactions (Mann et al 1990; Dahlbäck 1991; Bach 1988; Esmon 1989). Insights gained from the model membrane studies have been applied to analysis of cellular participation in coagulation. In particular, it is clear that the activation involves binding of the enzyme to a regulatory protein (commonly referred to as a cofactor) on the surface of the membrane where the complex then appears to activate membrane bound substrate. Whether this substrate comes from a pool that is membrane-associated (Nesheim et al 1984; Rosing et al 1980) and/or binds to the membrane-associated activation complex from the bulk phase (Pusey & Nelsestuen 1983) has been the subject of several investigations and remains unresolved. With either model, the substrate is membrane-associated.

Direct membrane interaction by the vitamin K-dependent clotting factors involves a Ca^{2+}-mediated interaction with negatively charged phospholipid

Table 1 Properties of coagulation factors and their regulators

	Synthesis[1]	Mol Wt	Post-translation modifications[2]	Plasma concentration (μg/ml)	Function
Anticoagulant factor					
Protein C	Liver	62,000	Gla,[3] β-OH[4]	4	Zymogen
Protein S	Liver, endothelium	80,000	Gla,[3] β-OH[4]	25	Cofactor
Thrombomodulin	Endothelium	≈74,000	β-OH[4] chondroitin sulfate	Bound to endothelium	Cofactor
Antithrombin III	Liver	58,000	—	150	Protease inhibitor
TFPI[5]	Liver, endothelium	39,000	—	0.120[6]	Protease inhibitor
Coagulation factor					
Prothrombin	Liver	72,000	Gla	100	Zymogen
Factor X	Liver	56,000	Gla, β-OH[4]	10	Zymogen
Factor IX	Liver	56,000	Gla, β-OH[4]	5	Zymogen
Factor VII	Liver	50,000	Gla, β-OH[4]	0.5	Zymogen
Fibrinogen	Liver	320,000	—	3,000	Clot/adhesion
Factor V	Liver	330,000	—	7	Cofactor
Factor VIII	Liver	330,000	—	0.1	Cofactor
Tissue factor	Monocytes, endothelium,[7] extravascular cells	37,000	—	—	Cofactor

[1] Several factors are synthesized at alternative sites also, usually at lower levels.
[2] All of these factors are glycoproteins.
[3] Gla = 4-carboxyglutamic acid. Synthesis is vitamin K-dependent.
[4] Hydroxylation of aspartic Asp or Asn.
[5] Tissue factor pathway inhibitor.
[6] Significant TFPI is membrane-associated.
[7] Inducible with cytokines and endotoxin on monocytes and endothelium.

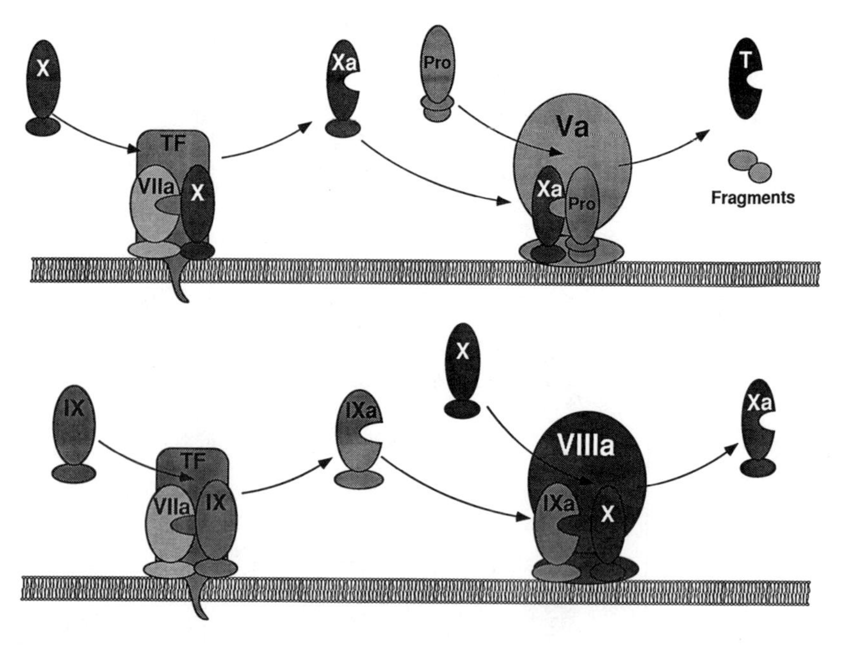

Figure 1 The assembly and function of coagulation complexes. Inactive coagulation factors (zymogens) interact with membrane-associated activation complexes involving a serine protease, a regulatory factor (cofactor), negatively charged membranes, and Ca^{2+}. The outer leaflet (top) of the membrane must contain a significant amount of aminophospolipid (phosphatidyl serine) to function optimally. In the cases of factors V and VIII, not shown in the figure to simplify presentation, the cofactors must be proteolytically activated to Va and VIIIa before they can serve their binding functions in the complexes. Even though factor VII binds to tissue factor, most investigators now believe that human factor VII must be activated to function. Note that factor Xa can be generated either by the tissue factor-factor VIIa or factor IXa-factor VIIIa complexes. Factor Xa is chemically and functionally equivalent in either case and can function in the prothrombin activation complex. See text for discussion. Abbreviations: pro, prothrombin; t, thrombin; TF, tissue factor; factor is deleted in front of all roman numerals for simplification.

surface that has been proposed to involve a Ca^{2+} bridge formation between Gla residues of the vitamin K-dependent and negatively charged membrane phospholipids of which phosphatidyl serine is the most active (Cutsforth et al 1989). Factors Va and VIIIa interact reversibly with negatively charged membranes in a process that does not require Ca^{2+}, while tissue factor is an integral membrane protein with a classical transmembrane region. The affinity of the enzymes and substrates for the membrane surface is moderate (K_d ~1 μM), which is too weak to allow effective membrane binding at the concentrations at which active coagulation enzymes function physiologically (nM and below) (Mann et al 1990). High affinity interaction occurs through additional direct binding interactions between the enzyme and the membrane-bound cofactor. Since the binding interactions involve independent sites, the overall affinity of the enzyme for the surface is increased at least 1000-fold. Experimentally, this results in specific binding of the protease to appropriate cells only when the cofactor is present (Miletich et al 1978; Tracy et al 1992).

The clotting cascade ultimately leads to thrombin formation. In recent years, it has become increasingly apparent that thrombin plays a major role in regulating activating cells, propagating inflammatory responses, controlling vascular tone, and both enhancing and inhibiting its own formation (Figure 3). It is apparent that this multiplicity of functions must be tightly controlled. Cells and the coagulation system interact through a variety of mechanisms to maintain thrombin formation at low levels and to restrict coagulant levels of thrombin to the wound site.

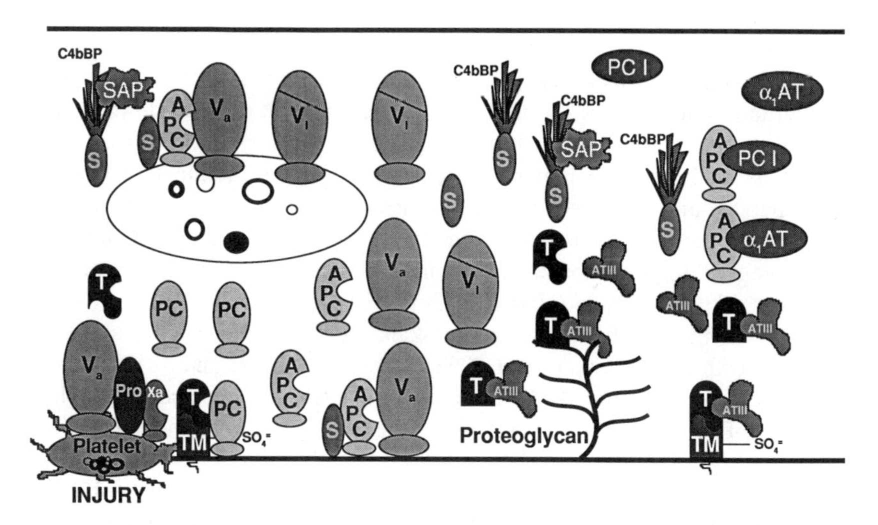

Figure 2 Protein C pathway—normal function. Natural anticoagulant proteins. The major known anticoagulant mechanisms in plasma are illustrated in Figure 2. They involve the protein C-protein S-TM system, the heparin-antithrombin system, and the tissue factor pathway inhibitor (not shown, see Figure 4). Thrombin activity can be neutralized by reaction with antithrombin III either on thrombomodulin or on heparin-like proteoglycans, shown as the tree-like structure with bound thrombin-antithrombin III. Abbreviations: T, thrombin; Pro, prothrombin; Va, factor Va; Xa, factor Xa; S, protein S; PC, protein C; APC, activated protein C; ATIII, antithrombin III; Vi, the activated protein C- protein S complex that converts factor Va to an inactive complex, illustrated by the slash through the larger part of the two-subunit-factor Va molecule; C4bBP, C4b binding protein; SAP, serum amyloid P; PCI, protein C inhibitor; α_1AT, α_1-antitrypsin. (Adapted from Esmon 1992; The protein C anticoagulant pathway. *Arterioscl. Thromb.* 12:135–45; with permission of the Am. Heart Assoc. Inc.)

NATURAL ANTICOAGULANT COMPLEXES

Blood contains three major anticoagulant protein systems that regulate blood coagulation. One involves antithrombin III (ATIII) and is potently enhanced by heparin; the second involves two vitamin K-dependent factors: protein C and protein S and an endothelial cell protein, thrombomodulin (TM); and the third involves the tissue factor pathway inhibitor (TFPI), a Kunitz-type inhibitor. A schematic representation of the three pathways is shown in Figure 2. The three anticoagulant mechanisms play distinct and apparently complementary roles. ATIII functions by inhibiting the serine protease components of blood coagulation, primarily thrombin, factor Xa and factor IXa (Rosenberg & Rosenberg 1984). Heparin functions to alter the conformation of ATIII and with some enzymes, thrombin in particular, to interact with both the enzyme and its inhibitor to form a ternary complex (Danielsson et al 1986) that facilitates formation of a stable, inactive complex. After complex formation of the enzyme with ATIII, heparin affinity is diminished and the heparin dissociates and begins to function catalytically.

In the protein C pathway, protein C circulates as an inactive zymogen. When thrombin is formed at the site of local injury, it diffuses downstream where it binds reversibly and with high affinity ($K_d \sim 0.5$ nM) to TM, which is an integral membrane protein inserted in the luminal surface of the endothelium. The thrombin-TM complex rapidly converts protein C to activated protein C (Esmon & Owen 1981). Activated protein C then binds to protein S to form membrane-bound complexes on model membranes (Walker 1980), platelets (Harris & Esmon 1985), and endothelium (Stern et al 1986) that inactivate factor Va (Walker et al 1979; Kisiel et al 1977; Suzuki et al 1983; Bakker et al 1992; Tracy et al 1992) and VIIIa (Vehar & Davie 1980; Fay et al 1991). Protein S circulates both free and in complex with C4bBP, an inhibitory protein of the classical complement pathway (Dahlbäck & Stenflo 1981; Dahlbäck 1991). Only the free form of protein S is functional as an anticoagulant (Dahlbäck 1986; Comp et al 1984), although both forms can associate with membrane surfaces (Furmaniak-Kazmierczak et al 1992; Schwalbe et al 1990). After complex formation, activated protein C is inactivated by either α-1-protease inhibitor (formerly referred to as α-1-antitrypsin) or the protein C inhibitor (Heeb et al 1989a,b). The reaction of activated protein C with the protein C inhibitor is enhanced by heparin and some other glycosaminoglycans (Suzuki et al 1984; Pratt & Church 1992).

Inhibition of the tissue factor-factor VIIa complex can occur via either of two mechanisms. One mechanism involves the tissue factor pathway inhibitor (TFPI), which has a complex mode of action. TFPI has three tandemly repeated Kunitz-type inhibitor domains. One of these binds to and

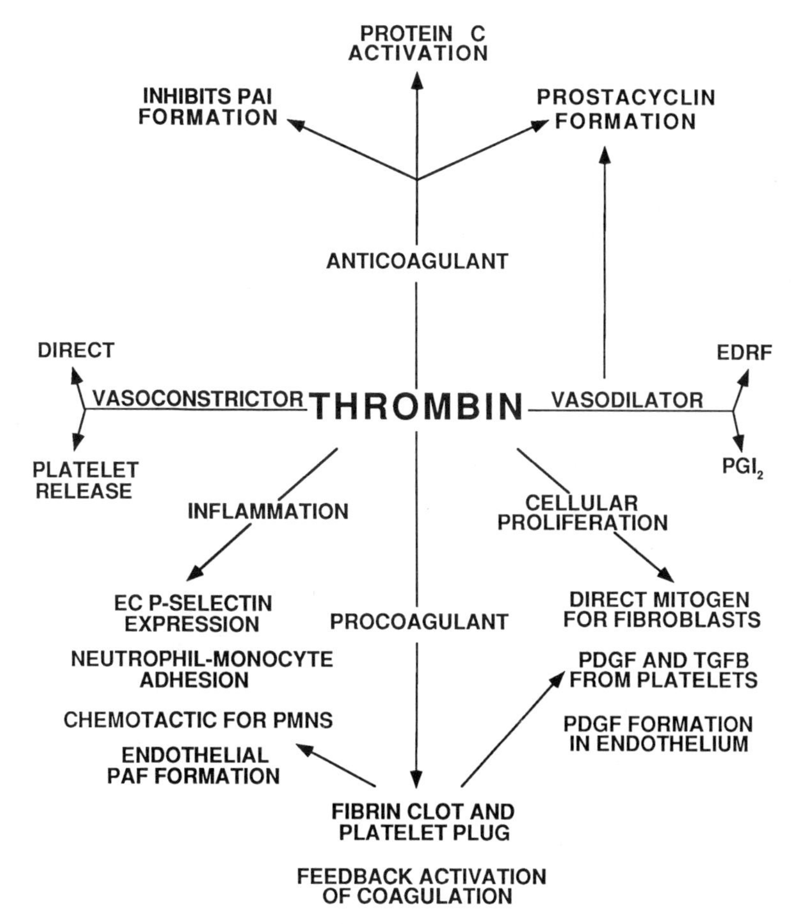

Figure 3 Cellular humoral responses to nM concentrations of thrombin. Biological functions of thrombin. All of the biological functions listed in this figure occur at thrombin concentrations easily obtained in the region of blood clots. See text for discussion. Abbreviations: EC, endothelial cell; EDRF, endothelial cell-derived relaxing factor; PDGF, platelet-derived growth factor; TGFβ, transforming growth factor β; PGI2, prostacyclin; PAF, platelet-activating factor.

inhibits factor Xa (Broze et al 1990). The resultant complex of factor Xa-TFPI interacts with the factor VIIa-tissue factor complex on membrane surfaces (Broze et al 1988, 1990; Rapaport 1989; Callander et al 1992a; Rapaport & Rao 1992). The fate of this complex in vivo is unknown. Inhibition of the tissue factor is reversible since active tissue factor can be regenerated by removing the Ca^{2+} from the inhibited complex, which dissociates upon Ca^{2+} removal (Rapaport & Rao 1992). Much of the TFPI in plasma is normally associated with lipoproteins, but the functional significance of this association remains unclear. Perhaps more importantly, an even larger pool of the inhibitor is associated with the vessel wall and can be liberated by infusion of heparin, which suggests that the inhibitor may be associated with vascular heparin-like molecules (Novotny et al 1991). TFPI constitutes a second anticoagulant mechanism that is apparently localized to the endothelium, but in this case the localization is incomplete and probably involves glycosaminoglycans.

A second mechanism for inhibition of the tissue factor-factor VIIa complex has recently been described. For years, it was recognized that factor VII or VIIa was insensitive to ATIII and heparin. Unlike free factor VIIa, the tissue factor-factor VIIa complex is inhibited by ATIII (Shigematsu et al 1992) in the presence of heparin. Whether vascular heparins can actually interact with ATIII to enhance inactivation of the cell associated tissue factor-factor VIIa complex is unknown.

A second heparin-dependent thrombin inhibitor, heparin cofactor II, exists in plasma. The physiological importance of this inhibitor is unknown, but it reacts with dermatan sulfate in the extravascular space (Van Deerlin & Tollefsen 1991), and peptides from heparin cofactor II are chemotactic for leukocytes (Church et al 1992). Thrombin itself has chemotactic activity for monocytes (Bar-Shavit & Wilner 1986). Thus thrombin formation in the extravascular space may serve as a mechanism for attracting inflammatory cells. This, coupled with the capacity of monocytes/macrophages to synthesize and assemble most, if not all, of the necessary coagulation components to generate thrombin (Altieri & Edgington 1989), may provide an important clue as to the mechanisms by which monocytes participate in vascular injury and disease.

At least two potential mechanisms for expression of heparin-like activity exist, one associated with the proteoglycans of the endothelium (de Agostini et al 1990) and the other, a chondroitin sulfate, covalently associated with TM (Bourin et al 1988; Parkinson et al 1992a). Recently, the heparin-like proteoglycans have been localized to the ablumuminal side of the endothelium with only a small concentration on the luminal surface (de Agostini et al 1990). Despite its relatively inaccessible location, it appears that this proteoglycan is important in accelerating the thrombin-antithrombin reaction. This conclusion is based primarily on the observation that heparinase essentially eliminates the acceleration of thrombin neutralization in the perfused hind limb of a rat (Marcum et al 1984). These observations offer an interesting apparent paradox with the observation that rapid clearance of thrombin from the circulation of rabbits relies on a high affinity thrombin receptor that could be occupied by thrombin (Lollar & Owen 1980). This bond likely involves TM. From the combined observations of the high capacity of the vasculature to bind thrombin (Esmon & Owen 1981), the relatively high concentration of TM on the surface of endothelium (~ 50 –100,000/cell) (Maruyama & Majerus 1985), and the ability of TM to accelerate thrombin inactivation by ATIII (Parkinson et al 1992b), it is apparent that TM must contribute to thrombin clearance and inactivation by ATIII. The relative contribution of TM and heparan proteoglycans to the inhibition of thrombin remains unknown. Regardless of which glycosaminoglycan is responsible for the inhibition, it is clear that the endothelial cell is responsible for elaboration of most of this anticoagulant signal.

The importance of the endothelial cell location for placement of the anticoagulants was noted by Busch et al (1982). The geometry of the circulation is such that when blood moves from the large vessels to the microcirculation, greater than 1000 cm^2 of endothelium can be exposed/ml of blood. If one assumes 100,000 proteoglycan or TM molecules/endothelial cell, then the concentration of these anticoagulants in the microcirculation is approximately 500 nM. Given that TM effectively anticoagulates blood when added at this concentration, there is little wonder that the hemostatic balance is far in favor of preventing coagulation when blood is in the microcirculation. This concept probably contributes significantly to the observation that blood flow is essential in preventing thrombosis. With impaired flow, blood resides in the larger vessels for extended periods of time. In the large vessels, the endothelial surface area, and hence anticoagulant concentration, is lower and thus coagulation can occur. Similar arguments may apply to the vascular heparan proteoglycans and perhaps to the TFPI associated with the endothelium.

INITIATION OF COAGULATION

How coagulation is initiated has been one of the major research questions since biochemical investigation of the process began. Two events can trigger the process. One involves exposure of blood to foreign surfaces, which initiates factor XII activation. Since patients deficient in factor XII have no bleeding complications, it is now generally believed that this pathway is not a critical initiation mechanism (Roberts & Lozier 1991).

The major alternative pathway is initiated by tissue factor. Tissue factor binds factor VII or VIIa (Rapaport & Rao 1992; Bach 1988). Factor VII is converted to its active form by several proteases, factor Xa and factor VIIa being the most likely physiological activators. The importance of the tissue factor pathway is suggested by the observation that factor VII-deficient patients do have bleeding complications (Roberts & Lozier 1991). Tissue factor is not normally present on vascular cell surfaces, but is found on extravascular cellular surfaces (Wilcox et al 1989; Callander et al 1992b;

Fleck et al 1990), especially adventitial fibroblasts (Drake et al 1989a). The cellular distribution of tissue factor provides a model for one level of control of coagulation. The vasculature endothelium lacks the major clot-promoting activity, but after injury, blood that reaches the subendothelium contacts cells constitutively expressing tissue factor and coagulation results.

Inflammation constitutes a major mechanism by which this normally passive vasculature can be altered to express tissue factor and initiate coagulation. Cytokines like IL-1 (Bevilacqua et al 1984; Nawroth et al 1986) and TNF (Herbert et al 1992; Ryan et al 1992) will stimulate tissue factor synthesis in endothelium in cell culture. Recent studies suggest that in primates, *E. coli*-induced septic shock leads to tissue factor expression, particularly on endothelium in the spleen and renal glomeruli and on splenic macrophages (Drake et al 1992). Endothelial cell tissue factor levels appear to be increased in preeclampsia (Labarrere et al 1990). Despite these observations, whether the endothelial cells actually contribute significantly to tissue factor expression and initiation of coagulation in vivo remains uncertain. While endothelial cells in culture can clearly respond to several stimuli by synthesizing tissue factor, human saphenous vein vessel segments appear to lack the capability to form tissue factor in response to endotoxin (Solberg et al 1990). These observations suggest that, except in the most extreme cases, inhibitory factors exist on the vessel wall that limit tissue factor production and aid in preventing vessel-associated fibrin formation.

Alternative mechanisms exist, however, for initiating coagulation in response to inflammation (Figure 4). Endotoxin will stimulate monocyte tissue factor synthesis (Herbert et al 1992; Osterud et al 1992), and monocyte/macrophage expression of tissue factor has been observed in atherosclerotic lesions (Wilcox et al 1989) and on monocytes of patients with inflammatory diseases such as septic shock and organ transplant rejection (Edwards et al 1987; Bach 1988). Monocytes might be envisioned as circulating loci for the initiation of coagulation, but monocytes and neutrophils can interact with adhesion molecules that are expressed on the surface of the endothelium in response to inflammatory mediators E selectin-responding to IL-1; TNF or endotoxin (Gimbrone et al 1990; Bevilacqua et al 1985), or P selectin responding to thrombin or histamine (McEver 1991; Larsen et al 1989). These endothelial cell adhesion molecules can potentially serve to attach the procoagulant monocytes to the endothelial cell surface. Presumably, platelets could adhere to bound neutrophils or monocytes through P selectin expressed on their surface. Thus whether the endothelium or the monocyte is the source of the tissue factor, the vascular surface would become more thrombogenic.

Anticoagulant pathways may be down-regulated by inflammation. Endotoxin or cytokines like TNF and IL-1 can lead to loss of TM (Moore et al

1987, 1989; Lentz et al 1991; Nawroth & Stern 1986; Conway & Rosenberg 1988; Nawroth et al 1986), a process that has been observed in patients with organ rejection and severe inflammation (Labarrere et al 1990; Hancock et al 1991), and circulating TM can be detected, presumably as a result of proteolysis by neutrophils (Takano et al 1990). Furthermore, C4bBP levels increase in inflammation (Dahlbäck 1991). This potentially decreases the free protein S levels (Taylor et al 1991) and further facilitates coagulation.

Although inflammation is one of the most potent triggers for intravascular coagulation, it is nonetheless true that even this process must be regulated. A first approach to understanding the complex interplay of the inflammatory mediators and the cytokines comes from observations with IL-4 (Kapiotis et al 1991; Herbert et al 1992). IL-4 prevents the down-regulation of TM on endothelium and the upregulation of tissue factor on both endothelium and monocytes by endotoxin, IL-1 and TNF. Undoubtedly, future studies will reveal more regulatory factors that will limit endothelial activation.

Monocytes may also be able to initiate coagulation via tissue factor-independent mechanisms. Recent studies have demonstrated that the monocyte can bind factor X to the Mac 1 receptor after monocyte activation by ADP and that the bound factor X is activated preferentially, apparently by a tissue factor-independent mechanism (Altieri et al 1988). Once activated, factor X can interact with monocyte factor V-like activity to generate thrombin from prothrombin. The factor V activity appears to be synthesized by the monocyte directly (Altieri & Edgington 1989) and expressed in response to cell activation with ADP or ionophore A23187. Monocytes can also bind and assemble prothrombinase with plasma coagulation factors, especially following endotoxin stimulation that results in the formation of monocyte microparticles (Robinson et al 1992) (see section on microparticle formation). Upon adhesion of monocytes to the selectins of activated endothelium, this activation pathway could potentially lead to thrombin or fibrin formation without significant participation of platelets. Another intriguing possibility is that the multiple mechanisms for thrombin generation with the monocyte may constitute a mechanism for extravascular thrombin generation. Thrombin could then serve cell signaling functions (see section on the thrombin receptor).

PLATELET PARTICIPATION IN THROMBIN FORMATION

For years it was recognized that cells, and in particular platelets, were required for rapid clotting in plasma. This platelet activity could be replaced by negatively charged artificial membranes, among which the most active are those containing phosphatidyl serine (Cutsforth et al 1989). Platelet activation is required for optimal expression of clotting activity (Zwaal et al 1989; Wiedmer et al 1986). Upon activation, factor V released from the platelet α granules (Ittyerah et al 1981) binds to the platelet surface and accelerates prothrombin activation (Tracy et al 1992; Miletich et al 1977). Platelet factor V concentration is sufficient to essentially saturate the critical binding sites on platelets (Kane & Majerus 1982). Factor V released from platelets is proteolytically cleaved, but not totally activated. The resultant factor V can be activated by factor Xa approximately 50 times faster than intact plasma factor V (Monkovic & Tracy 1990). Inside the platelets, and not released with normal platelet agonists, is a factor V-activating enzyme, the physiological function of which remains unknown (Kane et al 1982). Binding of factor Va to the platelet membrane is essential for the expression of high affinity binding sites for factor Xa. Platelets bind factor Va 1000 times tighter than model membranes (Tracy et al 1992). The platelet receptor shows considerable preference for the activated form of factor V since the precursor is not capable of functioning as an inhibitor of factor Va binding (Kane & Majerus 1982). If all of the factor X or factor V were activated in plasma, their concentrations would be more than sufficient to saturate the platelet binding sites (Kane & Majerus 1982). Hence, the maximum rate of prothrombin activation is determined not by the concentration of factor V or factor X, but by the number of potential binding sites that exist on the surface of platelets. The number of sites is dependent on the nature of the platelet agonist. Granule release per se is not the key element in forming prothrombinase sites since release of α granule contents is stimulated by all of the following agonists to essentially identical extents, but the ability to participate in prothrombin activation varies by more than tenfold (Zwaal et al 1989) with the potency of agonists increasing in the following order: ADP ≈ epinephrine, thrombin, collagen, collagen+thrombin, complement proteins C5b-9 (Sims et al 1989), and calcium ionophore.

The order of agonist effectiveness in promoting prothrombin activation sites correlates well with the ability of the agonist to translocate phosphatidyl serine to the outer leaflet of the membrane (Zwaal et al 1989). That the site could be composed, at least in part, of aminophospholipids was suggested by the observation that aminophospholipids are susceptible to phospholipases only after platelet activation (Zwaal et al 1989). That the negatively charged membrane that forms following activation is sufficient to actually bind proteins is supported by the observation that annexin V (also referred to as placental anticoagulant protein), a Ca^{2+}-dependent membrane-binding protein, binds to platelets only after activation, and the amount of binding observed with the membrane-binding protein follows the same agonist profile as the increase in prothrombinase (Thiagarajan & Tait 1990). Perhaps more importantly, annexin V blocks prothrombin activation on the activated platelets. Taken together the vast majority of the studies suggest that the platelet participation in prothrombin activation involves the formation of patches of negatively charged phospholipid on the cell surface. Whether a low level of hemostatically relevant surface exists on unperturbed platelets that is independent of aminophospholipid translocation remains unknown.

How the negatively charged surface is expressed remains the interest of several groups. A number of observations suggest that a major mechanism involves vesiculation. When platelets are stimulated with a variety of agonists, they release small "microparticles." Lentz's group was the first to demonstrate convincingly that these microparticles contributed to the increased prothrombin activation capacity associated with platelet activation (Sandberg et al 1985). More recent studies have shown a positive correlation between the ability of agonists to produce microparticles and the capacity to increase prothrombin activation. These microparticles bind factor Va, and residual membrane activity is found in the supernatant following separation of platelets from the microparticles (Sims et al 1989). Interesting differences exist between the remnant cells and the microparticles. Aminophospholipid

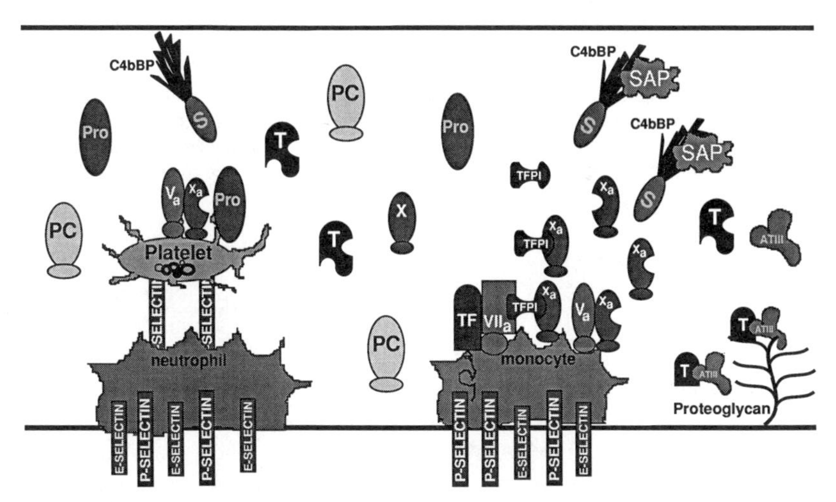

Figure 4 Protein pathway C after inflammation. Some of the influences of inflammation on the coagulation system. Inflammation can lead to TM down-regulation, decreases in free protein S, expression of leukocyte adhesion molecules, expression of tissue factor on endothelium and monocytes, neutrophil and monocyte attachment to the endothelium, and platelet microparticle formation. The net result can lead to vascular injury. See text for discussion. Abbreviations: PC, protein C; Pro, prothrombin; S, protein S; C4bBP, C4b binding protein; Va, factor Va; X, factor X; Xa, factor Xa; T, thrombin; TF, tissue factor; TFPI, tissue factor pathway inhibitor, heparin-like proteoglycan is depicted as in Figure 2.; VIIa, factor VIIa; SAP, serum amyloid P. (Adapted from Esmon 1992. The protein C anticoagulant pathway. *Arterioscl. Thromb.* 12:135–45; with permission of the Am. Heart Assoc. Inc.)

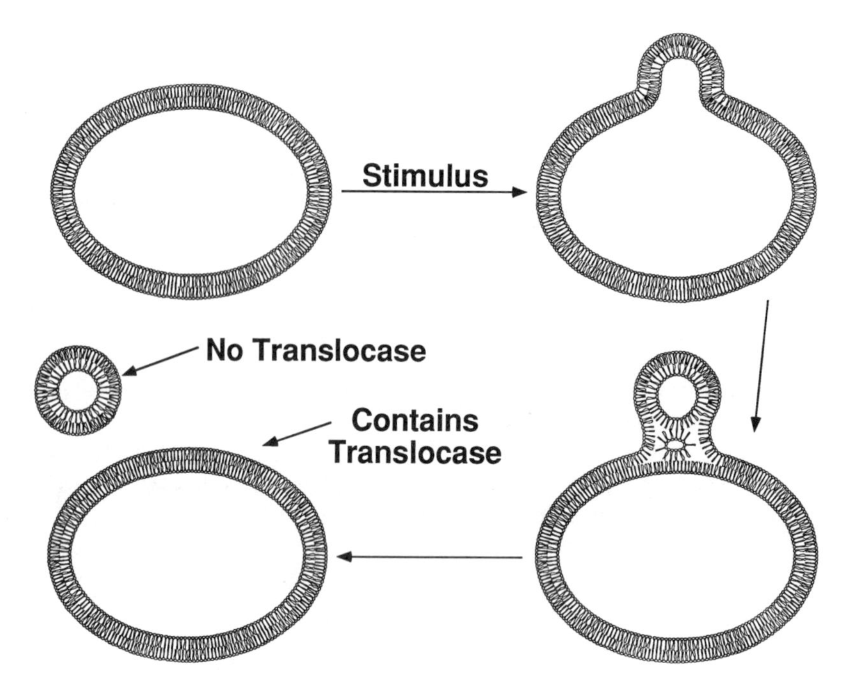

Figure 5 Microparticle formation on cell surfaces. As the cells vesiculate, the membrane bilayer cannot maintain its integrity as the vesicles are released which results in translocation of the aminophospholipids to the surface of the membrane. The resultant particles have a net negative surface charge and can therefore constitute sites for assembly of coagulation complexes as depicted in Figure 1.

translocase activity remains with the cells, but is not present in active form in the microparticles. As a result, with time the remnant cells reestablish the membrane phospholipid asymmetry by removing the aminophospholipids from the outer membrane leaflet, whereas the microparticles retain the negative charge characteristic of their release (Comfurius et al 1990). The transient loss of membrane asymmetry occurs when the microparticles bud from the membrane surface. At the budding interface, there is no possibility to maintain bilayer integrity and hence, at the budding interface, polar lipids from the inner leaflet becomes a part of the outer leaflet (Bevers et al 1991) (Figure 5).

What Patient Studies Tell Us About Platelet Participation in Prothrombin Activation

Differences between the platelet and model membranes are suggested by the observation that lupus anticoagulants that can inhibit coagulation on vesicle surfaces can be relatively ineffective in blocking platelet-dependent coagulation (Shapiro & Thiagarajan 1982). This discrepancy may be explained by the relatively slow association of antibodies with cell surface lipid as compared with coagulation factors (Dahlbäck et al 1983). This may account for the interesting observation that patients with lupus anticoagulant levels sufficient to prolong clinical assays dramatically almost never experience bleeding complications (Shapiro & Thiagarajan 1982). Perhaps this is related to the fact that the exposure of the negatively charged phospholipid occurs in concert with the release and processing of factor V. As suggested by Tracy et al (1984), this could work to help localize the clot. Indeed, these investigators found that deficiency of platelet factor V with only mildly reduced plasma factor V levels was associated with a bleeding complication in the patients. Whether this relates to factor V processing during release or to the temporal advantage in terms of receptor occupancy that occurs because of release from the platelet remains unclear. The importance of processing factor V to allow factor Xa binding is augmented by the fact that complex assembly is not only essential to rapid prothrombin activation, but the factor Xa-factor Va complex is resistant to inhibition by ATIII (Teitel & Rosenberg 1983).

Evidence that the platelet sites for assembly of prothrombinase are important for normal hemostasis comes from analysis of the Scott syndrome, which is characterized by the patient's inability to form factor X and prothrombin activation complexes (Bevers et al 1991). A probable cause for this deficiency is provided by the fact that the patient's platelets produce very few microparticles even when the platelets are challenged with strong agonists like the Ca^{2+} ionophore, A 23187 (Sims et al 1989). The exact cellular basis for this unusual resistance to microparticle formation is unknown, but it suggests that forming these particles may be a component of the hemostatic process. Recently, it was suggested, however, that microparticle formation is not an essential component of prothrombinase assembly on the platelet surface when the platelets are bound to von Willebrand factor (Swords et al 1992).

MEMBRANE SURFACES WITH OTHER COAGULATION REACTIONS

In general the requirements for the assembly of the factor VIIIa-factor IXa complex on platelet surfaces appears to be similar to the requirements for prothrombin activation complex assembly. These complexes form on microparticles and the activity increases with agonist stimulation (Gilbert et al 1991). Platelets and endothelial cells both appear to possess a factor IX/IXa binding site (Stern et al 1985; Ahmad et al 1989) that can participate in factor X activation, but no comparable site for factor X has been described on platelets. In the presence of factor VIIIa and factor X, these sites become relatively specific for factor IXa (Stern et al 1985; Ahmad et al 1989). In the Scott syndrome, the factor IX/IXa sites are present, but the addition of factor VIII and factor X does not convert these sites to factor IXa-specific sites, nor can the platelets effectively catalyze factor X activation (Ahmad et al 1989). On the endothelium, this site appears to be quite specific (Cheung et al 1992) since a small segment of the factor IX molecule appears to be sufficient for the binding activity.

Interaction of factor VIII with the activated platelet surface is somewhat more complicated than that of factor V. Factor VIII circulates bound to a high molecular weight polymeric protein, the von Willebrand factor, that inhibits the binding of factor VIII to the platelet surface (Nesheim et al 1991). Since the complex dissociates upon activation of factor VIII, activation may be essential for platelet interaction both by increasing factor VIII affinity and by overcoming the inhibitory activity associated with von Willebrand factor (Nesheim et al 1991). When the binding of isolated factor VIII to platelets was examined to determine whether factor V would compete for the factor VIII binding sites, complex patterns emerged, with factor V stimulating factor VIII binding at concentrations above the K_d for factor Va binding, but inhibiting factor VIII binding at still higher concentrations (Gilbert et al 1991). This appears to imply that some type of specificity exists above and beyond that imparted by negatively charged phospholipid. This specificity is consistent with the observation that factor Va binding to platelets appears to be much tighter than the binding to artificial membrane surfaces (Tracy et al 1992).

Membrane involvement in tissue factor activity is somewhat different than for the other complexes. Tissue factor-factor VIIa complexes catalyze three separate reactions: the activation of factor IX, the activation of factor X, and the conversion of factor VII to factor VIIa (Yamamoto et al 1992). In general, a negatively charged phospholipid surface is required for optimal activation of the target zymogens, factor X and factor IX (Ruf et al 1991; Rapaport & Rao 1992; Lawson & Mann 1991), but the activation of factor VII by the tissue factor-factor VIIa complex actually occurs more rapidly on neutral membranes composed of phosphatidyl choline than on phosphatidyl serine containing membranes (Neuenschwander & Morrissey 1992). Since the formation of factor VIIa may poise the system to clot, this priming event might have important regulatory functions prior to cell signaling events that lead to phosphatidyl serine exposure. Tissue factor is often found on microparticles released from cells and on the residual cell surface (Bona et al 1987; Rao et al 1992; Dvorak et al 1983) and many investigators have noted that full tissue factor activity is only detected after cell disruption (Drake et al 1989b;

Bach & Rifkin 1990). Interesting differences exist with tissue factor and the prothrombin activation complex. For instance, on tumor cells expressing both tissue factor and prothrombin activation sites, prothrombin activation is inhibited by annexin V, but tissue factor activation of factor X is not (Rao et al 1992). This suggests major differences in the aminophospholipid requirements for the two activation complexes.

In part, this can be explained by the fact that tissue factor and factor VII/VIIa bind effectively in the absence of phosphatidyl serine (Bach et al 1986), whereas factor Va-factor Xa complexes do not assemble with high affinity in the absence of a negatively charged phospholipid membrane surface (Mann et al 1990). Negatively charged phospholipids do, however, increase tissue factor activity (Ruf et al 1991; Bach & Rifkin 1990). Cellular disruption is well known to increase tissue factor activity, and recent studies have shown that this increase is not associated with significant increases in cell surface tissue factor antigen (Drake et al 1989b). Cell disruption per se is not critical since increases in cytosolic Ca^{2+} induced with Ca^{2+} ionophore enhance tissue factor activity, presumably by exposing phosphatidyl serine on the membrane surface (Bach & Rifkin 1990). Given that increased intracellular Ca^{2+} enhances vesiculation, it is likely that the increase in tissue factor activity in these cells is the result of a vesiculation process and that the control of tissue factor activity may be similar to the control of prothrombinase activity discussed above.

On the anticoagulant side, the thrombin-TM complex differs from most activators of vitamin K-dependent zymogens in that TM is constitutively expressed. Like tissue factor, TM-thrombin interaction is not dependent on negatively charged phospholipid. The activation of protein C proceeds effectively on neutral membranes with the catalytic properties of TM incorporated into neutral membranes similar to those of endothelial TM (Galvin et al 1987), although the activity is enhanced when incorporation is into phosphatidyl serine containing membranes (Freyssinet et al 1988). Thus in principle, vesiculation could lead to enhanced protein C activation through expression of phosphatidyl serine on the outer leaflet of the membrane. However, when endothelium is exposed to complement proteins C5b-9, the endothelial cells vesiculate, presumably because of a Ca^{2+} flux and vesicles are released, which catalyze prothrombin activation, but TM activity is essentially unaffected (Hamilton et al 1990). Whether some other stimulus can lead to exposure of aminophospholipids in the vicinity of TM to enhance activity is unknown.

The observation that complement proteins C5b-9 can interact with cells to initiate microparticle formation and enhance prothrombinase assembly constitutes still another example of the interplay between inflammation and coagulation. Earlier steps in the coagulation system are also enhanced by complement activation since C3b facilitate monocyte expression of tissue factor (Prydz et al 1977).

THROMBIN INFLUENCES ON CELLS

Thrombin has diverse influences on cells: (*a*) it is mitogenic for fibroblasts, macrophages, smooth muscle cells (when bound to extracellular matrix) (Bar-Shavit et al 1986); (*b*) it inhibits neurite outgrowth, a process specifically blocked by a protease inhibitor in the brain (Wagner et al 1989); (*c*) it triggers platelet activation with the concomitant release of vasoactive materials and growth factors; (*d*) it activates endothelial cell responses including von Willebrand factor release; (*e*) it stimulates growth factor production (V-sis) in endothelium (Daniel et al 1986); and (*f*) it initiates in endothelium the formation of platelet activating factor (PAF) (Lorant et al 1991), endothelial cell-derived relaxing factor (EDRF), and prostacyclin (Gimbrone 1986; Owen et al 1986). Thus thrombin modulates not only coagulation, but vascular tone and inflammation processes.

Thrombin activation of cells in general exhibits properties compatible with both binding and proteolytic events. Critical insights into the participation of thrombin in cell activation events came from the cloning and characterization of a cell surface thrombin receptor from platelets and endothelium (Vu et al 1991). In their landmark study, Coughlin's group described the structure of the receptor and delineated a feasible mechanism for signaling. The receptor spans the membrane seven times and contains a cleavage site. Upon cleavage by thrombin, the new amino terminus serves as a "tethered" ligand that binds to the receptor and appears to activate a G protein leading to Ca^{2+} flux, phosphatidyl inositol turnover, and cell activation (Hung et al 1992b). Interestingly, the receptor contains a peptide sequence with remote sequence similarity to hirudin, the leach thrombin inhibitor. This sequence in both the thrombin receptor and hirudin binds in a deep groove in thrombin known as the anion-binding exosite (Stubbs et al 1992; Liu et al 1991). These observations provide a molecular basis for the fact that TM can prevent platelet activation since, like the thrombin receptor, TM interacts with the anion-binding exosite (Hung et al 1992a; Ye et al 1992; Hofsteenge et al 1986). Most probably when blood passes through the microcirculation, the free thrombin concentration decreases rapidly because of the high affinity interaction with TM, the platelet surface is cleared of thrombin, and thrombin activation of the platelet is prevented. Consistent with this shared site hypothesis, thrombin mutants lacking the active site serine residue critical for proteolytic activity serve as competitive inhibitors of both thrombin-dependent TM activation of protein C and thrombin activation of platelets (Hung et al 1992a).

Identification of the cleavage site and the new amino terminus as the thrombin receptor agonist allowed the synthesis of peptides that could activate the receptor in the absence of thrombin (Scarborough et al 1992). By avoiding the potential complicating activity of thrombin binding and/or proteolysis of other receptors, these peptides have been useful in documenting that the cloned thrombin receptor is sufficient for many of the thrombin-dependent cellular biological responses (Vassallo et al 1992; Chao et al 1992; Simonet et al 1992) including enhanced wound healing and neovascularization (Carney et al 1992). In rat smooth muscle cells, the thrombin receptor level is low or absent when the cells are growth-arrested, or in the aorta, and it is elevated when the cells are stimulated to proliferate with basic fibroblast growth factor (Zhong et al 1992). This suggests that the thrombin receptor may be upregulated by vascular injury such as angioplasty and could play a role in the restenosis, a process of rapid smooth muscle cell proliferation. Eventually, the availability of thrombin receptor antagonist peptides may provide tools to test this possibility.

Exceptions to the active site-dependent cellular functions of thrombin include monocyte chemotaxis, macrophage (Bar-Shavit et al 1986), and smooth muscle cell growth factor activity (Bar-Shavit et al 1990). For monocyte chemotaxis, peptides derived from the anion-binding exosite of thrombin (residues 388–400 of prothrombin) will also elicit the chemotactic activity.

OTHER MODULATORS OF COAGULATION

Viral infection of endothelium may also modulate the cell coagulant function (Vicente et al 1991; Altieri et al 1991; Li et al 1992; Key et al 1990; Visser et al 1988) promoting fibrin deposition and the clinical manifestations of the infection. As one example, endothelium infected with herpes simplex virus 1 (HSV) increases the capacity to activate prothrombin (Visser et al 1988), synthesizes tissue factor and concomitantly down regulates TM (Key et al 1990). HSV infection leads to the expression of viral glycoprotein C, which binds factor X and aids in activation on the endothelium (Altieri et al 1991). The net result of all of these activities is the local facilitation of thrombin formation with its enhanced monocyte adherence and the potential to increase inflammatory damage at the wound site.

SUMMARY

Recent studies suggest that the generation of thrombin and other clotting factors plays a major role in cellular regulation and probably a variety of disease processes. Many of these cellular responses occur at thrombin levels lower than those required for fibrin clot formation. The availability of more specific reagents to intervene in the coagulation process will likely provide a better understanding of the role of the coagulation cascade and its inhibitors in normal and pathologic responses. In the immediate future, improved understanding of the role of inflammation and cellular responses in the control of coagulation offers many new and potentially safer interventions to prevent thrombosis.

Disclaimer: It was my intent in this chapter to provide some outlines of the role of membranes in the blood clotting process and the role of coagulation factors in cellular activation. The scope of the area is so extensive that many important contributions have been overlooked and space limitations forced citation of reviews rather than primary publications in many cases. I regret these oversights.

ACKNOWLEDGEMENTS

The research projects conducted by my group that have been discussed in this article were made possible by grants from the National Institutes of Health (R37 HL30340 and R01 HL29807). CTE is an investigator of the Howard Hughes Medical Institute. I would like to thank Karen Deatherage for helping in the preparation of the manuscript and also Jeff Box and Todd York for their help in preparing the figures.

Literature Cited

Ahmad, SS, Rawala-Sheikh, R, Ashby, B, Walsh, PN. 1989. Platelet receptor-mediated factor X activation by factor IXa. High-affinity factor IXa receptors induced by factor VIII are deficient on platelets in Scott Syndrome. *J. Clin. Invest.* 84:824–28

Altieri, DC, Edgington, TS. 1989. Sequential receptor cascade for coagulation proteins on monocytes: Constitutive biosynthesis and functional prothrombinase activity of a membrane form of factor V/Va. *J. Biol. Chem.* 264:2969–72

Altieri, DC, Etingin, OR, Fair, DS, Brunck, TK, Geltosky, JE, et al. 1991. Structurally homologous ligand binding of integrin Mac-1 and viral glycoprotein C receptors. *Science* 254:1200–2

Altieri, DC, Morrissey, JH, Edgington, TS. 1988. Adhesive receptor Mac-1 coordinates the activation of factor X on stimulated cells of monocytic and myeloid differentiation: An alternative initiation of the coagulation protease cascade. *Proc. Natl. Acad. Sci. USA* 85:7462–66

Bach, RR. 1988. Initiation of coagulation by tissue factor. *Crit. Rev. Biochem. Mol. Biol.* 23:339–68

Bach, R, Gentry, R, Nemerson, Y. 1986. Factor VII binding to tissue factor in reconstituted phospholipid vesicles: Induction of cooperativity by phosphatidylserine. *Biochemistry* 25:4007–20

Bach, R, Rifkin, DB. 1990. Expression of tissue factor procoagulant activity: Regulation by cytosolic calcium. *Proc. Natl. Acad. Sci. USA* 87:6995–99

Bakker, HM, Tans, G, Janssen-Claessen, T, Thomassen, MC, Hemker, HC, et al. 1992. The effect of phospholipids, calcium ions and protein S on rate constants of human factor Va inactivation by activated human protein C. *Eur. J. Biochem.* 208:171–78

Bar-Shavit, R, Benezra, M, Eldor, A, Hy-Am, E, Fenton, JW, Vlodavsky, I. 1990. Thrombin immobilized to extracellular matrix is a potent mitogen for vascular smooth muscle cells: nonenzymatic mode of action. *Cell Regul.* 1:453–63

Bar-Shavit, R, Kahn, AJ, Mann, KG, Wilner, GD. 1986. Identification of a thrombin sequence with growth factor activity on mac-

rophages. *Proc. Natl. Acad. Sci. USA* 83: 976–80
Bar-Shavit, R, Wilner, GD. 1986. Mediation of cellular events by thrombin. *Int. Rev. Exp. Pathol.* 29:213–41
Bevers, EM, Comfurius, P, Zwaal, RFA. 1991. Platelet procoagulant activity: Physiological significance and mechanisms of exposure. *Blood Rev.* 5:146–54
Bevilacqua, MP, Pober, JS, Majeau, GR, Cotran, RS, Gimbrone, MA, Jr. 1984. Interleukin-1 (IL-1) induces biosynthesis and cell surface expression of procoagulant activity in human vascular endothelial cells. *J. Exp. Med.* 160:618–23
Bevilacqua, MP, Pober, JS, Wheeler, ME, Cotran, RS, Gimbrone, MA, Jr. 1985. Interleukin 1 acts on cultured human vascular endothelium to increase the adhesion of polymorphonuclear leukocytes, monocytes, and related leukocyte cell lines. *J. Clin. Invest.* 76:2003–11
Bona, R, Lee, E, Rickles, F. 1987. Tissue factor apoprotein: Intracellular transport and expression in shed membrane vesicles. *Thromb. Res.* 48:487–500
Bourin, M-C, Öhlin, A-K, Lane, DA, Stenflo, J, Lindahl, U. 1988. Relationship between anticoagulant activities and polyanionic properties of rabbit thrombomodulin. *J. Biol. Chem.* 263:8044–52
Broze, GJ, Jr, Girard, TJ, Novotny, WF. 1990. Regulation of coagulation by a multivalent Kunitz-type inhibitor. *Biochemistry* 29:7539–46
Broze, GJ, Jr, Warren, LA, Novotny, WF, Higuchi, DA, Girard, JJ, Miletich, JP. 1988. The lipoprotein-associated coagulation inhibitor that inhibits the factor VII-tissue factor complex also inhibits factor Xa: Insight into its possible mechanism of action. *Blood* 71:335–43
Busch, C, Cancilla, P, DeBault, L, Goldsmith, J, Owen, W. 1982. Use of endothelium cultured on microcarriers as a model for the microcirculation. *Lab. Invest.* 47: 498–504
Callander, NS, Rao, LVM, Nordfang, O, Sandset, PM, Warn-Cramer, B, Rapaport, SI. 1992a. Mechanisms of binding of recombinant extrinsic pathway inhibitor (rEPI) to cultured cell surfaces: Evidence that rEPI can bind to and inhibit factor VIIa-tissue factor complexes in the absence of factor Xa. *J. Biol. Chem.* 267:876–82
Callander, NS, Varki, N, Rao, LVM. 1992b. Immunohistochemical identification of tissue factor in solid tumors. *Cancer* 70:1194–201
Carney, DH, Mann, R, Redin, WR, Pernia, SD, Berry, D, et al. 1992. Enhancement of incisional wound healing and neovascularization in normal rats by thrombin and synthetic thrombin receptor-activating peptides. *J. Clin. Invest.* 89:1469–77
Chao, BH, Kalkunte, S, Maraganore, JM, Stone, SR. 1992. Essential groups in synthetic agonist peptides for activation of the platelet thrombin receptor. *Biochemistry* 31: 6175–78
Cheung, W-F, Hamaguchi, N, Smith, KJ, Stafford, DW. 1992. The binding of human factor IX to endothelial cells is mediated by residues 3–11. *J. Biol. Chem.* 267: 20529–31
Church, FC, Pratt, CW, Hoffman, M. 1992. Leukocyte chemoattractant peptides from the serpin heparin cofactor II. *J. Biol. Chem.* 266:704–09
Comfurius, P, Senden, JMG, Tilly, RHJ, Schroit, AJ, Bevers, EM, Zwaal, RFA. 1990. Loss of membrane phospholipid asymmetry in platelets and red cells may be associated with calcium-induced shedding of plasma membrane and inhibition of aminophospholipid translocase. *Biochim. Biophys. Acta* 1026:153–60
Comp, PC, Nixon, RR, Cooper, MR, Esmon, CT. 1984. Familial protein S deficiency is associated with recurrent thrombosis. *J. Clin. Invest.* 74:2082–88
Conway, EM, Rosenberg, RD. 1988. Tumor necrosis factor suppresses transcription of the thrombomodulin gene in endothelial cells. *Mol. Cell. Biol.* 8:5588–92
Cutsforth, GA, Whitaker, RN, Hermans, J, Lentz, BR. 1989. A new model to describe extrinsic protein binding to phospholipid membranes of varying composition: Application to human coagulation proteins. *Biochemistry* 28:7453–61
Dahlbäck, B. 1986. Inhibition of protein Ca cofactor function of human and bovine protein S by C4b-binding protein. *J. Biol. Chem.* 261:12022–27
Dahlbäck, B. 1991. Protein S and C4b-binding protein: Components involved in the regulation of the protein C anticoagulant system. *Thromb. Haemostas.* 66:49–61
Dahlbäck, B, Nilsson, IM, Frohm, B. 1983. Inhibition of platelet prothrombinase activity by a lupus anticoagulant. *Blood* 1:218–25
Dahlbäck, B, Stenflo, J. 1981. High molecular weight complex in human plasma between vitamin K-dependent protein S and complement component C4b-binding protein. *Proc. Natl. Acad. Sci. USA* 78:2512–16
Daniel, TO, Gibbs, VC, Milfay, DF, Garovoy, MR, Williams, LT. 1986. Thrombin stimulates c-*cis* gene expression in microvascular endothelial cells. *J. Biol. Chem.* 261:9579–82
Danielsson, A, Raub, E, Lindahl, U, Bjork, I. 1986. Role of ternary complexes, in which heparin binds both antithrombin and

proteinase, in the acceleration of the reactions between antithrombin and thrombin or factor Xa. *J. Biol. Chem.* 261:15467–73

Davie, EW, Fujikawa, K, Kisiel, W. 1991. The coagulation cascade: Initiation, maintenance and regulation. *Biochemistry* 30: 10363–70

de Agostini, AI, Watkins, SC, Slayter, HS, Youssoufian, H, Rosenberg, RD. 1990. Localization of anticoagulantly active heparan sulfate proteoglycans in vascular endothelium: Antithrombin binding on cultured endothelial cells and perfused rat aorta. *J. Cell Biol.* 111:1293–304

Drake, TA, Cheng, J, Chang, A, Taylor, FB, Jr. 1993. Expression of tissue factor, thrombomodulin, and E-selectin in baboons with lethal *E. coli* sepsis. *Am. J. Pathol.* In press

Drake, TA, Morrissey, JH, Edgington, TS. 1989a. Selective cellular expression of tissue factor in human tissues: Implications for disorders of hemostasis and thrombosis. *Am. J. Pathol.* 134:1087–97

Drake, TA, Ruf, W, Morrissey, JH, Edgington, TS. 1989b. Functional tissue factor is entirely cell surface expressed on lipopolysaccharide-stimulated human blood monocytes and a constitutively tissue factor-producing neoplastic cell line. *J. Cell Biol.* 109:389–95

Dvorak, HF, Van DeWater, L, Bitzer, AM, Dvorak, AM, Anderson, D, et al. 1983. Procoagulant activity associated with plasma membrane vesicles shed by cultured tumor cells. *Cancer Res.* 43:4434–41

Edwards, RL, Levine, JB, Green, R, Duffy, M, Mathews, E, et al. 1987. Activation of blood coagulation in Crohn's disease: Increased plasma fibrinopeptide A levels and enhanced generation of monocyte tissue factor activity. *Gastroenterology* 92:329–37

Esmon, CT. 1989. The roles of protein C and thrombomodulin in the regulation of blood coagulation. *J. Biol. Chem.* 264:4743–46

Esmon, CT, Owen, WG. 1981. Identification of an endothelial cell cofactor for thrombin-catalyzed activation of protein C. *Proc. Natl. Acad. Sci. USA* 78:2249–52

Fay, PJ, Coumans, J-V, Walker, FJ. 1991. von Willebrand factor mediates protection of factor VIII from activated protein C-catalyzed inactivation. *J. Biol. Chem.* 266: 2172–77

Fleck, RA, Rao, LVM, Rapaport, SI, Varki, N. 1990. Localization of human tissue factor antigen by immunostaining with monospecific, polyclonal anti-human tissue factor antibody. *Thromb. Res.* 57:765–81

Freyssinet, J-M, Beretz, A, Klein-Soyer, C, Gauchy, J, Schuhler, S, Cazenave, J-P. 1988. Interference of blood-coagulation vitamin K-dependent proteins in the activation of human protein C: Involvement of the 4-carboxyglutamic acid domain in two distinct interactions with the thrombin-thrombomodulin complex and with phospholipids. *Biochem. J.* 256:501–7

Furie, B, Furie, BC. 1988. The molecular basis of blood coagulation. *Cell* 53:505–18

Furmaniak-Kazmierczak, E, Hu, CY, Esmon, CT. 1993. Protein S enhances C4b binding protein interaction with neutrophils. *Blood.* 81:405–11

Galvin, JB, Kurosawa, S, Moore, K, Esmon, CT, Esmon, NL. 1987. Reconstitution of rabbit thrombomodulin into phospholipid vesicles. *J. Biol. Chem.* 262:2199–205

Gilbert, GE, Sims, PJ, Wiedmer, T, Furie, B, Furie, BC, Shattil, SJ. 1991. Platelet-derived microparticles express high affinity receptors for factor VIII. *J. Biol. Chem.* 266:17261–68

Gimbrone, MA, Jr. 1986. Vascular endothelium: Nature's blood container In *Vascular Endothelium in Hemostasis and Thrombosis,* ed. MA. Gimbrone, Jr, pp. 1–13. Edinburgh: Churchill Livingstone

Gimbrone, MA, Jr, Bevilacqua, MP, Cybulsky, MI. 1990. Endothelial-dependent mechanisms of leukocyte adhesion in inflammation and atherosclerosis. *Ann. NY Acad. Sci.* 598:77–85

Hamilton, KK, Hattori, R, Esmon, CT, Sims, PJ. 1990. Complement proteins C5b-9 induce vesiculation of the endothelial plasma membrane and expose catalytic surface for assembly of the prothrombinase enzyme complex. *J. Biol. Chem.* 265:3809–14

Hancock, WW, Tanaka, K, Salem, HH, Tilney, NL, Atkins, R. C, Kupiec-Weglinski, JW. 1991. TNF as a mediator of cardiac transplant rejection, including effects on the intragraft protein C/protein S/thrombomodulin pathway. *Transplant. Proc.* 23:235–37

Harris, KW, Esmon, CT. 1985. Protein S is required for bovine platelets to support activated protein C binding and activity. *J. Biol. Chem.* 260:2007–10

Heeb, MJ, España, F, Griffin, JH. 1989a. Inhibition and complexation of activated protein C by two major inhibitors in plasma. *Blood* 73:446–54

Heeb, MJ, Mosher, D, Griffin, JH. 1989b. Activation and complexation of protein C and cleavage and decrease of protein S in plasma of patients with intravascular coagulation. *Blood* 73:455–61

Herbert, JM, Savi, P, Laplace, MC, Lale, A. 1992. IL-4 inhibits LPS-,1β-and TNFα-induced expression of tissue factor in endothelial cells and monocytes. *FEBS Lett.* 310:31–33

Hofsteenge, J, Taguchi, H, Stone, SR. 1986. Effect of thrombomodulin on the kinetics

of the interaction of thrombin with substrates and inhibitors. *Biochem. J.* 237:243–51
Hung, DT, Vu, T-KH, Wheaton, VI, Charo, IF, Nelken, NA, et al. 1992a. "Mirror image" antagonists of thrombin-induced platelet activation based on thrombin receptor structure. *J. Clin. Invest.* 89:444–50
Hung, DT, Wong, YH, Vu, T-KH, Coughlin, SR. 1992b. The cloned platelet thrombin receptor couples to at least two distinct effectors to stimulate phosphoinositide hydrolysis and inhibit adenylyl cyclase. *J. Biol. Chem.* 267:20831–34
Ittyerah, TR, Rawala, R, Colman, RW. 1981. Immunochemical studies of Factor V of bovine platelets. *Eur. J. Biochem.* 120:235–41
Kane, WH, Majerus, PW. 1982. The interaction of human coagulation Factor Va with platelets. *J. Biol. Chem.* 257:3963–69
Kane, WH, Mruk, JS, Majerus, PW. 1982. Activation of coagulation Factor V by a platelet protease. *J. Clin. Invest.* 70:1092–100
Kapiotis, S, Besemer, J, Bevec, D, Valent, P, Bettelheim, P, et al. 1991. Interleukin-4 counteracts pyrogen-induced downregulation of thrombomodulin in cultured human vascular endothelial cells. *Blood* 78:410–15
Key, NS, Vercellotti, GM, Winkelmann, JC, Moldow, CF, Goodman, JL, et al. 1990. Infection of vascular endothelial cells with herpes simplex virus enhances tissue factor activity and reduces thrombomodulin expression. *Proc. Natl. Acad. Sci. USA* 87: 7095–99
Kieffer, N, Phillips, DR. 1990. Platelet membrane glycoproteins: Functions in cellular interactions. *Annu. Rev. Cell Biol.* 6:329–57
Kisiel, W, Canfield, WM, Ericsson, LH, Davie, EW. 1977. Anticoagulant properties of bovine plasma protein C following activation by thrombin. *Biochemistry* 16:5824–31
Labarrere, CA, Esmon, CT, Carson, SD, Faulk, WP. 1990. Concordant expression of tissue factor and Class II MHC antigens in human placental endothelium. *Placenta* II:309–18
Larsen, E, Celi, A, Gilbert, GE, Furie, BC, Erban, JK, et al. 1989. PADGEM Protein: A receptor that mediates the interaction of activated platelets with neutrophils and monocytes. *Cell* 59:305–12
Lawson, JH, Mann, KG. 1991. Cooperative activation of human factor IX by the human extrinsic pathway of blood coagulation. *J. Biol. Chem.* 266:11317–27
Lentz, SR, Tsiang, M, Sadler, JE. 1991. Regulation of thrombomodulin by tumor necrosis factor-α: Comparison of transcriptional and posttranscriptional mechanisms. *Blood* 77:543–50
Li, C, Fung, S, Chung, S, Crow, A, Myers-Mason, N, et al. 1992. Monoclonal antiprothrombinase (3D4.3) prevents mortality from murine hepatitis virus (MHV-3) infection. *J. Exp. Med.* 176:689–97
Liu, L-W, Vu, T-KH, Esmon, CT, Coughlin, SR. 1991. The region of the thrombin receptor resembling hirudin binds to thrombin and alters enzyme specificity. *J. Biol. Chem.* 266:16977–80
Lollar, P, Owen, W. 1980. Clearance of thrombin from the circulation in rabbits by high-affinity binding sites on endothelium. *J. Clin. Invest.* 66:1222–30
Lorant, DE, Patel, KD, McIntyre, TM, McEver, RP, Prescott, SM, et al. 1991. Coexpression of GMP-140 and PAF by endothelium stimulated by histamine or thrombin: A juxtacrine system for adhesion and activation of neutrophils. *J. Cell Biol.* 115: 223–34
Mann, KG, Nesheim, ME, Church, WR, Haley, P, Krishnaswamy, S. 1990. Surface-dependent reactions of the vitamin K-dependent enzyme complexes. *Blood* 76:1–16
Marcum, JA, McKenney, JB, Rosenberg, RD. 1984. The acceleration of thrombin-antithrombin complex formation in hindquarters via naturally occurring heparin-like molecules bound to the endothelium. *J. Clin. Invest.* 74:341–50
Maruyama, I, Majerus, PW. 1985. The turnover of thrombin-thrombomodulin complex in cultured human umbilical vein endothelial cells and A549 lung cancer cells: Endocytosis and degradation of thrombin. *J. Biol. Chem.* 260:15432–38
McEver, RP. 1991. GMP-140: A receptor for neutrophils and monocytes on activated platelets and endothelium. *J. Cell. Biochem.* 45:156–61
Miletich, JP, Jackson, CM, Majerus, PW. 1977. Interaction of coagulation Factor Xa with human platelets. *Proc. Natl. Acad. Sci. USA* 74:4033–36
Miletich, JP, Jackson, CM, Majerus, PW. 1978. Properties of the factor Xa binding site on human platelets. *J. Biol. Chem.* 253:6908–16
Monkovic, DD, Tracy, PB. 1990. Functional characterization of human platelet-released factor V and its activation by factor Xa and thrombin. *J. Biol. Chem.* 265:17132–40
Moore, KL, Andreoli, SP, Esmon, NL, Esmon, CT, Bang, NU. 1987. Endotoxin enhances tissue factor and suppresses thrombomodulin expression of human vascular endothelium in vitro. *J. Clin. Invest.* 79:124–30
Moore, KL, Esmon, CT, Esmon, NL. 1989. Tumor necrosis factor leads to internaliza-

tion and degradation of thrombomodulin from the surface of bovine aortic endothelial cells in culture. *Blood* 73:159–65

Nawroth, PP, Handley, DA, Esmon, CT, Stern, DM. 1986. Interleukin-1 induces endothelial cell procoagulant while suppressing cell surface anticoagulant activity. *Proc. Natl. Acad. Sci. USA* 83:3460–64

Nawroth, PP, Stern, DM. 1986. Modulation of endothelial cell hemostatic properties by tumor necrosis factor. *J. Exp. Med.* 163: 740–45

Nesheim, M, Pittman, DD, Giles, AR, Fass, DN, Wang, JH, et al. 1991. The effect of plasma von Willebrand factor on the binding of human factor VIII to thrombin-activated human platelets. *J. Biol. Chem.* 266:17815–20

Nesheim, ME, Tracy, RP, Mann, KG. 1984. "Clotspeed," a mathematical simulation of the functional properties of prothrombinase. *Biol. Chem.* 259:1447–53

Neuenschwander, PF, Morrissey, JH. 1992. The factor VIIa-tissue factor complex exhibits strikingly different phospholipid requirements for activation of factor VII than for activation of factor X. 24th Int. Soc. Haematol., London, *Br. J. Haematol.* No. 214, p. 55 (Abstr.)

Novotny, WF, Palmier, M, Wun, T-C, Broze, GJ, Jr, Miletich, JP. 1991. Purification and properties of heparin-releasable lipoprotein-associated coagulation inhibitor. *Blood* 78: 394–400

Osterud, B, Olsen, JO, Wilsgard, L. 1992. Increased lipopolysaccharide-induced tissue factor activity and tumour necrosis factor production in monocytes after intake of aspirin: possible role of prostaglandin E2. *Blood Coag. Fibrinol.* 3:309–13

Owen, WG, Goeken, JA, Lollar, P. 1986. Thrombin-endothelium interactions. See Gimbrone, MA, Jr. pp. 57–69

Parkinson, JF, Koyama, T, Bang, NU, Preissner, KT. 1992a. Thrombomodulin: An anticoagulant cell surface proteoglycan with physiologically relevant glycosaminoglycan moiety. *Adv. Exp. Med. Biol.* 313:177–88

Parkinson, JF, Vlahos, CJ, Yan, SCB, Bang, NU. 1992b. Recombinant human thrombomodulin: Regulation of cofactor activity and anticoagulant function by a glycosaminoglycan side chain. *Biochem. J.* 283:151–57

Pratt, CW, Church, FC. 1992. Heparin binding to protein C inhibitor. *J. Biol. Chem.* 267:8789–94

Prydz, H, Allison, AC, Schorlemmer, HU. 1977. Further link between complement activation and blood coagulation. *Nature* 270: 173–74

Pusey, ML, Nelsestuen, GL. 1983. The physical significance of Km in the prothrombinase reaction. *Biochem. Biophys. Res. Commun.* 114:526

Rao, LVM, Tait, JF, Hoang, AD. 1992. Binding of Annexin V to a human ovarian carcinoma cell line (OC-2008). Contrasting effects on cell surface Factor VIIa/tissue factor activity and prothrombinase activity. *Thromb. Res.* 67:517–31

Rapaport, SI. 1989. Inhibition of factor VIIa/tissue factor-induced blood coagulation: With particular emphasis upon a factor Xa-dependent inhibitory mechanism. *Blood* 73:359–65

Rapaport, SI, Rao, LVM. 1992. Initiation and regulation of tissue factor-dependent blood coagulation. *Arterioscl. Thromb.* 12:1111–21

Roberts, HR, Lozier, JN. 1991. Other clotting factor deficiencies. In *Hematology: Basic Principles and Practice,* ed. R Hoffman, EJ Benz, Jr, SJ Shattil, B Furie, HJ Cohen, pp. 1332–42. New York: Churchill Livingstone

Robinson, RA, Worfolk, L, Tracy, PB. 1992. Endotoxin enhances the expression of monocyte prothrombinase activity. *Blood* 79: 406–16

Rosenberg, RD, Rosenberg, JS. 1984. Natural anticoagulant mechanisms. *J. Clin. Invest.* 74:1–6

Rosing, J, Tans, G, Govers-Riemslag, JWP, Zwaal, RFAZ, Hempker, HC. 1980. The role of phospholipids and factor Va in the prothrombinase complex. *J. Biol. Chem.* 255:274–83

Ruf, W, Rehemtulla, A, Morrissey, JH, Edgington, TS. 1991. Phospholipid-independent and dependent interactions required for tissue factor receptor and cofactor function. *J. Biol. Chem.* 266:2158–66

Ryan, J, Brett, J, Tijburg, P, Bach, RR, Kisiel, W, Stern, D. 1992. Tumor necrosis factor-induced endothelial tissue factor is associated with subendothelial matrix vesicles but is not expressed on the apical surface. *Blood* 80:966–74

Sandberg, H, Bode, AP, Dombrose, FA, Hoechli, M, Lentz, BR. 1985. Expression of coagulant activity in human platelets: Release of membranous vesicles providing platelet factor 1 and platelet factor 3. *Thromb. Res.* 39:63–79

Scarborough, RM, Naughton, M, Teng, W, Hung, DT, Rose, J, et al. 1992. Tethered ligand agonist peptides. Structural requirements for thrombin receptor activation reveal mechanism of proteolytic unmasking of agonist function. *J. Biol. Chem.* 267: 13146–49

Schwalbe, R, Dahlbäck, B, Hillarp, A, Nelsestuen, G. 1990. Assembly of protein S and C4b-binding protein on membranes. *J. Biol. Chem.* 265:16074–81

Shapiro, SS, Thiagarajan, P. 1982. Lupus anticoagulants. *Prog. Hemost. Thromb.* 6: 263–85

Shigematsu, Y, Miyata, T, Higashi, S, Miki, T, Sadler, JE, Iwanaga, S. 1992. Expression of human soluble tissue factor in yeast and enzymatic properties of its complex with factor VIIa. *J. Biol. Chem.* 267: 21329–37

Simonet, S, Bonhomme, E, Laubie, M, Thurieau, C, Fauchere, J-L, Verbeuren, TJ. 1992. Venous and arterial endothelial cells respond differently to thrombin and its endogenous receptor agonist. *Eur. J. Pharmacol.* 216:135–37

Sims, PJ, Wiedmer, T, Esmon, CT, Weiss, HJ, Shattil, SJ. 1989. Assembly of the platelet prothrombinase complex is linked to vesiculation of the platelet plasma membrane. *J. Biol. Chem.* 264:17049–57

Solberg, S, Osterud, B, Larsen, T, Sorlie, D. 1990. Lack of ability to synthesize tissue factor by endothelial cells in intact human saphenous veins. *Blood Coag. Fibrinol.* 1: 595–600

Stern, DM, Nawroth, PP, Harris, K, Esmon, CT. 1986. Cultured bovine aortic endothelial cells promote activated protein C-protein S-mediated inactivation of Factor Va. *J. Biol. Chem.* 261:713–18

Stern, DM, Nawroth, PP, Kisiel, WK, Vehar, GV, Esmon, CT. 1985. The binding of factor IXa to cultured bovine aortic endothelial cells: Induction of a specific site in the presence of factor VIII and X. *J. Biol. Chem.* 260:6717–22

Stubbs, MT, Oschkinat, H, Mayr, I, Huber, R, Angliker, H, et al. 1992. The interaction of thrombin with fibrinogen. A structural basis for its specificity. *Eur. J. Biochem.* 206:187–95

Suzuki, K, Nishioka, J, Kusumoto, H, Hashimoto, S. 1984. Mechanism of inhibition of activated protein C by protein C inhibitor. *J. Biochem. (Tokyo)* 95:187–95

Suzuki, K, Stenflo, J, Dahlbäck, B, Teodorsson, B. 1983. Inactivation of human coagulation factor V by activated protein C. *J. Biol. Chem.* 258:1914–20

Swords, NA, Tracy, PB, Mann, KG. 1992. Microvesicle release from adherent platelets is not essential for procoagulant activity. *Circulation* 86:2733 (Abstr.)

Takano, S, Kimura, S, Ohdama, S, Aoki, N. 1990. Plasma thrombomodulin in health and diseases. *Blood* 76:2024–29

Taylor, F, Chang, A, Ferrell, G, Mather, T, Catlett, R, et al. 1991. C4b-binding protein exacerbates the host response to *Escherichia coli*. *Blood* 78:357–63

Teitel, JM, Rosenberg, RD. 1983. Protection of factor Xa from neutralization by the heparin-antithrombin complex. *J. Clin. Invest.* 71:1383–91

Thiagarajan, P, Tait, JF. 1990. Binding of annexin V/placental anticoagulant protein I to platelets. Evidence for phosphatidylserine exposure in the procoagulant response of activated platelets. *J. Biol. Chem.* 265: 17420–23

Tracy, PB, Giles, AR, Mann, KG, Eide, LL, Hoogendoorn, H, Rivard, GE. 1984. Factor V (Quebec): a bleeding diathesis associated with a qualitative platelet factor V deficiency. *J. Clin. Invest.* 74:1221–28

Tracy, PB, Nesheim, ME, Mann, KG. 1992. Platelet factor Xa receptor. *Meth. Enzymol.* 215:329–60

Van Deerlin, VMD, Tollefsen, DM. 1991. The N-terminal acidic domain of heparin cofactor II mediates the inhibition of alpha-thrombin in the presence of glycosaminoglycans. *J. Biol. Chem.* 266:20223–31

Vassallo, RR, Jr, Kieber-Emmons, T, Cichowski, K, Brass, LF. 1992. Structure-function relationships in the activation of platelet thrombin receptors by receptor-derived peptides. *J. Biol. Chem.* 267:6081–85

Vehar, GA, Davie, EW. 1980. Preparation and properties of bovine factor VIII (antihemophilic factor). *Biochemistry* 19:401–10

Vicente, V, España, F, Tabernero, D, Estellés, A, Aznar, J, et al. 1991. Evidence of activation of the protein C pathway during acute vascular damage induced by Mediterranean spotted fever. *Blood* 78:416–22

Visser, MR, Tracy, PB, Vercellotti, GM, Goodman, JL, White, JG, Jacob, HS. 1988. Enhanced thrombin generation and platelet binding on herpes simplex virus-infected endothelium. *Proc. Natl. Acad. Sci. USA* 85:8227–30

Vu, T-KH, Hung, DT, Wheaton, VI, Coughlin, SR. 1991. Molecular cloning of a functional thrombin receptor reveals a novel proteolytic mechanism of receptor activation. *Cell* 64:1057–68

Wagner, SL, Geddes, JW, Cotman, CW. 1989. Protease nexin-I, an antithrombin with neurite outgrowth activity, is reduced in Alzheimer disease. *Proc. Natl. Acad. Sci. USA* 86:8284–88

Walker, FJ. 1980. Regulation of activated protein C by a new protein: A role for bovine protein S. *J. Biol. Chem.* 255:5521–24

Walker, FJ, Sexton, PW, Esmon, CT. 1979. Inhibition of blood coagulation by activated protein C through selective inactivation of activated factor V. *Biochim. Biophys. Acta* 571:333–42

Wiedmer, T, Esmon, CT, Sims, PJ. 1986. Complement proteins C5b-9 stimulate procoagulant activity through platelet prothrombinase. *Blood* 68:875–80

Wilcox, JN, Smith, KM, Schwartz, SM, Gordon, D. 1989. Localization of tissue factor in the normal vessel wall and in the atherosclerotic plaque. *Proc. Natl. Acad. Sci. USA* 86:2839–43

Yamamoto, M, Nakagaki, T, Kisiel, W. 1992. Tissue factor-dependent autoactivation of human blood coagulation factor VII. *J. Biol. Chem.* 267:19089–94

Ye, J, Liu, L-W, Esmon, CT, Johnson, AE. 1992. The fifth and sixth growth factor-like domains of thrombomodulin bind to the anion exosite of thrombin and alter its specificity. *J. Biol. Chem.* 267:11023–28

Zhong, C, Hayzer, DJ, Corson, MA, Runge, MS. 1992. Molecular cloning of the rat vascular smooth muscle thrombin receptor. *J. Biol. Chem.* 267:16975–79

Zwaal, RFA, Bevers, EM, Comfurius, P, Rosing, J, Tilly, R. HJ, Verhallen, PFJ. 1989. Loss of membrane phospholipid asymmetry during activation of blood platelets and sickled red cells; mechanisms and physiological significance. *Mol. Cell. Biochem.* 91:23–31

Annu. Rev. Cell Biol. 1993. 9:27–66

THE SPECTRIN-BASED MEMBRANE SKELETON AND MICRON-SCALE ORGANIZATION OF THE PLASMA MEMBRANE

Vann Bennett and Diana M. Gilligan

Howard Hughes Medical Institute and Departments of Biochemistry and Medicine, Duke University Medical Center, Durham North Carolina 27710

KEY WORDS: plasma membrane, cytoskeleton, ion channel, cell domains, cell adhesion molecule

---

CONTENTS

0743–4634/93/1115–0027$05.00

INTRODUCTION

Plasma membranes of metazoan cells are the site of convergence of disparate sets of molecules located in intercellular, transmembrane, and cytoplasmic compartments. Coordination between these compartments results in specialized cell domains and is essential for initial organization of cells into tissues during development, as well as physiological functions of cells in adult tissues. What is the structural basis for organization on the scale of cell domains? Conventional electron microscopy reveals an electron-dense undercoat as a common feature of many cell membranes, and such undercoats are especially well developed in specialized regions of the plasma membrane. Analysis of these multiprotein complexes at a molecular level is especially challenging due to the large number of components and their size, which can extend to microns. Moreover, these structures, although organized, are not sufficiently ordered for analysis by X-ray or electron diffraction.

Striking images of the polygonal spectrin-actin network of erythrocytes (Byers & Branton 1985; Liu et al 1987) suggest that, at least in some cases, the apparently amorphous material associated with membranes actually represents organized assemblies of proteins. Spectrin-based structures, perhaps closely related to the erythrocyte membrane skeleton, are a feature of many cells, since spectrin and associated proteins are ubiquitous components of metazoan plasma membranes. Functions of spectrin are likely to be diverse in different cells, but related in general to coupling of a variety of membrane-spanning cell surface proteins to cytoplasmic elements. This review deals with the current understanding of the spectrin skeleton in terms of its constituent proteins and the potential physiological and clinical relevance of these proteins in cells more complex than erythrocytes. Insights from this body of work may be useful in designing strategies for resolving other cell structures.

THE SPECTRIN-BASED SKELETON OF ERYTHROCYTES

The anucleate human erythrocyte represents an experiment of nature that provides a pure plasma membrane domain derived from a single cell type and has the added advantage (for the investigator) of mutations in the form of hereditary hemolytic anemias. The membrane skeleton of the erythrocyte was first detected as a dense meshwork of proteins that retained the original shape of the erythrocyte following solubilization of the membrane lipids with nonionic detergents (Yu et al 1973). High resolution electron microscopy of the stretched membrane skeleton reveals a regular lattice-like organization with 5–6 rod-shaped molecules 200 nm in length linked to vertices to form a sheet of five and six-sided polygons (Byers & Branton 1985; Liu et al 1987; Shen et al 1986). Legs of the polygons are comprised of spectrin, a flexible rod-shaped protein 200 nm in length (see below). Vertices of the network are formed by short actin filaments containing 13–15 actin monomers.

Spectrin and actin require accessory proteins to form a membrane-associated network (Figure 1). A major membrane attachment is provided by high affinity association of the beta subunit of spectrin with ankyrin at a site located in the midregion of spectrin tetramers (Bennett & Stenbuck 1979; Tyler et al 1979; Calvert et al 1980; Kennedy et al 1991). Ankyrin is a peripheral membrane protein which, in turn, is associated with the

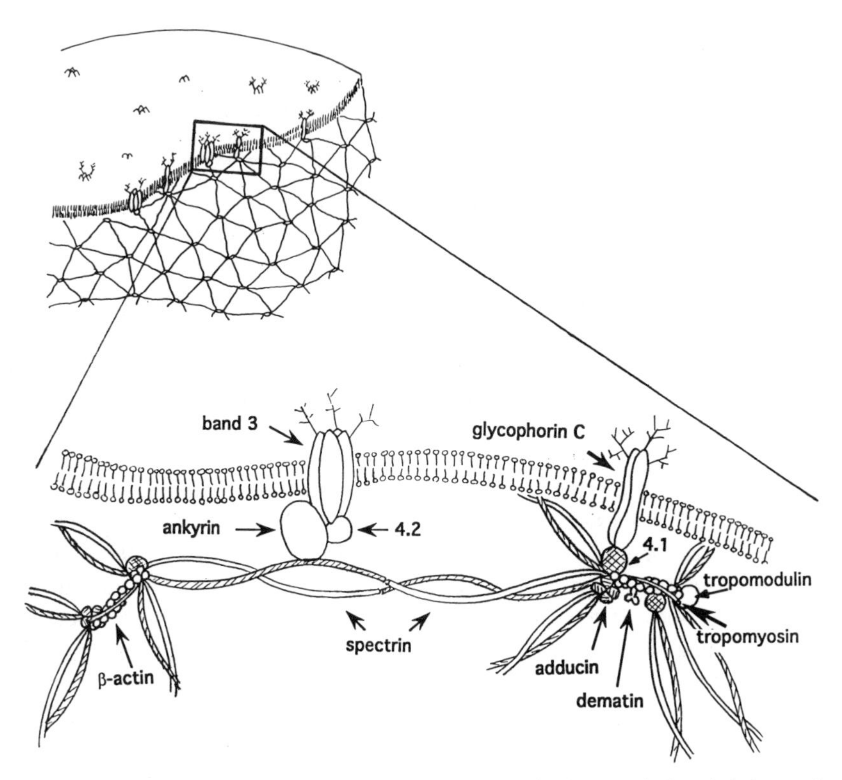

Figure 1 Schematic model of the organization of proteins in the spectrin-based skeleton of erythrocytes. The ensemble of proteins located at the spectrin/actin junctions are referred to as the junctional complexes.

cytoplasmic domain of the anion exchanger, a membrane-spanning integral protein (reviewed in Bennett 1990, 1992). Protein 4.2 associates with ankyrin as well as the cytoplasmic domain of the anion exchanger and may form a ternary complex with these proteins (Korsgren & Cohen 1986, 1988). Another association of spectrin with the membrane may be mediated by protein 4.1, which is located at the ends of spectrin molecules and recognizes membrane sites including glycophorin C and the anion exchanger (reviewed in Conboy 1993).

Association of multiple spectrin molecules with actin to form a two-dimensional meshwork involves a group of accessory proteins. The $(\text{spectrin})_{5-6}$ -actin oligomers, which form the vertices of the polygons, have been isolated from low ionic strength extracts of either erythrocyte ghosts (Beaven et al 1985), or detergent-extracted membrane skeletons (Matsuzaki et al 1985). Isolated spectrin-actin junctions contained spectrin and actin in the same molar ratios as in membrane skeletons (one spectrin dimer per two/three actin monomers). Spectrin-actin complexes contain additional proteins including protein 4.1 and dematin (protein 4.9). Other proteins may participate in spectrin-actin interactions, based on in vitro activities, and include adducin (Gardner & Bennett 1987), tropomyosin (Fowler & Bennett 1984), and a tropomyosin-binding protein, tropomodulin (Fowler 1987, 1990). A palmitoylated polypeptide (p55) copurifies with dematin (Ruff et al 1991) and may also be a junctional component. Protein 4.1, dematin, and adducin have been localized at spectrin-actin junctions by immunolabeling of membrane skeletons (Derick et al 1992). Accessory proteins of spectrin-actin junctions presumably are required for activities such as stabilization of spectrin-actin complexes, maintenance of actin filaments of a uniform length, and assembly of the membrane skeleton. Protein 4.1, adducin, and dematin are targets for protein kinases and thus may have a role in regulation of spectrin-actin interactions (Cohen & Gascard 1992).

Hereditary defects or deficiency in erythrocyte membrane skeletal proteins result in abnormally fragile red cells and mild to severe hereditary hemolytic anemias in humans and mice (reviewed by Gallagher & Forget 1993; Delaunay & Dhermy 1993; Conboy 1993; Palek & Lambert 1990; Cohen et al 1993). Mutations characterized at a molecular level result in reduced stability of spectrin tetramers and forms of protein 4.1 missing the spectrin-actin binding site. In addition, deficiencies of spectrin, ankyrin, protein 4.1, and protein 4.2 have been associated with abnormally fragile erythrocytes. These studies establish that the membrane skeleton is essential for normal survival of erythrocytes in the circulation and that defects in structural proteins can be the basis for disease. Mutations in skeletal proteins also have provided important insights into the structure and function of affected proteins and are discussed below in this context.

The influence of the spectrin skeleton on membrane mechanical properties and dynamic behavior of integral membrane proteins has been evaluated in mammalian erythrocytes without interference from other cytoskeletal components and organelles. The erythrocyte plasma membrane has the mechanical properties of an elastic semi-solid (Evans & Hochmuth 1977) that allows storage of energy during periods of brief deformation, with the consequence that the membrane will return to its original shape once the deformation stops. Direct evidence that elasticity of the membrane is due to spectrin is provided by studies with erythrocytes with varying degrees of spectrin-deficiency obtained from patients with hereditary spherocytosis (Waugh & Agre 1988). The shear modulus, a measure of energy required to deform the surface per unit area, varied in proportion to spectrin density over a twofold range. In addition, clinical severity of spherocytosis correlates well with the extent of spectrin deficiency and reduction in shear modulus (Agre et al 1985, 1986). Therefore, the ability of erythrocytes to sustain deformation and return to their previous shape is essential for prolonged life in the high shear environment of the vascular system.

The erythrocyte membrane skeleton limits diffusion of integral proteins such as the anion exchanger in the plane of the membrane bilayer and provides a model for how related membrane skeletons could define the topography of integral proteins in polarized cells. Measurements of translational diffusion reveal a fraction of anion exchangers that are immobile, and another fraction that is capable of slow diffusion at rates 50–200 times slower than expected in a lipid bilayer (Golan & Veatch 1980; Tsuji & Ohnishi 1986; Tsuji et al 1988). Spectrin appears to have a role in both types of restrictions because in spectrin-deficient erythrocytes, all of the anion exchangers exhibit rapid diffusion rates (Sheetz et al 1980).

A reduced diffusion rate of mobile anion exchangers could be the result of several mechanisms. One possibility is that these proteins are simply trapped within a spectrin-actin network that restricts long-range translation of anion exchangers, by analogy with a fence or corral (Tsuji et al 1988; Sheetz et al 1980). Direct interaction of the anion exchanger with ankyrin-spectrin complexes is responsible for the immobile fraction of anion exchangers, as shown by increasing the immobile fraction with addition of exogenous ankyrin (Tsuji & Ohnishi 1986) and decreasing the immobile fraction with addition of the spectrin-binding domain of ankyrin, which can displace intact ankyrin from spectrin (Fowler & Bennett 1978). Indirect and/or low affinity interactions limit the movements of those anion exchangers, in excess of ankyrin which are mobile at a reduced rate. Translational mobility of these molecules increased 10–50-fold following treatments that dissociate spectrin tetramers into dimers (Golan & Veatch 1980; Tsuji & Ohnishi 1986; Tsuji et al 1988). Another mechanism for immobilization of membrane proteins lacking direct connections on the cytoplasmic surface is through lateral associations with other integral membrane proteins. Glycophorin A, for example, can form complexes with the anion exchanger (Nigg et al 1980). In addition, associations of anion exchangers with proteins 4.2 and 4.1 (see below) may contribute to indirect interactions with spectrin or other membrane proteins. The spectrin skeleton, therefore, restricts the translational mobility of all of the anion exchangers either directly or indirectly, and potentially limits the mobility of other integral membrane proteins capable of association with anion exchangers.

SPECTRIN-BASED STRUCTURES IN MORE COMPLEX CELLS

Proteins closely related to spectrin and many of the associated proteins first discovered in erythrocytes are present in most tissues in vertebrates and invertebrates. The degree of similarity between erythrocyte proteins and more generally expressed forms varies from the same gene, with and without alternative exon usage, to closely related genes. In addition, several erythrocyte proteins are members of superfamilies of proteins (Table 1). Spectrins are associated with plasma membranes, in most cases, and are relatively abundant, comprising 3% of the total particulate protein in the case of vertebrate brain tissue. Spectrin thus can be viewed as the basic component of a system of plasma membrane-associated structural proteins, analogous

Table 1 Erythrocyte membrane skeletal proteins and related proteins expressed in other tissues

RBC protein	Same gene	Closely related genes	Gene family
$Alpha_{RBC}$ spectrin	—	$Alpha_G$ spectrin	Alpha-actinin, dystrophin
$Beta_R$ spectrin	Muscle, brain	$Beta_G$, $beta_{TW}$, $beta_H$ spectrins	Alpha-actinin, dystrophin
$Ankyrin_R$	Brain	$Ankyrin_B$, $ankyrin_{Node}$?$Ankyrin_G$	?
Protein 4.1	Many tissues	Ezrin, moesin, radixin	Talin; tyrosine phosphatase
Protein 4.2	Kidney, platelet, brain	?	Transglutaminases
Alpha adducin	Many tissues	?	?
Beta adducin	Many tissues	?	?
Dematin	Many tissues	?	?
Tropomodulin	Many tissues	?	?
p55	?	—	Dig-tumor suppressor; PSD-95

to the major cytoplasmic structures based on actin, tubulin, and intermediate filaments.

Spectrin and other constituents of the erythrocyte skeleton interact in tissues with a variety of proteins that are not expressed in mature erythrocytes, including various ion channels, cell adhesion molecules, tubulin, and intermediate filaments. Moreover, spectrin skeletons are likely to be polymorphic in various cell types and integrated with other cytoskeletal structures based on examples visualized in platelets (Hartwig & Desisto 1991) and hair cells of the cochlea (Holley & Ashmore 1990). Thus the simple paradigm of the erythrocyte skeleton is useful in terms of predicting some, but not all, potential components and provides only one possibility for how these components can be assembled into spectrin-based structures.

THE SPECTRIN FAMILY

Proteins closely related to erythrocyte spectrin are associated with plasma membranes in most vertebrate tissues (for reviews, see Goodman et al 1988; Bennett 1990; Mangeat 1988; Coleman et al 1989). Spectrins (also referred to as fodrin) have also been characterized in nonvertebrates including *Drosophila* (Byers et al 1992; Dubreuil et al 1989), echinoderms (Wessels & Chen 1993), *Dictyostelium* (Bennett & Condeelis 1988), and *C. elegans* (Loer 1992), as well as in higher plants (Maude et al 1991). Spectrins are extended, flexible molecules approximately 200–260 nm in length and 3–6 nm in width. They contain two subunits termed alpha (M_r 280,000) and beta (M_r 245,000–460,000) that are assembled into tetramers by the following alpha-beta interactions: (*a*) a relatively weak lateral association along the length of alpha and beta subunits oriented in an antiparallel configuration; (*b*) a tight lateral association beween the N-terminal domain of beta subunits and C-terminal domain of alpha subunits; and (*c*) head-head association between laterally associated heterodimers by linkage between the C-terminal domain of the beta and the N-terminal domain of alpha subunits (see Figure 2). Spectrins have binding sites for the following proteins: (*a*) for actin at the N-terminal domains of beta subunits that allow spectrins to cross-link actin filaments (Karinch et al 1990; Frappier et al 1992); (*b*) for calmodulin in the midregion of the alpha subunit of most spectrins (Harris et al 1988; Leto et al 1989). In the case of alpha spectrins of mammalian erythrocytes, calmodulin-binding sites are missing, and *Drosophila* spectrin contains a calmodulin-binding site at a different position than vertebrates (Dubreuil et al 1989). (*c*) For ankyrin, binding sites are located on the beta subunit of most spectrins at a site in the midregion of the tetramer (Tyler et al 1979; Davis & Bennett 1984; Kennedy et al 1991).

Physical properties of spectrin and interactions between subunits are best

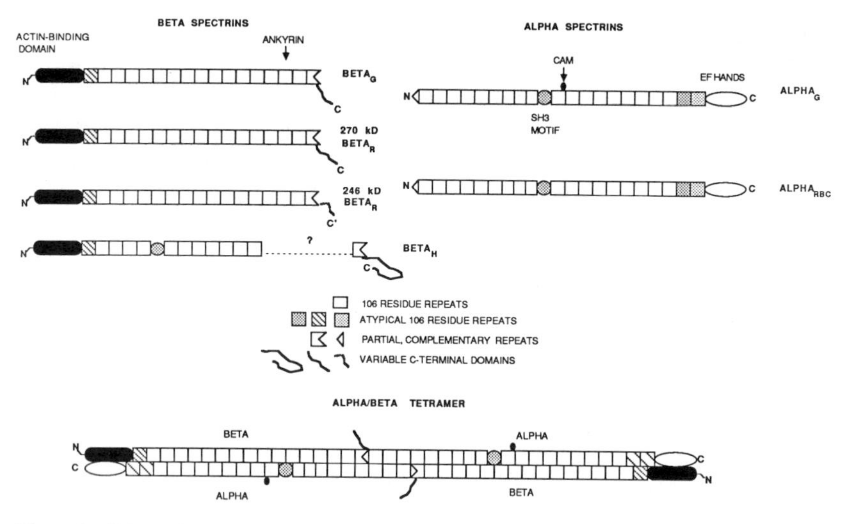

Figure 2 Schematic model for domain organization and subunit interactions of spectrins.

understood for erythrocyte spectrin. Fourier analysis of spectrin visualized in erythrocyte membrane skeletons indicates that the subunits twist about each other to form a helix (McGough & Josephs 1990). The pitch and diameter of the helix varies with the degree of extension of the spectrin molecules, which suggests that spectrin might function as an elastic spring. Flexiblility of spectrins is suggested by appearance in electron micrographs (Shotton et al 1979) and has been confirmed in solution by a variety of physical measurements (Elgsaeter 1978; Mikkelsen & Elgsaeter 1981; Mikkelsen et al 1984). Segmental motions with relaxation times on the order of 10 μs have been detected by dynamic light scattering (Budzynski et al 1992) and transient dichroism of spectrin labeled with the triplet probe eosin (Clague et al 1990). Interestingly, the flexibility of spectrin in solution also occurs in the context of a fully assembled membrane skeleton in erythrocytes (Clague et al 1990).

Alpha and beta subunits of the spectrin family contain a 106-amino acid repeating sequence first noted in erythroycte spectrin (Speicher & Marchesi 1984). Structural models have been proposed for the folding of 106-residue repeats of spectrin into a series of short alpha helical segments comprised of either three (Speicher & Marchesi 1984; Parry et al 1992) or four (Davison et al 1989) alpha helices. Evidence for independent folding of individual 106-residue repeats and determination of the starting residue for the repeated unit is based on bacterial expression of cDNAs encoding 106-residue segments that exhibit circular dichroism spectra and protease resistance nearly identical to native spectrin (Winograd & Branton 1991). Intron-exon boundaries do not correspond with the 106-amino acid repeated element (Kotula et al 1991), which is not surprising in view of the ancient origin of spectrins.

Diversity of Spectrins

Mammals express two types of alpha subunit encoded by distinct genes: a tissue invariant alpha subunit located on human chromosome 9, present in all mammalian tissues except for mature erythrocytes (Leto et al 1988), and the erythrocyte alpha subunit located on human chromosome 1 (Huebner et al 1985). Birds and other vertebrates have a single alpha subunit expressed in erythrocytes as well as other tissues.

Beta subunits of spectrin contain most of the recognition sites of spectrin for other proteins including ankyrin (Calvert et al 1980; Kennedy et al 1991; Davis & Bennett 1984), protein 4.1 (Coleman et al 1987), and actin (Karinch et al 1990; Frappier et al 1992), as well as the site for ankyrin-independent association of spectrin with membranes (Steiner et al 1989). Beta subunits also are a major source of diversity among spectrins, with a growing list of variants with specialized functions. The beta subunit family currently includes five isoforms that are likely to be encoded by distinct genes: (*a*)beta$_G$, a polypeptide expressed in most vertebrate tissues encoded by a gene located on human chromosome 2 (Hu et al 1992), and mouse chromosome 11 (Bloom et al 1992); (*b*)beta$_R$, first characterized in erythrocytes and also expressed as alternatively spliced forms in brain and skeletal muscle (Winkelmann et al 1990, 1991). The gene for beta$_R$ is located on human chromosome 14q23-q24.2 (Fukushima et al 1990) and mouse chromosome 12 (Laurila et al 1987); (*c*) beta$_{TW}$, a specialized subunit associated with terminal web in the apical domain of intestinal epithelial cells, found in birds but not in mammals (Glenney et al 1983); (*d*) beta$_{NM}$, a beta-type subunit identified at neuromuscular junctions based on cross-reactivity with antibodies (Bloch & Morrow 1989). The beta-related subunit at neuromuscular junctions may be an exception to the general rule that beta subunits are associated with alpha subunits. No alpha subunit could be detected by antibodies in one study (Bloch & Morrow 1989), although another group has reported alpha spectrin immunoreactivity at these sites (Vybiral et al 1992); (*e*) beta$_H$, a 430,000 dalton polypeptide initially discovered in *Drosophila* (Dubreuil et 1990) that forms tetramers with the alpha subunit 260 nm in length.

Beta spectrins exhibit functional differences in in vitro assays and have distinct patterns of expression and localization (Coleman et al 1989). Beta$_R$ and beta$_G$ spectrins have different affinities for ankyrins (Howe et al 1985; Davis & Bennett 1984) and for ankyrin-independent membrane sites (Steiner

& Bennett 1988), while beta$_{TW}$ spectrin lacks binding sites for protein 4.1 and ankyrin (Howe et al 1985; Coleman et al 1987). Several examples have been noted where two beta subunits are expressed in same cell, but are localized in specialized domains, for instance, avian intestinal epithelial cells have beta$_{TW}$ in their apical domains, while beta$_G$ is confined to the basolateral domains (Glenney et al 1983). Purkinje cells in the cerebellum (Lazarides & Nelson 1983; Riederer et al 1986) and neurons in the retina co-express beta$_R$ and beta$_G$ and exhibit differences in expression of these subunits in axons and cell bodies (Isayama et al 1991). A subset of neurons in forebrain expresses both beta$_R$ and beta$_G$ spectrins, although no obvious difference in localization could be detected (Lambert & Bennett 1993). Beta spectrins also exhibit different patterns of expression during brain development, with beta$_G$ appearing before birth, and beta$_R$ only after birth with maximal expression by day 10 (Riederer et al 1987; Hu et al 1992). Expression of a specialized beta subunit also occurs during differentiation of myoblasts (Nelson & Lazarides 1983). Cardiac muscle expresses beta$_G$ as well as several alternatively spliced forms of beta$_R$ spectrins, and these exhibit differences in localization (Vybiral et al 1992).

Spectrin Primary Structure and Domain Organization

Complete sequences have been determined for cDNAs encoding alpha subunits of spectrin from chicken brain (Wasenius et al 1989), human erythrocytes (Sahr et al 1990), human lung fibroblasts (Moon & McMahan 1990), and *Drosophila* (Dubreuil et al 1989). Alpha subunits of tissue spectrins are highly conserved with at least 90% identity between vertebrate species and 63% identity between *Drosophila* and chicken. The human erythrocyte alpha subunit, in contrast, has about 50–60% identity with the general human alpha subunit.

Alpha subunits contain 22 domains with the following features: domains 1–9 and 11–19 are comprised of 106-residue repeats; domain 10 is an SH3 (src homology domain 3) motif; domains 20 and 21 are related to 106-residue repeats with an 8 residue insertion; the C-terminal domain 22 includes two predicted EF-hand motifs and is likely to associate with the N-terminal domains of beta spectrins. Domain 11 of the generally expressed alpha subunit of vertebrates contains a 35-residue extension with the cleavage site for calcium-activated protease and a calmodulin-binding site (Harris et al 1998; Leto et al 1989). Diversity among vertebrate alpha spectrins results from several sites of alternate splicing. One site occurs between the SH3 motif and domain 11 and involves insertion of 60 bases; another results in an in-frame deletion of 18 bases within domain 21 near the C-terminus (Moon & McMahan 1990).

The SH3 motif of alpha spectrin domain 10 was first observed in the regulatory domains of several src-tyrosine kinases and the gamma isoform of phospholipase C (Koch et al 1991). Similar atomic structures have been resolved for SH3 domains derived from spectrin (Musacchio et al 1992) and src (Yu et al 1992), thereby indicating homology between these motifs at a conformational level. Other proteins associated with the plasma membrane cytoskeleton that also have similar SH3 motifs include a yeast actin-binding protein and myosin 1 of *Dictyostelium* (Drubin et al 1990), and the family of proteins related to p55 of erythrocyte membranes (Table 1; see below). Functions of spectrin SH3 domains may involve regulation or targeting of a class of yet to be characterized SH3 adaptor proteins to the plasma membrane. An interesting candidate for a SH3-binding protein has been identified that has homology with the GTPase activation domain of several proteins that regulate GTP-binding proteins (Cicchetti et al 1992).

Information regarding the primary structures of beta spectrins includes the complete sequences of the alternatively spliced 246 and 270-kd forms of human $beta_R$ spectrin (Winkelmann et al 1990, 1991), human (Hu et al 1992) and *Drosophila* (Byers et al 1992) $beta_G$ spectrins, and partial sequence for *Drosophila* $beta_H$ spectrin (Dubreuil et al 1990). $Beta_R$ and $beta_G$ spectrins contain 19 domains arranged in three distinct segments: the first segment contains a highly conserved N-terminal domain that associates with actin and neurofilaments (Frappier et al 1992; Karinch et al 1990) and with the C-terminal domains of alpha spectrins; segment two begins with an atypical domain related to the 106-residue repeats followed by 16 consecutive 106-residue repeat domains; segment three contains the C-terminal domain, which is the most variable region of beta spectrins, contains the site of alternative splicing in the case of $beta_R$ spectrins (Winkelman et al 1991), and is phosphorylated in 246-kd human erythrocyte $beta_R$ spectrin (Harris & Lux 1980). The ankyrin-binding site is located in domain 16 (repeat number 15) of $beta_R$ spectrin, which has a stretch of sequence that diverges from the 106-residue repeat motif (Kennedy et al 1991). $Beta_R$ and $beta_G$ spectrins are closely related in overall sequence with identical domain organization, and a sequence identity of 60%.

Head-head association between the N-terminus of alpha spectrin and C-terminal domain of beta spectrin subunits appears to result from pairing of incomplete but complementary 106-residue repeated motifs in each subunit to form a triple helical folding unit (Tse et al 1990; Speicher et al 1993). The partial repeat of the alpha subunit is located in the first 47 residues, while the repeat portion of the beta subunit is located internally in repeat number 17. C-terminal residues of beta spectrin extending beyond repeat 17 would be expected to be excluded from the triple helical domains and form a loop or globular domain near the midregion of spectrin tetramers.

Knobs have been oberved in the midregion of spectrin tetramers (Dubreuil et al 1990).

Spectrin Super-Family

The actin-binding domain of beta spectrin is a common feature of a group of proteins that cross-link actin filaments including alpha-actinin, dystrophin, ABP-280, ABP-120, and fimbrin (reviewed in Durbreuil 1991). Fimbrin contains two actin-binding domains in tandem without inserted sequence. ABP-120 and ABP-280 are elongated proteins with repeated elements, but these are not related to 106-residue repeats of spectrin. Alpha-actinins and dystrophins are more closely related to spectrin with additional shared features of 114–120-residue amino acid repeated sequences predicted to have alpha helical folding, as well as the EF hand domains of alpha spectrins. Repeats of alpha-actinin and dystrophin are most similar to the atypical repeats of beta spectrin (repeat 1) and alpha spectrin (domains 20 and 21) (Byers et al 1989).

Alpha-actinin and dystrophin polypeptides contain the C-terminal domain of alpha spectrin in addition to the N-terminal domain of beta spectrin. The fact that components of both alpha and beta spectrin are represented in the same polypeptides suggests that spectrins, alpha-actinin, and dystrophins evolved from a common progenitor containing both the N-terminal domain of beta spectrins and the C-terminal domain of alpha spectrins as well as a number of 106-residue repeats. Evolution of spectrin subunits from a common progenitor, possibly by unequal crossing over between homologous repetitive elements, has also been suggested based on comparison of 106-residue repeats of alpha and beta spectrin (Byers et al 1992).

Spectrin-Membrane Interactions

Spectrins are coupled to membrane-spanning proteins by both direct associations and through intermediary proteins such as ankyrins (see below). Ankyrin-independent binding of spectrin to high affinity protein sites has been measured in brain synaptosomal membranes (Steiner & Bennett 1988; Steiner et al 1989). Association of spectrin with these membrane sites was mediated by the beta subunit and was inhibited by calmodulin. One candidate for a spectrin-binding protein in brain is the cell adhesion molecule N-CAM180, which associated with spectrin in blot overlays (Pollerberg et al 1987). Interestingly, alternatively spliced forms of N-CAM missing the cytoplasmic domain did not associate with spectrin in these assays.

Spectrin in lymphocytes associates with CD45 (Lokeshwar & Bourguignon 1992; Bourguignon et al 1985), which is a member of a family of membrane-spanning glycoproteins with protein-tyrosine phosphatase activity. Spectrin binds to CD45 with a nanomolar affinity in an assay utilizing pure proteins and upregulates the tyrosine phosphatase activity of CD45 (Lokeshwar & Bourguignon 1992). The cGMP-gated cation channel of rod photoreceptor plasma membranes copurifies with a 240-kd polypeptide subsequently identified as spectrin based on immunoreactivity (Molday et al 1990). Direct linkage between spectrin and the cation channel is likely since antibodies against each protein co-immunoprecipitate the other from relatively pure preparations of the channel. The 240-kd polypeptide was localized to the inner surface of the rod-outer segment plasma membrane and excluded from the disc membranes, as would be expected for an association of this protein with the cGMP-gated cation channel in vivo.

Regulation of Spectrin Interactions

Spectrin exhibits dynamic behavior in cells that suggests mechanisms for regulated assembly and disassembly of spectrin-based structures. For example, gastric parietal cells redistribute the proton pump as well as spectrin to the apical domain following stimulation (Mercier et al 1989). Lymphocytes exhibit an activation-dependent redistribution of spectrin between a general plasma membrane pattern, a membrane-associated cap, and an intracellular aggregate (Gregorio et al 1992). During mitosis, spectrin is phosphorylated on the beta subunit and redistibutes from plasma membrane sites to the cytosol (Fowler & Adams 1992). Finally, epidermal growth factor stimulates phosphorylation of spectrin and redistribution of spectrin to membrane ruffles (Bretscher 1989).

Calcium acting through calmodulin (Steiner et al 1989) and calpain I (Harris et al 1988; Harris & Morrow 1990; Hu & Bennett 1991), and possibly association directly with alpha spectrin (Wallis et al 1992), abolishes several critical protein interactions of spectrin that could contribute to disassembly of spectrin-actin networks. In the presence of submicromolar Ca^{2+}, calmodulin is a competitive inhibitor of binding of spectrin to ankyrin-independent sites in brain membranes (Steiner et al 1989). Micromolar levels of calcium ion activate calpain I, which cleaves spectrin in the midregion of the alpha subunit close to the calmodulin-binding site (Harris at al 1988). Functional consequences of cleavage include a reduced affinity for brain membrane sites (Hu & Bennett 1991) and reduced ability to cross-link actin filaments (Harris & Morrow 1990). Higher levels of calcium result in cleavage of both alpha and beta subunits with loss of membrane binding and further reduction in ability to interact with actin. The combination of calmodulin, calpain, and calcium causes complete loss of actin-binding and dissociation of spectrin into dimers (Harris & Morrow 1990). Another level of regulation is provided by polyphosphoinositides, which cause activation of calpain and cleavage of spectrin at submicromolar concentrations of calcium (Saido et al 1992).

Evidence for a physiological role of calpain cleavage of spectrin in end-stage cells is based on observations of cleaved alpha spectrin following neutrophil degranulation (Jesaitis et al 1988), and platelet activation (Fox et al 1987). Spectrin degradation also accompanies neuronal degeneration (Seubert et al 1988a,b, 1990) and occurs in neonatal brain development (Najm et al 1991.

ANKYRINS

Ankyrins are a family of proteins initially discovered through the activity of erythrocyte ankyrin as the high affinity binding site for spectrin in erythrocyte membranes (reviewed in Bennett 1992). Ankyrins are expressed at especially high levels in vertebrate brain, where they comprises 0.5–1% of the total membrane protein. These proteins also are expressed in other cells and tissues including skeletal muscle, platelets, neutrophils, lymphocytes, cultured fibroblasts, and a variety of epithelial tissues (Bennett 1979; Nelson & Lazarides 1984; Drenckhahn & Bennett 1987). At least one ankyrin is expressed in *C. elegans* where a null mutation results in the unc44 phenotype with abnormal neuronal development (Otsuka et al 1992).

Ankyrins most likely function as adaptors between spectrin and the plasma membrane based on association of ankyrin with a diverse group of integral membrane proteins. Proteins currently known to associate with ankyrins include ion channels and cell adhesion molecules (Table 2). Ion channels that associate with ankyrin in in vitro assays and are co-localized with ankyrin in tissues include the anion exchanger in basolateral domains of

Table 2 Integral plasma membrane proteins linked to spectrin-based structures

Protein	Attachment site
Ionchannels	
Anion exchanger	Ankyrin
Na/K ATPase	Ankyrin
Voltage-sensitive Na channel	Ankyrin
Amiloride-sensitive Na channel	Ankyrin
cGMP-gated cation channel	Spectrin
Cell adhesion/signaling molecules	
CD45 (protein-tyrosine phosphatase)	Spectrin
N-CAM	Spectrin
CD44	Ankyrin
Nervous system CAMS related to L1/neurofascin	Ankyrin
E-cadherin	?

kidney collecting ducts, (Drenckhahn et al 1985; Schuster et al 1986), the Na^+/K^+ ATPase in basolateral domains of many epithelial tissues (Nelson & Veshnock 1987a,b; Koob et al 1990; Morrow et al 1989), and the voltage-dependent Na channel at nodes of Ranvier, axon initial segments, and the neuromuscular junction (Srinivasan et al 1988; Kordeli et al 1990; Kordeli & Bennett 1991; Flucher & Daniels 1989). In addition, the amiloride-sensitive Na channel in kidney, associates with ankyrin (Smith et al 1991) and is localized in apical membranes where one isoform of ankyrin also is present (Davis et al 1989a). Ankyrin-binding proteins involved in cell adhesion include lymphocyte adhesion antigen CD44 (Kalomiris & Bourguignon 1988, 1989) and ankyrin-binding glycoproteins that are members of the Ig-super family of nervous system cell adhesion molecules (Davis et al 1993).

Ankyrins isolated from mammalian erythrocytes and brain are monomers comprised of two highly conserved domains and a variable domain (Bennett & Stenbuck 1980; Davis & Bennett 1984; Lux et al 1990a,b; Lambert et al 1990; Otto et al 1991). The conserved domains are an N-terminal region of 89–95 kd, containing membrane-binding activity (Davis & Bennett 1984, 1990a) and a 62-kd spectrin-binding domain (Bennett 1978). C-terminal domains are the most variable and are sites of alternative splicing of pre-mRNA. The C-terminal domain of erythrocyte ankyrin has a regulatory role (Hall & Bennett 1987; Davis et al 1992). Ankyrins in solution exhibit moderate asymmetry reflected in a frictional ratio of about 1.5 (Bennett & Stenbuck 1980; Davis & Bennett 1984; Hall & Bennett 1987). The C-terminal domain provides the major contribution to the asymmetry of erythrocyte ankyrin, which becomes nearly globular following cleavage by calpain (Hall & Bennett 1987).

A striking feature shared by N-terminal domains of brain and erythrocyte ankyrins ($ankyrin_B$ and $ankyrin_R$, respectively) is a repeated 33-amino acid motif present in 24 contiguous copies (Lux et al 1990a; Lambert et al 1990; Otto et al 1991). Closely related repeats were originally noted in cell cycle control proteins CDC10, SWI6, and SWI4 of yeast and in proteins regulating tissue differentiation in *Drosophila* and *C. elegans* (Breeden & Nasmyth 1987). Proteins subsequently identified that contain 33-residue repeats include transcription factors (components of Nf-Kappa B, GABP-beta) and even a toxin from black widow spider venom (reviewed in Michael & Bennett 1992). The N-terminal domain of $ankyrin_R$ has a nearly spherical shape based on physical properties, which indicate that the repeats are not configured as an extended chain as is the case with 106-residue repeats of spectrin (see above). Six repeats represent the minimum folding unit, which suggests that the 89-kd domain with its 24 repeats is comprised of four subdomains (Michaely & Bennett 1992b).

The spectrin-binding domain of ankyrin$_R$ contains two subdomains: a N-terminal region of 80 residues that is acidic and enriched in proline residues, and a basic C-terminal portion of 488 residues. The acidic N-terminal portion is a likely candidate for contact with spectrin, since proteolytic fragments missing this region lose 95% of spectrin-binding activity (Davis & Bennett 1990a). The basic subdomain is highly conserved between ankyrin$_R$ and ankyrin$_B$, while the acidic region is more variable. The variable N-terminal subdomain may be responsible for ability of these ankyrins to preferentially associate with different spectrin isoforms.

The C-terminal domain of erythrocyte ankyrin modulates affinities of both the spectrin-binding and 89-kd domains (Hall & Bennett 1987). Much of the insight into activities of this domain is derived from study of an alternatively spliced variant (protein 2.2) with a 161-residue deletion (Lux et al 1990b). Protein 2.2 is an activated ankyrin with increased affinity for the cytoplasmic domain of the anion exchanger, spectrin, and tubulin compared to whole ankyrin (Davis et al 1992; Hall & Benett 1987). The alternatively spliced segment within the regulatory domain of erythrocyte ankyrin possibly has a repressor function and acts through an allosteric mechanism that involves interaction(s) at a site separate from the binding site for the anion exchanger (Davis et al 1992).

Ankyrin Genes and Patterns of Expression

Two ankyrins are currently characterized at the level of primary structure: ankyrin$_R$, which is expressed in erythrocytes and brain, and ankyrin$_B$, which is primarily expressed in brain. Ankyrin$_R$ is the product of the *ANK1* gene, located on human chromosome 8p11 (Lux et al 1990b), and ankyrin$_B$ is encoded by the *ANK2* gene, located on human chromosome 4 q25-q27 (Tse et al 1991).

Expression of ankyrin$_R$ in such different tissues as brain and erythrocytes is likely to involve distinct mechanisms for regulation of transcription and differences in exon usage. Different forms of ankyrin$_R$ are expressed in brain and erythrocytes because of alternative splicing of mRNA at sites located in the regulatory domain as well as the region between the N-terminal and spectrin-binding domains (Lambert & Bennett 1993). A physiological role for alternate forms of ankyrin$_R$ remains to be established.

A mutant mouse (*nb/nb*) is deficient in ankyrin$_R$ in erythrocytes as well as brain, and has provided a clear demonstration that the same gene is expressed in these different tissues (Peters et al 1991; Kordeli & Bennett 1991). Normoblastosis (*nb*) is a recessive mutation in a single gene linked to the erythrocyte ankyrin locus *Ank-1* on mouse chromosome 8 (Peters et al 1991). *Nb/nb* mice exhibit a 90% reduction in expression of ankyrin$_R$ in erythrocytes and brain with a phenotype of severe hemolytic anemia and progressive cerebellar ataxia. A subpopulation of Purkinje cell neurons in the cerebellum degenerates in these ankyrin-deficient animals within the first 3–4 months. An interesting question is the role of ankyrin in premature death of these neurons.

Ankyrin$_R$ is expressed late in the neonatal period of rat brain development, with about half of the synthesis occurring between day 15 and adult stages (Kunimoto et al 1991; Lambert & Bennett 1993). Ankyrin$_R$ is highly polarized in its cellular distribution and is confined to the plasma membrane in cell bodies and dendrites of neurons, but is missing from myelinated axons (Kordeli et al 1990; Kordeli & Bennett 1991; Lambert & Bennett 1993). Ankyrin$_R$ also is highly restricted in its expression and is limited to certain neurons: it is present in most motor neurons in the spinal cord and in both major types of neurons in the cerebellum, the Purkinje, and granule cells. However, in the forebrain ankyrin$_R$ is present in a minor subset of neurons whose function has not been identified (Lambert & Bennett 1993). The polarized distribution, the pattern of expression late in development, and restricted expression in a subset of neurons suggest that ankyrin$_R$ performs a specialized role in certain post-mitotic neurons.

Ankyrin$_B$ (220 kd) is the major isoform of ankyrin in adult brain tissue and has been localized by immunofluorescence to plasma membranes of neurons as well as glial cells throughout the brain (Kordeli & Bennett 1991). A 440-kd alternatively spliced variant of ankyrin$_B$ contains a sequence of a predicted size of 220 kd inserted between the regulatory domain and spectrin/membrane-binding domains (Otto et al 1991; Kunimoto et al 1991; W Chan et al, in preparation) (Figure 3). Properties of expressed portions of the inserted sequence suggest a configuration as an extended rod-shaped domain 200 nm in length (W Chan et al, in preparation). Thus 440-kd ankyrin$_B$ is predicted to have the shape of a ball, which comprises the membrane-binding and spectin-binding domains, and a chain comprised of the alternatively spliced sequence. Ankyrin$_B$ (440 kd) is expressed maximally in the neonatal period of rat development and is targeted to unmyelinated axons and possibly dendrites (Kunimoto et al 1991). Association of ankyrin$_B$ with a family of nervous system cell adhesion molecules (Davis et al 1993) suggests a possible function of ankyrin related to coupling of these membrane proteins to spectrin and the cytoskeleton.

Additional ankyrins are anticipated based on results with antibodies, but these proteins have yet to be cloned. A form of ankyrin distinct from ankyrin$_B$ is localized on the cytoplasmic surface of the axonal plasma membrane at nodes of Ranvier and axon initial segments (Kordeli et al 1990; Kordeli & Bennett 1991). A number of tissues express ankyrins

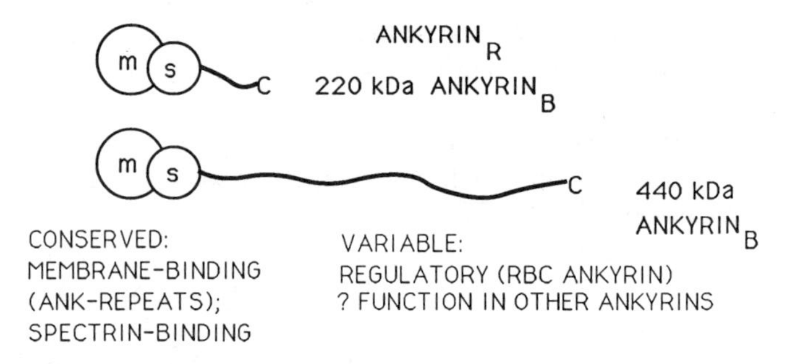

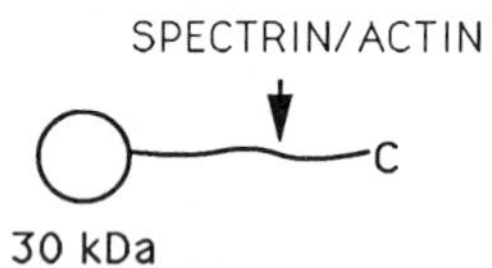

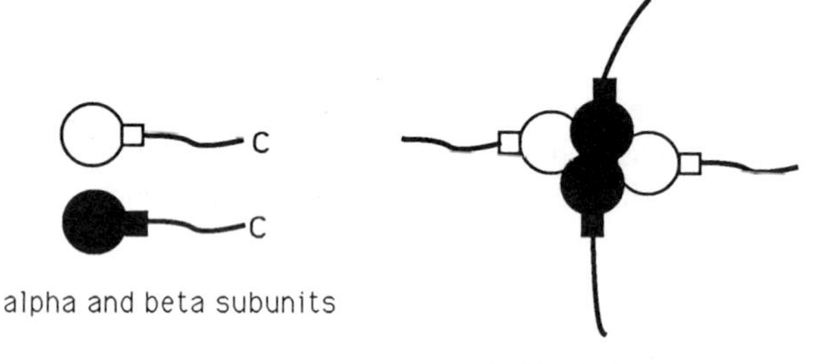

Figure 3 Schematic models for proteins associated with spectrin: ankyrins, protein 4.1, and adducin. The proposed configuration of 440-kd ankyrinB is based on analysis of expressed portions of the polypeptide chain (W Chan et al, in preparation), while that of adducin is based on studies with intact protein (Hughes et al 1992; Joshi & Bennett 1990; Joshi et al 1991). The shape of protein 4.1 is not based on physical measurements and should be considered speculative.

detected by radioimmunoassay and immunoblot techniques (Bennett 1979; Drenckhahn & Bennett 1987). These tissues may express a general form of ankyrin as a companion to the generally expressed alpha and beta spectrins. Alternatively, distinct ankyrin genes may be expressed in different cell types.

Recognition Between Ankyrin and Membrane Proteins

Thirty-three-residue repeats of ankyrin contain the binding site for the anion exchanger (Davis & Bennett 1990a; Davis et al 1991). The anion exchanger recognizes only certain repeats since constructs missing repeats 21 and 22 are inactive, and the comparable domain of ankyrin_B, which is comprised of closely related repeats, is about eightfold less active in binding to the anion echanger (Davis et al 1991). Other 33-amino acid repeats of ankyrin are likely to interact with proteins distinct from the anion exchanger. For example, ankyrin-binding proteins that are members of the Ig-superfamily of nervous system cell adhesion molecules associate with different repeats from the anion exchanger (P Michaely & V Bennett, unpublished data). Thirty-three-residue repeats of other proteins such as GABP beta, Nf-Kappa B precursor, Ikappa B, and alpha latrotoxin have been implicated in protein-protein interactions (reviewed Michaely & Bennett 1992a). Protein targets for repeat-bearing proteins are diverse in primary sequence, which suggests that 33-residue repeats in their folded conformation expose variable residues capable of interaction with a variety of protein ligands.

In addition to the potential for multiple interactions mediated by the N-terminal domain with its 33-residue repeats, ankyrin can also associate with membrane proteins through contacts with other domains. The Na^+/K^+ ATPase binds only weakly to the N-terminal domain and also associates with the spectrin-binding domain (Davis & Bennett 1990b). The fact that the anion exchanger and Na^+/K^+ ATPase associate with distinct sites on the ankyrin molecule suggests that these proteins independently evolved the capacity to interact with ankyrin. In support of this idea, primary sequences of the anion exchanger, and either alpha or beta subunits of the Na^+/K^+ ATPase are not obviously related. Moreover, the ankyrin-binding site of the anion exchanger requires tertiary structure involving at least 100 amino acids and contained in noncontiguous segments of the primary sequence (Davis et al 1989b; Willardson et al 1989), as would be expected if this protein evolved a conformation adapted to recognize ankyrin.

PROTEIN 4.2

Protein 4.2 is a N-myristylated polypeptide (Risinger et al 1992) that associates with the cytoplasmic domain of the erythrocyte anion exchanger as well as ankyrin (Korsgren & Cohen 1986, 1988; reviewed in Cohen et al 1993). The role of protein 4.2 in the membrane skeleton has not been defined, although deficiencies of protein 4.2 result in fragile erythrocytes and hemolytic anemia (Rybicki et al 1988; Ideguchi et al 1990). Protein

4.2 is expressed in brain, kidney, and platelets (Friedrichs et al 1988; White et al 1992). Protein 4.2 is homologous to transglutaminase enzymes in primary sequence and in genomic organization (Korsgren & Cohen 1991), although it lacks transglutaminase enzymatic activity.

Defects in protein 4.2 can result in tissue-specific diseases in addition to anemias. A platelet abnormality in secretory granules is linked to the protein 4.2 gene in mice with normal erythrocyte function (White et al 1992). The role of protein 4.2 in secretory vesicle assembly remains to be defined and may represent an activity distinct from the function of protein 4.2 in erythrocyte membranes.

PROTEIN 4.1

Protein 4.1 is associated with spectrin and actin in membrane skeletons and isolated spectrin-actin complexes in erythrocyte membranes (reviewed by Cohen & Gascard 1992; Conboy 1993). Protein 4.1 and a variety of splicoforms are expressed in many vertebrate tissues (Aster et al 1986; Anderson et al 1988). It is encoded by a gene located on human chromosome 1p33-p34.2 close to the Rh locus (Tang & Tang 1991). Protein 4.1 is a monomer in solution (Ohanian & Gratzer 1984) folded into four domains (Conboy et al 1986; Leto & Marchesi 1984): (*a*) a basic N-terminal domain of 30 kd that is homologous to domains of members of the protein 4.1 super-family (see below). This domain contains a calcium-dependent site for calmodulin (Kelly et al 1991), and is believed to mediate membrane associations of protein 4.1 (Leto & Marchesi 1984); (*b*) a 16-kd hydrophilic region that contains the site of phosphorylation by protein kinase C (Horne et al 1985); (*c*) a 10-kd domain that is phosphorylated by cyclic AMP-dependent protein kinase (Horne et al 1985) and contains a spectrin-binding region that promotes binding of spectrin to actin with nearly the same affinity as intact 4.1 (Correas et al 1986; Bennett & Davis 1982). A 21 residue cassette within the 10-kd domain has been identified as critical for interaction with spectrin and actin (Discher et al 1992); (*d*) a C-terminal acidic region of 12.6 kd (apparent M_r of 22 K on SDS-gels). Deamidation of Asn502 in this domain results in an increase in mobility of protein 4.1 and is a reaction that occurs slowly in circulating erythrocytes (Inaba et al 1992).

Multiple forms of protein 4.1 originate from a single 4.1 gene that produces multiple mRNAs by tissue-specific and developmentally regulated alternative splicing (Tang et al 1990; Conboy et al 1991; Chasis et al 1993). The multiple spliced forms of protein 4.1 differ in important functional domains. For example, a form of protein 4.1 missing the spectrin-actin binding domain is expressed in undifferentiated erythroid cells, and the spectrin-binding region is selectively inserted during differentiation of erythrocytes (Tang et al 1990; Conboy et al 1991). Another variant of protein 4.1 has an N-terminal addition of 209 residues resulting from an alternate start site for translation (Tang et al 1990). Protein 4.1 variants have been localized in nuclei (Correas 1991; Malek et al 1990), which suggests a new role(s) for this protein in addition to interaction with spectrin.

Protein 4.1-Spectrin-Actin Interactions

Protein 4.1 does not bind directly to actin, but does promote association of human erythrocyte spectrin with actin (Fowler & Taylor 1980; Ohanian et al 1984; Ungewickell et al 1979). Spectrin-actin complexes form prior to expression of competent protein 4.1 during maturation of erythrocytes (Tang et al 1990; Chasis et al 1993), which suggests that protein 4.1 acts to stabilize the spectrin-actin network after it has formed. Equilibrium measurements of spectrin-actin-protein 4.1 complex formation provide estimates of the standard free energy change of 17 kCal/mol (Ohanian et al 1984), thereby indicating a moderately stable ternary complex. Protein 4.1 also interacts directly with spectrin with a kd of $1–8 \times 10^{-7}$ M (Tyler et al 1980) at sites that have been localized in the N-terminal domain of beta spectrin based on photoaffinity-labeling (Becker et al 1990). A special role of the spectrin beta subunit is further suggested by studies with a spectrin isoform in avian intestinal brush borders, TW260/240, that has a variant beta subunit and does not exhibit 4.1-mediated association with actin (Coleman et al 1987).

Protein 4.1-Membrane Interactions

A major fraction of protein 4.1 remains associated with erythrocyte membranes following extraction of spectrin and actin. Membrane-associated protein 4.1 can promote spectrin-dependent interactions of actin with membranes (Cohen & Foley 1982). However, membrane-bound protein 4.1 cannot associate directly with spectrin in the absence of actin (Bennett & Stenbuck 1979; Cohen & Foley 1982). These observations suggest that the spectrin-binding domain of protein 4.1 is unavailable to spectrin, while protein 4.1 is associated with its membrane site(s). A ternary complex between spectrin, protein 4.1, and a membrane protein has not yet been demonstrated. It is possible that protein 4.1 fulfills distinct functions that involve association with membranes and stabilization of spectrin-actin complexes. It is noteworthy that the spectrin-actin binding domain of protein 4.1 alone is sufficient to restore mechanical stability to protein-4.1-deficient erythrocytes (Discher et al 1992).

Candidates for membrane-binding sites for protein 4.1 in erythrocytes include the anion exchanger (band 3) (Pasternack et al 1985) and glycophorin

C (Mueller & Morrison 1981; Sondag et al 1987). Evidence that glycophorin C interacts with protein 4.1 in vivo is based on the observation that erythrocytes deficient in protein 4.1 due to a defect in the gene for 4.1 have a 70% deficiency in glycophorin C, but normal amounts of glycophorin A and B. The association between isolated glycophorin C and protein 4.1 needs to be studied. Polyphosphoinositides have been implicated in interaction of protein 4.1 with glycophorins and could have a role in regulating these interactions (Anderson & Marchesi 1985).

The cytoplasmic domain of the anion exchanger (band 3) associates with protein 4.1 (Pasternack et al 1985) and represents the major class of membrane-binding sites for protein 4.1 in in vitro assays (Danilov et al 1990). The site of interaction between protein 4.1 and the anion exchanger has been localized to the N-terminal portion of the cytoplasmic domain, with secondary contacts at other sites (Lombardo et al 1992; Jons & Drenckhahn 1992). Protein 4.1 cannot form a ternary complex with spectrin and band 3, since the cytoplasmic domain of band 3 dissociates spectrin-protein 4.1 complexes (Pasternack et al 1985).

The relative affinity of protein 4.1 for its membrane sites is regulated by phosphorylation (Danilov et al 1990). Phosphorylation by protein kinase C, but not protein kinase A, preferentially abolishes binding to the anion exchanger without affecting association with other membrane sites (presumably due to glycophorin C).

Protein 4.1 Super-Family

A diverse group of proteins with the common feature of association with the plasma membrane share domains similar to the N-terminal 30-kd domain of protein 4.1. Current members of this super-family include ezrin (Gould et al 1989), radixin (Funayama et al 1991), moesin (Lankes & Furthmayr 1991), talin (Rees et al 1990), and a protein-tyrosine phosphatase (Gu et al 1991; Yang & Tonks 1991). Ezrin, radixin, and moesin are closely related with overall sequence identities in the range of 70%. These proteins contain, in addition to the 30-kd domain, a domain predicted to fold as an alpha helix, and a charged C-terminal domain. Ezrin was discovered based on high levels in microvilli and as a substrate for protein-tyrosine kinases (Bretscher 1989). Radixin was initially characterized as a component of adherens junctions and also is located at cleavage furrows of dividing cells (Funayama et al 1991).

The 30-kd domain of ezrin is sufficient to target ezrin constructs to the plasma membrane in cultured cells (Algrain et al 1993), and related domains may be involved in similar membrane targeting activies for other members of the family. Membrane site(s) interacting with 30-kd domains remain to be identified. C-terminal portions of these proteins contain a predicted alpha helical stretch and a highly charged region and are the most variable. In the case of ezrin, the C-terminal portion targets recombinant polypeptides to cytoskeletal structures in cultured cells (Algrain et al 1993).

ACCESSORY PROTEINS OF THE SPECTRIN-ACTIN-PROTEIN 4.1 JUNCTION

Beta-Actin

Evidence from cytochalasin binding and electron microscopy suggests that erythrocyte actin is organized into precisely defined short filaments (protofilaments) of about 30–40 nm (12–14 monomers) present in about 30,000 copies per cell (Byers et 1985; Brenner & Korn 1980; Pinder et al 1983). It is not yet understood how the length of these filaments is maintained, but it is likely that the specific isoforms of the proteins in the junctional complex (a term for the spectrin-actin complexes plus accessory proteins) have structural constraints that determine the configuration of the actin protofilament. Erythrocytes are unusual in expressing a single isoform of actin, beta-actin (Pinder et al 1978). This isoform may be uniquely suited for the formation of protofilaments, either based on its own secondary structure, or in combination with specific isoforms of actin-binding proteins. There is now evidence demonstrating membrane localization of beta-actin in non-erythroid cells in a region that is exclusive of F-actin (DeNofrio et al 1989; Hoock et al 1991). Other proteins that may contribute to the determination of protofilament length are tropomyosin and tropomodulin (see below).

Tropomyosin

Tropomyosin is a rod-like protein that self-associates in a head-to-tail fashion along the grooves of actin filaments in skeletal muscle. Tropomyosin has been purified from human erythrocyte membranes as a 60-kd dimer composed of two polypeptides of 29 and 27 kd, which resembles other tropomyosins in M_r and physical properties (Fowler & Bennett 1984). Tropomyosin is a good candidate for a structural role in stabilizing the short actin filaments since its length (33–34 nm) is close to that of 30-nm actin protofilaments, and it is present in sufficient amounts (70,000–80,000 copies per cell) to provide each short actin filament with two tropomyosin molecules lying along each groove of the actin helix.

Erythrocyte tropomyosin differs from muscle tropomyosin in that it exhibits only a weak head-to-tail association (Mak et al 1987). Peptide map analysis suggests that erythrocyte tropomyosin has a C-terminal domain similar to muscle isoforms, but an N-terminal domain similar to platelet tropomyosin (which also self-associates weakly). Erythrocyte tropomyosin may represent a hybrid isoform, with properties of both muscle and non-muscle isoforms. It will be of interest to determine if related tropomyosins are components of spectrin-based structures in other cells.

Tropomodulin

Tropomodulin is a 43-kd polypeptide identified in erythrocyte membranes that binds to tropomyosin (Fowler 1987; 1990). Immunoreactive forms of tropomodulin also are expressed in muscle, brain, lens, neutrophils, and endothelial cells (Fowler 1990). Purified tropomodulin binds to tropomyosin at a site located at one end of tropomysosin rods with an affinity of 0.5 μM (Fowler 1987; 1990). Its abundance in the erythrocyte (30,000 copies per cell) suggests association of one molecule per junctional complex. Tropomodulin is associated with the membrane skeleton after Triton extraction and removal of tropomyosin, which suggests additional protein interactions. Tropomodulin is an asymmetric monomer based on hydrodynamic properties (Fowler 1987). It is a non-competitive inhibitor of tropomyosin binding to F-actin (K_i of 0.7 μM), which suggests that tropomodulin and tropomyosin are not competing for the same binding site on F-actin (Fowler 1990). The affinity of the tropomodulin-tropomyosin complex for actin is reduced 20-fold compared to tropomyosin alone. However, if association of tropomodulin with tropomyosin greatly reduces affinity of tropomyosin for actin in vivo, it is not clear whether tropomyosin-tropomodulin complexes would be found on short actin filaments. Conditions on the erythrocyte membrane may differ from conditions of in vitro assays because of the different actin isoform and possibly other accessory proteins.

Tropomodulin is localized to the pointed (slow-polymerizing) ends of actin filaments in skeletal muscle (Fowler et al 1992) and therefore may be localized at the pointed ends of actin filaments in erythrocytes as well. Experiments with cytochalasin suggest that the slow-growing ends of the actin protofilaments are blocked (Pinder et al 1986). Tropomodulin, perhaps in concert with additional proteins, may cap the slowly polymerizing ends of actin filaments and thus contribute to stabilization of actin protofilaments in the erythrocyte skeleton.

The cDNA of human fetal liver tropomodulin encodes a polypeptide of 359 amino acids with no significant homology with known proteins (Sung et al 1992). The tropomyosin-binding domain has been localized to amino acids 39–138 by blot-binding of radiolabeled-tropomyosin to fusion proteins expressed in *E. coli*. Binding studies of erythroid and non-erythroid tropomyosins with erythrocyte tropomodulin demonstrate that erythrocyte tropomodulin preferentially binds erythrocyte tropomyosin compared to brain, platelet, and skeletal muscle tropomyosin (Sussman et al 1992). In addition, these binding assays combined with non-denaturing gels and oxidative cross-linking suggest that native tropomodulin is a dimer or tetramer.

Adducin

Adducin (adducere: to bring together), which contains two subunits with a M_r of 103 (alpha) and 97 kd (beta) (Gardner & Bennett 1986), was originally identified as a calmodulin-binding protein associated with the erythrocyte membrane skeleton. It appears to function as an assembly factor for construction of the spectrin-actin hexagonal array (Gardner & Bennett 1987). Adducin binds to spectrin-actin complexes with higher affinity than it binds to either spectrin or actin alone, and the formation of this ternary complex promotes binding of a second spectrin molecule to the complex. Recruitment of additional spectrin to spectrin-actin-adducin complexes is inhibited by calmodulin, which suggests a role for adducin in mediating Ca^{2+}-calmodulin dependent reorganization of the cytoskeleton. Adducin is expressed in early stages of erythropoiesis and switches from a Triton-soluble to a Triton-insoluble pool during erythrocyte maturation (Nehls et al 1991). This behavior is consistent with the idea that adducin promotes organization of the membrane lattice, which is then stabilized by protein 4.1 and ankyrin (Hanspal & Palek 1987). Adducin also bundles actin filaments at high concentrations (Mische et al 1987) although actin bundles have not been observed in erythrocytes. Immunoreactive forms of adducin are expressed in many cells and tissues including rat brain, kidney, liver, lung, and bovine lens (Bennett et al 1988). Adducin isolated from bovine brain has similar, but slightly different, properties compared to erythrocyte adducin (Bennett et al 1988).

Adducin, one of the major substrates for protein kinase C in erythrocytes, is also phosphorylated by protein kinase A (Ling et al 1986; Waseem et al 1988). Effects of phosphorylation on adducin have not yet been determined, but treatment of keratinocytes with phorbol ester results in phosphorylation of adducin and redistribution of adducin away from sites of cell-cell contact (Kaiser et al 1989).

Amino acid sequences of alpha and beta adducin are remarkably similar with 49% identity and 66% homology, which suggests evolution involving a gene duplication event (Joshi et al 1991). Adducin has a small region of similarity with the N-terminal actin-binding domains of beta spectrin and other actin filament-cross-linking proteins (see above). C-terminal ends of both alpha and beta adducin contain a highly basic segment similar to a domain in the MARCKS protein, which contains protein kinase C phosphorylation sites, exhibits calmodulin binding, and actin cross-linking activity (Blackshear 1993; Hartwig et al 1992). Calmodulin binding activity of adducin is located in the proposed neck region of the beta subunit (Scaramuzzino & Morrow 1991).

Adducin polypeptides are folded into three domains: a protease-resistant N-terminal globular head domain of about 350 residues, a second domain of undefined shape of about 100 residues, and a C-terminal tail domain configured as a random coil based on circular dichroism spectra (Joshi & Bennett 1990; Hughes et al 1992). Adducin was initially determined to be a dimer (Gardner & Bennett 1986), but more precise measurements suggest a dimer-tetramer equilibrium with tetramers representing the predominate form (Hughes et al 1992). The number of adducin molecules per erythrocyte therefore is between 15,000 and 30,000, depending on the ratio of dimers to tetramers. Adducin tetramers have been proposed to result from association of subunits through the globular head domains (Hughes et al 1992). The head domain is unable to cross-link spectrin and actin; these complex interactions may require contributions from head and tail domains simultaneously.

A flexible tail potentially could assist the function of adducin in simultaneous association with two large molecules such as spectrin and actin filaments. Separate binding sites of low affinity for spectrin and actin (one in the head and one in the tail) would result in the observed weak association with spectrin or actin alone (Gardner & Bennett 1987). However, participation of adducin in a ternary interaction with a previously formed spectrin-actin complex would be greatly favored because of the increase in effective concentration of binding sites. In addition, the tetrameric structure of adducin may provide a binding site for a second spectrin molecule, which is consistent with the observed recruitment of additional spectrin to the spectrin-actin-adducin complex (Gardner & Bennett 1987).

The gene encoding alpha adducin is located on human chromosome 4 p16.3, along with five other genes within the Huntington's disease locus (Taylor et al 1992), and the gene for beta adducin is on human chromosome 2 (D Gilligan et al, in preparation). Northern blots demonstrate an alpha adducin mRNA of 4 kb in rat brain, spleen, kidney, liver, human reticulocytes, and K562 cells (Joshi et al 1991). In contrast, beta adducin mRNAs show tissue-specific variability in size and abundance. Most striking is the relative amount of the two mRNAs in reticulocytes: alpha adducin mRNA is abundant, while beta adducin mRNA is rare. This suggests that synthesis of the beta subunit may be the limiting factor in assembly of adducin heteromers.

Alpha and beta adducin pre-mRNAs are subject to alternative splicing (Taylor et al 1992; D Gilligan et al, in preparation). Three alternatively spliced cDNAs have been identified that encode isoforms of beta adducin (D Gilligan et al, in preparation), and two forms of alpha adducin have been characterized (Taylor et al 1992). Beta adducin variants are designated in order of size of the predicted protein product as beta-1 (80 kd) (Joshi et al 1990), beta-2 (63 kd) (Tripodi et al 1991), and beta-3 (26 kd) (D Gilligan et al, in preparation). These three isoforms have identical N-terminal sequences, but different C-terminal regions. The highly basic C-terminus of beta-1 is deleted in both beta-2 and beta-3. The calmodulin binding site (Scaramuzzino & Morrow 1991) is present in beta-1 and beta-2, but deleted in beta-3. The beta-3 polypeptide is expressed in rat brain, heart and kidney, but not in erythrocyte ghosts or K562 cells (D Gilligan et al, in preparation). Multiple alpha and beta adducins suggest the potential for a variety of combinatorial possibilities for the assembled proteins.

Adducin is localized at lateral cell borders of epithelial cells including MDCK cells, keratinocytes, and intestinal epithelium (Kaiser et al 1989). Since adducin is one of the first proteins recruited to sites of calcium-dependent cell-cell contact in human keratinocytes, this suggests that it may play an important role in the organization of specialized cell junctions. Adducin is also abundant in all blood cells (which do not form junctions) and therefore is likely to have other roles such as local reorganization of the membrane skeleton for phagocytosis, cell-cell interactions, or motility. Alpha adducin has been identified without beta-1 adducin in fibroblasts (Waseem & Palfrey 1990) and may function independently in these cells or may associate with other beta subunits not recognized by the antibody that recognizes beta-1 adducin.

Adducin immunoreactivity has also been demonstrated in mouse oocytes and is localized by immunofluorescence to chromosomes in metaphase I and metaphase II oocytes (Pinto-Correia et al 1991). Following fertilization, maternal chromosomes lose their staining for adducin, and the fertilizing sperm chromatin does not stain for adducin. This unusual localization of adducin immunoreactivity suggests that adducin may have additional functions related to organization of metaphase chromosomes.

Band 4.9 (Dematin and p55)

The band 4.9 region on SDS-gels of human erythrocyte ghosts contains a major substrate for phosphorylation by protein kinase A and protein kinase C (Cohen et al 1993; Horne et al 1985; Palfrey & Waseem 1985). Initial studies on partially purified band 4.9 from human erythrocyte membranes (Siegel & Branton 1985) characterized an actin-bundling protein with apparent molecular weight of 145,000 and a tripartite globular structure with a radius of 5 nm. The relevance of actin bundling in adult erythrocytes is not known, and understanding the function of protein 4.9 in the junctional complex may require analysis of ternary molecular interactions. The name dematin (from the Greek dema, a bundle) has been proposed for the actin-bundling protein, consisting of 48- and 52-kd polypeptides (Husain-Chishti et al 1989). There are 43,000 trimers of dematin present per erythrocyte (Husain-Chisti et al 1988), which is similar to the number of short actin filaments. Actin-bundling activity of 48- and 52-kd polypeptides was inhibited by phosphorylation with protein kinase A and restored by dephosphorylation.

Immunoreactive polypeptides detected with antibody against human dematin are in the 47 to 52 kd range in bovine and avian brain and lens, and 50 and 54 kd in human platelets (Faquin et al 1988). Immunofluorescence showed crossreactivity in cortical fiber cells of avian lens and in neurons of avian brain. An antigenically related polypeptide, RET52, in mouse retina is a Triton-insoluble component of the rod inner segment and the outer synaptic layers (Roof et al 1991). Expression of RET52 in the synaptic region correlated with the formation of initial synaptic contacts and suggests a possible role for RET52 in synapse formation during development.

Expression of dematin during avian erythropoiesis includes five immunoreactive variants of 44, 47, 49, 50, and 52 kd expressed at different times during development (Faquin et al 1990). Forms synthesized at early times are degraded, whereas those synthesized at later times are not. Synthesis of the stable form of dematin late in erythropoiesis may be related to the stable assembly of spectrin, ankyrin, protein 4.1, and the anion transporter late in erythropoiesis (Cox et al 1987). Dematin may therefore play a role in stabilizing the membrane skeleton in differentiating erythrocytes.

A 55-kd polypeptide co-purified with dematin has subsequently been identified as a major palmitoylated protein in erythrocyte membranes (Maretzki et al 1990). The amino acid sequence of the human 55-kd polypeptide (Ruff et al 1991) reveals an SH3 motif (see above) and a guanylate kinase domain, which also are present in the discs-large tumor suppressor gene product located in septate junctions in *Drosophila* (Woods & Bryant 1991), and PSD-95, a protein associated with post-synaptic densities in vertebrate brain (Cho et al 1992). Post-synaptic densities and the septate junctions involved in the discs-large tumor suppressor mutation are special types of cell-cell junctions. Protein interactions of the conserved domains of p55 may have related roles in erythrocyte membranes and in these specialized membrane domains involved in cell-cell junctions.

FUNCTIONS OF SPECTRIN AND ASSOCIATED PROTEINS

Observations based on consequences of mutations have established that membrane skeletal proteins are important for cell viability and function and that alterations in proteins expressed in multiple cell types can result in tissue-specific defects. Hereditary hemolytic anemias in humans and mice have provided many examples of mutations involving spectrin, ankyrin, protein 4.1, and protein 4.2 that result in fragile or abnormally shaped erythrocytes (see above). *Nb/nb* mice have low expression of the ANK1 gene product in brain as well as in erythrocytes and exhibit degeneration of a population of Purkinje cells beginning a few months after birth (Peters et al 1991). In the nematode *C. elegans,* the gene associated with the *unc44* gene encodes an ankyrin with properties similar to $ankyrin_B$ (Otsuka et al 1992). *Unc44* mutants are uncoordinated and exhibit abnormal axon morphology for many classes of neurons. Mutations in the *hu-li tai shao* (*hts, too little nursing*) gene encoding adducin in *Drosophila* result in abnormal oogenesis (Yue & Spradling 1992). The Hts phenotype involves abnormalities in modified cleavage furrows known as ring canals and is consistent with a role of adducin in assembly of actin at these sites. Polymorphisms in adducin also have been linked to high blood pressure in the Milan strain of hypertensive rats (Tripodi et al 1991). Protein 4.2 is the target of the *Pallid* mutation in mice, which affects the ability of platelets to form functional secretory granules (White et al 1992). Protein 4.2 expressed in brain and erythroyctes is normal in these animals. Finally null mutations in the $beta_H$ spectrin gene in *Drosophila* have reduced viability, defective eye development, and leak hemolymph (Thomas & Kiehart 1992).

Targeted disruption of genes encoding membrane skeletal proteins in transgenic animals remains to be accomplished. However, expression of antisense RNA has demonstrated an important role for protein 4.1 in *Xenopus* development and a specific role in formation of the retina (Giebelhaus et al 1988).

Studies of localization and in vitro interactions suggest one function of spectrin-based structures involves organization of specialized domains of plasma membranes. Examples of domains enriched in isoforms of spectrin and ankyrin, as well as ankyrin-binding proteins, include basolateral domains of epithelial tissues (Morrow et al 1989; Drenckhahn et al 1985; Drenckhahn & Bennett 1987), neuromuscular junctions (Bloch & Morrow 1989; Flucher & Daniels 1989), and nodes of Ranvier and axon initial segments (Takeichi 1991; Kordeli et al 1990; Kordeli & Bennett 1991).

Sites of cell-cell contact in epithelial tissues are the best characterized example of a role of membrane skeletal proteins in a membrane domain (reviewed in Nelson 1992; Nelson et al 1991). Initiation of cell-cell contact by addition of calcium to cultured epithelial cells results in assembly of spectrin, ankyrin, and adducin in a detergent-resistant structure on the basolateral domains (Nelson et al 1987a; Kaiser et al 1989; Hammerton et al 1991). The Na^+/K^+ ATPase, which associates with ankyrin in in vitro assays (Nelson & Veshnock 1987b) also relocates to the same site as membrane skeletal proteins and is colocalized with ankyrin in basolateral domains of most epithelial cells (Koob et al 1990; Morrow et al 1989). These proteins also are colocalized in apical domains of the plasmalemma of retinal epithelial cells, which have a reversed polarity (Gundersen et al 1991).

A major question yet to be addressed is the role of spectrin-based structures in either targeting or possibly stabilization of the Na^+/K^+ ATPase. In cultured kidney cells, the Na^+/K^+ ATPase initially is randomly inserted on all membrane surfaces and targeting occurs by retention in basolateral domains (Hammerton et al 1991). This observation suggests spectrin-ankyrin complexes are not involved in initial targeting, although they may play a "fly-paper" role and adhere to membrane proteins once they arrive. An unresolved issue is how spectrin and ankyrin are themselves targeted to basolateral domains.

An early event in assembly of cell-cell junctions may be formation of cell-cell contacts mediated by interactions between external domains of the cadherin family of cell adhesion molecules (reviewed Takeichi 1991). Several observations suggest a linkage between cadherins and the spectrin system. Membrane skeletal proteins colocalize with E-cadherin in cultured epithelial cells and exhibit a cadherin-dependent assembly at sites of cell-cell contact (McNeil et al 1990). Moreover, a complex between E-cadherin, spectrin, and ankyrin can be isolated by immunoprecipitation with antibody against E-cadherin (Nelson et al 1990). Association of E-cadherin with either spectrin or ankyrin has not been demonstrated using purified proteins, and complexes of these proteins isolated from cell extracts may contain other molecules. Alpha, beta, and gamma catenins associate with cytoplasmic domains of cadherins (Takeichi 1991) and would be logical candidates to mediate connections with the spectrin skeleton.

Localization of ankyrin at nodes of Ranvier in the nervous system, basolateral domains of epithelial tissues, and basal infoldings of kidney distal tubule cells suggests some conserved structural features of these apparently unrelated cell domains. An interesting possibility is that ankyrin together with spectrin and associated proteins form the framework of a basic structure shared by many specialized plasma membrane domains. Specialized features that distinguish domains in different tissues would be provided by cell-specific expression of particular membrane proteins and isoforms of the structural proteins. Incorporation of specialized components into a general structural framework would not necessarily require high affinity protein interactions. The spectrin skeleton of erythrocytes, for example, restricts translational mobility of all of the anion exchangers even though only a fraction is directly coupled to the skeleton (see above).

CONCLUSIONS AND FUTURE PROSPECTS

Spectrin and a group of associated proteins defined by their roles in erythrocyte membrane skeletons are components of plasma membranes in metazoan organisms ranging from *C. elegans* to humans. Integral membrane proteins that interact with spectrin and ankyrin include a number of ion channels and cell adhesion molecules. An inferred function of spectrin and associated proteins, based on their localization in tissues and association with proteins in in vitro assays, is in assembly and/or maintenance of specialized domains on the cell surface. Recent progress in determination of primary structures for most of these proteins provides insight into protein domains and recognition sites and documents tissue-specific and developmental diversity through alternative splicing of pre-mRNAs and regulated expression of multiple related genes. In addition, several super-families have been defined, including actin filament cross-linking proteins related to spectrin, and the proteins (ezrin/moesin/radixin, protein tyrosine phosphatase, and talin) with a 30-kd domain related to protein 4.1. Defects or deficiencies of spectrin and associated proteins result in cell-type and tissue-specific abnormalities. Examples include defective oogenesis resulting from altered adducin in *Drosophila*, neuronal degeneration in ankyrin-deficient mice, defective secretory granules in mouse platelets with abnormal protein 4.2, as well as a number of hereditary hemolytic anemias in humans and mice involving spectrin, protein 4.1, ankyrin, and protein 4.2. Challenges for future work include understanding the mechanisms of targeting of membrane skeletal proteins to regions of the plasma membrane, and the general issue of resolving roles of membrane skeletal proteins and their protein interactions in cells organized into tissues.

ACKNOWLEDGMENTS

Brenda Sampson is gratefully acknowledged for help in preparation of the manuscript. Research from this laboratory has been supported in part by grants from the National Institutes of Health to VB and DMG.

Literature Cited

Agre, P, Asimos, A, Casella, J, McMillan, W. 1986. Inheritance pattern and clinical response to splenectomy as a reflection of erythrocyte spectrin deficiency in hereditary spherocytosis. *N. Engl. J. Med.* 315:1579–83

Agre, P, Casella, J, Zinkham, W, McMillan, C, Bennett, V. 1985. Partial deficiency of erythrocyte spectrin in hereditary spherocytosis. *Nature* 314:380–83

Algrain, M, Turunen, O, Vaheri, A, Louvard, P, Arpin, M. 1993. Ezrin contains cytoskeleton and membrane-binding domains accounting for its proposed role as a membrane-cytoskeletal linker. *J. Cell Biol.* 120: 129–39

Anderson, RA, Correas, I, Mazzucco, C, Castle, JD, Marchesi, VT. 1988. Tissue-specific analogues of erythrocyte protein 4.1 retain functional domains. *J. Cell. Biochem.* 37:269–84

Anderson, R, Marchesi, V. 1985. Regulation of the association of membrane skeletal protein 4.1 with glycophorin by a polyphosphoinositide. *Nature* 318:292–98

Aster, J, Brewer, G, Maisel, H. 1986. The 4.1-like proteins of the bovine lens: spectrin-binding proteins closely related in structure to red cell protein 4.1. *J. Cell Biol.* 103:115–22

Beaven, GH, Jean-Baptiste, L, Ungewickell, E, Baines, AJ, Shahbakhiti, F, et al. 1985. An examination of the soluble oligomeric complexes extracted from the red cell membrane and their relation to the membrane skeleton. *Eur. J. Cell Biol.* 36:299–306

Becker, P, Schwartz, M, Morrow, J, Lux, S. 1990. Radiolabel-transfer cross-linking demonstrates that protein 4.1 binds to the N-terminal region of beta spectrin and to actin in binding interactions. *Eur. J. Biochem.* 193:827–36

Bennett, V. 1978. Purification of an active proteolytic fragment of the membrane attachment site for human erythrocyte spectrin. *J. Biol. Chem.* 253:2292–99

Bennett, V. 1979. Immunoreactive forms of human erythrocyte ankyrin are present in diverse cells and tissues. *Nature* 281:597–99

Bennett, V. 1990. Spectrin-based membrane skeleton: a multipotential adaptor between plasma membrane and cytoplasm. *Phys. Rev.* 70:1029–65

Bennett, V. 1992. Ankyrins—Adaptors between diverse plasma membrane proteins and the cytoplasm. *J. Biol. Chem.* 267: 8703–6

Bennett, H, Condeelis, J. 1988. Isolation of an immunoreactive analogue of brain fodrin that is associated with the cell cortex of *Dictyostelium* amoebae. *Cell Motil. Cytoskeleton* 11:303–17

Bennett, V, Davis, J. 1982. Immunoreactive forms of human erythrocyte ankyrin are localized in mitotic structures in cultured cells and associated with microtubules in brain. *Cold Spring Harbor Symp. Quant. Biol.* 46:647–57

Bennett, V, Gardner, K, Steiner, J. 1988. Brain adducin: A protein kinase C substrate that may mediate site-directed assembly at the spectrin-actin junction. *J. Biol. Chem.* 263:5860–69

Bennett, V, Stenbuck, PJ. 1979. Identification and partial purification of ankyrin, the high affinity membrane attachment site for human erythrocyte spectrin. *J. Biol. Chem.* 254:2533–41

Bennett, V, Stenbuck, PJ. 1980. Human erythrocyte ankyrin. Purification and properties. *J. Biol. Chem.* 255:2540–48

Blackshear, P. 1993. The MARCKS family of cellular protein kinase C substrates. *J. Biol. Chem.* 268:1501–4

Bloch, RJ, Morrow, JS. 1989. An unusual beta-spectrin associated with clustered acetylcholine receptors. *J. Cell Biol.* 108:481–93

Bloom, ML, Lee, BK, Birkenmeier, CS, Ma, Y, Zimmer, WE, et al. 1992. Brain beta spectrin isoform 235 (Spnb-2) maps to mouse chromosome 11. *Mamm. Genet.* 3: 293–95

Bourguignon, LY, Suchard, SJ, Nagpal, ML, Glenney, JR. 1985. A T-lymphoma transmembrane glycoprotein (gp 180) is linked to the cytoskeletal protein, fodrin. *J. Cell Biol.* 101:477–87

Breeden, L, Nasmyth, K. 1987. Similarity between cell-cycle genes of budding yeast and the *Notch* gene of *Drosphila*. *Nature* 329:651–54

Brenner, SL, Korn, ED. 1980. Spectrin/actin complex isolated from sheep erythrocytes accelerates actin polymerization by simple nucleation. Evidence for oligomeric actin in the erythrocyte cytoskeleton. *J. Biol. Chem.* 255:1670–76

Bretscher, A. 1989. Rapid phosphorylation and reorganization of ezrin and spectrin accompany morphological changes induced in A-431 cells by epidermal growth factor. *J. Cell Biol.* 108:921–30

Budzynski, DM, Benight, AS, LaBrake, CC, Fung, LW. 1992. Dynamic light scattering investigations of human erythrocyte spectrin. *Biochemistry* 31:3653–60

Byers, TJ, Brandin, E, Lue, R, Winograd, E, Branton, D. 1992. The complete sequence of *Drosophila* beta spectrin reveals supramotifs comprising eight 106-residue repeats. *Proc. Natl. Acad. Sci. USA* 89:6187–91

Byers, TJ, Branton, D. 1985. Visualization of the protein associations in the erythrocyte membrane skeleton. *Proc. Natl. Acad. Sci. USA* 82:6153–57

Byers, TJ, Husain-Chishti, A, Dubreuil, RR, Branton, D, Goldstein, LSB. 1989. Sequence similarity of the amino-terminal domain of *Drosophila* beta spectrin to alpha actinin and dystrophin. *J. Cell Biol.* 109: 1633–41

Calvert, R, Bennett, P, Gratzer, W. 1980. Properties and structural role of the subunits of human spectrin. *Eur. J. Biochem.* 107: 355–61

Chasis, JA, Conlombel, L, Conboy, J, McGee, S, Andrews, K, et al. 1993. Differentiation associated switches in protein 4.1 expression. Synthesis of multiple structural isoforms during normal human erythropoiesis. *J. Clin. Invest.* 91:329–38

Cho, K, Hunt, C, Kennedy, M. 1992. The rat brain post-synaptic density fraction contains a homolog of the *Drosophila* discs-

large tumor suppressor protein. *Neuron* 9: 929–42
Cicchetti, P, Mayes, B, Theil, G, Baltimore, D. 1992. Identification of a protein that binds to the SH3 region of Abl and is similar to Bcr and GAP-rho. *Science* 257: 803–6
Clague, MJ, Harrison, JP, Morrison, IE, Wyatt, K, Cherry, RJ. 1990. Transient dichroism studies of spectrin rotational diffusion in solution and bound to erythrocyte membranes. *Biochemistry* 29:3898–904
Cohen, CM, Dotimas, E, Korsgren, C. 1993. Human erythrocyte membrane band 4.2 (pallidin). *Sem. Hematol.* 3012:119–37
Cohen, CM, Foley, SF. 1982. The role of band 4.1 in the association of actin with erythrocyte membranes. *Biochim. Biophys. Acta* 688:691–701
Cohen, CM, Gascard, P. 1992. Regulation and post-translational modification of erythocyte membrane and membrane skeletal proteins. *Sem. Hematol.* 29:244–92
Coleman, TR, Fishkind, DJ, Mooseker, MS, Morrow, JS. 1989. Contributions of the beta-subunit to spectrin structure and function. *Cell Motil. Cytoskeleton* 12:248–68
Coleman, T, Harris, A, Mische, S, Mooseker, M, Morrow, J. 1987. Beta spectrin bestows protein 4.1 sensitivity on spectrin-actin interactions. *J. Cell Biol.* 104:519–26
Conboy, JG. 1993. Stucture, function, and molecular genetics of erythroid membrane skeletal protein 4.1 in normal and abnormal red blood cells. *Sem. Hematol.* 30:58–73
Conboy, J, Chan, J, Chasis, J, Kan, Y, Mohandas, N. 1991. Tissue and development specific alternative RNA splicing regulates expression of multiple isoforms of erythroid membrane protein 4.1. *J. Biol. Chem.* 266:8273–80
Conboy, J, Kan, Y, Shohet, S, Mohandas, N. 1986. Molecular cloning of protein 4.1, a major structural element of the human erythrocyte membrane skeleton. *Proc. Natl. Acad. Sci. USA* 83:9512–16
Correas, I. 1991. Characterization of isoforms of protein 4.1 present in the nucleus. *Biochem. J.* 279:581–85
Correas, I, Leto, T, Speicher, D, Marchesi, V. 1986. Identification of the functional site of erythrocyte protein 4.1 involved in spectrin-actin associations. *J. Biol. Chem.* 261:3310–15
Cox, JV, Stack, JH, Lazarides, E. 1987. Erythroid anion transporter assembly is mediated by a developmentally regulated recruitment onto a preassembled membrane cytoskeleton. *J. Cell Biol.* 105:1450–56
Danilov, Y, Fennell, R, Ling, E, Cohen, C. 1990. Selective modulation of band 4.1 binding to erythrocyte membranes by protein kinase C. *J. Biol. Chem.* 265:2556–62
Davis, J, Bennett, V. 1984. Brain ankyrin—a membrane associated protein with binding sites for spectrin, tubulin and the cytoplasmic domain of the erythrocyte anion channel. *J. Biol. Chem.* 259:13550–59
Davis, J, Bennett, V. 1990a. Mapping the binding sites of human erythrocyte ankyrin for the anion exchanger and spectrin. *J. Biol. Chem.* 265:10589–96
Davis, L, Bennett, V. 1990b. The anion exchanger and Na^+K^+ATPase interact with distinct sites on ankyrin. *J. Biol. Chem.* 265:17252–56
Davis, J, Davis, L, Bennett, V. 1989a. Diversity in membrane binding sites of ankyrins: brain ankyrin, erythrocyte ankyrin, and processed erythrocyte ankyrin associate with distinct sites in kidney microsomes. *J. Biol. Chem.* 264:6417–26
Davis, J, McLaughlin, T, Bennett, V. 1993. Ankyrin-binding proteins related to nervous system cell adhesion molecules. Candidates to provide transmembrane and intercellular connections in adult brain. *J. Cell Biol.* 121:121–33
Davis, LH, Davis, JQ, Bennett, V. 1992. Ankyrin regulation: an alternatively spliced segment of the regulatory domain functions as an intramolecular modulator. *J. Biol. Chem.* 267:18966–72
Davis, L, Lux, SE, Bennett, V. 1989b. Mapping the ankyrin-binding site of the human erythrocyte anion exchanger. *J. Biol. Chem.* 264:9665–72
Davis, L, Otto, E, Bennett, V. 1991. Specific 33-residue repeat(s) of erythrocyte ankyrin associate with the anion exchanger. *J. Biol. Chem.* 266:11163–69
Davison, M, Baron, M, Wooton, J, Critchley, D. 1989. Structural analysis of homologous repeated domains in alpha-actinin and spectrin. *Int. J. Biol. Macromol.* 11:81–90
Delaunay, J, Dhermy, D. 1993. Mutations involving the spectrin heterodimer contact site: clinical expression and alterations in specific function. *Sem. Hematol.* 30:21–33
DeNofrio, D, Hoock, TC, Herman, IM. 1989. Functional sorting of actin isoforms in microvascular pericytes. *J. Cell Biol.* 109: 191–202
Derick, LH, Liu, S, Chishti, Athar H, Palek, J. 1992. Protein immunolocalization in the spread erythrocyte membrane skeleton. *Eur. J. Cell Biol.* 57:317–20
Discher, D, Parra, M, Conboy, J, Mohandas, N. 1992. Protein 4.1's alternatively-spliced spectrin-actin binding domain. *Mol. Biol. Cell* 3:269a
Drenckhahn, D, Bennett, V. 1987. Polarized distribution of M_r 210,000 and 190,000 analogs of erythrocyte ankyrin along the plasma membrane of transporting epithelia,

neurons and photoreceptors. *Eur. J. Cell Biol.* 43:479–86

Drenckhahn, D, Schluter, K, Allen, D, Bennett, V. 1985. Colocalization of band 3 with ankyrin and spectrin at the basal membrane of intercalated cells in the rat kidney. *Science* 230:1287–89

Drubin, D, Mulholland, J, Zhu, Z, Botstein, D. 1990. Homology of a yeast actin-binding protein to signal transduction proteins and myosin-1. *Nature* 343:288–90

Dubreuil, R. 1991. Structure and evolution of actin crosslinking proteins. *BioEssays* 13: 219–25

Dubreuil, RR, Byers, TJ, Sillman, AL, Bar-Zvi, D., Goldstein, LSB, Branton, D. 1989. The complete sequence of *Drosophila* alpha-spectrin: conservation of structural domains between alpha-spectrins and alpha-actinin. *J. Cell Biol.* 109:2197–205

Dubreuil, R, Byers, T, Stewart, C, Kiehart, D. 1990. A beta-spectrin isoform from *Drosophila* is similar in size to vertebrate dystrophin. *J. Cell Biol.* 111:1849–58

Elgsaeter, A. 1978. Human spectrin. I. A classcal light scattering study. *Biochim. Biophys. Acta* 536:235–44

Evans, EA, Hochmuth, RM. 1977. A solid-liquid composite model of the red cell membrane. *J. Membr. Biol.* 30:351–62

Faquin, WC, Husain-Chishti, A, Branton, D. 1990. Expression of dematin (protein 4.9) during avian erythropoiesis. *Eur. J. Cell Biol.* 53:48–58

Faquin, WC, Husain, A, Hung, J, Branton, D. 1988. An immunoreactive form of erythrocyte protein 4.9 is present in non-erythroid cells. *Eur. J. Cell Biol.* 46:168–75

Flucher, B, Daniels, M. 1989. Membrane proteins in the neuromuscular junction. Distribution of Na channels and ankyrin is complementary to acetylcholine receptors and the 43-kd protein. *Neuron* 3:163–75

Fowler, VM. 1987. Identification and purification of a novel M_r 43,000 tropomyosin binding protein from human erythrocyte membranes. *J. Biol. Chem.* 262:12792–800

Fowler, VM. 1990. Tropomodulin: A cytoskeletal protein that binds to the end of erythrocyte tropomyosin and inhibits tropomyosin binding to actin. *J. Cell Biol.* 111:471–82

Fowler, VM, Adams, EJH. 1992. Spectrin redistributes to the cytosol and is phosphorylated during mitosis in cultured cells. *J. Cell Biol.* 119:1559–72

Fowler, V, Bennett, V. 1978. Association of spectrin with its membrane attachment site restricts lateral mobility of human erythrocyte integral membrane proteins. *J. Supramolec. Struct.* 8:215–21

Fowler, V, Bennett, V. 1984. Erythrocyte membrane tropomyosin: Purification and properties. *J. Biol. Chem.* 259:5978–89

Fowler, VM, Sussman, MA, Miller, PG, Flucher, B, Flucher, BE, Daniels, MP. 1992. Tropomodulin is associated with the free (pointed) ends of the thin filaments in rat muscle. *J. Cell Biol.* 120:411–20

Fowler, V, Taylor, DL. 1980. Spectrin plus band 4.1 cross-link actin. Regulation by micromolar calcium. *J. Cell Biol.* 85:361–76

Fox, JE, Reynolds, CC, Morrow, JS, Phillips, DR. 1987. Spectrin is associated with membrane-bound actin filaments in platelets and is hydrolyzed by the Ca^{2+}-dependent protease during platelet activation. *Blood* 69: 537–45

Frappier, T, Derancourt, J, Pradel, LA. 1992. Actin and neurofilament binding domain of brain spectrin beta subunit. *Eur. J. Biochem.* 205:85–91

Friedrichs, B, Koob, R, Kraemer, D, Drenckhahn, D. 1988. Demonstration of immunoreactive forms of erythrocyte protein 4.2 in non-erythroid cells and tissues. *Eur. J. Cell Biol.* 48:121–27

Fukushima, Y, Byers, MG, Watkins, PC, Winkelmann, JC, Forget, BG, Shows, TB. 1990. Assignment of the gene for beta-spectrin (SPTB) to chromosome 14q23-q24.2 by in situ hybridization. *Cytogenet. Cell Genet.* 53:232–33

Funayama, N, Nagafuchi, A, Sato, N, Tsukita, S. 1991. Radixin is a novel member of the band 4.1 family. *J. Cell Biol.* 115:1039–48

Gallagher, PG, Forget, BG. 1993. Spectrin genes in health and disease. *Sem. Hematol.* 30:4–20

Gardner, K, Bennett, V. 1986. A new erythrocyte membrane-associated protein with calmodulin binding activity: identification and purification. *J. Biol. Chem.* 261:1339–48

Gardner, K, Bennett, V. 1987. Modulation of spectin-actin assembly by erythrocyte adducin. *Nature* 328:359–62

Giebelhaus, DH, Eib, DW, Moon, RT. 1988. Antisense RNA inhibits expression of membrane skeleton protein 4.1 during embryonic development of Xenopus. *Cell* 53:601–15

Glenney, J, Glenney, P, Weber, F. 1983. The spectrin-related molecule TW 260/240 cross-links actin bundles of the microvillus rootlets in the brush borders of intestinal epithelial cells. *J. Cell Biol.* 96:1491–96

Golan, DE, Veatch, W. 1980. Lateral mobility of band 3 in the human erythrocyte membrane studied by fluorescence photobleaching recovery: evidence for control by cytoskeletal interactions. *Proc. Natl. Acad. Sci. USA* 77:2537–41

Goodman, SR, Keith, KE, Whitfield, CF,

Riederer, BM, Zagon, IS. 1988. Spectrin and related molecules. *CRC Crit. Rev. Biochem.* 23:171–234

Gould, K, Bretcher, A, Esch, F, Hunter, T. 1989. cDNA cloning and sequencing of the protein-tyrosine kinase substrate, ezrin, reveals homology to band 4.1. *EMBO J.* 8:4133–52

Gregorio, CC, Kubo, RT, Bankert, RB, Repasky, EA. 1992. Translocation of spectrin and protein kinase C to a cytoplasmic aggregate upon lymphocyte activation. *Proc. Natl. Acad. Sci. USA* 89:4947–51

Gu, M, York, J, Warshawsky, I, Majerus, P. 2991. Identification, cloning and expression of a cytosolic megakaryocyte protein-tyrosine-phosphatase with sequence homology to cytoskeletal protein 4.1. *Proc. Natl. Acad. Sci. USA* 88:5867–71

Gundersen, D, Orlowski, J, Rodriquez-Boulan, E. 1991. Apical polarity of Na,K-ATPase in retinal pigment epithelium is linked to a reversal of the ankyrin-fodrin submembrane cytoskeleton. *J. Cell Biol.* 112:863–72

Hall, TG, Bennett, V. 1987. Regulatory domains of erythrocyte ankyrin. *J. Biol. Chem.* 262:10537–45

Hammerton, R, Krzeminski, K, Mays, R, Ryan, T, Wollner, D, Nelson, J. 1991. Mechanism for regulating cell surface distribution of Na^+, K^+-ATPase in polarized epithelial cells. *Science* 254:847–50

Hanspal, M, Palek, J. 1987. Synthesis and assembly of membrane skeletal proteins in mammalian red cell precursors. *J. Cell Biol.* 105:1417–24

Harris, AS, Corall, DE, Morrow, JS. 1988. The calmodulin binding site in alpha fodrin is near the calcium-dependent protease-I cleavage site. *J. Biol. Chem.* 263:15754–61

Harris, AS, Morrow, JS. 1990. Calmodulin and calcium-dependent protease I coordinately regulate the interaction of fodrin with actin. *Proc. Natl. Acad. Sci. USA* 87:3009–13

Harris, HW, Lux, SE. 1980. Structural characterization of the phosphorylation sites of human erythrocyte spectrin. *J. Biol. Chem.* 255:11512–20

Hartwig, JH, Desisto, M. 1991. The cytoskeleton of the resting human blood platelet: structure of the membrane skeleton and its attachment to actin filaments. *J. Cell Biol.* 112:407–25

Hartwig, JH, Thelen, M, Rosen, A, Janmey, PA, Nairn, AC, Aderem, A. 1992. MARCKS is an actin filament crosslinking protein regulated by protein kinase C and calcium-calmodulin. *Nature* 356:618–22

Holley, MC, Ashmore, JF. 1990. Spectrin, actin and the structure of the cortical lattice in mammalian cochlear outer hair cells. *J. Cell Sci.* 96:283–91

Hoock, TC, Newcomb, PM, Herman, IM. 1991. Beta actin and its mRNA are localized at the plasma membrane and the regions of moving cytoplasm during the cellular response to injury. *J. Cell Biol.* 112: 653–64

Horne, WC, Leto, TL, Marchesi, VT. 1985. Differential phosphorylation of multiple sites in protein 4.1 and protein 4.9 by phorbol ester-activated and cyclic AMP-dependent protein kinases. *J. Biol. Chem.* 260:9073–76

Howe, CL, Sacramone, LM, Mooseker, MS, Morrow, JS. 1985. Mechanisms of cytoskeletal regulation: modulation of membrane affinity in avian brush border and erythrocyte spectrins. *J. Cell Biol.* 101: 1379–85

Hu, R-J, Bennett, V. 1991. In vitro proteolysis of brain spectrin by calpain I inhibits association of spectrin with ankyrin-independent membrane-binding sites. *J. Biol. Chem.* 266:18200–5

Hu, R-J, Watanabe, W, Bennett, V. 1992. Characterization of human brain cDNA encoding the general isoform of beta-spectrin. *J. Biol. Chem.* 267:18715–22

Huebner, K, Palumbo, AP, Isobe, M, Kozak, CA, Monaco, S, et al. 1985. The alpha-spectrin gene is on chromosome I in mouse and man. *Proc. Natl. Acad. Sci. USA* 82:3790–93

Hughes, C, Gilligan, D, Bennett, V. 1992. Oligomeric structure and subunit association in erythrocyte adducin. *Mol. Biol. Cell* 3:366a

Husain-Chishti, A, Faquin, W, Wu, C-C, Branton, D. 1989. Purification of erythrocyte dematin (protein 4.9) reveals an endogenous protein kinase that modulates actin-bundling activity. *J. Biol. Chem.* 264: 8985–91

Husain-Chishti, A, Levin, A, Branton, D. 1988. Abolition of actin-bundling by phosphorylation of human erythrocyte protein 4.9. *Nature* 334:718–21

Ideguchi, H, Nishimura, J, Nawata, H, Hamasaki, N. 1990. A genetic defect of erythrocyte band 4.2 protein associated with hereditary spherocytosis. *Br. J. Haematol.* 74:347–53

Inaba, M, Gupta, K, Kuwabara, M, Takahoshi, T, Benz, E, Maede, Y. 1992. Deamidation of human erythrocyte protein 4.1: possible role in aging. *Blood* 79:3355–61

Isayama, T, Goodman, SR, Zagon, IS. 1991. Spectrin isoforms in the mammalian retina. *J. Neurosci.* 11:3531–38

Jesaitis, AJ, Bokoch, GM, Tolley, JO, Allen, RA. 1988. Lateral segregation of neutrophil chemotactic receptors into actin- and fodrin-rich plasma membrane microdomains depleted in guanyl nucleotide regulatory proteins. *J. Cell Biol.* 107:921–28

Jons, T, Drenckhahn, D. 1992. Identification of the binding interface involved in linkage of cytoskeletal protein 4.1 to the erythrocyte anion exchanger. *EMBO J.* 11:2863–67

Joshi, R, Bennett, V. 1990. Mapping the domain structure of human erythrocyte adducin. *J. Biol. Chem.* 265:13130–36

Joshi, R, Gilligan, DM, Otto, E, McLaughlin, T, Bennett, V. 1991. Primary structure and domain organization of human alpha and beta adducin. *J. Cell Biol.* 115:665–75

Kaiser, HW, O'Keefe, E, Bennett, V. 1989. Adducin: Ca^{2+}-dependent association with sites of cell-cell contact. *J. Cell Biol.* 109: 557–69

Kalomiris, EL, Bourguignon, LY. 1988. Mouse T lymphoma cells contain a transmembrane glycoprotein (GP85) that binds ankyrin. *J. Cell Biol.* 106:319–27

Kalomiris, EL, Bourguignon, LY. 1989. Lymphoma protein kinase C is associated with the transmembrane glycoprotein, GP85, and may function in GP85-ankyrin binding. *J. Biol. Chem.* 264:8113–19

Karinch, A, Zimmer, W, Goodman, S. 1990. The identification and sequence of the actin-binding domain of human red cell beta-spectrin. *J. Biol. Chem.* 265:11833–40

Kelly, G, Zelus, B, Moon, R. 1991. Identification of a calcium-dependent calmodulin-binding domain in *Xenopus* membrane skeleton protein 4.1. *J. Biol. Chem.* 266:12469–73

Kennedy, S, Warren,S, Forget, B, Morrow, J. 1991. Ankyrin binds to the 15th repetitive unit of erythroid and non-erythroid beta-spectrin. *J. Cell Biol.* 115:267–77

Koch, CA, Anderson, D, Moran, M, Ellis, C, Pawson, T. 1991. SH2 and SH3 domains: elements that control interactions of cytoplasmic signaling proteins. *Science* 252: 668–74

Koob, R, Kraemer, D, Trippe, G, Aebi, U, Drenckhahn, D. 1990. Association of kidney and parotid Na^+, K^+-ATPase microsomes with actin and analogs of spectrin and ankyrin. *Eur. J. Cell Biol.* 53:93–100

Kordeli, E, Bennett, V. 1991. Distinct ankyrin isoforms at neuron cell bodies and nodes of Ranvier resolved using erythrocyte ankyrin-deficient mice. *J. Cell Biol.* 114:1243–59

Kordeli, E, Davis, J, Trapp, B, Bennett, V. 1990. An isoform of ankyrin is localized at nodes of Ranvier in myelinated axons of central and peripheral nerves. *J. Cell Biol.* 110:1341–52

Korsgren, C, Cohen, CM. 1986. Purification and properties of human erythrocyte band 4.2. Association with the cytoplasmic domain of band 3. *J. Biol. Chem.* 261:5536–43

Korsgren, C, Cohen, C. 1988. Associations of human erythrocyte band 4.2. Binding to ankyrin and to the cytoplasmic domain of band 3. *J. Biol. Chem.* 263:10212–18

Korsgren, C, Cohen, CM. 1991. Organization of the gene for human erythrocyte membrane protein 4.2: structural similarities with the gene for the α subunit of factor XIII. *Proc. Natl. Acad. Sci. USA* 88:4840–44

Kotula, L, Laury-Kleintop, L, Showe, L, Sahr, K, Linnenbach, A, et al. 1991. The exon-intron organization of the human erythrocyte alpha-spectrin gene. *Genomics* 9:131–40

Kunimoto, M, Otto, E, Bennett, V. 1991. A new 440-kD isoform is the major ankyrin in neonatal rat brain. *J. Cell Biol.* 115: 1319–31

Lambert, S, Bennett, V. 1993. Post-mitotic expression of erythrocyte structural proteins in discrete neuronal populations of the rat brain. *J. Neurosci.* In press

Lambert, S, Yu, H, Prchal, J, Lawler, J, Ruff, P, et al. 1990. cDNA sequence for human erythrocyte ankyrin. *Proc. Natl. Acad. Sci. USA* 87:1730–34

Lankes, W, Furthmays, H. 1991. Moesin: a member of the protein 4.1-talin-ezrin family of proteins. *Proc. Natl. Acad. Sci. USA* 88:8297–301

Laurila, P, Cioe, L, Kozak, CA, Curtis, PJ. 1987. Assignment of mouse beta-spectrin gene to chromosome 12. *Somat. Cell Mol. Genet.* 13:93 97

Lazarides, E, Nelson, W. 1983. Erythrocyte and brain forms of spectrin in cerebellum: distinct membrane-cytoskeletal domains in neurons. *Science* 220:1295–96

Leto, TL, Fortugno-Erikson, D, Barton, D, Yang-Feng, TL, Francke, U, et al. 1988. Comparison of nonerythroid alpha-spectrin genes reveals strict homology among diverse species. *Mol. Cell. Biol.* 8:1–9

Leto, TL, Marchesi, VT. 1984. A structural model of human erythrocyte protein 4.1. *J. Biol. Chem.* 259:4603–8

Leto, TL, Pleasic, S, Forget, BG, Benz, EJ Jr, Marchesi, VT. 1989. Characterization of the calmodulin-binding site of nonerythroid alpha-spectrin. *J. Biol. Chem.* 264:5826–30

Ling, E, Gardner, K, Bennett, V. 1986. Protein kinase C phosphorylates a recently identified membrane skeleton-associated calmodulin-binding protein in human erythrocytes. *J. Biol. Chem.* 261:13875–78

Liu, S-C, Derick, LH, Palek, J. 1987. Visualization of the hexagonal lattice in the erythrocyte membrane skeleton. *J. Cell Biol.* 104:527–36

Loer, C. 1992. *C. elegans* spectrin/fodrin-like

clone free to a good home. *Worm Breeder's Gaz.* 12:54

Lokeshwar, V, Bourguignon, L. 1992. Tyrosine phosphatase activity of lymphoma CD45 (GP180) is regulated by a direct interaction with the cytoskeleton. *J. Biol. Chem.* 267:21551–57

Lombardo, C, Willardson, B, Low, P. 1992. Localization of the protein 4.1-binding site on the cytoplasmic domain of erythrocyte membrane band 3. *J. Biol. Chem.* 267: 9540–46

Lux, SE, John, K, Bennett, V. 1990a. Analysis of cDNA for human erythrocyte ankyrin indicates a repeated structure with homology to tissue-differentiation and cell-cycle control proteins. *Nature* 344:36–42

Lux, S, Tse, W, Menninger, J, John, K, Harris, P, et al. 1990b. Hereditary spherocytosis associated with deletion of human erythrocyte ankyrin gene on chromosome 8. *Nature* 345:736–39

Mak, AS, Roseborough, G, Baker, H. 1987. Tropomyosin from human erythrocyte membrane polymerizes poorly but binds F-actin effectively in the presence and absence of spectrin. *Biochim. Biophys. Acta* 912: 157–66

Malek, S, Katumuluwa, A, Pasternack, G. 1990. Identification and preliminary characterization of two related proliferation associated nuclear phosphoproteins. *J. Biol. Chem.* 265:13400–9

Mangeat, PH. 1988. Interaction of biological membranes with the cytoskeleton framework of living cells. *Biol. Cell* 64:261–81

Maretzki, D, Mariana, M, Lutz, H. 1990. Fatty acid acylation of membrane skeletal proteins in human erythrocytes. *FEBS Lett.* 259:305–10

Matsuzaki, F, Sutoh, K, Ikai, A. 1985. Structural unit of the erythrocyte cytoskeleton. Isolation and electron microscopic examination. *Eur. J. Cell Biol.* 39:153–60

Maude, D, Guillett, G, Rogers, PA, Charest, PM. 1991. Identification of a 220 kDa membrane-associated plant cell protein immunologically related to human b-spectrin. *FEBS Lett.* 294:77–80

McGough, AM, Josephs, R. 1990. On the structure of erythrocyte spectrin in partially expanded membrane skeletons. *Proc. Natl. Acad. Sci. USA* 87:5208–12

McNeil, H, Ozawa, M, Kemler, R, Nelson, WJ. 1990. Novel function of the cell adhesion molecule uvomorulin as an inducer of cell surface polarity. *Cell* 62:309–16

Mercier, F, Reggio, H, Devilliers, G, Bataille, D, Mangeat, P. 1989. Membrane-cytoskeleton dynamics in rat parietal cells: mobilization of actin and spectrin upon stimulation of gastric acid secretion. *J. Cell Biol.* 108:441–53

Michaely, P, Bennett, V. 1992a. ANK repeats of RBC ankyrin fold cooperatively in specific, ordered groups of six repeats. *Mol. Biol. Cell* 3:264a

Michaely, P, Bennett, V. 1992b. The ANK repeat: a ubiquitous motif involved in macromolecular recognition. *Trends Cell Biol.* 2:127-

Mikkelsen, A, Elgsaeter, A. 1981. Human spectrin. VA comparative electro-optic study of heterotetramers and heterodimers. *Biochim. Biophys. Acta* 668:74–80

Mikkelsen, A, Stokke, B, Elgsaeter, A. 1984. An electro-optic study of human erythrocyte spectrin dimers. The presence of calcium ions does not alter spectrin flexibility. *Biochim. Biophys. Acta* 786:95–102

Mische, SM, Mooseker, MS, Morrow, JS. 1987. Erythrocyte adducin: a calmodulin-regulated actin-bundling protein that stimulates spectrin-actin binding. *J. Cell Biol.* 105:2837–45

Molday, LL, Cook, NJ, Kaupp, UB, Molday, RS. 1990. The cGMP-gated cation channel of bovine rod photoreceptor cells is associated with a 240-kDa protein exhibiting immunochemical cross-reactivity with spectrin. *J. Biol. Chem.* 265:18690–95

Moon, R, McMahon, A. 1990. Generation of diversity of nonerythroid spectrins. *J. Biol. Chem.* 265:4427–33

Morrow, JS, Cianci, CD, Ardito, T, Mann, AS, Kashgarian, M. 1989. Ankyrin links fodrin to the alpha subunit of Na,K-ATPase in Madin-Darby canine kidney cells and in intact renal tubule cells. *J. Cell Biol.* 108: 455–65

Mueller, TJ, M. Morrison. 1981. Glyco-connectin (PAS2), a membrane attachment site for the human erythrocyte cytoskeleton. In *Erythrocyte Membranes 2: Recent Clinical and Experimental Advances,*ed. W Kruckeberg, J Eaton, G Brewer, pp. 95–112. New York: Liss

Musacchio, A, Noble, M, Pauptit, R, Wierenga, R, Saraste, M. 1992. Crystal structure of a SH3 domain. *Nature* 359: 851–55

Najm, I, Vanderklish, P, Lynch, G, Baudry, M. 1991. Effect of treatment with difluoromethylornithine on polyamine and spectrin breakdown levels in neonatal rat brain. *Dev. Brain Res.* 63:287–89

Nehls, V, Drenckhahn, D, Joshi, R, Bennett, V. 1991. Adducin in erythrocyte precursor cells of rats and humans: expression and compartmentalization. *Blood* 78:1692–96

Nelson, WJ. 1992. Regulation of cell surface polarity from bacteria to mammals. *Science* 258:948–55

Nelson, WJ, Hammerton, RW, McNeill, H. 1991. Role of the membrane-cytoskeleton in the spatial organization of the Na,K-ATPase in polarized epithelial cells. *Soc. Gen. Physiol. Ser.* 46:77–87

Nelson, WJ, Lazarides, E. 1983. Switching of subunit composition of muscle spectrin during myogenesis in vitro. *Nature* 304: 364–68

Nelson, WJ, Lazarides, E. 1984. Goblin (ankyrin) in striated muscle: identification of the potential membrane receptor for erythroid spectrin in muscle cells. *Proc. Natl. Acad. Sci. USA* 81:3292–96

Nelson, WJ, Shore, EM, Wang, AZ, Hammerton, RW. 1990. Identification of a membrane-cytoskeletal complex containing the cell adhesion molecule uvomorulin (E-cadherin), ankyrin, and fodrin in Madin-Darby canine kidney epithelial cells. *J. Cell Biol.* 110:349–57

Nelson, WJ, Veshnock, PJ. 1987a. Modulation of fodrin (membrane skeleton) stability by cell-cell contact in Madin-Darby canine kidney epithelial cells. *J. Cell Biol.* 104: 1527–37

Nelson, WJ, Veshnock, PJ. 1987b. Ankyrin binding to ($Na^+ + K^+$)ATPase and implications for the organization of membrane domains in polarized cells. *Nature* 328:533–36

Nigg, EA, Bron, C, Giradet, M, Cherry, RJ. 1980. Band 3-glycophorin A association in erythrocyte membranes demonstrated by combining protein diffusion measurements with antibody-induced cross-linking. *Biochemistry* 19:1887–93

Ohanian, V, Gratzer, W. 1984. Preparation of red cell membrane cytoskeletal constituents and characterisation of protein. *Eur. J. Biochem.* 144:375–79

Ohanian, V, Wolfe, LC, John, KM, Pinder, JC, Lux, SE, Gratzer, WB. 1984. Analysis of the ternary interaction of the red cell membrane skeletal proteins spectrin, actin, and 4.1. *Biochemistry* 23:4416–20

Otsuka, A, Boontrakulpoontawel, P, Zhang, Y, Tang, L, Sobery, A. 1992. A novel ankyrin is required for axon guidance in *C. elegans*. *Mol. Biol. Cell* 3:196a

Otto, E, Kunimoto, M, McLaughlin, T, Bennett, V. 1991. Isolation and characterization of cDNAs encoding human brain ankyrins reveal a family of alternatively spliced genes. *J. Cell Biol.* 114:241–53

Palek, J, Lambert, S. 1990. Genetics of the red cell membrane skeleton. *Sem. Hematol.* 27:290–332

Palfrey, H, Waseem, A. 1985. Protein kinase C in the human erythrocyte. *J. Biol. Chem.* 260:16021–29

Parry, DA, Dixon, TW, Cohen, D. 1992. Analysis of the three-alpha-helix motif in the spectrin superfamily of proteins. *Biophys. J.* 61:858–67

Pasternack, GR, Anderson, RA, Leto, TL, Marchesi, VT. 1985. Interactions between protein 4.1 and Band 3. *J. Biol. Chem.* 260:3676–83

Peters, LL, Birkenmeier, CS, Bronson, RT, White, RA, Lux, SE, et al. 1991. Purkinje cell degeneration associated with erythroid ankyrin deficiency in *nb/nb* mice. *J. Cell Biol.* 114:1233–41

Pinder, JC, Gratzer, WB. 1983. Structural and dynamic states of actin in the erythrocyte. *J. Cell Biol.* 96:768–75

Pinder, JC, Ungewickell, E, Bray, D, Gratzer, WB. 1978. The spectrin-actin complex and erythrocyte shape. *J. Supramolec. Struct.* 8:439–45

Pinder, JC, Weeds, AG, Gratzer, WB. 1986. Study of actin filament ends in the human red cell membrane. *J. Mol. Biol.* 191:461–68

Pinto-Correia, C, Goldstein, EG, Bennett, V, Sobel, JS. 1991. Immunofluorescence localization of an adducin-like protein in the chromosomes of mouse oocytes. *Dev. Biol.* 146:301–11

Pollerberg, GE, Burridge, K, Krebs, KE, Goodman, SR, Schachner, M. 1987. The 180 kD component of the neural cell adhesion molecule N-CAM is involved in a cell-cell contacts and cytoskeleton-membrane interactions. *Cell Tissue Res.* 250: 227–36

Rees, D, Ades, S, Singer, S, Hynes, R. 1990. Sequence and domain structure of talin. *Nature* 347:685–89

Riederer, B, Zagon, I, Goodman, S. 1987. Brain spectrin (240/235) and brain spectrin (240/235E): differential expression during mouse brain development. *J. Neurosci.* 7: 864–74

Riederer, BM, Zagon, IS, Goodman, SR. 1986. Brain spectrin (240/235) and brain spectrin (240/235E): two distinct spectrin subtypes with different locations within mammalian neural cells. *J. Cell Biol.* 102: 2088–97

Risinger, M, Dotimas, E, Cohen, C. 1992. Human erythrocyte protein 4.2, a high copy number membrane protein, is N-myristylated. *J. Biol. Chem.* 267:5680–85

Roof, D, Hayes, A, Hardenbergh, G, Adamian, M. 1991. A 52 kD cytoskeletal protein from retinal rod photoreceptors is related to erythrocyte dematin. *Invest. Ophthalmol. Vis. Sci.* 32:582–93

Ruff, P, Speicher, DW, Husain-Chishti, A. 1991. Molecular identification of a major palmitoylated erythrocyte membrane protein containing the src homology 3 motif. *Proc. Natl. Acad. Sci. USA* 88:6595–99

Rybicki, A, Heath, R, Wolf, J, Lubin, B, Schwartz, R. 1988. Deficiency of protein 4.1 in erythrocyte from a patient with

Coombs negative hemolytic anemia. *J. Clin. Invest.* 81:893–901

Sahr, K, Laurila, P, Kotula, L, Scarpa, A, Coupal, E, et al. 1990. The complete cDNA and polypeptide sequences of human erythroid alpha-spectrin. *J. Biol. Chem.* 265: 4434–43

Saido, T, Shibata, M, Takenawa, T, Murofushi, H, Suzuki, K. 1992. Positive regulation of micro-calpain action by polyphosphoinositide. *J. Biol. Chem.* 267: 24585–90

Scaramuzzino, DA, Morrow, JS. 1991. The calmodulin-binding domain of recombinant beta-adducin. *J. Cell Biol.* 115:42a

Schuster, V, Bonsib, S, Jennings, M. 1986. Two types of collecting direct mitochondrion-rich (intercalated) cells: lectin and band 3 cytochemistry. *Am. J. Physiol.* 251: 347–55

Seubert, P, Ivy, G, Larson, J, Lee, J, Shahi, K, et al. 1988a. Lesions of entorhinal cortex produce a calpain-mediated degradation of brain spectrin in dentate gyrus. I. Biochemical studies. *Brain Res.* 459:226–32

Seubert, P, Larson, J, Oliver, M, Junt, MW, Baudry, M, Lynch, G. 1988b. Stimulation of NMDA receptors induces proteolysis of spectrin in hippocampus. *Brain Res.* 460: 189–94

Seubert, P, Peterson, C, Vanderklish, P, Cotman, C, Lynch, G. 1990. Increased spectrin proteolysis in the brindled mouse brain. *Neurosci. Lett.* 108:303–8

Sheetz, MP, Schindler, M, Koppel, D. 1980. Lateral mobility of integral membrane proteins is increased in spherocytic erythrocytes. *Nature* 285:510–12

Shen, BW, Josephs, R, Steck, T. L. 1986. Ultrastructure of the intact skeleton of the human erythrocyte membrane. *J. Cell Biol.* 102:997–1006

Shotton, DM, Burke, BE, Branton, D. 1979. The molecular structure of human erythrocyte spectrin. *J. Mol. Biol.* 131:303–29

Siegel, DL, Branton, D. 1985. Partial purification and characterization of an actin-bundling protein, band 4.9, from human erythrocytes. *J. Cell Biol.* 100:775–85

Smith, PR, Saccomani, G, Joe, EH, Angelides, KJ, Benos, DJ. 1991. Amiloride-sensitive sodium channel is linked to the cytoskeleton in renal epithelial cells. *Proc. Natl. Acad. Sci. USA* 88:6971–75

Sondag, D, Allosio, N, Blanchard, D, Ducluzeau, MT, Colonna, P, et al. 1987. Gerbich reactivity in 4.1 (-) hereditary elliptocytosis and protein 4.1 level in blood group Gerbich deficiency. *Br. J. Haematol.* 65:43–50

Speicher, D, DeSilva, T, Speicher, K, Ursitti, J, Hemboch, P, Weglarz, L. 1993. Location of the human red cell spectrin tetramer binding site and detection of a related "closed" hairpin loop dimer using proteolytic footprinting. *J. Biol. Chem.* 268: 4227–35

Speicher, DW, Marchesi, VT. 1984. Erythrocyte spectrin is composed of many homologous triple helical segments. *Nature* 311:177–80

Srinivasan, Y, Elmer, L, Davis, J, Bennett, V, Angelides, K. 1988. Ankyrin and spectrin associate with voltage-dependent sodium channels in brain. *Nature* 333:177–80

Steiner, J, Bennett, V. 1988. Ankyrin-independent membrane protein binding sites for brain and erythrocyte spectrin. *J. Biol. Chem.* 263:14417–25

Steiner, JP, Walke, HT, Bennett, V. 1989. Calcium/calmodulin inhibits direct binding of spectrin to synaptosomal membranes. *J. Biol. Chem.* 264:2783–91

Sung, LA, Fowler, VM, Lambert, K, Sussman, MA, Karr, D, Chien, S. 1992. Molecular cloning and characterization of human fetal liver tropomodulin—A tropomyosin-binding protein. *J. Biol. Chem.* 267: 2616–21

Sussman, MA, Fowler, VM. 1992. Tropomodulin binding to tropomyosins Isoform-specific differences in affinity and stoichiometry. *Eur. J. Biochem.* 205:355–62

Takeichi, M. 1991. Cadherin cell adhesion receptors as a morphogenetic regulator. *Science* 251:1451–55

Tang, CJ, Tang, TK. 1991. Rapid localization of membrane skeletal protein 4.1 (EL1) to human chromosome 1p33——p34.2 by nonradioactive in situ hybridization. *Cytogenet. Cell Genet.* 57:119

Tang, T, Qin, Z, Leto, T, Marchesi, V, Benz, E. 1990. Heterogeneity of protein products arising from the protein 4.1 gene in erythroid and nonerythroid tissues. *J. Cell Biol.* 110:617–24

Taylor, S, Snell, R, Buckler, A, Ambrose, C, Duyao, M, et al. 1992. Cloning of the alpha-adducin gene from Huntington's disease candidate region of chromosome 4 by exon amplification. *Nat. Genet.* 2:223–27

Thomas, GH, Kiehart, DP. 1992. Mutation in the *karst* (*kst*) gene of *Drosophila melanogaster* alter the beta heavy-spectrin protein and disrupt development. *Mol. Biol. Cell* 3:p155a

Tripodi, G, Piscone, A, Borsani, G, Tisminetzky, S, Salardi, S, et al. 1991. Molecular cloning of an adducin-like protein: evidence of a polymorphism in the normotensive and hypertensive rats of the Milan strain. *Biochem. Biophy. Res. Comm.* 177:939–47

Tse, W, Lecomte, M, Costa, F, Gabarz, M,

Feo, C, et al. 1990. Point mutation in the beta-spectrin gene associated with alpha 1/74 hereditary elliptocytosis. *J. Clin. Invest.* 86:909–16

Tse, WT, Menninger, JC, Yang-Feng, TL, Francke, U, Sahr, KE, et al. 1991. Isolation and chromosomal localization of a novel nonerythroid ankyrin gene. *Genomics* 10: 858–66

Tsuji, A, Kawasaki, K, Ohnishi, S, Merkle, H, Kusumi, A. 1988. Regulation of band 3 mobilities in erythrocyte ghost membranes by protein association and cytoskeletal meshwork. *Biochemistry* 27:7447–52

Tsuji, A, Ohnishi, S. 1986. Restriction of the lateral motion of band 3 in the erythrocyte membrane by the cytoskeletal network: dependence on spectrin association state. *Biochemistry* 25:6133–39

Tyler, JM, Hargreaves, WR, Branton, D. 1979. Purification of two spectrin-binding proteins: biochemical and electron microscopic evidence for site-specific reassociation between spectrin and bands 2.1 and 4.1. *Proc. Natl. Acad. Sci. USA* 76:5192–96

Tyler, JM, Reinhardt, BN, Branton, D. 1980. Association of erythrocyte membrane proteins—Binding of purified bands 2.1 and 4.1 to spectrin. *J. Biol. Chem.* 255:7034–39

Ungewickell, E, Bennett, PM, Calvert, R, Ohanian, V, Gratzer, WB. 1979. In vitro formation of a complex between cytoskeletal proteins of the human erythrocyte. *Nature* 280:811–14

Vybiral, T, Winkelmann, JC, Roberts, R, Joe, E, Casey, DL, et al. 1992. Human cardiac and skeletal muscle spectrins: differential expression and localization. *Cell Motil. Cytoskeleton* 21:293–304

Wallis, C, Wenegieme, E, Babitch, J. 1992. Characterization of calcium binding to brain spectrin. *J. Biol. Chem.* 267:4333–37

Waseem, A, Palfrey, C. 1988. Erythrocyte adducin: Comparison of the alpha- and beta-subunits and multiple-site phosphorylation by protein kinase C and cAMP-dependent protein kinase. *Eur. J. Biochem.* 178:563–73

Waseem, A, Palfrey, HC. 1990. Identification and protein kinase C-dependent phosphorylation of alpha-adducin in human fibroblasts. *J. Cell Sci.* 96:93–98

Wasenius, VW, Saraste, M, Salven, P, Eramaa, M, Holm, L, Lehto, VP. 1989. Primary structure of the brain alpha-spectrin. *J. Cell Biol.* 108:79–93

Waugh, RE, Agre, P. 1988. Reductions of erythrocyte membrane viscoelastic coefficients reflect spectrin deficiencies in hereditary spherocytosis. *J. Clin. Invest.* 81:133–41

Wessels, Chen, SW. 1993. Transient localized accumulation of alpha spectrin during sea urchin morphogenesis. *Dev. Biol.* 155:161–71

White, RA, Peters, LL, Adkison, LR, Korsgren, C, Cohen, CM, Lux, SE. 1992. The murine pallid mutation is a platelet storage pool disease associated with the protein 4.2 (pallidin) gene. *Nat. Genet.* 2:80–83

Willardson, BM, Thevinin, BJ-M, Harrison, ML, Kuster, WM, Benson, MD, Low, PS. 1989. Localization of the ankyrin-binding site on erythrocyte membrane protein, band 3. *J. Biol. Chem.* 264:15893–99

Winkelmann, J, Chang, J, Tse, W, Scarpa, A, Marchesi, V, Forget, B. 1990. Full-length sequence of the cDNA for human erythroid beta-spectrin. *J. Biol. Chem.* 265: 11827–32

Winkelmann, J, Costa, F, Linzie, B. 1991. Beta spectrin in human skeletal muscle: tissue-specific differential processing of 3 beta spectrin pre-mRNA generates a beta spectrin isoform with a unique carboxy-terminus. *J. Biol. Chem.* 265:20449–54

Winograd, E, Branton, D. 1991. Phasing the conformational unit of spectrin. *Proc. Natl. Acad. Sci. USA* 88:10788–91

Woods, DF, Bryant, PJ. 1991. The discs-*large tumor suppressor* gene of Drosophila encodes a guanylate kinase homolog localized at septate junctions. *Cell* 66:451–64

Yang, Q, Tonks, N. 1991. Solation of a cDNA clone encoding a human protein-tyrosine phosphatase with homology to the cytoskeletal proteins band 4.1 ezrin, and talin. *Proc. Natl. Acad. Sci. USA* 88:5949–53

Yu, H, Rosen, MK, Shin, TB, Seidel-Dugan, C, Brugge, J, Schreiber, S. 1992. Solution identification of the SH3 domain of Src and identification of its ligand-binding site. *Science* 258:1665–68

Yu, J, Fischman, DA, Steck, TL. 1973. Selective solubilization of proteins and phospholipids from red blood cell membranes by nonionic detergents. *J. Supramolec. Struct.* 1:233–48

Yue, L, Spradling, A. 1992. *Hu-li tai shao,* a gene required for ring canal formation during *Drosophila* oogenesis, encodes a homolog of adducin. *Genes Dev.* 6:2443–54

Annu. Rev. Cell Biol. 1993. 9:67–99

FUNCTIONAL ELEMENTS OF THE CYTOSKELETON IN THE EARLY *DROSOPHILA* EMBRYO

Eyal D. Schejter and Eric Wieschaus
Department of Molecular Biology, Princeton University, Princeton, New Jersey 08544

KEY WORDS: cytoplasmic organization, nuclear migration, embryonic cortex, cytokinesis, *Drosophila* genetics

---

CONTENTS

INTRODUCTION

The combination of genetics and molecular biology has made the embryo of the fruitfly *Drosophila melanogaster* a mainstay for studies of pattern formation during early development. Similar approaches have demonstrated its potential for elucidating mechanisms that underlie changes in cell mor-

0743–4634/93/1115–0067$05.00

phology and behavior. In this review, we focus on one aspect of these studies: the emerging roles of the cytoskeleton as a mediator of morphogenesis in the early *Drosophila* embryo. The earliest stages of embryogenesis in *Drosophila* occur within a single, uninterrupted syncytium. During these stages, a series of diverse morphogenetic events takes place. These include rapid rounds of synchronous nuclear division, stereotypic nuclear migration through the common cytoplasm, dynamic rearrangements of the embryonic cortex, and finally, subdivision of the syncytium to form a uniform cellular blastoderm. The high degree of cytoplasmic organization required by this variety of processes is likely to be provided by the cytoskeleton. In *Drosophila,* as in all eukaryotes, the cytoskeleton performs a wide range of cellular functions. These include the determination of cell size, shape and motility properties, compartmentalization of the cytoplasm, transport and localization of macromolecules, and mediation of cell-cycle cues (Alberts et al 1989). Not surprisingly, the cytoskeleton is found to play a key role in the regulation and coordination of morphogenesis in the early *Drosophila* embryo.

The review is structured as follows: we first introduce the system by describing the major features of the early embryonic cytoskeleton. We survey the applicable molecular, cell biological, and genetic experimental methods used for identifying and studying cytoskeletal elements of the early embryo. We then take up in detail three salient issues that have emerged from such studies: the motion of nuclei and other organelles within the common cytoplasm; dynamic rearrangements of the surface under the influence of the cortical cytoskeleton; and the two events of cytokinesis that occur in the early embryo. Particular emphasis is placed on the functional roles of the cytoskeleton in these aspects of early embryonic development.

Utilization of *Drosophila* for the study of cytoskeletal processes is, of course, not confined to early embryonic stages. Those interested in a more general treatment of the subject should consult a recent comprehensive review (Fyrberg & Goldstein 1990) and additional material focusing on the use of genetic approaches to studies of the *Drosophila* cytoskeleton (Fyrberg 1989, 1993). Other reviews complement and extend some of the subjects covered here and address additional topics relevant to the early embryonic cytoskeleton (Warn 1986; Biessman & Walter 1989; Warn et al 1990; Bearer 1991; Glover 1991; Kiehart 1991; Schweisguth et al 1991; Foe et al 1993).

Synopsis of Early Embryonic Development and Major Features of the Embryonic Cytoskeleton

A number of thorough and detailed descriptions of early development of the *Drosophila* embryo using light microscopy are available (Rabinowitz 1941; Sonnenblick 1950; Zalokar & Erk 1976; Fullilove & Jacobson 1978;

Foe & Alberts 1983; Hartenstein & Campos-Ortega 1985). Readers interested in the specific techniques that have been developed for visualization and manipulation of *Drosophila* embryos are invited to consult recent comprehensive laboratory manuals (Roberts 1986; Ashburner 1989). Major features of both the microtubule- and microfilament-based cytoskeletons of the early *Drosophila* embryo have been determined primarily through immunofluorescent visualization of cytoskeletal elements in wholemount preparations. The pioneering work of Warn, Alberts, and their colleagues centered on studies of fixed specimens, stained with specific antibodies and with the F-actin-specific phallotoxins (summarized in Warn 1986; Karr & Alberts 1986). These studies have been augmented by the introduction of methods designed to detect cytoskeletal structures in the living embryo, using fluorescently labeled antibodies (Warn et al 1987), or fluorescent actin and tubulin derivatives (Kellogg et al 1988). Together these studies provide the basis for our understanding of cytoskeletal organization and dynamics in the early embryo, as summarized below.

The freshly laid *Drosophila* egg is composed primarily of yolk, except for a thin cortical layer of clear cytoplasm (periplasm), which underlies the embryonic plasma membrane (Figure 1*a,c*). Fertilization, followed by pronuclear fusion, takes place within the anterior of the oval-shaped egg. The major feature of the developmental stages that follow is the rapid, metasynchronous rounds of uninterrupted DNA synthesis and nuclear division, which proceed without intervening cytokinesis. The first seven divisions take place deep within the yolky interior. During division cycles 8 and 9, most nuclei, together with associated islands of clear cytoplasm (energids), migrate toward the periphery (Figure 1*a-d,f*). The few that do not migrate are termed yolk nuclei. They maintain their position in the center of the embryo, where they divide a few more times, become polyploid and remain acellular (Figure 1*d*).

During these early cleavage stages, two distinct cytoskeletal domains are recognized. One is associated with the clear cytoplasmic regions around the internal nuclei, the other with the cell cortex (Figure 1*c*). The microtubule-based cytoskeleton associated with the internal nuclei changes conformation during the nuclear division cycle. The mitotic spindle, which is the prominent structure at metaphase, gives way to an elongating astral array that interconnects neighboring nuclei during telophase and interphase (Baker et al 1993). Microfilament concentrations are also observed in the interior of the egg (Warn 1986; Hatanaka & Okada 1991a), but the exact nature of these structures requires more detailed description. In the cortex, microtubules form an extensive network, which is filamentous in nature and appears to emanate from numerous organizing centers. A homogeneous distribution of microfilaments, which is punctate in appearance, is also found in this cortical

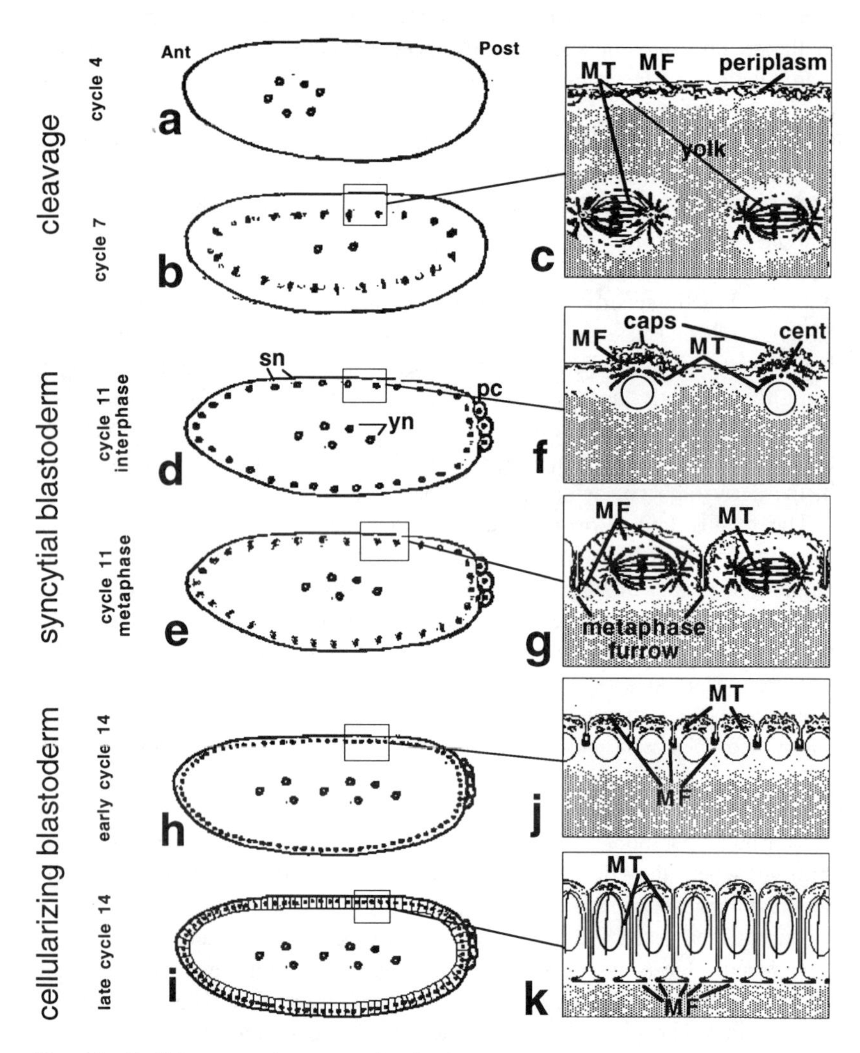

Figure 1 (*Left*) Representative stages of early cleavage (*a,b*), syncytial blastoderm (*d,e*), and cellularizing blastoderm (*h,i*) embryos. Note changes in the spatial arrangement of the nuclei, and formation of pole cells and somatic cells. Not all nuclei and cells are shown. Ant, anterior; Post, posterior; pc, pole cells; sn, somatic nuclei; yn, yolk nuclei. (*Right*) Enlargements of boxed regions, detailing major cytoskeletal domains characteristic of each stage. Yolk regions are stippled; cent, centrosomes; MF, microfilaments; MT, microtubules.

layer. In addition, microfilaments are found to underlie the many microprojections that cover the embryonic surface. Both cortical networks extend 3–4 μm below the surface and retain their mostly uniform appearance throughout the early cleavage cycles.

By early cycle 10, the nuclei complete migration and populate the thickened periplasm. Nuclei that reach the posterior pole are soon enclosed within pole cells, the germline progenitor cells. The remainder (referred to as the somatic nuclei) maintain a regularly spaced monolayer arrangement as they continue to divide within a syncytium (Figure 1*d-g*). The four cortical division cycles that ensue become progressively longer, from 9 min (at 25°C) for the early divisions, to 21 min at cycle 13. The developmental stage corresponding to cleavage cycles 10–13 is referred to as the syncytial blastoderm to distinguish it from the previous early cleavage cycles and the cellular blastoderm that forms afterwards (Figure 1*h-k*).

Nuclear migration into the cortex abolishes the distinction between the separate cytoskeletal domains of the early embryo. When the nuclei and their associated energids penetrate the periplasm, a dramatic reorganization of the cortical cytoskeleton takes place that results in the formation of discrete cortical cytoskeletal domains around each nucleus. These domains undergo cyclic rearrangements dictated by the nuclear division cycles of the syncytial blastoderm (Figure 1*f,g*). The vast majority of cortical microtubules become involved in the formation of short interphase arrays and mitotic spindle structures. Microfilaments are concentrated during interphase in cap-like structures just beneath bulges that form on the surface (Turner & Mahowald 1976; Warn & Magrath 1982). During mitosis they rearrange to underlie the entire, flattened surface, including transitory invaginations (termed metaphase furrows), which separate adjacent mitotic figures.

With the onset of interphase of cycle 14, the embryo undergoes a series of profound morphological rearrangements (Figure 1*h-k*). A considerable widening of the periplasmic layer, which began during the previous cycle, reaches its full extent. The approximately 6000 spherical nuclei, which constitute the cortical monolayer, change shape and elongate. Simultaneously, they are enclosed in individual cells by infoldings of the plasma membrane that descend through the periplasmic layer to the yolk. Immediately following the completion of blastoderm cellularization, the morphogenetic movements of gastrulation begin. The events of early cycle 14 are associated with a further reorganization of the cortical cytoskeleton (Figure 1*j,k*). Interphase microtubules form a basket-like structure around each elongating nucleus. Microfilaments are arranged primarily in an interlocking hexagonal array. This array is closely associated with the leading edge of the cellularization front as it descends between neighboring nuclei. Additional F-actin concentrations are found at the surface, within the lateral membrane extensions, and in aggregates within the deeper portions of the periplasm.

The role of intermediate filaments in early morphogenetic processes is uncertain. Proteins that are immunologically related to intermediate filament components, such as vimentin, have been identified in *Drosophila,* and their distribution patterns in the early embryo have been described (summarized in Biessmann & Walter 1989). Ultrastructural evidence of an intermediate filament cytoskeleton is lacking, however, and an exhaustive survey of invertebrate species specifically failed to identify intermediate filaments in Arthropods (Bartnik & Weber 1989). Given these ambiguities, in this review we have chosen to concentrate only on the features of the microfilament and microtubule based-cytoskeletons.

IDENTIFICATION AND PERTURBATION OF CYTOSKELETAL ELEMENTS

Biochemical and Molecular Approaches

Several experimental schemes have been used to identify cytoskeletal components and associated proteins found in the early *Drosophila* embryo. One approach is based on the isolation of conventional components of the *Drosophila* cytoskeleton, followed by an examination of the distributions of these elements during early embryogenesis. Purification of cytoskeletal elements from *Drosophila* cell extracts, by various fractionation techniques, led to the isolation of both the heavy and light chains of cytoplasmic (nonmuscle) myosin, α-spectrin, and the heavy chain of kinesin, followed by the eventual cloning of their structural genes (Kiehart & Feghali 1986; Byers et al 1987; Dubreuil et al 1987; Saxton et al 1988; Yang et al 1988; Kiehart et al 1989; Karess et al 1991). Cross-species homology at the nucleic-acid level enabled the identification of other conventional cytoskeletal elements including the genes for *Drosophila* actin and tubulin isoforms, some of which are expressed in the early embryo (Sanchez et al 1981; Kalfayan & Wensink 1982; Fyrberg et al 1983; Matthews et al 1989; Tobin et al 1990; Theurkauf 1992).

An alternative approach is aimed at identifying novel proteins specifically associated with the cytoskeleton of the early embryo. F-actin and microtubule affinity chromatography techniques, developed specifically for this purpose, have been used to purify some 40 actin-binding proteins and over 50 microtubule-associated proteins (Miller et al 1985, 1989, 1991; Kellogg et al 1989; Miller & Alberts 1989). Immunolocalization studies of many of these proteins have fostered an appreciation for the structural complexity of the early embryonic cytoskeleton. One promising extension of this work is the use of antibodies directed against these proteins for isolation, by affinity chromatography, of interacting elements (Kellogg & Alberts 1992). Such an approach should allow the identification and reconstruction of protein complexes that mediate various cytoskeletal functions.

Genetic Approaches

Mutational analysis provides a powerful approach to the study of functional contributions made by specific gene products to any biological process. Application of this experimental approach to morphogenesis in the early *Drosophila* embryo requires an appreciation of the genetic circumstances that prevail at this time. The freshly laid *Drosophila* egg is loaded with maternally derived gene products. Expression of the zygotic genome begins only during the syncytial blastoderm stage (e.g. Anderson & Lengyel 1979; Edgar & Schubiger 1986), at which time an interplay between the maternal and zygotic sets of genetic information determines the body plan of the organism. This course of events implies that morphogenesis in the early *Drosophila* embryo relies heavily on maternally provided gene products. Treatment of early embryos with inhibitors of RNA synthesis revealed that morphogenesis does, in fact, progress normally through the first 13 division cycles despite the block to zygotic gene transcription (Arking & Parente 1980; Edgar et al 1986). A similar conclusion was reached following a genetic screen designed to identify the early zygotic contributions to development (Merrill et al 1988; Wieschaus & Sweeton 1988). These findings demonstrate that maternally supplied factors provide all necessary functions during the early cleavage and syncytial blastoderm stages. Consequently, genetic analysis of the early embryonic stages requires elimination or modification of maternally rather than zygotically provided gene products.

Genetic screens that can accomplish this goal are designed to isolate maternal effect mutations. Such mutations represent a specific form of female sterility: affected females lay normal eggs, which can be fertilized, but embryonic development is prematurely arrested. Several screens for mutations causing female sterility in *Drosophila,* some quite extensive in nature, have been carried out for loci residing on the X chromosome (Gans et al 1975; Zalokar et al 1975; Mohler 1977; Perrimon et al 1986) and on the two major autosomes (Bakken 1973; Rice & Garen 1975; Tearle & Nusslein-Volhard 1987; Erdelyi & Szabad 1989; Schupbach & Wieschaus 1989; Szabad et al 1989; Spradling 1993). Maternal effect mutations uncovered in such screens provide the bulk of the genetic loci that have been studied in the context of the early embryonic cytoskeleton. Prominent examples are presented throughout this review.

Some inherent limitations place boundaries on the effectiveness of this approach. One major difficulty in female sterile screens arises from the requirement that affected females must survive to produce the mutant progeny. Elimination or modification of essential genes will result in zygotic lethality, so the maternal significance of such genes cannot be assessed. A promising alternative to the classical genetic approach is to selectively mutate genes that have been previously identified by molecular means. An example of this reverse genetic approach is provided by studies of the cytoplasmic (nonmuscle) myosin heavy chain gene. Ample evidence indicated that cytoplasmic myosin is both expressed and functional in the early embryo (Kiehart et al 1989, 1990; Young et al 1991). However, mutations in the gene had not been isolated in maternal effect screens, presumably due to a zygotic requirement. Only following cloning and cytogenetic mapping was it possible to establish that mutations in the cytoplasmic myosin gene lead to late embryonic lethality and correspond to a previously identified locus designated *zipper* (Young et al 1993). The early development of embryos homozygous for these mutations is not affected because substantial stores of cytoplasmic myosin are supplied maternally.

While the maternal contribution may mask the effects of zygotic mutations during early development, later zygotic requirement for the same function will result in lethality during advanced stages of development. This reasoning has prompted the search for genes affecting cell-cycle events among collections of larval and pupal lethal mutations (Ripoll et al 1987; Gatti & Baker 1989; Glover 1989) and may be applicable to genes controlling early cytoskeletal functions. For example, the locus *spaghetti-squash* was initially discovered since mutations result in late zygotic lethality because of faulty cytokinesis. This locus was found to encode the structural gene for the regulatory light chain of a cytoplasmic myosin (Karess et al 1991). The maternal requirement for zygotically essential genes can be assessed, in many cases, by analysis of progeny produced from chimeric female germlines (e.g. Chou & Perrimon 1992).

While maternally provided components govern most aspects of early embryogenesis, a screen specifically directed towards elucidation of early requirements for zygotic gene function has identified a handful of zygotic gene products required for proper morphogenesis prior to gastrulation (Merrill et al 1988; Wieschaus & Sweeton 1988). These genes contribute to various morphogenetic aspects of cycle 14, particularly to blastoderm cellularization, and are discussed in detail below.

Disruption and Modification of the Cytoskeleton by Epigenetic Means

Reagents that perturb the structural integrity of the cytoskeleton or release cytoskeletal elements from their natural subcellular environment have proven to be useful tools for functional studies. These reagents can be introduced into the embryo either by uptake from the external milieu following embryo permeabilization (Zalokar & Erk 1976) or by direct injection, a method that allows for greater control over the site and developmental stage of application. The two most widely used agents for cytoskeletal disruption are the cytochalasins, which inhibit microfilament polymerization, and the alkaloids colchicine/colcemid, which have a similar effect on microtubules. These drugs have been used to assess cytoskeletal function during all phases of early *Drosophila* development (Zalokar & Erk 1976; Foe & Alberts 1983; Edgar et al 1987; Hatanaka & Okada 1991a; Callaini et al 1992). Consequences of the application of the microfilament stabilizing agent phalloidin (Planques et al 1991) and microtubule disruption by cold treatment (Callaini & Marchini 1989; Callaini et al 1991) have also been examined. Additional treatments that indirectly affect the functioning of the cytoskeleton include the application of RNA synthesis inhibitors (Edgar et al 1986), the protein synthesis inhibitor cyclohexamide, which arrests the mitotic cycle of precellular embryos at interphase (Zalokar & Erk 1976; Edgar et al 1987; Baker et al 1993), and the DNA synthesis inhibitor aphidicolin, which results in the uncoupling of centrosomes from the nuclear chromatin (Raff & Glover 1988, 1989). An alternative method for generating such free centrosomes is by ultraviolet irradiation (Yasuda et al 1991), a treatment that is useful for delaying and studying nuclear migration as well (Okada et al 1974).

While modification or elimination of gene activity by mutation provides a particularly powerful method for assessing the functional roles of specific cytoskeletal elements, obtaining mutations in a gene of interest may not be a trivial matter. Alternative approaches focus on curtailment of the function of gene products, rather than mutation of the structural gene itself. One suitable method, in the context of the early *Drosophila* embryo, is injection of antisera to specific proteins into living embryos with the intent of disrupting functional activity. This approach has been used to study cytoplasmic myosin function throughout the pregastrulation stages (Kiehart et al 1990; D. Lutz & D. Kiehart, personal communication), to examine microtubule functions in the syncytial blastoderm by injection of anti α-tubulin (Warn et al 1987), and to assess the functions of a variety of actin-binding proteins (K. Miller, personal communication). Translational inhibition by introduction of antisense RNA transcripts, a relatively rare application in this field, has been used to mimic genetic disruption of the gene *germcell-less* (Jongens et al 1992), which is involved in pole cell formation.

NUCLEAR MIGRATION AND OTHER FORMS OF LARGE SCALE MOTION WITHIN THE CYTOPLASM

The dispersal and spatial redistribution of cytoplasmic contents in insect eggs are achieved by a variety of kinetic processes, both global and specific in their effects (reviewed in Counce 1973; van der Meer 1988). Participation of the cytoskeleton in controlling the spatial arrangement and movement of the nuclei and other organelles in the early *Drosophila* embryo has been primarily assessed following treatment with disrupting agents (Zalokar & Erk 1976; Foe & Alberts 1983; Raff & Glover 1989; Hatanaka & Okada 1991a). Such studies have led to the conclusion that both microtubule- and microfilament-based systems play important and distinct roles (see also discussion in Baker et al 1993). We discuss evidence for the involvement of cytoskeletal systems in three instances of cytoplasmic motion: the ordered migration of nuclei, periodic cortical cytoplasmic flows, and directed movement of particles within the cortex of the syncytial blastoderm.

Nuclear Migration

The nuclei are particularly suitable markers for cytoplasmic rearrangements, and the shifts in their spatial patterns during early *Drosophila* embryogenesis are well described (Zalokar & Erk 1976; Foe & Alberts 1983; Hatanaka & Okada 1991b; Baker et al 1993). During the early cleavage cycles, three temporal stages of nuclear migration are recognized. During the first four rounds of cleavage, the nuclei form an expanding sphere, within the anterior portion of the embryo (Figure 1*a*). The second stage of migration, during cycles 4–7, is characterized by preferential expansion of the sphere towards the poles. This stage ends with the nuclei positioned 30 μm from the surface, in an ellipsoidal arrangement that mirrors the contours of the egg (Figure 1*b*). Finally, during the next three cycles, most of the nuclei (except for the small group of internal yolk nuclei) migrate towards the periphery. This phase of migration is performed in an expanding ellipsoidal pattern. The nuclei complete their movements and penetrate the cortex nearly simultaneously by interphase of cycle 10 (Figure 1*d,f*).

Nuclear migration is a stepwise process. During the poleward expansion stage of cycles 4–7, it is confined primarily to prophase of the mitotic cycles (Baker et al 1993). The subsequent migration towards the cortex is restricted to telophase and early interphase (Foe & Alberts 1983). During division and migration the nuclei are accompanied by the surrounding cytoplasmic islands, and thus the spatial rearrangements apply to the entire energid. Involvement of cytoskeletal elements has been proposed in two specific instances. Microfilaments have been implicated in the poleward expansion during mid-cleavage cycles, while a microtubule-based mechanism appears to control migration towards the cortex during subsequent stages.

POLEWARD EXPANSION The role of a microfilament-based system in establishing an ellipsoidal arrangement of nuclei during mid-cleavage stages has been inferred from both cytochalasin treatments and from the study of a group of mutations that affect the migration process. Mutations in the maternal-effect loci *gs(1)N26*, *gs(1)N441* and *paralog*, as well as cytochalasin treatment during early cleavage stages, result in a common phenotype: the arrangement of nuclei prior to the cortical migration of cycles 8–9 remains spherical. As a result, following cortical migration, the nuclei populate the central cortex far in advance of the poles (Niki & Okada 1981; Thierry-Mieg 1982; Niki 1984, 1988; Hatanaka & Okada 1991a,b). However, while microfilament-based structures are severely disrupted in drug-treated embryos, such structures are only mildly affected in the mutant embryos. Abnormalities caused by the mutations include the appearance of aggregates in the normally smooth cortical layer of microfilaments and irregularities of F-actin distribution in the vicinity of the nuclei (Hatanaka & Okada 1991a). These findings suggest that the observed defect in nuclear migration is not simply a secondary consequence of general actin depolymerization, but may result from malfunction of specific actin-based structures. While the identity of these structures is not yet known, such findings underscore the utility of the genetic approach in refining observations obtained with chemical agents, which are less discriminating in their effects.

The primary functional roles of the three genes remain unknown. While mutations in the *gs(1)N26* and *paralog* genes exhibit effects in a variety of developmental stages, the *gs(1)N441* phenotype is restricted to the early embryo. This gene is thus a particularly attractive subject for studies of the nuclear migration process. At present, only a single mutant allele has been isolated in each gene. All three alleles are temperature-sensitive. This feature was used to determine the developmental period during which gene function is required for establishment of a normal pattern of nuclear migration. The temperature-sensitive period for *gs(1)N441* is during the late stages of oogenesis (Niki & Okada 1981), which suggests that aspects of the microfilament-based framework that operates in the embryo are set up well in advance.

MIGRATION TOWARDS THE CORTEX While the shift in the spatial arrangement of nuclei during mid-cleavage stages appears to involve a microfilament-based mechanism, migration towards the cortex during subsequent stages is likely to rely on a concerted and specific microtubule-dependent mechanism. Although genetic data regarding this process are still lacking, compelling observations have been made using various inhibitors. The specific involvement of microtubules is inferred from the sensitivity of the process to colchicine, but not to cytochalasin (Zalokar & Erk 1976; Hatanaka & Okada 1991a). Furthermore, the primary feature of the process appears to involve the microtubule-based migration of centrosomes, with the nuclei in the role of a passive hitchhiker. This striking conclusion was reached following studies on embryos treated with the DNA synthesis inhibitor aphidicolin (Raff & Glover 1988, 1989). Such treatment blocks chromatin replication, which leads to the dissociation of mitotic spindle components. Aphidicolin treatment of early cleavage stage embryos blocks the migration of the nuclei, but not that of centrosomes, which continue to divide and migrate in normal fashion. Colchicine treatment, however, will block centrosomal migration as well. The reliance of centrosomal replication and migration on cytoplasmic, rather than nuclear factors, has been demonstrated as well in the studies of a group of maternal-effect genes that are required for proper nuclear replication in the early embryo. This group includes the genetic loci *giant nuclei* (Freeman et al 1986); *fs*(1)Ya (Lin & Wolfner 1991); *plutonium* and *pan gu* (Shamanski & Orr-Weaver 1991). In many of the affected embryos, the mitotic cycles are blocked at a very early developmental stage. In spite of the block to nuclear replication, free centrosomes are generated, which continue to cycle and migrate to the surface.

The restriction of cortical nuclear migration to telophase and early interphase provides additional clues regarding the mechanistic basis of this process. Insight into this matter has been gained by examination of embryos treated with the protein synthesis inhibitor cyclohexamide, which results in cell cycle arrest at interphase (Baker et al 1993). Nuclear migration towards the cortex persists in these embryos, but is performed in a continuous rather than stepwise fashion so that the nuclei arrive at the cortex earlier than usual. These observations argue persuasively that the cell cycle dependency of the process results from a requirement for the particular configuration of microtubule structures during telophase and early interphase. Furthermore, all the required proteins are present at the onset of the migration process.

A mechanistic basis for the microtubule-dependent migration of nuclei to the cortex in early insect embryos has been proposed following investigation of this process in another Dipteran species, the gall midge *Wachtliella persicariae* (Wolf 1978). In gall midge embryos, cortical nuclear migration is also colchicine-sensitive and appears to be mediated by cytasters. These are enlarged mitotic asters, transiently attached via microtubule rays to cortical structures, that pull both nuclei and yolk particles through the egg interior. In *Drosophila,* however, no evidence exists for such long, surface-attached microtubules arrays. An elegant alternative explanation has been recently put forth by Baker and colleagues (1993), based on close examination of microtubule structures associated with nuclei within the interior of the embryo. During anaphase and prior stages of the nuclear division cycle, growing arrays of astral microtubules become prominent until an extensive microtubule network of interacting arrays from neighboring nuclei is formed. This network is proposed to maintain order in the spatial nuclear pattern by forcefully distancing neighboring nuclei from each other. During the migratory division cycles, the aster-derived arrays form asymmetrically, with considerably longer extensions of microtubules directed towards the interior. Asymmetric extension is proposed to translate the distancing force produced by the telophase network and propel nuclei toward the cortex. This mechanism accounts for the primary role of centrosomes and the cell-cycle dependency of the migratory process. It is particularly appealing since it relies only on intrinsic properties of the observed microtubule-based elements associated with the nucleus.

Motion and Transport of Particles in the Periplasm

The cortex of the early embryo is subject to a periodic contraction during each round of nuclear division (Fullilove & Jacobson 1978; Foe & Alberts 1983). This process begins during the initial stages of mitosis and consists of a strong inward contraction of the yolk mass at both poles of the embryo, coupled to a poleward flow of periplasm. As mitosis nears completion, both yolk and periplasm motions reverse direction and soon reestablish the original positional conformation. The periodic nature of the contractions suggests a role for the mitotic spindles in their initiation. Consistent with this notion is the finding that colchicine treatment abruptly halts both nuclear divisions and the flow of yolk and periplasm. A more direct role for microfilaments in bringing about the yolk contractions is suggested by the effects of cytochalasin, which arrests the contractions without disturbing the division cycle (Foe & Alberts 1983). The position and mechanistic properties of such an actin-based system are yet to be identified.

Motion of particles along the apical-basal axis of the periplasm during syncytial blastoderm stages is the final kinetic process discussed here. Following the arrival of nuclei at the cortex, large particles (possibly mitochondria) begin to move between the surface and the yolk mass (Foe & Alberts 1983). The particles move in saltatory fashion, with a strict confinement to motion that is perpendicular to the surface. This motion is sensitive to colchicine, but is unaffected by treatment with cytochalasin (Foe & Alberts 1983; Edgar et al 1987). A separate observation attesting to the presence of a radial system of microtubule tracks in the periplasm of syncytial blastoderms, comes from the analysis of expression patterns of an unusual "enhancer trap" element (Giniger et al 1993). The chimeric protein produced from this element contains both a kinesin heavy chain motor domain and bacterial β-galactosidase. When this element is expressed in early embryos, the protein first displays a general periplasmic distribution, then shifts to a tight localization at the border of the periplasm and the yolk mass (E. Giniger, personal communication; E. Wieschaus, unpublished). This result helps define the extent and orientation of the microtubule tracks extending between the cortex and the yolk. One possible function for a radial network of microtubules at this position is to aid in the clearing of the periplasm, which precedes cellularization of the blastoderm. Following the 13th nuclear division cycle, the clear cortical periplasm expands considerably as yolk and lipid particles are transported basally by a colchicine-sensitive mechanism (Zalokar & Erk 1976). A genetic inroad into this process is provided by the locus *halo,* which is required zygotically for periplasmic clearing (Merrill et al 1988; M. Postner & E. Wieschaus, in preparation). While yolk particles are removed normally from the periplasm of *halo* embryos, clearing of lipid droplets is defective and results in an unusually dark layer of cytoplasm below the cortical nuclei. The *halo* gene product may well function in the transport of lipid along a network of radial microtubule tracks. It should be stressed, however, that ultrastructural evidence for such a network is scant, with few reports in which appropriate microtubule structures were observed (e.g. Katoh & Ishikawa 1989).

Isolation of cytoskeletal components through biochemical means has produced the best example to date of a cytoplasmic motor protein active in the periplasm of the syncytial blastoderm. A 140 kd ATP-sensitive, actin-binding protein, originally isolated by F-actin affinity chromatography (Miller et al 1989), appears to be a novel unconventional myosin heavy chain (Kellerman & Miller 1992). Immunolocalization has shown that this protein is present in a punctate pattern throughout the periplasm, and in vivo motility studies have shown it to catalyze actin-based transport of associated particles to metaphase furrows (Kellerman & Miller 1992; K. Miller, personal communication). The nature of the associated particles and the functional significance of this microfilament-based motor activity are yet to be determined.

DYNAMIC REARRANGEMENTS AT THE CORTEX OF THE SYNCYTIAL BLASTODERM

In this section we discuss the roles of cytoskeletal elements in cortical events that take place during the syncytial blastoderm stage. Of particular significance is the functional interplay revealed between microtubule- and microfilament-based systems. Results from a variety of investigations argue that these cortical events involve a complex array of cytoskeletal-based interactions. This assessment is based not only on the obvious rearrangement of the major cytoskeletal components, but also on the involvement of other participants that have been identified by biochemical and genetic means. Immunocytological localization of more than a dozen actin-binding proteins isolated by F-actin affinity chromatography has revealed a unique dynamic pattern for each antigen examined during the syncytial stage mitotic cycles (Miller et al 1989). A genetic screen performed by Sullivan and colleagues (1993) has identified at least six loci that are first required specifically in the syncytial blastoderm, a feature shared by the *sponge* locus as well

(Postner et al 1992). Clearly, studies of syncytial *Drosophila* embryos bear the potential to be a rich and illuminating source on the nature of cortical cytoskeleton-based phenomena, and to add to the large body of information that has accumulated on this subject in other experimental systems (Bretscher 1991).

Correspondence of Surface and Cytoskeletal Dynamics

Coincident with the migration of nuclei and their associated cytoplasmic domains away from the yolky interior into the periphery of the embryo is a dramatic structural rearrangement of the embryonic cortex. Before the appearance of nuclei in the cortical region, the surface of the embryo is uniformly covered with a dense array of microprojections (Turner & Mahowald 1976; Warn et al 1984). At the onset of interphase of stage 10, simultaneous with the completion of nuclear migration, the surface is thrown into a series of protrusions (commonly referred to as buds) that directly overlie the nuclei (Warn & Magrath 1982, 1983; Foe & Alberts 1983; Stafstrom & Staehelin 1984). The surface of the buds is irregular and contains many microprojections, while the intervening membrane between buds is relatively smooth (Turner & Mahowald 1976). Following their initial appearance, the buds exhibit a cyclic behavior, in parallel with the nuclear division cycles of the syncytial blastoderm (Turner & Mahowald 1976; Foe & Alberts 1983; Stafstrom & Staehelin 1984). Upon entry into mitosis, the buds flatten and their margins give way to the formation of furrow-like indentations in the surface. These furrows separate all neighboring mitotic figures, reaching their full inward extent by metaphase, and retracting during the latter stages of mitosis. By interphase of the next cycle, new buds, occupying smaller individual areas, are reformed over all the somatic nuclei present in the cortex. Formation and interconversion of somatic buds and metaphase furrows continue through nuclear cycle 13. The buds that appear at the onset of cycle 14 precede the appearance of true cleavage furrows, which give rise to the cellular blastoderm.

The cyclic changes in surface structure of the syncytial blastoderm are closely mirrored by the rearrangement of the cortical cytoskeleton (Figure 1). The periplasm of the early cleavage stage embryo contains extensive networks of microtubules, microfilaments, myosin, and spectrin (Karr & Alberts 1986; Pesacreta et al 1989; Young et al 1991), which remain unaltered during the internal nuclear cleavage cycles (Figure 1*c*). The appearance of nuclei in the periplasm, however, results in profound reorganization of this cortical cytoskeleton (Figure 1*f,g*). The predominant microtubule structures become the interphase aster arrays nucleated by the centrosomes and the mitotic spindles (Karr & Alberts 1986; Warn & Warn 1986; Warn et al 1987; Kellogg et al 1988). The distribution of cortical microfilaments becomes subject to division cycle alterations as well (Warn et al 1984; Karr & Alberts 1986; Kellogg et al 1988). With the arrival of nuclei during interphase of cycle 10, and at interphase of subsequent cycles, F-actin forms a dome-like cap above each nucleus. This structure is closely associated with the cytoplasmic face of the surface buds and is excluded from the regions between them. With entry into mitosis and the flattening of the buds, the actin redistributes towards the cap margins and eventually lines the entire embryonic surface, including that of the metaphase furrows. As the daughter nuclei separate and surface buds reform, F-actin again concentrates into supranuclear cap structures. While the distribution of cortical myosin during the syncytial blastoderm stages coincides with the dynamic spatial organization of F-actin (Young et al 1991), the rearrangement of α-spectrin differs somewhat (Pesacreta et al 1989). Spectrin is incorporated into interphase caps during cycle 10 with a slight lag. Spectrin cap division takes place during mitosis, but does not include redistribution into metaphase furrows, so that the appearance of discrete caps is retained throughout.

The close correspondence between surface alterations and cytoskeletal rearrangements strongly suggests a causal relationship between the two. The dependency of surface dynamics on the normal rearrangement of the actin cytoskeleton is further inferred from the phenotype of embryos derived from females bearing the maternal effect mutation *sponge* (*spg;* Postner et al 1992). In these embryos, reorganization of cortical F-actin does not take place following nuclear migration and, as may be expected, somatic buds and metaphase furrows fail to form. A quantitative relationship is suggested by the shallow surface budding observed in aphidicolin-treated embryos, in which actin caps form only partially (Raff & Glover 1989). While these observations strongly suggest a mechanistic role for the actin-based cytoskeleton, models for the relevant mechanisms are conspicuously lacking. Initial notions involving furrowing induced by microfilament contraction have been abandoned as inconsistent with the localization of F-actin and myosin (Warn et al 1984; Young et al 1991) and the appearance of the embryonic surface, as observed by scanning electron microscopy (Turner & Mahowald 1976).

Functional roles for the cortical structures have been revealed following the examination of embryos in which they fail to form properly. A group of maternal effect mutations that displays this phenotype includes the loci *scrambled (sced), nuclear fallout (nuf), daughterless abo-like (dal),* and *spg* (Sullivan et al 1990, 1993; Postner et al 1992). A number of phenotypic features are shared among the four mutant loci, and these coincide with some of the effects of cytochalasin treatment on syncytial blastoderms (Zalokar & Erk 1976; Callaini et al 1992). Prominent defects include the partial or complete lack of metaphase furrows and the appearance of abnormal mitotic spindles. Consequently, nuclear divisions may fail or may result in nuclear fusions and other irregularities. Nuclei that participate in such abnormal events are found to recede into the interior of the embryo. The proportion of aberrant divisions increases with the progress of the syncytial cycles and the resultant decrease in internuclear distance. With the exception of *dal,* in which failure of centrosome separation appears to be the primary defect (see below), metaphase furrows frequently fail to separate morphologically normal spindles in the mutant embryos. This suggests that malformed spindles are a consequence rather than a cause of the missing furrows. These observations lead to the conclusion that a crucial function of metaphase furrows is to ensure the fidelity of nuclear division in the cortex of the syncytial blastoderm embryo by serving as a physical barrier between neighboring mitotic figures.

A functional role for the somatic buds that form over the interphase nuclei has been suggested following examination of the phenotype of embryos derived from *scrambled* mutant females (W. Theurkauf, personal communication). Despite a nearly complete inability to produce metaphase furrows, actin caps and their associated surface buds form and flatten in the usual cyclic fashion. While cytochalasin-treated embryos, which cannot form either surface structure, display a highly disordered interphase nuclear array, *scrambled* embryos display a relatively uniform and regular distribution of nuclei. This comparison suggests that the somatic buds, and perhaps other actin-based structures (Callaini et al 1993), restrict lateral movement of the nuclei and thus act to maintain a properly ordered nuclear array.

Centrosomes as Cytoplasmic Organizers

Simultaneous observation of microtubules and microfilaments offers persuasive evidence that the rearrangements of the two systems during the syncytial blastoderm stage are closely coupled, with the cortical microfilament network following the lead of the microtubule structures associated with the nucleus (Karr & Alberts 1986; Kellogg et al 1988). While the actin-based cytoskeleton may be mechanically responsible for the actual alterations in surface and cytoplasm structure, microtubule-based systems appear to act as the primary instructors. Centrosomes act as the major microtubule organizing centers of most animal cells (Vorobjev & Nadezhdina 1987; Kimble & Kuriyama 1992), nucleating microtubule arrays that vary in conformation through the cell cycle. Features of the centrosomal cycle in the early *Drosophila* embryo have been most closely followed during the division cycles of the syncytial blastoderm (Stafstrom & Staehelin 1984; Karr & Alberts 1986; Callaini & Riparbelli 1990). At interphase, a pair of centrosomes is positioned between the nucleus and the surface (Figure 1*f*).

Entering the mitotic cycle, they migrate to opposing sides of the nucleus, where they direct the formation of a spindle (Figure 1*g*). At the completion of mitosis, the centrosomes are restored to their apical position by a 90° rotation of the nucleus. A splitting of centrosomal material occurs during telophase and is separate from centriolar replication, which begins during interphase.

Many lines of evidence identify these organelles as key players in the interactions between cytoskeletal components during the progression of cortical events in the syncytial blastoderm. Surface changes and cyclic reorganization of cortical cytoskeletal networks are observed whenever centrosomes reach the cortex, and only in those regions occupied by centrosomes. Most significantly, cycles of surface budding and of actin and spectrin cap formation are initiated in embryos in which centrosomes migrate and populate the periplasm, free of associated nuclei (Raff & Glover 1988, 1989; Yasuda et al 1991). The complete reliance of the rearrangements of the cortical actin-based cytoskeleton on centrosomal function continues throughout the syncytial blastoderm stages. Treatment of syncytial blastoderms with disruptive agents such as high doses of anti-tubulin antibodies (Warn et al 1987) and low temperature (Callaini et al 1991) uncouples centrosomes from the associated nuclei. Such free centrosomes are sufficient to induce actin cap formation. Furthermore, the dimensions and shape of the caps are dictated by the number and spatial arrangement of the underlying centrosomal material. The distribution of free centrosomes in the treated embryos is irregular, with some cortical regions completely devoid of centrosomal components. Actin caps fail to form in these regions. A more subtle disturbance to centrosomal function is found in embryos from mothers bearing the mutation *daughterless abo-like (dal)*. The primary defect in these embryos is a failure of centrosome separation into inward migrating pairs of centrioles. A common consequence is an inability to form metaphase furrows (Sullivan et al 1990, 1993).

The mechanisms by which centrosomes direct cytoskeletal reorganization in syncytial embryos are unknown. Distinctions in the behavior of the cytoskeletal networks during different cortical cleavage cycles suggest that both direct and indirect mechanisms may be involved (Karr & Alberts 1986). During cycle 10, the F-actin distribution closely follows the movement of the centrosomes and the orientation of the mitotic spindle, splitting the interphase cap in two. A cap structure is then reformed as microfilaments follow the growth of astral microtubules late in the cycle. During mitosis of later cycles, the spatial correspondence between microfilament and microtubule concentrations is less apparent. F-actin is redistributed from interphase caps to underlie the entire surface, including the metaphase furrows, which implies an indirect influence by the centrosomes (Figure

1*g*). Phenotypic features of *dal* embryos suggest that metaphase furrow formation is directed by pairs of centrosomes from neighboring spindles (Sullivan et al 1990, 1993). Furrows fail to form only in the immediate vicinity of a nucleus in which centrosomal separation has failed and only on the side of the nucleus farthest away from the unseparated centrosome. These mutant features suggest an interaction among microtubules nucleated by neighboring asters in the determination of the site of metaphase furrow formation.

A glimpse into a possible pathway for transmission of instructions from the centrosomes to the actin-based cytoskeleton is provided by studies on antigen 13D2, an actin-binding protein isolated by F-actin affinity chromatography from embryonic extracts (Miller et al 1985). In wild-type embryos, 13D2 is found in interphase caps, which closely approximate the actin caps. The correspondence with actin distribution ceases during mitosis when 13D2 remains in slightly enlarged caps. In *spg* embryos, which do not reorganize the cortical actin network in response to the arrival of nuclei, actin-binding proteins generally fail to conform to their wild-type localization. Antigen 13D2 provides a striking exception, displaying a cap-like organization in the mutant embryos (Postner et al 1992). This finding identifies 13D2 as a potential intermediary between the two cytoskeletal networks, forming a cap structure in response to centrosomal cues, which in turn results in a similar rearrangement of cortical actin. The interaction between 13D2 and actin presumably requires a function provided by *spg*.

While genetic studies are providing inroads to the nature of the interactions between the two major cytoskeletal systems in the cortex of the syncytial blastoderm, convincing observations of a physical association are yet to be made. A possible beginning in this direction comes from a descriptive study of actin cap formation in the Mediterranean fruit fly *Ceratitis capitata,* whose early development closely resembles that of *Drosophila* (Callaini et al 1993). The larger size of the *Ceratitis* embryo has permitted the identification of novel actin structures in syncytial blastoderms of this species. These include an extensive perinuclear F-actin network, F-actin aggregates associated with spindle components and the centrosomes, and a heavy concentration of F-actin at the spindle pole during metaphase. Comparative studies with larger insects, which are more amenable to descriptive investigations, have been beneficial entry points for *Drosophila*-based research in such areas as nervous system structure (Thomas et al 1984) and may be essential for elucidation of the fine structure of the early embryonic cytoskeleton.

The instructive and mechanistic roles played by the centrosomes and their associated microtubule domains are not restricted to the syncytial blastoderm stages and have been inferred from a variety of observations throughout early *Drosophila* embryogenesis. This has led to a growing consensus on the primacy of centrosomes as organizers of cytoplasmic events (see also discussions in Karr & Alberts 1986; Raff & Glover 1988, 1989; Kellogg 1989; Yasuda et al 1991). As described elsewhere, centrosomes constitute the active moiety in cortical nuclear migration, function as inducers of pole cell formation, and act to determine the sites of cleavage furrow initiation during blastoderm cellularization. While descriptive observations are plentiful, a detailed molecular analysis is a clear priority for advancing the comprehension of centrosomal-based mechanisms. Mutations in *polo* (Sunkel & Glover 1988; Llamazares et al 1991), *merry-go-round* (Gonzalez et al 1988), and *dal* (Sullivan et al 1990, 1993) cause spindle abnormalities as a result of centrosomal malfunction and provide the beginning of a genetic approach to this issue. A biochemical approach has been made possible through the purification of antibodies specific to centrosomal antigens. These were isolated following screens for microtubule-associated proteins (Kellogg et al 1989) and for nuclear antigens (Frasch et al 1986) of the early embryo. Cloning of the structural genes encoding the centrosomal antigens (Whitfield et al 1988; Kellogg & Alberts 1992) now enables production of sufficient amounts of these proteins for functional characterization. This approach has been taken a step farther by using low-affinity antibodies to a centrosomal antigen as the basis for affinity chromatography purification of associated proteins. Such proteins presumably form functional complexes with the primary antigen. Using this method, a complex of at least ten proteins has been identified, and the individual components, as well as the complex as a whole, can now be characterized (Kellogg & Alberts 1992). An encouraging finding is that *Drosophila* γ-tubulin, which may function as a ubiquitous nucleation site for centrosome microtubules (Oakley 1992), is a component of this complex (Raff et al 1993).

CYTOKINESIS

Two separate events of cytokinesis take place during the early stages of *Drosophila* embryonic development. Shortly after the nuclear penetration of the cortical cytoplasm, a group of cells destined to establish the germline forms at the posterior end of the embryo (Figure 1*d,e,* Figure 2*a,b*). The major event of cell formation, in which several thousand somatic cells are formed simultaneously, takes place later, following four additional rounds of nuclear cleavage (Figure 1*h-k,* Figure 2*c-f*). The involvement of cytoskeletal elements in these processes is discussed primarily regarding two issues: the requirements for initiation, which are dependent on microtubule structures, and the actual mechanisms of cell formation, which rely heavily on microfilament-based systems.

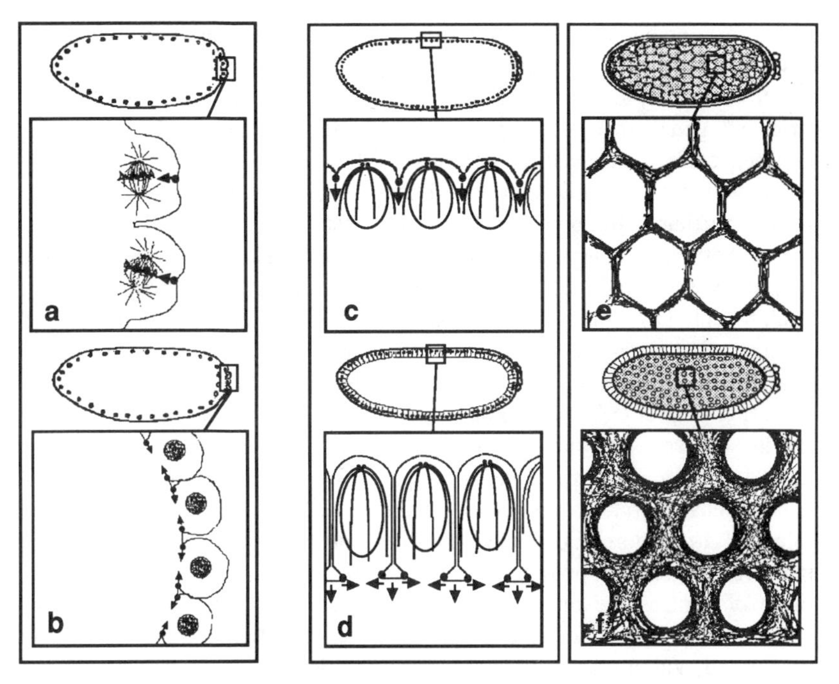

Figure 2 Cytokinesis in the early embryo. Detailed enlargements are shown beneath embryos of the corresponding stage. In the enlargements, closed circles mark sites of F-actin accumulation, while arrows point in the direction of cleavage. The two stages of pole bud cleavage are shown in panels (*a*) and (*b*). Two views of the early (*c,e*) and late (*d,f*) phases of cellularization are shown. Panels (*c*) and (*d*) are mid-sagittal sections. Panels (*e*) and (*f*) depict views from the surface of the cellularization front, demonstrating the changes in the structure of the actin-myosin array.

Pole Cell Formation

The first cytokinetic event in the early *Drosophila* embryo involves the formation of the progenitor germline cells at the posterior pole (Mahowald 1962; Turner & Mahowald 1976; Swanson & Poodry 1980; Foe & Alberts 1983). Arrival of the nuclei in the clear pocket of periplasm at the posterior is accompanied by pronounced surface blebbing, followed by the formation of large surface buds above each nucleus. During the ninth mitotic cycle, the polar buds divide and reform over the pole nuclei in a manner indistinguishable from somatic bud division. However, at the end of the tenth mitotic cycle, the polar buds enter into cytokinesis, during which they pinch off from the surface of the syncytial blastoderm and enclose the associated nuclei within spherical cells (Figure 2*a,b*).

Pole cell formation has been studied extensively in the context of posterior pattern establishment and germline determination. Although the exact molecular nature of the cytoplasmic pole cell determinants is still unknown, attention has centered on the polar granules (Mahowald 1962) and on a series of genes that affect pole cell formation and posterior patterning simultaneously (reviewed in Lehmann 1992). These cytoplasmic components, which are normally confined to the posterior pole of the embryo, are not sufficient for pole cell formation. Rather, they act in concert with nuclei that have migrated to the posterior periplasm. This interaction is temporally restricted. Delay in the arrival of nuclei at the posterior, as a result of physical blockage (Okada 1982), or genetic mutation (Niki & Okada 1981; Hatanaka & Okada 1991a), prevents the formation of pole cells. Additional insight into the nature of this interaction has been obtained by Raff & Glover (1989), who made the striking finding that the centrosomes are the active components that enable the processes of pole bud and subsequent pole cell formation to take place. In aphidicolin-treated embryos, centrosomes migrate to the cortex on their own and initiate the program of pole cell formation. Colchicine treatment alone, or in combination with aphidicolin, prevents pole cells from forming, presumably because of inhibition of centrosome migration. If colchicine is applied after nuclear migration and pole bud formation, the buds cease to divide and no pole cells are formed, although the buds remain intact (Raff & Glover 1989). Taken together, these results suggest that initiation of the process requires the presence of centrosomes and their ability to act as microtubule organizing centers.

The actual process of pole cell formation, which follows migration of nuclei into the polar plasm, has been described with an emphasis on the involvement of the actin-based cytoskeleton (Warn et al 1985). The first major change in F-actin distribution occurs during the tenth mitotic cycle (Figure 2*a*). Following reformation of the polar buds, a crescent-shaped accumulation of actin forms midway across their surface, above the mitotic spindle. This actin structure underlies a cleavage furrow, which invaginates and divides each of the buds in two. Once the buds are bisected, they are sealed off basally from the embryonic surface (Figure 2*b*). Both phases of pole cell formation are thought to be driven by contraction of an actin-myosin structure, a notion supported by the direct observation of myosin concentration at the bud edges during pole cell formation (Young et al 1991). Furthermore, cytochalasin treatment inhibits pole cell formation even if centrosomes and nuclei are present. If applied following initiation of the process, cytochalasin brings about the deterioration of the pole buds (Raff & Glover 1989). The initial partial cleavage of the pole buds resembles the action of the contractile arc that functions in various egg cleavages. The final pinching off of the pole cells is reminiscent of conventional cell division by a contractile ring of microfilaments (Schroeder 1990).

Blastoderm Cellularization

The transition from a syncytial to a cellular blastoderm is achieved by the simultaneous enclosure of all cortical somatic nuclei within individual cells (Mahowald 1963; Fullilove & Jacobson 1971; Simpson & Wieschaus 1990; Warn et al 1990). Cell divisions are initiated at the surface of the embryo, and membranes progressively extend through the periplasm towards the yolky interior, which leads to the formation of an epithelium of columnar cells, each 5 μm in diameter and nearly 35 μm in height (Figure 2*c,d*). Basal closure of the cells is incomplete since they maintain a narrow connection to the interior by way of small intercelluar bridges (Rickoll 1976). Cellularization is completed in just under an hour at 22°C and is followed immediately by the initial morphogenetic movements of gastrulation.

Blastoderm cellularization is coincident with a series of morphological events that take place as the embryo exits the 13th nuclear division cycle, the last of the rapid cycles that involve all the somatic nuclei. Additional features of this transition period include a marked change in the shape of the cortical nuclei (Mahowald 1963; Fullilove & Jacobson 1971), clearing of yolk and lipid particles from the periplasm, and a considerable increase in the size and density of the microvillar projections overlying the somatic buds (Turner & Mahowald 1976). Accompanying these developments is a new set of rearrangements in the structure of the major cytoskeletal systems. A distinctive perinuclear "basket" of microtubules is formed, nucleated by the pair of centrosomes positioned apically to the nuclei (Figure 2*c,d*). A restructuring of the actin-based cytoskeleton results in the formation of a hexagonal microfilament array, which is closely associated with the leading edge of the invaginating cleavage furrows (Figure 2*e*). Both these structures are thought to provide a mechanistic basis for some of the more prominent morphological events that commence during the early portion of cycle 14.

REARRANGEMENT OF THE MICROTUBULE-BASED CYTOSKELETON: NUCLEAR ELONGATION The early stage of cycle 14 is characterized in part by a dramatic alteration in the shape of the somatic nuclei (Figure 1*j,k,* Figure 2*c,d*), which assume a cylindrical appearance rather than the spherical morphology characteristic of the precellular stages (Foe & Alberts 1985). Nuclear growth, accompanied by an approximately 2.5-fold increase in nuclear volume, occurs almost entirely along the apical-basal axis (Mahowald 1963; Fullilove & Jacobson 1971). The process of nuclear elongation coincides with the appearance of a distinctive microtubule structure, the perinuclear "inverted basket" (Figure 2*c,d*), which has led to the suggestion that the two phenomena are functionally related.

The spatial distribution of microtubules during cycle 14 is well described (Fullilove & Jacobson 1971; Warn & Warn 1986; Callaini & Anselmi 1988; Wolf et al 1988; Katoh & Ishikawa 1989). Early in the cycle, small bundles and single microtubules emanate primarily from two microtubule-organizing centers positioned between the nucleus and the plasma membrane. The initial appearance of these microtubule formations is that of an astral array of short and randomly oriented microtubules, much like the structure surrounding syncytial blastoderm stage nuclei during interphase. However, as the nuclei begin to assume an elongated appearance, microtubules extending from the array are found to grow along and closely associate with the nuclear periphery (Figure 2*c,d*). Microtubule growth mirrors that of the nuclei in both direction and rate. Growth is confined to the axis perpendicular to the surface and just barely exceeds the base of the nuclei at all times. Both processes appear to come to a simultaneous halt toward the end of the slow phase of cleavage furrow formation. Since the microtubule array between the two centrosomes surrounds the nuclei apically, the resulting structure resembles a perinuclear basket that remains open from its basal aspect.

The suggestion that formation of the microtubule baskets and elongation of the nuclei are functionally linked is made on the basis of the close correspondence between the two processes and following the results of colchicine treatment of early cycle 14 embryos (Zalokar & Erk 1976; Foe & Alberts 1983; Edgar et al 1987). Such treatment inhibits formation of the perinuclear basket and prevents elongation of the nuclei, which remain rounded and swollen. Injection of embryos with colchicine after nuclear elongation is completed has no effect (Foe & Alberts 1983), thus prompting the proposal that the microtubule basket is resistant to this treatment. The notion that the inverted basket contains a unique population of microtubules is supported by the finding that acetylated microtubules, which are characteristic of relatively stable arrays, are exclusively observed in this particular structure prior to gastrulation (Wolf et al 1988). In an attempt to explain microtubule-dependent nuclear elongation, the microtubule basket can be viewed as a passive and rigid structure that serves to confine nuclear growth in the longitudinal direction (Fullilove & Jacobson 1971; Warn & Warn 1986; Wolf et al 1988). Growth can then be explained by an independent increase in nuclear volume, consistent with the observed swelling of nuclei in colchicine-treated embryos.

INITIATION OF CELLULARIZATION The general paradigm established for a variety of cortical processes described above, including formation of somatic buds, metaphase cleavage furrows, and pole cells, is that of an instructive function for microtubule-based structures and a mechanistic role for the actin-based cytoskeleton. This model is applicable to the process of cellularization as well. Two major principles are invoked in the description of conventional cytokinesis: cleavage furrow initiation as a result of stimulation by mitotic asters (Rappaport 1986); and progression of cell division as a result of the formation and action of a contractile ring of microfilaments (Mabuchi 1986; Salmon 1989; Schroeder 1990; Satterwhite & Pollard 1992). To a degree, both principles apply to cellularization of the *Drosophila* blastoderm.

In general, initiation of the cellularization process is compatible with the principles and observations described for initiation of cytokinesis. The key principle of cleavage furrow stimulation in conventional cytokinesis is positioning of the furrow at a point of influence overlap of the microtubule rays emanating from the pair of mitotic asters situated at either end of the spindle. The requirement for an intact microtubule-based structure to initiate cellularization has been shown by injection of colchicine or anti-tubulin antibodies into syncytial stage embryos, up to early cycle 14 (Zalokar & Erk 1976; Foe & Alberts 1983; Edgar et al 1987; Warn et al 1987). Furthermore, the cleavage furrows form at positions consistent with the influence overlap requirement (Figure 2*c*) at sites roughly equidistant from each of a pair of asters (Fullilove & Jacobson 1971). The precise molecular nature of the cleavage stimulus in cytokinesis remains an enigma. Speculation has centered on diffusible low molecular weight proteins, polyamines, and ions (Mabuchi 1986), while a variety of studies including careful ultrastructural examination (Asnes & Schroeder 1979) seem to rule out a direct role for the aster microtubules themselves. However, a recent study of cellularizing *Drosophila* blastoderms identified a population of microtubules that extend symmetrically from both spindle poles and form attachments to the plasma membrane at the site of furrowing (Katoh & Ishikawa 1989). This observation may point to a fundamental distinction in the nature of the stimulatory mechanism between cellularization and conventional cytokinesis.

REARRANGEMENT OF THE MICROFILAMENT-BASED CYTOSKELETON: A CONTRACTILE ACTIN-MYOSIN ARRAY DRIVES CELLULARIZATION The cellularization process can be divided into three distinct stages: a preliminary period during which cleavage furrows are established between the nuclei as they begin to change shape; a phase of relatively slow membrane extension, from the surface to the bases of the nuclei, which complete their elongation; and a final, rapid phase of membrane extension, which ends when the membranes reach the yolk and the cells are closed off basally. Following the completion of the 13th mitotic cycle, actin caps reform over the individual somatic nuclei, which are arranged in a cortical monolayer exhibiting a roughly hexagonal pattern. During the period when cleavage furrows are established, microfilaments display a novel dynamic behavior that departs from the sequence of events during earlier cycles, when the actin-based cytoskeleton closely follows the progress of the mitotic cycle (Warn & Magrath 1983; Simpson & Wieschaus 1990; Warn & Robert-Nicoud 1990). The interphase 14 caps first expand laterally, so that neighboring cap borders become juxtaposed. At these sites of overlap, a meshwork arrangement of F-actin rapidly gives way to the appearance of sharp actin lines, resulting in the formation of an array of interconnected hexagons, which are centered above the individual nuclei and are closely associated with the plasma membrane (Figure 2*c,e*). The hexagonal F-actin network marks the sites at which cleavage furrows will soon descend between neighboring nuclei.

The reorganization of cortical microfilaments into a network that lines the sites of cleavage furrow formation and remains associated with the cellularization front throughout the process strongly suggests that this network provides the structural basis for blastoderm cellularization. Additional support is provided by the continuous sensitivity of the cellularization process to treatment with cytochalasin or phalloidin (Zalokar & Erk 1976; Foe & Alberts 1983; Planques et al 1991). Furthermore, the colocalization of F-actin and cytoplasmic myosin in the hexagonal network (Warn et al 1980; Young et al 1991), the inhibition of membrane extension by application of antibodies directed against myosin prior to the onset of cellularization (Kiehart et al 1990; D. Lutz & D. Kiehart, personal communication), and the bipolar arrangement of microfilaments within the network (Katoh & Ishikawa 1989) all lend support to the notion that the network is contractile.

Subdivision of the syncytial periplasm into individual cells during blastoderm cellularization is a complex process in which two membrane partitions are formed. Cleavage furrows first descend from the surface, forming lateral divisions between neighboring nuclei (Figure 2*c*). A second division, perpendicular to the first, separates the nuclei and the periplasm from the underlying yolk sac. Construction of these divisions temporally overlaps, as the final, rapid phase of cleavage furrow invagination coincides with basal closure (Figure 2*d*). These changing roles are mirrored in alterations in the appearance of the actin-myosin array (Figure 2*e,f*). At the end of the early phase of membrane invagination, the hexagonal network breaks up into a series of rings that are connected by a relatively actin-poor mesh. As the membranes continue to extend toward the yolk, the circular units rapidly shrink, thereby sealing off the cells basally (Warn & Magrath 1983; Warn & Robert-Nicoud 1990; Young et al 1991).

The mechanistic contributions of the actin-myosin array to the different features of the cellularization process are a matter of debate. One viewpoint holds that a mechanism resembling the sliding filament contractile-ring mechanism, a classical explanation for conventional cytokinesis (Schroeder 1990; Satterwhite & Pollard 1992), underlies all aspects of blastoderm cellularization (Warn et al 1980; Warn & Magrath 1983; Kiehart 1991; Young et al 1991; Poodry 1992; Schejter et al 1992). Myosin-mediated shortening of microfilaments is posited to govern contraction of the individual units of the network, which are interlinked over the entire embryo and lead to the production of a net inward force along the interfaces of the hexagonal array. Presumed connections between the network and the plasma membrane result in the invagination of the cleavage furrows. A structural change in the network midway through the process allows for continued membrane extension, as well as basal closure. Others, notably Warn and colleagues, have argued that additional forces must be at play (Planques et al 1991; Warn & Robert-Nicoud 1990; Warn et al 1990). This view reasons that since limited disruptions to the network do not impede the overall progression of the process, global contraction is an unlikely mechanism for membrane invagination. According to this view, basal closure of the cells may well be driven by a conventional contractile ring mechanism. Lateral membrane extension may be achieved by incorporation of new membrane, which pushes on cleavage furrows that are kept taut by contractile forces provided by the individual actin-myosin units.

Genetic analysis of cellularization has focused on the roles of zygotic gene products during the cellularization process. Coincident with the morphological changes of the cycle 14 transition is a dramatic shift in the pattern of gene activity. Transcription rates from the somatic nuclei increase sharply (e.g. Anderson & Lengyel 1979; Edgar & Schubiger 1986), signaling the onset of an increasing reliance of the developmental program on zygotic gene products. The first morphological abnormalities resulting from lack of zygotic gene activity are detected at this stage (Merrill et al 1988; Wieschaus & Sweeton 1988). One of the earliest overt defects in embryo morphology due to zygotic mutation is in the rearrangement of the hexagonal actin-myosin network. Three genes contribute to this process: *nullo, serendipity*-α, and *bottleneck* (Schweisguth et al 1990; Simpson & Wieschaus 1990; Rose & Wieschaus 1992; Schejter et al 1992; E. Schejter & E. Wieschaus, in preparation). Loss of any one of these genes results in abnormalities of the hexagonal actin array formed at the cap borders and in aberrant cellularization because cleavage furrows fail to form at disrupted interfaces of the network. The initial defect observed in both *nullo* and *serendipity*-α mutant embryos is a large variability in the appearance of the hexagonal array, including unevenness in the thickness of the actin lines and occasional rupturing of the network. Consequently, cleavage furrows separating many neighboring nuclei are missing, resulting in the formation of multinucleate cells, which remain open basally. The hexagonal array of *bottleneck* mutant embryos forms normally, but rapidly takes on the appearance of a network undergoing basal closure. This premature and uneven closure results in the trapping of nuclei by the advancing cellularization front and in other distortions including the appearance of multinucleate cells. The structural genes for the three loci have been cloned (Vincent et al 1985; Schweisguth et al 1990; Rose & Wieschaus 1992; E. Schejter & E. Wieschaus, in preparation). Expression of all three during development is limited to the syncytial and cellularizing blastoderm, which suggests that they fulfill functions that are specific to the cellularization process. The three genes encode small, structurally novel proteins, two of which (nullo and bottleneck) possess an unusual excess of positively charged residues (pI ~11). The mutant phenotypes suggest that this group of genes participates in structural regulation of the hexagonal actin network during its formation and subsequent contraction. Immunolocalization of all three proteins to the leading edges of the advancing cellularization front is consistent with these roles (Schweisguth et al 1990; M. Postner & E. Wieschaus in preparation; E. Schejter & E. Wieschaus in preparation).

CONCLUSION

It has been our intention in this review to impart a sense of excitement regarding the potential of the early *Drosophila* embryo to serve as an experimental system for elucidation of features and mechanisms of the cytoskeleton. Descriptive, biochemical, and genetic techniques used in studying features of the early embryo cytoskeleton have been presented, some of which are novel and applicable to studies of the cytoskeleton in other systems. Aspects of organelle transport and movement, subdivision of the cytoplasm into functional compartments, interactions between cytoskeletal networks, and cell division have all been discussed, with an emphasis on the emerging roles played by the cytoskeleton in these processes. It appears certain that the introduction of a genetic approach complements the more classical methods used in the study of tissue and cell morphology and that this combination offers a powerful tool for elucidation of functional attributes of molecular components of the cell.

ACKNOWLEDGMENTS

We are indebted to many colleagues who provided us with invaluable advice and assistance. We would like to thank Jayne Baker, Giuliano Callaini, Ed Giniger, Dan Kiehart, Kathy Miller, Gerold Schubiger, Bill Sullivan, and Bill Theurkauf for kindly providing manuscripts prior to publication and discussing their experimental results with us. We are particularly grateful to Ken Irvine, Kathy Miller, Marya Postner, and Jeff Yuan for critical and thoughtful reading of this review. Supported by grant 5R01HD15587 from the National Institutes of Health to EW and a postdoctoral fellowship from the New Jersey Commission on Cancer Research to ES.

Literature Cited

Alberts, B, Bray, D, Lewis, J, Raff, M, Watson, JD. 1989. *Molecular Biology of the Cell*. pp. 613–80. New York/London: Garland. 1218 pp. 2nd ed.

Anderson, KV, Lengyel, JA. 1979. Rates of synthesis of major classes of RNA in *Drosophila* embryos. *Dev. Biol.* 70:217–31

Arking, R, Parente, A. 1980. Effects of RNA inhibitors on the development of *Drosophila* embryos permeabilized by a new technique. *J. Exp. Zool.* 212:183–94

Ashburner, MA. 1989. *Drosophila A Laboratory Manual*. Cold Spring Harbor: Cold Spring Harbor Lab. 434 pp.

Asnes, CF, Schroeder, TE. 1979. Cell cleavage: ultrastructural evidence against equatorial stimulation by astral microtubules. *Exp. Cell Res.* 122:327–38

Baker, J, Theurkauf, WE, Schubiger, G. 1993. Dynamic changes in microtubule configuration correlate with nuclear migration in the preblastoderm *Drosophila* embryo. *J. Cell Biol.* In press

Bakken, AH. 1973. A cytological and genetic study of oogenesis in *Drosophila melanogaster*. *Dev. Biol.* 33:100–22

Bartnik, E, Weber, W. 1989. Widespread occurence of intermediate filaments in invertebrates; common principles and aspects of diversion. *Eur. J. Cell Biol.* 50:17–33

Bearer, EL. 1991. Actin in the *Drosophila* embryo: is there a relationship to developmental cue localization? *BioEssays* 13:199–204

Biessmann, H, Walter, MF. 1989. Intermediate filaments during fertilization and early embryogenesis. In *The Cell Biology of Fertilization,* ed. H Schatten, G Schatten, pp. 189–224. San Diego: Academic. 404 pp.

Bretscher, A. 1991. Microfilament structure and function in the cortical cytoskeleton. *Annu. Rev. Cell Biol.* 7:337–74

Byers, TJ, Dubreuil, R, Branton, D, Kiehart, DP, Goldstein, LSB. 1987. *Drosophila* spectrin II. Conserved features of the alpha subunit are revealed by analysis of cDNA clones and fusion proteins. *J. Cell Biol.* 105:2103–10

Callaini, G, Anselmi, F. 1988. Centrosome splitting during nuclear elongation in the *Drosophila* embryo. *Exp. Cell Res.* 178: 415–25

Callaini, G, Dallai, R, Riparbelli, MG. 1991. Microfilament distribution in cold-treated *Drosophila* embryos. *Exp. Cell Res.* 194: 316–21

Callaini, G, Dallai, R, Riparbelli, MG. 1992. Cytochalasin induces spindle fusion in the syncytial blastoderm of the early *Drosophila* embryo. *Biol. Cell* 74:249–54

Callaini, G, Dallai, R, Riparbelli, MG. 1993. F-actin domains in the syncytial blastoderm of the dipteran *Ceratitis capitata*. *J. Cell Sci.* 104:97–104

Callaini, G, Marchini, D. 1989. Abnormal centrosomes in cold-treated embryos. *Exp. Cell Res.* 184:367–74

Callaini, G, Riparbelli, MG. 1990. Centriole and centrosome cycle in the early *Drosophila* embryo. *J. Cell Sci.* 97:539–43

Campos-Ortega, JA, Hartenstein, V. 1985. *The Embryonic Development of Drosophila Melanogaster*. Berlin: Springer-Verlag. 227 pp.

Chou, T, Perrimon, N. 1992. Use of a yeast site-specific recombinase to produce female germline chimeras in *Drosophila*. *Genetics*. 131:643–53

Counce, SJ. 1973. The causal analysis of insect embryogenesis. In *Developmental Systems: Insects,* ed. SJ Counce, CH Waddington, 2:1–156. London/New York: Academic. 304 pp.

Dubreuil, R, Byers, TJ, Branton, D, Goldstein, LSB, Kiehart, DP. 1987. *Drosophila* spectrin I. Characterization of the purified protein. *J. Cell Biol.* 105:2095–102

Edgar, BA, Kiehle, CP, Schubiger, G. 1986. Cell-cycle control by the nucleo-cytoplasmic ratio in early Drosophila development. *Cell* 44:365–72

Edgar, BA, Odell, GM, Schubiger, G. 1987. Cytoarchitecture and the patterning of *fushi tarazu* expression in the *Drosophila* blastoderm. *Genes Dev.* 1:1226–37

Edgar, BA, Schubiger, G. 1986. Parameters controlling transcriptional activation during early Drosophila development. *Cell* 44:871–7

Erdelyi, M, Szabad, J. 1989. Isolation and characterization of dominant female sterile mutations of *Drosophila melanogaster*. I. Mutations on the third chromosome. *Genetics* 122:111–27

Foe, VE, Alberts, BM. 1983. Studies of nuclear and cytoplasmic behavior during the five mitotic cycles that precede gastrulation in *Drosophila* embryogenesis. *J. Cell Sci.* 61:31–70

Foe, VE, Alberts, BM. 1985. Reversible chromosome condensation induced in *Drosophila* embryos by anoxia: visualization of interphase nuclear organization. *J. Cell Biol.* 100:1623–36

Foe, VE, Odell, GM, Edgar, BA. 1993. Mitosis and morphogenesis in the *Drosophila* embryo: point and counterpoint. In *The Development of Drosophila Melanogaster,* ed. M Bate, A Martinez-Arias. Cold Spring Harbor: Cold Spring Harbor Lab. In press

Frasch, M, Glover, D, Saumweber, H. 1986. Nuclear antigens follow different pathways into daughter nuclei during mitosis in early *Drosophila* embryos. *J. Cell Sci.* 82:155–72

Freeman, M, Nusslein-Volhard, C, Glover, DM. 1986. The dissociation of nuclear and centrosomal division in *gnu*, a mutation causing giant nuclei in Drosophila. *Cell* 46:457–68

Fullilove, SL, Jacobson, AG. 1971. Nuclear elongation and cytokinesis in *Drosophila montana. Dev. Biol.* 26:560–77

Fullilove, SL, Jacobson, AG. 1978. Embryonic development: descriptive. In *Genetics and Biology of Drosophila,* ed. M Ashburner, TRF Wright, 2C:105–227. New York: Academic. 617 pp.

Fyrberg, E. 1989. Study of contractile and cytoskeletal proteins using *Drosophila* genetics. *Cell Motil. Cytoskeleton* 14:118–27

Fyrberg, E. 1993. Genetic dissection of *Drosophila* cytoskeletal functions. *Adv. Cell Mol. Biol.* In press

Fyrberg, EA, Goldstein, LSB. 1990. The *Drosophila* cytoskeleton. *Annu. Rev. Cell Biol.* 6:559–96

Fyrberg, EA, Mahaffey, JW, Bond, BJ, Davidson, N. 1983. Transcripts of the six Drosophila actin genes accumulate in a stage- and tissue-specific manner. *Cell* 33: 115–23

Gans, M, Audit, C, Masson, M. 1975. Isolation and characterization of sex-linked female sterile mutants in *Drosophila melanogaster. Genetics* 81:683–704

Gatti, M, Baker, BS. 1989. Genes controlling essential cell-cycle functions in *Drosophila melanogaster. Genes Dev.* 3:438–53

Giniger, E, Wells, W, Jan LY, Jan, YN. 1993. Tracing neurons with a kinesin β-galactosidase fusion protein. *Roux's Arch. Dev. Biol.* 202:112–22

Glover, DM. 1989. Mitosis in *Drosophila. J. Cell Sci.* 92:137–46

Glover, DM. 1991. Mitosis in the *Drosophila* embryo - in and out of control. *Trends Genet.* 7:125–32

Gonzalez, C, Casal, J, Ripoll, P. 1988. Functional monopolar spindles caused by mutation in *mgr,* a cell division gene of *Drosophila melanogaster. J. Cell Sci.* 89: 39–47

Hatanaka, K, Okada, M. 1991a. Retarded nuclear migration in *Drosophila* embryos with aberrant F-actin reorganization caused by maternal mutations and by cytochalasin treatment. *Development* 111:909–20

Hatanaka, K, Okada, M. 1991b. Mutations affecting embryonic f-actin reorganization also affect separation of nuclei from their sisters and from the cortex in *Drosophila* cleavage embryos. *Dev. Growth Diff.* 33: 535–42

Jongens, TA, Hay, B, Jan, LY, Jan, YN. 1992. The *germ cell-less* gene product: a posteriorly localized component necessary for germ cell development in Drosophila. *Cell* 70:569–84

Kalfayan, L, Wensink, PC. 1982. Developmental regulation of Drosophila melanogaster alpha-tubulin genes. *Cell* 29: 91–98

Karess, RE, Chang, X-J, Edwards, KA, Kulkarni, S, Aguilera, I, Kiehart, DP. 1991. The regulatory light-chain of nonmuscle myosin is encoded by *spaghetti-squash,* a gene required for cytokinesis in Drosophila. *Cell* 65:1177–89

Karr, TL, Alberts, BM. 1986. Organization of the cytoskeleton in early *Drosophila* embryos. *J. Cell Biol.* 102:1494–1509

Katoh, K, Ishikawa, H. 1989. The cytoskeletal involvement in cellularization of the *Drosophila melanogaster* embryo. *Protoplasma* 150:83–95

Kellerman, KA, Miller, KG. 1992. An unconventional myosin heavy chain from *Drosophila melanogaster. J. Cell Biol.* 119: 823–34

Kellogg, DR. 1989. Organizing cytoplasmic events. *Nature* 340:99–100

Kellogg, DR, Alberts, BM. 1992. Purification of a multiprotein complex containing centrosomal proteins from the *Drosophila* embryo by chromatography with low-affinity polyclonal antibodies. *Mol. Biol. Cell.* 3:1–11

Kellogg, DR, Field, CM, Alberts, BM. 1989. Identification of microtubule-associated proteins in the centrosome, spindle, and kinetochore of the early *Drosophila* embryo. *J. Cell Biol.* 109:2977–91

Kellogg, DR, Mitchison, TJ, Alberts, BM. 1988. Behavior of microtubules and actin filaments in living *Drosophila* embryos. *Development* 103:675–86

Kiehart, DP. 1991. Contractile and cytoskeletal proteins in *Drosophila* embryogenesis. *Curr. Top. Memb.* 38:79–97

Kiehart, DP, Feghali, R. 1986. Cytoplasmic

myosin from *Drosophila melanogaster*. *J. Cell Biol.* 103:1517–25

Kiehart, DP, Ketchum, A, Young, P, Lutz, D, Alfentino, MRS, et al. 1990. Contractile proteins in *Drosophila* development. *Ann. NY Acad. Sci.* 582:233–51

Kiehart, DP, Lutz, MS, Chan, D, Ketchum, AS, Laymon, RA, et al. 1989. Identification of the gene for fly non-muscle myosin heavy chain: *Drosophila* myosin heavy chains are encoded by a gene family. *EMBO J.* 8:913–22

Kimble, M, Kuriyama, R. 1992. Functional components of microtubule-organizing centers. *Int. Rev. Cytol.* 136:1–50

Lehmann, R. 1992. Germ-plasm formation and germ-cell determination in *Drosophila*. *Curr. Opin. Genet. Dev.* 2:543–49

Lin, H, Wolfner, MF. 1991. The Drosophila maternal-effect gene *fs*(1)Ya encodes a cell cycle-dependent nuclear envelope component required for embryonic mitosis. *Cell* 64:49–62

Llamazares, S, Moriera, A, Tavares, A, Girdham, C, Spruce, BA, et al. 1991. *polo* encodes a protein kinase homolog required for mitosis in *Drosophila*. *Genes Dev.* 5: 2153–65

Mabuchi, I. 1986. Biochemical aspects of cytokinesis. *Int. Rev. Cytol.* 101:175–213

Mahowald, AP. 1962. Fine structure of pole cells and polar granules in *Drosophila melanogaster*. *J. Exp. Zool.* 151:201–7

Mahowald, AP. 1963. Electron microscopy of the formation of the cellular blastoderm in *Drosophila melanogaster*. *Exp. Cell Res.* 32:457–68

Matthews, KA, Miller, DFB, Kaufman, TC. 1989. Developmental distribution of RNA and protein products of the *Drosophila* alpha-tubulin gene family. *Dev. Biol.* 132: 45–61

Merrill, PT, Sweeton, D, Wieschaus, E. 1988. Requirements for autosomal gene activity during precellular stages of *Drosophila melanogaster*. *Development* 104:495–509

Miller, KG, Alberts, BM. 1989. F-actin affinity chromatography: Technique for isolating previously unidentified actin-binding proteins. *Proc. Natl. Acad. Sci. USA* 86: 4808–12

Miller, KG, Field, CM, Alberts, BM. 1989. Actin-binding proteins from *Drosophila* embryos: A complex network of interacting proteins detected by f-actin affinity chromatography. *J. Cell Biol.* 109:2963–75

Miller, KG, Field, CM, Alberts, BM, Kellogg, DR. 1991. Use of actin filament and microtubule affinity chromatography to identify proteins that bind to the cytoskeleton. *Meth. Enzymol.* 196:303–19

Miller, KG, Karr, TL, Kellogg, DR, Mohr, IJ, Walter, M., Alberts, BM. 1985. Studies on the cytoplasmic organization of early *Drosophila* embryos. *Cold Spring Harbor Symp. Quant. Biol.* 50:79–90

Mohler, JD. 1977. Developmental genetics of the *Drosophila* egg. I. Identification of 50 sex-linked cistrons with maternal effects on embryonic development. *Genetics* 85:259–72

Niki, Y. 1984. Developmental analysis of the grandchildless (*gs(1)N26*) mutation in *Drosophila melanogaster:* abnormal cleavage patterns and defects in pole cell formation. *Dev. Biol.* 103:182–89

Niki, Y. 1988. Developmental analysis of grandchildless (*gs(1)N441*) mutation in *Drosophila melanogaster;* abnormal formation of pole cells. *Jpn. J. Genet.* 63:23–32

Niki, Y, Okada, M. 1981. Isolation and characterization of grandchildless-like mutants in *Drosophila melanogaster*. *Roux's Arch. Dev. Biol.* 190:1–10

Oakley, BR. 1992. γ-tubulin: the microtubule organizer? *Trends Cell Biol.* 2:1–5

Okada, M. 1982. Loss of the ability to form pole cells in *Drosophila* embryos with artificially delayed nuclear arrival at the posterior pole. In *Embryonic Development, Part A: Genetic Aspects,* ed. MM Burger, G Weber, pp. 363–72. New York: Liss

Okada, M, Kleinman, IA, Schneiderman, HA. 1974. Restoration of fertility in sterilized *Drosophila* eggs by transplantation of polar cytoplasm. *Dev. Biol.* 37:43–54

Perrimon, N, Mohler, D, Engstrom, L, Mahowald, AP. 1986. X-linked female-sterile loci in *Drosophila melanogaster*. *Genetics* 113:695–712

Pesacreta, TC, Byers, TJ, Dubreuil, R, Kiehart, DP, Branton, D. 1989. *Drosophila* spectrin: the membrane skeleton during embryogenesis. *J. Cell Biol.* 108:1697–709

Planques, V, Warn, A, Warn, RM. 1991. The effects of microinjection of rhodamine-phalloidin on mitosis and cytokinesis in early stage *Drosophila* embryos. *Exp. Cell Res.* 192:557–66

Poodry, CF. 1992. Morphogenesis of *Drosophila*. In *Morphogenesis: An Analysis of the Development of Biological Form,* ed. EF Rossomando, S Alexander, pp. 143–88. New York: Dekker

Postner, MA, Miller, KG, Wieschaus, EF. 1992. Maternal effect mutations of the *sponge* locus affect actin cytoskeletal rearrangements in *Drosophila melanogaster* embryos. *J. Cell Biol.* 119:1205–18

Rabinowitz, M. 1941. Studies on the cytology and early embryology of the egg of *Drosophila melanogaster*. *J. Morphol.* 69:1–49

Raff, JW, Glover, DM. 1988. Nuclear and cytoplasmic mitotic cycles continue in *Drosophila* embryos in which DNA synthesis is inhibited with aphidicolin. *J. Cell Biol.* 107:2009–19

Raff, JW, Glover, DM. 1989. Centrosomes, not nuclei, initiate pole cell formation in Drosophila embryos. *Cell* 57:611–19

Raff, JW, Kellogg, DR, Alberts, BM. 1993. *Drosophila* γ-tubulin is part of a complex containing two previously identified centrosomal MAPs. *J. Cell Biol.* 121:823–35

Rappaport, R. 1986. Establishment of the mechanism of cytokinesis in animal cells. *Int. Rev. Cytol.* 105:245–81

Rice, TB, Garen, A. 1975. Localized defects of blastoderm formation in maternal effect mutants of *Drosophila*. *Dev. Biol.* 43:277–86

Rickoll, WL. 1976. Cytoplasmic continuity between embryonic cells and the primitive yolk sac during early gastrulation in *Drosophila melanogaster*. *Dev. Biol.* 49:304–10

Ripoll, P, Casal, J, Gonzalez, C. 1987. Towards the genetic dissection of mitosis in *Drosophila*. *BioEssays* 7:204–10

Roberts, DB, ed. 1986. *Drosophila a practical approach*. Oxford/Washington: IRL. 295 pp.

Rose, LS, Wieschaus, E. 1992. The *Drosophila* cellularization gene *nullo* produces a blastoderm-specific transcript whose levels respond to the nucleocytoplasmic ratio. *Genes Dev.* 6:1255–68

Salmon, ED. 1989. Cytokinesis in animal cells. *Curr. Opin. Cell Biol.* 1:541–47

Sanchez, F, Natzle, JE, Cleveland, DE, Kirschner, MW, McCarthy, BJ. 1981. A dispersed multigene family encoding tubulin in Drosophila melanogaster. *Cell* 22:845–54

Satterwhite, LL, Pollard, TD. 1992. Cytokinesis. *Curr. Opin. Cell Biol.* 4:43–52

Saxton, WM, Porter, ME, Cohn, SA, Scholey, JM, Raff, EC, McIntosh, JR. 1988. *Drosophila* kinesin: characterization of microtubule motility and ATPase. *Proc. Natl. Acad. Sci. USA* 85:1109–13

Schejter, ED, Rose, LS, Postner, MA, Wieschaus, E. 1992. Role of the zygotic genome in the restructuring of the actin cytoskeleton at the cycle-14 transition during *Drosophila* embryogenesis. Cold Spring Harbor Symp. Quant. Biol. 57:653–59

Schroeder, TE. 1990. The contractile ring and furrowing in dividing cells. *Ann. NY Acad. Sci.* 582:78–87

Schupbach, T, Wieschaus, E. 1989. Female sterile mutations on the second chromosome of *Drosophila melanogaster*. I. Maternal effect mutations. *Genetics* 121:101–17

Schweisguth, F, Lepesant, J-A, Vincent, A. 1990. The *serendipity alpha* gene encodes a membrane-associated protein required for the cellularization of the *Drosophila* embryo. *Genes Dev.* 4:922–31

Schweisguth, F, Vincent, A, Lepesant, J-A. 1991. Genetic analysis of the cellularization of the *Drosophila* embryo. *Biol. Cell* 72:15–23

Shamanski, FL, Orr-Weaver, TL. 1991. The Drosophila *plutonium* and *pan-gu* genes regulate entry into S phase at fertilization. *Cell* 66:1289–300

Simpson, L, Wieschaus, E. 1990. Zygotic activity of the *nullo* locus is required to stabilize the actin-myosin network during cellularization in *Drosophila* embryos. *Development* 110:851–63

Sonnenblick, BP. 1950. The early embryology of *Drosophila* melanogaster. In *Biology of Drosophila,* ed. M Demerec, pp. 62–167. New York: Wiley & Sons. 632 pp.

Spradling, AC. 1993. Molecular genetics of oogenesis. See Foe, VE 1993. In press

Stafstrom, JP, Staehelin, LA. 1984. Dynamics of the nuclear envelope and of nuclear pore complexes during mitosis in the *Drosophila* embryo. *Eur. J. Cell Biol.* 34:179–89

Sullivan, W, Fogarty, P, Theurkauf, W. 1993. Mutational analysis of the cytoskeletal organization in syncytial *Drosophila* embryos. *Development*. In press

Sullivan, W, Minden, JS, Alberts, BM. 1990. *daughterless-abo-like,* a *Drosophila* maternal effect mutation that exhibits abnormal centrosome separation during the late blastoderm divisions. *Development* 110:311–23

Sunkel, CE, Glover, DM. 1988. *polo,* a mitotic mutant of *Drosophila* displaying abnormal spindle poles. *J. Cell Sci.* 89:25–38

Swanson, MM, Poodry, CA. 1980. Pole cell formation in *Drosophila melanogaster*. *Dev. Biol.* 75:419–30

Szabad, J, Erdelyi, M, Hoffmann, G, Szidonya, J, Wright, TRF. 1989. Isolation and characterization of dominant female sterile mutations of *Drosophila melanogaster*. II. Mutations on the second chromosome. *Genetics* 122:823–35

Tearle, R, Nusslein-Volhard, C. 1987. Tubingen mutants and stock list. *Drosoph. Info. Ser.* 66:209–69

Theurkauf, WE. 1992. Behavior of structurally divergent α-tubulin isotypes during *Drosophila* embryogenesis: evidence for post-translational regulation of isotype abundance. *Dev. Biol.* 154:205–17

Thierry-Mieg, D. 1982. *Paralog,* a control mutant in *Drosophila melanogaster*. *Genetics* 100:209–37

Thomas, JB, Bastiani, MJ, Bate, M, Goodman, CS. 1984. From grasshopper to *Drosophila*: a common plan for neuronal development. *Nature* 310:203–7

Tobin, SL, Cook, PJ, Burn, TC. 1990. Transcripts of individual actin genes are differentially distributed during embryogenesis. *Dev. Genet.* 11:15–26

Turner, FR, Mahowald, AP. 1976. Scanning electron microscopy of *Drosophila* embryogenesis. *Dev. Biol.* 50:95–108

Van Der Meer, JM. 1988. The role of metabolism and calcium in the control of mitosis and ooplasmic movements in insect eggs: a working hypothesis. *Biol. Rev.* 63: 109–57

Vincent, A, Colot, HV, Rosbash, M. 1985. Sequence and structure of the *serendipity* locus of *Drosophila melanogaster*. A densely transcribed region including a blastoderm-specific gene. *J. Mol. Biol.* 186: 149–66

Vorobyev, IA, Nadezhdina, ES. 1987. The centrosome and its role in the organization of microtubules. *Int. Rev. Cytol.* 106:227–93

Warn, RM. 1986. The cytoskeleton of the early *Drosophila* embryo. *J. Cell Sci. Suppl.* 5:311–28

Warn, RM, Bullard, B, Magrath, R. 1980. Changes in the distribution of cortical myosin during the cellularization of the *Drosophila* embryo. *J. Embryol. Exp. Morph.* 57:167–76

Warn, RM, Flegg, L, Warn, A. 1987. An investigation of microtubule organization and functions in living *Drosophila* embryos by injection of a fluorescently labeled antibody to tyrosinated α-tubulin. *J. Cell Biol.* 105:1721–30

Warn, RM, Magrath, R. 1982. Observations by a novel method of surface changes during the syncytial blastoderm stage of the *Drosophila* embryo. *Dev. Biol.* 89:540–48

Warn, RM, Magrath, R. 1983. F-actin distribution during the cellularization of the *Drosophila* embryo visualized with FL-phalloidin. *Exp. Cell Res.* 143:103–14

Warn, RM, Magrath, R, Webb, S. 1984. Distribution of f-actin during cleavage of the *Drosophila* syncytial blastoderm. *J. Cell Biol.* 98:156–62

Warn, RM, Robert-Nicoud, M. 1990. F-actin organization during the cellularization of the *Drosophila* embryo as revealed with a confocal laser scanning microscope. *J. Cell Sci.* 96:35–42

Warn, RM, Smith, L, Warn, A. 1985. Three distinct distributions of f-actin occur during the divisions of polar surface caps to produce pole cells in *Drosophila* embryos. *J. Cell Biol.* 100:1010–15

Warn, RM, Warn, A. 1986. Microtubule arrays present during the syncytial and cellular blastoderm stages of the early *Drosophila* embryo. *Exp. Cell Res.* 163:201–10

Warn, RM, Warn, A, Planques, V, Robert-Nicoud, M. 1990. Cytokinesis in the early *Drosophila* embryo. *Ann. NY Acad. Sci.* 582:222–32

Whitfield, WGF, Miller, SE, Saumweber, H, Frasch, M, Glover DM. 1988. Cloning of a gene encoding an antigen associated with the centrosome in *Drosophila*. *J. Cell Sci.* 89:467–80

Wieschaus, E, Sweeton, D. 1988. Requirements for X-linked zygotic activity during cellularization of early *Drosophila* embryos. *Development* 104:483–93

Wolf, R. 1978. The cytaster, a colchicine-sensitve migration organelle of cleavage nuclei in an insect egg. *Dev. Biol.* 62:464–72

Wolf, N, Regan, CL, Fuller, MT. 1988. Temporal and spatial pattern of differences in microtubule behavior during *Drosophila* embryogenesis revealed by distribution of a tubulin isoform. *Development* 102:311–24

Yang, JT, Saxton, WM, Goldstein, LSB. 1988. Isolation and characterization of the gene encoding the heavy chain of *Drosophila* kinesin. *Proc. Natl. Acad. Sci. USA* 85:1864–8

Yasuda, GK, Baker, J, Schubiger, G. 1991. Independent roles of centrosomes and DNA in organizing the *Drosophila* cytoskeleton. *Development* 111:379–91

Young, PE, Pesacreta, TC, Kiehart, DP. 1991. Dynamic changes in the distribution of cytoplasmic myosin during *Drosophila* embryogenesis. *Development* 111:1–14

Young, PE, Richman, AM, Ketchum, AS, Kiehart, DP. 1993. Morphogenesis in *Drosophila* requires nonmuscle myosin heavy chain function. *Genes Dev.* 7:29–41

Zalokar, M, Audit, C, Erk, I. 1975. Developmental defects of female-sterile mutants of *Drosophila melanogaster*. *Dev. Biol.* 47: 419–32

Zalokar, M, Erk, I. 1976. Division and migration of nuclei during early embryogenesis of *Drosophila melanogaster*. *J. Microsc. Biol. Cell.* 25:97–106

Annu. Rev. Cell Biology. 1993. 9:101–28

SUPERANTIGENS: Bacterial and Viral Proteins that Manipulate the Immune System

Mark T. Scherer, Leszek Ignatowicz, and Gary M. Winslow
Howard Hughes Medical Institute, Division of Basic Immunology, Department of Medicine, National Jewish Center for Immunology and Respiratory Medicine, Denver, Colorado 80206

John W. Kappler
Departments of Microbiology and Immunology, and Medicine, University of Colorado Health Sciences Center, Denver, Colorado 80206

Philippa Marrack
Departments of Microbiology and Immunology, and Medicine, and Department of Biochemistry, Biophysics and Genetics, University of Colorado Health Sciences Center, Denver, Colorado 80206

KEY WORDS: mammary tumor virus, Mtv, staphylococcal enterotoxins, Mls, T cell receptor Vβ

CONTENTS

0743–4634/93/1115–0101$05.00

INTRODUCTION

Lymphocytes allow the immune system to recognize and respond specifically to a host of foreign antigens. Three known types of lymphocytes exist that are characterized by three different types of clonally distributed receptors, which they use to recognize antigen.

B lymphocytes have receptors consisting of immunoglobulins. These cells usually recognize foreign material in its native form; for example during a virus infection, B cells bind the free virus and produce serum immunoglobulins that have the same specificity as the membrane bound receptor form. These antibodies bind cell-free virus and inhibit its infectious activity, or stimulate its uptake by phagocytic cells such as macrophages, thereby causing the clearance of the infectious agent from the animal (Amzel et al 1974; Richards et al 1975).

Some T cells have receptors consisting of γ and δ chains. The specificities of $\gamma\delta$ receptors are not completely understood. However, it is likely that in at least some instances these receptors are able to recognize products produced by damaged tissues of their hosts. For example, the receptors on a major group of $\gamma\delta$-bearing T cells can react with peptide fragments of host heat shock proteins. Thus it is thought that $\gamma\delta^+$ T cells respond to the by-products of infection, such as host proteins made in damaged or shocked tissue (Allison & Havran 1991; Brenner et al 1986; Davis & Bjorkman 1988).

Finally, another set of T cells have receptors consisting of α and β chains. $\alpha\beta$ receptors recognize most antigens as peptide fragments bound in grooves on cell surface proteins, the products of genes mapping to the major histocompatibility complex (MHC) (Bjorkman et al 1987; Garrett et al 1989). Two different types of MHC proteins exist, Class I and Class II. Complex and interesting intracellular transportation pathways lead to the binding of peptides from different sources to the grooves of Class I and Class II MHC proteins. Peptides destined to bind to Class I are usually derived from proteins in the cytoplasm of the Class I-bearing cells (Townsend & Bodmer 1989). These can include viral matrix proteins. Peptides destined to bind to Class II MHC are usually derived from proteins in the endosomes, lysozomes, plasma membrane, or extracellular milieu of the cells (Brown et al 1988). Thus unlike B cells or $\gamma\delta$-bearing T cells, $\alpha\beta$-bearing T cells recognize foreign material as peptide fragments derived from almost any portion of the invading organism, and it is often suggested that the immune system has developed these three methods of recognizing the appearance of foreign substances in the body as fail-safe mechanisms. Thus some antigens, for example influenza virus, may be able to escape attack by B cells and antibodies by mutation of the major viral coat proteins. Such a stratagem would not, however, allow the virus to escape the attention of αβ- or γδ-bearing T cells, which recognize invading organisms by mechanisms that differ from those of B cells.

These three types of receptors share several characteristics. In each case the receptors are constructed of two clonally variable polypeptide chains. In each case the genes that will encode these receptors are assembled during lymphocyte development by rearrangement of one of several or many genetic elements. For example, a gene for a T cell receptor (TCR) β chain is formed during T cell development by rearrangement of 1 of 2 Dβ genes to 1 of about 12 Jβ genes. Subsequently 1 of about 20 (mouse) or 60 (human) Vβ genes rearranges to lie next to DβJβ and form the functional complete β gene in that T cell. Likewise, TCR α chains are formed by the rearrangement of 1 of about 50 Vαs to lie next to 1 of about 50 Jαs. Because so many combinations of Vβ, Dβ, Vα, etc exist, and because nucleotides that are not germ line-encoded can be introduced at the joining points of these different genetic elements, any animal can make many different αβ TCRs, perhaps as many as 10^{14} (Davis & Bjorkman 1988; Kronenberg et al 1986; Lewis & Gellert 1989). For conventional peptide antigens, all the variable elements of the α and β chains of the TCR are directly involved in the process of antigen/MHC recognition. Therefore, only about 1 in 105 or 1 in 106 of all αβ-bearing T cells in a naive individual is able to respond to a particular antigen/MHC combination.

For reasons that are not fully understood, in some cases microorganisms have co-opted the receptors of the immune system for their own purposes. Staphylococci, for example, produce a group of proteins, including protein A and protein G, that binds to antibody molecules at sites distant from those usually used to engage antigen. Microorganisms also produce a group of proteins that binds to MHC Class II molecules and can then engage particular TCR Vβ sequences. Because there are only a limited number of Vβ regions in any given animal, these Vβ-specific proteins can interact with a sizable percentage of all $\alpha\beta^+$ T cells. Toxic shock syndrome toxin 1 (TSST1), for example, engages nearly all T cells bearing Vβ2 in humans, a population that is about 7% of the mature $\alpha\beta^+$ cells in any given individual (Choi et al 1990b). The unusual ability of these antigens to stimulate such a high frequency of T cells has led to the name superantigens.

Superantigens and their stimulation of T cells have several other properties that distinguish them from the antigenic peptide/MHC combinations that normally engage αβ TCRs. Superantigens bind to a site on MHC Class II that is outside the peptide-binding groove described by Bjorkman et al (1987) (Dellabona et al 1990). Superantigens act on T cells as intact proteins, without the processing that is required to produce conventional, MHC-groove binding, antigenic peptides. Unlike conventional peptide antigens, superantigens bind and are presented to T cells by many alleles of MHC class II molecules (Fleischer & Schrezenmeier 1988; Herman et al 1990; White et al 1989). Finally, no superantigens have yet been discovered that bind to Class I MHC molecules, and all known forms of these protcins are specific for the Vβ regions of TCRs.

At present, two kinds of superantigens are known. Several types of bacteria, such as staphylococci or streptococci, make bacterial toxins with the properties of superantigens. The second sources of known superantigens are the retroviruses, mouse mammary tumor viruses (MTVs, Dyson et al 1991a; Frankel et al 1991; Marrack et al 1991; Woodland et al 1990b, 1991c). Bacterial and viral superantigens are not related in amino acid sequence, and therefore they appear to have evolved independently. However, the fact that both sources of superantigens are potentially pathogenic organisms that can infect the same host (*Mus. musculus*) suggests that the two groups could have influenced each other during evolution.

Here we review the effects of superantigens on the immune system. We then describe the structures and functions of the two known types of superantigens and finish with a discussion of the roles these proteins may play in the life cycles of the microorganisms that produce them.

SUPERANTIGENS AND THE IMMUNE SYSTEM

Superantigens react with a large proportion of all $\alpha\beta^+$ T cells in most hosts, therefore most superantigens have a powerful effect on the repertoire of T cells. Since superantigens react with T cell receptor Vβs after combination with Class II MHC rather than Class I MHC, in some cases the reaction between superantigen and T cell depends upon expression by the T cell of the Class II-binding protein, CD4. Well-expressed, high affinity superantigens react with T cells bearing either CD4 or CD8 because the avidity of the reaction between the T cell receptor Vβ and the superantigen/Class II complex is thought to be so high that additional binding activity from the Class II-specific adhesion molecule CD4 is not also required. On the other hand, poor superantigens bind to target T cell receptors with low avidity, and therefore successful engagement of T cells by such superantigens requires additional energy provided by the interaction between CD4 and Class II molecules. $CD8^+$, $CD4^-$ T cells are therefore usually not affected by such superantigens, regardless of the quality of their T cell receptors.

If mature animals are exposed to superantigens, mature T cells bearing target Vβs in these animals respond to the challenge by rapid proliferation and production of lymphokines. Attack by TSST1, for example, can cause the percentage of $V\beta2^+$ T cells in humans to rise from 7% (see above) to more than 50% (Choi et al 1989). In fact, it is thought that the high levels of lymphokines, such as tumor necrosis factor and interleukin 2, which result from this massive T cell stimulation, are the causative agents in the symptoms of toxic shock (Bohach et al 1990). Thus TSST1 is not in itself directly toxic, rather the disease it causes is an indirect consequence of T cell stimulation (Marrack et al 1990).

In some cases, however, superantigens are present constitutively in animals. Superantigens produced by integrated MTVs, for example, are produced in mice almost as soon as the immune system begins to develop. Consequently such superantigens are viewed by the immune system as self, and the animal is tolerant to them. Superantigens of this type have been invaluable, in fact, in dissecting the mechanisms whereby higher mammals become self-tolerant. Over the years, several hypotheses have been suggested to account for the fact that the T cells of a healthy animal do not respond to the tissues of their own hosts. Many years ago Lederberg (1959) suggested that this fact might be due to the death of self-reactive T cells during some immature, death-sensitive stage of their development. Later it was suggested that such cells might not die, but perhaps rather be inactivated by contact with self. Finally the idea that potentially self-reactive cells might be kept at bay by a lymphocyte network has also gained some support. There is evidence that each of these hypotheses might account for self-tolerance under different cirumstances. However, superantigens were instrumental in the discovery that clonal deletion, the death of immature self-reactive T cells upon contact with self-antigen, is a major feature of the mechanisms of tolerance. The superantigen encoded by *Mtv7,* for example, located on chromosome 1 of some strains of mice, reacts with T cells bearing Vβs 6, 7, 8.1 and 9, a population that can be up to 30% of all the αβ-bearing T cells in a given mouse (Kappler et al 1988; MacDonald et al 1988). In mice that express the correct alleles of Class II MHC, expression of vSAG7 causes the intrathymic deletion of nearly all T cells bearing the target Vβs.

Extrathymic superantigens can also cause the death or inactivation of mature T cells (Blackman et al 1990). This is surprising since until recently it was assumed that most encounters between mature T cells and antigen would lead to T cell proliferation and activation rather than T cell death. However, it is now clear that under some circumstances antigen or superantigen can cause the death of target T cells. This phenomenon, which is thought to mimic induction of tolerance to self-antigens that cannot reach the thymus, is mysterious. Immunologists are still struggling to find out why mature T cells sometimes respond to antigen by proliferation and differentiation, or may sometimes die or become inactivated in response to a similar challenge.

It is possible that the outcome is controlled by the nature of the antigen-presenting cells involved. Antigen-presenting B cells are frequently tolerogenic, whereas other types of "professional" antigen-presenting cells such as macrophages or dendritic cells may usually be stimulatory. Superantigens bind as intact proteins to naked Class II MHC molecules. Conventional antigens must be proteolyzed, converted into MHC-binding peptides, before they can be recognized by T cells. Therefore it is likely that most T cells will be confronted by superantigen on resting B cells, which are poor processors of intact protein, but bear abundant Class II MHC molecules. Conventional antigens, however, may usually be recognized by T cells after processing by stimulatory macrophages. Therefore tolerance induction of mature T cells may be easier to induce with superantigens rather than conventional protein antigens. In any event, superantigens have proved to be invaluable in studies of tolerance induction in either immature or mature T cells.

STAPHYLOCOCCAL AND STREPTOCOCCAL SUPERANTIGENS

These gram-positive bacteria are well-characterized toxins. They cause a number of diseases including food poisoning, toxic shock and scalded skin syndrome in humans and other mammals. The primary amino acid sequences of many of these toxins have been known for some time, in some cases from protein sequencing, and in other cases deduced from the sequences of the genes encoding them (Bergsdoll 1970; Huang & Bergsdoll 1970; Huang et al 1987; Marrack & Kappler 1990). Recently the structure of one of these toxins, staphylococcal enterotoxin B (SEB), has been solved by X-ray crystallography (Swaminathan et al 1992). It has also long been known that these toxins have potent T cell stimulatory activity (Peavy et al 1970; Smith & Johnson 1975). It has only recently been realized, however, that the toxins act as superantigens, binding to Class II MHC on presenting cells, and engaging TCRs via Vβ (Janeway et al 1989; White et al 1989).

The toxins are known to be relatively compact and protease-resistant, properties that might be expected of proteins that can act enterically to cause food poisoning. Extensive experiments involving toxins fragmented in various ways suggest that they function as intact proteins (Fischer et al 1989; Fraser 1989; Scholl et al 1989). Some clipping is allowed as long as it does not disrupt the tertiary structure of the protein. Staphylococcal enterotoxin B, for example, contains an intrachain disulfide bond. The protein may be cleaved within this disulfide loop (and indeed frequently is, in commercial preparations), and this cleavage does not appear to interfere with the function of the protein (Dalidowicz et al 1966; Spero et al 1973). Inspection of the solved structure of SEB suggests that the disulfide-bonded loop lies to one side of the protein, away from the sites thought to bind Class II molecules or TCR Vβ

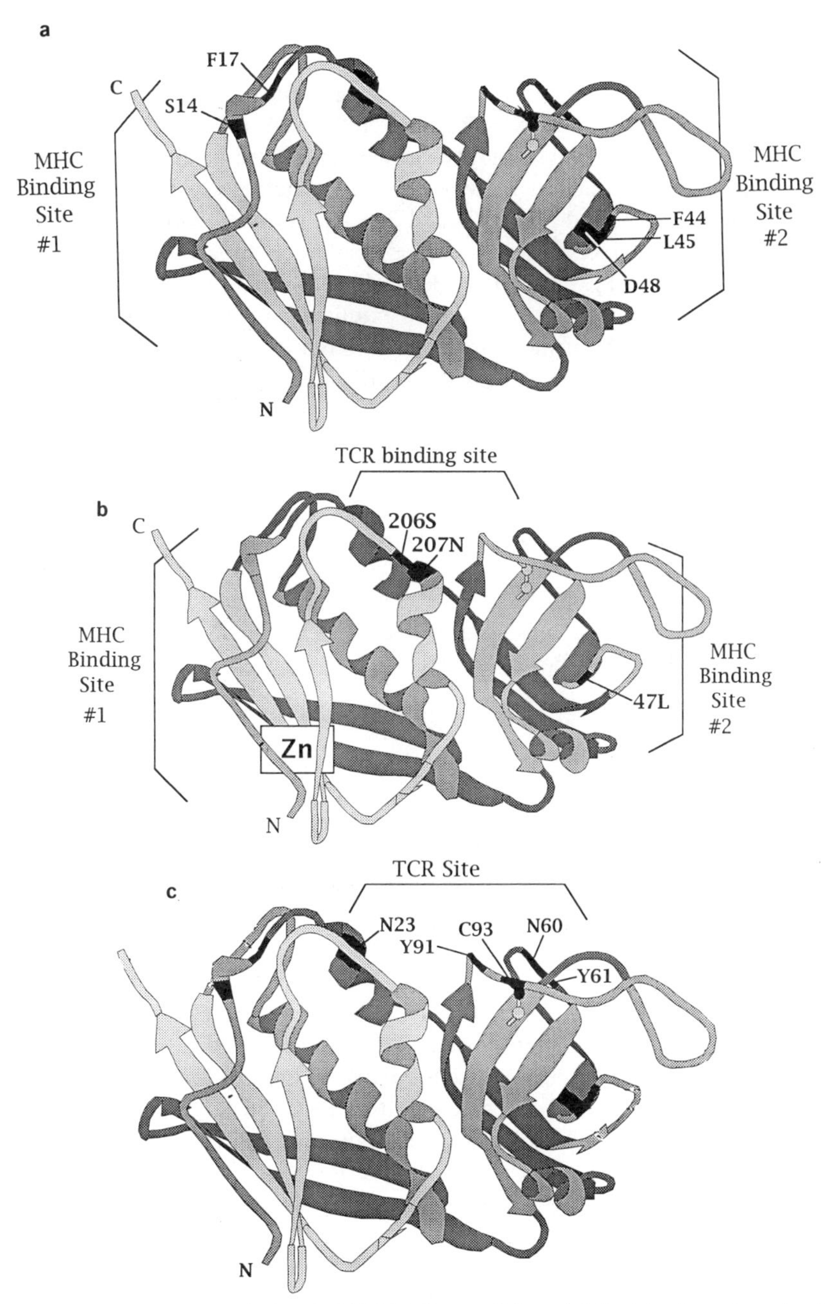

Figure 1 Sites of Vβ and Class II MHC binding on SEB and SEA. The three-dimensional structure of SEB is shown in Figures 1a and c (Swaminathan et al 1992). Figure 1b shows a structure for SEA modeled on that of SEB. The regions and amino acids of SEB and SEA known to be involved in Class II binding are indicated on Figures 1a and b. The regions and amino acids of SEA and SEB known to be involved in Vβ binding are shown in Figures 1b and c.

Many of the staphylococcal and streptococcal enterotoxins have obvious sequence similarities. Staphylococcal enterotoxin A (SEA) and SEE, for example, are about 90% homologous at the amino acid level. These toxins are also obviously, but less closely, related to another group of staphylococcal toxins, SEB, and SEC1, 2, and 3 (Callahan et al 1990; Marrack & Kappler 1990). Yet other toxins, for example streptococcal toxin C and staphylococcal TSST1, have barely detectable sequence similarities to those mentioned above, and the staphylococcal exfoliating toxins have no similarity to the others. In spite of the tremendous differences in primary amino acid sequence between the toxins, all these proteins share one property, their ability to bind MHC Class II molecules and stimulate T cells. They differ, however, in their Vβ specificities. Toxins with related amino acid sequences have related specificities for Vβ . Staphylococcal enterotoxin A and SEE, for example, both engage T cell receptors bearing mouse Vβ3, 7, and 17, albeit with different avidities. Staphylococcal enterotoxin A does, however, have an additional specificity for T cells bearing Vβ1 (Kappler et al 1989). Staphylococcal enterotoxin B and its close sequence relatives, SECs 1–3, share reactivity with T cells bearing members of the mouse Vβ8 family (Callahan et al 1990; Janeway et al 1989).

Sites on the Bacterial Toxins that Bind Class II Molecules and TCRβ and Vice Versa

The sites on MHC Class II molecules that bind the bacterial toxins are not fully understood. It is known that at least some of these toxins do not interfere with binding of conventional antigenic peptides to Class II molecules, and therefore the toxins do not bind in the groove of Class II molecules known to bind peptides (Dellabona et al 1990). This is confusing because the toxins almost certainly bind at more than one site on MHC molecules, as suggested by the fact that TSST1 and SEB do not compete for binding to Class II molecules, whereas SEA inhibits Class II engagement by either of these toxins (Chintagumpala et al 1991; Scholl et al 1989; See et al 1990). Data on SEB mutants (discussed below) suggest that two different sites on SEB are involved. Therefore, overall the data suggest that two separate sites on MHC molecules are engaged by different sites on the various toxins. For each toxin, the relative importance of the two sites in Class II binding may vary. Different sites may dominate in the binding of TSST1 and SEB, whereas SEA may bind equally well to both sites.

This idea may account for the contradictory data on the location of the toxin-binding sites on Class II. We have shown that a histidine residue on the Class II β chain, and on a side of Class II away from that thought to engage TCR or conventional peptides, is involved in binding SEA (Herman et al 1991). Changes in this residue do not, however, affect SEB binding.

By contrast amino acids on both the α and β chain of Class II are involved in binding TSST1. Moreover, for this toxin, it appeared that the α chain of Class II might have a high affinity binding site and the β chain of Class II a low affinity binding site (Braunstein et al 1992; Panina-Bordignon et al 1992).

The possibility that there are two binding sites on toxins for Class II molecules is supported by work on the amino acids in the toxins that affect Class II binding. We have reported a collection of SEB mutants which, because of amino acid changes, have lost the ability to bind either to Class II or to TCR Vβ (Kappler et al 1992). Analysis of the ability of these mutants to bind to Class II reveals two stretches of important amino acids. One stretch involves SEB phenyl alanine 44 and leucine 45, a pair of amino acids that are relatively conserved among all the bacterial superantigens, which suggests that this site might be common to most of the toxins. Another stretch of amino acids involves SEB 14–17, a region of the molecule that varies considerably from one bacterial superantigen to another. These two sites are mapped onto the reported structure of SEB in Figure 1a (Swaminathan et al 1992). It is clear that they lie on different parts of the molecule. They do, however, lie on the same surface, a fact that may be important in their function.

The notion of two Class II-binding sites is supported by work on SEA. Studies from Fraser's laboratory have implicated a zinc binding site (Fraser et al 1992) in the binding of SEA to Class II molecules. However, our own experiments have suggested that SEA leucine 47, the analogue of leucine 45 in SEB, is involved in the binding of SEA to Class II (J. Clements & J. Kappler, unpublished data). If SEA is modeled on the solved structure for SEB (Swaminathan et al 1992), as shown in Figure 1b, the two SEA Class II-binding sites lie in positions similar to those found on SEB, at distant sites on the same surface of the molecule.

Therefore, it seems that the bacterial superantigens bind Class II at two different sites. Although it is possible that the same toxin uses both sites to bind the same MHC molecule, it is more likely that the two binding sites are involved in crosslinking Class II proteins. Since T cell activation is known to be heavily dependent upon crosslinking T cell receptors and their accessory molecules (Meuer et al 1983), it is likely that this crosslinking activity is important to the strength of the effects of the bacterial superantigens on T cells.

The portion of Vβ that engages the bacterial toxins is relatively well understood. Although a formal structure for TCRs is not yet available, the structure of the variable regions of TCRs can probably be modeled with a fair degree of confidence on the known structure of immunoglobulin light and heavy chain variable regions (Chothia et al 1988; White et al 1993).

With such a model in hand, the binding site on Vβ for bacterial toxins is the TCR equivalent of immunoglobulin Hv4 (Choi et al 1990a; White et al 1993). This site consists of a pair of β strands and the loop between them, which lie on the surface of the T cell receptor exposed to the environment, away from the binding site of Vβ for Vα or conventional peptide plus MHC.

Likewise, the area on the bacterial toxins that engages Vβ is also relatively well established. Mutational and sequence exchange studies (Hudson et al 1993; Irwin et al 1992; Kappler et al 1992; Mollick et al 1993; Swaminathan et al 1992) have shown that the residues responsible for Vβ engagement lie in a pocket on one surface of both SEA and SEB (Figures 1b and c). It is not difficult to imagine that the predicted Hv4-like loop of Vβ could dip down and contact the amino acid residues lining this pocket.

In spite of the conclusion that the Vβ-binding site is confined to the pocket, as illustrated in Figure 1, it is still difficult to predict the Vβ specificity of a particular toxin from its primary amino acid sequence modeled onto the known structure of SEB. As mentioned above, toxins with similar Vβ specificities share amino acid sequences, but the shared residues are not noticeably concentrated on the residues that line the Vβ-binding pocket. In addition, the exfoliating toxins are so unlike the other bacterial products that we cannot even tell how to align the two sets of proteins in order to allow a search of this type.

MOUSE MAMMARY TUMOR VIRUS SUPERANTIGENS

Genetics of Endogenous Superantigens

Mouse mammary tumor viruses (MTVs) are retroviruses that are found either as free virus particles transmitted from mother to pup in milk or as integrants in mouse genomes, transmitted in the germ line from parents to their progeny (Kozak et al 1987). Many integrants have been identified by their chromosomal positions and presence or absence in various mouse strains (Frankel et al 1991; Lee & Eicher 1990; Peters et al 1986). In some cases the chromosomal location of particular MTVs has not been identified, but the existence of these viruses has been established by the unique size of the fragments containing their DNA after digestion with particular restriction enzymes. The MTVs each encode a superantigen (vSAG) in the 3 long terminal repeats of their DNA. Like the bacterial superantigens, the vSAGs act by binding to Class II molecules, then engaging T cell receptor Vβ, and thus activating T cells. Unlike the bacterial superantigens, however, the activity of vSAGs varies considerably, depending upon the Class II protein involved. vSAGs associated with any I-E Class II protein react well with T cells (Kappler et al 1987, 1988; MacDonald et al 1988). vSAGs are also stimulatory when associated with some I-A Class II proteins, for example those of the k, b, and d alleles, but not with others, in particular I-A^q. Interestingly, the patterns of Class II association are the same regardless of the vSAG involved. Since the different vSAGs are similar in sequence, except for their C terminal Vβ-binding sites (see below), this Class II specificity on the part of the vSAGs is thought to be because vSAGs may not bind well to particular Class II sequences (Winslow et al 1992). This conclusion is not certain, however, and it may be that there are other incompatibilities involved.

The integrated MTVs have been numbered approximately in the order of their discovery. Exogenous MTVs are known by letters designating their strain of origin. The superantigen encoded by each of these is known by the number or letters of the MTV from which it is derived. For example *Mtv7,* located on chromosome 1 of some strains of mice, codes for vSAG7. The milk-borne virus of C3H/HeJ mice is known as MTV(C3H), and the vSAG it encodes is called vSAG(C3H). Expression of these vSAGs in mice has profound effects on the T cells of their hosts (described above), and these effects have led to great interest in the genetics, structure, and function of the vSAGs.

Although the presence of MTVs in a few mouse strains, or the distribution of a particular MTV in several mouse strains has been widely reported, a list of all MTVs present in many strains has never been published. To resolve discrepancies and gaps in the literature, Scherer et al (our laboratory) examined the vSAGs expressed in 19 mouse strains. The presence of vSAGs was established by Southern blot (M. Scherer & L. Ignatowicz, unpublished results) (with a vSAG-specific probe) after digestion of mouse DNA with two restriction enzymes not previously used in such studies. These blots allowed the unambiguous identification of the vSAGs present in each mouse strain (Figure 2). A summary of the various vSAGs present in different mouse strains is shown in Table 1. The MTV vSAGs are extensively distributed throughout all the commonly used strains. Only a few strains have been found to have no endogenous MTVs, those derived from a breeding pair originally trapped in Czechoslovakia (Gallahan et al 1987), and a new strain recently derived by M. Scherer (unpublished data). As has been noted, and not surprisingly, since these viruses are transmitted in the germ line, related mouse strains share MTV integrants. Mice in the C57 group, for example, all have MTVs 8, 9, and 17. Some members of the group have additional MTVs, however, for example C57L mice have MTVs 11 and 29 as well.

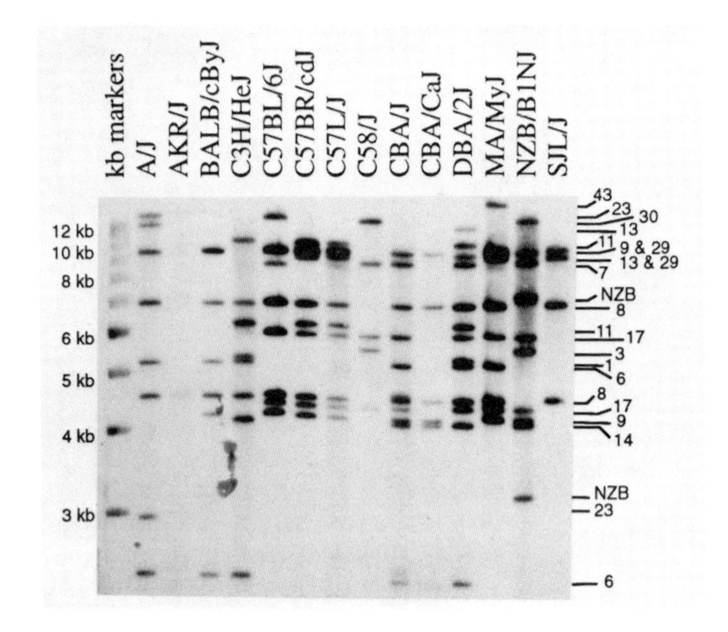

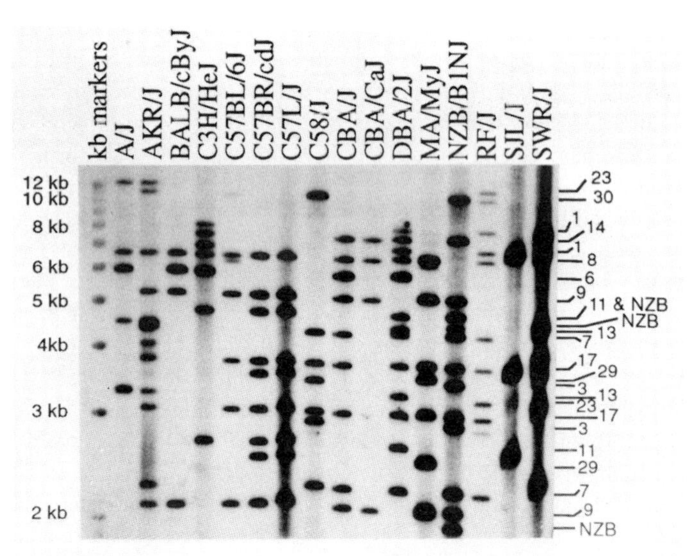

Figure 2 Selected Southern Blots showing Mtv distribution in various strains of mice. Genomic DNA was prepared from livers of female individuals from various stains of mice. The DNA was digested with either Bst XI (*top*) or Nco I (*bottom*), then separated by electrophoresis on an agarose gel. Southern blotting was performed using a 1 kb fragment of the vSAG gene from the exo(C3H) MTV as a probe (Pullen et al 1992b). The MTV assignments were made by comparing strain distributions of fragments to EcoR I and Pvu II digests and by checking sequences obtained by PCR.

Table 1 Stain distribution of some MTV proviruses

		MTV	1	3	6	7	8	9	11	13	14	17	23	29	30	31[a]	43
Chromosome			7	11	16	1	6	12	14	4	4	4	6	6	12	Y[a]	
cM from centromere			31	72	31	73	31	40	8	47	10	20	32	20–24	2		
Vβ Specificity			3, 17?	3, 17	3, 5, 17?	6,7, 8.1, 9	11, 12 17?	5, 11, 12 17?	11, 12 17	3, 17?	?	?	7	16?	5?	?	6,7 8.1 9
Strains	H2	IE															
A/J	a	k	−	−	+	−	+	−	−	+	−	−	+	−	−	−	−
AKR/J	k	k	−	−	−	+	+	+	−	−	−	+	+	−	+	+	−
BALB/cByJ	d	d	−	−	+	−	+	+	−	−	−	−	−	−	−	−	−
B10.BR	k	k	−	−	−	−	+	+	−	−	−	+	−	−	+	−	−
C3H/HeJ	k	k	+	−	+	−	+	−	+	−	+	−	−	−	−	−	−
C57BL/6J	b	-	−	−	−	−	+	+	−	−	−	+	−	−	+	−	−
C57BR/cdJ	k	k	−	−	−	−	+	+	+	−	−	+	−	+	−	?	−
C57L/J	b	-	−	−	−	−	+	+	+	−	−	+	−	+	−	?	−
C58/J	k	k	−	+	−	+	−	−	−	−	−	+	−	−	+	−	−
CBA/CaJ	k	k	−	−	−	−	+	+	−	−	+	−	−	−	−	−	−
CBA/J	k	k	−	−	+	+	+	+	−	−	+	+	−	−	−	−	−
DBA/1J	q	-	+	−	+	+	+	−	+	+	+	+	−	−	−	−	−
DBA/2J	d	d	+	−	+	+	+	−	+	+	+	+	−	−	−	−	−
LP/J	b	-	−	−	+	−	+	+	+	+	?	+	−	−	+	?	−
MA/MyJ	k	k	−	−	−	−	+	+	−	−	−	+	−	+	−	?	+
NOD/Lt[b]	nod	-	−	+	−	−	−	−	−	−	−	+	−	−	−	+	−

Table 1 (*Continued*)

		MTV	1	3	6	7	8	9	11	13	14	17	23	29	30	31[a]	43
Strains	H2	IE															
NZB/B1NJ[c]	d	d	–	+	–	+	–	+	–	–	+	+	–	–	+	–	–
NZW	z	u	–	+	+	–	+	–	–	–	–	+	+	+	–	+	–
RF/J[d]	k	k	+	–	–	+	+	–	–	–	–	+	+	–	+	?	–
RIIIS/J[d]	r	r	–	–	–	–	+	–	–	–	+	–	–	–	–	?	–
SJL/J	s	-	–	–	–	–	+	–	–	–	–	–	–	+	–	+	–
SM/J	v	v	–	–	+	+	+	–	–	–	+	+	–	–	–	?	–
SWR/J	q	-	–	–	–	+	+	–	–	–	+	+	–	–	–	+	–

The MTV proviruses found in bold-faced strains were determined by Southern blot in our laboratory, while italicized strains were from (Fairchild et al 1992; Frankel et al 1991). The chromosomal location of the MTVs is from (Peters 1992). The Vβ specificity is from (Bill et al 1989; Blackman et al 1992; Fairchild et al 1992; Gollob & Palmer 1992; Happ et al 1989; Jouvin-Marche et al 1992b; Kappler et al, 1988, 1987; MacDonald et al 1988; McDuffie et al 1992; Okada et al 1990; Pullen et al 1992b, 1988; Rudy et al 1992; Tomonari & Fairchild 1992; Tomonari et al 1992; Woodland et al 1990b).

[a] *Mtv31* is on the Y chromosome of *M. musculus domesticus* strains, and not found in *M. musculus musculus* strains (Prochazka & Leiter 1990).

[b] Some, but not all NOD mice also contain *Mtv-45* (Fairchild et al 1992)

[c] NZB mice also contain *Mtv27*, which deletes Vβ3 bearing T cells (Tomonari et al 1992), *Mtv28* (Tanaka & Matsuzawa 1990), and *Mtv44*, which deletes Vβ6, 8.1, and 9 bearing T cells (Fairchild et al 1992).

[d] RIII/S mice probably contain *Mtv21* (Peters et al 1986).

The Structure of Viral Superantigens

All viral superantigens (vSAGs) characterized to date are encoded in the 3 long terminal repeat (LTR) plus the last base (A) of the envelope genes of the MTVs (Acha-Orbea & Palmer 1992; Choi et al 1991; Dyson et al 1991b; Frankel et al 1991; Marrack et al 1991; Woodland et al 1990a, 1991b). The vSAGs consist of a family of proteins 317–321 amino acids in length, which is highly conserved except for the last 30 residues (Figure 3) (Beutner et al 1992; Crouse & Pauley 1989; Donehohower et al 1983; Fasel et al 1982; Pullen et al 1992b). The promoter for the vSAGs is contained within the 5 LTR and contains a glucocorticoid-inducible transcriptional enhancer element (Kuo et al 1988).

Experiments using cell-free systems (Choi et al 1992; Knight et al 1992; Korman et al 1992) and insect cells (Brandt-Carlson & Butel 1991) showed that vSAG proteins contain four glycosylation sites and have a molecular weight of about 37 K when translated in vitro in the absence of microsomes, and 45 K when translated in vitro in the presence of microsomes. The vSAGs are transported into microsomes as type II transmembrane proteins, i.e. with their N termini in the cytoplasm and an extracellular C terminus. There is a possible

```
Residue              290       300       310       320
                      |         |         |         |
Consensus  YIYLGTGMN-VWGKIFHYTKEGAVARILEHISADTFGMSYND*       Vβ specificity

vSAG7      ·········-F·····D··E···I·K··YNMKYTHG·RVGF·PF*     6,7,8.1,9
vSAG43     ·········-F·····D··E····K··YNMKYTHG·RVGF·PF*      6,7,8.1,9
vSAG(SW)   ·········-F·····D··E···I·K·IYN·KYTHG·RIGF·PF*     6,7,8.1,9

vSAG17     ····R·RI·--·K·············Q·········DIR··K*       ?
vSAG23     ····R·RI·--·K·············Q·········DIR··K*       7
vSAG(M12)  ·········-···R·LQ·········Q···········I*          7

vSAG1      ········I-H·--V·YNSR·E·KRH·I···K·LPLAF*           3,   17?
vSAG3      ········I-H·--V·YNSR·E·KRH·I···K·LPLTF*           3,   17
vSAG6      ········I-H·--V·YNSR·E·KRH·I···K·LPLAF*           3,5,17?
vSAG13     ········I-H·--V·YNSR·E·KRH·I···K·LPLAF*           3,   17?

vSAG8      ·········-··············Q··············G*           11,12,17?
vSAG9      ·········-··············L··············G*         5,11,12,17?
vSAG11     ·········-··············Q··············G*           11,12,17
vSAG30     ····R·EI·-··············L··············G*         5?

vSAG(C3H)  ········H-F·········-····T··GLI··Y··K·Y····YE*    14,15
vSAG(GR)   ········H-F···V·-·········GLI··Y··K·Y····Y·*      14,15
```

Figure 3 Comparison of amino acid sequences for selected MTV vSAGs. The sequences are aligned by family. A consensus sequence for each family was created by choosing the most commonly occurring amino acid at each location, then similarly creating a general consensus sequence using the consensus sequences for each family. The sequences of vSAG23 vSAG30, and vSAG(M12) are from L. Ignatowicz et al, submitted. The other sequences are from (Beutner et al 1992; Choi et al 1991, 1992; Crouse & Pauley 1989; Donehohower et al 1983; Fasel et al 1982; Held et al 1992; King et al 1990; Korman et al 1992; Pullen et al 1992b; Rudy et al 1992).

signal/transmembrane sequence at residues 45–63. Although there are five possible initiation codons, only the first two produce a functional superantigen capable of stimulating T cells in vitro (Choi et al 1992).

Using an antibody specific for the C terminus of vSAG7, Winslow et al (1992) were able to detect the superantigen on the surface of LPS-stimulated B cells, but not on resting or activated T cells, resting B cells, or thymocytes. They showed that although vSAG7 is synthesized as a 45-kd protein, the carboxy-terminus is found on the surface of LPS-stimulated B cells and transfected cells only as an 18.5-kd polypeptide. This 18.5-kd form is not glycosylated, and is presumed to be derived from the 45-kd form by proteolysis at a consensus protease site (Arg-Lys-Arg-Arg) near residue 170.

The C terminus is important for superantigen function as demonstrated by the fact that antibodies to the last 10–20 residues can inhibit superantigen stimulation of T cells (Acha-Orbea & Palmer 1992; Mohan et al 1993; Winslow et al 1992). In addition, there is evidence that a region varying somewhat between vSAGs around residues 190–200 is partly responsible for Vβ specificity (Y. Choi, unpublished observations).

Studies of the Vβ8.2 region of the T cell receptor showed that residues 17 or 18, 22, 70, 71, and 73 are important for binding vSAG7 (Pullen et al 1990, 1991). Similar results have been reported for Vβ17 (Cazenave et al 1990). Models of the TCR based on immunoglobulin would put these residues on the solvent-exposed face of the TCR, away from the binding site of TCR for MHC/peptide antigen and in a region of Vβ close to that used to engage the bacterial superantigens. The fact that mutant Vβ8.2 chains, which are unable to bind vSAGs, can still be stimulated by conventional antigenic peptides plus MHC (Pullen et al 1991) supports this hypothesis.

T Cell Receptor Specificity

The vSAG proteins can be grouped according to homology at the 3′ end, and these groupings correlate with the known Vβ specificities (Figure 3). So far, five different subfamilies of vSAG have been discovered, reacting with T cells bearing Vβ6, 7, 8.1, and 9; Vβ7; Vβ3; Vβ5, 11, and 12; or Vβ14 and 15. The discussion below is divided according to these different families.

VSAGS SPECIFIC FOR Vβ6, Vβ7, Vβ8.1, AND Vβ9 Four different vSAGs in inbred strains of mice (vSAG7, vSAG43, vSAG44, and vSAG(SW)) have been sequenced and reported to react (by deletion) with T cells bearing Vβs 6, 7, 8.1, and 9. Three of these vSAGs are encoded by endogenous MTVs (vSAG7, vSAG43, and vSAG44) and one is encoded by an exogenous MTV (vSAG(SW) transmitted in milk from mother to pup. vSAG7 has long been known as Mls-1[a] (Festenstein 1973) and is encoded by *Mtv7* (Frankel et al 1991) integrated in 9 of 19 tested strains of mice. vSAG43 has been found in Ma/My mice and differs in sequence from vSAG7 at only two amino acids (Rudy et al 1992). The exovirus encoding vSAG(SW) was found in a substrain of BALB/c mice and also has a sequence closely related to that of vSAG7 (Held et al 1992). In addition, vSAG44, in NZB mice, has been reported to cause deletion of Vβ6, Vβ8.1, and Vβ9-bearing T cells (Tomonari & Fairchild 1992). The specificity of this superantigen is so similar to those of others in this group that it is likely that it will also be similar in the amino acid sequence.

VSAGS SPECIFIC FOR Vβ7 Two superantigens have been found that are specific only for Vβ7-bearing T cells. The first is the product of *Mtv23*. vSAG23 appears to have low affinity for Vβ7 T cells and causes only minor deletion of Vβ7 T cells in vivo (L. Ignatowicz et al, submitted). The sequence of vSAG23 differs by only two amino acids from the recently published sequence of vSAG17 (Korman et al 1992), which is not expressed in vivo (see discussion below).

The second Vβ7-specific superantigen is vSAG(M12) found in the M12.4 lymphoma (L. Ignatowicz et al, submitted). This is a strong superantigen that can significantly stimulate normal T cells as well as some $V\beta7^+$ T cell hybrids in vitro.

VSAGS SPECIFIC FOR Vβ3 The mouse *Mls-2*[a] and *Mls-3*[a] genes were discovered to be responsible for clonal deletion of $V\beta3^+$ T cells (Abe et al 1988; Pullen et al 1988, 1989). Later, these genes were found to correspond to vSAGs encoded by *Mtvs1* and *6*. In addition, vSAGs produced by *Mtvs3* and *13* also deleted $V\beta3^+$ T cells (Frankel et al 1991; Pullen et al 1992a). The vSAG proteins of these MTVs have almost identical C terminal sequences, differing only in the last three residues (Figure 1). The similarity in sequences suggests that the same or closely related exogenous viruses at various times became established in the germline of mice four separate times. *Mtv13* is only found in strains that also have *Mtv6*. The superantigen encoded by *Mtv27* has also been reported to delete $V\beta3^+$ T cells (Tomonari et al 1992).

VSAGS SPECIFIC FOR Vβ5, Vβ11 AND Vβ12 A third group of vSAGs is capable of interacting with T cells bearing Vβ11, Vβ12, and/or Vβ5. The responsible genes have been designated *Mls*[f] (Ryan et al 1991). vSAGs from *Mtv8, 9,* and *11* all induce deletion of Vβ11 and Vβ12-bearing T cells in vivo. vSAG8 only deletes the Vβ11 T cells bearing CD4, and not those bearing CD8. This is thought to be an indication of a weak superantigen (see above), probably because of poor expression in vivo (Corley et al 1992). vSAG9 also deletes Vβ5-bearing T cells (Dyson et al 1991b; Foo-Phillips et al 1992; Woodland et al 1991a). This deletion is dependent upon the presence of the IE class II molecule as well (Woodland et al 1990b).

Interestingly, vSAG6 deletes Vβ5-bearing T cells in vivo, even though its C terminal amino acid sequence is not homologous to any other of the Vβ5, Vβ11, and Vβ12 reactive vSAGs (Gollob & Palmer 1992). However, vSAG6 has no effect on Vβ11 or Vβ12 (Foo-Phillips et al 1992). This apparent contradiction to the rules for vSAG specificity has yet to be resolved. In addition, the 60-kd *gag* gene product from replication-defective MuLV has been reported to expand Vβ5 and sometimes Vβ11-bearing $CD4^+$ T cells (Hugin et al 1991).

VΒ14 AND VΒ15-SPECIFIC VSAGS Two different vSAGs produced by exogenous MTVs have been found with similar Vβ specificity for Vβs 14 and 15. The first is the vSAG of exogenous MTV found in C3H/HeJ mice, which causes deletion of Vβ14 and Vβ15-bearing T cells (Choi et al 1991; Marrack et al 1991). A second is the vSAG encoded by an endogenous and exogenous MTV found in GR mice (*Mtv2*), which has been reported as a deletor of Vβ14-bearing T cells (Acha-Orbea et al 1991; Michalides et al 1981). Both of these exogenous MTVs are transmitted to the pups with mother's milk during the first days of life and can produce an infectious virus.

VΒ17A1 AND VΒ17A2-SPECIFIC VSAGS Most laboratory strains of mice contain a nonfunctional gene for Vβ17, Vβ17b. The functional Vβ17a1 allele, however, is expressed in four strains. This allele binds several vSAGs. $V\beta17a1^+$ T cells are eliminated in many strains of mice that express IE and various vSAGs (Abe et al 1988; Kappler et al 1987; Marrack & Kappler 1988). It is likely that Vβ17a1 reacts with vSAGs 1, 3, 6, 8, 9, and 13, and perhaps others as well (Blackman et al 1992; McDuffie et al 1992). Another allele of Vβ17 has been described, Vβ17a2. Vβ17a2 differs from Vβ17a1 in an amino acid thought to be a contact site for vSAGs (Cazenave et al 1990; Pullen et al 1990). Consequently Vβ17a2-bearing cells have a more restricted specificity for vSAGs. Vβ17a2 T cells are known to react with vSAG6 and a previously unknown vSAG from MAI mice (Jouvin-Marche et al 1992a). In unpublished studies, we have shown that a variant of Vβ17 found in Czech mice, Vβ17a3, also reacts with vSAG6, and probably other vSAGs in the vSAG6 family. This family also reacts with T cells bearing Vβ3 (see above), the mouse Vβ that is most like Vβ17 in amino acid sequence.

MTVS NOT EXPRESSING VSAGS Most of the Mtv proviruses express vSAG genes. Normally, the viral 5 LTR contains the MTV promoter, which drives the expression of the *vsag* gene in the 3 LTR (Gunzburg & Salmons 1992). On a Southern blot, both LTRs are visible as separate bands if the restriction enzyme used cuts within the virus genome. However, *Mtv14* yields only one LTR band, regardless of the restriction enzymes used (M. Scherer, unpublished results). Thus *Mtv14* may lack either the 5 or 3 LTR, and the *Mtv14* vSAG may not be expressed. However, although *Mtv14* is present in almost half of the strains we analyzed, we could not find any T cell deletion common to all the *Mtv14*-bearing strains.

The second vSAG that may not be expressed is that of *Mtv17*. *Mtv17* is present in 14 of 19 analyzed strains. The predicted sequence of vSAG17 differs from that of vSAG23 by only two amino acids (Korman et al 1992), which suggests that this *Mtv* might encode another Vβ7-specific vSAG. However, a recently obtained (C58/JxCBA/CaJ)F4 mouse, which contains only one *Mtv, Mtv17,* did not have reduced numbers of Vβ7-bearing T cells (M. Scherer, unpublished results). *Mtv17* may therefore not encode a functional vSAG. *Mtv17* contains both 5 and 3 LTRs, but it has been suggested that the *vsag17* might not be expressed because of a mutation in the glucocorticoid-inducible transcriptional enhancer element of the *Mtv17* LTR (Kuo et al 1988).

VSAGS WITH UNKNOWN Vβ SPECIFICITY The following Mtvs either do not express superantigen, or else have a superantigen of unknown specificity. The first, *Mtv29,* was found in four strains of mice. This virus has both LTRs and therefore its *vsag* might be expressed. Three of the $Mtv29^+$ strains have a deletion in their TCR β locus, which restricted our ability to find out whether vSAG29 affects the T cell repertoire. However, examination of (B10.BR × C57BR/cdJ)F1 mice, which contain a full complement of Vβs and *Mtv29,* revealed no evidence of reduced percentages of T cells bearing any Vβ tested other than Vβ5 (P. Marrack & L. Ignatowicz, unpublished results). $V\beta5^+$ T cells are probably deleted by vSAGs 8 and 9 expressed in these mice. However, it has been recently demonstrated that B cell lymphomas from SJL mice have an ability to stimulate T cells bearing Vβ16 and that this activity is linked to an Mtv vSAG, presumably that of Mtv29 (Tsiagbe et al 1993).

Mtv30 yields only one LTR-containing band in Southern blots of NcoI, BstXI, and PvuII restricted DNA. The bands are always at least 10 kb in size, however, so they could each contain the complete virus and both LTRs (Moore et al 1987). Therefore vSAG30 may be expressed and, indeed, experiments (M. Scherer, unpublished results) suggest that this may sometimes be the case. When vSAG30 is expressed, it appears to react with vβ5$^+$ T cells (M. Scherer, unpublished results).

The following *Mtvs* have been described, but their vSAGs have not been studied in any detail. They are mentioned here for completeness. *Mtv4* has only been found in SHN Swiss mice (Koizumi et al 1990). *Mtv5* was defined as a provirus that expresses a *gag,* but not an *env,* gene in EL/NiA mice (Imai & Hilgers 1979). *Mtv18* was defined as a new 9.2 kb EcoRI restriction fragment found in the recombinant inbred strain CxBJ (Hilgers & Michalides 1984). Similar EcoRI fragments of 9.2 and 5.5 kb found in the B6.C-KH-84 strain derived from a C57/BL6×BALB/c founder were named *Mtv22* (Colombo et al 1987). *Mtv20*, defined as 5.3 and 13kb EcoRI fragments, is found in some colonies of C57BL mice, but not in C57BL/6 or C57BL/10 (Vaidya et al 1983). *Mtv21* was defined as the 8.0 and 8.1 kb EcoRI fragments found in BR6 mice, but not in C57BL mice (Peters et al 1986). Since BR6 mice were created from a cross between C57BL and RIII, *Mtv21* is probably present in RIII mice also. *Mtv25* was defined as the 5.9 and 6.1 kb EcoRI fragments found in DDSio mice (Peters et al 1986). *Mtv28* was defined as the 5.8 kb EcoRI fragment found on the X chromosome of NZB mice (Eicher & Lee 1990). *Mtv31* is found on the Y chromosome of *M. musculus domesticus* strains and not found in *M. musculus musculus* strains (Prochazka & Leiter 1990).

OTHER SOURCES OF SUPERANTIGENS

Numerous publications have described superantigens produced by organisms other than those described above. It is well established that a rodent mycoplasma, *Mycoplasma arthritidis,* (Cole et al 1981) produces a soluble basic protein with classic superantigenic activity. This superantigen has not been completely characterized, but it may play some role in the induction of arthritis in rodents by the microorganism producing it. Other prokaryotes reported to produce superantigens include *Pseudomonas* (Legaard et al 1991) and *Chlostridium perfringens* (Bowness et al 1992).

Some viruses other than the MTVs have also been reported to produce superantigens. These include the rabies virus nucleocapsid protein (Lafon et al 1992), Herpesvirus saimiri (Nicholas et al 1990; Thomson & Nicholas 1991), Epstein Barr virus (J.D. Donahue et al, submitted), a virus responsible for immunodeficiencies in mice, the MAIDS virus (Hugin et al 1991), and HIV itself (Laurence et al 1992). The inclusion of the lymphotropic viruses, EBV, MAIDS, and HIV on this list is particularly interesting given the evidence, discussed below, that the viral superantigens may serve to provide activated lymphocytes in which invading viruses can grow.

DISCUSSION

Bacterial vs Viral Superantigens

While bacterial and viral superantigens can have similar effects on T cells and on the T cell repertoire, there are surprisingly few similarities between the proteins. There is no significant similarity in amino acid sequence between the two types of superantigens. Bacterial superantigens are soluble globular proteins that are functional in native form (Fischer et al 1989; Fraser 1989; Scholl et al 1989), while viral superantigens are type II membrane bound proteins that require proteolytic cleavage to produce the functional form (Winslow et al 1992). Nevertheless, they both are able to bind class II molecules and Vβ and subsequently activate and/or kill T cells. Their similar function seems to be an example of convergent evolution.

Binding of bacterial superantigens to class II MHC molecules is easy to demonstrate (Fraser 1989), while direct binding of vSAGs to class II molecules has been demonstrated only with some difficulty (G. Winslow, personal observation), which suggests that the vSAG class II interaction is of lower affinity. The ability to bind class II MHC rather than class I may be an advantage to the superantigens, since the professional antigen-presenting cells on which class II is found are able to provide any necessary co-stimulatory signals required to aid T cell activation.

Although both bacterial and viral superantigens bind the Vβ region of the T cell receptor, they seem to bind slightly different regions of the Vβ. The vSAGs bind at or near residues 17, 18, 22, 70, and 71, which are thought to lie on a β strand and variable loop distant from the part of the Vβ that binds MHC (Pullen et al 1990, 1991). SEB interacts with Vβ residues 66, 68, 72, and 74 (White et al 1993), which are thought to lie on two β strands close to the part of Vβ that binds MHC.

The Significance of Superantigens

SUPERANTIGENS ALLOW MICROORGANISMS TO EVADE THE IMMUNE SYSTEM

The bacteria that encode the superantigenic toxins may benefit from the large non-specific activation of all T cell bearing certain Vβs. The activated T cells would not be specific for the bacteria and might so overwhelm the system with cytokines and the ensuing shock that a specific response would be prevented. In addition, after the large proliferation of T cells, the activated cells later become anergic and/or die, thus being unable to later mount an immune response (McCormack et al 1993; Rellahan et al 1990; Webb et al 1990) In this way, the infected individuals could be immunocompromised.

THE FUNCTION OF VIRAL SUPERANTIGENS It has been suggested that vSAGs function to stimulate a pool of lymphocytes to replicate, providing a host for the virus (Acha-Orbea & Palmer 1991; Golovkina et al 1992; Hainaut et al 1990). The viral life cycle seems to require at least two stages: first the penetration by milk-borne virus of the gut, and later infection of mammary endothelial tissue to allow the spread of the virus to the next generation. The vSAG may exist on the surface of the MTV virions, bound to captured host class II MHC on the viral membrane. In this case the vSAG may be a ligand for the T cell receptor and allow the virus to directly infect T cells present in the gut. Concomitant activation of the T cell host would then increase the number of host cells and possibly increase the chance of successful integration of viral DNA. Alternately, a class II-bearing cell may be the initial target of the virus. If a B cell were infected, its ability to activate all T cells bearing certain Vβ would allow the T cell to activate the B cell in turn. T cells then could also be infected later. If antigen-presenting cells are the target for the initial infection, the eventual deletion of vSAG reactive T cells would not affect the viral life cycle.

POSSIBLE SELECTION FOR EXPRESSION OF MTV VSAGS It has been shown that the expression of a *vsag* transgene can prevent infection with the same exoMTV (Golovkina et al 1992) . Since the majority of endogenous Mtvs found in mouse strains are expressed, while only a few produce live virus, it is possible that the expression of the *vsag* gene of integrated Mtvs has been selected for in the mouse to prevent MTV infection, and the resultant tumors and other virally associated effects such as the release of lymphokines and the resulting flu-like symptoms. If that were the case, mice with at least one endogenous MTV from each vSAG family would be completely protected. This would have the presumably deleterious effect of narrowing the T cell repertoire significantly. However, both wild mice and laboratory mice strains have large genomic and/or MTV-mediated Vβ deletions (Pullen et al 1990; Huppi et al 1988), yet have no known deficiency in their ability to mount T cell responses to a wide variety of infectious agents. It should also be noted that endogenous MTV superantigens, coupled with deletions at the Vβ8 locus, have been shown to protect mice against disease caused by bacterial toxins such as staphylococcal enterotoxin B (Marrack & Kappler 1990). Perhaps endogenous retroviral superantigens can protect mice both against exogenous viruses and bacterial toxins.

CONCLUDING REMARKS

Superantigens appear to have independently evolved at least two times, presumably as a way for pathogens to avoid, or take advantage of, the host immune response. It remains to be discovered whether superantigens were derived by the pathogens from a preexisting host protein, perhaps one involved in regulating the immune repertoire, or whether the pathogen evolved a new function from a completely unrelated gene. It seems likely that other pathogens will also be found to use superantigens, and several candidates are already under investigation.

Superantigens have already been widely used as a tool in investigations of the development of the T cell repertoire, and as a way to uniformly activate or inactivate large numbers of T cells in normal mice. In the future, superantigens may also be used clinically as a powerful way to augment or prevent certain immune responses. Diagnostic tests of the Vβ T cell repertoire could indicate the presence of a certain pathogen or the likelihood of a particular form of autoimmune disease caused by a particular superantigen. Modified soluble forms of superantigens may be created that can act as T cell receptor agonists or antagonists. These could then serve to block certain unwanted responses, such as in the case of autoimmune disease, either by clonally deleting or by blocking the receptors of certain T cells. Suitably modified superantigens may also serve as specific adjuvants increasing the frequency of T cells able to respond to particular antigens by increasing the frequency of T cells bearing a certain β chain. Thus it is likely that agents originally used by pathogens to co-opt the immune system will be co-opted by humans to serve their own purposes.

Literature Cited

Abe, R, Vacchio, MS, Fox, B, Hodes, RJ. 1988. Preferential expression of the T-cell receptor V beta 3 gene by Mls^{c} reactive T cells. *Nature* 335:827–30

Acha-Orbea, H, Palmer, E. 1991. Regulated expression of mouse mammary tumor proviral genes in cells of the B lineage. *J. Exp. Med.* 174:1439–50

Acha-Orbea, H, Palmer, E. 1992. The open reading frames in the 3 long terminal repeats of several mouse mammary tumor virus integrants encode V beta 3-specific superantigens. *J. Exp. Med.* 175:41–47

Acha-Orbea, H, Shakhov, AN, Scarpellino, L, Kolb, E, Müller, V, et al. 1991. Clonal deletion of Vβ14-bearing T cells in mice transgenic for mammary tumour virus. *Nature* 350:207–11

Allison, JP, Havran, WL. 1991. The immunobiology of T-cells with invariant gammadelta antigen receptors. *Annu. Rev. Immunol.* 9:679–705

Amzel, L, Poljak, R, Saul F, Varga, J, Richards, F. 1974. The three-dimensional structure of a combining region-ligand complex of immunoglobulin NEW at 3.5Å resolution. *Proc. Natl. Acad. Sci. USA* 71:1427

Bergdoll, MS. 1970. Enterotoxins. In *Microbial Toxins III. Bacterial Protein Toxins.* pp. 265–326. New York: Academic

Beutner, U, Frankel, WN, Cote, MS, Coffin, JM, Huber, BT. 1992. Mls-1 is encoded by the long terminal repeat open reading frame of the mouse mammary tumor provirus Mtv-7. *Proc. Natl. Acad. Sci. USA* 89:5432–36

Bill, J, Kanagawa, O, Woodland, DL, Palmer, E. 1989. The MHC molecule I-E is necessary but not sufficient for the clonal deletion of Vβ11-bearing T cells. *J. Exp. Med.* 169:1405–19

Bjorkman, PJ, Saper, MA, Samraoui, B, Bennett, WS, Strominger, JL, Wiley, DC. 1987. The foreign antigen binding site and T cell recognition regions of class I histocompatibility antigens. *Nature* 329:512–8

Blackman, MA, Gerhard-Burgert, H, Woodland, DL, Palmer, E, Kappler, JW,

Marrack, P. 1990. A role for clonal inactivation in T cell tolerance to Mls-1[a]. *Nature* 345:540–42

Blackman, MA, Lund, FE, Surman, S, Corley, RB, Woodland, DL. 1992. Major histocompatibility complex-restricted recognition of retroviral superantigens by Vβ17+ T cells. *J. Exp. Med.* 176:275–80

Bohach, GA, Fast, DJ, Nelson, RD, Schlievert, PM. 1990. Staphylococcal and streptococcal pyrogenic toxins involved in toxic shock syndrome and related illnesses. *Crit. Rev. Microbiol.* 17:251–72

Bowness, P, Moss, PA, Tranter, H, Bell, JI, McMichael, AJ. 1992. Clostridium perfringens enterotoxin is a superantigen reactive with human T cell receptors Vβ6.9 and Vβ22. *J. Exp. Med.* 176:898–93

Brandt-Carlson, C, Butel, JS. 1991. Detection and characterization of a glycoprotein encoded by the mouse mammary tumor virus long terminal repeat gene. *J. Virol.* 65: 6051–60

Braunstein, NS, Weber, DA, Wang, XC, Long, EO, Karp, D. 1992. Sequences in both class II major histocompatibility complex alpha and beta chains contribute to the binding of the superantigen toxic shock syndrome toxin 1. *J. Exp. Med.* 175:1301–5

Brenner, MB, Dialynas, DP, Stroming, JL, Smith, JA, Owen, FL, et al. 1986. Identification of a putative second T cell receptor. *Nature* 322:145–49

Brown, JH, Jardetzky, T, Saper, MA, Samraoui, B, Bjorkman, PJ, Wiley, DC. 1988. A hypothetical model of the foreign antigen binding site of class II histocompatibility molecules. [Published erratum appears in *Nature* 1988. 23;333(6175):786]. *Nature* 332:845–50

Callahan, JE, Herman, A, Kappler, JW, Marrack, P. 1990. Stimulation of B10.BR T cells with superantigenic staphylococcal toxins. *J. Immunol.* 144:2473–79

Cazenave, PA, Marche, PN, Jouvin, ME, Voegtle, D, Bonhomme, F. et al. 1990. Vβ17 gene polymorphism in wild-derived mouse strains: two amino acid substitutions in the Vβ17 region greatly alter T cell receptor specificity. *Cell* 63:717–28

Chintagumpala, MM, Mollick, JA, Rich, RR. 1991. Staphylococcal toxins bind to different sites on HLA-DR. *J. Immunol.* 147: 3876–81

Choi, YW, Herman, A, DiGiusto, D, Wade, T, Marrack, P, Kappler, J. 1990a. Residues of the variable region of the T-cell-receptor beta-chain that interact with *S. aureus* toxin superantigens. *Nature* 346:471–73

Choi, Y, Kappler, JW, Marrack, P. 1991. A superantigen encoded in the open reading frame of the 3 long terminal repeat of mouse mammary tumour virus. *Nature* 350:203–7

Choi, Y, Kotzin, B, Herron, L, Callahan, J, Marrack, P, Kappler, J. 1989. Interaction of *Staphylococcus aureus* toxin "superantigens" with human T cells. *Proc. Natl. Acad. Sci. USA* 86:8941–45

Choi, Y, Lafferty J, Clements, J, Todd, J, Gelfand, E, et al. 1990b. Selective expansion of T cells expressing Vβ2 in toxic shock syndrome. *J. Exp. Med.* 172:981–84

Choi, Y, Marrack, P, Kappler, J. 1992. Structural analysis of a mouse mammary tumor virus superantigen. *J. Exp. Med.* 175:847–52

Chothia, C, Boswell, DR, Lesk, AM. 1988. The outline structure of the T-cell alpha beta receptor. *EMBO J.* 7:3745–55

Cole, BC, Daynes, RA, Ward, JR. 1981. Stimulation of mouse lymphocytes by a mitogen derived from *Mycoplasma arthritidis*. I. Transformation is associated with an H-2-linked gene that maps to the I-E/I-C subregion. *J. Immunol.* 127:1931–36

Colombo, MP, Melvold, RW, Wettstein, PJ. 1987. Inheritance of a mutant histocompatibility gene and a new mammary tumor virus genome in the B6.KH-84 mouse strain. *Immunogenetics* 26:99–104

Corley, RB, Lund, FE, Randall, TD, King, LB, Doerre, S, Woodland, DL. 1992. Mouse mammary tumor proviral gene expression in cells of the B lineage. *Sem. Immunol.* 4:287–96

Crouse, CA, Pauley, RJ. 1989. Molecular cloning and sequencing of the MTV-1 LTR: evidence for LTR sequence alteration. *Virus Res.* 12:123–38

Dalidowicz, JD, Silverman, SJ, Schantz, EJ, Stefanye, D, Spero, L. 1966. Chemical and biological properties of reduced and alkylated staphylococcal enterotoxin B. *Biochemistry* 5:2375–81

Davis, MM, Bjorkman, PJ. 1988. T-cell antigen receptor genes and T-cell recognition. [Published erratum appears in *Nature* 1988 335:744]. *Nature* 334:395–402

Dellabona, P, Peccoud, J, Kappler, J, Marrack, P, Benoist, C, Mathis, D. 1990. Superantigens interact with MHC class II molecules outside of the antigen groove. *Cell* 62:1115–21

Donehohower, LA, Fleurdelys, B, Hagher, GL. 1983. Further evidence for the protein coding potential of the mouse mammary tumor virus long terminal repeat: nucleotide suquence of an endogenous proviral long terminal repeat. *J. Virol.* 45:941–47

Dyson, PJ, Knight, AM, Fairchild, S, Simpson, E, Tomonari, K. 1991a. Expression of endogenous Mtv provirus transcripts in BALB/c splenic lymphocytes. *Proc. Soc. Exp. Biol. Med.* 196:316–20

Dyson, PJ, Knight, AM, Fairchild, S, Simpson, E, Tomonari, K. 1991b. Genes encoding ligands for deletion of Vβ11 T cells cosegregate with mammary tumour virus genomes. *Nature* 349:531–32

Eicher, EM, Lee, BK. 1990. The NXSM recombinant inbred strains of mice: genetic profile for 58 loci including the *Mtv* proviral loci. *Genetics* 125:431–46

Fairchild, S, Rosenwasser, OA, Dyson, J, Tomonari, K. 1992. Tcrb-V3+ T-cell deletion and a new mouse mammary tumor provirus, *Mtv-44*. *Immunogenetics* 36:189–94

Fasel, N, Pearson, K, Buetti, E, Diggelman, H. 1982. The region of MMTV DNA containing the LTR includes a long coding sequence and signals for hormonally regulated transcription. *EMBO J.* 1:4–7

Festenstein, H. 1973. Immunogenetic and biological aspects of in vitro lymphocyte allotransformation (MLR) in the mouse. *Transplant. Rev.* 15:62–88

Fischer, H, Dohlsten, M, Lindvall, M, Sjogren, HO, Carlsson, R. 1989. Binding of staphylococcal enterotoxin A to HLA-DR on B cell lines. *J. Immunol.* 142:3151–7

Fleischer, B, Schrezenmeier, H. 1988. T cell stimulation by staphylococcal enterotoxins. Clonally variable response and requirement for major histocompatibility complex class II molecules on accessory or target cells. *J. Exp. Med.* 167:1697–707

Foo-Phillips, M, Kozak, CA, Principato, MAC. 1992. Characterization of the Mls[f] system. II. Identification of mouse mammary tumor virus proviruses involved in the clonal deletion of self-Mls[f]-reactive T cells. *J. Immunol.* 149:3440–47

Frankel, WN, Rudy, C, Coffin, JM, Huber, BT. 1991. Linkage of Mls genes to endogenous mammary tumour viruses of inbred mice. *Nature* 349:526–28

Fraser, JD. 1989. High-affinity binding of staphylococcal enterotoxins A and B to HLA-DR. *Nature* 339:221–23

Fraser, JD, Urban, RG, Strominger, JL, Robinson, H. 1992. Zinc regulates the function of two superantigens. *Proc. Natl. Acad. Sci. USA* 89:5507–11

Gallahan, D, Kozak, C, Callahan, R. 1987. Mammary tumorigenesis in feral mice: identification of a new int locus in mouse mammary tumor virus (Czech II)-induced mammary tumors. *J. Virol.* 61:66–74

Garrett, TP, Saper, MA, Bjorkman, PJ, Strominger, JL, Wiley, DC. 1989. Specificity pockets for the side chains of peptide antigens in HLA-Aw68. *Nature* 342:692–96

Gollob, KJ, Palmer, E. 1992. Divergent viral superantigens delete Vβ5+ T lymphocytes. *Proc. Natl. Acad. Sci. USA* 89:5138–41

Golovkina, TV, Chervonsky, A, Dudley, JP, Ross, SR. 1992. Transgenic mouse mammary tumor virus superantigen expression prevents viral infection. *Cell* 69:637–45

Gunzburg, WH, Salmons, B. 1992. Factors controlling the expression of mouse mammary tumor virus. *Biochem. J.* 283:625–32

Hainaut, P, Castellazzi, M, Gonzales, D, Clausse, N, Hilgers, J, Crepin, M. 1990. Developmental and hormonal regulation of a mouse mammary tumour virus glycoprotein in normal mouse mammary epithelium. *J. Mol. Endocrinol.* 4:101–6

Happ, MP, Woodland, DL, Palmer, E. 1989. A third T-cell receptor b-chain variable region gene encodes reactivity to Mls-1[a] gene products. *Proc. Natl. Acad. Sci. USA* 86:6293–96

Held, W, Shakhov, AN, Waanders, G, Scarpelino, L, Luethy, R, et al. 1992. An exogenous mouse mammary tumor virus with properties of Mls-1a (*Mtv-7*). *J. Exp. Med.* 175:1623

Herman, A, Croteau, G, Sekaly, RP, Kappler, J, Marrack, P. 990. HLA-DR alleles differ in their ability to present staphylococcal enterotoxins to T cells. *J. Exp. Med.* 172:709–17

Herman, A, Labrecque, N, Thibodeau, J, Marrack, P, Kappler, J, Sekaly, R-P. 1991. Identification of the staphylococcal enterotoxin A superantigen binding site in the β1 domain of the human histocompatibility antigen HLA-DR. *Proc. Natl. Acad. Sci. USA* 88:9956–58

Hilgers, J, Michalides, R. 1984. Mouse news. *Mouse Newslett.* 70:63–64

Huang, IY, Bergsdoll, MS. 1970. The primary structure of Staphylococcal enterotoxin B. III. The cyanogen bromide peptides of reduced and aminoethylated enterotoxin B , and the complete amino acid sequence. *J. Biol. Chem.* 245:3518–25

Huang, IY, Hughes JL, Bergdoll MS, Schantz EJ. 1987. Complete amino acid sequence of Staphylococcal enterotoxin A. *J. Biol. Chem.* 262:7006–13

Hudson, KR, Robinson, H, Fraser, JD. 1993. Two adjacent residues in staphylococcal enterotoxins A and E determine T cell receptor Vβ specificity. *J. Exp. Med.* 177:175–84

Hugin, AW, Vacchio, MS, Morse, HC. 1991. A virus-encoded superantigen in a retrovirus-induced immunodeficiency syndrome of mice. *Science* 252:424–27

Huppi, KE, D'Hoostelaere, A, Mock, BA, Jouvin-Marche, E, Behlke, MA, et al. 1988. T-cell receptor VTβ genes in natural populations of mice. *Immunogenetics* 27:51–56

Imai, S, Hilgers, J. 1979. Levels of mammary tumor virus proteins (MTV p27 and MTV gp52) in the milk of low and high mammary cancer strains of Japanese origin compared with European and American strains. *Int. J. Cancer* 24:359–64

Irwin, MJ, Hudson, KR, Fraser, JD, Gascoigne, NR. 1992. Enterotoxin residues determining T-cell receptor V beta binding specificity. *Nature* 359:841–43

Janeway, CA, Yagi, J, Conrad, PJ, Katz, ME, Jones, B, et al. 1989. T-cell responses to Mls and to bacterial proteins that mimic its behavior. *Immunol. Rev.* 107:61

Jouvin-Marche, E, Cazenave, P-A, Voegtle, D, Marche, PN. 1992a. Vβ17 T-cell deletion by endogenous mammary tumor virus in wild-type-derived mouse strain. *Proc. Natl. Acad. Sci. USA* 89:3232–35

Jouvin-Marche, E, Cazenave, PA, Voegtle, D, Marche, PN. 1992b. Immunosuppression by the MMTV superantigen? *Immunol. Today* 13:77

Kappler, JW, Herman, A, Clements, J, Marrack, P. 1992. Mutations defining functional regions of the superantigen staphylococcal enterotoxin B. *J. Exp. Med.* 175: 387–96

Kappler, J, Kotzin, B, Herron, L, Gelfand, EW, Bigler, RD, et al. 1989. V beta-specific stimulation of human T cells by staphylococcal toxins. *Science* 244:811–13

Kappler, JW, Staerz, U, White, J, Marrack, PC. 1988. Self-tolerance eliminates T cells specific for Mls-modified products of the major histocompatibility complex. *Nature* 332:35–40

Kappler, JW, Wade, T, White, J, Kushnir, E, Blackman, M, et al. 1987. A T cell receptor V beta segment that imparts reactivity to a class II major histocompatibility complex product. *Cell* 49:263–71

King, LB, Lund, FE, White, DA, Sharma, S, Corley, RB. 1990. Molecular events in B lymphocyte differentiation. Inducible expression of the endogenous mouse mammary tumor proviral gene, Mtv-9. *J. Immunol.* 144:3218–27

Knight, AM, Harrison, GB, Pease, RJ, Robinson, PJ, Dyson, PJ. 1992. Transcription factor loading on the MMTV promoter: a bimodal mechanism for promoter activation. *Science* 255:1573–76

Koizumi, A, Tsukada, M, Wada, Y, Kamiyama, S. 1990. Demonstration of a new mouse mammary tumor virus locus in the genome of the mammary tumor prone strain SHN mouse. *Biochem. Biophys. Res. Comm.* 166:336–42

Korman, AJ, Bourgarel, P, Meo, T, Rieckof, GE. 1992. The mouse mammary tumor virus long terminal repeat encodes a type II transmembrane glycoprotein. *EMBO J.* 11:1901–5

Kozak, C, Peters, G, Pauley, R, Morris, V, Michalides, R, et al. 1987. A standarized nomenclature for endogenous mouse mammary tumor viruses. *J. Virol.* 61:1651–54

Kronenberg, M, Siu, G, Hood, LE, Shastri, N. 1986. The molecular genetics of the T-cell antigen receptor and T-cell antigen recognition. *Annu. Rev. Immunol.* 4:529–91

Kuo, W-L, Vilander, LR, Huang, M, Peterson, DO. 1988. A transcriptionally defective long terminal repeat within an endogenous copy of mouse mammary tumor virus proviral DNA. *J. Virol.* 62:2394–402

Lafon, M, Lafage, M, Martinez-Arends, A, Ramirez, R, Vuillier, F, et al. 1992. Evidence for a viral superantigen in humans. *Nature* 358:507–10

Laurence, J, Hodtsev, AS, Posnett, DN. 1992. Superantigen implicated in dependence of HIV-1 replication in T cells on TCR V beta expression. *Nature* 358:255–59

Lederberg, J. 1959. Genes and antibodies. *Science* 129:1649–53

Lee, BK, Eicher, EM. 1990. Segregation patterns of endogenous mouse mammary tumor viruses in five recombinant inbred strain sets. *J. Virol.* 64:4568–72

Legaard, PK, Legrand, RD, Misfeldt, ML. 1991. The superantigen Pseudomonas exotoxin A requires additional functions from accessory cells for T lymphocyte proliferation. *Cell. Immunol.* 135:372–82

Lewis, S, Gellert, M. 1989. The mechanism of antigen receptor gene assembly. *Cell* 59:585–88

MacDonald, HR, Schneider, R, Lees, RK, Howe, RC, Acha-Orbea, H, et al. 1988. T-cell receptor Vβ use predicts reactivity and tolerance to Mls^a-encoded antigens. *Nature* 332:40–45

Marrack, P, Blackman, M, Kushnir, E, Kappler, J. 1990. The toxicity of staphylococcal enterotoxin B in mice is mediated by T cells. *J. Exp. Med.* 171:455–64

Marrack, P, Kappler, J. 1988. T cells can distinguish between allogeneic major histocompatibility complex products on different cell types. *Nature* 332:840–42

Marrack, P, Kappler, J. 1990. The staphylococcal enterotoxins and their relatives. *Science* 248:705–11

Marrack, P, Kushnir, E, Kappler, J. 1991. A maternally inherited superantigen encoded by a mammary tumour virus. *Nature* 349:525–26

McCormack, JE, Callahan, JE, Kappler, J, Marrack, P. 1993. Profound deletion of mature T cells in vivo by chronic exposure to exogenous superantigen. *J. Immunol.* In press

McDuffie, M, Schweiger, D, Reitz, B, Ostrowska, A, Knight, AM, Dyson, PJ. 1992. I-E-independent deletion of Vβ17a+ T cells

by Mtv-3 from the nonobese diabetic mouse. *J. Immunol.* 148:2097–212
Meuer, S, Hodgdon, J, Hussey, R, Protentis, J, Schlossman, S, Reinherz, E. 1983. Antigen-like effects of monoclonal antibodies directed at receptors on human T cell clones. *J. Exp. Med.* 158:988
Michalides, R, van Nie, R, Nusse, R, Hynes, NE, Groner, B. 1981. Mammary tumor induction loci in GR and DBAf mice contain one provirus of the mouse mammary tumor virus. *Cell* 23:165–73
Mohan, N, Mottershead, D, Subramanyam, M, Beutner, U, Huber, BT. 1993. Production and characterization of an Mls-1-specific monoclonal antibody. *J. Exp. Med.* 177:351–58
Mollick, JA, McMasters, RL, Grossman, D, Rich, RR. 1993. Localization of a site on bacterial superantigens that determines T cell receptor B chain specificity. *J. Exp. Med.* 177:283–93
Moore, R, Dixon, M, Smith, R, Peters, G, Dickson, C. 1987. Complete nucleotide sequence of a milk-transmitted mouse mammary tumor virus: two frameshift suppression events are required for translation of *gag* and *pol*. *J. Virol.* 61:480–90
Nicholas, J, Smith, EP, Coles, L, Honess, R. 1990. Gene expression in cells infected with gamma herpes virus saimiri: properties of transcripts from two immediate-early genes. *Virology* 179:189–200
Okada, CY, Holzmann, B, Guidos, C, Palmer, E, Weissman, IL. 1990. Characterization of a rat monoclonal antibody specific for a determinant encoded by the Vβ7 gene segment. Depletion of Vβ7+ T cells in mice with Mls-1[a] haplotype. *J. Immunol.* 144:3473–77
Panina-Bordignon, P, Fu, XT, Lanzavecchia, A, Karr, RW. 1992. Identification of HLA-DR alpha chain residues critical for binding of the toxic shock syndrome toxin superantigen. *J. Exp. Med.* 176:1779–84
Peavy, DL, Adler, WH, Smith, RT. 1970. The mitogenic effects of endotoxin and staphylococcal enterotoxin B on mouse spleen cells and human peripheral lymphocytes. *J. Immunol.* 105:1453
Peters, G, Placzek, M, Brookes, S, Kozak, C, Smith, R, Dickson, C. 986. Characterization, chromosome assignment, and segregation analysis of endogenous proviral units of mouse mammary tumor virus. *J. Virol.* 59:535–44
Peters, J. 1992. Locus map of the mouse. *Mouse Genome* 90:8–21
Prochazka, M, Leiter, EH. 1990. Identification of a novel mouse mammary tumor proviral locus (*Mtv-31*) on chromosome Y. *Mouse Genome* 87:111
Pullen, AM, Bill, J, Kubo, RT, Marrack, P, Kappler, JW. 1991. Analysis of the interaction site for the self superantigen Mls-1(a) on T-cell receptor V beta. *J. Exp. Med.* 173:1183–92
Pullen, AM, Choi, Y, Kushnir, E, Kappler, J, Marrack, P. 1992a. Acquired Mls-1a-like clonal deletion in Mls-1b mice. *J. Exp. Med.* 175:453–60
Pullen, AM, Choi, Y, Kushnir, E, Kappler, J, Marrack, P. 1992b. The open reading frames in the 3 long terminal repeats of several mouse mammary tumor virus integrants encode Vβ3-specific superantigens. *J. Exp. Med.* 175:41–47
Pullen, AM, Marrack, P, Kappler, JW. 1988. The T-cell repertoire is heavily influenced by tolerance to polymorphic self-antigens. *Nature* 335:796–801
Pullen, AM, Marrack, P, Kappler, JW. 1989. Evidence that Mls-2 antigens which delete V beta 3+ T cells are controlled by multiple genes. *J. Immunol.* 142:3033–37
Pullen, AM, Wade, T, Marrack, P, Kappler, JW. 1990. Identification of the region of T cell receptor beta chain that interacts with the self-superantigen MIs-1a. *Cell* 61: 1365–74
Rellahan, BL, Jones, LA, Kruisbeek, AM, Fry, AM, Matis, LA. 1990. In vivo induction of anergy in peripheral Vβ8+ T cells by staphylococcal enterotoxin B. *J. Exp. Med.* 172:1091–100
Richards, F, Konigsberg, W, Rosenstein, RJV. 1975. On the specificities of antibodies. *Science* 189:130
Rudy, CK, Kraus, E, Palmer, E, Huber, BT. 1992. Mls-1-like superantigen in the MA/MyJ mouse is encoded by a new mammary tumor provirus that is distinct from *Mtv-7*. *J. Exp. Med.* 175:1613–21
Ryan, JJ, Lejeune, HB, Mond, JJ, Finkelman, FD. 1991. Allostimulatory analysis of a newly-defined and widely-distributed Mls superantigen. *Immunogenetics* 34:88–100
Scholl, PR, Diez, A, Geha, RS. 1989. Staphylococcal enterotoxin B and toxic shock syndrome toxin-1 bind to distinct sites on HLA-DR and HLA-DQ molecules. *J. Immunol.end_I 143:2583–8*
See, RH, Krystal, G, Chow, AW. 1990. Binding competition of toxic shock syndrome toxin 1 and other staphylococcal exoproteins for receptors on human peripheral blood mononuclear cells. *Infect. Immunity* 58: 2392–96
Smith, BG, Johnson HM. 1975. The effect of staphylococcal enterotoxins on the primary in vitro immune response. *J. Immunol.* 115:575
Spero, L, Warren, JR, Metzger, JF. 1973. Effect of single peptide bond scission by trypsin on the structure and activity of

staphylococcal enterotoxin B. *J. Biol. Chem.* 248:7289–94
Swaminathan, S, Furey, W, Pletcher, J, Sax, M. 1992. The crystal structure of Staphylococcal enterotoxin B, a superantigen. *Nature* 359:801–5
Tanaka, S, Matsuzawa, A. 1990. The NXSM recombinant inbred strains of mice: genetic profile for 58 loci including the Mtv proviral loci. *Genetics* 125:431–46
Thomson, BJ, Nicholas, J. 1991. Superantigen function *Nature* 351:530
Tomonari, K, Fairchild, S. 1992. Positive and negative selection of Tcrb-V6+ T cells. *Immunogenetics* 36:230–37
Tomonari, K, Fairchild, S, Rosenwasser, OA. 1992. Tcrb-V3+ T-cell deletion and a mouse mammary tumor provirus, *Mtv-27*. *Immunogenetics* 36:302–5
Townsend, A, Bodmer, H. 1989. Antigen recognition by class I-restricted T lymphocytes. *Annu. Rev. Immunol.* 7:601–24
Tsiagbe, V, Yoshimoto, T, Asakawa, J, Cho, S, Meruelo, D, Thorbecke, G. 1993. *EMBO J.* In press
Vaidya, AB, Taraschi, NE, Tancin, SL, Long, CA. 1983. Regulation of endogenous murine mammary tumor virus expression in C57BL mouse lactating mammary glands: transcription of functional mRNA with a block at the translational level. *J. Virol.* 47:818–28
Webb, S, Morris, C, Sprent, J. 1990. Extrathymic tolerance of mature T cells: clonal elimination as a consequence of immunity. *Cell* 63:1249–56
White, J, Herman, A, Pullen, AM, Kubo, R, Kappler, JW, Marrack, P. 1989. The Vβ-specific superantigen staphylococcal enterotoxin B: stimulation of mature T cells and clonal deletion in neonatal mice. *Cell* 56:27–35
White, J, Pullen, A, Choi, K, Marrack, P, Kappler, JW. 1993. Antigen recognition properties of mutant Vβ3+ T cell receptors are consistent with an immunoglobulin-like structure for the receptor. *J. Exp. Med.* 177:119–25
Winslow, GM, Scherer, MT, Kappler, JW, Marrack, P. 1992. Detection and biochemical characterization of the mouse mammary tumor virus-7 superantigen (Mls-1[a]). *Cell* 71:719–30
Woodland, D, Happ, MP, Bill, J, Palmer, E. 1990a. Molecular events in B lymphocyte differentiation. Inducible expression of the endogenous mouse mammary tumor proviral gene, Mtv-9. *J. Immunol* 144:3218–27
Woodland, D, Happ, MP, Bill, J, Palmer, E. 1990b. Requirement for cotolerogenic gene products in the clonal deletion of I-E reactive T cells. *Science* 247:964–67
Woodland, DL, Happ, MP, Gollob, KJ, Palmer, E. 1991a. An endogenous retrovirus mediating deletion of alpha beta T cells? *Nature* 349:529–30
Woodland, DL, Happ, MP, Gollob, KJ, Palmer, E. 1991b. Genes encoding ligands for deletion of V beta 11 T cells cosegregate with mammary tumour virus genomes. *Nature* 349:531–2
Woodland, DL, Lund, FE, Happ, MP, Blackman, MA, Palmer, E, Corley, RB. 1991c. Endogenous superantigen expression is controlled by mouse mammary tumor proviral loci. *J. Exp. Med.* 174:1255–58

Annu. Rev. Cell Biol. 1993. 9:129–61

SIGNAL-DEPENDENT MEMBRANE PROTEIN TRAFFICKING IN THE ENDOCYTIC PATHWAY

I. S. Trowbridge and J. F. Collawn
Department of Cancer Biology, The Salk Institute, P. O. Box 85800, San Diego, California 92186

C. R. Hopkins
MRC Laboratory for Molecular Cell Biology, University College London, Gower Street, London WC1E 6BT, England

KEY WORDS: receptor-mediated endocytosis, internalization signals, clathrin-coated pits, plasma membrane, endosome

---

CONTENTS

INTRODUCTION

The impressive diversity of function that typifies eukaryote cells depends upon the structural differentiation of their membrane boundaries. To generate

0743–4634/93/1115–0129$05.00

and maintain these structures, proteins must be transported from their site of synthesis in the endoplasmic reticulum to predetermined destinations throughout the cell. This requires that the trafficking proteins display sorting signals that are recognized by the molecular machinery responsible for route selection located at the access points to the main trafficking pathways. Sorting decisions for most proteins need to be made only once as they traverse their biosynthetic pathways since their final destination, the cellular location where they perform their function, becomes their permanent residence. However, there are also integral membrane proteins that traffic throughout their lifetimes. The main function of these proteins is to carry ligands between different cellular compartments, and they need to shuttle back and forth repeatedly, often making as many as 300 round trips during their 1–2 day lifetimes. In such circumstances, even a low frequency of error in recognizing a sorting signal will distort trafficking patterns unless the missorted protein can be returned to its original location. It is to be expected, therefore, that the accuracy of sorting will be high, either because the fidelity of the recognition mechanisms involved is of a high order, or because less selective recognition mechanisms are coupled to signal-independent recycling pathways.

The most thoroughly understood examples of signal-dependent membrane protein trafficking are found in receptor-mediated endocytosis, the process by which integral membrane proteins are selectively internalized from the plasma membrane to the endocytic pathway. Although receptor-mediated endocytosis is most commonly thought of as a mechanism for the efficient cellular uptake of nutrients bound to specific receptors, it is also important in regulating the protein composition of the plasma membrane. Thus while internalization plays a role in attenuating the responses of signal-transducing receptors by removing receptor-ligand complexes from the cell surface, externalization of proteins such as glucose transporters can transform the pattern of nutrient flux into the cell within minutes. Also, where a more generalized remodeling of the cell surface is required (e.g. following neurotransmitter release or establishment of new epithelial cell polarity), essentially the same processes of membrane protein relocation come into play.

During receptor-mediated endocytosis, itinerant membrane proteins are segregated from resident plasma membrane proteins and concentrated in specialized regions of the plasma membrane known as clathrin-coated pits. Invagination and pinching off of coated pits into the cytoplasm generates transport vesicles that deliver their cargo to the endosomal compartment. Some proteins then recycle back to the cell surface while others are sorted to a variety of destinations. Virtually all the current information about signal-dependent membrane protein sorting has been derived from studies of receptors that recycle constitutively within the endocytic pathway. Nevertheless, the molecular principles derived in these studies are likely to apply to the signal-dependent sorting steps in all of the intracellular pathways that derive their integral membrane proteins from the endoplasmic reticulum.

COMPARTMENTS WITHIN THE ENDOCYTIC PATHWAY

Recycling Pathway between the Cell Surface and the Endosome

Early studies of the endocytic pathway focused on the uptake and delivery of macromolecular materials to the lysosome. For studies using fluid tracers like horseradish peroxidase (HRP), this emphasis was well-justified (up to 60% of internalized HRP reaches the lysosome within 60 min) (Thilo 1985; Steinman et al 1976; Griffiths et al 1989), but for membrane proteins involved in receptor-mediated endocytosis, the route to the lysosome is much less significant. Recycling receptors such as receptors for transferrin (Tf), low density lipoprotein (LDL), α-2-macroglobulin, and asialoglycoprotein (ASGP) recycle from the endocytic pathway back to the plasma membrane with > 99% efficiency (Goldstein et al 1985; Salzman & Maxfield 1988), and overall, only ~3% of internalized surface-labeled plasma membrane proteins are delivered to the lysosome even after prolonged incubations (Haylett & Thilo 1986; Draye et al 1987). The volume of the intracellular compartments traversed by recycling receptors is not known, but it has been estimated that an amount of membrane equal to the entire cell surface (Steinman et al 1976; Griffiths et al 1989) is internalized and recycled every 1–2 hr. Compared with the membrane areas of other intracellular compartments quantified by morphometry, this suggests that at steady state the recycling pathway is comprised of a surface area roughly equivalent to that of the lysosomal compartment, i.e. ~400 μm^2 (Griffiths et al 1989; Steinman et al 1976).

Extensive morphological studies on a variety of recycling receptor populations have been made (reviewed by van Deurs et al 1989; Courtoy 1991), and it is clear that in most cell types the recycling pathway is extremely pleiomorphic and includes numerous tubular and vacuolar structures. Pulse-chase experiments with radiolabeled ligands demonstrate that the intracellular processing time for receptors such as transferrin receptor (TR), low density lipoprotein receptor (LDLR), and asialoglycoprotein receptor (ASGPR) can be as rapid as 5 min (Wileman et al 1985). With continuous incubations, however, linear uptake for periods of up to 45 min are observed, which indicates that among this heterogeneous collection of structures, some components provide a rapid return to the surface, whereas processing is more protracted within others. Nevertheless, double-label morphological studies consistently show that the various recycling receptors that enter the cell via coated pits are all processed within the same subset of endocytic elements (Stoorvogel et al 1987, 1989; Yamashiro et al 1984; Dickson et al 1983), and studies on tailless mutants (Jing et al 1990; Felder et al 1990) suggest that once receptors enter the pathway, they travel back to the cell surface constitutively, i.e. without the need for further trafficking information. Thus although the recycling pathway is morphologically heterogeneous and processes internalized receptors at widely varying rates of flux, it represents a single pathway uncomplicated by internal signal-dependent sorting mechanisms.

For receptors to exit from the recycling pathway and move to other intracellular destinations (e.g. towards the lysosome or synaptic vesicles) or to new domains on the cell surface (as in transcytosis) requires selective routing and presumably depends upon the possession of additional, specific sorting signals. A schematic diagram of the recycling pathway and its relationship to the main biosynthetic (exocytic) pathway is shown in Figure 1.

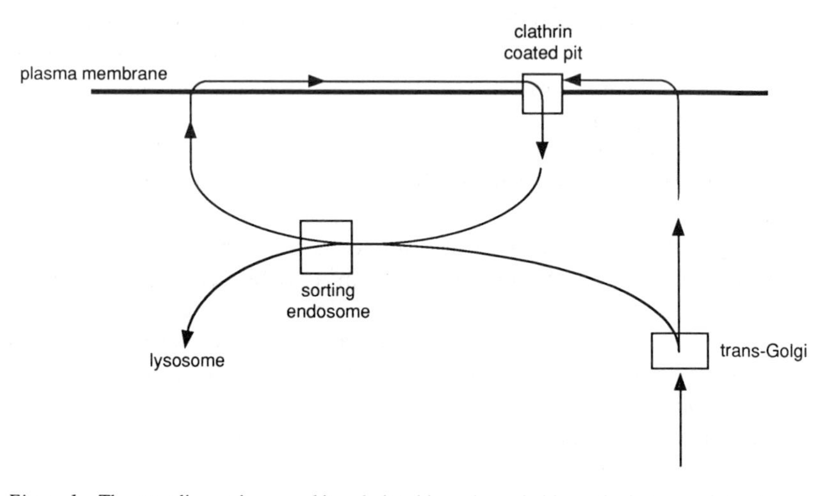

Figure 1 The recycling pathway and its relationship to the main biosynthetic (exocytic) pathway. Sorting on the plasma membrane into the recycling pathway takes place during internalization in the coated pit, sorting on the recycling pathway into the lysosomal pathway takes place in the sorting endosome. The recycling pathway also receives trafficking proteins from the exocytic pathway by an intracellular route after they have been sorted in the *trans* Golgi.

Sorting Endosome

Morphological studies suggest that the material that is carried in the fluid phase within the vacuoles, tubulo-vesicles and tubular networks enroute to the lysosome eventually accumulates within vacuoles roughly 0.5–1.0 μm in diameter. Maxfield and his colleagues, in their extensive light microscope studies on CHO cells (Yamashiro et al 1984; Salzman & Maxfield 1988; Dunn et al 1989; Dunn & Maxfield 1992), have termed these vacuoles collectively the sorting endosome. At the electron microscope level, these structures almost certainly correspond to the vacuoles that over the years have been variously described as receptosomes, CURL, multivesicular endosomes, early endosomes, and most recently, carrier vesicles (Courtoy 1991; Gruenberg & Howell 1989).

In situations where integral membrane proteins enter the endocytic pathway and, instead of recycling, are routed towards the lysosome, sorting endosomes accumulate 50–60 nm diameter vesicles within their lumena. The membrane of these internal vesicles, which is derived from the perimeter membrane of the vacuole, is highly enriched in internalized receptors. The multivesicular bodies that develop in this way thus probably provide a means of removing integral membrane proteins from the perimeter membrane to the lumen of the pathway where they are degraded. Selective removal of membrane proteins by this mechanism has been most convincingly demonstrated for the epidermal growth factor receptor (EGFR) (McKanna et al 1979; Felder et al 1990; Hopkins et al 1990).

Recycling Endosome

Networks of 60–80nm diameter tubules containing recycling receptors (e.g. TR) are distributed throughout the cytoplasm (Hopkins & Trowbridge 1983; Yamashiro et al 1984; Tooze & Hollinshead 1991, 1992) and are often concentrated in the pericentriolar area. These tubular elements, which frequently show direct continuity with the vacuoles of the sorting endosome, almost certainly represent the most distal compartment on the recycling pathway (Geuze et al 1983; Linderman & Lauffburger 1988). The relationship between the sorting and recycling endosomes is unclear since it is not known how many times recycling receptors pass through the sorting endosome, and nothing is known about their rate of flux or the efficiency with which their ligand dissociates. However, the efficiency with which a ligand such as LDL is removed from its receptor (less than 10% returns to the surface compared with 30% for a fluid phase tracer) (Greenspan & St. Claire 1984) suggests that most recycling receptors will traverse the vacuoles of the sorting endosome at least once during each cycle of intracellular processing.

Endosome Formation

Vesicles derived from plasma membrane-coated pits fuse with an interconnected system of irregularly shaped vacuoles and tubular elements which, when seen in conventional microscope thin sections, appear as separate tubulo-vesicles. In vivo studies suggest that free elements within this system have a high propensity to fuse with each other (Gruenberg & Howell 1989). The proportion of vacuolar-to-tubular elements varies, and with the influx of ligand, larger, more spherical vacuolar elements emerge (Hopkins et al 1990). These localized (0.5–1.0 μm diameter) expansions are seen to move along the 2–300 nm diameter tubules in which they develop and can be identified as sorting endosomes because they accumulate labeled tracers (e.g. EGFR) destined for the lysosome. The mechanisms that regulate these processes of fusion and shape change remain to be clarified, but a recent study suggests that the small GTP-binding protein, rab 4, plays an important role (van der Sluijs et al 1992).

Quantitative studies on the fluid phase uptake of HRP show that it becomes distributed throughout the endosomal compartment in <2 min (Griffiths et al 1989). Thereafter, no significant increase in the number of labeled elements containing HRP occurs, but the concentration within the system steadily increases over 15–20 min. These observations are compatible with the view of the endosome as a pre-existing system of extensively interconnected elements. They do not, however, preclude the possibility (discussed further below) that individual sorting endosome vacuoles can arise from within the endosome by a process of gradual maturation.

Fate of Sorting Endosomes

Sorting endosome vacuoles, which accumulate dissociated ligand from recycling receptors (and which, following the internalization of receptors like EGFR accumulate internal vesicles), are the primary site of entry into the pathway leading to the lysosomal compartment. Whether they mature into lysosomes or become carrier vesicles that fuse with pre-existing lysosomes is still debated (Griffiths & Gruenberg 1991; Roederer et al 1987; Stoorvogel et al 1991). Consequently, considerable confusion exists in the literature about where the endosomal compartment ends and the lysosomal compartment begins. There is general agreement, however, that recycling receptors reach sorting endosomes (Dunn et al 1989) and that they are still able to do so when the temperature is lowered to 20°C, or the microtubular cytoskeleton is depolymerized (Marsh et al 1983; Gruenberg & Howell 1989). Since these treatments also prevent fluid phase tracers from penetrating farther along the lysosomal pathway, they provide a useful means of identifying the limit to which rapidly recycling receptor populations like

TR migrate. The next compartment beyond this boundary is variously described as either the late endosome or the prelysosome; the former because it is believed that sorting events continue within this compartment, the latter because the compartment contains typical lysosomal constituents such as the acidic, highly glycosylated membrane proteins, the Lgps (lysosomal membrane glycoproteins). We favor restricting the term endosome to elements that contain rapidly recycling membrane proteins like TR and that can be defined operationally because they are accessible to endocytic tracers at 20°C. Because of the asynchrony of processing, which is a characteristic of the endosome system (Stoorvogel et al 1991), we will avoid using the terms early and late, and instead of late endosome, we will use the term prelysosome.

The prelysosome, when defined as the compartment beyond TR-containing elements, is characterized by a high content of Lgps, CI-M6PR and a pH significantly lower than the endosome (Maxfield & Yamashiro 1991; Kornfeld & Mellman 1989). It can be regarded as a separate compartment because it can be purified away from the TR-containing endosome compartment by a variety of preparative methods. These methods have demonstrated that the prelysosomal compartment has a greater bouyant density and carries a more cationic surface charge (as indicated by free flow electrophoresis). Most importantly, comparison of the proteins within TR-containing endosomes with those of prelysosomes separated by these and other methods demonstrates that the most abrupt change in protein composition along the endocytic pathway occurs at the transition between these two compartments (Schmid et al 1988; Beaumelle et al 1990).

Direct Entry into the Endocytic Pathway from the Golgi Complex

A variety of membrane proteins are delivered to the lysosome from the biosynthetic pathway by an intracellular route. The best documented is that used by the cation-independent mannose-6-phosphate receptor (CI-M6PR), which carries lysosomal enzymes from clathrin-coated buds on *trans* Golgi elements to the endocytic pathway (Kornfeld & Mellman 1989). As discussed below, the adaptor complexes within the clathrin coats of these buds are different from those of the pits on the plasma membrane and, while there is evidence to suggest they can recognize signals in the cytoplasmic domain of CI-M6PR, they have not yet been shown to identify signals carried by either TR or LDLR. Nevertheless, it is clear that when newly synthesized proteins like LDLR are delivered to the basolateral borders of polarized epithelial cells, they are identified by sorting mechanisms that are intracellular and recognize sorting signals located in their cytoplasmic domains. The LDLR signal recognized in this sorting step is not identical, but overlaps with the internalization signal in this domain (see below), and it is probable, therefore, that the *trans* Golgi clathrin-coated pits, which sort CI-M6PR, are also capable of sorting signal-bearing receptors directly into the endosome en route to the plasma membrane.

Initial studies on the distribution of CI-M6PRs in the endocytic pathway showed that they were concentrated in a distinctive subset of vacuolar elements (Griffiths et al 1988). Because mature lysosomes containing active hydrolases lack CI-M6PR, these elements were designated as prelysosomal (Kornfeld & Mellman 1989). They were thought to lie beyond the TR-containing endosome because they could not be labeled with fluid phase tracer at 20°C. Accumulating evidence now suggests, however, that CI-M6PRs are also present in significant amounts within TR-containing elements (Stoorvogel et al 1989; Beaumelle et al 1990) and, recently, several studies have shown (Ludwig et al 1991; Rijnboutt et al 1992; J Hirst & CR Hopkins, unpublished observations) that when newly synthesized CI-M6PRs leave the *trans* Golgi, they are delivered first to an early endocytic compartment, probably the TR-containing endosomal compartment. These studies confirm the existence of a direct pathway to the endosome from the *trans* Golgi and raise the important consideration that the sorting mechanism, which identifies lysosome targeting signals (discussed in detail below), is located within the endosome rather than the *trans* Golgi.

The delivery route out of the biosynthetic pathway in the Golgi for newly synthesized Lgps continues to be a matter of debate. Studies on avian and human Lgp-A (LEP100, LAMP-1) (Williams & Fukuda 1990; Carlsson & Fukuda 1992; Matthews et al 1992) suggest that they follow the same itinerary as lysosomal acid phosphatase (LAP), a hydrolase that travels as an integral membrane protein from the *trans* Golgi to the endocytic pathway via the cell surface (Waheed et al 1988; Braun et al 1989). In contrast, a recent study by Harter & Mellman (1992) on CHO cells found that newly synthesized Lgp-A was transported intracellularly from the *trans* Golgi directly to lysosomes. In this study, newly synthesized Lgp-A was only detected on the cell surface when high levels of the glycoprotein were expressed and, therefore, these workers concluded that the direct intracellular route is the major pathway.

The designation of the compartment beyond the endosome as a prelysosome is justified by its content of CI-M6PR and the absence of TR. It is clear that it receives membrane protein traffic directly from the endosome. There are, however, additional pathways between this compartment, the endosome, and the biosynthetic pathway in the *trans* Golgi. They remain to be outlined, but there is little doubt that they exist because there is good evidence to show that the CI-M6PR recycles from the prelysosome-lysosome segment of the endocytic pathway back to the Golgi (Duncan & Kornfeld

1988), and proteins such as class II MHC, which appear to travel well beyond the prelysosome, must also be retrieved to the cell surface (Peters et al 1991). However, there is no indication at the present time of where the sorting mechanisms associated with these pathways are located or how many sorting signals may be involved.

SORTING SIGNALS

Internalization Signals of Recycling Receptors

Substantial advances have been made in the last six years in understanding the sorting signals for receptors that are constitutively internalized from plasma membrane clathrin-coated pits to the endocytic pathway. A major factor contributing to this progress has been the extensive use of recombinant DNA techniques to create mutant receptors and the development of assays to analyze their function quantitatively. The internalization signals of three recycling receptors have been fully characterized. These data, together with more limited information about other receptors and membrane proteins, have permitted a model defining the important common features of this class of sorting signal to be formulated. Contrary to widely held expectations, internalization signals are not conformational determinants or "signal patches" dependent upon the overall structure of the cytoplasmic domain (Pfeffer & Rothman 1987); instead, they are a family of structural motifs specified by short linear arrays of amino acids that differ in specific sequence, but share a common three-dimensional conformation and chemistry.

The first clue to the location and structure of internalization signals came from the analysis of naturally occurring internalization-defective, mutant LDLRs from patients with familial hypercholesterolemia (Goldstein et al 1985; Davis et al 1986). Three independent mutant LDLRs had altered cytoplasmic domains and, in the case of one patient, JD, a single base change resulted in the modification of a tyrosine residue in the cytoplasmic domain to cysteine (Davis et al 1986). These results implied that LDLRs displayed a positive signal for internalization in the cytoplasmic domain and, as later confirmed by the analysis of other tyrosine substitution mutants generated in vitro (Davis et al 1987), that tyrosine was an important element of the signal. The importance of a tyrosine residue for rapid internalization was further emphasized by the fact that insertion of a tyrosine into the 10-residue cytoplasmic tail of influenza virus hemagglutinin (HA) dramatically increased its rate of endocytosis, presumably by generating an artificial internalization signal (Lazarovits & Roth 1988). Internalization activity was enhanced, however, only if the tyrosine was inserted at position 543, 5 residues from the transmembrane region, but not at positions 540 or 546, thus indicating that the structural environment of the tyrosine residue was also important. Interestingly, in light of what we now know about the sequence patterns of internalization signals, the endocytic rate of the Tyr-543 mutant HA was only ~4% per min, an order of magnitude less than the most efficiently internalized receptors such as TR (Hopkins & Trowbridge 1983; Lund et al 1990), CI-M6PR (Jadot et al 1992), or the polymeric immunoglobulin receptor (poly-IgR) (Breitfeld et al 1990), which implies that the artificial signal created by the substitution of a single residue was not optimal.

The concept of tyrosine-containing internalization signals was subsequently extended to other recycling receptors by functional analysis of mutant receptors generated in vitro (Mostov et al 1986; Rothenberger et al 1987; Miettinen et al 1989; Lobel et al 1989; Jing et al 1990). Invariably, receptors with extensive deletions in their cytoplasmic domains were poorly internalized, although in the cases in which the question was carefully examined, they were not entirely excluded from coated pits and were internalized at ~10% of wild-type rate (e.g. Jing et al 1990). Alteration of specific tyrosine residues within the cytoplasmic domains of many receptors (Lobel et al 1989; Jing et al 1990; McGraw & Maxfield 1990; Alvarez et al 1990; Breitfeld et al 1990; Fuhrer et al 1991) and other membrane glycoproteins (Williams & Fukuda 1990; Peters et al 1990) also severely impaired endocytosis. However, despite determination of the primary sequences of several recycling receptors, the structural requirements for rapid internalization remained a perplexing problem (Goldstein et al 1985) since there were no obvious sequence similarities in their cytoplasmic domains to account for their shared propensity to cluster in coated pits (Pearse & Robinson 1990). Consequently, initial efforts to predict internalization signals met with only limited success (Vega & Strominger 1989; Ktistakis et al 1990). Another confounding factor was that receptors known to cluster in coated pits differed in orientation with respect to the cell membrane, and their cytoplasmic domains varied widely in size (Goldstein et al 1985).

SEQUENCES OF INTERNALIZATION SIGNALS The complete sequences for the internalization signals of the three receptors, LDLR, TR, and CI-M6PR, have been elucidated. In each case, mutant receptors were expressed in recipient cells lacking functional endogenous receptors, and their internalization efficiencies were determined using quantitative assays. Signals were localized to a specific region of the cytoplasmic domain and individual residues important for high efficiency endocytosis identified. In this manner, the LDLR internalization signal was localized to the first 22 amino acids of the 50-residue carboxy-terminal domain, which included Tyr-807, the essential tyrosine residue identified by the JD mutation (Davis et al 1987).

Alteration of either Phe-802, Asn-804, Pro-805, or Tyr-807 to alanine reduced internalization efficiency by more than 50% (Chen et al 1990), which implied that the sequence pattern for the LDLR signal was FXNPXY. Attention was focused initially on the four carboxy-terminal residues, NPXY, because this sequence pattern was conserved in all members of the LDLR gene family, including LRP and GP330. The sequence was also found more frequently than expected by chance in the cytoplasmic domains of other cell surface receptors (Chen et al 1990). However, the functional LDLR signal is probably the 6-residue sequence FXNPXY, which itself is conserved in LDLRs from six species (Chen et al 1990). As described below, not only is this 6-residue sequence required for rapid internalization of the LDLR, but transplantation of FDNPVY (although not NPVY) into the TR cytoplasmic domain promotes rapid internalization (Collawn et al 1991).

Similarly, the CI-M6PR internalization signal was localized to the first 29 membrane-proximal amino acids of the 163-residue carboxy-terminal cytoplasmic tail (Canfield et al 1991). This region includes two tyrosine residues at positions 24 and 26 from the transmembrane region previously shown to be required for rapid endocytosis of the CI-M6PR (Lobel et al 1989). Their simultaneous replacement by alanine interfered with endocytosis. Mutation of each tyrosine independently indicated that Tyr-26 was more important than Tyr-24; further analysis identified the CI-M6PR internalization signal as the 4-residue sequence pattern YXKV, spanning residues 26–29, with Tyr-24 able to partially replace the requirement for Tyr-26 (Canfield et al 1991; Jadot et al 1992).

Characterization of the TR internalization signal was complicated by the fact that virtually all cultured cells must display TRs because iron uptake from Tf is required to sustain cell growth (Trowbridge et al 1991). To surmount this problem, two groups independently selected variant CHO cell lines lacking TRs capable of binding human Tf (McGraw et al 1987; Alvarez et al 1989). As an alternative approach, Jing et al (1990) expressed the human TR in chicken embryo fibroblasts (CEF). Although CEFs express endogenous receptors for ovatransferrin, they do not bind most mammalian Tfs, including human Tf. Consequently, endogenous chick TRs do not directly interfere with the functional analysis of mutant human TRs. Using this latter system, the TR internalization signal was localized to residues 19–28 by deletional analysis (Jing et al 1990). Insertion of these ten amino acids into a tailless receptor restored wild-type activity, which indicated unambiguously that this portion of the cytoplasmic tail was sufficient for rapid internalization. To define the minimum sequence required for rapid internalization, incremental 1-residue deletions were made from the carboxy-terminus of the 10-residue region inserted in the tailless mutant (Collawn et al 1990). The results of this analysis and alanine scanning mutations indicated that residues 24–28 are not specifically required, but function as a spacer region that separates the internalization signal, YTRF, from the cell membrane (Collawn et al 1990).

Although the YTRF sequence is now known to promote rapid internalization of the CI-M6PR (Jadot et al 1992), the conclusion that this 4-residue signal is sufficient for TR internalization has been questioned (Girones et al 1991; McGraw et al 1991). It was reported that Chinese hamster TR had a cysteine residue at position 20 and that YTRF signal was not, therefore, conserved in hamster (Alvarez et al 1989; Girones et al 1991). More recent work, however, indicates that the YTRF sequence is conserved in hamster, as it also is in mouse and chicken (DL Domingo & IS Trowbridge, unpublished observations; Gerhardt et al 1991). It was further argued that the TR internalization signal included residues in the amino-terminal 16-residues of the TR cytoplasmic tail because amino acid substitutions or deletions in this region reduced internalization activity (Girones et al 1991; McGraw et al 1991). However, as residues 19–28 are sufficient for activity (Jing et al 1990), the effect of mutations in the amino-terminal 18 residues of the TR tail must be indirect — either changing the conformation of the internalization signal or sterically blocking its interaction with coated pits. Recently, similar indirect effects of mutations on the activity of the LAP internalization signal have been reported (Lehmann et al 1992).

Extensive mutagenesis of the TR, CI-M6PR, and LDLR internalization signals has allowed generic sequence patterns of 4- and 6-residue signals to be identified (Figure 2). Analysis of the TR signal indicates that an amino-terminal aromatic residue and either a carboxy-terminal aromatic or large hydrophobic residue are required for activity. A positively charged residue is preferred at position 3, but substitution of other aliphatic residues only reduces activity by 50%. Although altering the residue at position 2 to either alanine or phenylalanine does not reduce activity, a change to

Figure 2 Mutagenesis of internalization signals. The figure shows the quantitative effects on internalization efficiency of single amino acid substitutions in the 4-residue internalization signals of TR and CI-M6PR and NPVY, the carboxy-terminal four amino acids of the 6-residue LDLR signal. TR data (*top panel*) displayed as closed circles are from Jing et al (1990); Collawn et al (1990); JF Collawn et al (unpublished observations). TR data displayed as open circles are from McGraw & Maxfield (1990). The data on mutagenesis of the CI-M6PR internalization signal (*middle panel*) are from Canfield et al (1991) and Jadot et al (1992). The data displayed as closed circles are from experiments in which single amino acid substitutions were introduced into a mutant receptor with a 29-residue truncated cytoplasmic domain, and in which Tyr-24 was altered to alanine giving a carboxy-terminal sequence of AKYSKV. The data displayed as open symbols represent point mutations introduced into the wild-type receptor. In this group, the receptor in which Tyr-26 was changed to Phe also had Tyr-24 altered to Phe. Mutagenesis data for the LDLR signal (*bottom panel*) are from Davis et al (1987) and Chen et al (1990).

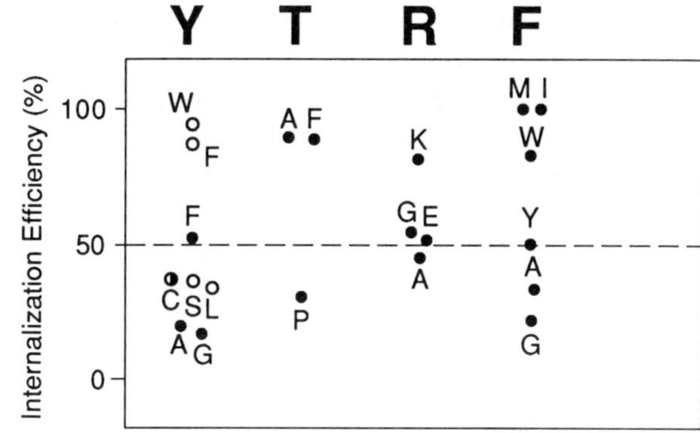

CI-M6PR Signal

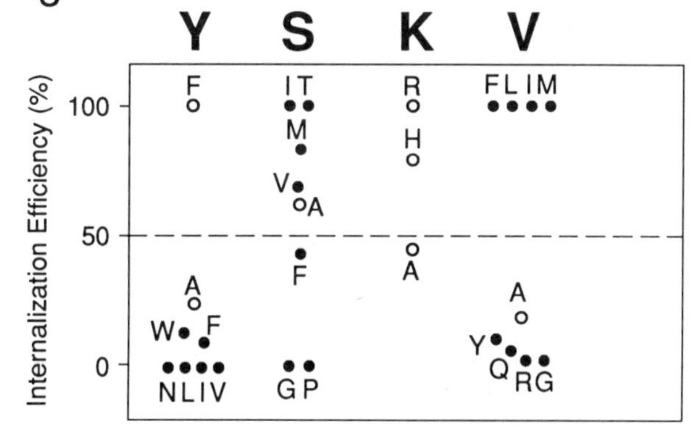

LDLR Signal

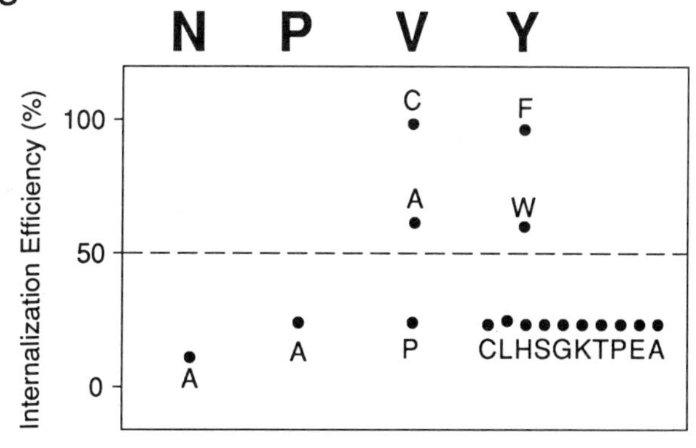

proline causes significant loss of activity. The structural requirements for activity of the CI-M6PR signal are similar to the TR signal except that they appear to be more stringent at positions 1 and 4. It should be noted, however, that most CI-M6PR mutations were analyzed in the context of the 29-residue truncated cytoplasmic domain, which places the YSKV signal at the carboxy-terminus, potentially destabilizing its structure. Overall, the data indicate that 4-residue internalization signals conform to a sequence pattern, aromatic-X-X-aromatic/large hydrophobic (where X stands for many but not all amino acids). The sequence pattern for 6-residue signals related to LDLR signal is less well-defined and fits the broad sequence pattern, aromatic-X-X-X-X-aromatic, with asparagine and proline possibly preferred at the third and fourth positions, respectively.

Putative internalization signals in the cytoplasmic domains of recycling receptors, Lgps, and other membrane proteins can be identified by searching for 4- or 6-residue sequence patterns similar to known signals (Figure 3; Canfield et al 1991; Trowbridge 1991). Although only one of the predicted sequences has been directly shown to help promote internalization of the protein in which it is found (Lehmann et al 1992), many of the 4-residue sequences from recycling receptors and Lgps contain aromatic residues known to be important for endocytosis and also have significant activity when transplanted into TR and/or CI-M6PR (Figure 3). Thus these sequences can serve as internalization signals in an appropriate structural context, which suggests that they function as internalization signals in their native membrane protcins. An exception appears to be the predicted 4-residue sequence YRHV from LAP. Although this sequence is sufficient to promote internalization of TR and M6PR (Figure 3), a 6-residue sequence, PGYRVH, is apparently required for rapid internalization of LAP (Lehmann et al 1992). The GY motif is found in other Lgps, but the glycine residue is not important for their internalization (Williams & Fukuda 1990; Hunziker et al 1991).

Other membrane proteins also display potential internalization signals that may be of functional significance (Figure 3). Internalization signals still cannot be identified unambiguously, however, on the basis of sequence alone. At least one potential signal, YSLL, in the FcRII receptor is apparently not required for internalization (Miettinen et al 1992). Further, some predicted signals are not active when transplanted into TR or CI-M6PR (Figure 3). In contrast, the CD3γ subunit sequence, YQPL (Letourneur & Klausner 1992), would not have been expected to have activity because of the proline residue at position 3 (Wilmot & Thornton 1988). Nevertheless, inspection of cytoplasmic tail sequences can provide important clues about whether a specific protein might be endocytosed. For example, Chen et al (1990) noted that the cytoplasmic domain of the β-amyloid precursor protein (β-APP) contains a 6-residue sequence similar to the LDLR internalization

	Cytoplasmic Domain	Internalization Efficiency %	
Receptors		TR	CI-M6PR
LDLR	12 FDNPVY 32	47	ND
CD M6PR	12 FPHLAF 26 YRGV 19	ND	26
Mannose R	12 FENTLY 27	24	ND
CI M6P/IGF-IIR	23 YKYSKV 134	107	100
poly IgR	81 YSAF 18	60	150
TR	19 YTRF 38	100	160
ASGPR	4 YQDL 32	62	ND
	4 FQDI 50	16	ND
Lysosomal Membrane Glycoproteins			
Lgp-A	7 YQTI	48	120
Lgp-B	6 YEQF	ND	110
LAP	6 PGYRHV 7	78	92
Other Membrane Proteins			
β-APP	33 YENPTY 8		
mIgG	19 YRNM 5		
CD3γ	21 YQPL 19		
FcγIIIγ	20 YTGL 7 YETL 7		
VSV G	18 YTDI 7		

Figure 3 Internalization signals of recycling receptors and other integral membrane proteins. The figure shows experimentally determined and putative internalization signals located in the cytoplasmic domains of membrane proteins. Amino acids that have been shown by alanine scanning to be important for endocytosis are in boldface type. The orientation of the membrane proteins and the size of their cytoplasmic tails are indicated with the start of the transmembrane region (shown as open boxes). The values for internalization efficiency shown at the right-hand side of the figure refer to the efficiency with which signals transplanted into the cytoplasmic domain of either TR or CI-M6PR promote internalization relative to the native signal. The 4-residue signal, YRGV, of the CD-M6PR, was transplanted into the CI-M6PR and the 4-residue sequence, YRHV, of the LAP signal, was transplanted into TR and CI-M6PR. N. D. is not determined. The data for the internalization signals and sequences are from Chen et al 1990 (LDLR); Johnson & Kornfeld, 1990 (CD-M6PR); Taylor et al 1990 (Mannose R); Canfield et al 1991 (CI-M6PR); Breitfeld et al 1990 (poly-IgR); Collawn et al 1990 (TR); Fuhrer et al 1991 (ASGPR); Williams & Fukuda 1990 (Lgp-A); Granger et al 1990 (Lgp-B); Lehmann et al 1992 (LAP); Kang et al 1987 (β-APP); Bensmana & LeFranc 1990 (mIgG); Letourneur & Klausner 1992 (CD3γ); Amigorena et al 1992 (Fcγ111γ); and Roth et al 1986 (VSVG). Data on the internalization efficiencies of signals transplanted into TR are from Collawn et al 1991; JC Collawn et al, unpublished observations; and for transplantation into CI-M6PR from Jadot et al 1992.

signal (Figure 3). Proteolytic processing of β-APP generates β-amyloid, which is deposited in the brains of Alzheimer's patients and is likely the causal agent of the disease (for reviews, see Selkoe 1991; Hardy & Allsop 1991). Recently, a chimeric TR has been constructed containing the β-APP cytoplasmic domain, and this molecule is efficiently internalized and recycled back to the cell surface, which strongly suggests that β-APP itself traffics along the endocytic pathway (A Lai & IS Trowbridge, unpublished observations). Altered trafficking of β-APP along the endocytic pathway may be involved in the generation of β-amyloid.

Notably absent from the list of integral membrane proteins with internalization signals are class I and II molecules of the major histocompatibility complex (MHC). Although some reports suggest class II MHC molecules are endocytosed (Guagliardi et al 1990; Reid & Watts 1990), the internalization rate is low (~2% per min), and there is general agreement that the major sites at which class I and II MHC molecules encounter antigen are different compartments along the biosynthetic pathway (Peters et al 1991).

Internalization Signals Are Tight Turns

There is increasing evidence that internalization signals share a common conformational determinant, a tight turn. A tight turn was first implicated as a structural feature of internalization signals by computational studies of the conformation preferences of the TR and LDLR internalization signals (Collawn et al 1990). The rationale for this work was based on the evidence from deletional analysis that the activity of the TR signal was relatively independent of adjacent residues (Jing et al 1990; Collawn et al 1990). This implied that the structure of the signal was self-determined, so it was hoped that the YTRF sequence might have a preferred conformation in proteins of known three-dimensional structure that could be identified by searching the Brookhaven Protein Data Bank. In practice, because of the limited number of protein structures in that database, it was necessary to search for analogues of the YTRF signal, which had at least 50% internalization activity. Strikingly, it was found that 8 of 10 of the most closely related sequences occurred as tight turns (either Type 1 β turns, or terminal or subterminal turns of an α-helix) that were independent of surrounding secondary structure and, with one possible exception, were surface-exposed. An independent search of the protein crystal database for analogues of the LDLR tetrapeptide sequence, NPVY, revealed that they also had a high propensity to be tight turns, which led to the conclusion that this structure may be a common feature of internalization signals (Collawn et al 1990).

Subsequently, direct evidence from two-dimensional NMR analysis of the structure of a synthetic nonapeptide corresponding to the LDLR signal

(Bansal & Gierasch 1991) indicated that the NPVY sequence in the LDLR internalization signal is a β-turn. Interestingly, structural analysis of synthetic peptides containing single amino acid substitutions in the NPVY sequence indicated that there was a strong correlation between the extent to which the alteration disrupted the turn conformation of the peptide and its effect on the internalization of mutant LDLR (Bansal & Gierasch 1991). It is also notable that the asparagine-proline combination in the NPVY sequence is the most statistically favored amino acid pair at the first and second positions of type I β-turns and the first turn of an α-helix (Wilmot & Thornton 1988; Richardson & Richardson 1988). NMR studies of a synthetic peptide corresponding to the cytoplasmic tail of LAP also showed that the tyrosine residue essential for rapid internalization is in a β-turn consisting of the sequence PPGY and that the turn was destabilized by amino acid substitutions that impaired internalization (Eberle et al 1991; Lehmann et al 1992). However, the interpretion of these data is complicated by the fact that the internalization signal of LAP appears to be PGYRHV (Lehmann et al 1992) and, in contrast to PPGY, the tetrapeptide, YRHV, is active as a signal when transplanted into TR as well as CI-M6PR (JF Collawn & IS Trowbridge, unpublished observations; Jadot et al 1992). Thus it is predicted that, by analogy to the LDLR and TR signals, tyrosine should be the first residue of a tight turn formed by the tetrapeptide, YRHV.

A prediction, based on the tight turn model, suggested that an internalization signal from one receptor could be expected to retain activity when transplanted into the cytoplasmic domain of another receptor (Collawn et al 1990). This prediction has now been confirmed in two independent studies (Collawn et al 1991; Jadot et al 1992). These studies also showed that internalization signals exchanged between type I and type II membrane proteins were active, thus indicating that their activity was independent of polypeptide polarity with respect to the cell membrane (Figure 3). However, despite all the concordant data supporting a tight turn model, the three-dimensional structure of a receptor cytoplasmic tail has not been determined, so that there is still no direct evidence that any internalization signal is a tight turn in its native environment.

This reservation notwithstanding, tyrosine-containing internalization signals are likely to be similar to signal sequences that promote the translocation of nascent polypeptide chains into the endoplasmic reticulum (Gierasch 1989) in that they differ in precise sequence, but provide exposed surface features with clearly defined physical characteristics that can be distributed at a variety of locations along the protein. They are thus quite unlike the short, highly conserved terminal sequences identified by the retention mechanisms of the endoplasmic reticulum.

Intracellular Sorting Signals

LYSOSOMAL SORTING SIGNALS As discussed earlier, Lgps can reach the lysosome either by an intracellular route or via the cell surface and, therefore, must have multiple sorting signals. Lgp-A and LAP, both type I membrane proteins with short cytoplasmic tails of 11 and 19 amino acids, respectively, are prototypes for this family of molecules that currently includes three other members Lgp-B (Lgp-2), Limp-1 and CD63 (Lamp 3) (reviewed in Fukuda 1991). Delivery of LAP and Lgp-A to the lysosome via the cell surface requires a tyrosine residue in the cytoplasmic domain that comprises part of an internalization signal (Peters et al 1990; Williams & Fukuda 1990; see Figure 3). Altering the tyrosine residue essential for rapid endocytosis of Lgp-A also blocks delivery of Lgp-A to the lysosome via the intracellular route (Harter & Mellman 1992), which indicates that a tyrosine-containing sorting signal is also required when Lgp-A travels along the intracellular route to the lysosome. If, as suggested earlier, there is a direct trafficking pathway from the *trans* Golgi to the endosome, this sorting signal may be involved in directing Lgp-A along this route. The glycine residue in the conserved GY motif of Lgps and adjacent to the amino-terminus of the predicted 4-residue internalization signal of Lgp-A appears to be an important element of the sorting signal recognized in the *trans* Golgi (Harter & Mellman 1992), which suggests that the *trans* Golgi sorting signal and internalization signals are related but not identical.

Two independent studies have identified a sorting signal distinct from tyrosine-containing signals that may also be recognized at the plasma membrane and in the *trans* Golgi. In one study, two sorting signals were identified in the cytoplasmic domain of the CD3γ chain of the T cell antigen receptor (Letourneur & Klausner 1992). One signal was a tyrosine-containing signal, YQPL, that also functioned as an internalization signal as well as a sorting signal on an intracellular route to the lysosome. The other lysosomal sorting signal consisted of six amino acids, DKQTLL, of which only the two leucine residues could be shown to be specifically required. This signal, which has been termed a di-leucine signal, was also active as an internalization signal in the context of the CD3γ chain cytoplasmic tail. Interestingly, this di-leucine signal is active as an internalization signal when transplanted into the cytoplasmic domain of human TR (JF Collawn et al, unpublished observations), which suggests that in the coated pits of the plasma membrane, tyrosine-containing signals and di-leucine signals are functionally equivalent.

A similar di-leucine lysosomal sorting signal has been identified in the CI-M6PR cytoplasmic domain (Johnson & Kornfeld, 1992). In these studies, it was shown that deletion of HLLPM at the carboxy-terminus of the CI-M6PR cytoplasmic tail abrogated sorting of a marker enzyme, cathepsin

D, from the *trans* Golgi to the lysosome. Of these amino acids, both leucines and the histidine residue were important for efficient sorting. Although the di-leucine signals identified in the two studies have similar properties with regard to their ability to mediate intracellular sorting to the lysosome, the CI-M6PR di-leucine signal does not appear to promote endocytosis and, therefore, differs in this respect from the CD3γ signal (see Lobel et al 1989).

The fact that di-leucine signals can promote endocytosis may explain why several membrane proteins that lack tyrosine signals, notably CD4 and the human FcII B_2 receptor, can participate in receptor-mediated endocytosis (Miettinen et al 1989; Pelchen-Matthews et al 1991, 1992). Both these molecules contain di-leucine sequence motifs in their cytoplasmic domains that could function as internalization signals. It is also interesting that the most recently characterized lysosomal membrane protein, LIMP II, lacks a tyrosine-containing signal in its cytoplasmic domain but has a leucine-isoleucine dipeptide close to its carboxy-terminus (Vega et al 1991).

BASOLATERAL SORTING SIGNALS An important, although not exclusive, mechanism by which polarized cells maintain the selective expression of integral transmembrane proteins on specific regions of the cell surface is by differential sorting as membrane proteins traverse the biosynthetic pathway. Of the different polarized cell types, the molecular basis of membrane protein trafficking has been studied most extensively in epithelial cells. Until recently, the most influential model for polarized sorting in epithelial cells proposed that trafficking of transmembrane proteins to the basolateral surface did not require a specific signal, whereas sorting to the apical surface was signal-dependent (e.g. Simons & Wandinger-Ness 1990). Now evidence is accumulating that basolateral sorting of at least some transmembrane proteins is signal-dependent and is directed by signals in their cytoplasmic tails. Tyrosine-containing signals in the cytoplasmic tail of several integral membrane proteins have been implicated in sorting to the basolateral surface of MDCK cells, although none has been fully characterized. The proteins studied include the Tyr-543 mutant influenza HA, which in contrast to wild-type HA, is sorted to the basolateral surface (Brewer & Roth 1991), and Lgp-A, which is not sorted to the basolateral surface if the tyrosine residue in its cytoplasmic tail is altered to alanine (Hunziker et al 1991). The cytoplasmic domain of the LDLR contains two independent basolateral sorting signals (Matter et al 1992). The region containing the membrane-proximal basolateral sorting signal includes the FXNPXY internalization signal. The tyrosine residue of the internalization signal is included in the basolateral sorting signal, but additional residues on the carboxy-terminal side of the internalization signal are also required. The distal basolateral sorting signal is localized to the carboxy-terminal 19 residues of the LDLR cytoplasmic domain. Although a tyrosine residue is important for activity of the distal basolateral sorting signal, the signal does not promote internalization.

Studies of the poly-IgR support the idea that basolateral targeting is signal-dependent since it was shown that residues 655–668 in the membrane-proximal region of the poly-IgR cytoplasmic tail mediated direct sorting to the basolateral surface (Casanova et al 1991). This region of the poly-IgR cytoplasmic tail was also sufficient to sort a chimeric molecule, in which the external domain of the poly-Ig receptor was replaced by that of placental alkaline phosphatase, to the basolateral surface of MDCK cells. The region in which the basolateral sorting signal is found includes a tyrosine residue (Tyr-668) and may form a tight turn. However, this tyrosine residue does not appear to be important for basolateral targeting, and the region containing the basolateral sorting signal does not promote rapid endocytosis (Okamoto et al 1992).

Modulation of Sorting Signals

Early indications that the internalization signals carried by TR were modified during recycling were not borne out, and it is believed, therefore, that the transit of receptors such as TR, LDLR, and ASGPR throughout the recycling pathway is constitutive. There are, however, trafficking proteins that migrate intermittently, usually in response to ligand binding, and in these instances, there is likely to be some means of switching their sorting signals on and off.

The most thoroughly studied receptors of this type are members of the tyrosine kinase growth factor receptor family, most notably the EGFR and insulin receptor. Mutant EGFRs lacking a cytoplasmic domain are poorly internalized (Prywes et al 1986), which raises the questions of how ligand binding induces internalization and what structural features of the cytoplasmic domain are required. An attractive hypothesis providing answers to both questions is that ligand binding induces a structural change in the cytoplasmic domain of the EGFR that exposes a recognition determinant for clathrin-coated pits similar to the internalization signals of constitutively recycling receptors. One mechanism by which such a conformational change could be induced is by autophosphorylation of tyrosine residues, which implies that internalization is dependent on the kinase activity of the receptor. Conflicting results have been obtained from studies of kinase-negative mutant EGF receptors generated either by insertional inactivation of the kinase domain or by altering Lys-721, a residue essential for catalytic activity. On the basis of biochemical assays and qualitative immunofluorescence, it was concluded that kinase-negative mutant receptors were not internalized

(Glenney et al 1988; Chen et al 1989). More recently, however, immunoelectron microscopic studies, together with detailed kinetic analysis of EGFR internalization, suggest that kinase-negative EGFRs are internalized normally but are rapidly recycled back to the cell surface (Felder et al 1990, 1992).

Both the EGFR and insulin receptor cytoplasmic domains contain multiple copies of the LDLR-related NPXY sequence pattern. However, the three NPXY sequences in the EGF receptor do not appear to be necessary for ligand-induced internalization because deletion of the carboxy-terminal region of the EGFR cytoplasmic tail containing them does not inhibit this process (Prywes et al 1986). Deletion of the membrane-proximal region of the insulin receptor cytoplasmic domain containing NPXY and related GPXY sequence does, however, appear to impair internalization (Thies et al 1990; Backer et al 1990). The alteration of the two tyrosine residues in these sequences also partially inhibits internalization, which suggests that an internalization motif is involved (Rajagopalan et al 1991; Backer et al 1992). It should be noted, however, that neither of the two NPXY-related sequences in the insulin receptor has an aromatic residue two residues upstream corresponding to the FXNPXY sequence pattern required for LDLR internalization. Deletional analysis of the cytoplasmic domain of the EGF receptor has also identified a region, the CaIn domain, distinct from the NPXY sequences that may be important for ligand-induced internalization (Chen et al 1989). Interestingly, the deletion that defines this region, spanning residues 966–1007, interrupts the tetrapeptide sequence, YRAL, which is located in a predicted turn, conforms to the sequence pattern of 4-residue internalization signals, and is active when transplanted into the human TR (JF Collawn & IS Trowbridge, unpublished observations). Whether ligand-induced internalization involves a similar signal to that of constitutively recycling receptors, which is only exposed after ligand binding, however, is still unclear.

It should also be noted that all the constitutively recycling integral membrane proteins with well-characterized internalization signals are type I and II membrane proteins that span the plasma membrane lipid bilayer once. However, the plasma membrane expression of several polytopic type III membrane proteins (Singer 1990) with multiple transmembrane regions, including the insulin-sensitive glucose transporter, GLUT 4 (Robinson et al 1992; Piper et al 1992), β2-adrenergic receptor (Valiquette et al 1990), and synaptophysin (Maycox et al 1992), are thought to be regulated by receptor-mediated endocytosis. The structural determinants that mediate clustering of these proteins in coated pits are unknown.

In polarized cells, transcytotic pathways that carry ligand-receptor complexes from the basolateral to the apical surface begin with internalization into the basolateral endosome (Hopkins 1991; Okamoto et al 1992). To be transcytosed, trafficking proteins, such as the poly-IgR, which follow this route, must therefore be sorted out of this recycling pathway (which contains TR) and into an apical exocytic route. Although the location at which this takes place has not been defined, the signals required for the complicated itinerary that the poly-IgR follows have been well-characterized, and it is clear that, while basolateral delivery and internalization depend upon well-defined sequences of amino acids in the cytoplasmic domain (discussed above), exit from the recycling pathway involves phosophorylation of a serine at position 664, i.e. between the two signals (Casanova et al 1990). The consequence of this modification at the molecular level is probably to switch on an apical signal because the phosphoserine can be replaced by an aspartate (which carries a negative charge mimicking phosphate). At the cellular level, it should allow the basolateral endosome to distinguish poly-IgRs traveling to the basolateral border from those that are returning after having been internalized.

In a recent study on the CI-M6PR (Meresse & Hoflack 1993), it was shown that two serines in the cytoplasmic domain become transiently phosphorylated during the time this receptor emerges from the biosynthetic pathway in the *trans* Golgi and enters the endocytic pathway. The recycling of this receptor between these two pathways is believed to be constitutive, and thus this modification could operate at any point in the cycle. As discussed above, the evidence available on the itinerary followed by CI-M6PR indicates that they need to be sorted in the *trans* Golgi, the sorting endosome, and the prelysosome (as well as the plasma membrane), so that it should not be surprising if either the tyrosine-containing or the di-leucine signals within their tails are supplemented at one of the intracellular steps by a regulatory phosphorylation-dephosphorylation modification.

The realization that motifs as different as those containing tyrosine and di-leucine can function as internalization signals and the indication that the affinities of these motifs for the various sorting mechanisms could be altered because of specific neighboring residues or by post-translational change such as phosphorylation clearly implies that the molecular interactions will undoubtedly require isolation and characterization of both signal and recognition machinery.

MOLECULAR SORTING MACHINERY

The only unequivocally identified mechanisms for signal-dependent sorting of trafficking membrane proteins are domains defined by clathrin lattices present on the cytoplasmic face of the plasma membrane and on a variety of cisternal elements in the *trans* Golgi area. The most abundant structural proteins of these domains have been thoroughly characterized biochemically and their primary structures determined. Several excellent reviews have recently appeared that describe these components in detail (Pearse & Robinson 1990; Keen 1990; Robinson 1992). High resolution electron microscopy suggests that the coat consists of an outer lattice of clathrin and an inner shell that contains brick-shaped adaptor (AP) complexes. The adaptor complexes thus lie immediately adjacent to the cytoplasmic face of the plasma membrane lipid bilayer and are probably able to interact directly with the cytoplasmic domains of its integral membrane proteins. Lattices are made by the three heavy and light clathrin chains forming triskelions, which then assemble into the characteristic polyhedral conformation of pentagons and hexagons. Two separate forms of light chains encoded by separate genes have been identified (Keen 1990). The adaptor complexes are composed of two 100 kd proteins and two smaller subunits. The 100 kd proteins of plasma membrane adaptor complexes (AP-2 or HA-2 complexes) are termed α- and β-adaptins and those of intracellular membranes (AP-1 or HA-1 complexes) γ- and β′-adaptins. There are extensive sequence similarities between them, although there is only 25% identity between the α- and γ-components.

Molecular genetic analysis suggests that in most tissues there are only two kinds of clathrin lattice, those of the plasma membrane characterized by α- adaptin and those of intracellular boundaries containing γ-adaptin. In clathrin coats isolated from brain, however, additional proteins, AP-3 (Murphy et al 1991) and auxillin (Ahle & Ungewickell 1990), have been identified. Two different α-adaptins (αA and αC) encoded by different genes have been identified, and a larger brain-specific isoform of αA is generated by alternative splicing (Robinson 1989). Larger isoforms of β-adaptin and clathrin light chains (Ponnambalam et al 1990; Jackson et al 1987) have also been described in the brain. Neurons are thus likely to have other additional lattice-based sorting mechanisms.

It is likely that the internalization signals of constitutively recycling receptors directly interact with coated pits, presumably by specific binding to adaptor complexes. However, evidence for a direct interaction between receptor cytoplasmic tails and adaptor complexes is limited and qualitative in nature (Pearse 1988; Glickman et al 1989; Beltzer & Speiss 1991; Pearse & Robinson 1990). Thus AP-2 complexes were shown to bind an affinity matrix of LDLR cytoplasmic tails prepared as bacterial fusion proteins (Pearse 1988). As expected for a specific interaction, AP-1 adaptors did not bind; however, maximum binding of AP-2 complexes is calculated to be only ~6% of the theoretical capacity of the affinity matrix. The binding of AP-2 complexes could be inhibited by soluble cytoplasmic tails of CI-M6PR, and poly-IgR, and a large molar excess (~10^3-fold) of the

Tyr-543 mutant HA cytoplasmic domain, but binding was only reduced ~50%. In subsequent studies, both AP-1 and AP-2 complexes were shown to bind to CI-M6PR cytoplasmic tails, and the addition of mutant CI-M6PR cytoplasmic tails with a defective internalization signal partially inhibited binding of AP-1, but not AP-2 complexes (Glickman et al 1989). These results suggest that internalization signals of different receptors bind to the same site on AP-2 adaptor complexes and that a structural determinant distinct from its internalization signal mediates binding of CI-M6PR to AP-1 complexes. Identification of a separate di-leucine lysosomal targeting signal at the carboxy-terminus of the CI-M6PR cytoplasmic tail supports the latter idea, but it would not agree with the accumulating evidence which suggests that tyrosine-based internalization signals may also be recognized at an intracellular location (Letourneur & Klausner 1992; Johnson & Kornfeld 1992). Thus although it remains likely that internalization and lysosomal sorting signals bind to both AP-1 and AP-2 adaptor complexes, they may do so with very different affinities. More detailed, quantitative studies of these interactions will be required for this issue to be resolved.

Binding of the ASGPR to the amino-terminal 60 kd region of β-adaptin has recently been reported using a ligand-blotting technique (Beltzer & Speiss 1991). The specificity of this interaction has not been firmly established, however, and its physiological significance is unclear (Robinson 1992). Intuitively, it would not be expected that the AP-2 binding site for internalization signals would be localized to a region of β-adaptin, which is so highly conserved (Kirchhausen et al 1989).

Some receptors, such as the CD-M6PR (Johnson et al 1990), contain more than one internalization signal, and the introduction of a second signal into the cytoplasmic domain of human TR can increase internalization efficiency (JF Collawn & IS Trowbridge, unpublished observations), which implies that two signals in the same polypeptide chain can increase binding to adaptors.

The sorting process that takes place in the clathrin-coated pits of the plasma membrane involves (a) the formation of the clathrin lattice in which trafficking receptors destined for uptake are concentrated, (*b*) invagination and pinching off of the membrane to form an endocytic vesicle, and (*c*) the rapid disassembly of the clathrin lattice. Morphological studies suggest that flat, planar lattices form before membrane invagination begins (Larkin et al 1986; Heuser & Anderson 1989; Lin et al 1991). Receptors like TR, which bear internalization signals, are actively involved in lattice formation since they have been shown to increase the area of lattice formed when expressed at high levels (Miller et al 1991). It is probable, however, that integral membrane proteins other than trafficking receptors are also involved in this process because lattices containing α- and β-adaptins only associate with these proteins when they are on the plasma membrane. It is thought, therefore, that a receptor protein in the plasma membrane acts as a nucleation site for adaptor complexes (Peeler et al 1993), and these complexes in turn recruit receptors with internalization signals. However, the possibility that the carboxy-terminal regions of the α- and β-adaptins (which are thought to form flexible ear-like appendages to the adaptor complexes) (Heuser & Keen 1988) may contain the binding site for such a membrane protein has now been discounted (Robinson 1992; Peeler et al 1993). With the association of trafficking receptors and adaptor complexes, clathrin triskelions are presumably recruited and the planar coat completed.

No correlation between the number of plasma membrane invaginations and the extent of lattice formation has been found (Miller et al 1991), and thus it is thought that vesiculation of the plasma membrane is probably independently regulated. The realignment of triskelions, which occurs during membrane invagination, is thus likely to be primarily concerned with accommodating the changes in membrane shape and holding recruited receptors in place while this process occurs.

Regulation of the vesiculation process in coated pits is currently being studied using semi-intact membrane systems (Moore et al 1989; Schmid & Smythe 1991; Schmid 1992), but these preparations are only just beginning to provide new insights into the molecular mechanisms involved in plasma membrane invagination (Carter et al 1993). Unlike similar work on other cellular pathways, this area has been hampered by a lack of genetic analysis. Well-characterized systems of receptor-mediated endocytosis in lower eukaryotes such as yeast have not been described, and it remains to be established if the internalization of alpha factor in *Saccharomyces cerevisiae* involves clathrin-coated pits (Chvatchko et al 1986; Payne et al 1988; Seeger & Payne 1992). Consequently, it is unclear whether the isolation of internalization-defective mutants in *S. cerevisiae* will yield information germane to receptor-mediated endocytosis in animal cells (Raths et al 1993). Ironically, although coated pits were first described in mosquito oocytes (Roth & Porter 1964), trafficking proteins analogous to mammalian receptors such as LDLR or TR have still not been described in *Drosophila* or other insects (Law et al 1992).

Recently, two GTP-binding proteins, Rab 5, a member of the Ras family, and dynamin, originally thought to be a microtubule-associated GTPase, have been directly implicated in endocytic vesicle formation (Bucci et al 1992; Chen et al 1991; van der Bliek & Meyerowitz 1991). These findings, together with the semi-intact in vitro systems which have recently been developed, are likely to bring new impetus to this area.

The only other sorting mechanism for which information is available is located in the sorting endosome. This mechanism separates receptors like EGFR destined for degradation in the lysosomes from recycling receptors. In studies comparing the movement of EGFR with a mutant receptor lacking endogenous tyrosine kinase activity, it was shown that the vesiculation process that removes EGFR to the internal vesicles of the sorting endosome requires the kinase activity in order to operate efficiently (Felder et al 1990). While the receptor is being processed in the sorting endosome, the most prominently phosphorylated substrate of the EGFR kinase is annexin 1, a 35 kd protein that interacts with both phospholipids and actin. Some forms of annexin 1 uncouple from these interactions when they become phosphorylated, and it is possible, therefore, that the kinase activity of clustered EGFR causes a local uncoupling that, as a consequence, allows the perimeter membrane of the sorting endosome to vesiculate inwards (Futter et al 1993).

Although the route taken by Lgps from the *trans* Golgi to the lysosome remains controversial, these proteins, as well as CI-M6PR, are known to be present in the endosome (Kornfeld & Mellman 1989). Therefore, like EGFR, they also need to be sorted away from the recycling pathway in order to move to the lysosome. This process, which presumably also takes place in the sorting endosome is, however, different from the removal of membrane proteins for degradation, since Lgps need to be delivered to the perimeter membrane of the lysosome. Regardless of whether sorting endosomes mature into lysosomes or form carrier vesicles, the main requirement of sorting signals that ensure that Lgps move towards the lysosome at this stage is that they prevent their proteins from being removed from the perimeter membrane. A variety of such retention mechanisms are known to operate along the main biosynthetic pathway, but although their signals are known, their anchorage mechanisms have yet to be defined. For Lgps, studies using VSVG-LEP 100 chimeras suggest that all the information for lysosomal targeting resides within the 11-residue cytoplasmic domain (Matthews et al 1992). Thus in the sorting endosome, these interactions must take place on the cytoplasmic side, rather than within the lipid bilayer of the perimeter membrane.

The information on CI-M6PR sorting in the endosome-prelysosome segment is less easy to understand because, although their sorting signals are also located within their cytoplasmic tails, morphological studies (Kornfeld & Mellman 1989) indicate that they become concentrated on internal rather than perimeter membranes. However, unlike the vesicles that sequester EGFR, these membranes would not be expected to lose continuity with the perimeter membrane since CI-M6PR needs to be retrieved for recycling back to the Golgi (Duncan & Kornfeld 1988).

CONCLUDING REMARKS

Characterization of tyrosine-containing internalization signals and progress in identifying other signals recognized by clathrin-coated domains are providing significantly new insight into the molecular mechanisms of signal-dependent membrane protein trafficking. All of the sorting signals identified to date in integral membrane proteins appear to be short sequences within the cytoplasmic domain. Like the tight turn of the tyrosine-containing signal of TR, they are likely to have a distinctive three-dimensional structure, and since some cytoplasmic domains contain more than one internalization signal (which are cooperative), it is probable that some trafficking proteins interact with more than one site on the sorting machinery. In addition, it seems that closely similar signals are recognized by the clathrin-coated domains of the plasma membrane and *trans* Golgi, but with different efficiency. Together, these observations suggest that the recognition mechanisms within clathrin-coated domains have a range of affinities for the signals on trafficking proteins; the higher the affinity, the more likely a protein is to be selected for transfer. Sorting mechanisms of this kind may have a relatively low level of discrimination for some proteins and would serve only to influence rather than open or close traffic flows. Nevertheless, in situations where such mechanisms of positive selection are coupled to recycling pathways, a high level of efficiency will be achieved at steady state. Clathrin-based sorting mechanisms are thus well-matched to recycling pathways, and it is to be expected that where, as in neurons, additional clathrin-based mechanisms exist, they too will be found to be coupled to a recycling pathway.

The view of clathrin-coated sorting mechanisms as sites at which proteins bearing internalization signals are selected for, but at which there is also a detectable flow of proteins that lack signals, is consistent with the low rates of internalization observed with tailless receptors. However, this view differs from the original concept of coated pits as molecular filters which, by some form of exclusion, were thought to prevent the transfer of proteins lacking the appropriate signals (Bretscher et al 1980). It seems likely, therefore, that cell surface proteins that are excluded entirely from coated pits are not freely mobile and are tethered in some way, either because they are anchored from below to the cortical cytoskeleton, or from above by their interactions with the extracellular matrix.

ACKNOWLEDGMENTS

This work was supported by grants CA34787 and CA37641 from the National Cancer Institute and by a NATO Collaborative Research Grant No. 880393 to IST, by a grant from the Arthritis Foundation to JFC, and by an MRC Program Grant to CH. We thank all our colleagues in the Department of Cancer Biology and the Laboratory for Molecular Cell Biology who participated in these studies and Joan Stewart for word processing. We would also especially like to acknowledge the important contributions of John Tainer and Leslie Kuhn in the Department of Molecular Biology at Scripps Research Institute in the computational analysis of the internalization signal structures.

Literature Cited

Ahle, S, Ungewickell, E. 1990. Auxilin, a newly identified clathrin-associated protein in coated vesicles from bovine brain. *J. Cell Biol.* 111:19–29

Alvarez, E, Girones, N, Davis, RJ. 1989. Intermolecular disulfide bonds are not required for the expression of the dimeric state and functional activity of the transferrin receptor. *EMBO J.* 8:2231–40

Alvarez, E, Girones, N, Davis, RJ. 1990. A point mutation in the cytoplasmic domain of the transferrin receptor inhibits endocytosis. *Biochem. J.* 267:31–35

Amigorena, S, Salamero, J, Davoust, J, Fridman, WH, Bonnerot, C. 1992. Tyrosine-containing motif that transduces cell activation signals also determines internalization and antigen presentation via type III receptors for IgG. *Nature* 358:337–41

Backer, JM, Kahn, CR, Cahill, DA, Ullrich, A, White, MF. 1990. Receptor-mediated internalization of insulin requires a 12-amino acid sequence in the juxtamembrane region of the insulin receptor β-subunit. *J. Biol. Chem.* 265:16450–54

Backer, JM, Shoelson, SE, Weiss, MA, Hua, QX, Cheatham, RB, et al. 1992. The insulin receptor juxtamembrane region contains two independent tyrosine/β-turn internalization signals. *J. Cell Biol.* 118: 831–39

Bansal, A, Gierasch, LM. 1991. The NPXY internalization signal of the LDLR adopts a reverse-turn confrontation. *Cell* 67:1195–201

Beaumelle, BD, Gibson, A, Hopkins, CR. 1990. Isolation and preliminary characterization fo the major membrane boundaries of the endocytic pathway in lymphocytes. *J. Cell Biol.* 111:1811–23

Beltzer, JP, Spiess, M. 1991. In vitro binding of the asialoglycoprotein receptor to the β adaptin of plasma membrane coated vesicles. *EMBO J.* 10:3735–42

Bensmana, M, LeFranc, M-P. 1990. Gene segments encoding membrane domains of the human immunoglobulin gamma 3 and alpha chains. *Immunogenetics* 32:321–30

Braun, M, Waheed, A, von Figura, K. 1989. Lysosomal acid phosphatase is transported to lysosomes via the cell surface. *EMBO J.* 8:3633–40

Breitfeld, PP, Casanova, JE, McKinnon, WC, Mostov, KE. 1990. Deletions in the cytoplasmic domain of the polymeric immunoglobulin receptor differentially affect endocytotic rate and postendocytotic traffic. *J. Biol. Chem.* 265:13750–57

Bretscher, MS, Thomson, JN, Pearse, BMF. 1980. Coated pits act as molecular filters. *J. Cell Biol.* 77:4156–59

Brewer, CB, Roth, MG. 1991. A single amino acid change in the cytoplasmic domain alters the polarized delivery of influenza virus hemagglutinin. *J. Cell Biol.* 114:413–21

Bucci, C, Parton, RG, Mather, IH, Stunnenberg, H, Simons, K, et al. 1992. The small GTPase rab5 functions as a regulatory factor in the early endocytic pathway. *Cell* 70:715–28

Canfield, WM, Johnson, KF, Ye, RD, Gregory, W, Kornfeld, S. 1991. Localization of the signal for rapid internalization of the bovine cation-independent mannose-6-phosphate/insulin-like growth factor-II receptor to amino acids 24–29 of the cytoplasmic tail. *J. Biol. Chem.* 266:5682–88

Carlsson, SR, Fukuda, M. 1992. The lysosomal membrane glycoprotein lamp-1 is transported to lysosomes by two alternative pathways. *Arch. Biochem. Biophys.* 296: 630–39

Carter, LL, Redelmeier, TE, Woollenweber, LA, Schmid, SL. 1993. Multiple GTP-binding proteins participate in clathrin-coated vesicle-mediated endocytosis. *J. Cell Biol.* 120:37–47

Casanova, JE, Apodaca, G, Mostov, KE. 1991. An autonomous signal for basolateral sorting in the cytoplasmic domain of the polymeric immunoglobulin receptor. *Cell* 66:65–75

Casanova, JE, Breitfeld, PP, Ross, SA, Mostov, KE. 1990. Phosphorylation of the polymeric immunoglobulin receptor re-

quired for its efficient transcytosis. *Science* 248:742–45
Chen, MS, Obar, RA, Schroeder, CC, Austin, TW, Poodry, CA, et al. 1991. Multiple forms of dynamin are encoded by *shibire,* a *Drosophila* gene involved in endocytosis. *Nature* 351:583–86
Chen, W-J, Goldstein, JL, Brown, MS. 1990. NPXY, a sequence often found in cytoplasmic tails, is required for coated pit-mediated internalization of the low density lipoprotein receptor. *J. Biol. Chem.* 265:3116–23
Chen, WS, Lazar, CS, Lund, KA, Welsh, JB, Chang, C-P, et al. 1989. Functional independence of the epidermal growth factor receptor from a domain required for ligand-induced internalization and calcium regulation. *Cell* 59:33–43
Chvatchko, Y, Howald, I, Riezman, H. 1986. Two yeast mutants defective in endocytosis are defective in pheromone response. *Cell* 46:355–64
Collawn, JF, Kuhn, LA, Liu, L-FS, Tainer, JA, Trowbridge, IS. 1991. Transplanted LDL and mannose-6-phosphate receptor internalization signals promote high-efficiency endocytosis of the transferrin receptor. *EMBO J.* 10:3247–53
Collawn, JF, Stangel, M, Kuhn, LA, Esekogwu, V, Jing, S, et al. Transferrin receptor internalization sequence YXRF implicates a tight turn as the structural recognition motif for endocytosis. *Cell* 63: 1061–72
Courtoy, PJ. 1991. Dissection of Endosomes. In *Trafficking of Membrane Proteins,* ed. JC Steer, J Harford, pp. 103–56. New York: Academic
Davis, CG, Lehrman, MA, Russell, DW, Anderson, RGW, Brown, M. S, Goldstein, JL. 1986. The J. D. mutation in familial hypercholesterolemia: amino acid substitution in cytoplasmic domain impedes internalization of LDL receptors. *Cell* 45:15–24
Davis, CG, van Driel, IR, Russell, DW, Brown, MS, Goldstein, JL. 1987. The low density lipoprotein receptor. Identification of amino acids in cytoplasmic domain required for rapid endocytosis. *J. Biol. Chem.* 262:4075–82
Dickson, RB, Hanover, JA, Willingham, MC, Pastan, I. 1983. Prelysosomal divergence of transferrin and epidermal growth factor during receptor-mediated endocytosis. *Biochemistry* 22:5667–74
Draye, JP, Quintart, J, Cowley, P, Baudheim, P. 1987. Relations between plasma membrane and lysosomal membrane of covalently labelled plasma membrane protein. *Eur. J. Biochem.* 170:395–403
Duncan, JR, Kornfeld, S. 1988. Intracellular movement of two mannose 6-phosphate receptors: return to the Golgi apparatus. *J. Cell Biol.* 106:617–28
Dunn, KW, Maxfield, FR. 1992. Delivery of ligands from sorting endosomes to late endosomes occurs by maturation of sorting endosomes. *J. Cell Biol.* 117:301–10
Dunn, KW, McGraw, TE, Maxfield, FR. 1989. Iterative fractionation of recycling receptors from lysosomally destined ligands in an early sorting endosome. *J. Cell Biol.* 109:3303–14
Eberle, W, Sander, C, Klaus, W, Schmidt, B, von Figura, K, Peters, C. 1991. The essential tyrosine of the internalization signal in lysosomal acid phosphatase is part of a β turn. *Cell* 67:1203–9
Felder, S, LaVin, J, Ullrich, A, Schlessinger, J. 1992. Kinetics of binding, endocytosis, and recycling of EGF receptor mutants. *J. Cell Biol.* 117:203–12
Felder, S, Miller, K, Moehren, G, Ullrich, A, Schlessinger, J, Hopkins, CR. 1990. Kinase activity controls the sorting of the epidermal growth factor receptor within the multivesicular body. *Cell* 61:623–34
Fuhrer, C, Geffen, I, Spiess, M. 1991. Endocytosis of the ASGP receptor H1 is reduced by mutation of tyrosine-5 but still occurs via coated pits. *J. Cell Biol.* 114: 423–31
Futter, CE, Felder, S, Schlessinger, J, Ullrich, A, Hopkins, CR. 1993. Annexin I is phosphorylated in the multivesicular body during the processing of the epidermal growth factor receptor. *J. Cell Biol.* 120:77–83
Fukuda, M. 1991. Lysosomal membrane glycoproteins. *J. Biol. Chem.* 266:21327–30
Gerhardt, EM, Chan, L-NL, Jing, S, Qi, M, Trowbridge, IS. 1991. The cDNA sequence and primary structure of the chicken transferrin receptor. *Gene* 102:249–54
Geuze, HJ, Slot, JW, Strous, GJ, Lodish, HF, Schwartz, AL. 1983. The pathway of the asialoglycoprotein-ligand during receptor-mediated endocytosis: a morphological study with colloidal gold/ligand in the human hepatoma cell line Hep G2. *Eur. J. Cell Biol.* 32:38–44
Gierasch, LM. 1989. Signal sequences. *Biochemistry* 28:923–30
Girones, N, Alvarez, E, Seth, A, Lin, I-M, Latour, DA, Davis, RJ. 1991. Mutational analysis of the cytoplasmic tail of the human transferrin receptor. *J. Biol. Chem.* 266: 19006–12
Glenney, JR, Chen, WS, Lazar, CS, Walton, GM, Zokas, LM, et al. 1988. Ligand-induced endocytosis of the EGF receptor is blocked by mutational inactivation and by microinjection of anti-phosphotyrosine antibodies. *Cell* 52:675–84
Glickman, JN, Conibear, E, Pearse, BMF. 1989. Specificity of binding of clathrin

adaptors to signals on the mannose-6-phosphate/insulin-like growth factor II receptor. *EMBO J.* 8:1041–47

Goldstein, JL, Brown, MS, Anderson, AGW, Russell, DW, Schneider, WJ. 1985. Receptor-mediated endocytosis: concepts emerging from the LDL receptor system. *Annu. Rev. Cell Biol.* 1:1–39

Granger, BL, Green, SA, Gabel, CA, Howe, CL, Mellman, I, Helenius, A. 1990. *J. Biol. Chem.* 265:12036–43

Greenspan, P, St Claire, RW. 1984. Retroendocytosis of LDL. Effect of lysosomal inhibitors on release of undegraded 125I LDL of altered composition from fibroblasts in culture. *J. Biol. Chem.* 259:1703–13

Griffiths, G, Back, R, Marsh, M. 1989. A quantitative analysis of the endocytic pathway in baby hamster kidney cells. *J. Cell Biol.* 109:2703–20

Griffiths, G, Gruenberg, J. 1991. The arguments for pre-existing early and late endosomes. *Trends Cell Biol.* 1:5–9

Griffiths, G, Hoflack, B, Simons, K, Mellman, I, Kornfeld, S. 1988. The mannose 6-phosphate receptor and the biogenesis of lysosomes. *Cell* 52:329–41

Gruenberg, J, Howell, KE. 1989. Membrane traffic in endocytosis: insights from cell-free assays. *Annu. Rev. Cell Biol.* 5:453–81

Guagliardi, LE, Koppelman, B, Blum, JS, Marks, MS, Cresswell, P, Brodsky, FM. 1990. Co-localization of molecules involved in antigen processing and presentation in an early endocytic compartment. *Nature* 343: 133–39

Hardy, J, Allsop, D. 1991. Amyloid deposition as the central event in the aetiology of Alzheimer's disease. *Trends Pharmacol. Sci.* 12:383–88

Harter, C, Mellman, I. 1992. Transport of the lysosomal membrane glycoprotein lgp120 (lgp-A) to lysosomes does not require appearance on the plasma membrane. *J. Cell Biol.* 117:311–25

Haylett, T, Thilo, L. 1986. Limited and selected transfer of plasma membrane glycoproteins to membrane of secondary lysosomes. *J. Cell Biol.* 103:1249–56

Heuser, JE, Anderson, RGW. 1989. Hypertonic media inhibit receptor-mediated endocytosis by blocking clathrin-coated pit formation. *J. Cell Biol.* 108:389–400

Heuser, JE, Keen, J. 1988. Deep-etch visualization of proteins involved in clathrin assembly. *J. Cell Biol.* 107:877–86

Hopkins, CR. 1991. Polarity signals. *Cell* 66:827–29

Hopkins, CR, Gibson, A, Shipman, M, Miller, K. 1990. Movement of internalized ligand-receptor complexes along a continuous endosomal reticulum. *Nature* 346:335–39

Hopkins, CR, Trowbridge, IS. 1983. Internalization and processing of transferrin and transferrin receptors in human carcinoma cells. *J. Cell. Biol.* 97:508–21

Hunziker, W, Harter, C, Matter, K, Mellman, I. 1991. Basolateral sorting in MDCK cells requires a distinct cytoplasmic domain determinant. *Cell* 66:907–20

Jackson, AP, Seow, H-F, Holmes, N, Drickamer, K, Parham, P. 1987. Clathrin light chains contain brain-specific insertion sequences and a region of homology with intermediate filaments. *Nature* 326: 154–59

Jadot, M, Canfield, WM, Gregory, W, Kornfeld, S. 1992. Characterization of the signal for rapid internalization of the bovine mannose-6-phosphate/insulin-like growth factor-II receptor. *J. Biol. Chem.* 267: 11069–77

Jing, S, Spencer, T, Miller, K, Hopkins, C, Trowbridge, IS. 1990. Role of the human transferrin receptor cytoplasmic domain in endocytosis: localization of a specific signal sequence for internalization. *J. Cell Biol.* 110:283–94

Johnson, KF, Chan, W, Kornfeld, S. 1990. Cation-dependent mannose-6-phosphate receptor contains two internalization signals in its cytoplasmic domain. *Proc. Natl. Acad. Sci. USA* 87:10010–14

Johnson, KF, Kornfeld, S. 1992. The cytoplasmic tail of the mannose 6-phosphate/insulin-like growth factor-II receptor has two signals for lysosomal enzyme sorting in the Golgi. *J. Cell Biol.* 119:249–57

Kang, J, Lemaire, H-G, Unterbeck, A, Salbaum, JM, Masters, CL, et al. 1987. The precursor of Alzheimer's disease amyloid A4 protein resembles a cell-surface receptor. *Nature* 325:733–36

Keen, JH. 1990. Clathrin and associated assembly and disassembly proteins. *Annu. Rev. Biochem.* 59:415–38

Kirchhausen, T, Nathanson, KL, Matsui, W, Vaisberg, A, Chow, EP, et al. 1989. Structural and functional division into two domains of the large (100- to 115-kDa) chains of the clathrin-associated protein complex AP-2. *Proc. Natl. Acad. Sci. USA* 86:2612–16

Kornfeld, S, Mellman, I. 1989. The biogenesis of lysosomes. *Annu. Rev. Cell Biol.* 5:483–525

Ktistakis, NT, Thomas, D, Roth, MG. 1990. Characteristics of the tyrosine recognition signal for internalization of transmembrane surface glycoproteins. *J. Cell Biol.* 111:1393–407

Larkin, JM, Donzell, WC, Anderson, RGW. 1986. Potassium-dependent assembly of coated pits: new coated pits form as planar clathrin lattices. *J. Cell Biol.* 103:2619–27

Law, JH, Ribeiro, JMC, Wells, MA. 1992. Biochemical insights derived from insect diversity. *Annu. Rev. Biochem.* 61:87–111
Lazarovits, J, Roth, M. 1988. A single amino acid change in the cytoplasmic domain allows the influenza virus hemagglutinin to be endocytosed through coated pits. *Cell* 53:743–52
Lehmann, LE, Eberle, W, Krull, S, Prill, V, Schmidt, B, et al. 1992. The internalization signal in the cytoplasmic tail of lysosomal acid phosphatase consists of the hexapeptide PGYRHV. *EMBO J.* 11:4391–99
Letourneur, F, Klausner, RD. 1992. A novel Di-Leucine motif and a tyrosine-based motif independently mediate lysosomal targeting and endocytosis of CD3 chains. *Cell* 69:1143–57
Lin, HC, Moore, MS, Sanan, DA, Anderson, RGW. 1991. Reconstitution of clathrin-coated pit budding from plasma membranes. *J. Cell Biol.* 114:881–91
Linderman, JL, Lauffburger, JA. 1988. Analysis of intracellular receptor/ligand sorting in the endosome. *J. Theor. Biol.* 132:203–45
Lobel, P, Fujimoto, K, Ye, RD, Griffiths, G, Kornfeld, S. 1989. Mutations in the cytoplasmic domain of the 275 kd mannose 6-phosphate receptor differentially alter lysosomal enzyme sorting and endocytosis. *Cell* 57:787–96
Ludwig, T, Griffiths, G, Hoflack, B. 1991. Distribution of newly synthesized lysosomal enzymes in the endocytic pathway of normal rat kidney cells. *J. Cell Biol.* 115:1561–72
Lund, KA, Opresko, LK, Starbuck, C, Walsh, BJ, Wiley, HS. 1990. Quantitative analysis of the endocytic system involved in hormone-induced receptor internalization. *J. Biol. Chem.* 265:15713–23
Marsh, M, Bolzau, E, Helenius, A. 1983. Penetration of Semliki Forest virus from acidic prelysosomal vacuoles. *Cell* 32:931–40
Matter, K, Hunziker, W, Mellman, I. 1992. Basolateral sorting of LDL receptor in MDCK cells the cytoplasmic domain contains two tyrosine-dependent targeting determinants. *Cell* 71:741–53
Matthews, PM, Martinie, JB, Fambrough, DM. 1992. The pathway and targeting signal for delivery of integral membrane glycoprotein LEP100 to lysosomes. *J. Cell. Biol.* 118:1027–40
Maxfield, FR, Yamashiro, JD. 1991. Acidification of organelles and the intracellular sorting of proteins during endocytosis. See Courtoy 1991, pp. 157–81
Maycox, PR, Link, E, Reetz, A, Morris, SA, Jahn, R. 1992. Clathrin-coated vesicles in nervous tissue are involved primarily in synaptic vesicle recycling. *J. Cell Biol.* 118: 1379–88
McGraw, TE, Greenfield, L, Maxfield, FR. 1987. Functional expression of the human transferrin receptor cDNA in Chinese hamster ovary cells deficient in endogenous transferrin receptor. *J. Cell. Biol.* 105:207–14
McGraw, TE, Maxfield, FR. 1990. Human transferrin receptor internalization is partially dependent upon an aromatic amino acid in the cytoplasmic domain. *Cell Regul.* 1:369–77
McGraw, TE, Pytowski, B, Arzt, J, Ferrone, C. 1991. Mutagenesis of the human transferrin receptor: two cytoplasmic phenylalanines are required for efficient internalization and a second-site mutation is capable of reverting an internalization-defective phenotype. *J. Cell Biol.* 112: 853–61
McKanna, JA, Haigler, HT, Cohen, S. 1979. Hormone receptor topology and dynamics: morphological analysis using ferritin-labeled epidermal growth factor. *Proc. Natl. Acad. Sci. USA* 76:5689–93
Meresse, S, Hoflack, B. 1993. Phosphorylation of the cation-independent mannose-6-phosphate receptor is closely associated with its exit from the *trans*-Golgi network. *J. Cell Biol.* 120:167–77
Miettinen, HM, Matter, K, Hunziker, W, Rose, JK, Mellman, I. 1992. Fc receptor endocytosis is controlled by a cytoplasmic domain determinant that actively prevents coated pit localization. *J. Cell Biol.* 116:875–88
Miettinen, HM, Rose, JK, Mellman, I. 1989. Fc receptor isoforms exhibit distinct abilities for coated pit localization as a result of cytoplasmic domain heterogeneity. *Cell* 58:317–27
Miller, K, Shipman, M, Trowbridge, IS, Hopkins, CR. 1991. Transferrin receptors promote the formation of clathrin lattices. *Cell* 65:621–32
Moore, MS, Mahaffey, DT, Brodsky, FM, Anderson, RGW. 1989. Assembly of clathrin-coated pits onto purified plasma membranes. *Science* 236:558–63
Mostov, KE, de Bruyn Kops, A, Deitcher, DL. 1986. Deletion of the cytoplasmic domain of the polymeric immunoglobulin receptor prevents basolateral localization and endocytosis. *Cell* 47:359–64
Murphy, J-E, Pleasure, IT, Puszkin, S, Prasad, K, Keen, JH. 1991. Clathrin assembly protein AP-3. *J. Biol. Chem.* 266: 4401–8
Okamoto, CT, Shia, S-P, Bird, C, Mostov, KE, Roth, MG. 1992. The cytoplasmic domain of the polymeric immunoglobulin receptor contains two internalization signals

that are distinct from its basolateral sorting signal. *J. Biol. Chem.* 267:9925–32

Payne, GS, Baker, D, van Tuinen, E, Schekman, R. 1988. Protein transport to the vacuole and receptor-mediated endocytosis by clathrin heavy chain-deficient yeast. *J. Cell Biol.* 106:1453–61

Pearse, BMF. 1988. Receptors compete for adaptors found in plasma membrane coated pits. *EMBO J.* 7:3331–36

Pearse, BMF, Robinson, MS. 1990. Clathrin, adaptors, and sorting. *Annu. Rev. Cell Biol.* 6:151–71

Peeler, JS, Donzell, WC, Anderson, RGW. 1993. The appendage domain of the AP-2 subunit is not required for assembly or invagination of clathrin-coated pits. *J. Cell Biol.* 120:47–54

Pelchen-Matthews, A, Armes, JE, Griffiths, G, Marsh, M. 1991. Differential endocytosis of CD4 in lymphocytic and nonlymphocytic cells. *J. Exp. Med.* 173: 575–87

Pelchen-Matthews, A, Boulet, I, Fagard, R, Littman, D, Marsh, M. 1992. CD4-p56lck interaction inhibits CD4 endocytosis. *J. Cell Biol.* 117:279–90

Peters, C, Braun, M, Weber, B, Wendland, M, Schmidt, B, et al. 1990. Targeting of a lysosomal membrane protein: a tyrosine-containing endocytosis signal in the cytoplasmic tail of lysosomal acid phosphatase is necessary and sufficient for targeting to lysosomes. *EMBO J.* 9:3497–506

Peters, PJ, Neeffes, JJ, Oorschot, V, Ploegh, HL, Geuze, HJ. 1991. Segregation of MHC class II molecules from MHC class I molecules in the Golgi complex for transport to lysosomal compartments. *Nature* 349: 669–76

Pfeffer, SR, Rothman, JE. 1987. Biosynthetic protein transport and sorting by the endoplasmic reticulum and Golgi. *Annu. Rev. Biochem.* 56:829–52

Piper, RC, Tai, C, Slot, JW, Hahn, CS, Rice, CM, et al. 1992. The efficient intracellular sequestration of the insulin-regulatable glucose transporter (GLUT-4) is conferred by the NH_2 terminus. *J. Cell Biol.* 117:729–43

Ponnambalam, S, Robinson, MS, Jackson, AP, Peiper, L, Parham, P. 1990. Conservation and diversity in families of coated vesicle adaptins. *J. Biol. Chem.* 265:4814–20

Prywes, R, Livneh, E, Ullrich, A, Schlessinger, J. 1986. Mutations in the cytoplasmic domain of EGF receptors affect EGF binding and receptor internalization. *EMBO J.* 5:2179–90

Rajagopalan, M, Neidigh, JL, McClain, DA. 1991. Amino acid sequences Gly-Pro-Leu-Tyr and Asn-Pro-Glu-Tyr in the submembranous domain of the insulin receptor are required for normal endocytosis. *J. Biol. Chem.* 266:23068–73

Raths, S, Rohrer, J, Crausaz, F, Riezman, H. 1993. *end3* and *end4:* Two mutants defective in receptor-mediated and fluid-phase endocytosis in *Saccharomyces cerevisiae*. *J. Cell Biol.* 120:55–65

Reid, PA, Watts, C. 1990. Cycling of cell-surface MHC glycoproteins through primaquine-sensitive intracellular compartments. *Nature* 346:655–57

Richardson, JS, Richardson, DC. 1988. Amino acid preferences for specific locations at the ends of α helices. *Science* 240:1648–52

Rijnboutt, S, Stoorvogel, W, Geuze, HJ, Strous, GJ. 1992. Identification of subcellular compartments involved in biosynthetic processing of cathepsin D. *J. Biol. Chem.* 267:15665–72

Robinson, LJ, Pang, S, Harris, DS, Heuser, J, James, DE. 1992. Translocation of the glucose transporter (GLUT4) to the cell surface in permeabilized 3T3-L1 adipocytes: Effects of ATP, insulin, and GTPγS and localization of GLUT4 to clathrin lattices. *J. Cell Biol.* 117:1181–96

Robinson, MS. 1989. Cloning of cDNAs encoding two related 100-kD coated vesicle proteins (α-adaptins). *J. Cell Biol.* 108: 833–42

Robinson, MS. 1992. Adaptins. *Trends Cell Biol.* 2:293–97

Roederer, M, Bowser, R, Murphy, R.F. 1987. Kinetics and temperature dependence of exposure of endocytosed material to proteolytic enzymes and low pH: evidence for a maturation model for the formation of lysosomes. *J. Cell. Physiol.* 131:200–9

Roth, MG, Doyle, C, Sambrook, J, Gething, M-J. 1986. Heterologous transmembrane and cytoplasmic domains direct functional chimeric influenza virus hemagglutinins into the endocytic pathway. *J. Cell Biol.* 102:1271–83

Roth, TF, Porter, KR. 1964. Yolk protein uptake in the oocyte of the mosquito *Aedes aegypti*. *J. Cell Biol.* 20:313–32

Rothenberger, S, Iacopetta, BJ, Kuhn, LC. 1987. Endocytosis of the transferrin receptor requires the cytoplasmic domain but not its phosphorylation site. *Cell* 49:423–31

Salzman, NH, Maxfield, FR. 1988. Intracellular fusion of sequentially formed endocytic compartments. *J. Cell Biol.* 106: 1083–91

Schmid, SL. 1992. The mechanism of receptor-mediated endocytosis: more questions than answers. *BioEssays* 14:589–96

Schmid, SL, Fuchs, R, Male, P, Mellman, I. 1988. Two distinct subpopulations of endosomes involved in membrane recycling and transport to lysosomes. *Cell* 52:73–83

Schmid, SL, Smythe, E. 1991. Stage-specific assays for coated pit formation and coated vesicle budding in vitro. *J. Cell Biol.* 114: 869–80
Seeger, M, Payne, GS. 1992. A role for clathrin in the sorting of vacuolar proteins in the Golgi complex or yeast. *EMBO J.* 11:2811–18
Selkoe, DJ. 1991. The molecular pathology of Alzheimer's disease. *Neuron* 6:487–98
Simons, K, Wandinger-Ness, A. 1990. Polarized sorting in epithelia. *Cell* 62:207–10
Singer, SJ. 1990. The structure and insertion of integral proteins in membranes. *Annu. Rev. Cell Biol.* 6:247–96
Steinman, RM, Brodie, SE, Cohn, ZA. 1976. Membrane flow during pinocytosis. *J. Cell Biol.* 68:665–87
Stoorvogel, W, Geuze, HJ, Griffith, JM, Schwartz, AL, Strous, GJ. 1989. Relations between the intracellular pathways of the receptors for transferrin, asialoglycoprotein, and mannose 6-phosphate in human hepatoma cells. *J. Cell Biol.* 108:2137–48
Stoorvogel, W, Geuze, HJ, Strous, GJ. 1987. Sorting of endocytosed transferrin and asialoglycoprotein occurs immediately after internalization in HepG2 cells. *J. Cell Biol.* 104:1261–68
Stoorvogel, W, Strous, GJ, Geuze, HJ, Oorschot, V, Schwartz, A. L. 1991. Late endosomes derive from early endosomes by maturation. *Cell* 65:417–27
Taylor, ME, Conary, JT, Lennartz, MR, Stahl, PD, Drickamer, K. 1990. Primary structure of the mannose receptor contains multiple motifs resembling carbohydrate-recognition domains. *J. Biol. Chem.* 265: 12156–62
Thies, RS, Webster, NJ, McClain, DA. 1990. A domain of the insulin receptor required for endocytosis in rat fibroblasts. *J. Biol. Chem.* 265:10132–37
Thilo, L. 1985. Quantification of endocytosis-derived membrane traffic. *Biochim. Biophys. Acta* 822:243–66
Tooze, J, Hollinshead, M. 1991. Tubular early endosomal networks in AtT20 and other cells. *J. Cell Biol.* 115:635–53
Tooze, J, Hollinshead, M. 1992. In AtT20 and HeLa cells Brefeldin A induces the fusion of tubular endosomes and changes their distribution and some of their endocytic properties. *J. Cell Biol.* 118:813–30
Trowbridge, IS. 1991. Endocytosis and signals for internalization. *Curr. Opin. Cell Biol.* 3:634–41; Erratum 3:1062
Trowbridge, IS, Collawn, J, Jing, S, White, S, Esekogwu, V, Stangel, M. 991. Structure-function analysis of the human transferrin receptor: effects of anti-receptor monoclonal antibodies on tumor growth. In *Biotechnology of Plasma Proteins: Haemostasis, Thrombosis and Iron Proteins,* ed. HC Hemker, 58:139–47, Basel, Switzerland: Karger
Valiquette, M, Bonin, H, Hnatowich, M, Caron, MG, Lefkowitz, RJ, Bouvier, M. 1990. Involvement of tyrosine residues located in the carboxyl tail of the human β2-adrenergic receptor in agonist-induced down-regulation of the receptor. *Proc. Natl. Acad. Sci. USA* 87:5089–93
van der Bliek, AM, Meyerowitz, EM. 1991. Dynamin-like protein encoded by *Drosophila shibire* gene associated with vesicular traffic. *Nature* 351:411–14
van Deurs, B, Peterson, OW, Olsnes, S, Sandvig, K. 1989. The ways of endocytosis. *Int. Rev. Cytol.* 117:131–77
van der Sluijs, P, Hull, M, Webster, P, Male, P, Goud, B, Mellman, I. 1992. The small GTP-binding protein rab4 controls an early sorting event on the endocytic pathway. *Cell* 70:729–40
Vega, MA, Rodriguez, F, Segui, B, Cales, C, Alcalde, J, Sandoval, IV. 1991. Targeting of lysosomal integral membrane protein LIMP II. *J. Biol. Chem.* 266:16269–72
Vega, MA, Strominger, JL. 1989. Constitutive endocytosis of HLA class I antigens requires a specific portion of the intracytoplasmic tail that shares structural features with other endocytosed molecules. *Proc. Natl. Acad. Sci. USA* 86:2688–92
Waheed, A, Gottschalk, S, Hille, A, Krentler, C, Pohlmann, R, et al. 1988. Human lysosomal acid phosphatase is transported as a transmembrane protein to lysosomes in transfected baby hamster kidney cells. *EMBO J.* 7:2351–58
Wileman, T, Harding, C, Stahl, P. 1985. Receptor mediated endocytosis. *Biochem. J.* 232:1–14
Wilmot, CM, Thorton, JM. 1988. Analysis and prediction of the different types of β-turn in proteins. *J. Mol. Biol.* 203:221–32
Williams, MA, Fukuda, M. 1990. Accumulation of membrane glycoproteins in lysosomes requires a tyrosine residue at a particular position in the cytoplasmic tail. *J. Cell Biol.* 111:955–66
Yamashiro, DJ, Tycko, B, Fluss, SR, Maxfield, FR. 1984. Segregation of transferrin to a mildy acidic (ph 6.5) para-Golgi compartment in the recycling pathway. *Cell* 37:789–800

Annu. Rev. Cell Biol. 1993. 9:163–206

PEPTIDE BINDING TO MAJOR HISTOCOMPATIBILITY COMPLEX MOLECULES

Linda D. Barber and Peter Parham
Departments of Cell Biology and Microbiology and Immunology, Stanford University, Stanford, California 94305

KEY WORDS: MHC, class I, class II, peptide generation, assembly

---

CONTENTS

INTRODUCTION

Antigen-specific recognition in the vertebrate immune response is a characteristic of both T and B lymphocytes. This property is conferred by the presence of specific receptors that are clonally distributed on the surface of these cells. Antigen is recognized by B cells in its native conformation, via antigen-specific immunoglobulin receptor molecules. Although the antigen-specific T cell receptors (TCR) have structural and genetic similarities to immunoglobulins (Davis & Bjorkman 1988), the ligands they recognize consist of a fragment from the antigenic protein, known as the epitope, bound to a self protein—an antigen-presenting molecule—encoded by the major histocompatibility complex (MHC) (reviewed by Unanue 1984; Townsend & Bodmer 1989). Such MHC-restricted recognition of processed antigen facilitates discrimination of self from nonself, in that expression of

0743–4634/93/1115–0163$05.00

fragments of self proteins maintains tolerance, while antigen-specific T cell responses are triggered by the recognition of MHC bound peptides derived from foreign proteins (Schwartz 1985).

Mature T cells can be divided into two groups based on their cell surface expression of either CD8 or CD4 glycoproteins. Each group interacts with a different set of MHC molecules called class I and class II, respectively. Class I heavy chain transmembrane glycoproteins are present on the surface of almost all nucleated somatic cells and are associated with β_2-microglobulin (β_2-m). Class II molecules consist of two transmembrane glycoprotein chains, α and β, and exhibit a more restricted tissue distribution. They are expressed primarily on B cells, some activated T cells, members of the macrophage series, and dendritic cells.

Recognition of a cell expressing antigenic peptide in association with class I generally results in cytolysis of the cell presenting the antigen, hence CD8 T cells are cytotoxic T cells. In contrast, CD4 T cells are known as helper T cells because they perform a regulatory role in the immune response to an antigen. Correlating with these functional distinctions, class I and class II molecules bind peptides derived from proteins found in different cellular locations. In general, class I molecules bind peptides from intracellular proteins, while class II molecules bind peptides from extracellular proteins that gain access to acidic compartments associated with the endocytic pathway (Morrison et al 1986). Segregation is not absolute; there are reports of class I molecules presenting extracellular antigen (Staerz et al 1987) and of intracellular antigen being presented in association with class II (Jacobson et al 1989; Nuchtern et al 1990). However, the prevailing view is that functional distinction between T cell recognition of antigen presented by class I and class II molecules is maintained by use of different antigen processing and presentation pathways (Germain 1986; reviewed by Brodsky & Guagliardi 1991).

To facilitate effective surveillance for antigenic peptides, the MHC proteins expressed by a cell must possess the capacity to present a wide range of peptides. Diversity of the MHC proteins, from the expression of several different MHC genes (three class I and three class II loci in humans) and the extensive polymorphism exhibited by most of these gene loci within a population (Klein & Figueroa 1986), contributes to this requirement. However, a single allele must still be able to bind a large, yet finite number of structurally diverse peptides. Consequently, the complex of MHC and bound peptide represents an association quite unlike that of the classical lock and key mechanism typified by the specific interaction between an antibody and its antigen (Davies et al 1990).

Significant advances in our understanding of the molecular basis for peptide binding to MHC molecules have recently been made, primarily as a result of the determination of the three-dimensional structure of class I molecules, the isolation and characterization of naturally processed peptides bound by MHC proteins, and establishment of peptide-binding assays for class I and class II molecules. It has also become evident that several additional factors influence the assembly of the complexes in vivo. Chaperones appear to play a role in the folding and intracellular transport of both class I and class II molecules, and components that may be involved in the generation of peptides and their transport to the site of MHC binding have also been identified. These studies suggest that peptide binding plays an integral role in the assembly, stability, and intracellular transport of functional MHC molecules.

Unraveling the basis for the association of peptides with MHC proteins not only has relevance to our understanding of the immune response and consequent attempts to manipulate it for therapeutic purposes, it may also offer insights into the mechanisms underlying other similar interactions such as peptide binding by heat shock proteins (HSPs). In this review, recent experiments and observations that have led to our current understanding of the binding of peptides to MHC molecules are presented.

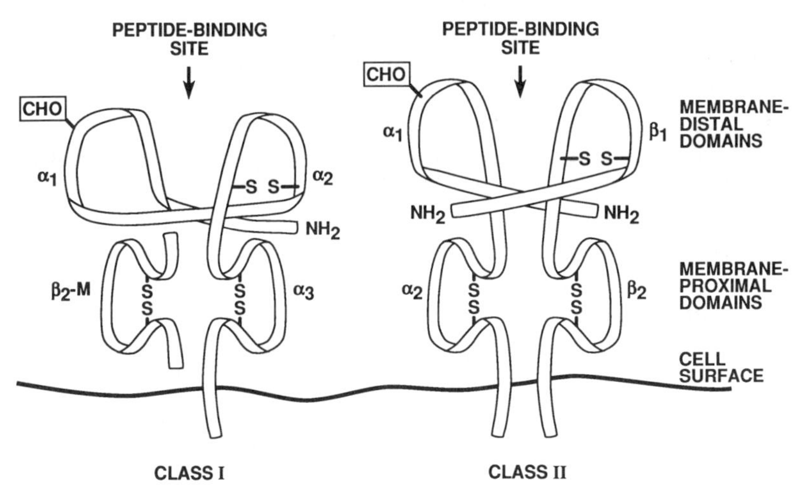

Figure 1 A schematic representation of MHC class I and class II molecules highlighting homologies. Both molecules comprise two membrane-distal and two membrane-proximal extracellular domains. The alignment of class I α1 to class II α1 and class I α2 to class II β1 is suggested by an N-linked glycosylation site (indicated by CHO), which is completely conserved in all class I and class II α1 domains, and by the disulfide-linked cysteines (indicated by -S S-) found in all class I α and class II β domains.

THE STRUCTURE OF MHC MOLECULES

The extracellular portions of class I and class II proteins are divided into four domains (Figure 1), each comprising approximately 90 amino acid residues. Two domains of the class I heavy chain, α1 and α2, are membrane-distal while a third heavy chain domain, α3, and β_2-m form membrane-proximal immunoglobulin-like domains. For class II molecules, the four domains are distributed symmetrically between the two chains such that each contributes one membrane-proximal (α2 and β_2) and one membrane-distal (α1 and β1) domain.

Prior to the determination of three-dimensional structures, the clustering of allelic polymorphism in the two membrane-distal domains of both class I and class II molecules and sequence conservation over the rest of the protein (Kaufman & Strominger 1982; Parham et al 1988) led to speculation that these domains represented the site of interaction with peptide and the TCR. This hypothesis was vindicated when Bjorkman et al (1987a) described the three-dimensional structure of the human class I molecule HLA-A2 (reviewed by Bjorkman & Parham 1990). A single peptide-binding site could be defined at the membrane-distal surface of the molecule that comprised a groove between two parallel α-helices supported by a floor of eight-strands of anti-parallel β-pleated sheet. The α1 and α2 domains each contribute four of the anti-parallel β-strands and one of the helices. Most polymorphic residues are located at or near the peptide-binding site, thus highlighting the functional importance of this region (Bjorkman et al 1987b). Subsequent refinement of the HLA-A2 structure (Saper et al 1991) and determination of the structures of the human class I molecules HLA-A68 (Garrett et al 1989; Silver et al 1992), HLA-B27 (Madden et al 1991, 1992) and the murine class I molecule H-$2K^b$ (Fremont et al 1992; Zhang et al 1992) confirmed these findings. In all these class I molecules, coming from different loci and species, the carbon backbone is preserved. However, residue differences between class I molecules have profound effects on the contours of the peptide-binding site, altering its shape, hydrophobicity, and electrostatic charge, which suggests a basis for the mechanism by which polymorphisms in MHC molecules affect the specificity of peptide binding.

A feature of the HLA-A2 and HLA-A68 crystal structures was the presence of electron-dense material in the peptide-binding site (Bjorkman et al 1987a, Garrett et al 1989). Although of approximately equal intensity to the density generated by the class I protein, the material was poorly resolved, which suggested that a mixture of peptides occupies the peptide-binding site and has co-crystallized with the class I molecules. Consistent with the proposal that allelic polymorphisms within the peptide-binding site modify the range of peptides bound, the form of the electron density appeared different in

HLA-A2 and HLA-A68, which indicated that different sets of peptides were bound by each allele.

Structures at higher resolution revealed subsites or pockets extending from the peptide-binding site that appear to admit the peptide since electron-dense material extends into them. Of the six pockets (denoted A through F) defined in HLA-A2 (Saper et al 1991), pockets A and F at the ends of the peptide-binding site are highly conserved between class I alleles and may serve to anchor the ends of a peptide in the binding site; the other pockets (B,C,D, and E) are polymorphic and have been postulated to determine peptide-binding specificity.

The peptide-binding site is approximately 25Å long, 10Å wide and 11Å deep, and is thus capable of accommodating peptides of 8–20 amino acids in an α-helical or extended chain conformation (Bjorkman et al 1987b). The first model of a peptide bound to a class I molecule was derived from combining the analysis of HLA-B27 crystals (Madden et al 1991) with information on the amino acid sequence of naturally processed peptides bound by this allele (Jardetzky et al 1991). It was proposed that a nonamer peptide adopted an extended conformation with its N-terminus at the left of the site in the conventional view, and predictions were made as to the direction in which each of the peptide side chains pointed. The crystals of HLA-B27 comprise molecules in which the bound peptide is more ordered than in other alleles examined. Consequently, resolution at 2.1Å (Madden et al 1992) permitted direct visualization of the nonamer peptide backbone in the binding site, and the conserved arginine residue identified at the second position in peptides bound by HLA-B27 (Jardetzky et al 1991) clearly extended into the B pocket. HLA-B27-specific features of the B pocket formed by histidine at 9, threonine at 24, negatively charged glutamic acid at 45, and cysteine at 67 create a site complementary in size, shape, and charge to the long, positively charged side chain of arginine.

Site-specific mutagenesis of class I proteins supports the proposal that the structure of allele-specific pockets determines the peptide side chains they can accommodate. Point mutation of residues forming the B pocket in HLA-B27 and HLA-A2 show that this structure plays a critical role in peptide binding. Rojo et al (1993) found that a nonconservative substitution at position 45 abolished the preference for arginine at position 2 in the endogenous peptides bound by HLA-B27, presumably reflecting loss of a favorable electrostatic interaction. Similarly, introduction of polar residues into the hydrophobic B pocket of HLA-A2, which is believed to accommodate the conserved isoleucine/leucine at position 2 of peptides bound by this allele (Falk et al 1991), inhibits presentation of three different peptides to T cells (Utz et al 1992). Thus residue changes within allele-specific pockets alter the repertoire of peptides bound.

The first crystallographic structures were generated from HLA-A2, A68, and B27 proteins isolated from human cells after papain cleavage from the membrane. The heterogeneous sets of endogenous peptides they contained prevented precise visualization of the peptide in its binding site. Information on the molecular contacts between bound peptide and the class I protein required the production of crystals from a source of homogeneous complexes. This advance was facilitated by in vitro reconstitution of denatured class I proteins in the presence of a single peptide species (Silver et al 1991), and crystals of the influenza virus nucleoprotein peptide 91-99 (KTGGPIYKR) bound to HLA-A68 were subsequently described (Silver et al 1992). The advent of methods for producing MHC proteins in milligram quantities has afforded an abundant source of material for the production of reconstituted class I molecules. Class I proteins expressed as insoluble aggregates in bacterial cells have been used to generate crystals of H-2K^b complexed to vesicular stomatitis virus (VSV) nucleoprotein peptide 52–59 (RGYVYQGL) (Zhang et al 1992). In an alternative approach, Matsumura et al (1992b) produced soluble H-2K^b molecules without bound peptide in transfected *Drosophila melanogaster* cells. In vitro loading of these molecules with either the octamer VSV nucleoprotein peptide or the nonamer Sendai virus (SEV) nucleoprotein peptide 324–332 (FAPGNYPAL) yielded a large supply of homogeneous complexes for crystallographic analysis (Fremont et al 1992).

Comparison of these various complexes of class I and specific bound peptide reveals that peptides of very different sequences adopt essentially the same conformation within the binding site (Figure 2). The peptides form an extended β structure in which the N and C termini are held at the ends of the site, in the A and F pockets, respectively, while residues in the middle of the peptide are less constrained. The binding of peptide to class I primarily involves side chain atoms of the class I protein and main chain atoms of the peptide, with networks of hydrogen bonds between the peptide termini and the conserved A and F pockets providing the common basis for high affinity binding.

Demonstration that single amino acid additions or deletions to the N or C termini of peptides decreases binding efficiency by orders of magnitude emphasizes the contribution of interactions involving the peptide termini (Cerundolo et al 1991; Tsomides et al 1991; Matsumura et al 1992b). The importance of hydrogen bonding to the N terminus of bound peptide was confirmed by the creation of site-specific mutations at conserved sites in the A pocket of HLA-A2 (Latron et al 1992). Substitutions that result in loss of a hydroxyl group at either position 7 or 171 cause a significant decrease in the presentation of two viral peptides to T cells. Mutations at conserved sites in the F pocket did not appear to alter the efficiency of

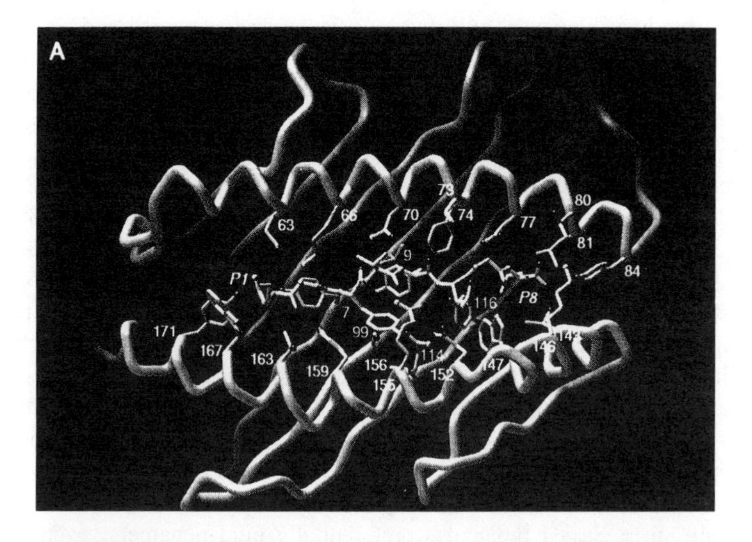

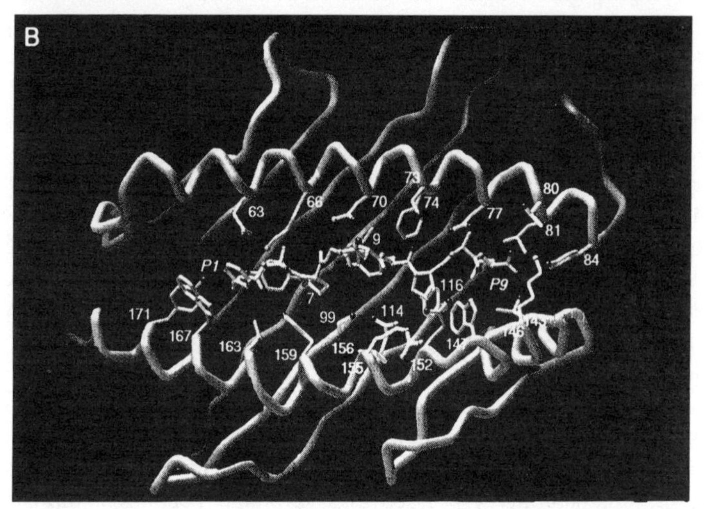

Figure 2 Surface views of the MHC class I molecule H-2K^b showing two viral peptides from VSV (*A*) and SEV (*B*), occupying the peptide-binding site. The carbon-α backbone structure of the membrane-distal domains of H2-K^b is represented as a ribbon diagram showing the eight-stranded β sheet topped by two H-2K^b-helices. The H-2K^b residues in van der Waals contact with the octamer VSV and nonamer SEV peptides are indicated. The figure is taken with permission from Fremont et al (1992), copyright 1992 by AAAS.

peptide binding; however, only two positions were examined and each was studied in isolation. A full assessment of the role of conserved sites in the A and F pockets in formation of a hydrogen bond network to the N and C termini of bound peptide will require mutagenesis of all eight conserved sites in multiple combinations.

Fremont et al (1992) reported that 21–36 hydrogen bonds and 88–119 van der Waal contacts and hydrophobic interactions involving 604–681Å^2 of the buried surface of peptide contribute to peptide binding by H-2K^b. This large number of interactions involving the main chain atoms of the peptide presumably accounts for the high affinity binding of a diverse range of peptide sequences to class I. Interactions involving peptide side chains are limited to the conserved anchoring residues. In H-2K^b molecules, the C pocket accommodates the tyrosine anchor residue at position 5 of the VSV peptide and position 6 of the SEV peptide, while the small B pocket appears not to be important (Matsumura et al 1992a).

Since both peptide termini are buried within the class I peptide-binding site, and since class I molecules preferentially bind nonamers, even when they form a minor component ($< 0.1\%$) in preparations of longer peptides (Schumacher et al 1991), the question arises as to the maximum peptide length that can be accommodated. Comparison of H-2K^b complexed with octamer VSV and nonamer SEV peptides shows that the N and C terminal interactions are similar (Figure 2). The extra residue in the nonamer is accommodated by a bulge in the middle of the peptide, such that glycine 4 and asparagine 5 of the SEV peptide kink out of the binding site and effectively occupy the same position as valine 4 of the VSV peptide (Fremont et al 1992). Guo et al (1992) have similarly suggested that the heterogeneity in length (9–11 residues) of naturally processed peptides bound by HLA-A68 is facilitated by the central region of the peptide sequence protruding from the binding site. In this central portion of the peptide, where direct contacts with class I are sparse, the presence of bound water molecules provides bridging hydrogen bonds to stabilize the interaction (Fremont et al 1992; Silver et al 1992). The maximum length that can be accommodated is unknown, but naturally processed class I bound peptides comprising 13 residues have been identified (Henderson et al 1992; Wei et al 1992).

Crystallographic studies of MHC class II molecules are in progress, but a structure has yet to be reported. Meanwhile, a hypothetical model of the peptide-binding site of the class II molecule has been proposed based on the structure of HLA-A2 (Brown et al 1988). In this model, class II α1 domain is aligned with class I α1 domain, and class II β1 domain is aligned with class I α2 domain. This alignment is suggested by the presence of a conserved glycosylation site at the same relative position in all class I and class II α1 sequences, and of disulfide-linked cysteines in all class I α2 and class II β1 sequences, but not in the α1 domains of both molecules (Kaufman et al 1984). The structural homologies of class I and class II molecules are schematically represented in Figure 1.

With this alignment, sequence comparison shows that the structural features that characterize the class I peptide-binding site, namely the regions of α helix and β sheet, can also be identified at equivalent positions in class II molecules. A comparison of class I and class II molecules by Fourier transform infrared and circular dichroism spectroscopy showed that they possess similar amounts of α helix and β sheet secondary structure (Gorga et al 1989). This supports the proposal that class II molecules possess a peptide-binding site similar to class I molecules in which the N terminal portions of α1 and β1 fold into β-pleated sheets and form the floor of the binding site, while the C terminal regions of these domains adopt an α-helical conformation and form the sides of the site. The results of site-directed mutagenesis experiments assessing the serologic and functional importance of specific residues within the class II molecule are consistent with this hypothetical model. Antigen presentation by class II molecules is sensitive to many mutations at residues predicted to form the peptide-binding site (Ronchese et al 1987; Davis et al 1989), whereas serologic determinants map predominantly to exposed portions of the helices (Buerstedde et al 1988).

The molecular basis for peptide binding to class II appears similar to that described for class I. Using peptides in which most residues are mutated to either glycine, proline, or alanine, it was found that only a few specific residue side chains were necessary for peptides to bind to class II (Jardetzky et al 1990) and class I (Maryanski et al 1990; Romero et al 1991) molecules. Thus interactions between the peptide backbone and the MHC molecule form the basis for binding, and specificity is conferred by interactions involving one or two peptide residue side chains.

However, some differences are evident. The naturally processed peptides bound by class II molecules are longer, and they exhibit considerable length variation: between 13–25 residues for human HLA-DR1 molecules (Chicz et al 1992) and 12–22 residues for the murine I-E^b, I-A^b, I-A^d, I-A^k and I-A^s alleles (Hunt et al 1992b; Nelson et al 1992; Rudensky et al 1992). This variation can reflect the presence of overlapping sets of peptides containing the same core determinant, but with ragged N and C termini. Class II bound peptides are predicted to adopt an extended conformation in the binding site (Sette et al 1989b; Krieger et al 1991). In this form, accommodation of peptides longer than 15 residues probably requires that the termini extend beyond the binding site. Demonstration that N or C terminal truncations do not prevent antigen presentation by class II molecules (Hunt et al 1992b; Rudensky et al 1992) supports the proposal that the

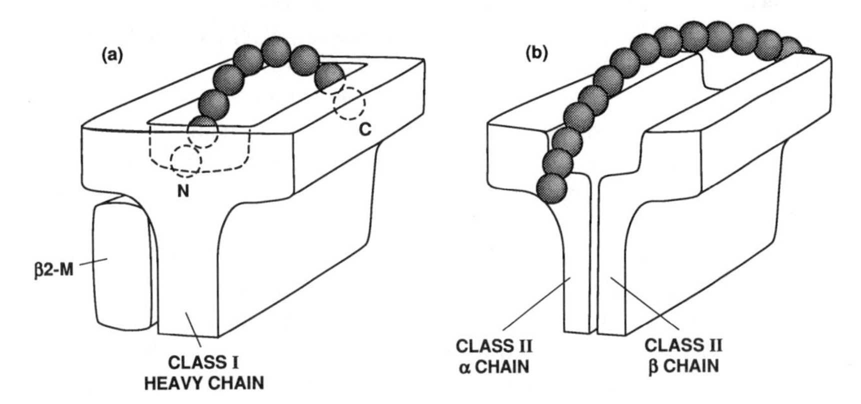

Figure 3 Schematic representation of (*a*) a 9-residue peptide occupying the binding site of a class I molecule, and (*b*) a 15-residue peptide occupying the binding site of a class II molecule.

peptide termini are not anchored within the binding site. A schematic representation of this model for peptides bound in the class I and class II peptide-binding sites is shown in Figure 3.

Margulies (1992) examined the model of the class II peptide-binding site for indications that the ends were open to enable accommodation of longer peptides. The closed nature of the class I peptide-binding site is primarily determined by the conserved tryptophan at 167 and tyrosine at 171 of the A pocket, and tyrosine at position 84 in the F pocket (Matsumura et al 1992a). These and several other conserved residues in the class I A and F pockets are absent in analogous regions of class II. Significantly, the tyrosine at position 171, whose important role in anchoring the peptide N terminus was highlighted by Latron et al (1992), can be replaced by several different amino acids in class II molecules.

The MHC peptide-binding site is adapted to interact with a diverse range of peptides. Therefore, it is not unexpected that apparently analogous structures with different functions can be identified in other proteins whose physiological role requires the binding of a wide range of ligands. Heat shock proteins of the HSP70 family catalyse ATP-dependent protein assembly and disassembly: reactions that require their interaction with diverse polypeptide sequences. Peptide binding by several members of this family has been demonstrated (Chiang et al 1989; Flynn et al 1989; VanBuskirk et al 1989; DeLuca-Flaherty et al 1990), and a binding site is predicted to reside within the C terminal portion of the proteins (Milarski & Morimoto 1989). MHC class I molecules and HSP70 proteins exhibit similarities in the relative amount and location of regions of secondary structure within their ligand-binding domains. Consequently, the class I structure has been used to generate hypothetical models of the HSP70 ligand-binding site (Flajnik et al 1991; Rippmann et al 1991).

Molecules that comprise a class I-like heavy chain associated with β_2-m include an Fc receptor expressed on the intestinal epithelium of newborn rodents, which mediates uptake of maternally derived immunoglobulins from milk (Simister & Mostov 1989), and the CD1 family (Calabi & Milstein 1986). The CD1 proteins are expressed predominantly on the intestinal epithelium (Bleicher et al 1990), and they can be recognized by a subset of T cells that are both CD4- and CD8-negative (Porcelli et al 1989). A functional peptide-binding site has yet to be demonstrated for these class I-like molecules, but Porcelli et al (1992) recently reported that T cells specific for *Mycobacterium tuberculosis* can be restricted by CD1, which suggests a role for CD1 in antigen presentation.

THE CHARACTERIZATION OF PEPTIDES BOUND BY MHC MOLECULES

Since the demonstration that synthetic peptides can be presented by MHC molecules to T cells (Thomas et al 1981; Townsend et al 1986), thereby substituting for naturally produced peptides, these reagents have been extensively used to define T cell epitopes (reviewed by Livingstone & Fathman 1987). The feasibility of mapping T cell epitopes by screening large numbers of synthetic peptides has been enhanced by the development of techniques enabling the rapid synthesis of many peptides simultaneously. For example, the multipin peptide synthesis strategy, in which peptides are produced on polyethylene pins and then cleaved for use in functional studies, has been utilized by Maeji et al (1990) to localize precisely a T cell epitope.

Minimum peptide sequences required for T cell recognition have been identified using sequentially truncated synthetic peptides (Sette et al 1987; Reddehase et al 1989). Testing peptide analogues containing sequential amino acid substitutions for their ability to stimulate T cells and for competition for presentation with native peptide can enable the designation of TCR and MHC contacts (Allen et al 1987). Comparisons of synthetic peptide epitopes have identified sequence similarities among peptides bound by particular MHC alleles, which may reflect functional specialization. For example, the murine class I molecule H-2M3 specifically binds N terminal formylated peptides (Shawar et al 1991) and thus, since bacteria initiate protein translation with N formyl methionine, H-2M3 may be adapted for a role in host defenses against intracellular prokaryotic organisms. In support of this proposal, H-2M3 has recently been shown to present N terminal formylated peptides derived from the intracellular pathogen *Listeria monocytogenes* to T cells (Kurlander et al 1992; Pamer et al 1992).

Demonstration that MHC restricted presentation of antigen to T cells involves direct binding of peptides to MHC molecules was first performed using labeled synthetic peptides and purified MHC proteins (Babbitt et al 1985; Chen & Parham 1989). In subsequent studies, peptides tagged by the incorporation of radioactive, photoreactive, fluorescent, or biotin moieties have been used to monitor the kinetics of binding to MHC molecules (Buus et al 1986; Tsomides et al 1991) and the effects of parameters such as pH (Harding et al 1991c; Jensen 1991a; Tampé & McConnell 1991) on the interaction.

The physiological relevance of these in vitro assay systems to peptide binding by class II is supported by the correlation between known MHC restriction patterns and the ability of purified class II molecules to bind peptide (Buus et al 1987; O'Sullivan et al 1990). In contrast, synthetic peptides known to be presented to T cells by class I molecules fail to bind appreciably to appropriate purified molecules in similar assay systems (Chen & Parham 1989). With hindsight, the small number of sites available for peptide binding (<0.3%) probably reflects prior occupancy with naturally processed peptides that co-purify with the class I molecules. Substantial binding of synthetic peptides in a physiologically relevant manner was first demonstrated using class I molecules on living cells (Benjamin et al 1991; Christinck et al 1991; Luescher et al 1991). The allele-specific binding of peptides to purified class I has now been reported using recombinant class I proteins produced by transfected insect cells (Godeau et al 1992; Matsumura et al 1992b).

Until recently, the relationship between the epitopes defined by synthetic peptides and the naturally occurring peptides bound by MHC proteins in vivo was uncertain. This has been resolved by the development of techniques for the isolation and characterization of naturally processed peptides bound by MHC molecules. Buus et al (1988) first described how acid denaturation of purified class II molecules could be used to elute the naturally processed peptides bound by MHC molecules. The low molecular weight material isolated from I-A^d and I-E^d could compete with synthetic peptide for binding to the class II molecules from which they were eluted. Demotz et al (1989) similarly isolated I-E^d-binding peptides from cells exposed to intact hen egg lysozyme (HEL) and demonstrated the presence of a naturally processed lysozyme peptide capable of stimulating HEL-specific T cells. Fractionation of the eluted material by reverse-phase high performance liquid chromatography (HPLC) was subsequently applied to purify the naturally processed peptides. Fractions retained biological activity (Wallny & Rammensee 1990), and the naturally occurring peptides were similar in apparent size and hydrophobicity to synthetic peptides previously shown to substitute for natural T cell epitopes (Rötzschke et al 1990b).

Application of these methods to virally infected cells led to the identification of two naturally processed nonamer peptide epitopes from influenza nucleoprotein presented by H-2K^d and H-2D^b (Rötzschke et al 1990a). A third viral epitope was isolated from H-2K^b molecules expressed by cells infected with VSV (Van Bleek & Nathenson 1990). Identification of this peptide as an octamer from the nucleocapsid protein was made possible by obtaining partial sequence data via the incorporation of specific radiolabeled amino acids during synthesis of the natural peptide by infected cells. In both studies, the isolated viral peptides corresponded to part of a synthetic peptide with previously characterized biological activity, thus establishing a direct link between epitopes defined by synthetic peptides and naturally processed antigens.

Determination of the sequences of naturally processed peptides by Edman degradation is technically challenging due to their diversity and consequent low abundance of any single species. However, by eluting and fractionating peptides from approximately 10 nmol quantities of MHC proteins, sequences for some of the dominant peptides bound by class I (Jardetzky et al 1991; Guo et al 1992) and class II molecules (Rudensky et al 1991, 1992; Chicz et al 1992; Nelson et al 1992) have been obtained.

The prescribed length of most class I bound peptides and the presence of conserved residues at certain positions in peptides bound by a particular class I allele enabled unfractionated peptide mixtures eluted from purified class I molecules to be analyzed by Edman degradation (Falk et al 1991). Based on the assumption that a strong signal for a residue at a certain cycle in the sequencing reaction indicates the occurrence of that amino acid in a majority of the peptides, the results obtained have been used to identify motifs that are predicted to confer allele-specific binding. The class I binding motifs defined to date, either by pool sequencing or by alignment of individual peptide sequences bound by the same allele, are shown in Table 1. The relevance of these putative motifs to the peptide repertoire bound by class I alleles is supported by several lines of evidence. Romero et al (1991) showed that a simple polyproline peptide analogue incorporating the H-2K^d binding motif specifically interacts with H-2K^d molecules since it is a potent competitor for peptide binding to this allele. Similarly, Corr et al (1992) demonstrated that a polyalanine peptide analogue containing the H-2L^d binding motif could effectively bind H-2L^d as assessed by induction of a conformation-dependent antibody epitope. Significantly, an examination of the sequences of the T cell epitopes presented by HLA-B27, H-2K^d and HLA-B8 reveals that the appropriate motif is contained within all of them (Jardetzky et al 1991; Romero et al 1991; Sutton et al 1993).

Table 1 Class I allele-specific peptide-binding motifs

Allele	Peptide residue position (from N to C terminus) 1	2	3	4	5	6	7	8	9	Reference
H-2D^b	—	—	—	—	N	—	—	—	M	Falk et al (1991)
H-2K^b	—	—	—	—	Y/F	—	—	L	—	Falk et al (1991)
H-2K^d	—	Y	—	—	—	—	—	—	I/L	Falk et al (1991)
H-2L^d	—	P	—	—	—	—	—	—	L/F/M	Corr et al (1992)
HLA-A2(01)	—	L	—	—	—	—	—	—	V	Falk et al (1991)
HLA-A2(05)	—	—	—	—	—	—	—	—	L	Rötzschke et al (1992)
HLA-A68	—	V	—	—	—	—	—	—	R/K	Guo et al (1992)
HLA-B8	—	—	R/K	—	R/K	—	—	—	I/L	Sutton et al (1993)
HLA-B27	—	R	—	—	—	—	—	—	R/K	Jardetzky et al (1991)
HLA-B35	—	P	—	—	—	—	—	—	Y	Hill et al (1992)
HLA-B53	—	P	—	—	—	—	—	—	—	Hill et al (1992)

Since these motifs are good indicators of whether a peptide will bind to a particular class I allele, their predictive power has begun to be explored by using them as algorithms for searching sequence databases to identify potential T cell epitopes. The H-2K^d-binding motif has been used to identify potential antigenic sequences within the listeriolysin protein produced by the bacteria *Listeria monocytogenes* (Pamer et al 1991). By screening synthetic versions of the potential epitopes for recognition by *Listeria monocytogenes*-specific T cells, an antigenic peptide comprising residues 91–99 was identified. Immunization of mice with the peptide elicited T cells that protected against infection with the bacteria, thus demonstrating the biological relevance of the epitope (Harty & Bevan 1992). Similar application of the HLA-B53 binding motif has identified a malarial epitope that may contribute to the resistance to severe malaria observed in individuals who express HLA-B53 (Hill et al 1992).

This new approach to the identification of T cell epitopes has obvious implications for vaccine development. However, extensive assessment of the importance of the HLA-A2-binding motif comprising leucine at position 2 and valine at position 9 (Parker et al 1992) suggests a note of caution in the use of simple motifs as predictive algorithms. Although peptides containing the motif were shown to form stable complexes with HLA-A2, only two out of five known T cell epitopes possess the motif, and its presence in a polyglycine peptide analogue was not sufficient to confer binding. Thus although many class I-bound peptides contain an appropriate motif, algorithms based on them may not predict the complete peptide repertoire.

Despite this limitation, their usefulness as a starting point for identifying potential antigenic peptides is evident from the studies of Pamer et al (1991) and Hill et al (1992).

Application of predictive algorithms to the identification of antigenic epitopes presented by class II molecules is hampered by the absence of easily identifiable motifs in peptides bound by class II alleles. Pool sequencing can not be usefully applied since the ragged nature of the N and C termini of class II bound peptides means that a shift in the N terminus of the peptide by one or two positions is often required to obtain an alignment of functionally equivalent residues. Those putative motifs identified are based on alignments of peptide epitopes characterized by in vitro studies or by sequencing individual naturally processed peptides. However, in some cases, the motifs defined have failed to yield consistent results. For example, by screening a random peptide library expressed on the surface of M13 phage for binding to purified HLA-DR1, Hammer et al (1992) postulated a binding motif comprising an aromatic amino acid at the N terminus, a methionine or leucine at position 4, and a small amino acid at position 6. In contrast, binding studies with polyalanine substituted analogues of a naturally processed peptide led Kropshofer et al (1992) to suggest that the presence of two bulky hydrophobic residues with a relative spacing of 1 and 8 within the peptide conferred binding to HLA-DR1. Yet another HLA-DR1-binding motif was identified by alignment of several naturally processed peptides with the sequences of 35 other peptides known to bind this allele. It comprises a positively charged group, a hydrogen bond donor, and a hydrophobic residue with a relative spacing of 1, 6, and 10, respectively, within the peptide (Chicz et al 1992). Evidently, the accuracy of the class II-binding motifs predicted to date is questionable.

The ease with which individual MHC bound peptides can be characterized and binding motifs identified is likely to be enhanced by the application of mass spectrometry. Edman sequencing is currently limited to the most abundant and clearly resolved peptides. Failure to obtain complete sequences, because of rapidly declining yields as each amino acid is sequentially removed, is often encountered for peptides with hydrophobic C termini; it may not be fortuitous that those peptide sequences most successfully analyzed by Edman degradation, namely those bound by HLA-B27 and HLA-A68 (Jardetzky et al 1991; Guo et al 1992), possess hydrophilic C terminal anchors. In an attempt to circumvent these problems, the technique of microcapillary reverse-phase HPLC coupled to electrospray ionization-tandem mass spectroscopy has been applied to the characterization of MHC bound peptides (Hunt et al 1992a,b; Henderson et al 1992; Sette et al

Table 2 Naturally processed peptides bound by class I alleles for which a potential protein source can be identified

Allele	Peptide sequence[a]	Potential protein source[b]	Reference
H-2L[d]	XPQKAGGFLM	Mouse phosphoglycerate kinase	Corr et al (1992)
HLA-A2	ILDKKVEKV	HSP84	Parker et al (1992)
HLA-A2	SLLPAIVEL	Human protein phosphatase 2A / Human transformation-associated protein p61	Hunt et al (1992a)
HLA-A2	YLLPAIVHI	Human ATP-dependent RNA helicase	Hunt et al (1992a)
HLA-A2	TLWVDPYEV	TIS21	Hunt et al (1992a)
HLA-A2	LLDVPTAAV	Human IP-30 leader peptide	Henderson et al (1992) Hunt et al (1992a) Wei & Cresswell (1992)
HLA-A2	LLLDVPTAAVQA	Human IP-30 leader peptide	Henderson et al (1992)
HLA-A2	LLLDVPTAAVQ	Human IP-30 leader peptide	Wei & Cresswell (1992)
HLA-A2	LLLDVPTAAV	Human IP-30 leader peptide	Henderson et al (1992)
HLA-A2	MLLSVPLLLG	Calreticulin leader peptide	Henderson et al (1992)
HLA-A68	DVFRDPALK	Ribosomal 60S homologue	Guo et al (1992)
HLA-A68	TVFDAKRLIGR	Human HSP70 protein B	Guo et al (1992)
HLA-B27	RRYQKSTEL	Human histone H3, H3.3	Jardetzky et al (1991)
HLA-B27	RRIKEIVKK	Human HSP89α	Jardetzky et al (1991)
HLA-B27	RRVKEVVKK	Human HSP89β	Jardetzky et al (1991)
HLA-B27	RRWLPAGDA	Human elongation factor 2	Jardetzky et al (1991)
HLA-B27	RRSKEITVR	Human ATP-dependent RNA helicase	Jardetzky et al (1991)
HLA-B27	FRYNGLIHR	Rat 60S ribosomal protein L28	Jardtezky et al (1991)
HLA-B53	YPAEITLYW	Human class I $\alpha 3$ domain	Hill et al (1992)

[a] Peptide positions for which a residue assignment could not be confidently made are designated with an X. [b] Only peptides with complete homology to a known protein sequence are listed.

Table 3 Naturally processed peptides bound by class II alleles for which a potential protein source can be identified

Allele	Peptide sequence[a]	Potential protein source[b]	Reference
I-A^b	HNEGFYVCPGPHRP	MuLV[c] envelope protein 145–158	Rudensky et al (1991)
I-A^b	HNEGFYVCPGPHR	MuLV envelope protein 145–157	Rudensky et al (1991)
I-A^b	ASFEAQGALANIAVDKA	Mouse class II I-E α chain 52–68	Rudensky et al (1991)
I-A^b	KPVSQMRMATPLLMR	Mouse invariant chain 86–100	Rudensky et al (1991)
I-A^b	RPDAEYWNSQPE	Mouse class II I-A β chain 55–66	Rudensky et al (1992)
I-A^b	XNADFKTPATLTVDKP	Immunoglobulin G heavy chain 59–74	Rudensky et al (1992)
I-A^d	WANLMEKIQASVATNPI	Mouse apolipoprotein-E 268–284	Hunt et al (1992b)
I-A^d	WANLMEKIQASVATNP	Mouse apolipoprotein-E 268–283	Hunt et al (1992b)
I-A^d	DAYHSRAIQVVRARKQ	Rat cystatin-C 40–55	Hunt et al (1992b)
I-A^d	ASFEAQGALANIAVDKA	Mouse class II I-E α chain 52–68	Hunt et al (1992b)
I-A^d	ASFEAQGALANIAVDK	Mouse class II I-E α chain 52–67	Hunt et al (1992b)
I-A^d	EEQTQQIRLQAEIFQAR	Mouse apolipoprotein-E 236–252	Hunt et al (1992b)
I-A^d	EQTQQIRLQAEIFQAR	Mouse apolipoprotein-E 237–252	Hunt et al (1992b)
I-A^d	KPVSQMRMATPLLMRPM	Mouse invariant chain 86–102	Hunt et al (1992b)
I-A^d	VPQLNQMVRTAAEVAGQ	Rat tranferrin receptor 442–459	Hunt et al (1992b)
I-A^k	IIANDQGNRTTPSY	HSP70 28–41	Nelson et al (1992)
I-A^k	TPRRGEVYTCHVEHP	Mouse class II I-A^k β chain 165–179	Nelson et al (1992)
I-A^k	KVHGSLARAGKVRGQTPKVAKQ	Rat S30 ribosomal protein 75–96	Nelson et al (1992)
I-A^k	AGKVRGQTPKVAKQEKKKKKT	Rat S30 ribosomal protein 83–103	Nelson et al (1992)
I-A^k	EPLVPLDNHIPENAQPG	Mouse ryudocan 84–100	Nelson et al (1992)
I-A^k	DGSTDYGILQINSR	Hen egg lysozyme 48–61	Nelson et al (1992)
I-A^k	DGSTDYGILQINS	Hen egg lysozyme 48–60	Nelson et al (1992)
I-A^k	DGSTDYGILQINSRW	Hen egg lysozyme 48–62	Nelson et al (1992)
I-A^k	DYGILQINSRWW	Hen egg lysozyme 52–63	Nelson et al (1992)
I-A^s	IRLKITDSGPRVPIGPN	MuLV envelope protein 255–269	Rudensky et al (1992)
I-A^s	WPSQSITCNVAHPASST	Immunoglobulin G2a 194–210	Rudensky et al (1992)
I-A^s	NVEVHTAQTQTHREDY	Immunoglobulin G2a 281–296	Rudensky et al (1992)

Table 3 *(Continued)*

Allele	Peptide sequence[a]	Potential protein source[b]	Reference
I-A^s	KPTEVSGKLVHANFGT	Transferrin receptor 203–218	Rudensky et al (1992)
I-E^b	SPSYVYHQFERRAKYK	MuLV envelope protein 454–469	Rudensky et al (1991)
I-E^b	SPSYVYHQFERRAKY	MuLV envelope protein 454–468	Rudensky et al (1991)
I-E^b	SPSYVYHQFERRAK	MuLV envelope protein 454–467	Rudensky et al (1991)
I-E^b	GKYLYEIARRHPYFYAP	Bovine serum albumin 141–157	Rudensky et al (1992)
HLA-DR1	VGSDWRFLRGYHQYAYDG	Human class I HLA-A2 103–120	Chicz et al (1992)
HLA-DR1	VGSDWRFLRGYHQYA	Human class I HLA-A2 103–117	Chicz et al (1992)
HLA-DR1	VGSDWRFLRGYHQY	Human class I HLA-A2 103–116	Chicz et al (1992)
HLA-DR1	GSDWRFLRGYHQYA	Human class I HLA-A2 104–117	Chicz et al (1992)
HLA-DR1	SDWRFLRGYHQYA	Human class I HLA-A2 105–117	Chicz et al (1992)
HLA-DR1	LPKPPKPVSKMRMATPLLMQALPMG	Human invariant chain 81–105	Chicz et al (1992)
HLA-DR1	LPKPPKPVSKMRMATPLLMQALPM	Human invariant chain 81–104	Chicz et al (1992)
HLA-DR1	LPKPPKPVSKMRMATPLLMQALP	Human invariant chain 81–103	Chicz et al (1992)
HLA-DR1	PKPPKPVSKMRMATPLIMQALPMG	Human invariant chain 82–105	Chicz et al (1992)
HLA-DR1	PKPPKPVSKMRMATPLIMQALPM	Human invariant chain 82–104	Chicz et al (1992)
HLA-DR1	PKPPKPVSKMRMATPLLMQALP	Human invariant chain 82–103	Chicz et al (1992)
HLA-DR1	KPPKPVSKMRMATPLLMQALPM	Human invariant chain 83–104	Chicz et al (1992)
HLA-DR1	KPPKPVSKMRMATPLIMQALP	Human invariant chain 83–103	Chicz et al (1992)
HLA-DR1	PPKPVSKMRMATPLLMQALP	Human invariant chain 84–103	Chicz et al (1992)
HLA-DR1	KMRMATPLLMQALPM	Human invariant chain 90–104	Chicz et al (1992)
HLA-DR1	KMRMATPLLMQALP	Human invariant chain 90–103	Chicz et al (1992)
HLA-DR1	IPADLRIISANGCKVDNS	Sodium and potassium ATPase 199–216	Chicz et al (1992)

HLA-DR1	RVEYHFLSPYVSPKESP	Transferrin receptor 680–696	Chicz et al (1992)
HLA-DR1	YKHTLNQIDSVKVWPRRPT	Bovine fetuin 56–74	Chicz et al (1992)
HLA-DR1	YKHTLNQIDSVKVWPRRP	Bovine fetuin 56–73	Chicz et al (1992)
HLA-DR3	LPKPPKPVSKMRMATPLLMQALPM	Human invariant chain 81–104	Riberdy et al (1992) Sette et al (1992a)
HLA-DR3	LPKPPKPVSKMRMATPLLMQALP	Human invariant chain 81–103	Riberdy et al (1992) Sette et al (1992a)
HLA-DR3	LPKPPKPVSKMRMATPL	Human invariant chain 81–97	Riberdy et al (1992)
HLA-DR3	PKPPKPVSKMRMATPLLMQA	Human invariant chain 82–101	Riberdy et al (1992)
HLA-DR3	PKPPKPVSKMRMATPL	Human invariant chain 82–97	Riberdy et al (1992)
HLA-DR3	KPPKPVSKMRMATPLLMQALPM	Human invariant chain 83–104	Riberdy et al (1992) Sette et al (1992b)
HLA-DR3	KPPKPVSKMRMATPLLMQALP	Human invariant chain 83–103	Sette et al (1992b)
HLA-DR3	KPPKPVSKMRMATPLLMQ	Human invariant chain 83–100	Riberdy et al (1992)

[a] See footnote a in Table 2. [b] See footnote b in Table 2. [c] Mouse leukemia virus

1992a). This potentially highly sensitive analytical method determines the molecular mass of each peptide component and the approximate number of individual peptides, and it is reported that sequence information can be obtained from sub-pmol amounts of individual peptides.

By a combination of this approach and conventional Edman degradation sequencing, an increasing number of naturally processed peptides bound by MHC molecules are being characterized. A preliminary assessment by mass spectrometry indicates that the range of peptides bound by both class I and class II molecules is broad. In excess of 200 different species, with no dominant peptide, are bound by class I HLA-A2 molecules (Hunt et al 1992a), and between 650 and 2000 different peptides appear to be associated with class II I-A^d (Hunt et al 1992b). Candidate proteins from which these peptides may derive have been identified by screening gene and protein sequence databases, and a list of these naturally processed class I and class II bound peptides is presented in Tables 2 and 3, respectively. The origin of these peptides exemplifies the proposal that most peptides bound by class II are derived from extracellular or integral membrane proteins that can gain access to endosomal compartments, while those of intracellular origin, from the cytoplasm and nucleus, associate with class I (Germain 1986).

ASSEMBLY OF MHC: PEPTIDE COMPLEXES

Although peptide binding to MHC molecules is influenced by residues comprising the binding site and the peptide sequence itself, the difference in the origins of the peptides bound by class I and class II molecules indicates that other parameters also regulate the interactions. Foremost is the utilization of at least two distinct pathways for peptide generation and association with MHC molecules (Figure 4) (reviewed by Braciale & Braciale 1991; Brodsky & Guagliardi 1991). Our current understanding of the assembly of class I molecules is that peptides generated in the cytoplasm are transported into the endoplasmic reticulum (ER) and then bind to class I in a pre-Golgi compartment. Once peptides are bound, class I molecules migrate to the cell surface via the exocytic pathway common to secretory and other membrane bound molecules. In contrast, peptides bound by class II molecules are generated in an acidic compartment associated with the endocytic pathway. Nascent class II molecules are actively targeted to this locality for association with peptide prior to expression at the cell surface. Although the exact molecular basis for the segregation of class I and class II assembly is not fully understood, association of MHC proteins with chaperone molecules has an influential role in determining the site of peptide binding.

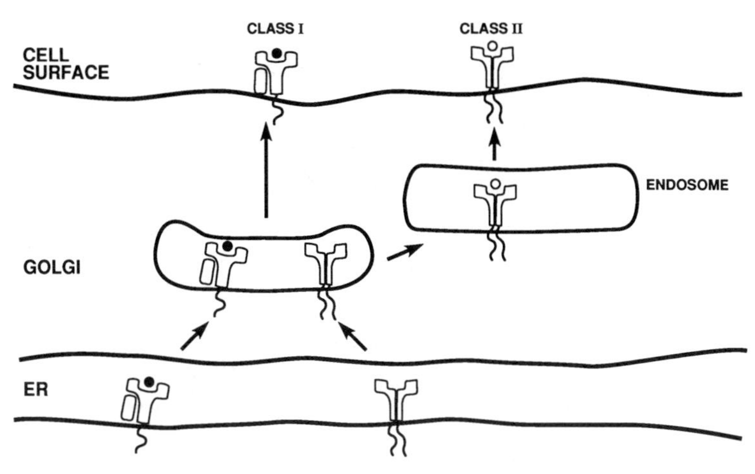

Figure 4 A schematic representation of the intracellular assembly and transport of the MHC class I and class II molecules. Most class I molecules bind peptides (*closed circle*) derived from intracellular proteins in the ER and are then transported to the cell surface via the exocytic pathway. In contrast, most class II molecules are sorted from the Golgi to the endocytic pathway where they bind (*open circle*) peptides derived from the extracellular proteins prior to transport to the cell surface.

In Vivo Binding of Peptide to MHC Class I Molecules

The current understanding of the in vivo pathway for peptide binding to class I molecules is illustrated in Figure 5. Most class I bound peptides are generated by limited proteolytic degradation of proteins in the cytoplasm. Peptides targeted to the cytoplasm using recombinant vaccinia virus expression vectors are presented to T cells by class I molecules (Gould et al 1989; Whitton & Oldstone 1989). Demonstration that enhanced cytoplasmic degradation of influenza nucleoprotein by fusion of ubiquitin to its N terminus facilitated presentation of nucleoprotein-derived peptides to T cells by class I molecules (Townsend et al 1988) provides evidence that peptides are generated in the cytoplasm.

The proteases involved in peptide generation have not been characterized, but the involvement of ubiquitin conjugation, which targets proteins for degradation by proteasomes, suggests that these protease complexes may play a role. Proteasomes are constitutively expressed in the cytoplasm and comprise 20–30 subunits, each of molecular weight 15–30 K (reviewed by Goldberg & Rock 1992). They account for most of the proteolytic activity within the cell and exhibit many distinct protease activities, which suggests a potential for generating a wide range of peptides. A link between

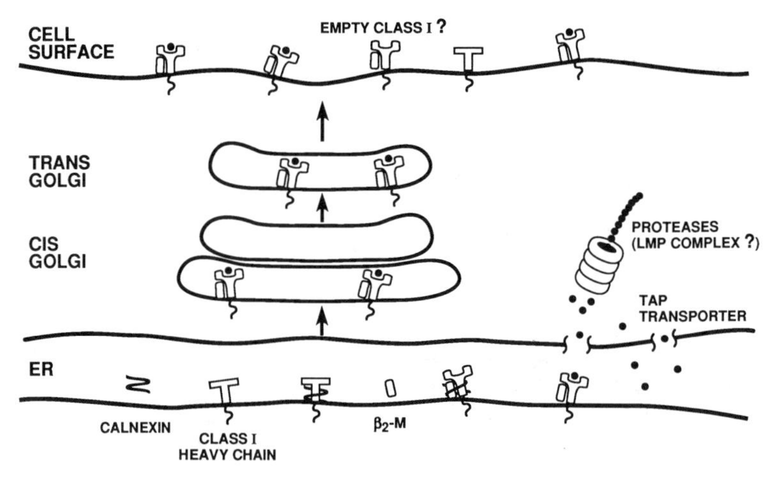

Figure 5 In vivo pathway for peptide generation and binding to class I molecules. Intracellular proteins (*row of closed circles*) are degraded into peptides (*closed circle*) by partial proteolysis in the cytoplasm, which may involve a proteasome like LMP complex. Peptides are transported into the ER by a mechanism involving TAP protein complexes where they bind nascent class I molecules. Some peptides bound by class I are derived from signal sequences cleaved from proteins following their translocation into the ER. Peptide binding confers stability to the complex and facilitates transport through the Golgi to the cell surface. Calnexin retains free class I heavy chains and empty class I molecules in the ER.

proteasomes and antigen processing is suggested by the presence of two genes in the class II region of the MHC that encode proteins with sequence homology to proteasome subunits (Glynne et al 1991; Kelly et al 1991; Martinez et al 1991). However, although these proteins form part of a low molecular weight polypeptide (LMP) complex closely related to proteasomes (Brown et al 1991; Ortiz-Navarrete et al 1991; Arnold et al 1992), there is no direct evidence linking them to the generation of peptides for presentation by class I molecules. Indeed, studies with mutant cell lines defective in their ability to supply peptides for binding to class I have shown that peptide presentation can be restored despite the absence of the two MHC-encoded subunits LMP-2 and LMP-7 (Arnold et al 1992; Momburg et al 1992). Evidently there is no absolute requirement for these subunits, but the possibility remains that they may increase the efficiency of peptide generation.

There is speculation that in vivo protease activity may influence the repertoire of peptides bound by class I molecules. The observed preference for hydrophobic amino acids at the C terminus of several naturally processed peptides bound by class I alleles (Table 1) could reflect the specificity of the peptide-generating enzymes, rather than complementation with the F pocket of the class I peptide-binding site. This possibility is supported by the failure to observe a residue preference at the C terminus when a random mixture of synthetic peptides was offered for binding to H-2K^{b} molecules in vitro (Matsumura et al 1992b). Evidence that the proteolytic machinery may exhibit preferential cutting sites is also suggested by a report that insertion of an epitope from murine cytomegalovirus at different sites in a carrier protein altered the efficiency of peptide generation (Del Val et al 1991). However, it should be noted that in a similar study, repositioning of an epitope from hemagglutinin within the protein failed to influence peptide generation (Hahn et al 1991), and mutagenesis of epitope-flanking residues in the protein also had little effect (Hahn et al 1992).

Class I proteins are co-translationally inserted into the ER (Dobberstein et al 1979), and thus peptide binding to class I does not occur at the site of peptide generation. The precise location where class I molecules associate with peptide has not been established, but several studies implicate the early portion of the exocytic pathway; either the ER lumen or the *cis* Golgi. Treatment of cells with brefeldin A, which prevents protein movement out of the ER by disrupting compartmentalization of the *cis/trans* Golgi, blocks presentation of antigenic proteins by class I molecules (Nuchtern et al 1989; Yewdell & Bennick 1989). Expression of adenovirus E3/19K or cytomegalovirus E proteins, which specifically bind to nascent class I proteins and retain them in the ER, also interferes with antigen presentation (Jefferies et al 1990; Del Val et al 1992). That assembly of the class I heavy chain, β_2-m, and peptide can occur in the ER compartment immediately after protein synthesis has been demonstrated by Kvist & Hamann (1990) using an in vitro translation system and microsomal vesicles to mimic the ER environment.

Peptides generated in the cytoplasm must therefore cross a membrane to associate with class I molecules in the ER. Transmembrane passage may involve the signal recognition particle-dependent pathway (Walter & Lingappa 1986). Indeed, HLA-A2 molecules expressed by the antigen processing mutant cell line T2 predominantly bind peptides derived from N terminal signal sequences (Henderson et al 1992; Wei & Cresswell 1992). These peptides are also bound by HLA-A2 molecules expressed by normal cells (Henderson et al 1992), but they form a minor component of the peptide repertoire, perhaps because they are generally longer than the optimal size for binding to class I. Since peptides bound by class I do not require signal sequences (Townsend et al 1985), an alternative mechanism presumably exists for supplying the ER with suitable peptides for binding to class I. Increasing circumstantial evidence invokes two proteins, TAP1 and TAP2, encoded within the class II region of the MHC in the transport of peptides (Deverson et al 1990; Monaco et al 1990; Spies et al 1990; Trowsdale et al 1990; Bahram et al 1991).

Sequence homology indicates both TAP proteins belong to the ATP-binding cassette (ABC) transporter superfamily, whose members share a conserved ATP-binding domain and a hydrophobic transmembrane domain and include molecules known to transport peptides across membranes (reviewed by Higgins 1992). TAP1 and TAP2 physically associate (Kelly et al 1992; Spies et al 1992), and thus the putative class I peptide transporter appears to be dimeric. Immunostaining shows the TAP complexes are confined to the ER and *cis* Golgi membranes (Kleijmeer et al 1992); a location consistent with a role in transporting peptides into this compartment for binding to class I. A role for TAP proteins in peptide transport is further supported by studies of mutant cell lines in which naturally processed peptides fail to be presented to T cells by class I molecules (Townsend et al 1989; Cerundolo et al 1990). The mutant phenotype results from the absence of functional TAP proteins and can be corrected by transfection of appropriate TAP genes (Spies & DeMars 1991; Attaya et al 1992; Arnold et al 1992; Momburg et al 1992).

The TAP proteins clearly have an important role in peptide binding by class I molecules. However, since it has not been formally demonstrated that they transport peptides, their function remains speculative. A provocative report by Lévy et al (1991) showed that peptide transport across ER membranes appears to be intact in microsomes prepared from the mutant cell line T2, which lacks both TAP proteins. The mutant phenotype did not manifest itself until the ATP-dependent assembly of class I with peptide in the lumen of T2 microsomes. Thus TAP proteins may not transport the peptides themselves but rather a co-factor required for their efficient binding to class I molecules. Such a co-factor could be involved, for example, in generating optimally sized peptides. Resolving the function of TAP proteins will require analysis of functional TAP complexes in vitro.

The MHC linked LMP and TAP genes exhibit polymorphism (Kelly et al 1991; Colonna et al 1992; S. H. Powis et al 1992), and sequence variation may influence the peptides supplied. Comparison of naturally processed peptides bound by the rat class I molecule RT1.A^a expressed in the context of two different TAP2 proteins revealed a dramatic difference in the hydrophobicity of the peptide repertoire (S. J. Powis et al 1992). This difference influences recognition of RT1.A^a by both T cells and antibodies (Livingstone et al 1989). The polymorphisms in the rat TAP2 protein are located in the transmembrane domains where changes in other related transporter proteins are known to alter transport specificity (Higgins 1992). As yet, extensive polymorphisms in structurally pertinent positions have not been found in murine or human TAP proteins (Colonna et al 1992; S. H. Powis et al 1992). However, a phenomenon akin to that seen in the rat exists in a family where an MHC-linked polymorphism alters the ability of HLA-B27 to present antigen to T cells (Pazmany et al 1992). Similarly, an MHC-linked factor alters recognition of the murine class I molecule Qa-1 by T cells (Aldrich et al 1988). The cause of these effects is unknown, but it could reflect influences from the peptide-generating and transport machinery.

The sequence of events in peptide binding to nascent class I heavy chains and β_2-m in the ER is poorly understood. Assembly could proceed via an empty class I and β_2-m heterodimer or via a peptide and free class I heavy chain intermediate (Elliott 1991). Evidence indicates both pathways are possible. In cell lysates, class I heavy chains can bind peptide in the absence of β_2-m (Elliott et al 1991), and peptide can stabilize apparently empty, thermodynamically labile, class I molecules (Townsend et al 1990). Irrespective of the assembly route, once formed, complexes of class I with bound peptide are very stable. Half-lives of 200–600 hr for purified complexes of a nonamer peptide from human immunodeficiency virus bound to HLA-A2 (Tsomides et al 1991) and >110 hr for a nonamer peptide from influenza nucleoprotein bound to H-2D^b at 4°C (Cerundolo et al 1991) imply that the binding of optimally sized peptides is an irreversible event. Small increases in length significantly reduce stability (Cerundolo et al 1991), thus reflecting the size constraints imposed by the anchoring of both peptide termini in the binding site.

In most cases, only class I molecules associated with peptide leave the ER for completion of N-linked glycosylation in the Golgi and transit, via the exocytic pathway, to the cell surface. Immunostaining shows that class I molecules expressed by the peptide supply-deficient mutant cell line T2 accumulate in the ER (Baas et al 1992). In normal cells, marked differences in the delay between class I synthesis and transport to the cell surface are observed between different alleles; with export from the ER being the rate-limiting step (Neefjes & Ploegh 1988). The variable assembly rates probably reflect differences in the supply of appropriate peptides for binding to each allele and differences in the relative affinities of the heavy chain for β_2-m and peptides.

Peptide binding appears to facilitate cell surface expression by promoting the assembly of stable, properly folded class I complexes. Expression of natively conformed H-2L^d by normal mouse cells can be enhanced (Lie et al 1990) and class I expression by peptide supply-deficient mutant cells induced (Townsend et al 1989; Cerundolo et al 1990) by providing appropriate peptides. Peptide-induced stabilization is associated with a conformational change in the class I molecule that can be detected serologically.

H-2L^{d} exists in two serologically distinguishable conformations (Lie et al 1991), and addition of peptide to cell lysates facilitates conversion of H-2L^{d}alt to properly folded H-2L^{d} molecules (Smith et al 1992). Similarly, binding of exogenous peptide by detergent-solubilized class I molecules from peptide supply-deficient mutant cell lines induces a conformation detectable by antibodies that recognize epitopes displayed only by properly folded class I heavy chains (Townsend et al 1990).

Although most class I molecules appear to require bound peptide for stability, H-2K^{b} and H-2D^{p} molecules are stably expressed without bound peptides at the surface of human T2 cells (Wei & Cresswell 1992). A thermostability comparison of H-2K^{d} molecules reconstituted with and without bound peptide showed that the empty form was less resistant to heat denaturation, but peptide binding only conferred marginal stability at physiological temperatures (Fahnestock et al 1992). However, the in vivo relevance of these findings is questionable because in both studies murine class I heavy chains are associated with human β_2-m, and this high affinity interaction (Hochman et al 1988) may confer greater stability.

Since most evidence indicates that peptide binding is a prerequisite for stable expression of class I at the cell surface, a mechanism probably exists for retaining partially assembled class I molecules in the ER. Evidence that a quality control mechanism may operate is suggested by the observation that nascent murine class I molecules transiently associate with an ER-resident protein (Degen & Williams 1991). This p88 protein remains associated during and after the class I heavy chain assembles with β_2-m, but dissociates prior to completion of N-linked glycosylation in the Golgi. Dissociation may correlate with peptide binding, as indicated by the prolonged association of p88 with class I molecules in peptide supply-deficient cell lines (Degen et al 1992). The p88 protein and its human homologue IP90 are products of the calnexin gene (Ahluwalia et al 1992; Galvin et al 1992; Hochstenbach et al 1992), which appears to function as a chaperone for several proteins including TCRs and immunoglobulins. The correct folding and assembly of multimeric protein complexes in vivo is often facilitated by chaperones (reviewed by Gething & Sambrook 1992). Therefore, although the function of class I-associated calnexin is currently unknown, it may either retain class I molecules in the ER until peptide is bound, or directly assist in assembly.

Despite intracellular quality control, there are reports of free heavy chains and class I molecules without bound peptide at the cell surface. H-2D^{b} and H-2D^{dm6} can be transported and expressed at the cell surface without associating with β_2-m (Allen et al 1986; Rubocki et al 1991; Bix & Raulet 1992). Most studies indicate that these heavy chains are nonfunctional.

However, H-2D^b heavy chains expressed by cells from β_2-m knockout mice can be recognized by a conformation-dependent antibody and by alloreactive T cells (Bix & Raulet 1992), although its peptide-binding capabilities have yet to be assessed. Other instances where free heavy chains have been found (Schnabl et al 1990; Smith & Barber 1990) probably result from denaturation of previously assembled class I molecules.

The possibility that class I molecules are exported to the cell surface without bound peptide is a controversial issue. Demonstration that class I molecules are stably expressed by the peptide supply-deficient mutant RMA-S if the cells are grown at reduced temperature was interpreted as evidence that empty molecules are exported to the cell surface, where they rapidly denature unless stabilized by reduced temperature or peptide binding (Ljunggren et al 1990). However, these molecules may not be devoid of bound peptide. RMA-S cells maintain limited ability to present naturally processed antigen to T cells (Esquivel et al 1992; Hosken & Bevan 1992), and several peptide-dependent alloreactive and minor histocompatibility antigen-specific T cells recognize RMA-S cells (Rötzschke et al 1991). The instability of the class I molecules expressed by RMA-S may result from an inadequate supply of appropriate peptides forcing the binding of low affinity, perhaps sub-optimally sized peptides, which dissociate more readily.

The physiologically important site for peptide binding to class I molecules is the ER and *cis* Golgi, but binding can also occur at the cell surface since fixed cells present exogenous peptides in association with class I to T cells (Hosken et al 1989). The mechanism of peptide binding at the cell surface is uncertain; it may involve empty class I molecules, re-assembly from free heavy chains and β_2-m, or peptide exchange. In many cases, addition of exogenous β_2-m is required (Vitiello et al 1990; Rock et al 1992) and thus, since the concentration of free β_2-m in body fluids is low (Poulik 1975), peptide binding to cell surface class I in vivo is probably limited.

In Vivo Binding of Peptide to MHC Class II Molecules

The current understanding of the in vivo pathway for peptide binding to class II molecules is illustrated in Figure 6. Class II α and β chains are co-translationally inserted into the ER where they rapidly associate (Kvist et al 1982). Unlike class I and most other membrane-bound proteins where the rate-limiting step in cell surface expression is exit from the ER, class II molecules do not arrive at the cell surface until 2–3 hr after they have undergone carbohydrate modification (Neefjes et al 1990). During this post-Golgi delay, class II molecules digress from the default exocytic pathway and intersect the endocytic pathway, as shown by the intracellular

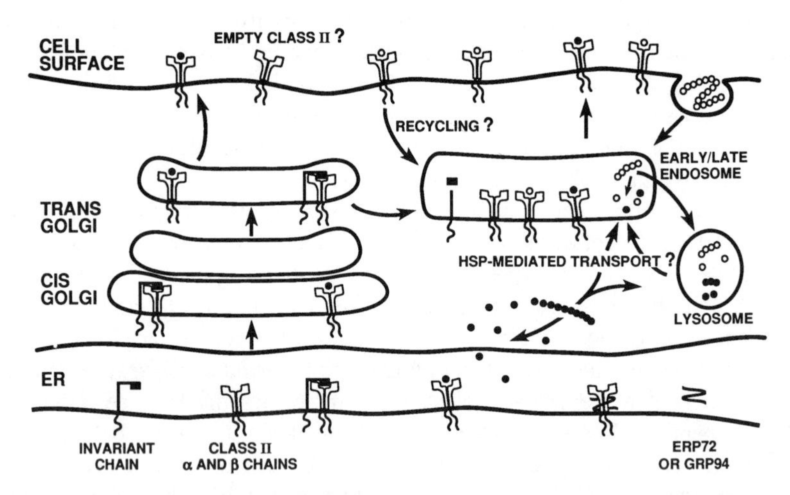

Figure 6 In vivo pathway for peptide generation and binding to class II molecules. A limited number of nascent class II molecules may bind peptides (*closed circle*) in the ER. However, most molecules associate with the invariant chain, transit through the Golgi and are targeted to the endocytic pathway by the invariant chain. In a poorly defined endosomal compartment, the invariant chain is proteolytically degraded and peptide receptive class II molecules released. Internalized extracellular proteins (*open circle*) are partially degraded in endosomes/lysosomes, and the peptides (*open circle*) generated bind class II molecules, which are then transported to the cell surface. Transport of peptides from lysosomes to endosomes and import of cytoplasmic proteins (*row of closed circles*)/peptides (*closed circle*) into the endocytic pathway may be HSP-mediated.

co-localization of nascent class II molecules and proteins internalized by receptor-mediated (Cresswell 1985) and fluid-phase endocytosis (Neefjes et al 1990).

Targeting of class II molecules into the endocytic pathway is a consequence of their transient intracellular association with a membrane-bound chaperone protein known as the invariant chain. This association begins soon after synthesis (Kvist et al 1982) and involves a complex of three class II molecules linked to an invariant chain trimer (Roche et al 1991). Only in this form is class II efficiently transported out of the ER and directed to the endocytic pathway by a signal sequence located in the cytoplasmic tail of the invariant chain (Bakke & Dobberstein 1990; Lotteau et al 1990). Improperly conformed multimeric units are retained in the ER by an invariant chain-encoded signal (Lotteau et al 1990), and most class II molecules not associated with the invariant chain are retained by the ER resident proteins ERp72 and GRP94 (Schaiff et al 1992). In the absence of the invariant chain, class II molecules can still be expressed at the cell surface (Miller

& Germain 1986; Sekaly et al 1986), but they are transported via the exocytic route (Lotteau et al 1990).

Class II molecules complexed to the invariant chain are unable to bind peptides (Roche & Cresswell 1990; Teyton et al 1990), perhaps because the binding site is either sterically blocked or a conformation is adopted that prevents peptide binding. Both class I and class II can bind the same peptides (Perkins et al 1989; Hickling et al 1990), and class II molecules are assembled in the ER environment where class I molecules efficiently bind peptides. However, there are few reports of class II associating with peptides in the early exocytic pathway (Chen et al 1990; Nuchtern et al 1990; Thomas et al 1990; Brooks et al 1991; Weiss & Bogen 1991). Association with the invariant chain is presumably responsible for preventing access to this source of peptides, and therefore it will be of interest to determine which, if any, peptides are bound by class II molecules expressed in the absence of the invariant chain.

Dissociation of the invariant chain releases class II molecules competent to bind peptides (Roche & Cresswell 1990). In vitro proteolysis by cathepsin B, a protease abundant in the endocytic pathway, releases peptide-receptive class II molecules from their association with the invariant chain (Roche & Cresswell 1991), which suggests that this is the likely in vivo mechanism for invariant chain removal. Class II molecules predominantly bind peptides derived from extracellular proteins that have been internalized by fluid-phase or receptor-mediated uptake into acidic vesicles associated with the endocytic pathway. Immunostaining studies showing co-localization of internalized antigen, class II, and invariant chain in the endocytic pathway (Guagliardi et al 1990; Lamb et al 1991) suggest that the proteolytic environment, which releases class II from its association with the invariant chain, is also the site of peptide generation and binding.

The acidic environment of endosomes and lysosomes is believed to be important for antigen processing because reagents that raise the pH in these compartments, such as ammonium chloride and chloroquine, inhibit presentation of peptides to T cells by class II molecules (Ziegler & Unanue 1982). Studies with drugs are often inconclusive because of their pleiotropic effects, but demonstration that incubation of antigenic protein with endosomal and lysosomal extracts generates peptides capable of binding to class II molecules (Collins et al 1991) supports this proposal.

Peptide generation is a two-step process; there is a requirement for protein unfolding by reduction of disulfide bonds (Collins et al 1991; Jensen 1991b), followed by partial proteolytic degradation (Buus & Werdelin 1986; Puri & Factorovich 1988; Takahashi et al 1989). The intense proteolytic activity within the endocytic pathway raises the question of how total protein degradation is prevented. The protease inhibitor, leupeptin, can augment the generation of ovalbumin peptides bound by class II molecules (Vidard et al 1991), which suggests that in vivo protease activity can destroy potential epitopes. Peptides may be protected from total degradation by rapidly binding to class II molecules, a view supported by the in vitro demonstration that peptides bound to class II molecules are shielded from proteolysis by cathepsin B and pronase E (Mouritsen et al 1992a). Interestingly, this study also showed that class II bound peptide could still be modified by the exopeptidase, aminopeptidase N. Both peptide termini appear to be exposed, unlike class I bound peptides where they are buried within the binding site, and thus in vivo trimming may be an explanation for the ragged termini of naturally processed class II bound peptides (Table 3). An extreme interpretation is that class II may bind intact protein, which is subsequently trimmed; a feasible possibility given that unfolded whole antigens can bind to class II molecules (Lee et al 1988; Sette et al 1989a).

The exact site where class II molecules bind peptide has not been established. Endocytosed proteins first encounter early endosomes, progress through late endosomes, and finally end up in lysosomes. Each site has distinct acidic properties, with conditions becoming increasingly harsh during progression along the pathway (Mellman et al 1986). Studies in which several antigenic proteins were targeted for processing to different sites within the pathway (by encapsulation in liposomes of differing resistance to breakdown by acid pH) showed that peptides were generated most effectively within lysosomes (Harding et al 1991a,b). Ultrastructural studies aimed at localizing class II molecules in the pathway have yielded divergent results. Guagliardi et al (1990) identified class II molecules in early endosomes, whereas Peters et al (1991) found them to be most predominant in vesicles related to lysosomes. However, class II molecules are not present in lysosomes and therefore, if these vesicles are an important site for peptide generation, a mechanism must exist for transporting peptides to the site of association with class II.

Peptide transport is also implied by instances of antigen being routed from the cytoplasm into the endosomal pathway for processing and presentation by class II molecules (Jaraquemada et al 1990). The transport mechanism is not the same as that utilized by peptides bound by class I molecules since presentation of cytoplasmic antigen by HLA-DR1 is possible in a mutant cell line lacking TAP proteins (Malnati et al 1992). HSPs are possible candidates. Protein transfer to lysosomes can be mediated by members of the HSP70 family (Chiang et al 1989), and the peptide-binding protein PBP 72/74, which plays a role in antigen presentation by class II molecules (Lakey et al 1987), is a HSP (Van Buskirk et al 1989). The precise function of PBP 72/74 has not been established, but the protein is present in endocytic vesicles, and it binds naturally processed peptides in vivo (DeNagel & Pierce 1992).

Location of the gene encoding PBP 72/74 is unknown, but at least three HSP70 genes map within the class III region of the MHC (Sargent et al 1989; Milner & Campbell 1990). It has recently emerged that an MHC-encoded factor influences antigen presentation by class II molecules. Several mutant cell lines fail to present peptides to T cells when processing is required, despite the expression of normal class II and invariant chain proteins (Mellins et al 1990), and the defect maps to the class II region of the MHC (Mellins et al 1991). The same phenotype is also exhibited by products of transfected class II genes expressed by the cell line 721.174 and its derivative T2, which possess a homozygous deletion within the MHC including all TAP, LMP, and functional class II genes (Ceman et al 1992; Riberdy & Cresswell 1992). The phenotype is reminiscent of that manifested by defective TAP proteins in class I peptide supply-deficient cells, but involvement of TAP genes is not implicated. Cells in which the defect results from a point mutation express functional class I molecules (Mellins et al 1991), and a cell line lacking TAP1 does not exhibit the mutant phenotype (Ceman et al 1992). The defect may involve LMP proteins, but it is more probably connected to a HSP or an as yet unidentified factor.

Class II molecules are presumed to leave the endocytic compartment rapidly after acquiring peptides, although the route to the cell surface and possible existence of a retention mechanism to prevent class II molecules leaving the endocytic pathway without bound peptide have not been determined. Rapid transit from the endocytic pathway is implicated to reduce the risk of class II degradation in this proteolytic environment. Indeed, their proteolysis probably explains the presence of class II-derived fragments in the pool of naturally processed peptides bound by class II molecules (Table 3). Speculation that peptide binding to class II is rapid is at odds with the slow association kinetics observed in vitro using purified class II molecules at neutral pH conditions (Babbitt et al 1985; Buus et al 1986). Peptide binding by class II molecules expressed on live cells is more rapid (Roosneck et al 1988; Ceppellini et al 1989), which suggests that a mechanism exists in vivo to facilitate association.

Since peptide binding occurs in an acidic environment, attention has focused on whether pH enhances efficiency. Jensen (1990) showed class II molecules expressed by fixed cells stimulate T cells more efficiently after incubation with peptide at lower pH (4.5–5.5). Direct positive influence of acidic pH on peptide binding has been confirmed by in vitro binding assays using various combinations of purified class II molecules and peptides (Harding et al 1991c; Jensen 1991a; Sadegh-Nasseri & Germain 1991; Tampé & McConnell 1991; Wettstein et al 1991).

No consensus exists on the molecular basis for promotion of peptide binding by class II molecules at acidic pH. The rate of peptide association is enhanced between pH 5–6 (Harding et al 1991c; Jensen 1991a), but there is controversy over whether the number of available binding sites also increases. Sadegh-Nasseri & Germain (1991) reported increased dissociation of class II at the same acidic pH as that required for the enhanced association of peptide. Similarly, analysis of ovalbumin peptide bound to I-A^d showed that complexes became unstable at pH 5, and this correlated with an increase in the amount and rate of peptide association (Tampé & McConnell 1991). Thus class II may undergo acid-induced destabilization to release pre-bound peptides and thereby generate free binding sites. However, other reports show complexes of class II and peptide are stable for many hours at pH values sufficient to enhance peptide binding (Lee & Watts 1990; Jensen 1992). Destabilization required exposure to pH <4.5, an environment unlikely to be encountered by class II molecules in the endocytic pathway, and resulted in irreversible denaturation (Harding et al 1991c; Jensen 1991a). Therefore, Jensen (1992) speculated that mildly acidic pH enhances peptide binding to class II molecules already devoid of bound peptide.

Most in vitro peptide-binding studies use purified class II proteins that already contain pre-bound peptides, thus they are of limited relevance to the in vivo association of peptides with nascent empty class II molecules. Such studies do show that optimum pH for binding varies with the combination of class II allele and peptide, which may have physiological implications for the site of peptide binding. Peptides bind I-A molecules at a pH found in early endosomes, while I-E molecules bind peptides optimally at a pH associated with more acidic endosomes (Jensen 1991a; Sette et al 1992b). Perhaps surprisingly, in vitro binding of myelin basic protein-derived peptides to I-A^k and I-A^s is sub-optimal at acidic pH (Nag et al 1992) and, so far, no preferential association of peptides at low pH has been observed for most human class II alleles (Sette et al 1992b). Thus although the class II bound peptide repertoire is evidently biased towards peptides with enhanced ability to bind at low pH, acidic conditions may not be a requirement. Peptide binding to class II molecules merely has to be acid-tolerant given the physiological environment in which it occurs.

Factors other than low pH may contribute to the rapid association of peptides with class II molecules observed in vivo. Demonstration that, at optimal pH, 20 hr or more are required to achieve maximum peptide binding by detergent-solubilized class II molecules compared to 1–2 hr for class II molecules expressed at the surface of fixed cells (Jensen 1991a) suggests involvement of membrane-associated factors. A report that cell membrane-derived lipids enhance the rate of peptide association, but not dissociation, to purified mouse class II molecules by 10–50-fold (Roof et al 1990) shows plasma membrane components can facilitate the interaction.

At neutral pH, complexes of class II and bound peptide are highly stable. Half-lives of several hours to several days are reported for a variety of purified class II and bound peptide combinations (Buus et al 1986; Sadegh-Nasseri & McConnell 1989; Stern & Wiley 1991; Tampé & McConnell 1991). Lanzavecchia et al (1992) confirmed that the observations made in vitro are reflected in vivo; complexes of class II and bound peptide are expressed at the cell surface with a half-life of approximately 25 hr, the same as the half-life of the class II molecule, which indicates an irreversible association ended only by class II degradation.

Peptide binding appears to confer stability to class II molecules by inducing a conformational change; a situation analogous to the role of peptide in determining the structure of class I molecules. Sadegh-Nasseri & Germain (1991) identified two class II molecular conformations by SDS-PAGE. Most molecules migrate in a SDS-stable compact form, but acid treatment generates an unstable floppy form that can be converted back to the compact molecule if exposed to peptides and neutralized. Physiological relevance of this peptide-induced structural change was established in a parallel study in which acquisition of SDS stability by nascent class II molecules was shown to correlate with dissociation from the invariant chain in an endocytic compartment, by inference the site of peptide binding (Germain & Hendrix 1991). Further support for a link between peptide binding, a conformational change, and conference of stability is provided by studies of class II peptide supply-deficient mutant cell lines (Mellins et al 1990). The cells express SDS-unstable HLA-DR3 molecules that are serologically distinguishable, and thus conformationally different, from HLA-DR3 expressed by wild-type parental cells.

Instability of the class II molecules was interpreted as reflecting reduced occupancy of the binding site with peptides. Since expression of class II at the cell surface is not impeded in the peptide supply-deficient mutant cells (Mellins et al 1990), peptide binding does not appear to be a prerequisite for surface expression, unlike for class I molecules. In support of this, numerous class II molecules in the SDS-unstable form have been identified at the surface of normal spleen cells (Germain & Hendrix 1991). However, it was recently shown that unstable class II molecules are not necessarily devoid of bound peptides. Transfected HLA-DR3 genes expressed by the peptide supply-deficient mutant cell lines 721.174 and its derivative T2 exhibit the unstable, conformationally altered phenotype (Ceman et al 1992; Riberdy & Cresswell 1992). Yet most of the molecules possess bound peptides which, unlike HLA-DR3 from normal cells, are predominantly derived from within residues 81–104 of the invariant chain (Riberdy et al 1992; Sette et al 1992a). Peptides derived from this portion of the invariant chain have been isolated from several class II alleles expressed by normal cells (Table 3), which raises the possibility that this sequence may not occupy the conventional binding site, but instead interacts with a conserved portion of the class II molecule.

The invariant chain-derived peptides are associated with approximately 70% of the HLA-DR3 molecules expressed at the surface of 721.174 cells, the remaining molecules appear to be genuinely empty (Sette et al 1992a). Peptide binding to empty class II molecules at the cell surface offers an explanation for the ability of fixed cells to present exogenously added peptides to T cells (Shimonkevitz et al 1983). There is also evidence that peptide binding can occur via peptide exchange. Recycling of class II molecules is suggested by the ability of mouse cells to generate and present class II bound peptides in the absence of protein synthesis (Harding & Unanue 1989) and the observation that murine class II molecules are endocytosed and re-appear at the cell surface (Machy et al 1990; Salamero et al 1990). However, recycling is at odds with the proposal that internalized class II molecules do not encounter a pH low enough to induce destabilization and so release pre-bound peptides (Jensen 1991a). In human cells, recycling seems to play a negligible role; processed antigen binds primarily to nascent class II molecules (Davidson et al 1991), and the peptide acquired during biosynthesis is retained throughout the life of the molecule (Lanzavecchia et al 1992).

SUMMARY AND PERSPECTIVES

It is now evident that bound peptide forms an integral third component of a MHC molecule. The stability conferred to MHC structure by peptide binding presumably ensures the heterotrimeric complex is expressed at the cell surface for sufficient time to allow T cell surveillance for pathogen-derived peptides. The functional dichotomy between T cell recognition of peptides presented by class I and class II molecules is maintained by use of different assembly pathways. The retention of empty class I molecules in the ER and limited peptide binding at the cell surface effectively confines class I to binding those peptides encountered early in biosynthesis. These peptide-binding properties are designed to focus class I on presenting antigens of intracellular origin to T cells such as peptides derived from viral proteins synthesized within host cells, or peptides from intracellular parasites. If this distinction is not preserved, binding of peptides from the extracellular environment would result in the killing of uninfected cells by cytotoxic T cells. Association of nascent class II molecules with invariant chain in the

ER and during their intracellular transport targets them into the endocytic pathway and confines them to binding peptides generated predominantly in endocytic compartments. These in vivo peptide-binding properties ensure class II molecules present peptides of extracellular origin to helper T cells, and thereby initiate an appropriate response such as immunoglobulin production.

Recent advances in understanding the molecular basis for interactions between bound peptides and MHC proteins have afforded explanations for the capacity of a given MHC molecule to bind a diverse, but not infinite, range of peptides with high affinity. The advent of methods for predicting which antigenic peptides possess the potential to bind MHC molecules has implications for vaccine development. In addition, characterization of properties that confer MHC binding may enable design of MHC-specific antagonists for use in preventing adverse triggering of self-reactive T cells, which probably contributes to autoimmune diseases, and for blockade of T cell responses to foreign MHC molecules, which can induce rejection of therapeutically transplanted tissues.

ACKNOWLEDGMENTS

We thank John Domena, Jenny Gumperz, and William Hildebrand for helpful comments on the manuscript. Linda D. Barber is an American Heart Association, California Affiliate postdoctoral research fellow.

Literature Cited

Ahluwalia, N, Bergeron, JJM, Wada, I, Degen, E, Williams, DB. 1992. The p88 molecular chaperone is identical to the endoplasmic reticulum membrane protein, calnexin. *J. Biol. Chem.* 267:10914–18

Aldrich, CJ, Rodgers, JR, Rich, RR. 1988. Regulation of Qa-1 expression and determinant modification by an H-2D-linked gene, Qdm. *Immunogenetics* 28:334–44

Allen, H, Fraser, J, Flyer, D, Calvin, S, Flavell, R. 1986. β_2-microglobulin is not required for cell surface expression of the murine class I histocompatibility antigen H-2D^b or of a truncated H-2D^b. *Proc. Natl. Acad. Sci. USA* 83:7447–51

Allen, PM, Matsueda, GR, Evans, RJ, Dunbar, JB, Marshall, GR, Unanue, ER. 1987. Identification of the T-cell and Ia contact residues of a T-cell antigenic epitope. *Nature* 327:713–15

Arnold, D, Driscoll, J, Androlewicz, M, Hughes, E, Cresswell, P, Spies, T. 1992. Proteasome subunits encoded in the MHC are not generally required for the processing of peptides bound by MHC class I molecules. *Nature* 360:171–74

Attaya, M, Jameson, S, Martinez, CK, Hermel, E, Aldrich, C, et al. 1992. Ham-2 corrects the class I antigen processing defect in RMA-S cells. *Nature* 355:647–49

Baas, EJ, van Santen, H, Kleijmeer, MJ, Geuze, HJ, Peters, PJ, Ploegh, HL. 1992. Peptide-induced stabilization and intracellular localization of empty HLA class I complexes. *J. Exp. Med.* 176:147–56

Babbitt, BP, Allen, PM, Matsueda, G, Haber, E, Unanue, ER. 1985. Binding of immunogenic peptides to Ia histocompatibility molecules. *Nature* 317:359–61

Bahram, S, Arnold, D, Bresnahan, M, Strominger, JL, Spies, T. 1991. Two putative subunits of a peptide pump encoded in the human major histocompatibility complex class II region. *Proc. Natl. Acad. Sci. USA* 88:10094–98

Bakke, O, Dobberstein, B. 1990. MHC class II-associated invariant chain contains a sort-

ing signal for endosomal compartments. *Cell* 63:707–16
Benjamin, RJ, Madrigal, JA, Parham, P. 1991. Peptide binding to empty HLA-B27 molecules of viable human cells. *Nature* 351:74–77
Bix, M, Raulet, D. 1992. Functionally conformed free class I heavy chains exist on the surface of β_2 microglobulin negative cells. *J. Exp. Med.* 176:829–34
Bjorkman, PJ, Parham, P. 1990. Structure, function, and diversity of class I major histocompatibility complex molecules. *Annu. Rev. Biochem.* 59:253–88
Bjorkman, PJ, Saper, MA, Samraoui, B, Bennett, WS, Strominger, JL, Wiley, DC. 1987a. Structure of the human class I histocompatibility antigen HLA-A2. *Nature* 329:506–12
Bjorkman, PJ, Saper, MA, Samraoui, B, Bennett, WS, Strominger, JL, Wiley, DC. 1987b. The foreign antigen binding site and T cell recognition regions of class I histocompatibility antigens. *Nature* 329:512–18
Bleicher, PA, Balk, SP, Hagen, SJ, Blumberg, RS, Flotte, TJ, Terhorst, C. 1990. Expression of murine CD1 on gastrointestinal epithelium. *Science* 250:679–82
Braciale, TJ, Braciale, VL. 1991. Antigen presentation: structural themes and functional variations. *Immunol. Today* 12:124–29
Brodsky, FM, Guagliardi, LE. 1991. The cell biology of antigen processing and presentation. *Annu. Rev. Immunol.* 9:707–44
Brooks, A, Hartley, S, Kjer-Nielsen, L, Perera, J, Goodnow, CC, et al. 1991. Class II-restricted presentation of an endogenously derived immunodominant T-cell determinant of hen egg lysozyme. *Proc. Natl. Acad. Sci. USA* 88:3290–94
Brown, JH, Jardetzky, T, Saper, MA, Samraoui, B, Bjorkman, PJ, Wiley, DC. 1988. A hypothetical model of the foreign antigen binding site of class II histocompatibility molecules. *Nature* 332:845–50
Brown, MG, Driscoll, J. Monaco, JJ. 1991. Structural and serological similarity of MHC-linked LMP and proteasome (multicatalytic proteinase) complexes. *Nature* 353:355–57
Buerstedde, JM, Pease, LR, Bell, MP, Nilson, AE, Buerstedde, G, et al. 1988. Identification of an immunodominant region on the I-A β chain using site-directed mutagenesis and DNA-mediated gene transfer. *J. Exp. Med.* 167:473–87
Buus, S, Sette, A, Colon, SM, Grey, HM. 1988. Autologous peptides constitutively occupy the antigen binding site on Ia. *Science* 242:1045–47
Buus, S, Sette, A, Colon, SM, Jenis, DM, Grey, HM. 1986. Isolation and characterization of antigen-Ia complexes involved in T cell recognition. *Cell* 47:1071–77
Buus, S, Sette, A, Colon, SM, Miles, C, Grey, HM. 1987. The relationship between major histocompatibility complex (MHC) restriction and the capacity of Ia to bind immunogenic peptides. *Science* 235:1353–58
Buus, S, Werdelin, O. 1986. A group-specific inhibitor of lysosomal cysteine proteinases selectively inhibits both proteolytic degradation and presentation of the antigen dinitrophenyl-poly-L-lysine by guinea pig accessory cells to T cells. *J. Immunol.* 136: 452–58
Calabi, F, Milstein, C. 1986. A novel family of human major histocompatibility complex-related genes not mapping to chromosome 6. *Nature* 323:540–43
Ceman, S, Rudersdorf, R, Long, EO, DeMars, R. 1992. MHC class II deletion mutant expresses normal levels of transgene encoded class II molecules that have abnormal conformation and impaired antigen presentation ability. *J. Immunol.* 149:754–61
Ceppellini, R, Frumento, G, Ferrara, GB, Tosi, R, Chersi, A, Pernis, B. 1989. Binding of labelled influenza matrix peptide to HLA-DR in living B lymphoid cells. *Nature* 339:392–94
Cerundolo, V, Alexander, J, Anderson, K, Lamb, C, Cresswell, P, et al. 1990. Presentation of viral antigen controlled by a gene in the major histocompatability complex. *Nature* 345:449–52
Cerundolo, V, Elliott, T, Elvin, J, Bastin, J, Rammensee, HG, Townsend, A. 1991. The binding affinity and dissociation rates of peptides for class I major histocompatability complex molecules. *Eur. J. Immunol.* 21: 2069–75
Chen, BP, Madrigal, A, Parham, P. 1990. Cytotoxic T cell recognition of an endogenous class I HLA peptide presented by a class II HLA molecule. *J. Exp. Med.* 172: 779–88
Chen, BP, Parham, P. 1989. Direct binding of influenza peptides to class I HLA molecules. *Nature* 337:743–45
Chiang, HL, Terlecky, SR, Plant, CP, Dice, JF. 1989. A role for a 70-kilodalton heat shock protein in lysosomal degradation of intracellular proteins. *Science* 246:382–85
Chicz, RM, Urban, RG, Lane, WS, Gorga, JC, Stern, LJ, et al. 1992. Predominant naturally processed peptides bound to HLA-DR1 are derived from MHC-related molecules and are heterogeneous in size. *Nature* 358:764–68
Christinck, ER, Luscher, MA, Barber, BH, Williams, DB. 1991. Peptide binding to

class I MHC on living cells and quantitation of complexes required for CTL lysis. *Nature* 352:67–69

Collins, DS, Unanue, ER, Harding, CV. 1991. Reduction of disulphide bonds within lysosomes is a key step in antigen processing. *J. Immunol.* 147:4054–59

Colonna, M, Bresnahan, M, Bahram, S, Strominger, JL, Spies, T. 1992. Allelic variants of the human putative peptide transporter involved in antigen processing. *Proc. Natl. Acad. Sci. USA* 89:3932–36

Corr, M, Boyd, LF, Frankel, SR, Kozlowski, S, Padlan, EA, Margulies, DH. 1992. Endogenous peptides of a soluble major histocompatibility complex class I molecule, H-2$L^d{}_s$: Sequence motif, quantitative binding, and molecular modeling of the complex. *J. Exp. Med.* 176:1681–92

Cresswell, P. 1985. Intracellular class II HLA antigens are accessible to transferrin-neuraminidase conjugates internalized by receptor-mediated endocytosis. *Proc. Natl. Acad. Sci. USA* 82:8188–92

Davidson, HW, Reid, PA, Lanzavecchia, A, Watts, C. 1991. Processed antigen binds to newly synthesized MHC class II molecules in antigen-specific B lymphocytes. *Cell* 67: 105–16

Davies, DR, Padlan, EA, Sheriff, S. 1990. Antibody-antigen complexes. *Annu. Rev. Biochem.* 59:439–73

Davis, CB, Buerstedde, JM, McKean, DJ, Jones, PP, McDevitt, HO, Wraith, DC. 1989. The role of polymorphic I-A^k β chain residues in presentation of a peptide from myelin basic protein. *J. Exp. Med.* 169: 2239–44

Davis, MM, Bjorkman, PJ. 1988. T-cell antigen receptor genes and T-cell recognition. *Nature* 334:395–402

Degen, E, Cohen-Doyle, MF, Williams, DB. 1992. Efficient dissociation of the p88 chaperone from major histocompatibility complex class I molecules requires both β_2-microglobulin and peptide. *J. Exp. Med.* 175:1653–61

Degen, E, Williams, DB. 1991. Participation of a novel 88-kD protein in the biogenesis of murine class I histocompatibility molecules. *J. Cell Biol.* 112:1099–115

DeLuca-Flaherty, C, McKay, DB, Parham, P, Hill, BL. 1990. Uncoating protein (hsc70) binds a conformationally labile domain of clathrin light chain LC_a to stimulate ATP hydrolysis. *Cell* 62:875–87

Del Val, M, Hengel, H, Häcker, H, Hartlaub, U, Ruppert, T, et al. 1992. Cytomegalovirus prevents antigen presentation by blocking the transport of peptide-loaded major histocompatibility complex class I molecules into the medial-Golgi compartment. *J. Exp. Med.* 176:729–38

Del Val, M, Schlicht, H, Ruppert, T, Reddehase, MJ, Koszinowski, UH. 1991. Efficient processing of an antigenic sequence for presentation by MHC class I molecules depends on its neighboring residues in the protein. *Cell* 66:1145–53

Demotz, S, Grey, HM, Appella, E, Sette, A. 1989. Characterization of a naturally processed MHC class II-restricted T cell determinant of hen egg lysozyme. *Nature* 342: 682–84

DeNagel, DC, Pierce, SK. 1992. A case for chaperones in antigen processing. *Immunol. Today* 13:86–89

Deverson, EV, Gow, IR, Coadwell, WJ, Monaco, JJ, Butcher, GW, Howard, JC, 1990. MHC class II region encoding proteins related to the multidrug resistance family of transmembrane transporters. *Nature* 348: 738–41

Dobberstein, B, Garoff, H, Warren, G, Robinson, PJ. 1979. Cell-free synthesis and membrane insertion of mouse H-$2D^d$ histocompatibility antigen and β_2-microglobulin. *Cell* 17:759–69

Elliott, T. 1991. How do peptides associate with MHC class I molecules? *Immunol. Today* 12:386–88

Elliott, T, Cerundolo, V, Elvin, J, Townsend, A. 1991. Peptide-induced conformational change of the class I heavy chain. *Nature* 351:402–6

Esquivel, F, Yewdell, J, Bennick, J. 1992. RMA-S cells present endogenously synthesized cytosolic proteins to class I-restricted cytotoxic T lymphocytes. *J. Exp. Med.* 175: 163–68

Fahnestock, ML, Tamir, I, Narhi, L, Bjorkman, PJ. 1992. Thermal stability comparison of purified empty and peptide-filled forms of a class I MHC molecule. *Science* 258:1658–62

Falk, K, Rötzschke, O, Stevanović, S, Jung, G, Rammensee, HG. 1991. Allele-specific motifs revealed by sequencing of self-peptides eluted from MHC molecules. *Nature* 351:290–96

Flajnik, MF, Camilo, C, Kramer, J, Kasahara, M. 1991. Which came first, MHC class I or class II? *Immunogenetics* 33:295–300

Flynn, GC, Chappell, TG, Rothman, JE. 1989. Peptide binding and release by proteins implicated as catalysts of protein assembly. *Science* 245:385–90

Fremont, DH, Matsumura, M, Stura, EA, Peterson, PA, Wilson, IA. 1992. Crystal structures of two viral peptides in complex with murine MHC class I H-$2K^b$. *Science* 257:919–27

Galvin, K, Krishna, S, Ponchel, F, Frohlich, M, Cummings, DE, et al. 1992. The major histocompatibility complex class I antigen-binding protein p88 is the product of the calnexin gene. *Proc. Natl. Acad. Sci. USA* 89:8452–56

Garrett, TPJ, Saper, MA, Bjorkman, PJ, Strominger, JL, Wiley, DC. 1989. Specificity pockets for the side chains of peptide antigens in HLA-Aw68. *Nature* 342:692–96

Germain, RN. 1986. The ins and outs of antigen processing and presentation. *Nature* 322:687–89

Germain RN, Hendrix, LR. 1991. MHC class II structure, occupancy and surface expression determined by post-endoplasmic reticulum antigen binding. *Nature* 353:134–39

Gething, MJ, Sambrook, J. 1992. Protein folding in the cell. *Nature* 355:33–45

Glynne, R, Powis, SH, Beck, S, Kelly, A, Kerr, L, Trowsdale, J. 1991. A proteasome-related gene between the two ABC transporter loci in the class II region of the human MHC. *Nature* 353:357–60

Godeau, F, Luescher, IF, Ojcius, DM, Saucier, C, Mottez, E, et al. 1992. Purification and ligand binding of a soluble class I major histocompatibility complex molecule consisting of the first three domains of $H\text{-}2K^d$ fused to β_2-microglobulin expressed in the baculovirus-insect cell system. *J. Biol. Chem.* 267:24223–29

Goldberg, AL, Rock, KL. 1992. Proteolysis, proteasomes and antigen presentation. *Nature* 357:375–79

Gorga, JC, Dong, A, Manning, MC, Woody, RW, Caughey, WS, Strominger, JL. 1989. Comparison of the secondary structures of human class I and class II major histocompatibility complex antigens by Fourier transform infrared and circular dichroism spectroscopy. *Proc. Natl. Acad. Sci. USA* 86:2321–25

Gould, K, Cossins, J, Bastin, J, Brownlee, GG, Townsend, A. 1989. A 15 amino acid fragment of influenza nucleoprotein synthesized in the cytoplasm is presented to class I-restricted cytotoxic T lymphocytes. *J. Exp. Med.* 170:1051–56

Guagliardi, LE, Koppelman, B, Blum, JS, Marks, MS, Cresswell, P. Brodsky, FM. 1990. Co-localization of molecules involved in antigen processing and presentation in an early endocytic compartment. *Nature* 343: 133–39

Guo, H, Jardetzky, TS, Garrett, TPJ, Lane, WS, Strominger, JL, Wiley, DC. 1992. Different length peptides bind to HLA-Aw68 similarly at their ends but bulge out in the middle. *Nature* 360:364–66

Hahn, YS, Braciale, VL, Braciale, TJ. 1991. Presentation of viral antigen to class I major histocompatibility complex-restricted cytotoxic T lymphocytes. Recognition of an immunodominant influenza hemagglutinin site by cytotoxic T lymphocytes is independent of the position of the site in the hemagglutinin translation product. *J. Exp. Med.* 174:733–36

Hahn, YS, Hahn, CS, Braciale, VL, Braciale, TJ, Rice, CM. 1992. CD8+ T cell recognition of an endogenously processed epitope is regulated primarily by residues within the epitope. *J. Exp. Med.* 176:1335–41

Hammer, J, Takacs, B, Sinigaglia, F. 1992. Identification of a motif for HLA-DR1 binding peptides using M13 display libraries. *J. Exp. Med.* 176:1007–13

Harding, CV, Collins, DS, Kanagawa, O, Unanue, ER. 1991a. Liposome-encapsulated antigens engender lysosomal processing for class II MHC presentation and cytosolic processing for class I presentation. *J. Immunol.* 147:2860–63

Harding, CV, Collins, DS, Slot, JW, Geuze, HJ, Unanue, ER. 1991b. Liposome-encapsulated antigens are processed in lysosomes, recycled, and presented to T cells. *Cell* 64:393–401

Harding, CV, Roof, RW, Allen, PM, Unanue, ER. 1991c. Effects of pH and polysaccharides on peptide binding to class II major histocompatibility complex molecules. *Proc. Natl. Acad. Sci. USA* 88:2740–44

Harding, CV, Unanue, ER. 1989. Antigen processing and intracellular Ia. Possible roles of endocytosis and protein synthesis in Ia function. *J. Immunol.* 142:12–19

Harty, JT, Bevan, MJ. 1992. CD8+ T cells specific for a single nonamer epitope of *Listeria monocytogenes* are protective in vivo. *J. Exp. Med.* 175:1531–38

Henderson, RA, Michel, H, Sakaguchi, K, Shabanowitz, J, Appella, E, et al. 1992. HLA-A2.1-associated peptides from a mutant cell line: a second pathway of antigen presentation. *Science* 255:1264–66

Hickling, JK, Fenton, CM, Howland, K, Marsh, SGE, Rothbard, JB. 1990. Peptides recognized by class I restricted T cells also bind to MHC class II molecules. *Int. Immunol.* 2:435–41

Higgins, CF, 1992. ABC transporters: from microorganisms to man. *Annu. Rev. Cell Biol.* 8:67–113

Hill, AVS, Elvin, J, Willis, AC, Aidoo, M, Allsopp, CEM, et al. 1992. Molecular analysis of the association of HLA-B53 and resistance to severe malaria. *Nature* 360: 434–39

Hochman, JH, Shimizu, Y, DeMars, R, Edidin, M. 1988. Specific associations of fluorescent β_2-microglobulin with cell surfaces. The affinity of different H-2 and HLA antigens for β_2-microglobulin. *J. Immunol.* 140:2322–29

Hochstenbach, F, David, V, Watkins, S, Brenner, MB. 1992. Endoplasmic reticulum resident protein of 90 kilodaltons associates with the T- and B- cell antigen receptors and major histocompatibility complex antigens during their assembly. *Proc. Natl. Acad. Sci. USA* 89:4734–38

Hosken, NA, Bevan, MJ. 1992. An endogenous antigenic peptide bypasses the class I antigen presntation defect in RMA-S. *J. Exp. Med.* 175:719–29

Hosken, NA, Bevan, MJ, Carbone, FR. 1989. Class I-restricted presentation occurs without internalization or processing of exogenous antigenic peptides. *J. Immunol.* 142: 1079–83

Hunt, DF, Henderson, RA, Shabanowitz, J, Sakaguchi, K, Michel, H, et al. 1992a. Characterization of peptides bound to the class I MHC molecule HLA-A2.1 by mass spectrometry. *Science* 255:1261–63

Hunt, DF, Michel, H, Dickinson, TA, Shabanowitz, J, Cox, AL, et al. 1992b. Peptides presented to the immune system by the murine class II major histocompatibility complex molecule I-A^d. *Science* 256: 1817–20

Jacobson, S, Sekaly, RP, Jacobson, CL, McFarland, HF, Long, EO. 1989. HLA class II-restricted presentation of cytoplasmic measles virus antigens to cytotoxic T cells. *J. Virol.* 63:1756–62

Jaraquemada, D, Marti, M, Long, EO. 1990. An endogenous processing pathway in vaccinia virus-infected cells for presentation of cytoplasmic antigens to class II-restricted T cells. *J. Exp. Med.* 172:947–54

Jardetzky, TS, Gorga, JC, Busch, R, Rothbard, JB, Strominger, JL, Wiley, DC. 1990. Peptide binding to HLA-DR1: a peptide with most residues substituted to alanine retains MHC binding. *EMBO J.* 9:1797–803

Jardetzky, TS, Lane, WS, Robinson, RA, Madden, DR, Wiley, DC. 1991. Identification of self peptides bound to purified HLA-B27. *Nature* 353:326–29

Jefferies, WA, Burgert, HG. 1990. E3/19K from adenovirus 2 is an immunosubversive protein that binds to a structural motif regulating the intracellular transport of major histocompatibility complex class I proteins. *J. Exp. Med.* 172:1653–64

Jensen, PE. 1990. Regulation of antigen presentation by acidic pH. *J. Exp. Med.* 171:1779–84

Jensen, PE. 1991a. Enhanced binding of peptide antigen to purified class II major histocompatibility glycoproteins at acidic pH. *J. Exp. Med.* 174:1111–20

Jensen, PE. 1991b. Reduction of disulphide bonds during antigen processing: evidence from a thiol-dependent insulin determinant. *J. Exp. Med.* 174:1121–30

Jensen, PE. 1992. Long-lived complexes between peptide and class II major histocompatibility complex are formed at low pH with no requirement for pH neutralization. *J. Exp. Med.* 176:793–98

Kaufman, JF, Auffray, C, Korman, AJ, Shackelford, DA, Strominger, JL. 1984. The class II molecules of the human and murine major histocompatibility complex. *Cell* 36:1–13

Kaufman, JF, Strominger, JL. 1982. HLA-DR light chain has a polymorphic N-terminal region and a conserved immunoglobulin like C-terminal region. *Nature* 297:694–97

Kelly, A, Powis, SH, Glynne, R, Radley, E, Beck, S, Trowsdale, J. 1991. Second proteasome-related gene in the human MHC class II region. *Nature* 353:667–68

Kelly, A, Powis, SH, Kerr, L, Mockridge, I, Elliott, T, et al. 1992. Assembly and function of the two ABC transporter proteins encoded in the human major histocompatability complex. *Nature* 355:641–44

Kleijmeer, MJ, Kelly, A, Geuze, HJ, Slot, JW, Townsend, A, Trowsdale, J. 1992. Location of MHC-encoded transporters in the endoplasmic reticulum and *cis*-Golgi. *Nature* 357:342–44

Klein, J, Figueroa, F. 1986. Evolution of the major histocompatibility complex. *CRC Crit. Rev. Immunol.* 6:295–386

Krieger, JI, Karr, RW, Grey, HM, Yu, W, O'Sullivan, D, et al. 1991. Single amino acid changes in DR and antigen define residues critical for peptide-MHC binding and T cell recognition. *J. Immunol.* 146: 2331–40

Kropschofer, H, Max, H, Müller, CA, Hesse, F, Stevanovic, S, et al. 1992. Self-peptide released from class II HLA-DR1 exhibits a hydrophobic two-residue contact motif. *J. Exp. Med.* 175:1799–803

Kurlander, RJ, Shawar, SM, Brown, ML, Rich, RR. 1992. Specialized role for a murine class I-b MHC molecule in prokaryotic host defences. *Science* 257:678–79

Kvist, S, Hamann, U. 1990. A nucleoprotein peptide of influenza A virus stimulates assembly of HLA-B27 class I heavy chains and β_2-microglobulin translated in vitro. *Nature* 348:446–48

Kvist, S, Wiman, K, Claesson, L, Peterson, PA, Dobberstein, B. 1982. Membrane insertion and oligomeric assembly of HLA-DR histocompatibility antigens. *Cell* 29:61–69

Lakey, EK, Margoliash, E, Pierce, SK. 1987. Identification of a peptide binding protein that plays a role in antigen presentation. *Proc. Natl. Acad. Sci. USA* 84:1659–63

Lamb, CA, Yewdell, JW, Bennick, JR, Cresswell, P. 1991. Invariant chain targets HLA class II molecules to acidic endosomes containing internalized influenza virus. *Proc. Natl. Acad. Sci. USA* 88:5998–6002

Lanzavecchia, A, Reid, PA, Watts, C. 1992.

Irreversible association of peptides with class II MHC molecules in living cells. *Nature* 357:249–52

Latron, F, Pazmany, L, Morrison, J, Moots, R, Saper, MA, et al. 1992. A critical role for conserved residues in the cleft of HLA-A2 in presentation of a nonapeptide to T cells. *Science* 257:964–67

Lee, JM, Watts, TH. 1990. On the dissociation and reassociation of MHC class II-foreign peptide complexes: evidence that brief transit through an acidic compartment is not sufficient for binding site regeneration. *J. Immunol.* 144:1829–34

Lee, P, Matsueda, GR, Allen, PM. 1988. T cell recognition of fibrinogen. A determinant on the A α-chain does not require processing. *J. Immunol.* 140:1063–68

Lévy, F, Gabathuler, R, Larsson, R, Kvist, S. 1991. ATP is required for in vitro assembly of MHC class I antigens but not for transfer of peptides across the ER membrane. *Cell* 67:265–74

Lie, WR, Myers, NB, Connolly, JM, Gorka, J, Lee, DR, Hansen, TH. 1991. The specific binding of peptide ligand to L^d class I major histocompatibility complex molecules determines their antigenic structure. *J. Exp. Med.* 173:449–59

Lie, WR, Myers, NB, Gorka, J, Rubocki, RJ, Connolly, JM, Hansen, TH. 1990. Peptide ligand-induced conformation and surface expression of the L^d class I MHC molecule. *Nature* 344:439–41

Livingstone, AM, Fathman, CG. 1987. The structure of T cell epitopes. *Annu. Rev. Immunol.* 5:477–501

Livingstone, AM, Powis, SJ, Diamond, AG, Butcher, GW, Howard, JC. 1989. A *trans*-acting major histocompatibility complex-linked gene whose alleles determine gain and loss changes in the antigenic structure of a classical class I molecule J. Exp. Med. 170:777–95

Ljunggren, HG, Stam, NJ, Öhlén, C, Neefjes, JJ, Höglund, P, et al. 1990. Empty MHC class I molecules come out in the cold. *Nature* 346:476–80

Lotteau, V, Teyton, L, Peleraux, A, Nilsson, T, Karlsson, L, et al. 1990. Intracellular transport of class II MHC molecules directed by invariant chain. *Nature* 348:600–5

Luescher, IF, Romero, P, Cerottini, JC, Maryanski, JL. 1991. Specific binding of antigenic peptides to cell-associated MHC class I molecules. *Nature* 351:72–74

Machy, P, Bizozzero, JP, Reggio, H, Leserman, L. 1990. Endocytosis and recycling of MHC-encoded class II molecules by mouse B lymphocytes. *J. Immunol.* 145: 1350–55

Madden, DR, Gorga, JC, Strominger, JL, Wiley, DC. 1991. The structure of HLA-B27 reveals nonamer self peptides bound in an extended conformation. *Nature* 353:321–25

Madden, DR, Gorga, JC, Strominger, JL, Wiley, DC. 1992. The three-dimensional structure of HLA-B27 at 2.1Å resolution suggests a general mechanism for tight peptide binding to MHC. *Cell* 70:1035–48

Maeji, NJ, Bray, AM, Geysen, HM. 1990. Multi-pin peptide synthesis strategy for T cell determinant analysis. *J. Immunol. Methods* 134:23–33

Malnati, MS, Marti, M, LaVaute, T, Jaraquemada, D, Biddison, W, et al. 1992. Processing pathways for presentation of cytosolic antigen to MHC class II-restricted T cells. *Nature* 357:702–04

Margulies, DH. 1992. Peptides tailored to perfection? *Curr. Biol.* 2:211–13

Martinez, CK, Monaco, JJ. 1991. Homology of proteasome subunits to a major histocompatibility complex-linked LMP gene. *Nature* 353:664–67

Maryanski, JL, Verdini, AS, Weber, PC, Salemme, FR, Corradin, G. 1990. Competitor analogs for defining T cell antigens: peptides incorporating a putative binding motif and polyproline or polyglycine spacers. *Cell* 60:63–72

Matsumura, M, Fremont, DH, Peterson, PA, Wilson, IA. 1992a. Emerging principles for the recognition of peptide antigens by MHC class I molecules. *Science* 257:927–34

Matsumura, M, Saito, Y, Jackson, MR, Song, ES, Peterson, PA. 1992b. In vitro peptide binding to soluble empty class I major histocompatibility complex molecules isolated from transfected *Drosophila melanogaster* cells. *J. Biol. Chem.* 267:23589–95

Mellins, E, Kempin, S, Smith, L, Monji, T, Pious, D. 1991. A gene required for class II-restricted antigen presentation maps to the major histocompatibility complex. *J. Exp. Med.* 174:1607–15

Mellins, E, Smith, L, Arp, B, Cotner, T, Celis, E, Pious, D. 1990. Defective processing and presentation of exogenous antigens in mutants with normal HLA class II genes. *Nature* 343:71–74

Mellman, I, Fuchs, R, Helenius, A. 1986. Acidification of the endocytic and exocytic pathways. *Annu. Rev. Biochem.* 55:663–700

Milarski, KL, Morimoto, RI. 1989. Mutational analysis of the human HSP70 protein: distinct domains for nuclear localization and adenosine triphosphate binding. *J. Cell Biol.* 109:1947–62

Miller, J, Germain, RN. 1986. Efficient cell surface expression of class II MHC molecules in the absence of associated invariant chain. *J. Exp. Med.* 164:1478–89

Milner, CM, Campbell, RD. 1990. Structure

and expression of the three MHC-linked HSP70 genes. *Immunogenetics* 32:242–51

Momburg, F, Ortiz-Navarrete, V, Neefjes, J, Goulmy, E, van de Wal, Y, et al. 1992. Proteasome subunits encoded by the major histocompatibility complex are not essential for antigen presentation. *Nature* 360:174–77

Monaco, JJ, Cho, S, Attaya, M. 1990. Transport protein genes in the murine MHC: possible implications for antigen processing. *Science* 250:1723–26

Morrison, LA, Lukacher, AE, Braciale, VL, Fan, DP, Braciale, TJ. 1986. Differences in antigen presentation to MHC class I- and class II-restricted influenza virus-specific cytotoxic T lymphocyte clones. *J. Exp. Med.* 163:903–21

Mouritsen, S, Meldal, M, Werdelin, O, Stryhn Hansen, A, Buus, S. 1992a. MHC molecules protect T cell epitopes against proteolytic destruction. *J. Immunol.* 149: 1987–93

Mouritsen, S, Stryhn Hansen, A, Petersen, B, Buus, S. 1992b. pH dependence of the interaction between immunogenic peptides and MHC class II molecules. Evidence for an acidic intracellular compartment being the organelle of interaction. *J. Immunol.* 148:1438–44

Nag, B, Passmore, D, Deshpande, SV, Clark, BR. 1992. In vitro maximum binding of antigenic peptides to murine MHC class II molecules does not always take place at the acidic pH of the in vivo endosomal compartment. *J. Immunol.* 148:369–72

Neefjes, JJ, Ploegh, HL. 1988. Allele and locus-specific differences in cell surface expression and the association of HLA class I heavy chain with β_2-microglobulin: differential effects of inhibition of glycosylation on class I subunit association. *Eur. J. Immunol.* 18:801–10

Neefjes, JJ, Stollorz, V, Peters, PJ, Geuze, HJ, Ploegh, HL. 1990. The biosynthetic pathway of MHC class II but not class I molecules intersects the endocytic route. *Cell* 61:171–83

Nelson, CA, Roof, RW, McCourt, DW, Unanue, ER. 1992. Identification of the naturally processed form of hen egg white lysozyme bound to the murine histocompatibility complex class II molecule I-A^k. *Proc. Natl. Acad. Sci. USA* 89:7380–83

Nuchtern, JG, Biddison, WE, Klausner, RD. 1990. Class II MHC molecules can use the endogenous pathway of antigen presentation. *Nature* 343:74–76

Nuchtern, JG, Bonifacino, JS, Biddison, WE, Klausner, RD. 1989. Brefeldin A implicates egress from endoplasmic reticulum in class I restricted antigen presentation. *Nature* 339:223–26

Ortiz-Navarrete, V, Seelig, A, Gernold, M, Frentzel, S, Kloetzel, PM, Hämmerling, GJ. 1991. Subunit of the '20S' proteasome (multicatalytic proteinase) encoded by the major histocompatibility complex. *Nature* 353:662–64

O'Sullivan, D, Sidney, J, Appella, E, Walker, L, Phillips, L, et al. 1990. Characterization of the specificity of peptide binding to four DR haplotypes. *J. Immunol.* 146:1799–808

Pamer, EG, Harty, JT, Bevan, MJ. 1991. Precise prediction of a dominant class I MHC-restricted epitope of *Listeria monocytogenes*. *Nature* 353:852–55

Pamer, EG, Wang, CR, Flaherty, L, Fischer Lindahl, K, Bevan, MJ. 1992. H-2M3 presents a *Listeria monocytogenes* peptide to cytotoxic T lymphocytes. *Cell* 70:215–23

Parham, P, Lomen, CE, Lawlor, DA, Ways, JP, Holmes, N, et al. 1988. Nature of polymorphism in HLA-A, -B, and -C molecules. *Proc. Natl. Acad. Sci. USA* 85: 4005–9

Parker, KC, Bednarek, MA, Hull, LK, Utz, U, Cunningham, B, et al. 1992. Sequence motifs important for peptide binding to the human MHC class I molecule, HLA-A2. *J. Immunol.* 149:3580–87

Pazmany, L, Rowland-Jones, S, Huet, S, Hill, A, Sutton, J, et al. 1992. Genetic modulation of antigen presentation by HLA-B27 molecules. *J. Exp. Med.* 175:361–69

Perkins, DL, Lai, M, Smith, JA, Gefter, ML. 1989. Identical peptides recognized by MHC class I- and class II-restricted T cells. *J. Exp. Med.* 170:279–89

Peters, PJ, Neefjes, JJ, Oorschot, V, Ploegh, HL, Geuze, HJ. 1991. Segregation of MHC class II molecules from MHC class I molecules in the Golgi complex for transport to lysosomal compartments. *Nature* 349: 669–76

Porcelli, S, Brenner, MB, Greenstein, JL, Balk, SP, Terhorst, C, Bleicher, PA. 1989. Recognition of cluster of differentiation 1 antigens by human CD4- CD8- cytolytic T lymphocytes. *Nature* 341:447–50

Porcelli, S, Morita, CT, Brenner, MB. 1992. CD1b restricts the response of human CD4–8- T lymphocytes to a microbial antigen. *Nature* 360:593–97

Poulik, MD. 1975. In *Trace Components of Plasma: Isolation and Clinical Significance*, ed. GA Jamieson, TJ Greenwalt, 155–78. New York: Liss. 428 pp

Powis, SH, Mockridge, I, Kelly, A, Kerr, L, Glynne, R, et al 1992. Polymorphism in a second ABC transporter gene located within the class II region of the human major histocompatibility complex. *Proc. Natl. Acad. Sci. USA* 89:1463–67

Powis, SJ, Deverson, EV, Coadwell, WJ, Ciruela, A, Huskisson, NS, et al. 1992. Effect of polymorphism of an MHC-linked transporter on the peptides assembled in a class I molecule. *Nature* 357:211–15

Puri, J, Factorovich, Y. 1988. Selective inhibition of antigen presentation to cloned T cells by proteases inhibitors. *J. Immunol.* 141:3313–17

Reddehase, MJ, Rothbard, JB, Koszinowski, UH. 1989. A pentapeptide as minimal antigenic determinant for MHC class I-restricted T lymphocytes. *Nature* 337:651–53

Riberdy, JM, Cresswell, P. 1992. The antigen-processing mutant T2 suggests a role for MHC-linked genes in class II antigen presentation. *J. Immunol.* 148:2586–90

Riberdy, JM, Newcomb, JR, Surman, MJ, Barbosa, JA, Cresswell, P. 1992. HLA-DR molecules from an antigen-processing mutant cell line are associated with invariant chain peptides. *Nature* 360:474–77

Rippmann, F, Taylor, WR, Rothbard, JB, Green, NM. 1991. A hypothetical model for the peptide binding domain of hsp70 based on the peptide binding domain of HLA. *EMBO J.* 10:1053–59

Roche, PA, Cresswell, P. 1990. Invariant chain association with HLA-DR molecules inhibits immunogenic peptide binding. *Nature* 345:615–18

Roche, PA, Cresswell, P. 1991. Proteolysis of the class II-associated invariant chain generates a peptide binding site in intracellular HLA-DR molecules. *Proc. Natl. Acad. Sci. USA* 88:3150–54

Roche, PA, Marks, MS, Cresswell, P. 1991. Formation of a nine-subunit complex by HLA class II glycoproteins and the invariant chain. *Nature* 354:392–94

Rock, KL, Rothstein, L, Benacerraf, B. 1992. Analysis of the association of peptides of optimal length to class I molecules on the surface of cells. *Proc. Natl. Acad. Sci. USA* 89:8918–22

Rojo, S, García, F, Villadangos, JA, López de Castro, JA. 1993. Changes inside the repertoire of peptides bound to HLA-B27 subtypes and to site-specific mutants in and outside pocket B. *J. Exp. Med.* 177:613–20

Romero, P, Corradin, G, Luescher, IF, Maryanski, JL. 1991. H-2K^d-restricted antigenic peptides share a simple binding motif. *J. Exp. Med.* 174:603–12

Ronchese, F, Brown, MA, Germain, RN. 1987. Structure-function analysis of the A_β^{bm12} mutation using site-directed mutagenesis and DNA-mediated gene transfer. *J. Immunol.* 139:629–38

Roof, RW, Luescher, IF, Unanue, ER. 1990. Phospholipids enhance the binding of peptides to class II major histocompatibility molecules. *Proc. Natl. Acad. Sci. USA* 87: 1735–39

Roosneck, E, Demotz, S, Corradin, G, Lanzavecchia, A. 1988. Kinetics of MHC-antigen complex formation on antigen-presenting cells. *J. Immunol.* 140:4079–82

Rötzschke, O, Falk, K, Deres, K, Schild, H, Norda, M, et al. 1990a. Isolation and analysis of naturally processed viral peptides as recognised by cytotoxic T cells. *Nature* 348: 252–54

Rötzschke, O, Falk, K, Faath, S, Rammensee, HG. 1991. On the nature of peptides involved in alloreactivity. *J. Exp. Med.* 174: 1059–71

Rötzschke, O, Falk, K, Stevanović, S, Jung, G, Rammensee, HG. 1992. Peptide motifs of closely related HLA class I molecules encompass substantial differences. *Eur. J. Immunol.* 22:2453–56

Rötzschke, O, Falk, K, Wallny, HJ, Faath, S, Rammensee, HG. 1990b. Characterization of naturally occuring minor histocompatibility peptides including H-4 and H-Y. *Science* 249:283–87

Rubocki, RJ, Connolly, JM, Hansen, TH, Melvold, RW, Kim, BS, et al. 1991. Mutation at amino acid position 133 of H-2D^d prevents β_2m association and immune recognition but not surface expression. *J. Immunol.* 146:2352–57

Rudensky, AY, Preston-Hurlburt, P, Al-Ramadi, BK, Rothbard, J, Janeway, CA. 1992. Truncated variants of peptides isolated from MHC class II molecules suggest sequence motifs. *Nature* 359:429–31

Rudensky, AY, Preston-Hurlburt, P, Hong, S, Barlow, A, Janeway, CA. 1991. Sequence analysis of peptides bound to MHC class II molecules. *Nature* 353:622–27

Sadegh-Nasseri, S, Germain, RN. 1991. A role for peptide in determining MHC class II structure. *Nature* 353:167–70

Sadegh-Nasseri, S, McConnell, HM. 1989. A kinetic intermediate in the reaction of an antigenic peptide and I-E^k. *Nature* 337:274–76

Salamero, J, Humbert, M, Cosson, P, Davoust, J. 1990. Mouse B lymphocyte specific endocytosis and recycling of MHC class II molecules. *EMBO J.* 9:3489–96

Saper, MA, Bjorkman, PJ, Wiley, DC. 1991. Refined structure of the human histocompatibility antigen HLA-A2 at 2.6Å resolution. *J. Mol. Biol.* 219:277–319

Sargent, CA, Dunham, I, Trowsdale, J, Campbell, RD. 1989. Human major histocompatibility complex contains genes for the major heat shock protein HSP70. *Proc. Natl. Acad. Sci. USA* 86:1968–72

Schaiff, WT, Hruska, KA, McCourt, DW, Green, M, Schwartz, BD. 1992. HLA-DR associates with specific stress proteins and is retained in the endoplasmic reticulum in invariant chain negative cells. *J. Exp. Med.* 176:657–66

Schnabl, E, Stockinger, H, Majdic, O,

Gaugitsch, H, Lindley, IJD, et al. 1990. Activated human T lymphocytes express MHC class I heavy chains not associated with β_2-microglobulin. *J. Exp. Med.* 171: 1431–42

Schumacher, TNM, DeBruijn, MLH, Vernie, LN, Kast, WM, Melief, CJM, et al. 1991. Peptide selection by MHC class I molecules. *Nature* 350:703–6

Schwartz, RH. 1985. T-lymphocyte recognition of antigen in association with gene products of the major histocompatibility complex. *Annu. Rev. Immunol.* 3:237–61

Sekaly, RP, Tonnelle, C, Strubin, M, Mach, B, Long, EO. 1986. Cell surface expression of class II histocompatibility antigens occurs in the absence of the invariant chain. *J. Exp. Med.* 164:1490–504

Sette, A, Adorini, L, Colon, SM, Buus, S, Grey, HM. 1989a. Capacity of intact proteins to bind to MHC class II molecules. *J. Immunol.* 143:1265- 67

Sette, A, Buus, S, Colon, S, Smith, JA, Miles, C, Grey, HM. 1987. Structural characteristics of an antigen required for its interaction with Ia and recognition by T cells. *Nature* 328:395–99

Sette, A, Ceman, S, Kubo, RT, Sakaguchi, K, Appella, E, et al. 1992a. Invariant chain peptides in most HLA-DR molecules of an antigen-processing mutant. *Science* 258:1801–4

Sette, A, Lamont, A, Buus, S, Colon, SM, Miles, C, Grey, HM. 1989b. Effect of conformational propensity of peptide antigens in their interaction with MHC class II molecules. Failure to document the importance of regular secondary structure. *J. Immunol.* 143:1268–73

Sette, A, Southwood, S, O'Sullivan, D, Gaeta, FCA, Sidney, J, Grey, HM. 1992b. Effect of pH on MHC class II-peptide interactions. *J. Immunol. 148* :844–51

Shawar, SM, Vyas, JM, Rodgers, JR, Cook, RG, Rich, RR. 1991. Specialized functions of major histocompatibility complex class I molecules. II. Hmt binds N-formylated peptides of mitochondrial and prokaryotic origin. *J. Exp. Med.* 174:941–44

Shimonkevitz, R, Kappler, J, Marrack, P, Grey, H. 1983. Antigen recognition by H-2-restricted T cells. I. Cell-free antigen processing. *J. Exp. Med.* 158:303–16

Silver, ML, Guo, HC, Strominger, JL, Wiley, DC. 1992. Atomic structure of a human MHC molecule presenting an influenza virus peptide. *Nature* 360:367–69

Silver, ML, Parker, KC, Wiley, DC. 1991. Reconstitution by MHC-restricted peptides of HLA-A2 heavy chain with β_2-microglobulin in vitro. *Nature* 350:619–22

Simister, NE, Mostov, KE. 1989. An Fc receptor structurally related to MHC class I antigens. *Nature* 337:184–87

Smith, JD, Lie, WR, Gorka, J, Kindle, CS, Myers, NB, Hansen, TH. 1992. Disparate interaction of peptide ligand with nascent versus mature class I major histocompatibility complex molecules: comparisons of peptide binding to alternative forms of L^d in cell lysates and the cell surface. *J. Exp. Med.* 175:191–202

Smith, MH, Barber, BH. 1990. The conformational flexibility of class I H-2 molecules as revealed by anti-peptide antibodies specific for intracytoplasmic determinants: differential reactivity of β_2-microglobulin "bound" and "free" H-$2K^b$ heavy chains. *Mol. Immunol.* 27:169–80

Spies, T, Bresnahan, M, Bahram, S, Arnold, D, Blanck, G, et al. 1990. A gene in the human major histocompatibility complex class II region controlling the class I antigen presentation pathway. *Nature* 348:744–47

Spies, T, Cerundolo, V, Colonna, M, Cresswell, P, Townsend, A, DeMars, R. 1992. Presentation of viral antigen by MHC class I molecules is dependent on a putative peptide transporter heterodimer. *Nature* 355:644–46

Spies, T, DeMars, R. 1991. Restored expression of major histocompatibility class I molecules by gene transfer of a putative peptide transporter. *Nature* 351:323–24

Staerz, UD, Karasuyama, H, Garner, AM. 1987. Cytotoxic T lymphocytes against a soluble protein. *Nature* 329:449–51

Stern, LJ, Wiley, DC. 1992. The human class II MHC protein HLA-DR1 assembles as empty $\alpha\beta$ heterodimers in the absence of antigenic peptide. *Cell* 68:465–77

Sutton, J, Rowland-Jones, S, Rosenberg, W, Nixon, D, Gotch, F, et al. 1993. A sequence pattern for peptides presented to cytotoxic T lymphocytes by HLA-B8 revealed by analysis of epitopes and eluted peptides. *Eur. J. Immunol.* 23:447–54

Takahashi, H, Cease, KB, Berzofsky, JA. 1989. Identification of proteases that process distinct epitopes on the same protein. *J. Immunol.* 142:2221–29

Tampé, R, McConnell, HM. 1991. Kinetics of antigenic peptide binding to the class II major histocompatibility molecule I-A^d. *Proc. Natl. Acad. Sci. USA* 88:4661–65

Teyton, L, O'Sullivan, D, Dickson, PW, Lotteau, V, Sette, A, et al. 1990. Invariant chain distinguishes between the exogenous and endogenous antigen presentation pathways. *Nature* 348:39–44

Thomas, DB, Hodgson, J, Riska, PF, Graham, CM. 1990. The role of the endoplasmic reticulum in antigen processing: N-glycosylation of influenza hemagglutinin abrogates CD4+ cytotoxic T cell recogni-

tion of endogenously processed antigen. *J. Immunol.* 144:2789–94
Thomas, JW, Danho, W, Bullesbach, E, Föhles, J, Rosenthal, A. S. 1981. Immune response gene control of determinant selection. III Polypeptide fragments of insulin are differentially recognized by T but not by B cells in insulin immune guinea pigs. *J. Immunol.* 126:1095–100
Townsend, A, Bastin, J, Gould, K, Brownlee, G, Andrew, M, et al. 1988. Defective presentation to class I-restricted cytotoxic T lymphocytes in vaccinia-infected cells is overcome by enhanced degradation of antigen. *J. Exp. Med.* 168:1211–24
Townsend A, Bodmer H. 1989. Antigen recognition by class I-restricted T lymphocytes. *Annu. Rev. Immunol.* 7:601–24
Townsend, A, Elliott, T, Cerundolo, V, Foster, L. Barber, B, Tse, A. 1990. Assembly of MHC class I molecules analyzed in vitro. *Cell* 62:285–95
Townsend, A, Öhlén C, Bastin, J, Ljunggren, HG, Foster, L, Kärre, K. 1989. Association of class I major histocompatibility heavy and light chains induced by viral peptides. *Nature* 340:443–48
Townsend, ARM, Gotch, FM, Davey, J. 1985. Cytotoxic T cells recognize fragments of the influenza nucleoprotein. *Cell* 42:457–67
Townsend, ARM, Rothbard, J, Gotch, FM, Bahadur, G, Wraith, D, McMichael, AJ. 1986. The epitopes of influenza nucleoprotein recognized by cytotoxic T lymphocytes can be defined by short synthetic peptides. *Cell* 44:959–68
Trowsdale, J, Hanson, I, Mockridge, I, Beck, S, Townsend, A, Kelly, A. 1990. Sequences encoded in the class II region of the MHC related to the 'ABC' superfamily of transporters. *Nature* 348:741–44
Tsomides, TJ, Walker, BD, Eisen, HN. 1991. An optimal viral peptide recognized by CD8+ T cells binds very tightly to the restricting class I major histocompatibility complex protein on the intact cells but not to the purified class I protein. *Proc. Natl. Acad. Sci. USA* 88:11276–80
Unanue, ER. 1984. Antigen-presenting function of the macrophage. *Annu. Rev. Immunol.* 2:395–428
Utz, U, Koenig, S, Coligan, JE, Biddison, WE. 1992. Presentation of three different viral peptides, HTLV-1 TAX, HCMV gB, and influenza virus M1, is determined by common structural features of the HLA-A2.1 molecule. *J. Immunol.* 149:214–21
Van Bleek, GM, Nathenson, SG. 1990. Isolation of an endogenously processed immunodominant viral peptide from the class I H-2K^b molecule. *Nature* 348:213–16
VanBuskirk, A, Crump, BL, Margoliash, E, Pierce, SK. 1989. A peptide binding protein having a role in antigen presentation is a member of the HSP70 heat shock family. *J. Exp. Med.* 170:1799–809
Vidard, L, Rock, KL, Benacerraf, B. 1991. The generation of immunogenic peptides can be selectively increased or decreased by proteolytic enzyme inhibitors. *J. Immunol.* 147:1786–91
Vitiello, A, Potter, TA, Sherman, LA. 1990. The role of β_2-microglobulin in peptide binding by class I molecules. *Science* 250:1423–26
Wallny, HJ, Rammensee, HG. 1990. Identification of classical minor histocompatibility antigen as cell-derived peptide. *Nature* 343:275–78
Walter, P, Lingappa, VR. 1986. Mechanism of protein translocation across the endoplasmic reticulum membrane. *Annu. Rev. Cell Biol.* 2:499–516
Wei, ML, Cresswell, P. 1992. HLA-A2 molecules in an antigen-processing mutant cell contain signal sequence-derived peptides. *Nature* 356:443–46
Weiss, S, Bogen, B. 1991. MHC class II-restricted presentation of intracellular antigen. *Cell* 64:767–76
Wettstein, DA, Boniface, JJ, Reay, PA, Schild, H, Davis, MM. 1991. Expression of a class II major histocompatibility complex (MHC) heterodimer in a lipid-linked form with enhanced peptide/soluble MHC complex formation at low pH. *J. Exp. Med.* 174:219–28
Whitton, JL, Oldstone, MBA. 1989. Class I MHC can present an endogenous peptide to cytotoxic T lymphocytes. *J. Exp. Med.* 170: 1033–38
Yewdell, JW, Bennick, JR. 1989. Brefeldin A specifically inhibits presentation of protein antigens to cytotoxic T lymphocytes. *Science* 244:1072–75
Zhang, W, Young, ACM, Imarai. M, Nathenson, SG, Sacchettini, JC. 1992. Crystal structure of the major histocompatibility complex class I H-2K^b molecule containing a single viral peptide: Implications for peptide binding and T-cell receptor recognition. *Proc. Natl. Acad. Sci. USA* 89:8403–7
Ziegler, HK, Unanue, ER. 1982. Decrease in macrophage antigen catabolism caused by ammonia and chloroquine is associated with inhibition of antigen presentation of T cells. *Proc. Natl. Acad. Sci. USA* 79:175–78

Annu. Rev. Cell Biol. 1993. 9:207–35

ASSEMBLY AND INTRACELLULAR TRANSPORT OF MHC CLASS I MOLECULES

Michael R. Jackson and Per A. Peterson
Department of Immunology, IMM8, The Scripps Research Institute, 10666 North Torrey Pines Road, La Jolla, California 92037

KEY WORDS: antigen presentation, crystal structure

CONTENTS

0743–4634/93/1115–0207$05.00

INTRODUCTION

The immune system has a variety of means by which it can eliminate pathogens. However, some parts of the immune system are primarily focused on pathogens with a certain life style. For example, cytolytic T cells, which usually express the CD8 molecule, are effector cells that can recognize and destroy host cells that have become infected with a pathogen. Therefore, such CD8+ T cells are mainly responsible for combating intracellular pathogens that commonly utilize the host biosynthetic machinery, e.g. viruses. The other major subset of T cells, the T helper cells, which express the cell surface protein CD4, are central in the defense against bacterial infections. Such cells rarely act directly on bacteria, but recruit other components of the immune sytem, such as antibody-producing B cells, to eliminate the invader.

In contrast to the antibody-bearing B cells, T cells do not recognize foreign proteins directly. Instead, antigen-specific T cell receptors interact with antigenic peptides bound by specialized presenting molecules. Two major types of these specialized molecules, called major histocompatibility complex (MHC) proteins exist; class I molecules, which consist of one transmembrane chain associated with the small subunit β2microglobulin, and class II molecules, which are heterodimers formed from two transmembrane chains. The MHC molecules, which are genetically polymorphic and encoded by several closely linked loci, occur as multiple isotypes. Thus most humans express at least three class I isotypes, denoted HLA-A, -B, and -C, and typically two allelic forms of each isotype are expressed in an individual. An analogous situation exists for class II molecules. The reason for this extensive polymorphism was explained when the crystal structure of the first MHC class I molecule was determined (Bjorkman et al 1987). The polymorphic residues were found to be clustered around the peptide-binding pocket, thereby altering its shape and presumably the spectrum of peptides that can be bound. Thus to mount an effective immune response against most pathogens, the immune system has evolved many allelic forms of the class I molecules so as to be able to accomodate as many different antigenic peptides as possible.

The existence of two structurally similar classes of MHC molecules reflects their distinct roles (Bevan 1987). MHC class I molecules are loaded in the exocytic pathway (ER to Golgi to cell surface) with peptides derived from the pool of proteins synthesized in the cytoplasm of a cell. In contrast, MHC class II molecules are targeted from their site of synthesis in the ER to the endosomal compartments by the binding of an accessory protein, the invariant chain; this chain has the second important function of preventing them from binding peptide until after they arrive at the endosome (Teyton

& Peterson 1992). In keeping with this division of labor between MHC class I and class II molecules, CD8-expressing T cells recognize antigenic peptides in association with MHC class I molecules, while CD4-expressing T helper cells recognize peptides in the context of MHC class II molecules.

The role of MHC class I molecules is to provide an up-to-date report at the cell surface of the type and quantities of proteins being synthesized in a cell. Throughout its life time, a cell sends a continuous stream of newly synthesized class I molecules from the endoplasmic reticulum (ER) to the cell surface. En route these molecules bind peptides derived from newly synthesized proteins manufactured by the cell. It is this peptide-class I complex displayed at the plasma membrane that is surveyed by patrolling CD8 T-cells. In an uninfected cell, the class I molecules report that all is normal by displaying peptides derived from endogenous proteins (self-peptides) (Heath et al 1989; Jardetzky et al 1991; Corr et al 1992). If, however, the cell is infected by a virus, peptides derived from the newly synthesized viral proteins will be rapidly displayed in the context of the class I molecules on the cell surface, thus leaving the cell at the mercy of the circulating killer T cells.

The circulating CD8-positive T cells have been educated to discriminate MHC/peptide complexes that contain self-peptides from those containing peptides derived from non-self proteins. Education of naive T cells occurs in the thymus by the processes of positive and negative selection (Bevan 1977; von Boehmer et al 1978; Zinkernagel et al 1978). The details of these processes are poorly understood, but it is widely accepted that MHC molecules containing self-peptides are used in the teaching process (Kappler et al 1987; Nikolic-Zugic & Bevan 1990; Van Kaer et al 1992).

STRUCTURE OF MHC CLASS I MOLECULES

Fully assembled MHC class I molecules are composed of a polymorphic heavy chain noncovalently associated with β2 microglobulin and a short peptide of typically 8–9 amino acids. Heavy chains are encoded in the class I region of the MHC (called HLA in humans and H-2 in mouse). They consist of three extracellular domains (α1,α2, and α3) followed by a single transmembrane domain and a relatively short (~35 residues) cytoplasmic tail. Crystallographic studies have shown that the two N terminal domains (α1 and α2) fold together to form the sides of a peptide-binding groove composed of two α helices, which sit on a platform of eight anti-parallel β-sheets (Bjorkman et al 1987). The peptide is bound in the groove in an extended conformation, and a few side chains of the peptide are accomodated in specific pockets in the groove (Madden et al 1991; Fremont et al 1992; Matsumura et al 1992a). Beneath the antigen-binding region, the α3 domain folds into an immunoglobulin-like domain that forms extensive contacts with β2microglobulin. Almost all of the allelic polymorphism is accounted for by amino acids that are located in the α1 and α2 domains of the heavy chains.

Most of the allelic residues occupy positions that would affect the shape of the antigen-binding groove (Bjorkman et al 1987; Garrett et al 1989). It is the variability created by these highly polymorphic amino acids of the α1 and α2 domains of the heavy chains that determines the shape of the antigen-binding groove (Fremont et al 1992; Matsumura et al 1992a) and thereby the spectrum of peptides that can be bound by each allelic protein.

Crystal structure analyses of two viral peptides in complex with murine MHC class I H-2K^b showed that the antigen-binding site is more of a pocket than a groove (see Figure 1A) (Fremont et al 1992; Zhang et al 1992). In the case of K^b, this pocket best accomodates peptides of 8 amino acids. However, a 9-amino acid long peptide may also be accomodated. This is accomplished by the nonapeptide maintaining similar interactions at the amino and carboxy termini as the octapeptide, which allows the central residues of the nonapeptide to bulge out from the pocket (Figure 1B) (Fremont et al 1992; Guo et al 1992). The amino terminus of both the 8- and 9-amino acid long peptides are deeply buried relative to the carboxy terminus. Indeed, most (>75%) of the peptide is sufficiently buried to escape direct contact with a T cell receptor (Figure 1A) and, consequently, the peptide contributes suprisingly little to the surface area of the class I/peptide complex (Fremont et al 1992). This has led to the suggestion that T cells are sensing what little exposed peptide is available plus subtle alterations in the structure of the class I heavy chain resulting from accomodation of a specific peptide in the binding groove. Structural alterations in the heavy chain have been detected upon peptide binding, and some of these changes depend on the exact sequence of the peptide bound (Fremont et al 1992; Bluestone et al 1992; Catipović et al 1992).

NATURE OF ANTIGENIC PEPTIDE BOUND BY MHC CLASS I MOLECULES

Although it has been recognized for some time that antigen presented by MHC molecules is in the form of peptide fragments (Townsend et al 1986), the recent development of techniques to isolate and sequence the pool of naturally processed peptides bound by a given class I isotype has provided new insights into the nature of the antigens (Van Bleek et al 1990; Falk et al 1991). Analysis of the self-peptides extracted from different MHC molecules showed that the peptides were typically 8- or 9-amino acids long depending on the particular MHC molecule analyzed, and they contained

Table 1 Allele-specific consensus motifs

Murine class I			Human class I		
	Peptide length	Amino acid position		Peptide length	Amino acid position
		1 2 3 4 5 6 7 8 9			1 2 3 4 5 6 7 8 9
K^b	8	– – Y – F – – L	HLA A2.1	9	– L – – – – – – V
		F Y	HLA B27	9	L
					I
K^d	9	– Y – – – – – – L	HLA B27	9	– R – – – – – – –
		F I			
D^b	9	– – – – N – – – M	HLA Aw68	9–11	– V – – – – – – R
					T K
L^d	9	– P – – – – – – L			
		F			
		M			

Allele-specific consensus motifs determined by sequencing the pool of naturally processed peptides bound by various human (Falk et al 1991; Jardetzky et al 1991; Guo et al 1992), and murine class I molecules (Falk et al 1991; see Rotzschke & Falk 1991; Corr et al 1992). These data suggest that one to three residues in a peptide are critical for high affinity binding to a given MHC class I molecule. Assuming that little restriction exists with regard to the side chains of residues in other positions, it can be calculated that a given class I molecule can accomodate well over one million discrete peptides.

class-I-specific motifs of amino acid residues putatively involved in binding to the MHC molecule (Table 1). For example, HLA-A2 molecules prefer to bind nonapeptides with leucine in position 2 and a valine in position 9, whereas H-2K^b molecules preferentially bind octapeptides with a tyrosine in position 3, tyrosine or phenylalanine in position 5, and leucine in position 8 (Falk et al 1991; Matsumura et al 1992b). A comparison of high resolution crystal structures of class I molecules containing a single peptide species provides much of the explanation for these characteristics (Fremont et al 1992; Silver et al 1992; Zhang et al 1992). Conserved pockets at both ends of the peptide-binding groove were found to accomodate the NH_2 and COOH termini of the peptide, whereas a deep polymorphic pocket in the middle of the groove binds the main anchor residues of a peptide.

ASSEMBLY OF MHC CLASS I MOLECULES IN VITRO

Analysis of the biochemical properties of MHC class I molecules has been severely hampered by the fact that molecules purified from the cell surface of a normal cell already contain peptides. Such complexes appear to be extremely stable and have low affinity for added peptides (Chen & Parham

1989). Initial studies therefore utilized detergent lysates prepared from mutant mammalian cells (producing heavy chains and β2m in the absence of peptide; see below for details) as a source of empty complexes (Schumacher et al 1990; Elliott et al 1991; Townsend et al 1990; Cerundolo et al 1991). In the absence of peptide, complexes of heavy chains and β2m were found to be extremely unstable and readily dissociated, especially when the concentration of the subunits was low (Townsend et al 1989; Ljunggren et al 1990; Elliott et al 1991). Addition of peptide to the lysates, however, led to thermo-stabilization of the complex (Elliott et al 1991; Schumacher et al 1991). The degree of stabilization was dependent on the sequence and length of the peptide (Schumacher et al 1991). However, the presence of detergent in these analyses questions the significance of the findings.

In order to establish a direct method for quantitatively measuring the physical interactions between synthetic peptides and class I molecules, a source of soluble empty class I molecules was required. It was found that soluble empty class I could be purified from the culture media of *Drosophila* cells transfected with cDNAs encoding a truncated heavy chain and β2m (Jackson et al 1992; Matsumura et al 1992b). Using purified soluble empty class I molecules, it was shown that in contrast to similar studies carried out in the presence of detergent (Townsend et al 1990; Elliott et al 1991; Cerundolo et al 1991), the interactions between the heavy chain and β2m and between the heterodimer and peptide are reversible processes that obey the laws of mass action (Matsumura et al 1992b). Thus only the concentration of the three reactants, heavy chain, β2m, and peptide, determine the degree of complex formation. Evaluation of the class I peptide dissociation kinetics in the presence and absence of detergent showed that free heavy chain most

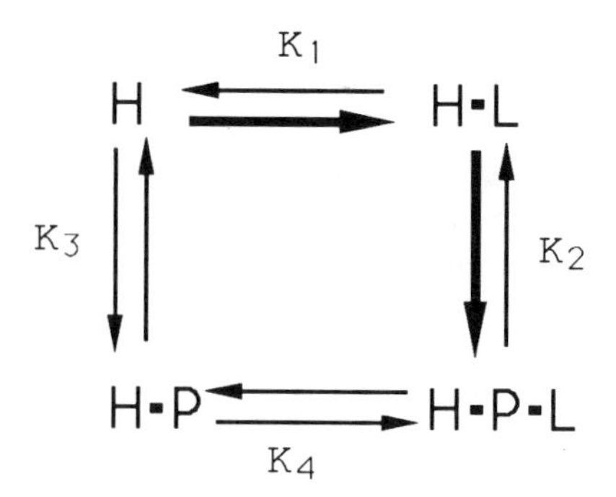

Figure 2 Schematic assembly pathway for class I MHC molecules in vitro. H, L, and P denote the heavy chain, β2m, and peptide, respectively. The thick arrows indicate the proposed major pathways in vitro.

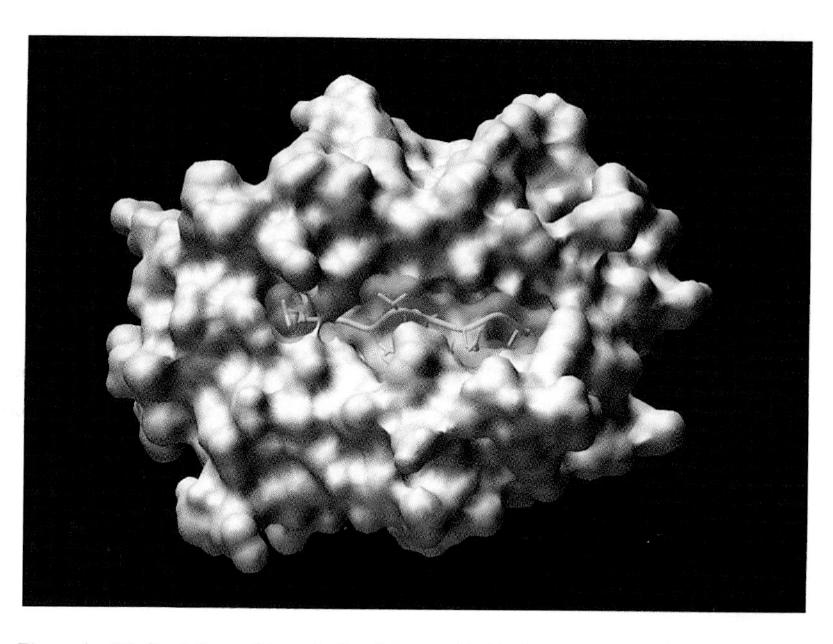

Figure 1 (*A*) Crystallographic analysis of the peptide binding site of H-2K^b. The solvent accessible surface of the VSV-8 peptide complexed with H-2K^b is shown in yellow. Surfaces of the peptide are displayed transparently while surfaces of the $\alpha 1\alpha 2$ domain of the heavy chain are shown solid.

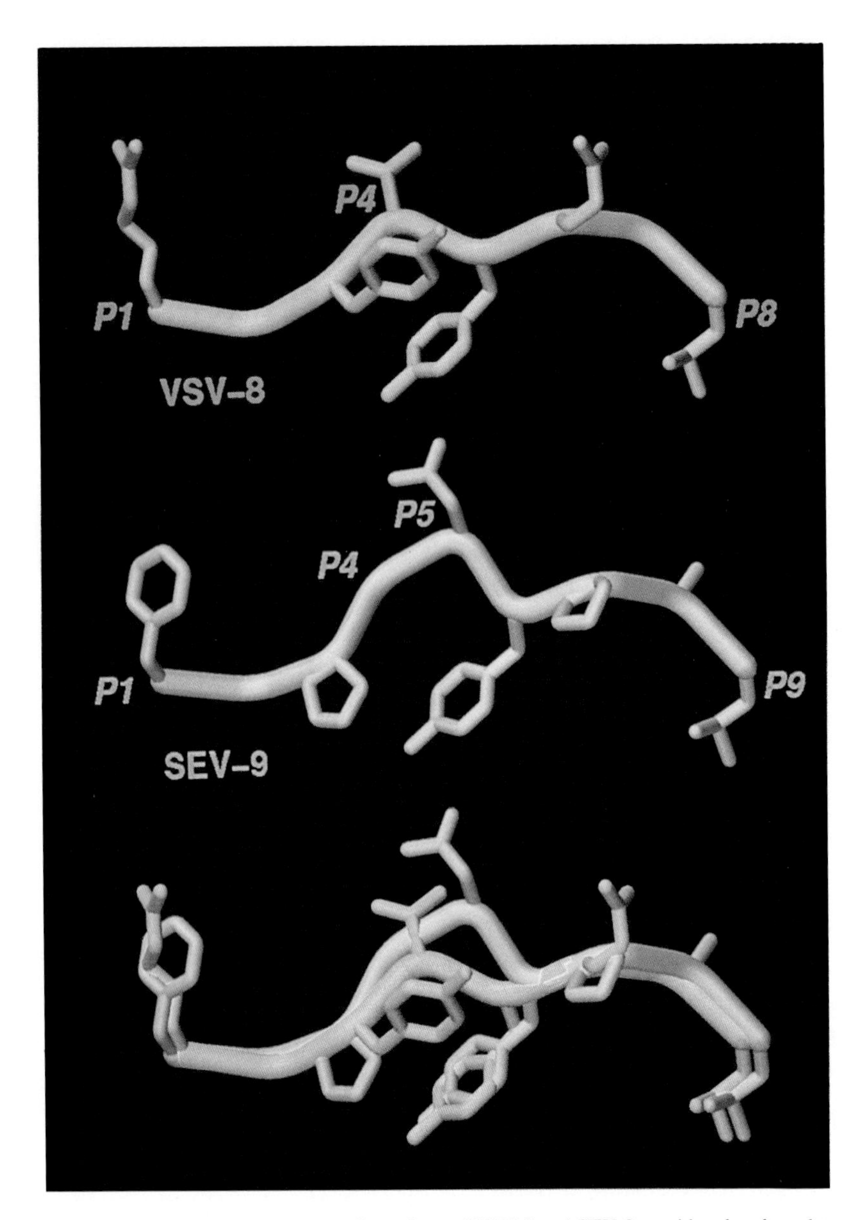

Figure 1 (*B*) Comparison of the conformations of VSV-8 and SEV-9 peptide when bound to H-2K^b. The overlap of the two peptides shows that the NH_2- and COOH-terminal residues have almost identical conformations, while the extra residue in the nonamer is accommodated in a bulge. *Figure 1 A* and *B* have been reproduced with permission from *Science* (Fremont et al 1992).

likely interacts with detergent that prevents its reassociation with β2m. Consequently the initial studies in detergent lysates (Townsend et al 1990) overestimated the role of peptide in the class I assembly process (Elliott 1991). Indeed, only heavy chains already associated with β2m could be demonstrated to bind peptide at the nanomolar level (Elliott et al 1991; Matsumura et al 1992b). One hundred to 1000-fold higher peptide concentrations were found to be necessary for the binding of free heavy chains to peptide (Elliott et al 1991; Matsumura et al 1992b). The assembly pathway for class I peptide complexes in vitro is therefore best described by the sequential binding of heavy chains first to β2m and then to peptide (K1 and K2 in Figure 2). The alternate route where the existence of heavy chain/peptide complexes precedes the binding of β2m (K3 and K4) (Elliott 1991) requires much higher levels of peptide and must therefore be a relatively minor assembly pathway in vitro.

Soluble empty K^b molecules have also been very useful for carrying out affinity measurements of a variety of peptides (Table 2). Using a competitive binding assay and a radioiodinated peptide it was shown (Matsumura et al 1992b) that peptides corresponding to naturally processed epitopes displayed the highest affinities and their K_d values were remarkably similar (2.7–4.1 nM) consistent with observations in detergent lysates (Schumacher et al 1991; Cerundolo et al 1991) and on living cells (Christinck et al 1991). Peptides slightly shorter or longer by just 1 or 2 amino acids had affinities lower by a factor of 2–100, and iodinating the standard peptide resulted in a tenfold reduction in its affinity for K^b molecules. Two residue extensions in the NH_2-terminus of naturally processed octapeptides reduced their affinities by a factor of ~100. By contrast, two residue extensions at the COOH terminus only reduced the affinities by a factor of ~5. These results suggest that precise positioning of the NH_2-terminal residue of a peptide is more important than that of the COOH-terminal residue for high affinity binding (see above).

Interestingly, peptides encoding epitopes that are naturally processed, but presented by other haplotypes, i.e. not restricted to H2-K^b, were in some cases found to be only marginally worse at binding K^b molecules than the radioiodinated standard. Furthermore, analysis of the binding of a library of random octamer peptides to the K^b molecule showed that all peptides could bind. However, their affinities represented a complete continuum from $1\times10^{-1}\times10^{-3}$ M (Y. Saito, unpublished data), which suggests that although haplotype-specific anchor residues may be important for very high affinity binding, their absence from a peptide does not result in a drastic drop in affinity and does not necessarily preclude a peptide from interacting with a class I molecule.

Table 2 Peptide dissociation constants (K_d) from H2-K^b

VSV-8	RGYVYQGL	3.7×10^{-9}
VSV-7	GYVYQGL	5.3×10^{-8}
VSV-9N	LRGYVYQGL	7.3×10^{-9}
VSV-10N	DLRGYVYQGL	3.9×10^{-7}
VSV-9C	RGYVYQGLK	6.9×10^{-9}
VSV-10C	RGYVYQGLKS	2.1×10^{-8}
^{125}IVSV-8	RG**Y**V**Y**QGL	3.3×10^{-8}
OVA-8	SIINFEKL	4.1×10^{-9}
OVA-9N	LSIINFEKL	8.9×10^{-8}
OVA-10N	LESIINFEKL	2.8×10^{-7}
OVA-9C	SIINFEKLT	4.1×10^{-9}
OVA-10C	SIINFEKLTE	4.1×10^{-9}
OVA-24	EQLESIINFEKLTEWTSSNVMEEER	7.1×10^{-5}
SEV-9	FAPGNYPAL	2.7×10^{-9}
INF NP	ASNENMETM	2.3×10^{-5}
P1A	KYQAVTTTL	1.1×10^{-6}
Pol (HIV)	ILKEPVHGV	5.3×10^{-6}
Histone H3	RRYQKSTEL	5.2×10^{-7}

The dissociation constant for various peptides for H2-K^b as determined using a competitive binding assay (Matsumura et al 1992b). Peptides encode the following peptide epitopes: VSV-8, vesicular stomatitis virus nucleocapsid protein (residues 52–59) (Van Bleek & Nathenson 1990); OVA-8, ovalbumin (residues 257–264) (Carbone & Bevan 1989); SEV-9, Sendai virus nucleoprotein (residues 324–332) (Schumacher et al 1991); INF NP, H2-D^b restricted influenza viral nucleoprotein (residues 366–374), (Rotzschke & Falk 1991); P1A, H2-K^d restricted peptide, which is a P815 tumor antigen (Sibille et al 1990); Pol (HIV), HLA A2.1 restricted HIV reverse transcriptase (residues 461–485) (Walker et al 1989); and HLA-B27 restricted peptide derived from human histone H3 protein (Jardetzky et al 1991).

ASSEMBLY AND TRANSPORT OF MHC CLASS I MOLECULES IN CULTURED CELLS

Overview

MHC class I heavy chains and β2microglobulin are synthesized with signal sequences, directing the nascent polypeptides to the ER. After translocation, the signal sequences are cleaved and the two subunits fold and subsequently associate. These events are restricted to the lumen of the ER. Details of the location of peptide generation are more sketchy. However, it is now generally accepted that peptide generation occurs primarily in the cytoplasm and that peptides are subsequently transported into the ER where they associate with the class I heavy chain and β2m. The fully assembled class I molecule is then transported to the cell surface, via the exocytic pathway.

It is this trimolecular complex displayed on the cell surface that is the functional molecule recognized by CD8+ T cells. The requirement for transport of the newly assembled class I/peptide complexes for presentation of antigen is nicely demonstrated by the ability of the fungal metabolite, brefeldin A, to totally inhibit this process (Nuchtern et al 1989; Yewdell & Bennick 1989). As with many multi-subunit proteins, the assembly and intracellular transport of class I molecules seems tightly regulated (Hurtley & Helenius 1989) with the apparent overall goal of maximizing the number of peptide-containing class I complexes displayed on the cell surface.

How and Where are Antigenic Peptides Generated?

The evidence suggesting that the cytoplasm is the source of peptides for loading onto class I molecules is substantial, and circumstantial evidence indicates that it is primarily nascent polypeptides that serve as the main source of antigenic peptides; for example, cyclohexamide efficiently blocks the loading of naturally processed peptides onto class I molecules (E. Song & P. Peterson, unpublished observation). However, the demonstration that exogenous proteins introduced into the cytoplasm can enter the class I processing pathway (Moore et al 1988) shows that nascent polypeptides are not the exclusive source of antigen.

The discovery that two subunits, LMP2 and LMP7 in mouse (Martinez & Monaco 1991; Brown et al 1991) and RING12 and RING10 in humans (Glynne et al 1991; Kelly et al 1991), of the multi-catalytic protein protease or proteasome are encoded by genes located in the MHC region, and in tandem with the *TAP1* and *TAP2* genes (see below), has led to much speculation that the proteasome generates the peptides for class I molecules.

The proteasome is abundant, ubiquitous, and located in the cytoplasm as well as in the nucleus (Orlowski 1990). In addition, this peptide generator has multiple protease specificities that allow cleavage at almost any peptide bond of the substrate, thereby generating a great diversity of peptides. Unlike other proteases, its affinity for its substrate appears to be affected by peptide length (Orlowski 1990). The end products of proteasome-mediated digestion are typically short peptides of different lengths (Dick et al 1991). In vitro studies with purified proteasome have shown that it is capable of generating peptides from whole denatured antigen that can be bound by class I molecules and elicit a T cell response (J. Waters & P. Peterson, unpublished data). More direct evidence for involvement of the proteasome in the generation of the peptides loaded onto class I has been difficult to obtain.

The proteasome subunits encoded in the MHC region are not essential for the generation of antigenic peptides since a cell line deficient in these subunits presents peptides to T cells (Arnold et al 1992; Momburg et al

1992). However, the MHC-encoded subunits of the proteasome may turn out to facilitate the proteasome's processing of antigen. For example, they are induced by INFγ (Yang et al 1992b), a treatment that increases the peptide supply to class I molecules. Perhaps they alter the proteolytic activity of the proteasome so that it produces the octa- and nonapeptides that class I molecules prefer instead of the shorter peptides that it normally creates. Alternatively, they could couple the proteasome directly to the peptide transporter and thereby facilitate their passage across the ER membrane to the class I molecules in the ER lumen (Falk et al 1990).

While the details of the pathway of generation and delivery of peptides to assembling class I molecules remain to be worked out, in the absence of contrary evidence it seems likely that the proteasome generates the majority of peptides that are subsequently loaded onto class I molecules.

Putative Peptide Transporters

The identification of two genes, *TAP1* and *TAP2*, located in the MHC region that share substantial homologies with ATP-dependent transporter proteins led to the immediate speculation that their gene products may be responsible for transporting peptides from the cytoplasm to the lumen of the ER (Monaco et al 1990; Spies et al 1990; Trowsdale et al 1990). These genes were previously refered to as *HAM1* and *HAM2* in mouse (Monaco et al 1990), *MTP1* and *MTP2* in rats (Deverson et al 1990), and *PSF1* and *PSF2* (Spies et al 1990) or *RING4* and *RING11* (Trowsdale et al 1990; S.H. Powis et al 1992) in humans. They encode two INFγ-inducible proteins of approximately 75 kd, each thought to consist of six hydrophobic segments that span the membrane, followed by a large cytoplasmically located domain that contains the ATP-binding cassette. Since there is evidence that other members of this protein family (Higgins et al 1990) such as oppD/F and STE6 can transport short peptide sequences, the role of TAP1 and TAP2 as antigen transporters seems quite plausible.

Many of the class I assembly-deficient mutant cell lines result from mutations in either one (Spies & DeMars 1991; Powis et al 1991; Attaya et al 1992) or both of the putative transporter genes. For example, analysis of the *TAP2* gene in the RMA/S mutant showed that a premature stop codon has been introduced by a point mutation early in the coding sequence (Yang et al 1992a). Transfection of RMA/S with murine or rat *TAP2* cDNA restored its antigen presentation and class I assembly capabilities (Powis et al 1991; Attaya et al 1992). The T2 mutant (Salter et al 1985), which has a 1 megabase pair chromosomal deletion in the MHC class II region encompassing the *TAP1* and *TAP2* genes, required co-transfection of cDNAs encoding TAP1 and TAP2 (Arnold et al 1992; Momburg et al 1992) to restore its antigen presentation capacity. These results showed that, like other members of this family of proteins, the functional transporter consists of either a heterodimer or at least some combination of TAP1 and TAP2 proteins. Immunoprecipitation data indicate that TAP1 and TAP2 associate (Spies et al 1992); however, in normal cells the degree of association is surprisingly low (Y. Yang & P. Peterson, unpublished observation), which raises the possibility that the transport process may be regulated by dimerization of these two proteins.

The proposed role of TAP1 and TAP 2 in transporting peptide from the cytoplasm to the ER has been challenged by the finding that short peptides can enter microsomal vesicles in vitro in the absence of ATP and that this occurs in microsomes obtained from the T2 mutant cell line (Levy et al 1991). It is unclear whether this in vitro assay faithfully reflects in vivo conditions, especially because the peptide used was 11-amino acids long and biotinylated. Furthermore, the functional arguments in favor of a role for TAP1 and 2 in peptide transport are very compelling. First, functional loss of either transporter chain by mutation causes an arrest in class I assembly and antigen presentation (Powis et al 1991; Spies & DeMars 1991; Kelly et al 1992; Spies et al 1992; Attaya et al 1992). Second, although such mutants are virtually unable to load peptides generated in the cytosol either by infection or transfection, the defect can be bypassed if a known peptide epitope is provided at the cell surface (Townsend et al 1989) or is introduced into the ER lumen by transfection with an added leader sequence that can target it independently to the lumen of the ER by the signal sequence translocation pathway (Anderson et al 1991). Thirdly, a polymorphism in the rat *TAP2* genes appears to determine the spectrum of peptides loaded by a class I molecule and thus presentation of antigen (S. J. Powis et al 1992), thereby establishing a firm relationship between direct interaction between the *TAP* gene products and the peptides loaded onto class I molecules. However, it is important to remember that there is no direct evidence that the *TAP* gene products actually transport the peptide across the ER membrane; they could, for example, be involved in chaperonin-like peptide binding during the delivery process.

Folding and Assembly of Class I Molecules In Vivo

Analysis of the folding and assembly and transport of class I molecules has centered around the use of monoclonal antibodies that in most instances recognize discontinuous determinants on the class I molecule. Even though the exact residues involved in these determinants are rarely known, these antibodies provide some measure of the conformation of the class I molecule. Antibodies whose binding is influenced by polymorphic residues on the $\alpha1$ and $\alpha2$ domains are typically the most sensitive measure that all three domains ($\alpha1$, $\alpha2$, and $\alpha3$) of the heavy chain are correctly folded. However, in some cases the antibody affinity may be strongly influenced by the sequence of the peptide bound or the source of the β2m subunit (Bluestone et al 1992; Catipović et al 1992; Jackson et al 1992). Heavy chains folded in the absence of the β2m subunit are often only partially folded (α1-α2-, α3+) and consequently are not bound by these so called β2m-dependent antibodies; however, they can, in some instances, be identified by monoclonal antibodies that recognize the α3 domain of class I molecules.

Rabbit antisera against either denatured class I molecules (Krangel et al 1979; Kissonerghis et al 1980; Sege et al 1981) or to synthetic peptides corresponding to portions of the cytoplasmic tails of the heavy chain (Smith et al 1986) have been used to detect unfolded class I molecules.

The Role of β2m in Transport and Assembly

Early events in the folding and assembly of class I molecules were first analyzed using a monoclonal antibody W6/32 (which recognizes a determinant expressed on most human class I molecules but only when associated with β2m) and a rabbit serum that reacts only with denatured heavy chains (Krangel et al 1979). With these two antibodies, it was found that the halftime for the conversion of the unfolded to the folded β2m-associated form of the heavy chain was about 5 min. The N-linked oligosaccharides on the W6/32-reactive class I molecules became endo H-resistant with a halftime of approximately 40 min, while endo H-resistant free heavy chains could not be detected and were most likely degraded in the ER. The W6/32-reactive class I molecules were found to be relatively resistant to proteolysis, while the free heavy chains were easily digested. The role of β2m in promoting transport and a conformational alteration of the heavy chain was clearly demonstrated by the finding that in the human Daudi cell line, which is deficient in β2m (Arce-Gomez et al 1978), heavy chains did not react with the W6/32 antibody, and their N-linked oligosaccharides did not become resistant to endo H, which indicated that they were not transported out of the ER (Kissonerghis et al 1980). Similarly in mouse cell lines deficient in β2m, e.g. R1E, it was reported that little or no transport of some heavy chains was detected (Allen et al 1986), while for others, e.g. H2-D^b or L^d, transport occurs, but at much lower efficiency, and the surface-expressed molecules are not folded properly (Allen et al 1986; Williams et al 1989; Hansen et al 1988).

These data have been challenged recently by the findings that functional H2-D^b molecules are transported, albeit at low levels, to the surface of T cells isolated from β2m-deficient mice and that with highly sensitive staining procedures, very low levels of properly folded D^b molecules can be detected on the surface of transfected R1E cells (Bix & Raulet 1992). These data show that the binding of peptide to the heavy chain and transport of this complex to the cell surface can occur in vivo in the absence of β2m; however, the low levels of heavy chain expressed on the cell surface in these cells indicate that in normal cells such a means of peptide presentation is most likely a minor pathway.

The lack of transport of heavy chains in the absence of β2m undoubtably reflects the failure of such malfolded heavy chains to negotiate the quality control machinery of the ER. Retention of misfolded or unassembled membrane and secretory proteins in the ER is now a widely recognized phenomenon (Hurtley & Helenius 1989; Rose & Doms 1988). In the case of class I heavy chains, retention in the ER in the absence of β2m is conserved from *Xenopus* oocytes (Severinsson & Peterson 1984) to humans (Sege et al 1981), most likely indicating that it is based on recognition of general features of misfolded proteins, e.g. exposed hydrophobic domains rather than the recognition of any class I-specific signals (Gething & Sambrook 1992). The leakage of low levels of heavy chain to the cell surface in β2m-deficient cells is presumably a result of transient acquistion of a properly folded form of the heavy chain rather than an indication of poor quality control (Hurtley & Helenius 1989).

The Role of Peptide in Transport and Assembly

In the past few years our understanding of the assembly of the peptide/class I complex has been revolutionized by the characterization of mutant cell lines that synthesize normal amounts of class I heavy chains and β2m, but are abnormal in their ability to assemble class I molecules and present antigen. Two cell lines mentioned previously, the murine T lymphoma, RMA/S (Ljunggren et al 1989), and the human B lymphoma, T2 (Salter et al 1985), which efficiently present synthetic peptide antigens, but not (or at least only at low levels) biosynthetic antigens to CD8 T cells, have been most informative (Townsend et al 1989; Cerundolo et al 1990).

The first indications that it was peptide deficiency that resulted in the phenotype of the above assembly-deficient mutants came from work by Townsend and colleagues on the RMA/S cell line (Townsend et al 1989). Incubating RMA/S cells in the presence of peptides known to bind to the class I molecules increased the surface expression of properly folded class I molecules. But there was also an enhancement in the association of endo H-sensitive heavy chains with β2m, suggesting that the increased surface expression might be the result of some of this peptide reaching the ER via an uncharacterized retrograde transport route (Townsend et al 1989). Surface class I expression on RMA/S cells was subsequently found to be induced by simply lowering the incubation temperature, or by adding anti-class I antibodies to the culture medium (Ljunggren et al 1990; Ortiz-Navarrete 1991). These cold-induced, surface-expressed molecules behaved like no other class I molecule previously characterized. They had all the properties appropriate for unstable, peptide-free dimers of heavy chain and β2m. Ljunggren et al (1990) called these class I molecules empty. Manipulations that led to an increase in the class I expression at the surface of RMA/S cells could now be interpreted as simply the result of stabilizing the heavy chain/β2m heterodimer that had already reached the cell surface.

Biosynthetic pulse chase labeling experiments in the RMA/S mutant indicated that heavy chains and β2m associated at very low levels, if at all, in these cells. However, these findings were contradicted by cross-linking studies (Y. Yang & P. Peterson, unpublished data) that showed that the association of the class I heavy chain and β2m was as rapid in RMA/S as in normal cell lines. The inability to identify these empty class I molecules without cross-linking stems from their instability in detergent lysates, especially when the concentrations of heavy chain and β2m are low. Under these conditions, the empty class I molecules easily dissociate, but can be stabilized by very low concentrations of peptide, or driven to reassociate with higher concentrations of peptide. Indeed, the initial observation that exposing RMA/S cells to peptide results in stabilization of endo H-sensitive heavy chains/β2m complexes is now believed to result from post-lysis stabilization by low levels of peptide (Elliott et al 1991).

From these analyses it may be concluded that peptide is not essential for the association of class I heavy chains with β2m, but in the absence of peptide, such complexes are unstable. At body temperature, the empty class I molecules are, however, stable enough that some remain associated sufficiently long to be transported past the quality control machinery of the ER (Hurtley & Helenius 1989). These escapees most likely continually disassemble en route to and at the cell surface such that only very low levels accumulate at the plasma membrane unless peptide is provided in the cell medium. The observation that reduced temperature and peptide act synergistically in inducing surface expression of class I molecules in RMA/S (Ljunggren et al 1990) indicates that they play different roles. Whereas peptides are more potent in stabilizing surface expressed molecules, reduced temperature actually increases the number of empty molecules that successfully navigate the exocytic pathway. Thus surface-expressed class I molecules induced by reduced temperature are still fragile and readily disintegrate if the cells are returned to 37°C (Ljunggren et al 1990).

Initial characterization of the mutant cell line T2 (Salter et al 1985), which displays a complete loss of surface expression of HLA B5 and an 80% reduction in the expression of HLA-A2 (Salter & Cresswell 1986), suggested that it may have a similar defect as RMA/S (Cerundolo et al 1990). Incubation of T2 cells with peptide increased the surface expression of HLA-A2 (Baas et al 1992). This was not found for other HLA isotypes, e.g. B7, Bw58, or B5 (Alexander et al 1989). Furthermore, in contrast to RMA/S, reduced temperature did not increase the amount of class I molecules reaching the cell surface. However, pulse-chase experiments indicated that heavy chain/β2m complexes formed in the ER reacted with conformation-specific monoclonal antibodies (Baas et al, 1992). Such complexes were empty because they were temperature-sensitive and could be stabilized by addition of the appropriate peptides. Curiously, while HLA molecules were in general maintained in the ER of T2 cells, transfection of T2 with H2 genes showed that murine molecules were transported, albeit poorly, to the cell surface (Alexander et al 1989; Baas et al 1992).

Based on these observations, T2 cells seem to more stringently control the exit of properly folded, but empty, class I molecules from the ER than do RMA/S. The reason why HLA-A2 molecules managed, at least in part, to evade this control system became clear when it was shown (Wei & Cresswell 1992; Henderson et al 1992) that at the surface of T2 cells, these molecules contained peptides derived from used signal peptides, cleaved-off of proteins translocated into the ER lumen. Most interestingly, two of the peptides identified were 11- and 13-amino acids long, which contrasts with the previous finding that peptides associated with HLA-A2 in normal cells are 9-amino acids long. The higher affinity of HLA-A2 for nonamers presumably results in an insignificant representation of these longer signal sequence peptides in the repertoire of peptides bound to HLA-A2 in a normal cell. These findings do, however, suggest that in normal cells antigen may already be processed into peptides of typically 8 or 9 amino acids prior to delivery into the ER, and question the existence of proposed ER-located proteases that might trim longer peptides (Falk et al 1990; Eisenlohr et al 1992).

The existence of empty class I molecules is not restricted to mutant cell lines. In normal cells, thermo-sensitive, class I heavy chain/β2m dimers, which can be stabilized by addition of peptide, can be identified at early pulse-chase times, which indicates that such complexes are indeed intermediates in the biogenesis of functional class I molecules (E. Song & P. Peterson, unpublished data). In addition, incubation of many cell lines in appropriate peptides has been shown to result in increased surface expression of functional class I molecules (Lie et al 1990; Ortiz-Navarrete et al 1991; Benjamin et al 1991; Smith et al 1992). Whether these molecules reach the surface empty or containing low affinity peptides, which are subsequently displaced, is not clear. In most cell types it does seem that the number of assembling class I molecules is in excess over high affinity peptides (Lie et al 1990; Ortiz-Navarrete et al 1991). This might well result in the formation of complexes containing relatively low affinity peptides, which might subsequently be transported to the cell surface. Such complexes are probably short-lived at the cell surface, although they might survive long enough to be recognized by T cells. This could be one way a cell could extend the repertoire of peptides it is able to present to T cells.

Accessory Molecules Controlling the Transport of Class I Molecules

In a normal cell it has been shown that different class I molecules sharing up to 80% sequence identity are exported from the ER at different rates (Williams et al 1985; Degen & Williams 1991). In general, these differences could not be attributed to differences in the rates of folding of the class I heavy chain or its association with β2m, although weak association with β2m might be important for some isotypes (Beck et al 1986). Cross-linking studies showed that newly synthesized class I heavy chains associated rapidly and quantitatively with an 88-kd protein (Degen & Williams 1991). The class I-p88 complex was shown to exist transiently, and the length of the association correlated directly with the rate of ER to Golgi transport of each heavy chain. Based on immunological criteria, p88 has been identified as calnexin or p90 (Ahluwalia et al 1992; Galvin et al 1992), a molecular chaperone known to bind transiently to a variety of nascent proteins in the lumen of the ER, e.g. subunits of the T cell receptor (TCR) and MHC class II molecules (Hochstenbach et al 1992). The observation that dissociation of p88 from the heavy chain was not triggered by binding of β2m (Degen & Williams 1991) and that p88 remains associated with intracellular heavy chains throughout their life time in β2m-deficient cells suggests that p88 retains class I molecules in the ER until formation of the trimolecular complex (Degen et al 1992), a suggestion that is further supported by the observation that in RMA/S cells there is a prolonged association between p88 and empty class I molecules (Degen et al 1992).

Further evidence for the involvement of accessory molecules in the assembly of class I has come from the observation that immediately after synthesis the cytoplasmic tail of the heavy chain is unreactive to antibodies. Interestingly, while this is the case for the tail of wild-type K^b molecules, the cytoplasmic tail of a mutant K^b heavy chain with two point mutations in the α2 domain is fully reactive throughout the pulse/chase experiment (E. Song & P. Peterson, unpublished observation). The molecular basis for these observations is unknown; however, they could be the result of steric effects from the juxtapositioning of the calnexin cytoplasmic tail. Recent data (see Ahluwalia et al 1992) indicate that it is the transmembrane domain plus a few flanking residues of class I molecules that interact with calnexin because, for example, an interaction between calnexin and a soluble, GPI anchored class I molecule, which lacks only the cytoplasmic and the transmembrane domains, could not be detected.

Analysis of the sub-cellular distribution of unassembled class I molecules in a mutant murine cell line CMT 64.5 (which expresses little or no surface class I molecules unless induced with INFγ) (Klar & Hammerling 1989) suggested that although their steady-state distribution is primarily the ER, they may recirculate between the ER and the Golgi (Hsu et al 1991). In particular, at 16°C, the class I molecules were observed to redistribute from the ER to Golgi-like structures and, upon warming up to 37°C, they apparently returned to the ER. However, it is not clear whether the so-called unassembled molecules identified in these experiments were empty, properly folded class I/β2m dimers, or partially folded heavy chains, or a combination of both. Given the thermo-sensitive nature of empty class I molecules (Ljunggren et al 1990), such temperature shift experiments must be interpreted with caution.

The suggestion that empty class I molecules are maintained in the ER by retrieval from post-ER compartments is tempting given that the sequence of the cytoplasmic tail of canine and human calnexin (Wada et al 1991; Galvin et al 1992) reveals an ER targeting motif (Jackson et al 1990) that has been shown to be capable of directing the retrieval of marker proteins back to the ER from the Golgi complex (Jackson et al 1993).

Although all the data point to calnexin as being the protein responsible for maintaining empty class I molecules in the ER, a molecular explanation as to why different class I molecules are transported with such disparate rates remains unclear. If we assume that release of the class I heavy chain from calnexin is triggered by attaining a specific conformation that is stabilized by binding peptide, then these transport rates might reflect the availability of high affinity peptides that can bind to a specific heavy chain. Alternatively, calnexin may have different affinities for each class I heavy chain. Interspecies differences in calnexin might then explain the observed variations in the ability of human cells to effectively maintain empty human and murine class I molecules in the ER (Alexander et al 1989).

Mutations in the heavy chains of class I molecules, as simple as a single amino acid substitution, have been shown to result in significant reductions in their transport rates (Miyazaki et al 1986; Williams et al 1988). In most instances the mutant heavy chain assumed an altered conformation, in agreement with the essential requirement for a properly folded molecule to attain a transport-competent state (Rose & Doms 1988).

Where is Peptide Loaded onto Class I Molecules In Vivo?

All the available evidence points to the ER as the site where peptide is bound by class I molecules; however, this is not formally proven. Biosynthetic pulse-chase analyses show that soon after synthesis a significant proportion of the class I molecules are associated with β2m. However, such complexes are thermo-sensitive, which indicates that they are empty or contain poorly binding peptides (E. Song & P. Peterson, unpublished data). With time, and before the N-linked mannose on the heavy chain are trimmed, thermo-stable class I molecules are observed. These data point to a loading station before the protein reaches the medial Golgi. As supporting data, class I molecules can be maintained in the ER either by association with the adenoviral E3/E19 protein (Andersson et al 1985), or if the cytoplasmic tail of the class I molecule is replaced with that from the E19 protein (Jackson et al 1993). In these cases, the class I molecules become thermo-stable, which indicates that they acquire peptide (E. Song & P. Peterson, unpublished data). However, retention of such complexes in the ER is via retrieval from post-ER compartments (Jackson et al 1993), so we cannot exclude early Golgi compartments as loading sites. Indeed, the observation that empty class I molecules can be found in Golgi compartments has led to much speculation that it is here, rather than in the ER, that peptides are loaded (Hsu et al 1991; Baas et al 1992; Kleijmeer et al 1992). However, immunolocalization of the putative peptide transporters shows that most are localized in the ER membrane (Kleijmeer et al 1992; Yang et al 1992a), while only low levels are found in early Golgi compartments. Given the much larger volume of the ER than the early Golgi compartments, the ER is most likely to be the primary site for peptide loading onto class I molecules.

Loading of Peptide onto Class I Molecules at the Cell Surface

In vivo, the binding of peptide at the cell surface is likely to be of minor importance in antigen presentation by class I molecules. Indeed, it has long been recognized that cell surface class I molecules are extremely poor binders of extracellularly provided peptides (see Chen & Parham 1989), a feature that presumably reflects the need for an individual cell to maintain its own immunological identity in its class I display in order to avoid being killed as an innocent bystander. Nonetheless, peptide binding to the surface of living cells has attracted a great deal of attention and generated much controversy (Townsend et al 1989; Christinck et al 1991; Kozlowski et al 1991; Lie et al 1990; Luescher et al 1991; Rock et al 1990, 1991a; Vitello et al 1990; Sherman et al 1992).

Basically two pathways for peptide binding to class I molecules have been proposed. Either they bind to empty class I molecules and/or they are involved in a peptide exchange reaction with already assembled class I/peptide complexes. It should be realized that peptide binding to class I molecules at the cell surface is likely to follow similar rules to those determined for soluble class I molecules in vitro (see above). Thus it is important to consider the length and concentration of the peptides used, whether the peptide is modified or unlabeled, the source and concentration of the β2m, and the haplotype of the heavy chain. Many of the experiments described in the literature have unfortunately used peptides of incorrect length, which we now know to have relatively low affinity (see Table 2) for the class I molecules in combination with xenogeneic β2m (Christinck et al 1991; Luescher et al 1991; Kozlowski et al 1991; Rock et al 1990, 1991a). In the case of murine molecules, it has been concluded that β2m is essential for both the loading of peptide onto empty class I molecules and in exchange reactions. However, such experiments must be interpreted with caution, because these findings could reflect either the need for proteases in the fetal calf serum to trim the peptide to the correct length (Sherman et al 1992) and/or the requirement for a higher affinity β2m (e.g. bovine or human) (Bernabeu et al 1984; Hochman et al 1988) to stabilize the low affinity of the heavy chain for peptides that are too long.

Similar caution must also be applied to interpretations of experiments analyzing peptide exchange or peptide binding where the peptide has been labeled, e.g. biotinylated or iodinated, in light of the observation that such peptides may have a significantly reduced affinity for the class I molecule (Matsumura et al 1992b). Thus peptide exchange, which appears to occur in the absence of β2m exchange (Smith et al 1992; Matsumura et al 1992b) is most likely important if the class I molecule expressed at the cell surface complexed with relatively low affinity peptides (Ortiz-Navarrete & Hammerling 1991; Benjamin et al 1991).

The relatively low levels of β2m in culture media (relative to the levels in exocytic pathway) most likely result in little reassembly at the cell surface of class I molecules once the heavy chain and β2m have dissociated (Matsumura et al 1992b). Indeed, high levels of denatured nonfunctional heavy chains can be detected at the cell surface of normal (cultured) cells (Rock et al 1991b, Ortiz-Navarrete & Hammerling 1991; Madrigal et al 1991). They are presumably spent heavy chains destined for degradation, which may be part of the wasteful but necessary cycle of events needed to continually update the class I display.

MODEL SYSTEMS TO STUDY CLASS I ASSEMBLY AND TRANSPORT

Our understanding of MHC class I presentation may now be sufficiently detailed that it is worthwhile to reconstitute the complete presentation pathway in a model system. The ideal cell for this type of analysis should have all of the essential components necessary for folding, translocating, glycosylating, assembling and transporting of proteins, but be devoid of

MHC molecules and MHC-dedicated accessory proteins required in the presentation pathway. Attempts to express MHC class I molecules in heterologous systems have, however, met with mixed success. For example, expression of MHC in bacteria resulted in little assembly of the heterodimer (Garboczi et al 1992; Parker et al 1992). Perhaps the most promising host cell for these types of studies are insect cells (Jackson et al 1992), which do not have an MHC-based immune sytem.

Transfection of *Drosophila* Schneider cells with a variety of class I heavy chains and β2m showed that both human and murine class I molecules were readily assembled and transported to the cell surface (Jackson et al 1992). However, the assembled class I molecules had all the properties of empty class I molecules, which indicated that these cells were unable to load peptide intracellularly (note that insect cells are cultured at 27°C). The surface-expressed class I molecules loaded extracellularly with peptide were fully functional in terms of presentation to T cells; however, as expected, the *Drosophila* cells were incapable of presenting antigen derived from an intracellular source to T cells (Jackson et al 1992). Thus it seems that specialized proteins are not required for the assembly and transport of empty MHC class I molecules. However, insect cells are missing certain MHC-dedicated accessory molecules, for example, TAP1 and TAP2, LMP2 and LMP7, which are presumably required to provide antigenic peptides to the assembling heterodimers.

The relative transport rates of various murine class I molecules in these insect cells were distinctly different from those previously determined in either normal mammalian cells or in the peptide loading-deficient RMA/S or T2 mammalian cells. Thus for example the $t_{1/2}$ transport rates of D^b, K^b, and L^d were 15, 25, and 60 min, respectively, in insect cells, whereas in normal mammalian cells (Degen & Williams 1991), they are 55, 20, 55 min, respectively. What appeared to primarily account for the differences in transport rate in the insect cells were differences in the thermostability of the empty complexes (M. Jackson & P. Peterson, unpublished data; Jackson et al 1992). A direct relationship between thermostability of the empty complexes and transport was observed. Co-expression of murine class I molecules with human β2m (murine heavy chains have a higher affinity for human β2m) (Hochman et al 1988) resulted in complexes that were more thermo-stable and more rapidly transported (M. Jackson & P. Peterson, unpublished data). These results suggest that in mammalian cells specific accessory molecules (e.g. calnexin), rather than the inherent properties of the class I molecules, control the transport of class I molecules out of the ER.

In contrast to mammalian cells (Allen et al 1986; Williams et al 1989), all murine and human heavy chains expressed in *Drosophila* cells in the absence of β2m were found to be transported to the cell surface where they could reassociate with β2m and/or peptide (Jackson et al 1992). Although the relative transport rates of the free murine heavy chains were much slower and the molecules more susceptible to degradation than when they were co-expressed with β2m, the relative rank order of transport was maintained. These results suggest that the retention and degradation of free heavy chains in the ER of mammalian cells is primarily carried out by non-MHC dedicated quality control machinery. However, the finding that this process is more complete in mammalian cells than in *Drosophila* cells suggests that mammalian cells may exert a second level of control to prevent free heavy chains from leaving the ER.

Co-transfection of murine heavy chains and β2m along with a cDNA encoding calnexin resulted in drastic changes in the transport rates (M. Jackson & D. Williams, unpublished observation), such that they became similar to those observed in the peptide-deficient cell line RMA/S. Furthermore, transport of free heavy chains in insect cells co-transfected with calnexin was undetectable. These data provide clear evidence for the role of calnexin in retaining both free class I heavy chains (Degen et al 1992) and those associated with β2m (Degen & Williams 1991) in the ER. As heavy chains and β2m efficiently assembled in the absence of calnexin in insect cells, the role of calnexin is implicated in retaining the empty molecules until a peptide is bound rather than in maintaining the conformation of the heavy chain (Degen et al 1992). However, we do not know if the association with calnexin is essential in the loading of peptide in vivo or whether it only plays a role in improving the efficiency of this process.

In order to reconstitute the full class I presentation pathway, *Drosophila* cells have been transfected with cDNAs encoding murine TAP1 and TAP2 along with the MHC class I subunits. Preliminary data suggest that the expression of TAP1 and TAP2 is not sufficient to result in peptide addition to the assembled class I molecules (M. Jackson et al, unpublished data). Both TAP1 and TAP2 are, however, expressed at high levels in the insect cells, which suggests that additional components, e.g. LMP2 and LMP7, calnexin, and or other as yet unidentified components, are required to fully reconstitute this system.

SCHEMATIC REPRESENTATION OF THE ASSEMBLY AND TRANSPORT OF MHC CLASS I MOLECULES

The basic MHC class I assembly pathway as presently understood can be described as follows (Figure 3): The heavy chain and β2m are targeted to the ER via signal sequences. Once cleaved, the molecules fold and associate

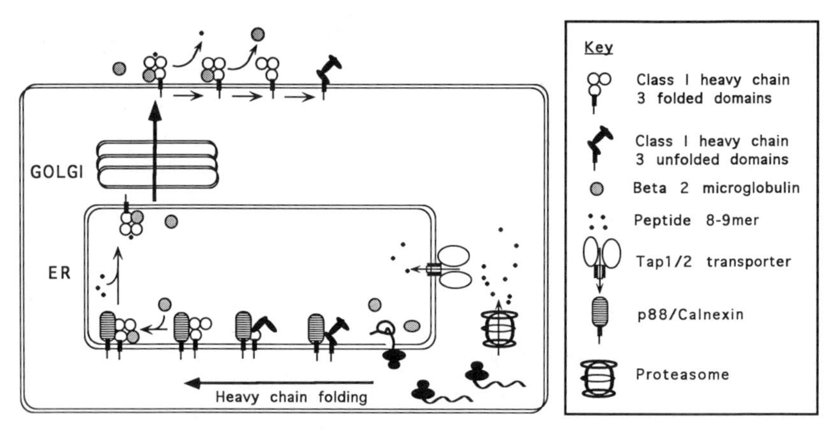

Figure 3 Schematic representation of MHC class I assembly and transport in vivo.

with one another. In the absence of peptide, the weakly associated heterodimer is in equilibrium with partially folded free heavy chains and a pool of free correctly folded β2m. Both partially folded free heavy chains and the majority of the heterodimer are maintained in the ER by association with calnexin. Denatured heavy chains may be retained in the ER by other chaperonins, and if they cannot be folded into productive molecules, they are targeted for degradation. Peptides, generated primarily from nascent polypeptides by the proteasome system in the cytoplasm, are supplied to the assembling class I molecules in the ER via a TAP1 TAP2 dimer. Release of the heavy chain from calnexin requires it to attain a specific transport-competent conformation, which is highly favored by the presence of peptide and β2m, but not absolutely dependent upon their presence. The calnexin-binding and the transport-competent conformations of the heavy chain are in equilibrium, and this equilibrium is most likely different for each allele. Under normal conditions, release of the heavy chain from calnexin is typically achieved by peptide addition; however, empty class I molecules, which transiently attain the transport-competent conformation, are able to escape and be transported along the exocytotic pathway to the cell surface. Residency at the cell surface depends on peptide occupancy; once the peptide is lost the heavy chain rapidly denatures into a nonfunctional form.

REGULATION OF CLASS I PRESENTATION IN VIVO

Many of the components involved in the class I presentation pathway, e.g. class I heavy chain, β2m, TAP1 and TAP2, and the MHC-encoded subunits of the proteasome, are INFγ-inducible. This feature presumably provides an organism with a way to upregulate its class I surveillance system at localized sites of infection. INFγ-treated cells are effectively in a state of "red alert," which results in an improved efficiency in the reporting process and speed with which an infected cell is identified and killed. Similarly, cells that do not regenerate or are essential for survival, for example neurons (Joly et al 1991) and germ cells (Bikoff et al 1991), have down-regulated the expression of both MHC class I heavy chains and TAP1 and TAP2, presumably to prevent lysis by cytolytic T cells upon viral infection.

The class I presentation system has undoubtedly been under much evolutionary selection due to viruses and bacteria, and there are many examples of interference of the class I presentation pathway by viral proteins to gain a selective advantage. In turn, it seems that some of the idiosyncrasies of the class I assembly and transport pathway may have evolved to combat evasion of the presentation system by viruses or bacteria. For example, in both human and mice there are typically both slowly transported and more rapidly tranported MHC class I isotypes. These differences may be important to ensure that the cell has the ability to rapidly report the presence of a virus, as well as having a reserve of class I molecules that are slowly released to the cell surface. The latter would be especially useful to detect viruses, which upon infection, switch off host protein synthesis and thus production of nascent class I molecules.

CONCLUSIONS AND PERSPECTIVE

In the last few years our understanding of the molecular basis of MHC class I antigen processing and presentation has made impressive progress. The basic outline of the pathway is now defined: antigen is degraded in the cytoplasm, passed to waiting MHC class I molecules in the ER, and then rapidly transported to the cell surface for inspection by T cells. The most ill-defined part of this pathway is the generation of antigen in the cytoplasm. Attention is focused on the proteasome; however, it seems likely that other accessory molecules will be involved in both the unfolding of proteins prior to degradation and protection/delivery of peptide fragments once formed. While the role of such accessory molecules is purely speculative, Hsp70, a major cytoplasmic chaperone (Morimoto & Milarski 1990) appears well suited to carry out one or both of these tasks. Indeed, a role for Hsp70 in antigen presentation is implicated by the discovery of three Hsp70 genes in the MHC region (Sargent et al 1989). As to the primary source of antigen, nascent polypeptide maintained in an unfolded state by Hsp70, or stuck in an unproductive folding pathway and bound by GroEL-like mammalian proteins (Ellis & van der Vies 1991), represents a prime candidate. Direct transfer of protein from mammalian GroEL to the proteasome could be facilitated by the fact that GroEL and the proteasome are both composed of cylinders consisting of seven subunit rings.

Direct proof that the proteasome is utilized in vivo in the generation of antigenic peptides may be difficult to obtain. Mutagenesis studies of conserved proteasomal subunits are lethal to the cell (Orino et al 1991). Perhaps our best course is to examine the role of the MHC-encoded subunits of the proteasome. Although these subunits are not essential for production of antigen, it seems highly likely that they in some way aid in this process, perhaps in redirecting the efforts of the proteasome away from the ubiqitin-dependent pathway (in which proteins are degraded to amino acids) and towards the degradation of nascent polypeptides into peptides.

Although the role of the transporter proteins TAP1 and TAP2 remains to be formally proven, the evidence seems overwhelming that they are peptide transporters. What is now required is an in vitro assay that directly measures the activity of these transporters. Once this is available, we should expect answers to the remaining questions concerning this part of the pathway. For example, what is the specificity of the pumps in terms of sequence of peptides transported, length of peptides transported, etc? Do peptides have to be delivered bound by a molecular chaperone? Is the pump regulated by, for example, phosphorylation?

The assembly and intracellular transport component of the MHC class I presentation pathway is perhaps the best understood. The evidence from both the in vitro and in vivo studies identifies the primary pathway of assembly as proceeding via peptide binding to preformed classI/β2m dimers. Peptide is delivered to these complexes in the ER and the complexes are rapidly transported to the cell surface. Nevertheless, some questions remain. For example, what happens to peptides not bound by the MHC, are they secreted or are they degraded before they get to the cell surface? What triggers the release of heavy chain from calnexin? Is calnexin essential for the peptide addition to heavy chains? How is the turnover of class I molecules regulated at the cell surface?

With the basic outline of the pathway secure, we can expect that many of the missing details will become evident in the not too distant future. Given the enormous selection pressure exerted on the MHC by ever-evolving viruses and bacteria, there is little doubt that these details will provide fascinating insights into the tricks of the trade of antigen presentation.

ACKNOWLEDGMENTS

We are especially grateful to our colleagues Y. Yang, E. Song, M. Matsumura, Y. Saito, A. Brumark, J. Waters, K. Fruh, and L. Karlsson for helpful comments and for allowing us to cite unpublished data. We thank D. Fremont and I. Wilson for Figure 1. Work in the authors' laboratory was supported by the National Institutes of Health.

Literature Cited

Ahluwalia, N, Bergeron, JJM, Wada, I, Degen, E, Williams, DB. 1992. The p88 molecular chaperone is identical to the endoplasmic reticulum membrane protein, calnexin. *J. Biol. Chem.* 267:10914–18

Alexander, J, Payne, JA, Murray, R, Frelinger, JA, Cresswell, P. 1989. Differential transport requirements of HLA and H-2 class I glycoproteins. *Immunogenetics* 29: 380–88

Allen, H, Fraser, J, Flyer, D, Calvin, S, Flavell, R. 1986. β2-microglobulin is not required for the cell surface expression of the murine class I histocompatibility antigen H-2D^b or of a truncated H-2D^b. *Proc. Natl. Acad. Sci. USA* 83:7447–51

Anderson, K, Cresswell, P, Gammon, M, Hermes, J, Williamson, A, Zweerink, H. 1991. Endogenously synthesized peptide with an endoplasmic reticulum signal sequence sensitizes antigen processing mutant cells to class-I restricted cell-mediated lysis. *J. Exp. Med.* 174:489–92

Andersson, M, Paabo, S, Nilsson, T, Peterson, PA. 1985. Impaired intracellular transport of class I MHC antigens as a possible means for adenovirus to evade immune surveillance. *Cell* 43:215–22

Arce-Gomez, B, Jones, EA, Barnstable, CJ, Solomon, E, Bodmer, WF. 1978. The genetic control of HLA-A and B antigens in somatic cell hybrids : Requirement for β2-microglobulin. *Tissue Antigens* 11:96–104

Arnold, D, Driscoll, J, Androlewicz, M, Hughes, E, Cresswell, P, Spies, T. 1992. Proteasome subunits encoded in the MHC are not generally required for the processing of peptides bound by MHC class I molecules. *Nature* 360:171–73

Attaya, M, Jameson, S, Martinez, CK, Hermel, E, Aldrichs, C, et al. 1992. Ham-2 corrects the class I antigen-processing defect in RMA-S cells. *Nature* 355:647–49

Baas, EJ, van Santen, H-M, Kleijmeer, MJ, Geuze, HJ, Peters, PJ, Ploegh, HL. 1992. Peptide-induced stabilization and intracellular transport of empty HLA class I complexes. *J. Exp. Med.* 176:147–56

Beck, S, Hansen, TH, Cullen, SE, Lee, DR. 1986. Slower processing and weaker β2microglobulin association, and lower surface expression of H-2Ld are influenced by its amino terminus. *J. Immunol.* 137:916–23

Benjamin, RJ, Madrigal, JA, Parham, P. 1991. Peptide binding to empty HLA-B27 molecules of viable human cells. *Nature* 351:74–77

Bernabeu, C, Van de Rijn, M, Lerch, P, Terhorst, C. 1984. β2-microglobulin from serum associates with MHC class I antigens on the surface of cultured cells. *Nature* 308:642–45

Bevan, MJ. 1977. In a radiation chimaera, host H-2 antigens determine immune responsiveness of donor cytotoxic cells. *Nature* 269:417–18

Bevan, MJ. 1987. Class discrimination in the world of immunology. *Nature* 325:192–94

Bikoff, EK, Jaffe, L, Ribaudo, RK, Otten, GR, Germain, RN, Robertson, EJ. 1991. MHC class I surface expression in embryo-derived cell lines inducible with peptide or interferon. *Nature* 354:235–38

Bix, M, Raulet, D. 1992. Functionally conformed free class I heavy chains exist on the surface of β2-microglobulin negative cells. *J. Exp. Med.* 176:829–34

Bjorkman, PJ, Saper, MA, Samraoui, B, Bennett, WS, Strominger, JL, Wiley, DC. 1987. Structure of the human histocompatibility antigen, HLA-A2. *Nature* 329:506–12

Bluestone, JA, Jameson, S, Miller, S, Dick, R. 1992. Peptide-induced conformational changes in class I heavy chains alters MHC recognition. *J. Exp Med.* 176:1757–61

Brown, MG, Driscoll, J, Monaco, JJ. 1991. Structural and serological similarity of MHC-linked LMP and proteasome (multicatalytic proteinase) complexes. *Nature* 353:355–60

Catipović, B, Porto, JD, Mage, M, Johansen, TE, Schneck, JP. 1992. Major histocompatibility complex conformational epitopes are peptide specific. *J. Exp. Med.* 176:1611–18

Carbone, FR, Bevan, MJ. 1989. Induction of ovalbumin-specific cytotoxic T cells by in vivo peptide immunization. *J. Exp. Med.* 169:603–12

Cerundolo, V, Alexander, J, Lamb, C, Cresswell, P, McMichael, A, et al. 1990. Presentation of viral antigen controlled by a gene in the MHC. *Nature* 345:449–52

Cerundolo, V, Elliott, T, Elvin, J, Bastin, J, Rammensee, H-G, Townsend, A. 1991. The binding affinity and dissociation rates of peptides for class I major histocompatibility complex molecules. *Eur. J. Immunol.* 21:2069–75

Chen, BP, Parham, P. 1989. Direct binding of influenza peptides to class I HLA molecules. *Nature* 337:743–45

Christinck, ER, Luscher, MA, Barber, BH, Williams, DB. 1991. Peptide binding to class I MHC on living cells and quantitation of complexes required for CTL lysis. *Nature* 352:67–70

Corr, M, Boyd, LF, Frankel, SR, Kozlowski,

S, Padlan, EA, Margulies, DH. 1992. Endogenous peptides of a soluble major histocompatibility complex class I molecule, H-2Lds: Sequence motif, quantitative binding and molecular modeling of the complex. *J. Exp. Med.* 176:1681–92

Degen, E, Cohen-Doyle, MF, Williams, DB. 1992. Efficient dissociation of the p88 chaperone from MHC class I molecules requires both β2-microglobulin and peptide. *J. Exp. Med.* 175:1653–61

Degen, E, Williams, DB. 1991. Participation of a novel 88-kD protein in the biogenesis of murine class I histocompatibility molecules. *J. Cell Biol.* 112:1099–115

Deverson, ER, Gow, IR, Coadwell, WJ, Monaco, JJ, Butcher, GW, Howard, JC. 1990. MHC class II region encoding proteins related to the multidrug resistance family of membrane transporters. *Nature* 348:738–41

Dick, LR, Moomaw, CR, Dematino, GN, Slaughter, CA. 1991. Degradation of oxidized insulin B chain by the multiproteinase complex macroplain (Proteasome). *Biochemistry* 30:2725–34

Eisenlohr, LC, Bacik, I, Bennik, JR, Bernstein, K, Yewdell, JW. 1992. Expression of a membrane protease enhances presentation of endogenous antigens to MHC class I-restricted T lymphocytes. *Cell* 71:963–72

Elliott, T. 1991. How do peptides associate with MHC class I molecules ? *Immunol. Today* 12:386–88

Elliott, T, Cerundolo, V, Elvin, J, Townsend, A. 1991. Peptide-induced conformational change of the class I heavy chain. *Nature* 351:402–6

Ellis, RJ, van der Vies, SM. 1991. Molecular chaperones. *Annu. Rev. Biochem.* 60:321–47

Falk, K, Rotzschke, O, Rammensee, H-G. 1990. Cellular peptide composition governed by major histocompatibility complex class I molecules. *Nature* 348:248–51

Falk, K, Rotzschke, O, Stevanovic, S, Jung, G, Rammensee, H-G. 1991. Allele-specific motifs revealed by sequencing of self-peptides eluted from MHC molecules. *Nature* 353:290–96

Fremont, DH, Matsumura, M, Stura, EA, Peterson, PA, Wilson, IA. 1992. Crystal structures of two viral peptides in complex with murine class I H-2K^b MHC. *Science* 257:919–27

Galvin, K, Krishna, S, Ponchel, F, Frohlich, M, Cimmings, DE, et al. 1992. The major histocompatibility complex class I antigen-binding protein p88 is the product of the calnexin gene. *Proc. Natl. Acad. Sci. USA* 89:8452–56

Garboczi, DN, Hung, DT, Wiley, DC. 1992. HLA-A2-peptide complexes. Refolding and crystallization of molecules expressed in *Escherichia coli* and complexed with single antigenic peptides. *Proc. Natl. Acad. Sci. USA* 89:3429–33

Garrett, TPJ, Saper, MA, Bjorkman, PJ, Strominger, JL, Wiley, DC. 1989. Specific pockets for the side chains of peptide antigens in HLA-Aw68. *Nature* 342:692–96

Gething, M-J, Sambrook, J. 1992. Protein folding in the cell. *Nature* 355:33–44

Glynne, R, Powis, SH, Beck, S, Kelly, A, Kerr, L-A, Trowsdale, J. 1991. A proteasome-related gene between the two ABC transporter loci in the class II region of the human MHC. *Nature* 353:357–60

Guo, H-C, Jardetzky, TS, Garrett, TPJ, Lane, WS, Strominger, JL, Wiley, DC. 1992. Different length peptides bind to HLA-Aw68 similarly at their ends but bulge out in the middle. *Nature* 360:364–66

Hansen, TH, Myers, NB, Lee, DR. 1988. Studies of two antigenic forms of Ld with disparate β2-microglobulin (β2m) association suggests that β2m facilitates the folding of the α1 and α2 domains during de novo synthesis. *J. Immunol.* 140:3522–27

Heath, WR, Hurd, ME, Carbone, FR, Sherman, LA. 1989. Peptide-dependent recognition of H-2K^b alloreactive cytotoxic T lymphocytes. *Nature* 341:749–52

Henderson, RA, Michel, H, Sakaguchi, K, Shabanowitz, E, Apella, E, et al. 1992. HLA-A2.1-associated peptides from a mutant cell line: a second pathway of antigen presentation. *Science* 255:1264–67

Higgins, CF, Hyde, SC, Mimmack, MM, Gildeadi, V, Gill, DR, Gallagher, MP. 1990. Binding protein-dependent transport systems. *J. Bioenerg. Biomembr.* 22:571–92

Hochman, JH, Shimizu, Y, DeMars, R, Edidin, M. 1988. Specific association of fluorescent β2-microglobulin with cell surfaces: The affinity of different H-2 and HLA antigens for β2-microglobulin. *J. Immunol.* 140:2322–29

Hochstenbach, F, David, V, Watkins, S, Brenner, MB. 1992. Endoplasmic reticulum resident protein of 90 kilodaltons associates with the T- and B-cell antigen receptors and major histocompatibility complex antigens during their assembly. *Proc. Natl. Acad. Sci. USA* 89:4734–38

Hsu, VW, Yuan, LC, Nuchtern, JG, Lippincott-Schwartz, J, Hammerling, GJ, Klausner, RD. 1991. A recycling pathway between the endoplasmic reticulum and the Golgi apparatus for retention of unassembled MHC class I molecules. *Nature* 352:441–44

Hurtley, SM, Helenius, A. 1989. Protein oligomerization in the endoplasmic reticulum. *Annu. Rev. Cell Biol.* 5:277–307

Jackson, MR, Nilsson, T, Peterson, PA.

1990. Identification of a consensus motif for retention of transmembrane proteins in the endoplasmic reticulum. *EMBO J.* 9: 3153–62

Jackson, MR, Nilsson, T, Peterson, PA. 1993. Retrieval of transmembrane proteins to the endoplasmic reticulum. *J. Cell Biol.* 121: 317–33

Jackson, MR, Song, ES, Yang, Y, Peterson, PA. 1992. Empty and peptide containing conformers of class I major histocompatibility complex molecules expressed in *Drosophila melanogaster* cells. *Proc. Natl. Acad. Sci. USA* 89:12117–21

Jardetzky, TS, Lane, WS, Robinson, RA, Madden, DR, Wiley, DC. 1991. Identification of self-peptides bound to purified HLA-B27. *Nature* 353:326–29

Joly, E, Mucke, L, Oldstone, MBA. 1991. Viral persistence in neurons explained by lack of major histocompatibility class I expression. *Science* 253:1283–85

Kappler, JW, Roehm, N, Marrack, P. 1987. T cell tolerance by clonal elimination in the thymus. *Cell* 49:273–80

Kelly, A, Powis, SH, Glynne, R, Radley, E, Beck, S, Trowsdale, J. 1991. Second proteasome-related gene in the human MHC class II region. *Nature* 353:667–68

Kissonerghis, A-M, Owen, MJ, Lodish, HF. 1980. Biosynthesis of HLA-A and HLA-B antigens in vivo. *J. Biol. Chem.* 255:9678–84

Klar, D, Hammerling, GJ. 1989. Induction of the assembly of MHC class I heavy chains with β2-microglobulin by interferon-g. *EMBO J.* 8:475–81

Kleijmeer, MJ, Kelly, A, Geuze, HJ, Slot, JW, Townsend, A, Trowsdale, J. 1992. Location of MHC-encoded transporters in the endoplasmic reticulum and *cis*-Golgi. *Nature* 357:342–44

Kozlowski, S, Takeshita, T, Boehncke, W-H, Takahashi, H, Boyd, LF, et al. 1991. Excess β2-microglobulin promoting functional peptide association with purified soluble class I MHC molecules. *Nature* 349:74–77

Krangel, MS, Orr, HT, Strominger, JL. 1979. Assembly and maturation of HLA-A and HLA-B antigens in vivo. *Cell* 18:979–91

Levy, F, Gabathuler, R, Larsson, R, Kvist, S. 1991. ATP is required for in vitro assembly of MHC class I antigens but not for transfer of peptides across the ER membrane. *Cell* 67:265–74

Lie, W-R, Myers, NB, Gorka, J, Rubocki, RJ, Connolly, JM, Hansen, TH. 1990. Peptide ligand-induced conformation and surface expression of Ld class I MHC molecules. *Nature* 344:439–41

Ljunggren, H-G, Paabo, S, Cochet, M, Kling, G, Kourilsky, P, Karre, K. 1989. Molecular analysis of H-2 deficient lymphoma lines; distinct defects in biosynthesis and association of MHC class I/beta-2microglobulin observed in cells with increased sensitivity to NK cell lysis. *J. Immunol.* 142:2911–17

Ljunggren, H-G, Stan, NJ, Ohlen, C, Neefjes, JJ, Hoglund, P, et al. 1990. Empty MHC class I molecules come out in the cold. *Nature* 346:476–80

Luescher, IF, Romero, P, Cerottini, J-C, Maryanski, JL. 1991. Specific binding of antigenic peptides to cell-associated MHC class I molecules. *Nature* 351:72–74

Madden, DR, Gorga, JC, Strominger, JL, Wiley, DC. 1991. The structure of HLA-B27 reveals nonamer self-peptides bound in an extended conformation. *Nature* 353:321–27

Madrigal, JA, Belich, MP, Benjamin, RJ, Little, A-M, Hildebrand, D, et al. 1991. Molecular definition of a polymorphic antigen (LA45) of free HLA-A and -B heavy chains found on the cell surface of activated B and T cells. *J. Exp. Med.* 174:1085–95

Martinez, CK, Monaco, JJ. 1991. Homology of the proteasome subunits to a major histocompatibility complex-linked LMP gene. *Nature* 353:664–67

Matsumura, M, Fremont, D, Peterson, PA, Wilson, IA. 1992a. Emerging principles for the recognition of peptide antigens by MHC class I molecules. *Science* 257:927–34

Matsumura, M, Saito, Y, Jackson, MR, Song, ES, Peterson, PA. 1992b. In vitro peptide binding to soluble empty class I major histocompatibility complex molecules isolated from transfected *Drosophila melanogaster* cells. *J. Biol. Chem.* 267:23589–95

Miyazaki, J-I, Appella, E, Ozato, K. 1986. Intracellular transport blockage caused by disruption of the disulfide bridge in the third external domain of a major histocompatibility complex class I antigen. *Proc. Natl. Acad. Sci. USA* 83:757–61

Momburg, F, Ortiz-Navarette, V, Neefjes, J, Goulmy, E, van der Wal, Y, et al. 1992. Proteasome subunits encoded by the major histocompatibility complex are not essential for antigen presentation. *Nature* 360:174–77

Monaco, JJ, Cho, S, Attaya, M. 1990. Transport protein genes in the murine MHC: possible implications for antigen processing. *Science* 250:1723–26

Moore, MW, Carbone, FR, Bevan, MJ. 1988. Introduction of soluble proteins into the class I pathway of antigen processing and presentation. *Cell* 54:777–85

Morimoto, RI, Milarski, KL. 1990. Expression and function of vertebrate HSP 70 genes. In *Stress Proteins in Biology and Medicine,* ed. RI Morimoto, A Tissieres, C Georgopoulos. pp. 322–59. New York: Cold Spring Habor Lab. Press

Nikolic-Zugic, J, Bevan, MJ. 1990. Role of

self-peptides in positively selecting the T-cell repertoire. *Nature* 344:65–67
Nuchtern, JG, Bonifacino, JS, Biddison, WE, Klausner, RD. 1989. Brefeldin A inhibits egress from the endoplasmic reticulum in class I restricted antigen presentation. *Nature* 339:223–26
Orino, E, Tanaka, K, Tamura, T, Sone, T, Ogura, T, Ichihara, A 1991. ATP dependant reversible association of proteasomes with multiple protein components to form 26S complexes that degrade ubiquinated proteins in human HL-60 cells. *FEBS Lett.* 284:206–10
Ortiz-Navarrete, V, Hammerling, GJ. 1991. Surface appearance and instability of empty H-2 class I molecules under physiological conditions. *Proc. Natl. Acad. Sci. USA* 88:3594–97
Ortiz-Navarrete, V, Seelig, VA, Gernold, M, Frentzel, S, Kloetzel, PM, Hammerling, GJ. 1991. Subunit of the '20S' proteasome (multicatalytic proteinase) encoded by the major histocompatibility complex. *Nature* 353:662–64
Orlowski, M. 1990. The multicatalytic proteinase complex, a major extralysosomal proteolytic system. *Biochemistry* 29:10289–97
Parker, KC, Carreno, BM, Sestak, BM, Utz, U, Biddison, WE, Coligan, JE. 1992. Peptide binding to HLA-A2 and HLA-B27 isolated from *Escherichia coli*. *J. Biol. Chem.* 267:5451–59
Powis, SH, Mockridge, I, Kelly, A, Kerr, L-A, Glynne, R, et al. 1992. Polymorphism in a second ABC transporter gene located within the class II region of the human major histocompatibility complex. *Proc. Natl. Acad. Sci. USA* 89:1463–67
Powis, SJ, Deverson, EV, Coadwell, WJ, Ciruela, A, Huskisson, NS, et al. 1992. Effect of polymorphism of an MHC-linked transporter on the peptides assembled in a class I molecule. *Nature* 357:211–15
Powis, SJ, Townsend, ARM, Deverson, EV, Bastin, J, Butcher, GW, Howard, JC. 1991. Restoration of antigen presentation in the mutant cell line RMA-S by an MHC-linked transporter. *Nature* 354:528–31
Rock, KL, Gamble, SR, Rothstein, LE, Benaceraff, B. 1991a. Reassociation with β2-microglobulin is necessary for Db class I major histocompatibility complex binding of an exogenous influenza peptide. *Proc. Natl. Acad. Sci. USA* 88:301–4
Rock, KL, Gamble, SR, Rothstein, L, Gramm, C, Benaceraff, B. 1991b. Dissociation of β2-microglobulin leads to the accumulation of a substantial pool of inactive class I MHC heavy chains on the cell surface. *Cell* 65:611–20
Rock, KL, Rothstein, LE, Gamble, SR, Benaceraff, B. 1990. Reassociation with β2-microglobulin is necessary for Kb class I major histocompatibility complex binding exogenous peptides. *Proc. Natl. Acad. Sci. USA* 87:7517–21
Rose, JK, Doms, RW. 1988. Regulation of protein export from the endoplasmic reticulum. *Annu. Rev. Cell. Biol.* 4:257–88
Rotzchke, O, Falk, K. 1991. Naturally occuring peptide antigens derived from the MHC class I restricted processing pathway. *Immunol. Today* 12:447–55
Salter, RD, Cresswell, P. 1986. Impaired assembly and transport of HLA-A and -B antigens in TxB cell hybrids. *EMBO J.* 5:943–49
Salter, RD, Howell, DN, Cresswell, P. 1985. Genes regulating HLA class I antigen expression in T-B lymphoblast hybrids. *Immunogenetics* 21:235–46
Sargent, CA, Dunham, I, Trowsdale, J, Campbell, RD. 1989. Human major histocompatibility complex contains genes for the major heat shock protein HSP 70. *Proc. Natl. Acad. Sci. USA* 86:1968–72
Schumacher, TNM, De Bruijn, MLH, Vernie, LN, Kast, WM, Melief, CJM, et al. 1991. Peptide selection by MHC class I molecules. *Nature* 350:703–6
Schumacher, TNM, Heemels, M-T, Neefjes, JJ, Kast, WM, Melief, CJM, Ploegh, HL. 1990. Direct binding of peptide to empty MHC class I molecules on intact cells and in vitro. *Cell* 62:563–67
Sege, K, Rask, L, Peterson, PA. 1981. Role of β2-microgobulin in the intracellular processing of HLA antigens. *Biochemistry* 20: 4523–30
Severinsson, L, Peterson, PA. 1984. β2-microglobulin induces intracellular transport of human class I transplantation antigen heavy chains in *Xenopus laevis* oocytes. *J. Cell Biol.* 99:226–32
Sherman, LA, Burke, TA, Biggs, JA. 1992. Extracellular processing of antigens that bind class I major histocompatibility molecules. *J. Exp. Med.* 175:1221–26
Sibille, C, Chomez, P, Wildmann, C, Van Pel, A, Deplaen, E, et al. 1990. Structure of the gene of tum-transplantation antigen p198: A point mutation generates a new antigenic peptide. *J. Exp. Med.* 172:35–45
Silver, ML, Guo, H-G, Strominger, JL, Wiley, DC. 1992. Atomic structure of a human MHC molecule presenting an influenza virus peptide. *Nature* 360:367–69
Smith, JD, Lie, W-R, Gorka, J, Myers, NB, Hansen, TH. 1992. Extensive peptide ligand exchange by surface class I major histocompatibility complex molecules independent of exogenous β2-microglobulin. *Proc. Natl. Acad. Sci. USA* 89:7767–71
Smith, MH, Parker, JMR, Hodges, RS, Bar-

ber, BH. 1986. The preparation and characterisation of anti-peptide hetero antisera recognizing subregions of the intracytoplasmic domain of class I H-2 antigens. *Mol. Immunol.* 23:1077–92

Spies, T, Bresnahan, M, Bahram, S, Arnold, D, Blanck, E, et al. 1990. A gene in the human major histocompatibility complex class II region controlling the class I antigen presentation pathway. *Nature* 348:744–47

Spies, T, Cerundolo, V, Colonna, M, Cresswell, P, Townsend, A, DeMars, R. 1992. Presentation of viral antigen by MHC class I molecules is dependent on a putative peptide transporter heterodimer. *Nature* 355:644–46

Spies, T, DeMars, R. 1991. Restored expression of major histocompatibility class I molecules by gene transfer of a putative peptide transporter. *Nature* 351:323–24

Teyton, L, Peterson, PA. 1992. Assembly and transport of MHC class II molecules. *New Biol.* 4:441–47

Townsend, ARM, Bastin, J, Gould, K, Brownlee, GG. 1986. Cytotoxic T lymphocytes recognize influenza hemagglutinin that lacks a signal sequence. *Nature* 324: 575–77

Townsend, A, Elliot, T, Cerundolo, V, Foster, L, Barber, B, Tse, A. 1990. Assembly of MHC class I molecules analyzed in vitro. *Cell* 62:285–95

Townsend, A, Ohlen, C, Bastin. J, Ljunggren, H-G, Foster, L, Karre, K. 1989. Association of class I major histocompatibility heavy and light chains induced by viral peptides. *Nature* 340:443–48

Trowsdale, J, Hanson, I, Mockridge, I, Beck, S, Townsend, A, Kelly, A. 1990. Sequences encoded in the class II region of the MHC related to the "ABC" superfamily of transporters. *Nature* 348:741–44

Van Bleek, GM, Nathenson, SG. 1990. Isolation of an endogenously processed immunodominant viral peptide from class I H-$2K^b$ molecule. *Nature* 348:213–16

Van Kaer, L, Ashton-Rickardt, PG, Ploegh, HL, Tonegawa, S. 1992. TAP1 mutant mice are deficient in antigen presentation, surface class I molecules and CD4-CD8+ T cells. *Cell* 71:1205–14

Vitiello, A, Potter, TA, Sherman, LA. 1990. The role of β2-microglobulin in peptide binding by class I molecules. *Science* 250:1423–26

von Boehmer, H, Haas, W, Jerne, NK. 1978. Major histocompatibilty complex-linked immune responsiveness is acquired by lymphocytes of low responder mice differentiating in thymus of high responder mice. *Proc. Natl. Acad. Sci. USA* 75:2439–42

Wada, I, Rindress, D, Cameron, PH, Ou, W-J, Doherty, JJ, et al. 1991. SSRα and associated calnexin are major calcium binding proteins of the endoplasmic reticulum membrane. *J. Biol. Chem.* 266: 19599–610

Walker, BD, Flexner, C, Birch-Limberger, K, Fisher, L, Paradis, TJ, et al. 1989. Long-term culture and fine specificity of human cytotoxic T-lymphocyte clones reactive with human immunodeficiency virus type 1. *Proc. Natl. Acad. Sci. USA* 86:9514–18

Wei, ML, Cresswell, P. 1992. HLA-A2 molecules in an antigen processing mutant cell contain signal sequence-derived peptides. *Nature* 356:443–46

Williams, DB, Barber, BH, Flavell, RA, Allen, H. 1989. Role of β2-microglobulin in the intracellular transport and surface expression of murine class I histocompatibility molecules. *J. Immunol.* 142:2796–806

Williams, DB, Borriello, F, Zeff, RA, Nathenson, SG. 1988. Intracellular transport of class I histocompatibility molecules. Influence of protein folding on transport to the cell surface. *J. Biol. Chem.* 263:4549–60

Williams, DB, Swiedler, SJ, Hart, GW. 1985. Intracellular transport of membrane glycoproteins: Two closely related histocompatibility antigens differ in their rates of transit to the cell surface. *J. Cell Biol.* 101:725–34

Yang, Y, Fruh, K, Chambers, J, Waters, JB, Wu, L, et al. 1992a. Major histocompatibility complex (MHC)-encoded HAM2 is necessary for antigenic peptide loading onto class I MHC molecules. *J. Biol. Chem.* 267:11669–72

Yang, Y, Waters, JB, Fruh, K, Peterson, PA. 1992b. Proteasomes are regulated by Interferon γ: Implications for antigen processing. *Proc. Natl. Acad. Sci. USA* 89:4928–32

Yewdell, JW, Bennick, JR. 1989. Brefeldin A specifically inhibits presentation of protein antigens to cytotoxic T lymphocytes. *Science* 244:1072–75

Zhang, W, Young, ACM, Imarai, M, Nathenson, SG, Sacchettini, JC. 1992. Crystal structure of the major histocompatibility complex class I H-$2K^b$ molecule containing a single viral peptide: Implications for peptide binding and T-cell recognition. *Proc. Natl. Acad. Sci. USA* 89:8403–7

Zinkernagel, RM, Callahan, GN, Klein, J, Dennert, G. 1978. Cytotoxic T cells learn specificity for self H-2 during differentiation in the thymus. *Nature* 271:251–53

Annu. Rev. Cell Biol. 1993. 9:237–64

BIOLOGY OF ANIMAL LECTINS

Kurt Drickamer

Department of Biochemistry and Molecular Biophysics, Columbia University, 630 West 168th Street, New York, New York 10032

Maureen E. Taylor

Institute of Glycobiology, Department of Biochemistry, University of Oxford, South Parks Road, Oxford OX1 3QU England

KEY WORDS: lectins, carbohydrates, endocytosis, cell adhesion, innate immunity

---

CONTENTS

INTRODUCTION

Biological Functions of Carbohydrates

For nearly 100 years, lectin research focused on proteins from plants. These nonenzymatic, sugar-binding proteins have been invaluable tools in the

0743–4634/93/1115–0237$05.00

structural and functional analysis of animal cell glycoconjugates because of their ability to discriminate among the myriad of complex carbohydrate structures that are found on the surface of cells, in extracellular matrices, and attached to soluble glycoproteins. As recently as 1986, a review on lectins devoted only one page to animal lectins (Lis & Sharon 1986). However, the field of animal lectins is expanding rapidly. These proteins have the ability to recognize either carbohydrates endogenous to the animal or those that are presented to it by microbial invaders. The description of the asialoglycoprotein receptor in mammalian liver provided the first model for how an animal lectin might discriminate in a useful way between various glycoproteins (summarized in Ashwell & Harford 1982). In the ensuing years, a large number of animal lectins have been isolated and in many cases have been ascribed roles in biological recognition events.

The existence of animal lectins establishes that the carbohydrate portions of glycoconjugates can present biologically important information and suggests that this may be one of their most significant functions. This function may be contrasted with the many effects of the carbohydrate portions of glycoconjugates that result from the physical properties of the sugars themselves and from interaction of the saccharides with each other and with the proteins and lipids to which they are covalently attached. For example, the properties of proteoglycans in various biological matrices are dominated by the interactions of the constituent glycosaminoglycans with water. Other effects of carbohydrate conjugated to proteins can be more subtle and varied (Parekh 1991). The fact that diverse properties of carbohydrates have been exploited by nature is analogous to the variety of ways in which other biological polymers such as proteins and RNA function.

As novel, sugar-specific interactions are discovered, it is reasonable to hypothesize that lectins mediate the interactions by decoding the information present in these selected saccharides. We are still in the early stages of identifying situations in which sugar-lectin interactions play a critical role. The list of lectin-mediated processes already includes diverse biological phenomena, from intracellular routing of glycoproteins, to cell-cell adhesion and phagocytosis. The state of our understanding of the lectin components of carbohydrate-lectin interactions varies widely. In some cases, the existence of the lectins is known largely by inference; specific binding of natural glycoproteins or neoglycoproteins to various cellular fractions can provide the first evidence for the presence of endogenous lectins (Facy et al 1990). Similarly, the abundance and variety of animal lectins is demonstrated by the large number of polypeptides that can be isolated by affinity chromatography on immobilized sugars (Gabius 1991). This review emphasizes vertebrate lectins that have been isolated and characterized at the level of primary structure.

Classification of Animal Lectins

Animal lectins have been grouped into classes based on the nature of their carbohydrate ligands, the biological processes in which they participate, their subcellular localization, and their dependence on divalent cations. Since the primary structures of approximately 100 animal lectins have been determined, classification based on shared sequence characteristics is now one of the most useful ways to sort the lectins. While the overall architecture of the lectin proteins varies widely, carbohydrate-binding activity can often be ascribed to a limited portion of a given lectin. This active segment can be designated the carbohydrate-recognition domain (CRD). Comparison of the sequences of these CRDs reveals that they fall into relatively few groups (Drickamer 1988). CRDs in each group share a pattern of invariant and highly conserved amino acid residues at a characteristic spacing.

CRDs in several of the major lectin groups share properties beyond similarity of primary structure. The C-type CRDs derive their name from the fact that they require calcium ions for activity. In addition, they are all extracellular, although they bind a diversity of sugars. In contrast, although the S-type lectins are found both inside and outside cells, they often are dependent on reducing agents (thiols) for full activity. They display no requirement for divalent cations, and they all bind β-galactosides. P-type CRDs bind mannose 6-phosphate as their primary ligand. The relationship between these shared properties and the sequence similarities in each group are discussed in more detail below.

Ca^{2+}-DEPENDENT (C-TYPE) LECTINS

C-type animal lectins are found in serum, extracellular matrix, and membranes (Table 1). These proteins share a common sequence motif consisting of 14 invariant and 18 highly conserved amino acid residues in the 115–130 amino acid segment that constitutes the C-type CRD (Drickamer 1988, 1993b). Although each of these CRDs binds ligand in a Ca^{2+}-dependent manner, the overall architecture of a C-type lectin is defined by the way in which the CRD is combined with other domains. The additional domains determine many of the functions of the lectins.

Type II Endocytic Receptors

Many C-type lectins are membrane-bound receptors that mediate endocytosis of glycoproteins. Apart from the macrophage mannose receptor, which is described in the next section, these proteins are structurally similar to each other. Each is a type II transmembrane protein with a short NH_2-terminal cytoplasmic tail, an internal, uncleaved signal sequence, and an extracellular

Table 1 Groups of C-type animal lectins

Functions	Evolutionary group	Location	Organization	Ligands	Examples
Endocytic receptors	II	Plasma membrane	Type II transmembrane	Endogenous	Asialoglycoprotein receptor
				Endogenous?	Kupffer cell receptor
				Endogenous	Chicken hepatic lectin
				Endogenous	Lymphocyte IgE F_c receptor
	V			Exogenous?	Natural killer cell receptors
	VI	Plasma membrane	Type I transmembrane	Exogenous	Mannose receptor
Adhesion molecules	IV	Plasma membrane	Type I transmembrane	Endogenous	Selectins
Humoral defense	III	Extracellular	Soluble, collagenous	Exogenous	Mannose-binding proteins Pulmonary surfactant apoproteins
Proteoglycans	I	Matrix	Extended	Endogenous	Aggrecan Versican Neurocan
Free CRDs	VII	Secreted	Soluble	Unknown	Pancreatic stone protein Pancreatitis/hepatoma protein

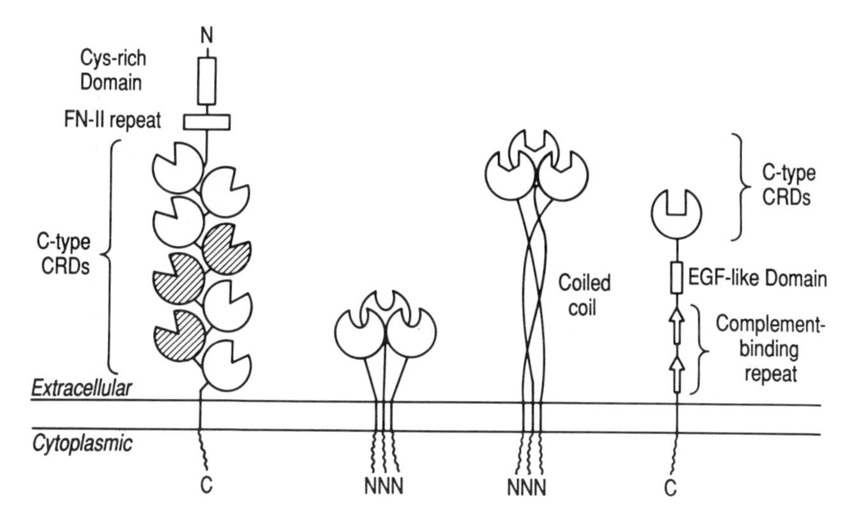

Figure 1 Organization of membrane-bound C-type animal lectins. The domain organization of three endocytic receptors and the selectin cell adhesion molecules is shown, along with the oligomeric structure, where it is known. From left to right: the macrophage mannose receptor; two examples of Type II endocytic receptors (the chicken hepatic lectin and the Kupffer cell receptor); and L-selectin.

portion consisting of a neck region and a COOH-terminal CRD. Examples of these proteins are represented by the middle two structures in Figure 1.

MAMMALIAN ASIALOGLYCOPROTEIN RECEPTOR The asialoglycoprotein receptor is found on hepatocytes and has specificity for galactose and N-acetylgalactosamine (Spiess 1990). It mediates clearance of serum glycoproteins with complex oligosaccharides from which the terminal sialic acid residues have been removed, leaving exposed terminal galactose residues. Glycoproteins bound by this receptor are internalized and transported to lysosomes for degradation. A neuraminidase exposed to the circulation is presumed to be responsible for the removal of sialic acid residues, although such an enzyme has not been identified. It has been proposed that the half-life of a serum glycoprotein may be controlled by the number of complex oligosaccharides that it bears and their accessibility to neuraminidase (Drickamer 1991).

The asialoglycoprotein receptor consists of two different polypeptides, both of which are similar in overall structure to the chicken hepatic lectin shown in Figure 1 (McPhaul & Berg 1986). In all species examined, one of the constituent polypeptides is predominant. Expression studies in fibroblasts have indicated that both polypeptides are required to form a functional receptor (McPhaul & Berg 1986). The purified receptor is a hexamer, although the stoichiometry of the different polypeptides in the oligomer has not been clearly defined (Halberg et al 1987; Herzig & Weigel 1990). Oligomerization of polypeptides each with a single CRD leads to an increased affinity of the receptor for multivalent ligands, as demonstrated using synthetic ligands containing clusters of terminal galactose or N-acetyl-galactosamine residues (Lee et al 1984). Such increased affinity brought about by clustering of CRDs is a common feature of many C-type lectins. The two different polypeptides of the rat asialoglycoprotein receptor bind different terminal galactose residues of a triantennary glycopeptide (Rice et al 1990), which indicates that clustering of CRDs may determine the specificity as well as the affinity of interaction.

CHICKEN HEPATIC LECTIN The chicken hepatic lectin is the avian homologue of the asialoglycoprotein receptor (Kawasaki & Ashwell 1977). In the membrane, this receptor is a trimer of a single type of subunit similar in overall structure to the polypeptides of the asialoglycoprotein receptor (Verrey & Drickamer 1993). The chicken receptor binds N-acetylglucosamine rather than galactose or N-acetylgalactosamine. It mediates clearance of glycoproteins from the serum, although binding of this receptor requires that N-acetylglucosamine residues be exposed. The mechanism by which this occurs is not clear. Like the asialoglycoprotein receptor, the chicken hepatic lectin binds multivalent ligands with much higher affinity than it binds monovalent ligands (Lee et al 1989). Again, this is likely to be the result of clustering of multiple CRDs in the native oligomer, since a single CRD produced by proteolysis shows reduced affinity for a multivalent glycoprotein ligand (Loeb & Drickamer 1988).

MACROPHAGE GALACTOSE RECEPTOR Rat peritoneal macrophages express a galactose-specific receptor that is similar in structure to the hepatic asialoglycoprotein receptor, although it is an oligomer of a single type of polypeptide (Ii et al 1990). While physiological functions and ligands of this receptor are unknown, there is some evidence that the homologous mouse protein binds tumor cells (M Sato et al 1992).

KUPFFER CELL RECEPTOR A fucose- and galactose-specific receptor is found on Kupffer cells, the resident macrophages of the liver (Hoyle & Hill 1988). The polypeptide of this receptor is much longer than the chicken hepatic lectin and the asialoglycoprotein receptor because of the presence of repeated heptad sequences in the neck region between the membrane and the CRD (Figure 1). This neck region is likely to form a coiled coil of α-helices (Beavil et al 1992). Physiological ligands and functions for the Kupffer cell receptor have not been determined, although it displays selective binding to certain globosides (Tiemeyer et al 1992). Like the peritoneal macrophage receptor, the Kupffer cell receptor may be involved in interactions with tumor cells since inhibition of attachment of lymphoma cells to the liver by fucose has been reported (Vavasseur et al 1990).

PLACENTAL MANNOSE RECEPTOR A type II receptor with specificity for mannose and fucose has recently been identified in human placenta (Curtis et al 1992). This receptor was cloned based on its ability to bind human immunodeficiency virus (HIV) envelope glycoprotein gp120. The receptor binds to mannose-containing oligosaccharides of gp120 with high affinity. Cells transfected with the receptor can bind and internalize HIV, but do not become infected. This receptor may play a role in CD4-independent internalization of HIV, although the cell type of origin in the placenta has not been identified.

LYMPHOCYTE RECEPTORS Several proteins, related in structure to the endocytic receptors, are found on the surfaces of lymphocytes. They are type II transmembrane proteins with COOH-terminal domains containing many of the residues characteristic of C-type CRDs. One of these proteins is the lymphocyte low affinity IgE F_c receptor, CD23 (Ludin et al 1987). Binding of IgE is mediated by the CRD-like portion of the receptor and requires Ca^{2+}, although deglycosylation of IgE does not affect binding to the receptor (Bettler et al 1989, 1992). Binding is inhibited by specific peptides from the F_c region, but not by sugars (Vercelli et al 1989; Richards & Katz 1990). A second protein, CD21, has recently been identified as a ligand for CD23 (Aubry et al 1992). Interaction of CD23 with this ligand is reduced by tunicamycin treatment, which suggests that sugars are involved (Pochon et al 1992). The mouse form of CD23 contains all of the residues shown to interact with mannose in the crystal structure of a mannose-binding CRD (discussed below), thus suggesting that this receptor should be able to bind carbohydrate.

Other type II proteins with putative C-type CRDs are found on natural killer cells. These proteins are encoded by a multigene family (Houchins et al 1991; Wong et al 1991). Ligands for these receptors have not been identified. Since the sequences of the putative CRDs in these proteins are highly divergent from other C-type CRDs, it is unclear whether these domains bind Ca^{2+}, much less carbohydrates.

Macrophage Mannose Receptor

The mannose receptor of macrophages and hepatic endothelial cells mediates binding and internalization of glycoconjugates terminating in mannose, fucose or, N-acetylglucosamine (Stahl 1990). These sugar residues are not normally found at the termini of mammalian cell-surface or serum glycoproteins, but are common components of the cell surfaces of bacteria, fungi, and parasites. Recognition of these carbohydrates by the mannose receptor allows discrimination of self from non-self, so the receptor is believed to play a role in the innate immune response against pathogenic microorganisms. There is evidence for involvement of the mannose receptor in phagocytosis of several different organisms, including yeasts (Ezekowitz et al 1990), *Pneumocystis carinii* (Ezekowitz et al 1991), and *Leishmania donovani* (Chakraborty & Das 1988). Endogenous glycoproteins bearing high mannose oligosaccharides, including lysosomal enzymes (Stahl & Schlesinger 1980) and tissue plasminogen activator (Otter et al 1991), are also ligands for the mannose receptor. These potentially harmful proteins are often released from cells in response to pathological events. Ligands internalized via the mannose receptor are transported to the lysosomes for degradation.

In contrast to other endocytic C-type lectins, the mannose receptor is a type I transmembrane protein (Figure 1, far left) (Taylor et al 1990). The extracellular portion of the receptor consists of an N-terminal cysteine-rich domain, a fibronectin type II repeat and eight C-type CRDs. The mannose receptor is the only protein known to have more than one C-type CRD within a single polypeptide.

Expression of portions of the receptor in vitro, in fibroblasts, bacteria, and insect cells, has been used to determine which of the extracellular domains are involved in binding and endocytosis of ligand (Taylor et al 1992; Taylor & Drickamer 1993). The N-terminal cysteine-rich domain and the fibronectin type II repeat are not required for endocytosis of glycoproteins. The function of these domains is unclear. Only one CRD shows detectable carbohydrate-binding activity when expressed in isolation. This single CRD can mimic the monosaccharide-binding properties of the receptor. However, several CRDs are necessary to achieve high affinity binding to multivalent glycoprotein ligands. Two of the CRDs form a protease-resistant, ligand-binding core, sufficient to bind some glycoproteins with high affinity, but three additional CRDs are required to reproduce the affinity of the intact receptor for yeast mannan. These results indicate that both valency and geometry of glycoconjugates are important determinants of binding affinity.

Like the asialoglycoprotein receptor, the chicken hepatic lectin, and the collectins (discussed below), the mannose receptor achieves high affinity binding to multivalent oligosaccharides by clustering of several CRDs, each with weak affinity for monosaccharides. However, clustering of CRDs in the mannose receptor occurs within a single polypeptide, rather than through the formation of oligomers. The different forms of clustering seen in different C-type lectins probably allows selection of different types of ligand. The arrangement of CRDs in the hepatic lectins may provide optimal binding to clusters of terminal residues that are found in desialylated complex chains of serum glycoproteins cleared by these receptors. In contrast, the linear arrangement of CRDs in the mannose receptor may be ideal for binding repeated polymers such as yeast mannan. While the hepatic lectins bind endogenous ligands with a limited set of structures, the mannose receptor must recognize a diversity of foreign ligands. The presence of multiple CRDs in the mannose receptor may provide the flexibility needed to interact with this diversity of foreign ligands.

Selectin Cell Adhesion Molecules

The initial phase of adhesion between leukocytes and endothelia is a weak, transient adhesion mediated by a group of Ca^{2+}-dependent adhesion molecules, designated selectins. The selectins are considered in more detail elsewhere (M. Bevilacqua, this volume), but a few features relevant to the present discussion are summarized here. L-selectin on T cells targets these cells to peripheral lymph nodes, while E- and P-selectins on endothelium interact with neutrophils and monocytes (Lasky 1992). In addition, P-selectin on platelets mediates their binding to endothelium. The selectins are type I transmembrane proteins. The extracellular domain of each, shown diagrammatically in Figure 1 (far right), consists of an NH_2-terminal C-type CRD, an epidermal growth factor-like domain, and variable numbers of complement-binding repeats. The oligomeric structure of the selectins has not yet been established.

The range of endogenous ligands that can be bound by the selectins remains to be fully established, but it is clear that all three bind structures related to the sialyl Lewis X antigen, in which both terminal sialic acid and fucose are present (Lasky 1992). A related structure, in which sialic acid is replaced by a sulfate group, is also a ligand for E-selectin (Yuen et al 1992). Sulfatides appear to be candidate cell surface ligands for at least P-selectin (Aruffo et al 1991). These results suggest that a portion of the binding site on the selectins must accommodate a negatively charged ligand (see below). Not surprisingly, the portion of the selectin polypeptide required for ligand binding includes the CRD segment at the NH_2 terminus (Watson et al 1990; Kansas et al 1991). The complement-binding segments can influence ligand binding (Watson et al 1991), but there is no evidence that domains other than the C-type CRD interact directly with sugars.

Collectins

The collectins are a group of soluble proteins that consist of oligomers (9–27 subunits) of polypeptides with COOH-terminal CRDs in association with NH_2-terminal collagen-like domains (Table 2). They fall into two morphological subgroups; those with shorter collagenous domains form bouquet-like

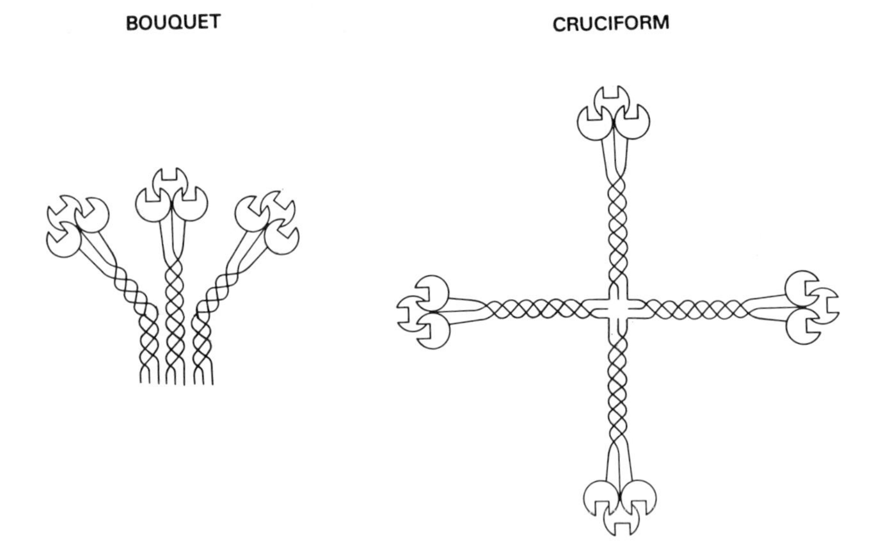

Figure 2 Summary of subunit and oligomeric structure of collagen-containing lectins. The native molecular weights, shown in Table 2, of some of the lectins in each subgroup indicate that the oligomers shown here must associate into larger structures.

structures, while those with more Gly-X-Y repeats are cruciform (Figure 2). The collectins include mannose-binding proteins found in mammalian serum and liver (Drickamer et al 1986; Ezekowitz et al 1988), the bovine serum protein conglutinin (Y-M Lee et al 1991), and pulmonary surfactant apoproteins A (Benson et al 1985) and D (Rust et al 1991; Lu et al 1992). Like the macrophage mannose receptor, these proteins bind carbohydrates found on the surfaces of pathogenic microorganisms and are believed to play a role in the innate immune response.

MANNOSE-BINDING PROTEINS Two forms of rat mannose-binding protein, mannose-binding protein A (predominant in serum) and mannose-binding protein C (predominant in liver), have been cloned and sequenced (Drickamer et al 1986). The two forms are similar in overall structure, but have different carbohydrate specificities. Mannose-binding protein A binds to terminal mannose, fucose, and N-acetylglucosamine residues, while mannose-binding protein C recognizes the trimannosyl core of complex N-linked oligosaccharides (Childs et al 1990). In humans, only one form has been sequenced (Ezekowitz et al 1988; Taylor et al 1989). The human form is believed to correspond to rat mannose-binding protein A because of its

Table 2 Subgroups of collectins

Subgroup morphology	Examples	Gly-X-Y repeats	Central interruption	M_r	Possible oligomer	Ligands
Bouquet	Mannose-binding protein-A	18	+	650–700,000	$(\alpha_3)_6$	Man, GlcNAc, Fuc
	Mannose-binding protein-C	20	+	200–220,000	$(\alpha_3)_3$	Man, Fuc
	Surfactant SP-A	24	+	>400,000	$(\alpha_3)_6$	Man, Glc, Fuc, Gal
Cruciform	Conglutinin	57	–	400,000	$(\alpha_3)_4$	Man, GlcNAc, Fuc
	Surfactant SP-D	59	–		$(\alpha_3)_4$	Glc

predominance in serum and its specificity for terminal mannose, fucose, and N-acetylglucosamine residues (Taylor & Summerfield 1987).

Serum mannose-binding protein mediates defense either via complement fixation or by direct opsonization. The similarity in overall structural organization of mannose-binding proteins and complement component C1q suggests that serum mannose-binding protein might be able to act in place of C1q and fix complement via the classical pathway. Such activation would occur after binding of mannose-binding protein to carbohydrates on the surfaces of pathogens. Consistent with this hypothesis, activation of complement in vitro by rat mannose-binding protein A, in an antibody- and C1q-independent manner, has been demonstrated (Ikeda et al 1987; Lu et al 1990). Killing of bacteria mediated by mannose-binding protein and complement through the classical pathway has also been demonstrated (Kawasaki et al 1989). Although mannose-binding protein can activate the $C1r_2C1s_2$ complex in vitro, recent evidence suggests that it uses a novel C1s-like serine protease to initiate the complement cascade in vivo (Matsushita & Fujita 1992). This protease can be isolated, under certain conditions, in complex with human mannose-binding protein, and has both C4- and C2-consuming activities. As well as activating complement, human MBP may have a direct opsonic effect, since it can induce phagocytosis of a gram-negative bacterium *Salmonella montivideo* in the absence of serum (Kuhlman et al 1989), probably by interaction with leukocyte C1q receptors (Malhotra et al 1988).

Analysis of the structure and expression of the human mannose-binding protein gene has provided further evidence for the role of this protein in host defense. Serum mannose-binding protein is synthesized in the liver and secreted into serum. Synthesis and secretion are regulated as part of the acute phase response (Ezekowitz et al 1988). Thus serum levels can be increased rapidly in response to trauma when the risk of infection is high. The upstream region of the mannose-binding protein gene contains several control elements, including sequences resembling heat shock response elements and glucocorticoid receptor binding sites (Taylor et al 1989) that may be responsible for regulating expression.

Clinical evidence for the importance of mannose-binding protein has come from the identification of a mannose-binding protein deficiency syndrome (Super et al 1989). Low levels of mannose-binding protein in serum are associated with a defect in the ability to opsonize yeast. Infants who have this defect suffer from recurrent, severe infections caused by a range of microorganisms. The molecular basis of this defect is a point mutation in the first exon of the mannose-binding protein gene, which results in the replacement of glycine with aspartic acid (Sumiya et al 1991). This amino acid change disrupts one of the Gly-X-Y repeats of the collagen-like domains, which presumably prevents stable formation of the triple helix. Heterozygotes as well as homozygotes have low serum levels of mannose-binding protein since the presence of mutant polypeptides suppresses the formation of oligomers from normal polypeptides. Although the variant form of mannose-binding protein can be secreted from cells in culture, it is unclear how this observation reflects on the levels of the protein in vivo (Super et al 1992). Since adults carrying this mutation do not suffer from increased numbers of infections, it is believed that mannose-binding protein function is most important in infancy during the period between disappearance of maternal antibody and the acquisition of a full antibody repertoire. The frequency of this mutation in both Caucasian and Chinese populations (approximately 7%) suggests that it may confer selective advantage under certain circumstances (Lipscombe et al 1992).

PULMONARY SURFACTANT APOPROTEINS Pulmonary surfactant apoproteins A and D (SP-A and SP-D) are two components of the surfactant that line alveoli in the lung. SP-A interacts with several monosaccharides, including mannose, glucose, fucose, and galactose, but not with amino sugars (Haagsman et al 1987; Childs et al 1992), and SP-D interacts with glucose-containing structures (Persson et al 1990). SP-A has been shown to recognize several natural glycoproteins and glycolipids (Childs et al 1992). Both the carbohydrate and lipid moieties of glycolipids seem to be involved in SP-A binding. SP-D binds strongly to phosphatidylinositol (Persson et al 1992).

Like mannose-binding protein, both SP-A and SP-D can bind pathogenic organisms through their surface carbohydrates. SP-A can bind *Pneumocystis carinii,* a common cause of pneumonia in immunocompromised patients (Zimmerman et al 1992), while SP-D has been shown to bind *Escherichia coli* and other gram-negative bacteria (Kuan et al 1992). These interactions are inhibitable by the appropriate sugars. Complement fixation by these two proteins has not yet been demonstrated, although SP-A does interact with the leukocyte C1q receptor (Malhotra et al 1988).

While considerable evidence suggests that SP-A and SP-D have a role in defending the lung against infection, alternative functions have been proposed. Interaction of SP-A and SP-D with endogenous glycolipids in lung fluid and on the surfaces of alveolar cells are consistent with proposed roles in the uptake and recycling of surfactant protein-lipid complexes (Wright et al 1989). It has also been proposed that interaction of SP-A with both lipids and carbohydrates may be critical for formation of tubular myelin in the lung (Childs et al 1992).

Proteoglycans

C-type CRDs have been identified in three proteoglycans: aggrecan, the primary proteoglycan of cartilage matrix (Doege et al 1991), versican, a high molecular weight proteoglycan synthesized by fibroblasts (Zimmermann

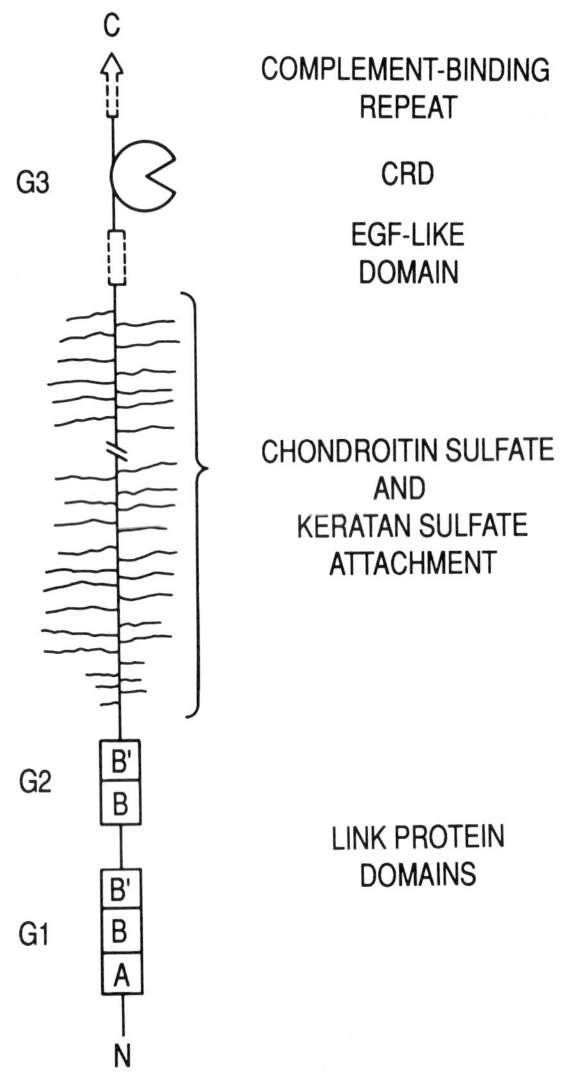

Figure 3 Domain organization of the proteoglycans that contain C-type CRDs. The G2 domain, found in cartilage aggrecan, is absent from versican and neurocan. Alternative splicing of aggrecan mRNA generates forms in which either the adjacent epidermal growth factor-like domain or the complement regulatory repeat are not present. Two copies of the epidermal growth factor-like domain are found in versican and neurocan.

& Ruoslahti 1989), and neurocan, a proteoglycan found in brain (Rauch et al 1992). The primary structures of the core proteins from these proteoglycans are summarized in Figure 3. As in the selectins, the CRDs are adjacent to epidermal growth factor-like sequences and complement-regulatory repeats, but the order of the domains is different. Alternative splicing of aggrecan can generate forms in which some of these domains are missing.

The CRD of aggrecan expressed in an in vitro translation system binds densely substituted sugar resins. A solid phase binding assay employing the domain expressed in bacteria reveals that this CRD has unusually broad saccharide-binding specificity and that the affinity for monosaccharides is as much as tenfold weaker than the affinity of other C-type CRDs such as those from the asialoglycoprotein receptor and mannose-binding protein (Saleque et al 1993). The best ligands are fucose and galactose. The COOH-terminal portion of aggrecan, including the CRD, has been visualized in the electron microscope as a globular domain designated G3 (Paulsson et al 1987). The G3 domain appears to be absent in older tissue, which suggests that it may serve a transient function during cartilage development.

Isolated Carbohydrate-Recognition Domains

Based on the presence of a shared sequence motif similar to that seen in C-type lectins, CRDs without flanking accessory domains have been described. Only lower vertebrate and invertebrate proteins of this type (Drickamer 1993a) have been shown to have sugar-binding activity, but experimental evidence for the function of the proteins with known sugar-binding activity is lacking. The most coherent cluster of free CRDs in vertebrates consists of proteins associated with the pancreas. Two distinct human proteins fall into this recently described group (C-type animal lectins group VII). The first is the protein component of pancreatic stones (deCaro et al 1987); the second was initially identified as a cDNA, designated HIP, expressed in hepatocellular carcinomas, but not in normal hepatic tissue (Lasserre et al 1992). This second protein is found in normal pancreas as well. Expression of the rat homologue of the stone protein is increased in regenerating pancreas (Watanabe et al 1990). A rat protein associated with pancreatitis is the homologue of the HIP protein (Iovanna et al 1991). The pancreatic thread protein, which forms filaments at neutral pH, is the bovine homologue of HIP and pancreatitis-associated protein (Iovanna et al 1991).

Two additional mammalian proteins consist of CRDs alone, although they are not closely related in sequence to the group VII proteins or to CRDs in other C-type lectins. These are tetranectin, a serum protein that binds to the fourth kringle domain of plasminogen (Fuhlendorff et al 1987), and the major basic protein of eosinophil granules, which has also been described as a T cell-derived immunoregulatory factor (Yoshimatsu et al 1992). Two different antifreeze proteins from arctic fish serum also have many of the sequence features of C-type CRDs (Ng & Hew 1992).

Mechanism of Carbohydrate Recognition

CARBOHYDRATE-RECOGNITION DOMAIN STRUCTURE The structure of the CRD from rat serum mannose-binding protein, shown in Figure 4, has been deduced by analysis of crystals produced in the presence and absence of a mannose-containing oligosaccharide (Weis et al 1991a,b, 1992). It is characterized by the presence of a modest amount of regular secondary structure, which is largely confined to the lower two thirds of the domain, while the upper third consists of a continuous segment of polypeptide folded into a series of four loops. These loops form binding sites for two Ca^{2+}s. The domain is stabilized by two disulfide bonds.

The structure is consistent with many aspects of the biochemistry of the C-type lectins. For instance, the mannose-binding CRD shows second order dependence on Ca^{2+} for ligand binding, a fact that is explained by the presence of two cations in the structure. The arrangement of the NH_2- and COOH-termini near each other explains how CRDs at the NH_2-terminal end

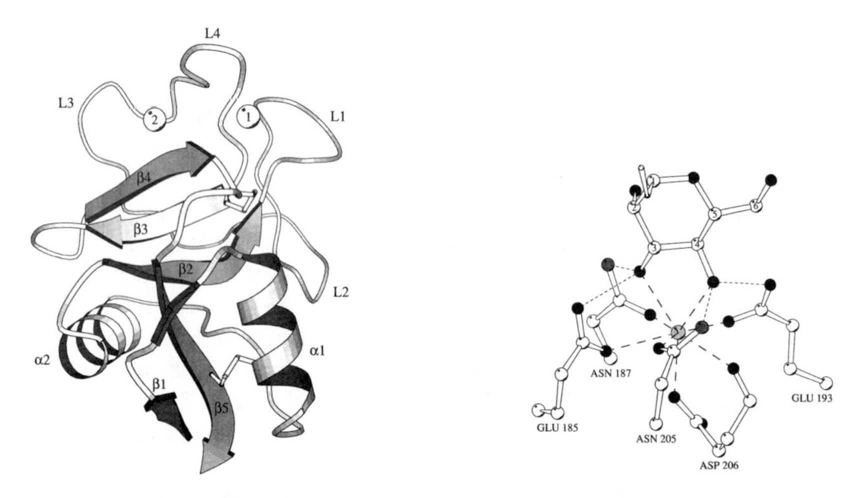

Figure 4 Structure of C-type CRD from rat serum mannose-binding protein. (*left*) Ribbon diagram showing secondary structure elements of the mannose-binding CRD. Spheres 1 and 2 represent calcium ions. (*right*) Detailed view of region surrounding Ca^{2+} 2 in complex with mannose-containing ligand. The calcium ion is shown as a light grey sphere. White, dark grey, and black spheres represent carbon, nitrogen, and oxygen, respectively. Ca^{2+} coordination bonds are denoted by long thick dashes, while short dashes represent hydrogen bonds. Numbers on the mannose carbon atoms represent ring positions. Reprinted from Weis et al (1992) with permission.

of type I transmembrane proteins, such as the selectins, and at the COOH-terminal end of type II transmembrane proteins, such as the asialoglycoprotein receptor, can be projected from the cell surface in a similar orientation. Finally, the compactness of the domain explains its protease resistance, which is lost upon removal of Ca^{2+}, presumably because the upper loops become exposed to proteases in the absence of bound cation.

The pattern of conserved residues that forms the sequence motif characteristic of C-type CRDs is also explained by the structure. The highly conserved residues include four cysteines that form the two disulfide bonds. In addition, many of the Ca^{2+} ligands, consisting of glutamic acid, aspartic acid, and asparagine residues, are conserved in all the C-type CRDs. Several glycine and proline residues found at critical turn positions are invariant, while the remainder of the conserved positions are occupied by aliphatic and aromatic residues that constitute the hydrophobic core of the domain.

LIGAND BINDING As shown in Figure 4, interaction between the CRD and the sugar is limited to a small portion of the protein near Ca^{2+} 2. The local structure of the protein-sugar complex reveals an intimate association between equatorial hydroxyl groups 3 and 4 of the terminal mannose residue, and Ca^{2+} 2. These two hydroxyl groups serve as coordination ligands for the Ca^{2+}. The complex is stabilized by two hydrogen bonds to each hydroxyl group, involving asparagine and glutamic acid residues that also serve as ligands for the Ca^{2+}.

From the perspective of sugar recognition, the structure of the complex is entirely consistent with the known specificity of the mannose-binding CRD. Other sugars with equatorial 3 and 4 hydroxyl groups in a similar orientation, such as N-acetylglucosamine, are also ligands for the protein. The importance of these two groups was emphasized in studies using various modified forms of the sugar ligand (RT Lee et al 1991). In addition, L-fucose can apparently bind because its 2 and 3 hydroxyl groups are oriented in the same way as the 3 and 4 hydroxyl groups of mannose.

Comparison of the sequences of CRDs with demonstrated sugar-binding activity reveals the conserved presence of three of the five ligands for Ca^{2+} 2: Asp206, Asn205, and Glu193. The remaining two ligands, found in the sequence Glu-Pro-$Asn^{185-187}$, are always present in lectins that bind ligands similar to those bound by mannose-binding protein, but the sequence is invariably changed to Gln-Pro-Asp in lectins that bind galactose. Changing the mannose-binding CRD to include this alternative sequence changes the preferred binding selectivity from mannose to galactose (Drickamer 1992). These results indicate that the binding sites in these diverse C-type CRDs probably involve the same region around Ca^{2+} 2 and that the residues at positions corresponding to residues 185–187 in the mannose-binding CRD are likely to be critical determinants of ligand-binding specificity.

REVERSIBILITY OF BINDING A pH-dependent change in ligand-binding activity allows C-type CRDs in endocytic receptors to release their ligands in the slightly acidic luminal environment of endosomes so that the receptors can be recycled to the cell surface while the ligands are routed to lysosomes for degradation (Mellman et al 1986). Studies of the CRD from the chicken hepatic lectin show that as the pH is reduced, the affinity of the CRD for Ca^{2+} declines (Loeb & Drickamer 1988). This loss of affinity means that the domain does not have bound Ca^{2+} at endosomal pH. In the absence of bound Ca^{2+}, no sugar-binding site is formed.

A novel class of mutations in the mannose-binding CRD, isolated by random mutagenesis, manifests loss of sugar binding activity under physiological conditions because of slighly decreased affinity for Ca^{2+} (Quesenberry & Drickamer 1992). In this respect, they mimic the low pH conformation of the CRD. The residues changed in these mutants are generally amino acids that form part of the hydrophobic core and are at some distance from the Ca^{2+}-binding loops. Their phenotype suggests that a very slight alteration in the arrangement of the more rigid lower portion of the CRD disposes the upper loops in such a way that the Ca^{2+}-binding sites do not form as readily as in the wild-type domain. Using these mutated domains as a model, it has been proposed that lowering the pH results in titration of groups in the lower portion of the CRD, which indirectly causes loss of sugar-binding activity by reducing the affinity of the domain for Ca^{2+}.

VALENCY As noted above, clustering of CRDs is one mechanism by which affinity and specificity for complex oligosaccharide ligands is increased. Recent analysis of E-selectin suggests that incorporation of secondary binding sites within a single CRD is an alternative way to achieve this result. Modeling suggests that the terminal fucose of the sialyl Lewis X ligand may bind in a manner analogous to the way that it binds to the mannose-binding protein. Mutational analysis indicates that positively charged residues located at either end of strand β5 play an important role in ligand binding (Erbe et al 1992). It is possible that these residues contribute to formation of a sialic acid-binding subsite.

EVOLUTION OF C-TYPE CRDS The evolution of C-type animal lectins has been discussed in detail in two recent reviews (Drickamer 1993a,b). The diversity in protein architecture found in the lectins is presumed to result from shuffling of exons encoding the CRD and others domains. Consistent with the structural studies, relatively minor changes near the sugar-binding site have allowed the CRDs to accommodate a variety of ligands, some of which are probably nonsaccharide structures.

SOLUBLE β-GALACTOSIDE-BINDING (S-TYPE) LECTINS

Organization of S-Type Lectins

The galactose-specific S-type lectins bind β-galactosides in a Ca^{2+}-independent manner (Wang et al 1991). S-type lectins have been characterized from many different species. They show a wide tissue distribution and are usually found intracellularly. S-type lectins for which primary structures have been determined fall into three distinct groups: L-14, L-30 and L-36 (following the nomenclature of Wang et al 1991, and Oda et al 1993). The organization of their constituent polypeptides is summarized in Figure 5.

The CRDs in L-14 and L-30 lectins from various mammalian and bird

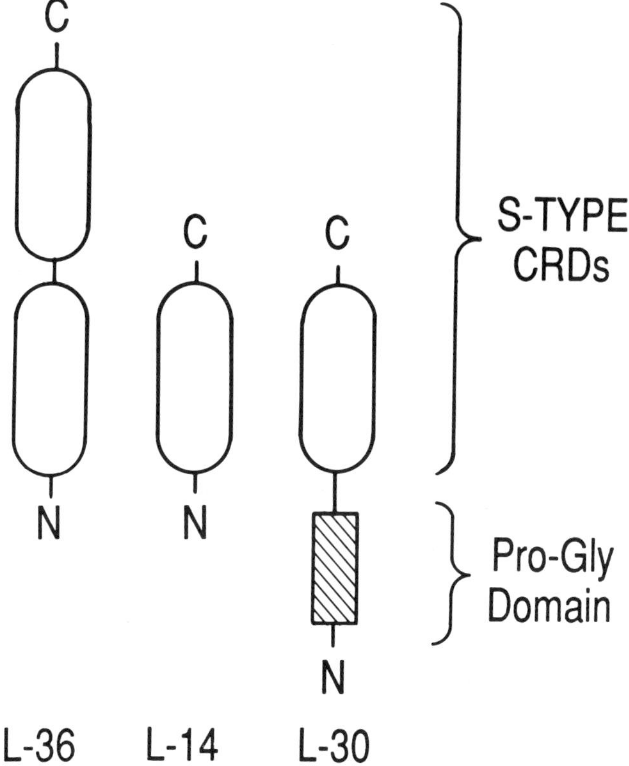

Figure 5 Structure of three groups of S-type animal lectins. The S-type CRDs are denoted by ovals.

species share 19 invariant and 36 conserved residues, which define the S-type CRD (Abbott & Feizi 1991; Hirabayashi & Kasai 1991; Wang et al 1991). Homologous L-14 lectins from fish and amphibians share many of the same residues (Marschal et al 1992). No cysteine residues are invariant in the S-type motif, and all the cysteine residues in these proteins appear to be present as sulfhydryl groups rather than as disulfides. The designation of these lectins as S-type was based on the fact that many are stabilized by the presence of thiols. However, the proteins are not inactivated by alkylation; both of the cysteine residues in a chicken S-type lectin can be changed to serine residues without affecting sugar binding (Hirabayashi & Kasai 1991), and none is present in L-36 (Oda et al 1993). Thus it seems unlikely that sulfhydryl groups per se are required for activity. However, oxidation of sulfhydryls to form disulfides does result in inactivation of bovine L-14, thus explaining the thiol dependence of this lectin (Tracey et al 1992).

L-14 Lectins

L-14 S-type lectins contain subunits of approximately 14,000 daltons and are found as homodimers. Expression of a truncated L-14 lectin shows that almost the entire polypeptide is required for carbohydrate binding (Abbott & Feizi 1991). Two distinct L-14 lectins have been described in humans (Gitt et al 1992) and in chickens (Sasakura et al 1990).

Several different functions have been proposed for L-14 S-type lectins, but none has been convincingly proven. For example, these lectins can bind with high affinity to the polylactosamine structures of laminin, which suggests a role in cell-matrix interactions (Sato & Hughes 1992). The bivalent nature of these lectins, resulting from their dimerization, would allow them to link together glycoconjugates. L-14 lectins can be found extracellularly as well as in the cytoplasm, but the mechanism by which they are released from cells is unclear, since the cDNAs do not encode a signal sequence. Mechanisms involving protein pumps or intracellular vesiculation have been proposed (Cooper & Barondes 1990).

Expression of L-14 is developmentally regulated. For example, during myoblast differentiation, the localization of L-14 changes from the cytoplasm to the extracellular fluid, which suggests that these lectins may have a role in cell-matrix interactions during muscle development (Cooper & Barondes 1990). A role in cellular differentiation is also supported by the appearance of L-14 in the trophoectoderm of the preimplantation blastocyst, later in myotomes, and in specific subsets of neurons (Hynes et al 1990; Poirier et al 1992). Since L-14 lectins are found in high concentrations in many tissues, even apparently pure preparations may be contaminated with lower abundance proteins that confer upon the preparation various biological activities. This problem must be dealt with before association of L-14 lectins with activities such as growth promotion or suppression can be considered proven (Sanford & Harris-Hooker 1990; Wells & Mallucci 1991).

L-30 Lectins

L-30 S-type lectins have been isolated only from mammalian species. They have molecular weights of approximately 30 K and usually exist as monomers. L-30 polypeptides consist of a C-terminal CRD, homologous to that of the L-14 lectins, coupled to another domain, which is rich in proline and glycine (Wang et al 1991).

Like the L-14 lectins, the L-30 lectins can be found both intracellularly and extracellularly, although the L-30 cDNA also does not encode a signal sequence. Two different types of function have been described for intracellular and extracellular forms of L-30. In its extracellular form, L-30 has been independently identified as a macrophage cell-surface antigen (Mac-2; Cherayil et al 1989), as a tumor cell-surface lectin (Raz et al 1989), and as an IgE-binding protein in basophilic leukemia cells (Albrandt et al 1987). Thus there are hints that it may be involved in modulation of an immune function. A possible explanation for finding L-30 associated with the cell surface is that it binds to exposed galactose residues on plasma membrane glycoconjugates. A proposed role of L-30 in cell-matrix adhesion would require the protein to be multivalent, although most studies have shown that it exists as a monomer (Wang et al 1991). However, under some conditions, L-30 can form dimers that are stabilized either by interchain disulfide bonds (Woo et al 1991), or by association of the proline and glycine-rich domain (Hsu et al 1992).

The L-30 lectin of mouse fibroblasts is located predominantly in the nucleus, in the form of a ribonucleoprotein complex. The lectin moves from the cytoplasm to the nucleus as part of the RNA-protein complex as the cell changes from the proliferating to the quiescent state (Agrwal et al 1989). These observations and the homology of the NH_2-terminal domain of L-30 to other proteins of ribonuclear complexes are consistent with a possible role of the lectin in RNA processing (Jia & Wang 1988; Wang et al 1991). It is possible that L-30 interacts with O-linked sugars found on proteins of transcription complexes and nuclear pore complexes (Hart et al 1989). So far only N-acetylglucosamine, which is not a ligand for L-30, has been found associated with these proteins, but it is possible that other sugars remain to be identified.

L-36 Lectins

L-36 S-type lectins from rat (Oda et al 1993) and from the nematode, *Caenorhabditis elegans* (Hirabayashi et al 1992), have recently been characterized. The polypeptides of these lectins consists of two S-type CRDs in tandem. Each of the CRDs contains most of the invariant and conserved residues found in the vertebrate L-14 and L-30 lectins, and each of the CRDs has saccharide-binding activity.

OTHER GROUPS OF ANIMAL LECTINS

Mannose 6-phosphate receptors and their role in intracellular targeting of lysosomal enzymes have been considered in detail in a recent review (Kornfeld 1992). They share a common sequence motif in their extracytoplasmic domains. One P-type CRD is present in the cation-dependent receptor, while the motif is repeated 15 times in the cation-independent receptor. The P-type CRD motif has not been seen in any other proteins and is not related to the C-type or S-type CRD motifs. The importance of clustered CRDs in the recognition of multivalent ligands is reminiscent of the situation with C-type animal lectins. The finding that the cation-independent mannose 6-phosphate receptor is also the receptor for insulin-like growth factor II suggests that some of the domains may be active in binding other ligands.

The pentraxins, C-reactive protein and serum amyloid P are serum proteins that share the property of forming oligomers based on an annular pentamer. Although their sequences are not related to C-type CRDs, these proteins show Ca^{2+}-dependent binding properties, including binding to bacterial cell surfaces. Serum amyloid P shows affinity for specific phosphorylated and sulfated saccharides (Loveless et al 1992). C-reactive protein, which is an acute phase reactant, has been identified as a galactose-specific particle receptor on liver macrophages (Kempka et al 1990). The properties of these proteins are in some respects parallel to those of the mannose-binding proteins.

SUMMARY AND PERSPECTIVE

In spite of the great diversity of animal lectins, it is striking that the ability to bind saccharides often resides in a discrete domain (the CRD) that functions somewhat independently of the rest of the molecule in which it is found. The division of the lectins into carbohydrate-binding and effector domains is of great experimental value and provides a useful conceptual approach to lectin structure. This division may also reflect an evolutionary coming together of the effector functions with sugar-binding domains able to target a desired activity with greater precision. The existence of domains that are clearly homologous with the P or C-type CRDs, but that do not have demonstrable saccharide-binding activity, suggests that the CRDs have evolved from more general binding domains. One challenge will be to show

how the unique biological properties of the lectins derive from the molecular details that are now being deciphered.

Additional animal proteins with sugar-binding activity have been isolated, but have not yet been characterized at the primary structure level. A striking example of such a lectin is sialoadhesin, a macrophage surface protein that mediates sialic acid-dependent interactions with specific subpopulations of hematopoietic cells and lymphocytes (Crocker et al 1991). Isolation of a cDNA for sialoadhesin will reveal whether it is related to other sialic acid-binding lectins such as influenza virus hemagglutinin and will provide an important tool for defining its natural ligands. Characterization of other lectin-like activities, such as the β-glucan receptor of macrophages, has just reached the point where sugar-binding activity can be associated with specific polypeptides (Czop & Kay 1991).

Finally, in some systems a role for specific sugars is evident, but there is no consensus about what lectins recognize these sugars and mediate their biological activities. Particular interest is directed toward the initial interaction between sperm and egg, which in mammalian systems seems to involve α-galactosyl residues in the zona pelucida (Florman & Wassarman 1986). Clear identification of a sperm receptor for these sugars is a task for the future. The best understood examples of saccharide-mediated cell adhesion involve dynamic systems such as leukocyte-endothelium interaction. Examples of carbohydrate-lectin interactions in more static systems, such as those between slowly migrating cells in development, also remain to be defined at the molecular level.

ACKNOWLEDGMENTS

KD is supported by grant GM42628 from the National Institutes of Health and a Faculty Salary Award from the American Cancer Society. We thank David Ashford for help with transatlantic communication. We apologize to those whose relevant work is not discussed and/or cited because of space limitations.

Literature Cited

Abbott, WM, Feizi, T. 1991. Soluble 14-kDa β-galactoside-specific bovine lectin: evidence from mutagenesis and proteolysis that almost the complete chain is necessary for integrity of the carbohydrate-recognition domain. *J. Biol. Chem.* 266:5552–57

Agrwal, N, Wang, JL, Voss, PG. 1989. Carbohydrate-binding protein 35: Levels of transcription and mRNA accumulation in quiescent and proliferating cells. *J. Biol. Chem.* 264:17236–42

Albrandt, K, Orida, NK, Liu, FT. 1987. An IgE-binding protein with a distinctive repetitive sequence and homology with an IgG receptor. *Proc. Natl. Acad. Sci. USA* 84: 6859–63

Aruffo, A, Kolanus, W, Walz, G, Fredman, P, Seed, B. 1991. CD62/P-selectin recognition of myeloid and tumor cell sulfatides. *Cell* 67:35–44

Ashwell, G, Harford, J. 1982. Carbohydrate-specific receptors of the liver. *Annu. Rev. Biochem.* 51:531–54

Aubry, J-P, Pochon, S, Graber, P, Jansen,

KU, Bonnefoy, J-Y. 1992. CD21 is a ligand for CD23 and regulates IgE production. *Nature* 358:505–7

Beavil, AJ, Edmeades, RL, Gould, HJ, Sutton, BJ. 1992. α-Helical coiled-coil stalks in the low-affinity receptor for IgE (F_cεRII/CD23) and related C-type lectins. *Proc. Natl. Acad. Sci. USA* 89:753–57

Benson, B, Hawgood, S, Schilling, J, Clements, J, Damm, D, et al. 1985. Structure of canine pulmonary surfactant apoprotein: cDNA and complete amino acid sequence. *Proc. Natl. Acad. Sci. USA* 82:6379–83

Bettler, B, Maier, R, Ruegg, D, Hoffstetter, H. 1989. Binding site for IgE of the human lymphocyte low-affinity F_cε receptor (F_cε/CD23) is confined to the domain homologous with animal lectins. *Proc. Natl. Acad. Sci. USA* 86:7118–22

Bettler, B, Texido, G, Raggini, S, Rüegg, D, Hofstetter, H. 1992. Immunoglobulin E-binding site in F_cε receptor (F_cεRII/CD23) identified by homolog-scanning mutagenesis. *J. Biol. Chem.* 267:185–91

Chakraborty, P, Das, PK. 1988. Role of mannose/N-acetylglucosamine receptors in blood clearance and cellular attachment of *Leishmania donovani*. *Mol. Biochem. Parasitol.* 28:55–62

Cherayil, BJ, Weiner, SJ, Pillai, S. 1989. The Mac-2 antigen is a galactose-specific lectin that binds IgEJ. *Exp. Med.* 170:1959–72

Childs, RA, Feizi, T, Yuen, CT, Drickamer, K, Quesenberry, MS. 1990. Differential recognition of core and terminal portions of oligosaccharide ligands by carbohydrate-recognition domains of two mannose binding proteins. *J. Biol. Chem.* 265:20770–77

Childs, RA, Wright, JR, Ross, GF, Yuen, C-T, Lawson, AM, et al. 1992. Specificity of lung surfactant protein SP-A for both the protein and the lipid moieties of certain neutral glycolipids. *J. Biol. Chem.* 267: 9972–79

Cooper, DN, Barondes, SH. 1990. Evidence of export of a muscle lectin from cytosol to extracellular matrix and for a novel secretory mechanism. *J. Cell Biol.* 110:1681–91

Crocker, PR, Kelm, S, Dubois, C, Martin, B, McWilliam, AS, et al. 1991. Purification and properties of sialoadhesin, a sialic-binding receptor of murine tissue macrophages. *EMBO J.* 10:1661–69

Curtis, BM, Scharnowske, S, Watson, AJ. 1992. Sequence and expression of a membrane-associated C-type lectin that exhibits CD4-independent binding of human immunodeficiency virus envelope glycoprotein gp120. *Proc. Natl. Acad. Sci. USA* 89: 8356–60

Czop, JK, Kay, J. 1991. Isolation and characterization of β-glucan receptors on human mononuclear phagocytes. *J. Exp. Med.* 173: 1511–20

deCaro, AM, Bonicel, JJ, Rouimi, P, deCaro, JD, Sarles, H, Rovery, M. 1987. Complete amino acid sequence of an immunoreactive form of human pancreatic stone protein isolated from pancreatic juice. *Eur. J. Biochem.* 168:201–7

Doege, KJ, Sasaki, M, Kimura, T, Yamada, Y. 1991. Complete coding sequence and deduced primary structure of the human cartilage large aggregating proteoglycan, aggrecan: human-specific repeats, and additional alternatively spliced forms. *J. Biol. Chem.* 266:894–902

Drickamer, K. 1988. Two distinct classes of carbohydrate-recognition domains in animal lectins. *J. Biol. Chem.* 263:9557–60

Drickamer, K. 1991. Clearing up glycoprotein hormones. *Cell* 67:1029–32

Drickamer, K. 1992. Engineering galactose-binding activity into a C-type mannose-binding protein. *Nature* 360:183–86

Drickamer, K. 1993a. Ca^{2+}-dependent carbohydrate-recognition domains in animal proteins. *Curr. Opin. Struct. Biol.* 3:393–400

Drickamer, K. 1993b. Evolution of Ca^{2+}-dependent animal lectins. *Prog. Nucleic Acid Res. Mol. Biol.* 45:207–32

Drickamer, K, Dordal, MS, Reynolds, L. 1986. Mannose-binding proteins isolated from rat liver contain carbohydrate-recognition domains linked to collagenous tails. *J. Biol. Chem.* 261:1034–46

Erbe, DV, Wolitzky, BA, Presta, LG, Norton, CR, Ramos, RJ, et al. 1992. Identification of an E-selectin region critical for carbohydrate recognition and cell adhesion. *J. Cell Biol.* 119:215–27

Ezekowitz, RAB, Day, LE, Herman, GA. 1988. A human mannose-binding protein is an acute phase reactant that shares sequence homology with other vertebrate lectins. *J. Exp. Med.* 167:1034–46

Ezekowitz, RAB, Sastry, K, Bailly, P, Warner, A. 1990. Molecular characterization of the human macrophage mannose receptor: Demonstration of multiple carbohydrate recognition-like domains and phagocytosis of yeasts in Cos-1 cells. *J. Exp. Med.* 172:1785–94

Ezekowitz, RAB, Williams, DJ, Koziel, H, Armstrong, MYK, Warner, A, et al. 1991. Uptake of *Pneumocystis carinii* mediated by the macrophage mannose receptor. *Nature* 351:155–58

Facy, P, Seve, A-P, Hubert, M, Monsigny, M, Hubert, J. 1990. Analysis of nuclear sugar-binding components in undifferentiated and in vitro differentiated human promyelocytic leukemia cells (HL60). *Exp. Cell Res.* 190:151–60

Florman, HM, Wassarman, PM. 1986. O-Linked oligosaccharides of mouse egg ZP3 account for its sperm receptor activity. *Cell* 41:313–24

Fuhlendorff, J, Clemmensen, I, Magnusson, S. 1987. Primary structure of tetranectin, a plasminogen kringle 4 binding plasma protein: homology with asialoglycoprotein receptor and cartilage proteoglycan core protein. *Biochemistry* 26:6757–64

Gabius, H-J. 1991. Detection and functions of mammalian lectins, with emphasis on membrane lectins. *Biochim. Biophys. Acta* 1071:1–18

Gitt, MW, Massa, SM, Leffler, H, Barondes, SH. 1992. Isolation and expression of a gene encoding L-14-II, a new human soluble lactose-binding lectin. *J. Biol. Chem.* 267:10601–6

Haagsman, HP, Hawgood, S, Sargeant, T, Buckley, D, White, RT, et al. 1987. The major lung surfactant protein, SP 28–36, is a calcium-dependent, carbohydrate-binding protein. *J. Biol. Chem.* 262:13877–80

Halberg, DF, Wager, RE, Farrell, DC, Hildreth, J, Quesenberry, MS, et al. 1987. Major and minor forms of the rat liver asialoglycoprotein receptor are independent galactose-binding proteins: primary structure and glycosylation heterogeneity of minor receptor forms. *J. Biol. Chem.* 262:9828–38

Hart, GW, Haltiwanger, RS, Holt, GD, Kelly, WG. 1989. Glycosylation in the nucleus and cytoplasm. *Annu. Rev. Biochem.* 58: 841–74

Herzig, M. CS, Weigel, PH. 1990. Surface and internal galactosyl receptors are heterooligomers and retain this structure after ligand internalization or receptor modulation. *Biochemistry* 29:6437–47

Hirabayashi, J, Kasai, K. 1991. Effect of amino acid substitution by site-directed mutagenesis on the carbohydrate-recognition and stability of human 14-kDa β-galactoside-binding lectin. *J. Biol. Chem.* 266: 23648–53

Hirabayashi, J, Satoh, M, Kasai, K. 1992. Evidence that *Caenorhabditis elegans* 32-kDa β-galactoside-binding protein in homologous to vertebrate β-galactoside-binding lectins: cDNA cloning and deduced amino acid sequence. *J. Biol. Chem.* 267: 15485–90

Houchins, JP, Yabe, T, McSherry, C, Bach, FH. 1991. DNA sequence analysis of NKG2, a family of related cDNA clones encoding type II integral membrane proteins on human natural killer cells. *J. Exp. Med.* 173:1017–20

Hoyle, GW, Hill, RL. 1988. Molecular cloning and sequencing of a cDNA for a carbohydrate binding receptor unique to rat Kupffer cells. *J. Biol. Chem.* 263:7487–92

Hsu, DK, Zuberi, RI, Liu, FT. 1992. Biochemical and biophysical characterization of human recombinant IgE-binding protein, an S-type animal lectin. *J. Biol. Chem.* 267: 14167–74

Hynes, MA, Gitt, M, Barondes, SH, Jessell, TM, Buck, LB. 1990. Selective expression of an endogenous lactose-binding lectin gene in subsets of central and peripheral neurons. *J. Neurosci.* 10:1004–13

Ii, M, Kurata, H, Itoh, N, Yamashina, I, Kawasaki, T. 1990. Molecular cloning and sequence analysis of cDNA encoding the macrophage lectin specific for galactose and N-acetylgalactosamine. *J. Biol. Chem.* 265: 11295–98

Ikeda, K, Sannoh, T, Kawasaki, N, Kawasaki, T, Yamashina, I. 1987. Serum lectin with known structure activates complement through the classical pathway. *J. Biol. Chem.* 262:7451–54

Iovanna, J, Orelle, B, Keim, V, Dagorn, J-C. 1991. Messenger RNA sequence and expression of rat pancreatitis-associated protein, a lectin-related protein overexpressed during acute experimental pancreatitis. *J. Biol. Chem.* 266:24664–69

Jia, S, Wang, JL. 1988. Carbohydrate binding protein 35: Complementary DNA sequence reveals homology with proteins of the heterogeneous nuclear RNP. *J. Biol. Chem.* 263:6009–11

Kansas, GS, Spertini, O, Stoolman, LM, Tedder, TF. 1991. Molecular mapping of functional domains of the leukocyte receptor for endothelium, LAM-1. *J. Cell Biol.* 114: 351–58

Kawasaki, N, Kawasaki, T, Yamashina, I. 1989. A serum lectin (mannan-binding protein) has complement-dependent bactericidal activity. *J. Biochem.* 106:483–89

Kawasaki, T, Ashwell, G. 1977. Isolation and characterization of an avian hepatic binding protein specific for N-acetylglucosamine-terminated glycoproteins. *J. Biol. Chem.* 252:6536–43

Kempka, G, Roos, PH, Kolb-Bachoven, V. 1990. A membrane-associated form of C-reactive protein is the galactose-specific particle receptor on rat liver macrophages. *J. Immunol.* 144:1004–9

Kornfeld, S. 1992. Structure and function of the mannose 6-phosphate/insulin-like growth factor II receptors. *Annu. Rev. Biochem.* 61:307–30

Kuan, S-F, Rust, K, Crouch, E. 1992. Interactions of surfactant protein D with bacterial lipopolysaccharides: surfactant protein D is an *Escherichia coli*-binding protein in bronchoalveolar lavage. *J. Clin. Invest.* 90: 97–106

Kuhlman, M, Joiner, K, Ezekowitz, RAB. 1989. The human mannose-binding protein functions as an opsonin. *J. Exp. Med.* 169: 1733–45

Lasky, LA. 1992. Selectins: interpreters of cell-specific carbohydrate information during inflammation. *Science* 258:964–69

Lasserre, C, Christa, L, Simon, M-T, Vernier, P, Brechot, C. 1992. A novel gene (HIP) activated in human primary liver cancer. *Cancer Res.* 52:5089–95

Lee, RT, Ichikawa, Y, Fay, M, Drickamer, K, Shao, M-C, Lee, YC. 1991. Ligand-binding characteristics of rat serum-type mannose-binding protein (MBP-A): homology of binding site architecture with mammalian and chicken hepatic lectins. *J. Biol. Chem.* 266:4810–15

Lee, RT, Lin, P, Lee, YC. 1984. New synthetic cluster ligands for galactose/N-acetylgalactosamine-specific lectin of mammalian liver. *Biochemistry* 23:4255–61

Lee, RT, Rice, KG, Rao, N. BN, Ichikawa, Y, Barthel, T, et al. 1989. Binding characteristics of N-acetylglucosamine-specific lectin of the isolated chicken hepatocytes: similarities to mammalian hepatic galactose/N-acetylgalactosamine-specific lectin. *Biochemistry* 28:8351–58

Lee, Y-M, Leiby, KR, Allar, J, Paris, K, Lerch, B, Okarma, TB. 1991. Primary structure of bovine conglutinin, a member of the C-type animal lectin family. *J. Biol. Chem.* 266:2715–23

Lipscombe, RJ, Lau, YL, Levinsky, RJ, Sumiya, M, Summerfield, JA, Turner, MW. 1992. Identical point mutation leading to low levels of mannose binding protein and poor C3b mediated opsonization in Chinese and Caucasian populations. *Immunol. Lett.* 32:253–58

Lis, H, Sharon, N. 1986. Lectins as molecules and as tools. *Annu. Rev. Biochem.* 55:35–67

Loeb, JA, Drickamer, K. 1988. Conformational changes in the chicken receptor for endocytosis of glycoproteins: modulation of binding activity by Ca^{2+} and pH. *J. Biol. Chem.* 263:9752–60

Loveless, RW, Floyd-O'Sullivan, G, Raynes, JG, Yuen, C-T, Feizi, T. 1992. Human serum amyloid P is a multispecific adhesive protein whose ligands include 6-phosphorylated mannose and the 3-sulfated saccharides galactose, N-acetylgalactosamine and glucuronic acid. *EMBO J.* 11:813–19

Lu, J, Thiel, S, Wiedemann, H, Timpl, R, Reid, KBM. 1990. Binding of the pentamer/hexamer forms of a mannan-binding protein to zymosan activates the proenzyme $C1r_2C1s_2$ complex of the classical pathway of complement, without involvement of C1q. *J. Immunol.* 144:2287–94

Lu, J, Willis, AC, Reid, KBM. 1992. Purification, characterization and cDNA cloning of human lung surfactant protein D. *Biochem. J.* 284:795–802

Ludin, C, Hofstetter, H, Sargati, M, Levy, CA, Suter, U, et al. 1987. Cloning and expression of the cDNA coding for a human lymphocyte IgE receptor. *EMBO J.* 6:109–14

Malhotra, R, Howell, CJ, Spur, BW, Youlten, LJ, Clark, TJ, et al. 1988. Human leukocyte C1q receptor binds other soluble proteins with collagen domains. *J. Exp. Med.* 172: 955–59

Marschal, P, Herrmann, J, Leffler, H, Barondes, SH, Cooper, DNW. 1992. Sequence and specificity of a soluble lactose-binding lectin from *Xenopus laevis* skin. *J. Biol. Chem.* 267:12942–49

Matsushita, M, Fujita, T. 1992. Activation of the classical complement pathway by mannose-binding protein in association with a novel C1s-like serine protease. *J. Exp. Med.* 176:1497–502

McPhaul, M, Berg, P. 1986. Formation of functional asialoglycoprotein receptor after transfection with cDNAs encoding the receptor proteins. *Proc. Natl. Acad. Sci. USA* 83:8863–67

Mellman, I, Fuchs, R, Helenius, A. 1986. Acidification of the endocytic and exocytic pathways. *Annu. Rev. Biochem.* 55:663–700

Ng, NFL, Hew, CL. 1992. Structure of an antifreeze polypeptide from the sea raven: disulfide bonds and similarity to lectin-binding proteins. *J. Biol. Chem.* 267:16069–75

Oda, Y, Herrmann, J, Gitt, MA, Turck, CW, Burlingame, AL, et al. 1993. Soluble lactose-binding lectin from rat intestine with two different carbohydrate-binding domains in the same peptide chain. *J. Biol. Chem.* 268:5929–39

Otter, M, Barrett-Bergshoeff, MM, Rijken, DC. 1991. Binding of tissue-type plasminogen activator by the mannose receptor. *J. Biol. Chem.* 266:13931–35

Parekh, RB. 1991. Effects of glycosylation on protein function. *Curr. Opin. Struct. Biol.* 1:750–54

Paulsson, M, Mörgelin, M, Wiedmann, H, Beardmore-Gray, M, Dunham, D, et al. 1987. Extended and globular protein domains in cartilage proteoglycans. *Biochem. J.* 245:763–72

Persson, A, Chang, D, Crouch, E. 1990. Surfactant protein D is a divalent cation-dependent carbohydrate-binding protein. *J. Biol. Chem.* 265:5755–60

Persson, AV, Gibbons, BJ, Shoemaker, JD, Moxley, MA, Longmore, WJ. 1992. The major glycolipid recognized by SP-D in surfactant is phosphatidylinositol. *Biochemistry* 31:12183–89

Pochon, S, Graber, P, Yeager, M, Jansen, K, Bernrad, AR, et al. 1992. Demonstration of a second ligand for the low affinity receptor for immunoglobulin E (CD23) using recombinant CD23 reconstituted into fluorescent microsomes. *J. Exp. Med.* 176: 389–97

Poirier, F, Timmons, PM, Chan, C. TJ, Guénet, JL, Rigby, PW. J. 1992. Expression of the L-14 soluble lectin during mouse embryogenesis suggests multiple roles during pre- and post-implantation development. *Development* 115:143–55

Quesenberry, MS, Drickamer, K. 1992. Role of conserved and nonconserved residues in the Ca^{2+}-dependent carbohydrate-recognition domain of a rat mannose-binding protein: analysis by random cassette mutagenesis. *J. Biol. Chem.* 267:10831–41

Rauch, U, Karthikeyan, L, Maurel, P, Margolis, RU, Margolis, RK. 1992. Cloning and primary structure of neurocan, a developmentally regulated, aggregating chondroitin sulfate proteoglycan of brain. *J. Biol. Chem.* 267:19536–47

Raz, A, Pazerininin, G, Carmi, P. 1989. Identification of the metastasis-associated galactoside-binding lectin as a chimeric gene product with homology to an IgE-binding protein. *Cancer Res.* 49:3489–93

Rice, KG, Weisz, OA, Barthel, T, Lee, RT, Lee, YC. 1990. Defined stoichiometry of binding between triantennary glycopeptide and the asialoglycoprotein receptor of rat hepatocytes. *J. Biol. Chem.* 265: 18429–34

Richards, ML, Katz, DH. 1990. The binding of IgE to murine F_c RII is calcium-dependent but not inhibited by carbohydrate. *J. Immunol.* 144:2638–46

Rust, K, Grosso, L, Zhang, V, Chang, D, Persson, A, et al. 1991. Human surfactant protein D: SP-D contains a C-type lectin carbohydrate recognition domain. *Arch. Biochem. Biophys.* 290:116–26

Saleque, S, Ruiz, N, Drickamer, K. 1993. Expression and characterization of a carbohydrate-binding fragment of rat aggrecan. *Glycobiology* 3:185–90

Sanford, GL, Harris-Hooker, S. 1990. Stimulation of vascular cell proliferation by β-galactoside specific lectins. *FASEB J.* 4: 2912–18

Sasakura, Y, Hirabayashi, J, Oda, Y, Ohyama, Y, Kasai, K. 1990. Structure of chicken 16-kDa β-galactoside-binding lectin: complete amino acid sequence, cloning of cDNA, and production of recombinant lectin. *J. Biol. Chem.* 265:21573–79

Sato, M, Kawakami, K, Osawa, T, Toyoshima, S. 1992. Molecular cloning and expression of cDNA encoding a galactose/N-acetylgalactosamine-specific lectin on mouse tumoricidal macrophages. *J. Biochem.* 111:331–36

Sato, S, Hughes, RC. 1992. Bind specificity of a baby hamster kidney lectin for H type I and II chains, polylactosamine glycans, and appropriately glycosylated forms of laminin and fibronectin. *J. Biol. Chem.* 267:6983–90

Spiess, M. 1990. The asialoglycoprotein receptor: a model for endocytic transport receptors. *Biochemistry* 29:10008–19

Stahl, PD. 1990. The macrophage mannose receptor: current status. *Am. J. Respir. Cell. Mol. Biol.* 2:317–18

Stahl, PD, Schlesinger, PH. 1980. Receptor-mediated pinocytosis of mannose/N-acetylglucosamine-terminated glycoproteins and lysosomal enzymes by macrophages. *Trends Biochem. Sci.* 5:194–96

Sumiya, M, Super, M, Tabona, P, Levinsky, RJ, Takayuki, A, et al. 1991. Molecular basis of opsonic defect in immunodeficient children. *Lancet* 337:1569–70

Super, M, Gillies, SD, Foley, S, Sastry, K, Schweinle, J-E, et al. 1992. Distinct and overlapping function of allelic forms of human mannose binding protein. *Nature Genet.* 2:50–55

Super, M, Thiel, S, Lu, J, Levinsky, RJ, Turner, MW. 1989. Association of low levels of mannan-binding protein with a common defect of opsonization. *Lancet* ii:1236–39

Taylor, ME, Bezouska, K, Drickamer, K. 1992. Contribution to ligand binding by multiple carbohydrate-recognition domains in the macrophage mannose receptor. *J. Biol. Chem.* 267:1719–26

Taylor, ME, Brickell, PM, Craig, RK, Summerfield, JA. 1989. Structure and evolutionary origin of the gene encoding a human serum mannose-binding protein. *Biochem. J.* 262:763–71

Taylor, ME, Conary, JT, Lennartz, MR, Stahl, PD, Drickamer, K. 1990. Primary structure of the mannose receptor contains multiple motifs resembling carbohydrate-recognition domains. *J. Biol. Chem.* 265: 12156–62

Taylor, ME, Drickamer, K. 1993. Structural requirements for high affinity binding of complex ligands by the macrophage mannose receptor. *J. Biol. Chem.* 268:399–404

Taylor, ME, Summerfield, JA. 1987. Carbohydrate-binding proteins of human serum: isolation of two mannose/fucose-specific lectins. *Biochim. Biophys. Acta* 915:60–67

Tiemeyer, M, Brandley, BK, Ishihara, M, Swiedler, SJ, Greene, J, et al. 1992. The binding specificity of normal and variant rat Kupffer cell (lectin) receptors expressed in COS cells. *J. Biol. Chem.* 267:12252–57

Tracey, BM, Feizi, T, Abbott, WM, Carruthers, RA, Green, BN, Lawson, AM. 1992. Subunit molecular mass assignment of 14,654 to the soluble β-galactoside-binding lectin from bovine heart muscle and demonstration of intramolecular disulfide bonding associated with oxidative inactivation. *J. Biol. Chem.* 267:10342–47

Vavasseur, F, Berrada, A, Heuze, F, Jotereau, F, Meflah, K. 1990. Fucose and galactose receptor and liver recognition by lymphoma cells. *Int. J. Cancer* 248:744–51

Vercelli, D, Helm, B, Marsh, P, Padlan, E, Geha, RS, Gould, H. 1989. The B-cell binding site on human immunoglobulin E. *Nature* 338:649–51

Verrey, F, Drickamer, K. 1993. Determinants of oligomeric structure in the chicken liver glycoprotein receptor. *Biochem. J.* 292: 149–55

Wang, JL, Laing, JG, Anderson, RL. 1991. Lectins in the cell nucleus. *Glycobiology* 1:243–52

Watanabe, T, Yonekura, H, Terazono, K, Yamamoto, H, Okamoto, H. 1990. Complete nucleotide sequence of human *reg* gene and its expression in normal and tumoral tissues: the *reg* protein, pancreatic stone protein, and pancreatic thread protein are one and the same product of the gene. *J. Biol. Chem.* 265:7432–39

Watson, SR, Imai, Y, Fennie, C, Geoffroy, JS, Rosen, SD, Lasky, LA. 1990. A homing receptor-IgG chimera as a probe for adhesive ligands of lymph node high endothelial venules. *J. Cell Biol.* 110:2221–29

Watson, SR, Imai, Y, Fennie, C, Geoffrey, J, Singer, M, et al. 1991. The complement-binding-like domains of the murine homing receptor facilitate lectin activity. *J. Cell Biol.* 115:235–43

Weis, WI, Crichlow, GV, Murthy, HMK, Hendrickson, WA, Drickamer, K. 1991a. Physical characterization and crystallization of the carbohydrate-recognition domain of a mannose-binding protein from rat. *J. Biol. Chem.* 266:20678–86

Weis, WI, Drickamer, K, Hendrickson, WA. 1992. Structure of a C-type mannose-binding protein complexed with an oligosaccharide. *Nature* 360:127–34

Weis, WI, Kahn, R, Fourme, R, Drickamer, K, Hendrickson, WA. 1991b. Structure of the calcium-dependent lectin domain from a rat mannose-binding protein determined by MAD phasing. *Science* 254:1608–15

Wells, V, Mallucci, L. 1991. Identification of an autocrine negative growth factor: mouse β-galactoside-binding protein is a cytostatic factor and cell growth regulator. *Cell* 64:91–97

Wong, S, Freedman, JD, Kelleher, C, Mager, D, Takei, F. 1991. Ly-49 multigene family: new members of a superfamily of type II membrane proteins with lectin-like domains. *J. Immunol.* 147:1417–23

Woo, H-J, Lotz, M, Jung, JU, Mercurio, AM. 1991. Carbohydrate-binding protein 35 (Mac-2), a laminin-binding lectin, forms functional dimers using cysteine 186. *J. Biol. Chem.* 266:18419–22

Wright, JR, Borchelt, JD, Hawgood, S. 1989. Lung surfactant apoprotein SP-A (26–36 kDa) binds with high affinity to isolated type II cells. *Proc. Natl. Acad. Sci. USA* 86:5410–14

Yoshimatsu, K, Ohya, Y, Shikata, Y, Seto, T, Hasegawa, Y, et al. 1992. Purification and cDNA cloning of a novel factor produced by a human T-cell hybridoma: sequence homology with animal lectins. *Mol. Immunol.* 29:537–46

Yuen, C-T, Lawson, AM, Chai, W, Larkin, M, Stoll, MS, et al. 1992. Novel ligands for the cell adhesion molecule E-selectin revealed by the neoglycolipid technology among O-linked oligosaccharides on an ovarian cystadenoma glycoprotein. *Biochemistry* 31:9126–31

Zimmerman, PE, Voelker, DR, McCormack, FX, Paulsrud, JR, Martin, WJ. 1992. 120 kDa Surface glycoprotein of *Pneumocystis carinii* is a ligand for surfactant protein A. *J. Clin. Invest.* 89:143–49

Zimmermann, DR, Ruoslahti, E. 1989. Multiple domains of the large fibroblast proteoglycan, versican. *EMBO J.* 8:2975–81

Annu. Rev. Cell Biol. 1993. 9:265–315

MACROMOLECULAR DOMAINS WITHIN THE CELL NUCLEUS

David L. Spector
Cold Spring Harbor Laboratory, New York 11724

KEY WORDS: nucleus, nuclear structure, nuclear domains

CONTENTS

0743–4634/93/1115–0265$05.00

INTRODUCTION

The nucleus is the main repository of genetic information in the eukaryotic cell and is commonly referred to as the control center of the cell. At one time the nucleus was thought to be a rather amorphous structure, however, numerous descriptive electron microscopic studies have revealed a variety of particulate and fibrillar structures (for a review see Busch, 1974–1984). Recently, the advent of antibody and recombinant DNA technologies, allowing for the development of specific molecular probes, has fueled a renaissance in the quest to understand the molecular organization of the cell nucleus. By using such probes, it has become possible to ask more specific questions to unravel the intricacies of nuclear organization. In this review, I concentrate on recent advances in identifying macromolecular domains within the mammalian cell nucleus and relate these findings to other organisms and to the earlier studies. Since the entire literature cannot be reviewed here, the reader is referred to more detailed reviews, whenever possible, on each aspect of nuclear structure.

NUCLEAR-ENVELOPE LAMINA COMPLEX

Nuclear Envelope

The nuclear contents are spatially separated from the cytoplasm by the nuclear envelope, a double-membrane structure that plays a role in regulating the nucleocytoplasmic exchange of molecules and macromolecular complexes (Maul 1977; Feldherr et al 1984) as well as perhaps serving as an anchoring site for interphase chromatin. The outer nuclear membrane is contiguous with the endoplasmic reticulum and is often studded with ribosomes, which are involved in protein synthesis. The inner nuclear membrane is adjacent to the nuclear lamina and chromatin. The space between the two nuclear membranes, known as the perinuclear space, is directly contiguous with the lumen of the endoplasmic reticulum. The inner and outer nuclear membranes are fused together at various locations, forming nuclear pores. These pores are occupied by complex macromolecular assemblies called nuclear pore complexes (NPC), which control the passage of molecules into and out of the nucleus (Paine et al 1975).

The number of nuclear pores per unit area varies among different cell types (Maul 1977) and can range from 2–4pores/μm^2 (mammalian lymphocytes) to over 60 pores/μm^2 (mature *Xenopus* oocyte). A common value for higher eukaryotic cells is 10–20 pores/μm^2 (2,000–4,000 pores per nucleus) (Gerace & Burke 1988). The NPC is thought to contain 100–200 polypeptides and to have a total protein mass of 124 ± 11 MDa as determined by scanning transmission electron microscopy (STEM) (Reichelt et al 1990). When examined in negative stain preparations, the diameter of the NPC is typically 120 nm (Franke 1974; Unwin & Milligan 1982; Milligan 1986).

Since the early structural studies on the NPC (Gall 1967; Franke & Scheer 1970), numerous models have been reported (reviewed by Gerace & Burke (1988). The NPC has been characterized by three-dimensional reconstruction of negatively stained preparations and, based on this study, a consensus model of the NPC was derived (Unwin & Milligan 1982). This model emphasizes the symmetrical structure and subunit composition of the NPC. It has octagonal symmetry about a central axis, which is perpendicular to the plane of the nuclear membrane. The NPC is composed of two widely separated coaxial rings (with inner diameters of 80 nm), one facing the cytoplasm and one facing the nucleoplasm. Eight elongated radial spokes extend between each of the rings and contact a centrally located particle termed the plug (35 nm diameter). The central channel of the NPC is thought to be involved in active transport, and eight peripheral channels, each measuring approximately 10 nm in diameter, are proposed to function in passive exchange of small molecules (Hinshaw et al 1992). The plug might represent an endogenous substrate caught in transit between the nucleoplasm and cytoplasm (Jarnik & Aebi 1991). However, Akey & Goldfarb (1989) have implicated the central plug as the transporter for nucleocytoplasmic exchange. In a cryoelectron microscopy study of frozen-hydrated NPCs, the central plug was observed in at least four distinct configurations: open, closed, docked, and open-in-transit (Akey 1990). Transport of proteins and RNAs between the nucleus and cytoplasm has been discussed in several recent reviews (Dingwall & Laskey 1986; Goldfarb 1989; Maquat 1991; Izaurralde & Mattaj 1992).

Akey (1989) verified and extended the Unwin & Milligan model (1982) by visualizing new structural details of NPCs in amorphous ice using image analysis of edge-on and en face projections of detergent-extracted NPCs. Using this approach, the radial spokes have been resolved into three domains including inner spokes that form an inner spoke ring, outer spokes that abut the membrane, and vertical supports that connect the central spokes to the nucleoplasmic and cytoplasmic coaxial rings. In addition, radial arms (Akey 1989) or knobs (Jarnik & Aebi 1991) were identified that may anchor the NPCs to the nuclear envelope. A candidate protein, which may be represented in the knobs, is gp210 (Gerace et al 1978). This protein was localized to the region of the nuclear envelope where the knobs have been visualized (Greber et al 1990).

Recently, several studies (Ris 1989, 1990, 1991; Jarnik & Aebi 1991; Goldberg & Allen 1992) revealed new aspects of the NPC that have dramatically altered the concept of the NPC as a structure whose inner and outer faces are the same. In the first group of studies, Ris (1989, 1990, 1991) employed high-resolution low-voltage scanning electron microscopy (LVSEM) to visualize critical point-dried/platinum-sputtered nuclear envelopes from *Xenopus laevis* and *Notophthalmus viridescens*. In these studies, the NPC appeared to be an hour glass-shaped structure with a ring measuring 120 nm in diameter on both the nuclear and cytoplasmic sides (Figure 1). Eight filaments were observed projecting from each of these rings. The filaments on the nuclear side were long and thin, ending in a 60 nm distal ring that may consist of eight granules. This structure gives the appearance of a fish-trap (Figure 1). Jarnik & Aebi (1991) observed a similar structure in NPCs after quick freezing/freeze drying/rotary metal shadowing. In this case, the fish-trap extended 50–100 nm beyond the plane of the nuclear envelope into the nuclear space. Furthermore, the fish-traps were destabilized in response to the removal of divalent cations (Jarnik & Aebi 1991). The filaments that project from the ring on the cytoplasmic side of the membrane are highly twisted and are shorter than those projecting from the nuclear ring (Ris 1991). These filaments may represent the eight particles described by Unwin & Milligan (1982) that decorate the cytoplasmic ring of negatively stained NPCs.

A nonmuscle isoform of myosin heavy chain has been shown to be associated with the NPC (Berrios & Fisher 1986; Berrios et al 1991), and it would be informative to map precisely this protein to determine if it localizes to the nuclear and/or cytoplasmic filaments that extend from the NPC. High resolution scanning electron microscopy (HRSEM) of nuclear

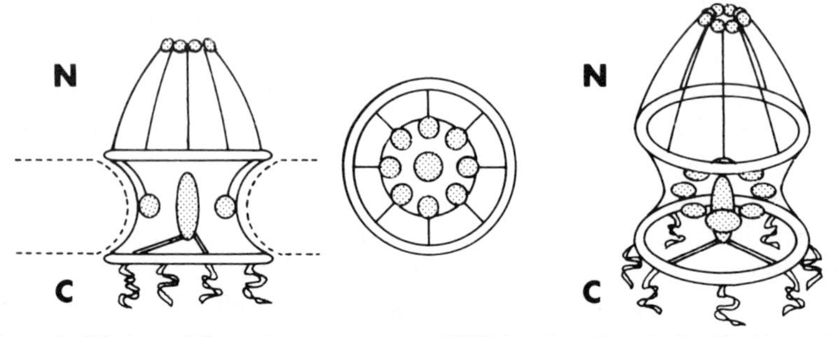

Figure 1 Diagram of the nuclear pore complex (NPC) based on data obtained by high voltage transmission electron microscopy and low voltage scanning electron microscopy. The NPC consists of three parts. (1) The hour-glass-like central component inserted into the nuclear pore, limited by a ring 120 nm in diameter both on the intranuclear side and the cytoplasmic side. In the midplane is a ring of eight particles that are connected by spokes to the intranuclear ring. A central cylinder is placed in the axis of the central component. (2) Attached to the ring on the cytoplasmic side (C) are eight short, twisted filaments that in the scanning microscope appear as cylinders twice as long as they are thick. (3) On the intranuclear side (N), eight long thin filaments project from the 120 nm ring into the nuclear space with their ends attached to a smaller ring, forming the fishtrap (courtesy of H. Ris, Univ. Wisconsin).

envelopes from the salamander *Triturus cristatus* recently confirmed the fish-trap structure of the intranuclear part of the NPC and revealed a new complex fibrillar network forming a canopy over and attached to the fish-traps (Goldberg & Allen 1992). Future studies are needed to elucidate the sublocalization of various NPC proteins and the dynamics of the pore complex in transport.

Nuclear Lamina

Internal and adjacent to the inner membrane of the nuclear envelope is an electron-opaque layer termed the nuclear lamina. In most mammalian cells, the lamina measures about 10 nm in thickness and is composed of three major polypeptides, lamins A (M_r= 70 K), B (M_r= 68 K), and C (M_r= 60 K), which are present in equal amounts in the interphase nucleus (Gerace & Blobel 1980). The B lamins are constitutively expressed in all mammalian somatic cells. However, the A and C lamins are not expressed early in development, but appear in most differentiated cell types (Lebel et al 1987; Stewart & Burke 1987; Rober et al 1989). Comparison of the cDNA sequences of both lamins A and C shows that the two proteins are identical for the first 566 amino acids from the N terminus. These two lamins are thought to arise from the same gene by alternative splicing (McKeon et al 1986). Lamins A and C purified from rat liver have been examined after low-angle shadowing, and both proteins appear as asymmetric rods consisting of a 50-nm tail (corresponding to the alpha-helical rod domain) and two globular heads (corresponding to the non-alpha helical extension at the carboxyl terminus) at one end (Aebi et al 1986).

Electron microscopic examination of detergent-extracted nuclear envelopes of *Xenopus* oocytes indicates that the lamina is a meshwork containing orthogonally oriented filaments (Aebi et al 1986; Scheer et al 1976). The filaments have a diameter of 10 nm and an axial repeat of 25 nm, which is similar to intermediate filaments (Weber & Geisler 1985). However, the lamins of mammalian cell nuclei, observed by electron microscopy, appear as an amorphous fibrillar meshwork (Aaronson & Blobel 1975; Dwyer & Blobel 1976). Under suitable ionic conditions, isolated mammalian lamins assemble into higher order paracrystals with an axial repeat of 25 nm (Aebi et al 1986; Goldman et al 1986). Lamins possess all the major structural properties that have been described for intermediate filament polypeptides at the levels of amino acid sequence, of protomers, and of assembled filaments.

The role of the nuclear lamins in nuclear envelope assembly has been examined using *Xenopus* cell-free extracts (Newport et al 1990) and CHO mitotic extracts (Burke & Gerace 1986). Depletion of lamins from a CHO mitotic cell-free system was shown to result in the inhibition of nuclear envelope assembly (Burke & Gerace 1986). However, depleting lamin LIII from a *Xenopus* cell-free nuclear extract did not block the initial assembly of the nuclear membrane or the nuclear pores around a chromatin substrate (Newport et al 1990). Further studies are needed to address the specific role of the nuclear lamins in nuclear envelope formation and chromosome decondensation.

Developmental regulation of lamin expression has been studied in early mouse embryos and in embryonal carcinoma stem cell lines (Stewart & Burke 1987; Lebel et al 1987; Schatten et al 1985). A single major B-type lamin is expressed in preimplantation embryos and in undifferentiated teratocarcinoma stem cell lines. Lamins A/C are not synthesized until the commencement of organogenesis.

In mammalian cells, the association of the lamina with the nuclear envelope appears to be mediated, at least in part, by lamin B (Worman et al 1988, 1990; Appelbaum et al 1990) through a receptor protein present in the inner nuclear membrane. In avian erythrocyte nuclear envelopes, a 58-kd major integral membrane protein is thought to be the lamin B receptor. Since the protein has three DNA-binding motifs, it was suggested that the lamin B receptor may also have a role in chromatin organization. The B lamins are also more resistant to extraction from the nuclear envelope compared to the A/C lamins (Gerace et al 1984). Appelbaum et al (1990) have shown that the association of lamin B with the p58 receptor is dependent upon the phosphorylation of p58 by protein kinase A.

During lamina depolymerization at mitosis, the B lamins remain associated with cytoplasmic membrane vesicles thought to be remnants of the nuclear envelope, whereas the A/C lamins become soluble (Gerace & Blobel 1980). The p34 cdc2-cyclin B complex has been shown to phosphorylate the lamins on mitosis-specific sites, which in vitro result in lamin disassembly (Peter et al 1990; Dessev et al 1991).

In order to study the assembly of the nuclear lamina, Goldman et al (1992) recently injected biotin labeled lamin A into the cytoplasm of mouse 3T3 cells. The protein was rapidly transported to the nucleus, where in some cells it was found not only at the nuclear periphery, but also in a series of intranuclear spots. Upon further investigation of uninjected cells for native lamin A, similar intranuclear structures were observed and were most prominent during the G1 and S phases of the cell cycle (Goldman et al 1992). The intranuclear lamin-enriched structures are thought to represent initial accumulation sites of newly synthesized lamin proteins or sites of lamin processing. Future studies at the electron microscopic level will be useful in determining the structural composition of these nuclear regions and their potential overlap with other nuclear domains.

The lamins are thought to function in regulating nuclear envelope structure and anchoring interphase chromatin at the nuclear periphery (Aaronson & Blobel 1975; Dwyer & Blobel 1976; Gerace et al 1978; Gerace et al 1982). Purified lamin A protein has been shown to bind to polynucleosomes in a saturable and specific fashion (Yuan et al 1991). Recently, Paddy et al (1990) used three-dimensional reconstruction analyses to suggest that the lamins do not form a continuous layer around the nuclear periphery, but that they are organized as a highly discontinuous network leaving large voids at the nuclear periphery that contain little or no lamin. In addition, it was found using light microscopy techniques that although a large fraction of the most peripheral chromatin loci were aligned with lamin fibers, most of these loci were too distant from the nearest lamin fiber (0.15–0.26 μm) to be in direct contact. The lamin-staining pattern was found to occupy about half (48% in HeLa cells) of the surface area of the nuclear periphery (Paddy et al 1990). It has been suggested that the organization of the nuclear periphery may contribute to controlling the template activity of chromatin in transcription or replication (Peter et al 1990). In support of this possibility, Newport et al (1990) found that quantitative immunodepletion of *Xenopus* lamin LIII from a nuclear assembly extract inhibits subsequent growth of the nuclear envelope and blocks DNA replication. Based on these data, it was suggested that the lamins may act as a substrate for the formation of replication complexes, or that they may regulate DNA condensation or formation of nuclear matrix elements, which are required for efficient replication to occur. Additional studies are needed to more closely examine the role of the lamins in chromatin organization and function.

CHROMATIN ORGANIZATION

Overview

Eukaryotic chromatin has been divided into two classes, heterochromatin and euchromatin, based on its state of condensation during interphase. Heterochromatin is a form that is condensed during interphase and is therefore generally considered to be transcriptionally inactive. It is commonly located in an irregular band around the nuclear periphery and around the nucleolus as well as in patches throughout the nucleoplasm. Heterochromatin is further divided into constitutive and facultative forms. Constitutive heterochromatin is highly enriched in repetitive sequences (for example, α-satellite DNA sequences) which are present in long tandem arrays and comprise approximately 10% of the mammalian genome (for a review, see Haaf & Schmid 1991). The function of this type of chromatin in the interphase cell is unknown. Facultative heterochromatin consists of potentially active chromatin and involves the interphase condensation of one chromosome or a set of chromosomes of a homologous pair. The best studied case of facultative heterochromatin is the Barr body, the condensation of one X chromosome in female placental mammals (Barr & Bertram 1949).

The amount of heterochromatin present in the nucleus varies with transcriptional activity; little heterochromatin is present in some transcriptionally active cancer cells, whereas the nuclei of orthochromatic erythroblasts and mature spermatozoa, both transcriptionally inactive, are practically filled with condensed chromatin. In an average eukaryotic cell approximately 90% of the chromatin is transcriptionally inactive at any given time, but all of the inactive chromatin may not be in a condensed state (for a review, see Haaf & Schmid 1991; Manuelidis 1990). However, the 10% of the chromatin that is transcriptionally active is always in the form of decondensed euchromatin. The higher order structure of chromatin has been addressed in a recent review (Manuelidis & Chen 1990).

Localization of DNA Sequences

CHROMOSOMES AND EXPRESSED GENES Recent developments in the ability to detect DNA sequences efficiently by fluorescence in situ hybridization (FISH) have made it possible to identify and localize specific sequences in interphase nuclei even when present in low copy number (for a review, see McNeil et al 1991). The idea that interphase chromatin may not be randomly distributed throughout the interphase nucleus, but may occupy distinct territories, was first expressed in the classical papers by Rabl (1885) and Boveri (1909). Rabl (1885) suggested that each chromosome in plant cells occupies a distinct domain throughout interphase that reflects its mitotic orientation. Boveri (1909) confirmed these studies by showing that chromosomes maintained relatively fixed positions in the nuclei of *Ascaris* eggs. Furthermore, these studies suggested that telomeres were attached to the nuclear envelope on one side of the nucleus and centromeres were attached on the opposite nuclear side. More recently, numerous studies have readdressed the initial questions asked by Rabl and Boveri in a variety of systems at significantly higher resolution. In *Drosophila*, polytene chromosomes have been differentially stained with vital dyes, and in conjunction with optical sectioning methods, it has been possible to examine chromosomes in living cells. These studies have revealed that polytene chromosomes are closely associated with the inner surface of the nuclear membrane and contact the membrane at specific sites (Agard & Sedat 1983; Hochstrasser & Sedat 1987; Mathog et al 1984). Furthermore, it was shown that chromosomes occupy distinct territories within diploid and polytene nuclei and spiral with the same handedness through the nucleus (Hilliker 1985; Mathog et al 1984; Hochstrasser et al 1986). Cremer et al (1982) showed that laser-UV-microirradiation of specific interphase nuclear areas in Chinese hamster cells damaged discrete chromosomal regions, which suggests that the genome is organized during interphase. Using probes specific for individual human chromosomes, several groups have shown that each chromosome occupies compact interphase domains and that homologous chromosome domains are often not adjacent (Trask et al 1988; Cremer et al 1988; Borden & Manuelidis 1988; Pinkel et al 1986, 1989). Manuelidis (1985b) examined the arrangement of individual human chromosomes in mouse-human cell hybrids and found that the human chromosome showed a reproducible position in the nuclei.

Several studies have provided data to support the concept that chromosomes are dynamic in the interphase nucleus and that their position is cell cycle-dependent. Using a composite probe to chromosome 8, the interphase position of the chromosome was observed to change during the cell cycle (Ferguson & Ward 1992). In G1 cells, chromosome 8 centromeres localized adjacent to the nuclear periphery and the chromosomal arms extended in toward the nuclear interior. However, in G2 the chromosome reoriented itself and the centromeres were internal and the chromosomal arms extended toward the nuclear periphery (Ferguson & Ward 1992). A similar redistribution was observed in brain tumor cells where centromeres were dispersed during G1 and S-phase and became clustered toward the nuclear interior during G2 (Manuelidis 1985a). These data suggest that the positioning of chromsomes in the interphase nucleus is not fixed but is dynamic and changes during the cell cycle perhaps to reflect gene expression.

The most provocative study demonstrating a correlation in chromosome position and cell physiology comes from work on human epileptic foci (Borden & Manuelidis 1988). In normal male cortical neurons, the X chromosome was localized to the nuclear periphery. However, when cells in an electrophysiologically defined seizure focus were observed, there was a dramatic increase from approximately 7 to 45% in the number of cells exhibiting internal nuclear localization of the X chromosome (Borden & Manuelidis 1988). A similar observation was previously reported in neurons after 8 hr of electrical stimulation (Barr & Bertram 1949).

Somatic pairing of homologous chromosomes has been reported in a number of plant and Diptera species (for a review, see Hilliker & Appels 1989). However, somatic pairing does not seem to occur in mammalian cell systems (Cremer et al 1982; Lawrence & Singer 1991; Huang & Spector 1991). Pinkel et al (1986) used probes specific to the X or Y chromosomes to show that the sex chromosomes did not exhibit somatic pairing in interphase human cells. Furthermore, examination of the localization of homologous sequences of several single copy genes including dystrophin, α-cardiac myosin, and the neu oncogene leads to the conclusion that there is no homologous pairing of these sequences in lymphocytes or fibroblasts (Lawrence et al 1990).

Specific gene sequences appear to reproducibly occupy nonrandom domains within the interphase nucleus. In examining single copy gene localization, Lawrence et al (1988) showed that the Epstein Barr virus genome, which is integrated into the Namalwa cell genome on chromosome 1, was not randomly localized in the nucleus, but was positioned to the inner 50% of the nuclear volume (Lawrence et al 1988). Furthermore, the neu oncogene was localized to internal regions of the nucleus, whereas the dystrophin genes were found close to the nuclear periphery (Lawrence & Singer 1991). Whether individual genes occupy specific X, Y, and Z nuclear coordinates at specific times during the cell cycle is not yet clear. However, based on the data currently available, it appears more likely that genes will occupy domains or nuclear regions, rather than absolute points, and their location may change with respect to the cell cycle, cell physiology, and/or their expression. It remains to be elucidated whether the specific location of a gene in the interphase nucleus directly effects its expression. However, the possibility that a gene can affect its regulation by binding factors that determine its spatial localization in the cell nucleus is an important consideration that needs to be tested by experimentation.

CENTROMERES The centromere is defined as the primary constriction of the mitotic chromosome. It is composed of chromosome-specific α-satellite DNA sequences plus a series of proteins (for a review, see Brinkley et al 1992). During interphase, centromeres have been identified using antibodies or nucleic acid probes (Brenner et al 1981; Manuelidis 1985b; Ochs & Press 1992). There appear to be cell-type specific patterns of centromere localization. In two rat-kangaroo cell lines and in Indian muntjac cells 20–36% of the cells examined showed a pair-wise arrangement of centromeres, which suggests that in these cell types, homologous chromosomes may occupy adjacent regions of the interphase nucleus (Hadlaczky et al 1986). In addition, in some cells, centromeres were clustered at one pole of the nucleus or formed symmetrical nonrandom patterns. However, other earlier reports have not found this to be the case in the same cell types (Cohen et al 1972; Sperling & Ludtke 1981). In a human carcinoma cell line, centromeres were localized adjacent to the nucleoli or the nuclear membrane in G1 and appeared to form necklace-like structures in S and G2 (Haaf & Schmid 1989). These structures may result from unraveling of centromeres during DNA replication, as expected from the repetitive subunit structure of the centromere recently proposed by Zinkowski et al (1991). In numerous plant species, centromeres have been reported to be grouped together near the nuclear envelope (for a review, see Hilliker & Appels 1989).

Rearrangements of centromere positions also appear to be related to developmental stages (Manuelidis 1985b). Centromeres in post-mitotic (non-dividing) Purkinji cells were localized in three large clusters similar to those observed in adult neurons. However, unlike the adult Purkinje neurons in which centromeres are associated with the nucleoli, centromeres in these cells were localized to various regions between the nucleoli and the nuclear envelope (Manuelidis 1985b). As nucleoli became more prominent in the Purkinji cells, the centromeres began to associate with the nucleolar rim as they do in adult cells. This study clearly showed a rearrangement of centromeres as a function of development. Different patterns of centromere arrangement were observed in small granule neurons and astrocytes. In differentiated dorsal root ganglia neuronal cells, centromeres were primarily associated with the nucleolus and positions between the nucleolus and the nuclear envelope (Billia & De Boni 1991). Recently, S. Henderson & D. Spector (submitted) found that in myoblasts, chromosome-specific α-satellite DNA sequences appeared to be randomly arranged with respect to each other, although they were preferentially localized to the nuclear periphery or the nucleolar surface. However, as myoblasts converged prior to cell fusion, homologous centromeres aligned parallel to the long axis of each cell. A similar arrangement was observed in myotubes and sections of striated muscle. Therefore, it appears that centromeres are arranged in different nonrandom patterns that are cell-type and cell-cycle-specific and that relate to the state of differentiation of the cell being examined.

TELOMERES Telomeres are present at the ends of chromosomes and are characterized by highly repetitive DNA sequences that are synthesized in many organisms by a unique enzyme called telomerase (Greider & Blackburn 1987). While by classical definition, telomeres were associated with the nuclear envelope (Rabl 1885), this does not appear to be the case in all cell types examined. Rawlins et al (1991) reported telomeres in two plant species to be associated primarily with the nuclear envelope and to a lesser extent with the nucleolar surface. In both cases, telomeres were clustered, and in *Vicia faba,* more than two-thirds of the telomeres were adjacent to an area of the nuclear envelope that represented less than 10% of the total surface area, which supported the Rabl configuration. In normal human fibroblasts, telomere sequences were also reported to be associated with the nuclear periphery (Lichter et al 1991). However, in some brain cell types, telomeres have been identified in internal portions of the nucleus, whereas in other cell types, they have been shown to be associated with the nuclear membrane (Borden & Manuelidis 1988; Ferguson & Ward 1992) or in intermediate positions (Billia & De Boni 1991). Therefore, the localization of telomeres in many animal cells does not follow the classical Rabl configuration observed in *Drosophila* polytene nuclei (Hochstrasser et al 1986) and in nuclei of several plant species (Avivi & Feldman 1980).

These differences demonstrate that interphase chromosome organization varies in different cell lineages. Centromeres and telomeres appear to be dynamic structures that change their nuclear position during the cell cycle. The dynamic organization of the genome in the interphase nucleus may be essential for the correct expression of genes during the cell cycle, differentiation, or for the abberant expression of the genome during certain pathological conditions. Future studies will determine the significance of the nonrandom arrangement of DNA sequences in the interphase nucleus and its effect on gene expression.

DNA REPLICATION

During S-phase of the mammalian cell cycle, the cell must copy the millions of nucleotides that form its genome with precise fidelity in order to maintain its genetic integrity (Jackson 1990). DNA replication initiates at origins of replication that occur in clusters or replication units (Cairns 1966; Huberman & Riggs 1968; Hand 1978), which are activated at different times throughout S-phase (Taylor 1960; Hand 1978). Reports that specific DNA sequences (sections of chromosomes) are replicated at discrete times during S-phase first suggested that each chromosome consists of multiple units of replication (Taylor 1960). The presence of subchromosomal replication units in mammalian cells was demonstrated using in situ autoradiography by Cairns (1966) and subsequently by Huberman & Riggs (1968). In addition, Schildkraut and co-workers (Calza et al 1984; Hatton et al 1988) have shown that chromosomal rearrangement of a gene can dramatically change the replication time of the gene.

In theory, a defined spatial organization of DNA replication and factors associated with replication in the mammalian cell nucleus would seem to be a logical possibility in order for the correct replication of the genome to take place. Numerous studies have used [^{3}H]-thymidine incorporation combined with in situ autoradiography to localize the sites of DNA replication in the cell nucleus (Comings & Okada 1973; Fakan & Hancock 1974; Hay & Revel 1966). These studies have generally identified two patterns of replication: one is localized to the perinuclear and perinucleolar regions and is thought to represent the replication of condensed heterochromatin; a second pattern is localized to internal nuclear sites that are thought to represent the replication of euchromatin. However, the data are limited by the level of resolution these techniques have provided.

The production of antibodies specific for 5-bromodeoxyuridine (Gratzner

1982) and the development of biotin-labeled nucleic acid affinity probes (Langer et al 1981) have made it possible for the sites of DNA replication to be localized at higher resolution. Bromodeoxyuridine (BrdU), a thymidine analogue, can be specifically incorporated into replicating DNA of cultured cells and visualized by immunofluorescence or immunoelectron microscopy. Using this approach with immunofluorescence microscopy, Nakamura et al (1986) detected intranuclear ring-like structures that they proposed to represent replicon clusters. However, these structures were only observed in cells grown in the presence of BrdU for long periods of time (>11 hr). Nakayasu & Berezney (1989) used fluorescence microscopy to map DNA replication sites in permeabilized cells that have incorporated biotinylated dUTP and in nonpermeabilized cells grown in the presence of BrdU. Using this approach, these investigators reported that DNA replication occurs at several hundred discrete granular sites termed replication granules. The distribution of these replication granules was observed in three distinct S-phase-dependent patterns defined as types I, II, and III (Nakayasu & Berezney 1989). The type I pattern occurred early in S-phase. This pattern was proposed to correspond to regions where the chromatin of R-bands (euchromatin) replicates. Cells exhibiting a type II pattern showed immunofluorescence over perinuclear, perinucleolar, and internal heterochromatic regions. This distribution pattern was prominent in mid to late S-phase and was thought to represent a transition where both sites of R and G bands (non-centromeric heterochromatin) were replicating. Finally, cells exhibiting a type III pattern were characterized by the exclusive occurrence of replication at heterochromatic regions corresponding to centromeric heterochromatin of C bands (Pardue & Gall 1970). Each of these DNA replication patterns appears, by light microscopy, to be staining different nuclear regions.

O'Keefe et al (1992) used a similar approach to Nakamura et al (1986), but examined the patterns of DNA replication in synchronized cells that were pulse-labeled with BrdU for 10 min at various times throughout S-phase. Using this approach, five different patterns of DNA replication were identified by immunofluorescence and immunoelectron microscopy, and their timing during S-phase was determined. The overall findings of O'Keefe et al (1992) are in agreement with previous studies of others who have found different patterns of DNA replication during S-phase (Nakayasu & Berezney 1989; van Dierendonck et al 1989; Mazzotti et al 1990; Fox et al 1991). However, the finding that the bulk of replication at the periphery of the nucleus occurs primarily during mid S-phase differs from that of Fox et al (1991), who suggested that peripheral chromatin replicates at the end of S-phase. Other studies have found less distinct patterns of BrdU (or dUTP) incorporation and report that regions of replicating DNA appear to vary only in size (and not pattern) of replication clusters during S-phase (Nakamura et al 1986; Banfalvi et al 1990).

In vitro, the sites of replication of *Xenopus* sperm nuclei, chicken erythrocyte nuclei, *Drosophila* polytene nuclei, and bacteriophage λ DNA induced by *Xenopus* cell-free extracts occur as discrete foci, which are maintained during replication (Mills et al 1989; Cox & Laskey 1991; Leno & Laskey 1991; Sleeman et al 1992). These differences raise the question of whether replication sites are dynamic (and move from one region of the nucleus to another), or whether they are fixed, with the DNA "spooling" through them (Pardoll et al 1980). Several studies have shown that DNA replication factors (Bravo & Macdonald-Bravo 1985, 1987; Celis & Celis 1985; Madsen & Celis 1985; Raska et al 1989a) display temporal changes in distribution similar to those that were observed by BrdU labeling (Nakayasu & Berezney 1989; van Dierendonck et al 1989; Mazzotti et al 1990; Fox et al 1991; O'Keefe et al 1992). Immunolabeling of DNA cruciforms (regions proposed to be located at or near replication origins) also shows a similar change in distribution during S-phase (Ward et al 1990, 1991).

These results suggest that DNA replication factors move to the DNA being replicated during S-phase. However, this does not exclude the possibility that the sites of replication are fixed. It has been proposed that replication occurs at discrete sites (Mills et al 1989; Cox & Laskey 1991), the positions of which are maintained via an attachment to an underlying nuclear skeleton (or nuclear matrix) (Pardoll et al 1980; Berezney et al 1982; Dijkwel et al 1986; Jackson & Cook 1986; Razin 1987; Foster & Collins 1985; Tubo & Berezney 1987; Vaughn et al 1990, Nakayasu & Berezney 1989). These results suggest a possible model in which replication factors, capable of moving throughout the nucleus, bind to specific DNA sequences at (or prior to) their initiation of replication. These complexes then temporarily associate with the nuclear matrix where DNA is replicated.

By combining BrdU labeling with in situ hybridization of DNA sequences, O'Keefe et al (1992) were able to demonstrate that α-satellite DNA sequences, which localize to either the nucleolar surface and/or the nuclear periphery, replicate during mid S-phase. These findings are in agreement with another recent study (Bartholdi 1991), which followed centromeres labeled with antibodies throughout the cell cycle and found that the intensity of centromere fluorescence increased during mid S-phase. It is noteworthy that the DNA replication pattern during mid-S phase delineates the previously established positions of centromeres (i.e. perinucleolar and/or peripheral) (Manuelidis 1984, 1985a; Rappold et al 1984; Bourgeois et al 1985; Borden & Manuelidis 1988; Haaf & Schmid 1989, 1991; Mukherjee & Parsa 1990;

Popp et al 1990; van Dekken et al 1990). Furthermore, these results are supported by previous biochemical studies that found that centromeres replicated in mid S-phase, but not in the first three (Dooley & Ozer 1977) or last two hours (Bostock & Prescott 1971) of S-phase. However, the results of O'Keefe et al (1992) differ with previous studies that suggested that centromeric DNA replicates very late in (or at the end of) S-phase (Lima-de-Faria & Jaworska 1968; Camargo & Cervenka 1982; Goldman et al 1984; Ten Hagen et al 1990).

Selig et al (1992) recently used in situ hybridization to determine the replication timing of specific DNA sequences. In diploid cells, unreplicated DNA regions result in singlet hybridization signals after in situ hybridization with a specific probe, and replicated loci are represented by doublets. Using this approach, it was possible to identify already replicated DNA sequences in random cell populations. Genes that replicate early in S-phase show a high percentage of doublets, whereas for late replicating genes, most nuclei show singlet hybridization signals (Selig et al 1992).

In principle, the technique of combining BrdU labeling with in situ hybridization and examination by confocal microscopy allows for the determination of the timing of replication of specific genes. Furthermore, changes in the timing of replication can be determined based upon whether a cell expresses a gene, cell differentiation, or cell transformation. Additionally, adjacent DNA probes along the length of a chromosome can be hybridized in situ to BrdU-labeled cells to map the differences in timing of contiguous DNA sequences (i.e. to determine if there is a smooth wave of replication within specific regions of a chromosome, or whether there are abrupt starts and stops in replication of sequences along the chromosome).

In summary, the replication of DNA appears to follow a dynamic order in the mammalian cell nucleus, with specific regions of DNA replicating at defined times and in defined locations of the nucleus during S-phase. These studies support the hypothesis that the nucleus is a highly organized system of defined structural and functional activities.

TRANSCRIPTION AND PRE-mRNA PROCESSING

Localization of Transcriptionally Active Nuclear Regions by [^{3}H]-Uridine Incorporation or In Situ Nick Translation

Changes in chromatin structure as well as interactions of *cis*-DNA sequences with *trans*-acting factors are required for the activation of genes. High resolution electron microscopic in situ autoradiography has been used by several groups to localize sites of active transcription within the cell nucleus (for a review, see Fakan & Puvion 1980). After pulses of [^{3}H]-uridine for as short as 2 min, non-nucleolar labeling was first observed over perichromatin fibrils, which are distributed throughout the nucleoplasm (Fakan & Bernhard 1971; Fakan & Nobis, 1978; Fakan et al 1976). These structures were shown to be RNAase-sensitive, and their appearance was inhibited by actinomycin D (Miyawaki 1974) or α-amanitin (Petrov & Sekeris 1971) pre-treatment. Based on these data, it was suggested that these fibrils may represent pre-mRNA (Fakan et al 1976). When RNAs were extracted from nuclear fractions enriched in perichromatin fibrils, a heterogeneous, high molecular weight distribution of RNA was observed that was distinct from ribosomal RNA (Bachellerie et al 1975). Based on chain length distribution and actinomycin D treatment, these RNAs were reported to represent hnRNA or pre-mRNA (Bachellerie et al 1975). The localization of pre-mRNA splicing factors to these fibrils (Fakan et al 1984; Spector et al 1991) strongly supports their role as pre-mRNA.

Several reports using in situ nick translation to visualize DNAase I-sensitive nuclear DNA (active genes) suggest that active genes are preferentially localized around the nuclear periphery (Hutchison & Weintraub 1985; Krystosek & Puck 1990). However, the possibilities that nucleotides or other components of the reaction mix are sequestered or trapped by the peripheral chromatin, or that DNAase I is bound to perinuclear actin, were not addressed in these studies. These possibilities are supported by the fact that when DNAase I incubation was performed in a separate reaction, prior to the incubation with polymerase and labeled nucleotide, it was possible to demonstrate a decrease in staining of the nuclear periphery and a concomitant increase in staining of an internal subset of chromatin (Krystosek & Puck 1990). In addition, other studies have shown heterochromatin or inactive chromatin (Comings 1980) including non-transcribed α-satellite DNA sequences (Manuelidis 1984; O'Keefe et al 1992) to be localized to the nuclear periphery. Based on [^{3}H]-uridine incorporation and the localization of nascent RNA (see below), it appears that active genes are distributed throughout the nucleoplasm and are not restricted to the nuclear periphery.

Pre-mRNA Splicing

During or after the transcription of pre-mRNA molecules, these transcripts are processed and then transported to the cytoplasm where they are translated into proteins. For most RNA polymerase II transcripts, this processing includes, not necessarily in this order, addition of a 7-methyl-guanosine cap structure at the 5′ end, hnRNP assembly, splicing of noncoding intron regions, and subsequent ligation of exons, polyadenylation, and the exchange of hnRNP proteins for mRNP proteins. Splicing occurs in a multicomponent complex termed a spliceosome. Many of the detailed biochemical steps involved in the pre-mRNA splicing reaction have been extensively studied in vitro and are well understood (for a review, see Green 1991). The earliest components shown to be essential for pre-mRNA splicing were small nuclear ribonucleoprotein particles (snRNPs). Each of the snRNP particles (U1, U2, U5) contains a single snRNA species, with the exception of the U4/U6 particle, which contains two snRNAs. In addition, each snRNP particle contains a common set of core proteins, as well as unique snRNA-specific proteins (for a review, see Zieve & Sauterer 1990).

The sub-localization of splicing components in the nucleus of mammalian cells was first suggested in the 1980's by immunofluorescent localization studies using human autoantibodies, which recognize protein or RNA components of snRNPs (Perraud et al 1979; Lerner et al 1981; Spector et al 1983). Localization studies using anti-Sm antibodies directed against a variety of snRNP-specific proteins (Spector et al 1983; Nyman et al 1986; Spector 1984; Reuter et al 1984; Verheijen et al 1986; Habets et al 1989), or antibodies specifically directed against U1 (Spector 1984; Verheijen et al 1986), U2 snRNP (Spector 1984; Habets et al 1989), or the m_3G cap structure of snRNAs (Reuter et al 1984) have shown these components to be concentrated in 20–50 nuclear speckles in addition to being distributed diffusely in the nucleoplasm of mammalian cells. The diffuse nuclear staining of snRNPs may represent an excess soluble population, snRNPs in transit to or from nascent transcripts, or it may represent snRNPs in transit to speckles from their assembly sites in the cytoplasm (Zieve & Sauterer 1990). In situ hybridization studies using probes to several of the snRNAs have shown a similar localization (Carmo-Fonseca et al 1992; Huang & Spector 1992a; Spector et al 1992). During mitosis the speckled pattern breaks up and snRNPs are distributed diffusely throughout the cytoplasm (Spector & Smith 1986). Several non-snRNP splicing factors (SC-35, SF2) have also been localized to speckled nuclear regions (Fu & Maniatis 1990; Spector et al 1991; A. Krainer & D. Spector, unpublished).

The protein U2AF, which facilitates U2 snRNP binding to the branch point (Ruskin et al 1988; Zamore & Green 1989, 1991), is the only splicing factor thus far identified that does not localize in a speckled distribution (Zamore & Green 1991). U2AF was shown to localize in coiled bodies (see section below on nuclear bodies) as well as in a diffuse distribution throughout the nucleoplasm (Zamore & Green 1991). The significance of the unique nuclear localization of U2AF, as compared to other splicing factors, remains to be determined. In addition to their speckled localization pattern, several studies have also shown snRNPs to be concentrated in coiled bodies in a percentage of cells in a population (Fakan et al 1984; Eliceiri & Ryerse 1984; Carmo-Fonseca et al 1991a,b, 1992; Raska et al 1991; Spector et al 1992; for a review, see Huang & Spector 1992b). However, the lack of DNA (Monneron & Bernhard 1969), [^{3}H]-uridine incorporation (Moreno Diaz de la Espina et al 1980), hnRNP proteins (Carmo-Fonseca et al 1991b; Raska et al 1991), and several non-snRNP splicing factors (SC-35, SF2) (Raska et al 1991; Huang & Spector 1992a; Carmo-Fonseca et al 1992; A. Krainer & D. Spector, unpublished) in coiled bodies supports the idea that these inclusions are not essential for pre-mRNA splicing.

At the electron microscopic level, the speckled distribution of splicing factors has been shown to represent nuclear regions enriched in interchromatin granules and perichromatin fibrils (Perraud et al 1979; Spector et al 1983; Fakan et al 1984; Puvion et al 1984; Spector et al 1991) (Figure 2). The interchromatin granule regions contain particles with a mean diameter of 20–25, nm which are linked together by thin fibrils (Monneron & Bernhard 1969). In situ autoradiographic studies following [^{3}H]-uridine incorporation have shown little to no labeling over internal regions of interchromatin granule clusters (Fakan & Bernhard 1971; Fakan & Nobis 1978). Based on these data and other studies, interchromatin granule clusters are thought to contain RNA species with a slow turnover rate (for a review, see Fakan & Puvion 1980). These studies are consistent with the findings that snRNPs are associated with these nuclear regions (Spector et al 1983; Fakan et al 1984; Puvion et al 1984; Carmo-Fonseca et al 1992; Huang & Spector 1992a). Perichromatin fibrils are found at the periphery of regions of condensed chromatin and dispersed throughout the interchromatin space (Fakan & Puvion 1980) including on the surface of interchromatin granule clusters (Figure 2). These fibrils have a diameter of 3–5 nm, but can measure up to 20 nm in diameter. In contrast to interchromatin granules, perichromatin fibrils are rapidly labeled with [^{3}H]-uridine, which suggests that they correspond to nascent pre-mRNA (Bachellerie et al 1975; Fakan et al 1976). Monneron & Bernhard (1969) first suggested a relationship between these fibrils and extranucleolar RNA synthesis. SnRNPs, SC-35 (Figure 2), and hnRNP antigens have been localized to these fibrils (Fakan et al 1984; Puvion et al 1984; Spector et al 1991).

In addition to the speckled localization of pre-mRNA splicing factors, several other proteins have been localized in a speckled nuclear distribution (Smith et al 1985, 1986; Spector et al 1987; Turner & Franchi 1987). Of particular interest are studies that have identified a protein of approximately 236 kd, which has been named centrophilin (Tousson et al 1991), or the nuclear mitotic apparatus protein (NuMA) (Lyderson & Pettijohn 1980; Compton et al 1992; Yang et al 1992). During interphase, NuMA localizes in a punctate nuclear distribution (Compton et al 1992). However, during mitosis NuMa associates with microtubules and progressively accumulates at the spindle poles before chromosome segregation at anaphase. Since analysis of the predicted protein sequence of NuMA is similar to myosins

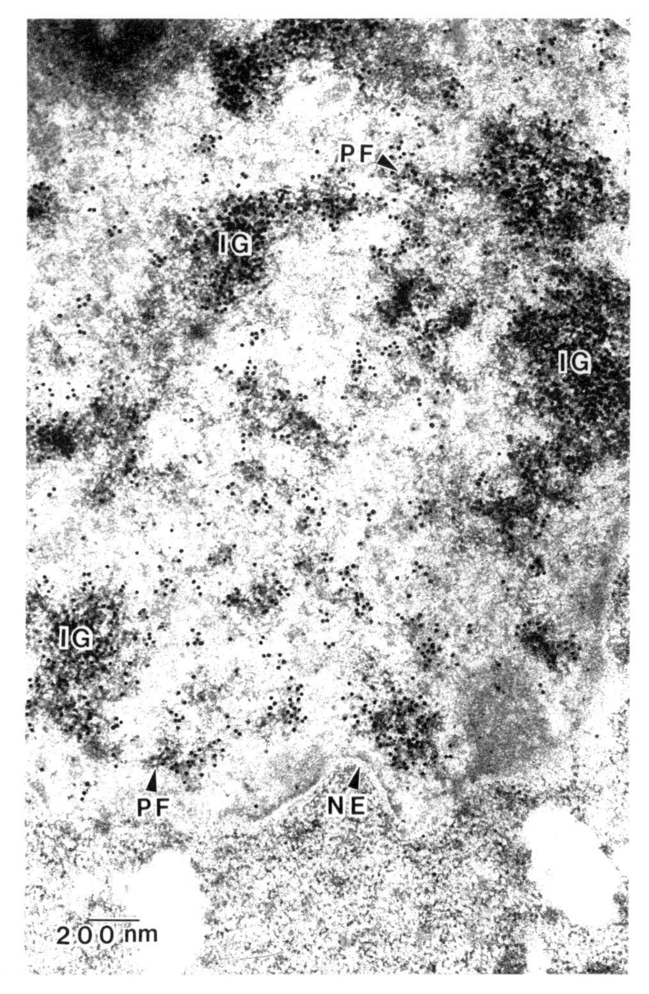

Figure 2 Immunogold labeling of a HeLa cell section with the anti-SC35 antibody. Pre-mRNA splicing factors are localized to clusters of interchromatin granules (IG) and perichromatin fibrils (PF). PFs are on the border of chromatin-enriched regions (lighter areas) and in many cases form connections that extend between the IG clusters and other nuclear regions. Numerous connections can be observed between innumostained regions (post-stained by the EDTA-regressive method) (Bernhard 1969). NE = nuclear envelope.

and intermediate filaments, which are capable of forming coiled-coil dimers, NuMa is thought to be a structural component of the nucleus (Lyderson & Pettijohn 1980; Yang et al 1992) and may be important in the organization of the speckled pattern.

Three-dimensional reconstruction techniques have shown that the snRNPs are not present in isolated islands; instead, portions of the speckled pattern are connected to form a latticework that extends between the nucleolar surface and the nuclear lamina-envelope (Spector 1990; Spector et al 1991). By immunofluorescence techniques, the latticework appears to be composed of two components, larger more intensely stained regions and less intensely stained regions, which in many cases appear to connect two of the larger regions. Others have previously identified a ribonucleoprotein network or interchromatin net in two-dimensional images by cytochemical staining (Puvion & Bernhard 1975), or in nuclear matrix preparations (Smetana et al 1963). For detailed elaborations of the nuclear matrix the reader is referred to several reviews (Berezney 1984; Nickerson et al 1990; Stuurman et al 1992b).

The most convincing data to demonstrate that portions of regions enriched in splicing factors connect to form a nuclear latticework come from immunogold localization studies (Spector et al 1991), which provides the highest resolution currently available. The organization of splicing factors in a latticework is probably dynamic and reflects the physiological state of the cell. Therefore, at any given time, the shape, connections between, and organization of, speckles may vary.

An Arginine/Serine-Rich Domain Targets Proteins to Nuclear Speckles

The sub-localization of splicing components to speckled nuclear regions has recently been attributed to a specific targeting signal (Li & Bingham 1991). An arginine/serine- (RS) rich domain composed of approximately 120 amino acids from two different *Drosophila* pre-mRNA splicing regulators, *suppressor-of-white-apricot* (*su*(w^a)) and *transformer (tra)*, was determined to target proteins into the nucleus and direct them to the speckled regions. Furthermore, the fusion of an RS domain to β-galactosidase directs this protein to the speckle region, which suggests that the RS domain is essential and sufficient to target proteins to this subnuclear region. Multiple splicing factors including *tra2* (Amrein et al 1988) in *Drosophila* and SC-35 (Fu & Maniatis 1992), SF2/ASF (Krainer et al 1991; Ge et al 1991), U2AF (Zamore et al 1992; Zhang et al 1992) and the U1 snRNP 70-kd polypeptide (Theissen et al 1986) in mammalian cells also contain an RS-rich domain. In fact, Roth and co-workers (Roth et al 1990; Zahler et al 1992; Mayeda et al 1992) had previously identified a family of nuclear phosphoproteins that contain an arginine- and serine-rich carboxy-terminal domain. A monoclonal antibody, which recognizes the entire family of proteins, labels mammalian cells in a speckled distribution (Roth et al 1990). The RS domain may serve to concentrate splicing factors in a sub-compartment of the nucleus (speckles) in order to favor interactions between splicing factors and pre-mRNA substrates, thus increasing the efficiency of spliceosome assembly, RNA cleavage, and ligation (Roth et al 1990).

Effect of RNA Polymerase II Transcription on the Localization of Splicing Factors

Several studies have used RNA polymerase II inhibitors to examine whether the nuclear organization of splicing factors was dependent upon active transcription. In one such study (Spector et al 1983) cells were incubated in medium containing the adenosine analogue 5,6-dichloro-1-β-D-ribofuranosylbenzimidazole (DRB). DRB is known to decrease pre-mRNA synthesis in HeLa cells by 70% and to inhibit the appearance of mRNA in the cytoplasm by >95% (Sehgal et al 1976). Upon incubation of cells with DRB at a concentration of 25 μg/ml for 2 hr, the nuclear speckles round up and the connections between speckles are lost (Spector et al 1983). This is similar to the change observed with α-amanitin treatment under conditions that inhibit RNA polymerase II (Carmo-Fonseca et al 1992; D. Spector et al, unpublished observations). However, unlike α-amanitin, the effect of DRB on the organization of splicing factors was found to be reversible within 30 min of removal of the drug (D. Spector et al, unpublished observations). Taken together, these data demonstrate that the organization of splicing factors in the cell nucleus is dependent upon RNA polymerase II activity. When α-amanitin-treated cells were examined at the electron microscopic level, a significant decrease in the number of perichromatin fibrils was observed as compared to control cells (Marinozzi & Fiume 1971; Petrov & Sekeris 1971; Sinclair & Brasch 1978; Kedinger & Simard 1974). In addition, Carmo-Fonseca et al (1992) showed that snRNPs are no longer concentrated in coiled bodies after α-amanitin or actinomycin D treatment. The decrease in the number of perichromatin fibrils after α-amanitin treatment supports the idea that the connections observed between the larger speckles represent snRNPs associated with perichromatin fibrils or sites of nascent transcripts. These findings are in agreement with previous studies that showed [^{3}H]-uridine incorporation as well as hnRNP, snRNP, and SC-35 antigens to be associated with perichromatin fibrils (Bachellerie et al 1975; Fakan et al 1976, 1984; Puvion et al 1984; Spector et al 1991). Furthermore, these data support the possibility that pre-mRNA splicing occurs cotranscriptionally as has been previously suggested for *Drosophila* embryo genes (Beyer & Osheim 1988), genes in mouse and *Drosophila* somatic cells (Fakan et al 1986), as well as puffs and Balbiani rings in *Chironomus* polytene chromosomes (Sass & Pederson 1984). Therefore, the organization of splicing factors in the nucleus is a reflection of the transcriptional activity of the cell.

Recently, Jiménez-García and Spector (1993) used adenovirus 2 as a model system to study the spatial and temporal relationship of transcription and pre-mRNA processing in the mammalian cell nucleus. The localization of viral RNA sequences was examined throughout the infection process and compared to the distribution of several host cell factors that are involved in transcription, pre-mRNA splicing, and in packaging RNA transcripts. Ad2 RNA was first visualized in the cell nucleus as 6–12 dots at 7 hr post-infection. The dots were shown to develop, over a 24 hr time course, into an elaborate series of rings (replication foci)(Puvion-Dutilleul & Puvion 1991; Puvion-Dutilleul et al 1992) and large dots that occupied a majority of the nuclear volume. At 10–14 hr post-infection, RNA polymerase II, pre-mRNA splicing factors, and hnRNP proteins were shuttled to the sites of Ad2 RNA where they colocalized. In contrast, the La protein, which is not involved in RNA polymerase II transcription or splicing (Stefano 1984), remained diffusely distributed throughout the nucleoplasm. As infection progressed, the sites of RNA localization increased in both size and number, and splicing factors continued to be shuttled to these new sites of Ad2 RNA localization. In Cos-1 cells transiently transfected with a plasmid containing a portion of the rat β-tropomyosin gene (Guo et al 1991), a similar shuttling of splicing factors to sites of new RNA synthesis was observed (Jiménez-García & Spector 1993).

At least two potential mechanisms, a scanning mechanism or a recruiting mechanism, were proposed to account for the transcription-dependent shuttling of both transcription and pre-mRNA processing factors upon adenovirus infection (Jiménez-García and Spector 1993). In the scanning mechanism, factors would continuously diffuse throughout the nucleus either in a soluble form or move on components of a nuclear matrix. When these factors reached a potential active site they would dock and transcription and processing would occur. Evidence has been provided that active sites of transcription (Berezney 1984), splicing components (Vogelstein & Hunt 1982; Spector et al 1983; Smith et al 1986; Zeitlin et al 1987), pre-mRNA (Ciejek et al 1982; Mariman et al 1982), and poly(A)$^+$ RNA (Xing & Lawrence 1991) are all associated with the nuclear matrix. However, the continuous movement of factors in living cells has not yet been demonstrated.

In the recruiting mechanism model, factors would be associated with specific storage and/or assembly sites in the nucleus. Prior to or at the initiation of transcription, these factors would be recruited to the active sites of transcription by another factor or chaperone molecule. Evidence supporting this model comes from previous studies that showed that a subpopulation of splicing factors is localized to interchromatin granule clusters in mammalian cell nuclei (Fakan et al 1984; Spector et al 1991). Since these clusters contain little labeled RNA, after short pulses with [^{3}H]-uridine (Fakan &

Nobis 1978), they may not represent active sites of transcription. Therefore, in uninfected cell nuclei, splicing factors are localized to both sites of active transcription (perichromatin fibrils) and storage and/or assembly sites (interchromatin granule clusters). Furthermore, upon introduction of new transcription sites into the cell nucleus, a concomitant decrease in the signal intensity of splicing factors was observed at host cell speckles with an increase at new active sites of viral transcription, while the overall level of snRNP proteins remained constant throughout the infection process (Jiménez-García & Spector 1993). These findings suggest that there are signals generated in the nucleus that regulate the compartmentalization of factors to nuclear regions where they will be functioning. Identification of these signaling mechanisms will be key to understanding the integration of a variety of functional events that occur within the boundaries of the nuclear envelope.

Distribution of hnRNP Proteins

Pre-mRNA or heterogeneous nuclear RNA (hnRNA) is present in the cell nucleus in the form of a complex that contains a discrete set of proteins. By electron microscopy, these hnRNP particles can be visualized as a linear array of 20–25 nm particles along each pre-mRNA molecule (reviewed in Chung & Wooley 1986; Dreyfuss 1986). Individual monoparticles, 40S structures, can be generated by cleavage of the linker RNA. Each monoparticle contains approximately 20 different proteins including the core hnRNP proteins (Beyer et al 1977; Piñol-Roma et al 1988) (for a review, see Dreyfuss 1986). HnRNP proteins are thought to be involved in the packaging and post-transcriptional processing of all pre-mRNAs (reviewed in Chung & Wooley 1986; Dreyfuss 1986). Using chromatin spreading methods, Beyer et al (1980) found RNP particles, averaging 24 nm in diameter, to be associated with the majority of hnRNA transcription units of *Drosophila melanogaster* embryos.

Antibodies have been generated to several hnRNP proteins and their organization within the nucleus has been examined by immunofluorescence and immunoelectron microscopy. The hnRNP core proteins were shown to be distributed diffusely throughout the nucleoplasm, but absent from the nucleoli in a variety of cell types examined (Jones et al 1980; Choi & Dreyfuss 1984; Dreyfuss et al 1984; Leser et al 1984; Piñol-Roma et al 1989). In nuclear matrix preparations, these proteins were found to be associated with fibrogranular material enmeshed in the core filament network (He et al 1991). The abundant hnRNP K (66 kd) and J (64 kd) proteins, which are the major poly(C) binding proteins in HeLa cells (Matunis et al 1992), also localized in a diffuse nuclear distribution. This distribution of hnRNP proteins appears to overlap with a portion of the speckled pattern observed with anti-splicing factor antibodies. Fakan et al (1984) have shown, by immunogold electron microscopy, that the hnRNP C proteins (41, 43 kd) are localized to perichromatin fibrils, but not to interchromatin granule clusters. The remainder of the diffuse staining observed with these antibodies may represent the localization of transcripts in transit to the nuclear envelope and/or a pool of hnRNP proteins in transit to new sites of transcription.

Two other hnRNP proteins, the L and I proteins, have been shown to localize in novel nuclear distributions (Piñol-Roma et al 1989; Ghetti et al 1992). The L protein is a 64–68 kd protein that stains the majority of non-nucleolar nascent transcripts from the loops of lampbrush chromosomes in the newt, being the first hnRNP protein found to be associated with these giant loops. In mammalian cells, this protein is distributed diffusely throughout the nucleoplasm. However, in addition, it stains 1–3 discrete non-nucleolar regions. The identity of these regions is unknown at this time. The 58-kd hnRNP I protein is identical to the previously reported polypyrimidine tract-binding protein (PTB) (Gil et al 1991; Patton et al 1991). In addition to its general nucleoplasmic distribution, a high concentration of the I protein appears to form a cap-like structure in HeLa cells that is associated with the surface of a portion of the nucleolus (Ghetti et al 1992). The identity of this structure is also unknown.

During mitosis, the C and A1 proteins were shown to be distributed throughout the cell. Different hnRNP proteins were found to return to cell nuclei after mitosis by different mechanisms: one transcription-dependent and the other transcription-independent (Piñol-Roma & Dreyfuss 1991). In recently divided cells, the C proteins accumulated in the daughter nuclei as soon as the nuclei were formed, while most of the A1 protein remained in the cytoplasm. Therefore, hnRNP complexes appear to dissociate from each other prior to their transport back to the nucleus after mitosis. Inhibition of RNA polymerase II transcription did not affect the return of the C proteins, but inhibited the transport of A1 to the nucleus (Piñol-Roma & Dreyfuss 1991). Therefore, transcription by RNA polymerase II is required for the return of the A1 protein, but not the C proteins, to the nucleus after mitosis. Other hnRNP proteins including A2, B1, B2, E, H, and L are also transported to the nucleus in a transcription-dependent mode (Piñol-Roma & Dreyfuss 1991).

Recently, the hnRNP A1 protein was found to shuttle between the nucleus and cytoplasm of interphase cells (Piñol-Roma & Dreyfuss 1992). In interphase cells in which transcription of RNA polymerase II was inhibited by actinomycin D, the A1 protein accumulated in the cytoplasm, whereas the C proteins and the U protein remained restricted to the nucleus. Cross-linking studies have shown that some of the A1 that accumulates in the cytoplasm is bound to poly(A)$^+$ RNA (Piñol-Roma & Dreyfuss 1992).

In addition to A1, other hnRNP proteins such as A2 and E also accumulate in the cytoplasm in the presence of RNA polymerase II inhibitors. Cells treated with translation inhibitors, or with actinomycin D at a concentration that would only inhibit RNA polymerase I transcription, showed no effect on the intracellular distribution of the A1 or C proteins. Similar results demonstrating a shuttling of the A1 protein were observed in cells treated with the RNA polymerase II inhibitor DRB and in human/*Xenopus laevis* interspecies heterokaryons in the absence of inhibitors of transcription (Piñol-Roma & Dreyfuss 1992). Based on these studies, it was proposed that hnRNP proteins may play a more active role in the transport of mRNA from the nucleus to the cytoplasm, in addition to packaging pre-mRNA molecules (Piñol-Roma & Dreyfuss 1992).

Localization of Splicing Factors in the Amphibian Germinal Vesicle

Gall and co-workers have examined the localization of a variety of splicing components in amphibian germinal vesicles (Gall & Callan 1989; Wu et al 1991; see Gall 1991 for a review). These studies have shown that lampbrush chromosome loops from germinal vesicles of the newt *Notophthalmus viridescens* are uniformly stained with antibodies against snRNPs, SC-35, and hnRNP proteins. Thus the loops appear to be packaged into a ribonucleoprotein complex that includes all of the components of the spliceosome for which probes are available. This situation would be comparable to the staining of perichromatin fibrils in mammalian cells by anti-snRNP (Fakan et al 1984; Puvion et al 1984), anti-hnRNP (Fakan et al 1984), and anti-SC-35 (Spector et al 1991) antibodies. Therefore, it is likely that pre-mRNA splicing occurs in close proximity to the sites of transcription, in mammalian cells and amphibian germinal vesicles, as previously suggested (Beyer & Osheim 1988; Fakan et al 1986; Sass & Pederson 1984).

In the germinal vesicle, anti-spliceosome antibodies also stain large extrachromosomal particles, designated snurposomes (Gall 1992). Snurposomes have been divided into three classes, designated A, B, and C. The A snurposomes vary in size from 1 to 4 μm in diameter and appear to contain exclusively U1 snRNPs, while the B snurposomes are 4 μm in diameter and contain U1, U2, U4, U5, and U6 snRNPs, as well as SC-35. C snurposomes (spheres) can be as large as 20 μm in diameter. They appear to contain granular inclusions, and B snurposomes are found on their surface. Gall and co-workers proposed that spliceosome components may be preassembled into macromolecular complexes in the B snurposome, analogous to the assembly of ribosomes in the nucleolus (Wu et al 1991). The complexes assembled in the snurposomes, rather than free snRNPs or hnRNPs, would then associate with nascent pre-mRNA. However, Amero et al (1992) presented evidence using *Drosophila* polytene chromosomes to show that hnRNPs and snRNPs are independently deposited at sites of transcription, which suggests that factors do not move to active transcription sites as macromolecular complexes. While B snurposomes may be equivalent to clusters of interchromatin granules in mammalian cells, counterparts for A and C snurposomes in mammalian cells have thus far not been identified. Future studies will shed more light on the similarities and/or differences in the association of factors and substrates in the above systems.

Localization of Splicing Factors in Yeast

The immunolocalization of several nuclear constituents associated with pre-mRNA processing has been reported in the budding yeast *Saccharomyces cerevisiae*. Last & Woolford (1986) produced antibodies against fusion proteins that contain portions of the precursor RNA processing (PRP)2 or PRP3 open reading frames. The PRP2 protein is thought to be associated with spliceosomes (Lin et al 1987; King & Beggs 1990) and the PRP3 protein with the U4/U6 snRNP (Ruby & Abelson 1991). These antibodies were used with light microscopy to show that polypeptides expressed from high-copy number plasmids were localized to the cell nucleus in *S. cerevisiae*. However, the precise subnuclear localization could not be determined in this study.

The PRP11 protein in *S. cerevisiae* has been shown to be specifically associated with the 40S spliceosome and a 30S complex (Chang et al 1988). By immunogold labeling, the PRP11 protein was localized to the non-nucleolar portion of the *S. cerevisiae* nucleus in a predominantly perinuclear distribution. Potashkin et al (1990) used an antibody against the m_3G cap structure of snRNAs and showed that 85% of the m_3G-labeled colloidal gold particles were localized to the nucleolar portion of the nucleoplasm of *Schizosaccharomyces pombe*. Brennwald et al (1988) have shown that this same m_3G antibody immunoprecipitates snRNPs U1 to U5 from *S. pombe* extracts, and U1 and U2 snRNPs are the most predominant snRNPs in this organism. A similar nucleolar localization was observed with an antibody that recognizes a protein component of U1 snRNP (Potashkin et al 1990). Based on these data, it was suggested that the classically designated nucleolar portion of the yeast nucleus is organized differently from the mammalian cell nucleolus and contains functional domains in addition to those associated with rRNA transcription and processing.

Recently, the localization of the PRP6 protein has been examined in *S. cerevisiae* (Elliott et al 1992). PRP6 is thought to be associated with the U4/U6 snRNP. The localization of this protein was found to be predominantly nucleoplasmic (Elliott et al 1992). Deconvolved immunofluorescent images have shown PRP6 to be localized in what appears to be a speckled nuclear distribution pattern with both more intense and weaker sites of staining within the nuclei (Elliott et al 1992). In many cases the less intensely stained regions appear to connect regions that are more intensely stained. This type of nuclear organization is similar to that reported in mammalian cells for the distribution of several pre-mRNA splicing components (Spector 1990; Spector et al 1991). More work needs to be done before a general scheme for the nuclear organization of pre-mRNA processing components in yeast begins to emerge.

Distribution of Nuclear Poly(A)$^+$ RNA

Since almost all messenger RNAs contain a polyadenylate tail, localization of nuclear poly(A)$^+$ RNA provides an easy means of identifying the majority of this RNA population in the cell. Carter et al (1991) examined the localization of endogenous poly(A)$^+$ RNA in interphase nuclei by fluorescence in situ hybridization. This study revealed a speckled distribution of poly(A)$^+$ RNA, which colocalized with the speckled pattern of snRNPs. Carter et al (1991) referred to these nuclear regions as transcript domains. The localization of poly(A)$^+$ RNA to the entire speckled pattern, which includes both interchromatin granules and perichromatin fibrils, is not consistent with both of these regions representing nascent transcripts (for a review, see Fakan & Puvion 1980). In previous studies, which examined [^{3}H]-uridine incorporation, a portion of the speckled pattern (interchromatin granule clusters) was found to contain little to no newly synthesized RNA (for a review, see Fakan & Puvion 1980). Therefore, the poly(A)$^+$ RNA localization observed may represent both nascent and stable species. It is likely that the portion of the poly(A)$^+$ RNA associated with perichromatin fibrils will be spliced and exported, while other RNA with a potentially longer half-life may remain in the nucleus and serve a structural role in interchromatin granule clusters. Carter et al (1993) did not observe connections between transcript domains that have been observed for the localization of splicing factors at the electron microscope level (Spector et al 1991). In addition, since the majority of transcript domains do not contact the nuclear envelope, the path that poly(A)$^+$ RNA follows to exit the nucleus remains unknown. However, the transport of several specific mRNAs will be discussed below. Future studies at the electron microscopic level are needed to examine the above inconsistencies.

Distribution of Specific Cellular RNA Transcripts

Several studies have examined the localization of specific cellular RNA transcripts. Wang et al (1991) microinjected fluorescently tagged β-globin pre-mRNA into interphase nuclei and showed that these RNA molecules localize in a speckled distribution coincident with the speckled pattern enriched in splicing factors. In contrast, microinjection of transcripts lacking an intron, or with a deleted polypyrimidine tract and 3′ splice site, resulted in a diffuse distribution of the transcripts. These results have shown that intron-containing transcripts have the ability to associate with nuclear speckles enriched in pre-mRNA splicing factors.

To localize endogenous RNA transcripts in the cell nucleus, Lawrence et al (1989) took advantage of Namalwa cells, which contain two copies of the Epstein Barr virus (EBV) genome closely integrated on chromosome 1 (Lawrence et al 1988). EBV RNA transcripts were detected as a "track" in the nuclei of Namalwa cells, which averaged 5 μm in length. The tracks appeared to extend between the nuclear interior and the nuclear envelope. Cells stably transfected with a plasmid containing the neu oncogene also showed a highly localized concentration of neu RNA. However, the distribution in many cases appears as clusters of dots rather than an elongated track as seen for the EBV RNA (Lawrence et al 1989). A dot-like distribution has also been shown for the *string* RNA within cells of a cycle-14 *Drosophila* embryo (O'Farrell et al 1989).

To elucidate the localization of specific nascent cellular transcripts and the pathways that these transcripts take from their sites of synthesis to the nuclear envelope, nascent RNA transcripts of an inducible gene, *c-fos*, were localized in the interphase nucleus by in situ hybridization (Huang & Spector 1991). Observations by confocal laser scanning microscopy and high voltage electron microscopy demonstrated that *c-fos* RNA transcripts extend as an elongated path, through the depth of the nucleus, from their sites of synthesis to the nuclear envelope. Electron microscopic in situ hybridization showed these transcripts to exit the nucleus at a very limited area that could be related to a group of nuclear pores and thus support the gene gating hypothesis (Blobel 1985). Double labeling of *c-fos* RNA transcripts and SC-35 revealed a close association between the RNA transcripts and the speckled regions (Huang & Spector 1991). Most recently, Xing et al (1993) localized fibronectin and neurotensin pre-mRNAs and found them to be highly concentrated in 1–2 sites per nucleus. Fibronectin RNA frequently accumulated in tracks up to 6 μm long. The gene was positioned at or near one end of the track. However, elongated tracks for neurotensin RNA were not observed. Similar to what was previously observed for *c-fos* RNA (Huang & Spector 1991), the site of the fibronectin gene was usually associated with larger transcript domains (speckles) (Xing et al 1993). Since hybridization to cDNA probes produced longer tracks than with intron probes, it was suggested that splicing occurs within a portion of the tracks. Intron probes also localized as a dispersed signal throughout the nucleoplasm and did not concentrate around the nuclear periphery (Xing et al 1993), as was reported previously for the acetylcholine receptor intron (Berman et al 1990). Taken together, the above studies showing a close association of RNA substrates or products with nuclear regions enriched in splicing factors (speckles) provide the strongest support to date that these nuclear regions are involved in pre-mRNA splicing.

Perhaps the best example of RNP transport comes from the work done with Balbiani ring (BR) granules in *Chironomus tentans*. BR granules are synthesized in BRs 1 and 2, two giant puffs on chromosome IV in the larval salivary glands. The particles are released from the chromatin axis and are found in the nucleoplasm with a higher frequency in the vicinity of the nuclear envelope, in particular close to the entrance of the nuclear pores (Mehlin et al 1992). The BR granules represent mRNP particles that encode information for secretory polypeptides with molecular masses around 10^6 daltons (for a review, see Daneholt 1982). Since BR granules found free in the nucleoplasm do not immunolabel with anti-snRNP antibodies, it was suggested that splicing takes place prior to the formation of these mature particles (Vazquez-Nin et al 1990). The 37 kb transcripts (75S RNA) (Wurtz et al 1990) are packaged into 50 nm RNP particles (Lamb & Daneholt 1979; Olins et al 1980), which have been characterized as a ribbon bent into a ring (Skoglund et al 1986). Recently, Olins et al (1992) demonstrated an RNA-rich substructure to BR granules. Each granule was found to contain approximately 10–12 RNA-rich particles (average particle diameter 9–10 nm). Electron microscopic tomographic studies have shown that the BR RNP particle positions itself in front of the central channel of the nuclear pore complex (Mehlin et al 1992; Olins et al 1992). The particle then becomes elongated and finally rod shaped with a diameter of 25 nm and a length of approximately 135 nm (Stevens & Swift 1966; Mehlin et al 1991), or cone-shaped with a leading string of particles passing through the nuclear envelope (Olins et al 1992). The unfolded particle is translocated through the nuclear pore in an oriented fashion with the 5′ end of the transcript in the lead (Mehlin et al 1992). The RNP particle is not reformed as a globular structure on the cytoplasmic side of the pore complex. Instead, the RNP fiber appears to associate with a row of putative ribosomes. It remains to be determined if this mode of transport is specialized for the BR granule, or if it is ubiquitous to all mRNP particles.

NUCLEOLAR ORGANIZATION

One of the most dramatic and noteworthy examples of the relationship between spatial organization and cell function is represented by the nucleolus, which is a distinct biochemical and structural entity within which ribosomal genes and their products are naturally sequestered from the rest of the genome and nucleoplasm. Ribosomal gene transcription, ribosomal RNA processing, and preribosomal particle formation all occur within this highly specialized region of the nucleus (for a review, see Busch & Smetana 1970; Hadjiolov 1985). Although the nucleolus is not membrane-bound, specific proteins have been localized to discrete functional regions within this structure (for a review, see Raska et al 1990; Hernandez-Verdun 1991) and a nucleolar localization signal for proteins has been identified (reviewed in Hatanaka 1990). Ultrastructurally, the nucleolus is composed of five organizational areas (Figure 3): (*a*) a dense fibrillar region, (*b*) fibrillar center(s), (*c*) a granular region, (*d*) nucleolar vacuoles, and (*e*) condensed nucleolar

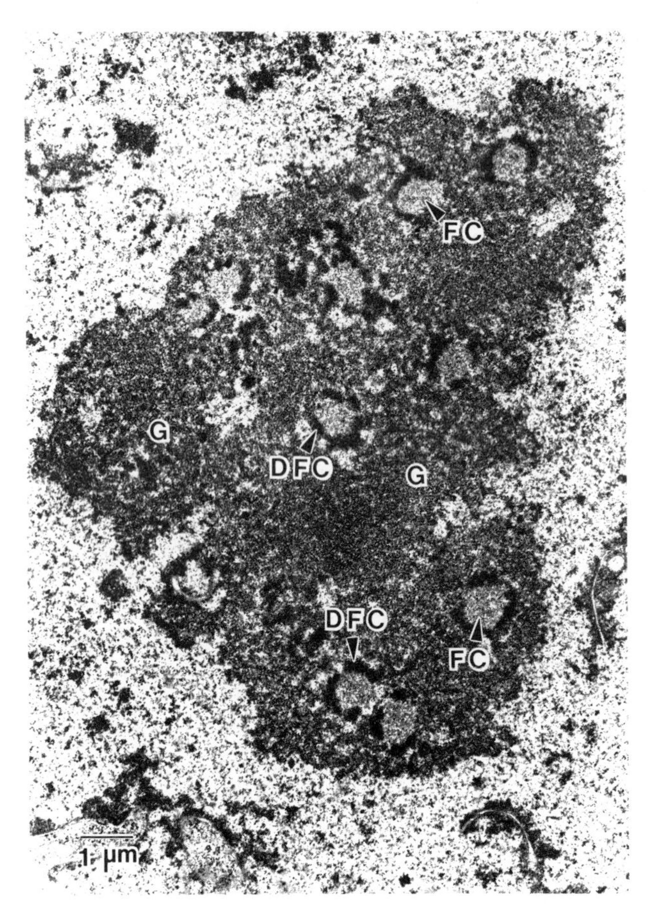

Figure 3 Thin section transmission electron micrograph of a nucleolus showing the relationship between the fibrillar centers (FC), dense fibrillar components (DFC), and granular region (G).

chromatin. DNA emanating from the chromosomal nucleolar-organizing-region (NOR) (McClintock 1934) is thought to form and organize the nucleolus (reviewed in Goessens 1984).

The fibrillar centers are thought to be the interphase equivalent of the NORs. However, there is no direct correspondence between the number of NORs and the number of fibrillar centers (Mirre & Knibiehler 1982; Vagner-Capodano & Stahl 1982). Three-dimensional reconstructions based on serial sections of the nucleolus have shown that the fibrillar centers in mouse oocytes are distinct structures that are separated from each other by the dense fibrillar component (Mirre & Stahl 1981). The granular region of the nucleolus is made up of particles 15–20 nm in diameter. These particles represent preribosomes at different stages of maturation as well as large and small ribosomal subunits. The nucleolus is usually surrounded by a shell of perinucleolar chromatin that is connected with regions of intranucleolar chromatin and is continuous with the fibrillar centers (Goessens 1979; Ashraf & Godward 1980). Recently Meier & Blobel (1992) identified a protein, Nopp140, which was shown to shuttle between the nucleus and the cytoplasm. Immunoelectron microscopy showed that Nopp140 localized along a limited number of curvilinear tracks that extend from the dense fibrillar region of the nucleolus to the nuclear envelope. The protein was proposed to function in import of ribosomal proteins and/or export of preribosomal subunits.

The rDNA of higher eukaryotes occurs in multiple copies and is organized in tandem repeats that are separated from each other by nontranscribed spacer regions of DNA. Human cells contain approximately 250 copies of rDNA (Lewin 1980) and the length of a single rDNA repeat unit is 44 kb (for a review, see Scheer & Benavente 1990). Recently, the sites of rDNA localization and of rRNA transcription have become somewhat controversial (for a review, see Jordan 1991). Early electron microscopic studies, which used [^{3}H]-uridine incorporation and in situ autoradiography, showed that the first labeled nucleolar structures were the dense fibrillar regions (Granboulan & Granboulan 1965), which suggested that these regions of the nucleolus were the sites of active rDNA transcription (reviewed in Fakan & Puvion 1980). However, in situ hybridization studies (Thiry & Thiry-Blaise 1989; Thiry 1992b) with rDNA specific probes, cytochemical studies with anti-DNA antibodies (Scheer et al 1987; Thiry et al 1988), and DNA labeling of cell sections with terminal deoxynucleotidyltransferase (Thiry 1992a) showed that DNA (rDNA) is not present in detectable amounts in the dense fibrillar regions of the nucleolus. In contrast, rDNA seems to be preferentially localized at the peripheral regions of the fibrillar centers (Thiry & Thiry-Blaise 1989) or throughout the fibrillar centers (Puvion-Dutilleul et al 1991; Thiry et al 1991; Thiry 1992b) and in association with intranucleolar condensed chromatin. Interestingly, this localization coincides with the localization of RNA polymerase I in the fibrillar centers (Scheer & Rose 1984; Scheer & Raska 1987), although others have localized RNA polymerase I (Raska et al 1989b), topoisomerase I (Raska et al 1989b), and the RNA polymerase I transcription factor UBF (Rendon et al 1992) to both the fibrillar centers and the dense fibrillar region of the nucleolus.

Recently, several in situ hybridization studies provided results that are contradictory to those described above. Wachtler et al (1990) localized rDNA exclusively in the dense fibrillar component of human lymphocytes, spermatocytes (Wachtler et al 1992), and Sertoli cells (Stahl et al 1991). In spermatocytes, rDNA was also found in the condensed chromatin associated with the nucleolus (Wachtler et al 1992). These data agree with the original [^{3}H]-uridine incorporation studies, and the authors conclude that rDNA is primarily located and transcribed in the dense fibrillar component, with a minor portion being present at the periphery of the fibrillar centers. It is unclear, at this time, why the in situ hybridization studies have resulted in different localizations of rDNA.

In an attempt to address the above discrepancies, several groups have reevaluated the localization of rRNA in nucleoli. Using [^{3}H]-uridine incorporation to localize the sites of nucleolar RNA synthesis, Thiry & Goessens (1991) found that the first sites of grains were in the dense fibrillar component as originally reported by Granboulan & Granboulan (1965). However, with prolonged exposure time, label did not increase over the dense fibrillar component, but it did become increasingly apparent over the fibrillar centers. The authors interpret these data to indicate that transcription occurs in the fibrillar centers and then rapidly moves to, and accumulates within, the dense fibrillar component. Therefore, based on quantity, initial autoradiographic labeling was observed in the dense fibrillar component.

However, an alternative explanation is that transcription primarily occurs within the dense fibrillar component, and perhaps, the transcripts move through the fibrillar centers to the granular region of the nucleolus where they are present in the form of pre-ribosomal particles. As a second approach to examine the localization of rRNA, an in situ in vitro transcription assay was developed. In this assay, sites were identified where biotinylated ribonucleotides were incorporated into RNA by an exogenous polymerase on the surface of cell sections (Thiry & Goessens 1991). These experiments, as well as others (Raska et al 1989b; Scheer & Benavente 1990; Derenzini et al 1990; Puvion-Dutilleul et al 1991), suggest that initiation of transcription in the nucleolus occurs preferentially at the periphery of the fibrillar centers, which results in the accumulation of ribosomal transcripts and the formation of the dense fibrillar component of the nucleolus.

In yet another study, Thiry (1992a) used polyadenylate nucleotidyl transferase and biotinylated ATP to label the ends of RNA at the surface of cell sections, which were then visualized by immunogold methods. This experiment demonstrated that RNA was present in the granular region, dense fibrillar component, and also in the fibrillar centers of the nucleoli. Identical results were observed in HeLa, HEp-2, and Ehrlich tumor cells, as well as in human lymphocytes and Sertoli cells. These results were confirmed by labeling with anti-RNA antibodies. In situ hybridization experiments specifically identified rRNA throughout the fibrillar centers as well as in the granular and dense fibrillar components of the nucleolus (Thiry 1992a). Since the only nucleolar region that labeled for both rRNA and rDNA was the fibrillar center, these regions were postulated to represent the sites of rDNA transcription.

Puvion-Dutilleul et al (1991) used a set of biotinylated probes from various regions along the rDNA gene to examine the specific localization of precursor and mature rRNA. The 5′ external transcribed spacer region of pre-rRNA was localized to the dense fibrillar region and the periphery of the fibrillar centers. In contrast, 18S and 28S mature rRNA were localized in both the dense fibrillar region and the granular region. Based on these data, it was suggested that the dense fibrillar region contains elongating 47S-45S precursor rRNA, whereas rRNA processing intermediates (34S) are mainly located in the granular region. These data are consistent with the localization of U3 small nuclear ribonucleoprotein particle (snRNP), which is involved in the processing of 45S pre-rRNA (Kass et al 1990), in the dense fibrillar component of the nucleolus (Ochs et al 1985). Since the periphery of the fibrillar centers is the only region where both rDNA and newly synthesized pre-rRNA colocalized, Puvion-Dutilleul et al (1991) have suggested that transcription of rRNA occurs in this region.

Although a significant amount of recent data has been generated on the localization of rDNA and rRNA in the nucleolus, a consensus has not yet been achieved. However, since the nucleolar fibrillar centers or the periphery of the fibrillar centers are the sites where rDNA, rRNA, RNA polymerase I, and topoisomerase I, as well as the RNA polymerase I transcription factor UBF colocalize, these sites are the most likely to represent the nucleolar regions where rDNA transcription occurs. The dense fibrillar component is likely to represent elongating transcripts and precursor molecules. Finally, the granular region represents the site of RNA processing intermediates and the localization of assembling and mature ribosomal subunits.

NUCLEAR BODIES

Nuclear bodies are round structures measuring between 0.3 and 1.5 μm in diameter (de The et al 1960). They represent a class of structures (Bouteille et al 1967) that are generally found in the nuclei of hyperstimulated or malignant cells (for a recent review, see Brasch & Ochs 1992); the most well-studied example being the increase of nuclear bodies in estrogen-stimulated tissues (for a review, see Brasch & Ochs 1992). In rooster liver cells, Brasch et al (1989) showed that a single dose of estrogen resulted in a significant increase in the number of nuclear bodies. The average number of nuclear bodies increased from 1.0 in control cells to 3.3 per nucleus 48 hr after induction. Furthermore, the percentage of cells containing nuclear bodies rose from 70% to 94%. Although most nuclear bodies appear fibrillar or granular in appearance, others are lipid-like. Based on cytochemical staining and autoradiographic studies, their composition is thought to be primarily protein and RNA. Recently Fusconi et al (1991) identified two autoantibodies that recognized proteins of 78–92 and 96–100 kd, which localized to nuclear bodies. Five to fifteen nuclear dots were observed in each cell of a sample derived from a human oviduct carcinoma. In a second study, Stuurman et al (1992a) identified a 126-kd antigen that localized to nuclear bodies in rat and human cells. In a human bladder carcinoma cell line, 21 ± 10 nuclear bodies were observed per cell nucleus. The observed structures were tightly bound to the nuclear matrix and did not contain centromere antigens or U1 snRNPs. These antibodies should provide valuable probes to learn more about the protein composition and function of nuclear bodies.

Perhaps the most extensively studied nuclear body is the coiled body. Coiled bodies were first identified by Cajal (1903), who characterized them as nucleolar "accessory bodies" by nucleolar silver staining at the light microscopic level. These structures are generally round and measure 0.5 to 1.0 μm in diameter and consist of coiled fibrillar strands (Monneron & Bernhard 1969; Moreno Diaz de la Espina et al 1982). Cytochemical staining of ribonucleoproteins by the EDTA regressive method (Bernhard 1969) demonstrates that coiled bodies contain RNPs (Monneron & Bernhard 1969). In addition, coiled bodies have been shown to contain orthophosphate ions and acid phosphatase activity (Moreno Diaz de la Espina et al 1982). Since snRNPs have been localized to these nuclear bodies (Fakan et al 1984; Elicieri & Ryerse 1984; Leser et al 1989; Carmo-Fonseca et al 1991; Huang & Spector 1992a), it has been suggested that they may have a role in pre-mRNA metabolism (Carmo-Fonseca et al 1992). However, the lack of DNA (Monneron & Bernhard 1969), [^{3}H]-uridine incorporation (Moreno Diaz de la Espina et al 1980), hnRNP proteins (Carmo-Fonseca et al 1991b; Raska et al 1991), and several non-snRNP splicing factors (see splicing section) in coiled bodies supports the idea that these inclusions are not essential for pre-mRNA splicing.

Recent studies on the coiled body have been fueled by the identification of a coiled-body-specific protein (Raska et al 1991). The 80-kd autoantigen, called p80-coilin, has been localized using sera from individuals with various autoimmune disorders (Raska et al 1990, 1991; Andrade et al 1991). Partial cDNA clones coding for the p80-coilin antigen have been characterized (Andrade et al 1991). The sequence of these clones did not reveal any of the traditional RNA-recognition motifs (Chan et al 1989; Dreyfuss et al 1988). However, the protein contains two stretches rich in arginine and lysine, which could be involved in RNA binding (Zamore et al 1990). p80-coilin is the first antigen identified that is restricted to the coiled body. Using the p80-coilin antibody, Raska et al (1991) determined that the number of coiled bodies varies among different cell types. However, in cells that grow rapidly, the number of coiled bodies per cell ranges from zero to seven as compared to between zero and two per nucleus for slower growing cells. While the number and size of coiled bodies appear to change in HeLa cells, according to the cell cycle, the total amount of coilin protein appears to remain constant (Andrade et al 1993). These data showed that coilin redistributes from an insoluble form in coiled bodies to a relatively soluble form during mitosis (Andrade et al 1993). In another recent study, Spector et al (1992) identified differences in the presence of coiled bodies in transformed cells, immortal cells, and cells of defined passage number. Coiled bodies were found in a low percentage (2 to 3%) of cells with defined passage number, in intermediate numbers of immortal cells (4 to 40%), but were most abundant in transformed cells (81 to 99%). Therefore, a direct correlation was found between the percentage of cells containing coiled bodies and transformation. Based on these findings it was proposed that the presence of coiled bodies is a reflection of potentially increased metabolic activity.

Several studies have supported the original findings of Cajal (1903), which showed a close association of coiled bodies with nucleoli (Monneron & Bernhard 1969; Hardin et al 1969; Seite et al 1982; Schultz 1989; Raska et al 1991). At the cytochemical level, Seite et al (1982) and Raska et al (1990) have shown that coiled bodies are stained by the nucleolar-organizing region (NOR) silver staining technique in a way similar to the fibrillar regions of the nucleolus. The association between coiled bodies and nucleoli was investigated further in cycling cells that were treated with actinomycin D or 5,6-dichloro-1-β-D-ribofuranosylbenzimidazole (DRB) at a dose that results in nucleolar segregation (Raska et al 1990). Nucleolar segregation results from the redistribution of nucleolar components into separated areas containing fibrillar or granular constituents. These studies demonstrated that when nucleoli segregate, the p80-coilin antigen colocalizes with fibrillarin-positive nucleolar components (Raska et al 1990). Fibrillarin is a 34-kd antigen associated with the nucleolar U3 snRNP particle (Lischwe et al 1985; Ochs et al 1985; Tyc & Steitz 1989), which is involved in pre-rRNA processing (Kass et al 1990). In cells incubated with lower concentrations of actinomycin D (0.02 μg/ml), nucleoli did not segregate. However, p80-coilin was observed to be associated with paranucleolar structures similar to what was observed in primary neuron cultures that had received no drug treatment. Furthermore, Lafarga et al (1991) recently showed an increase in the number of coiled bodies in supraoptic neurons of the rat after stimulation of nucleolar transcription. Based on these data, it was suggested that coiled bodies may be involved in the processing, transport, and storage of nucleolar metabolites (Raska et al 1990). However, the presence of other nuclear components not involved in nucleolar function suggests a more general role for this nuclear organelle (Raska et al 1991). The identification of coiled-body-specific antigens and of other proteins of known function to the coiled body have stimulated research on this nuclear organelle. The response of these bodies to changes in nuclear physiology makes them an interesting target for future functional studies.

OTHER DOMAINS

In addition to the domains described in the preceding sections, several other nuclear domains have been identified for which there is less functional and structural data available. The identification of each of these domains was the result of screening cells with human autoantibodies or monoclonal antibodies derived from immunizing mice with nuclear matrix preparations.

A series of antigens (peripherin, I1, PI1, and PI2) have been identified by Chaly et al (1984) that localize to different nuclear domains. One of these antigens, peripherin, also known as P1 (Chaly et al 1984), localizes to the nucleoplasmic side of the nuclear envelope and appears to extend into the nucleoplasm beyond the localization of the nuclear lamins. This protein was proposed to interact with the lamins and other nuclear proteins to mediate attachment of chromatin to the nuclear lamina (Chaly et al 1985). McKeon et al (1984) identified a 33-kd protein, which they called perichromin, that gives a similar interphase and mitotic localization as peripherin (Chaly et al 1984, 1989). However, it has not been determined if these two proteins are the same.

Ascoli & Maul (1991) identified a 55-kd protein that is specifically associated with a series of nuclear dots measuring 0.2–0.3 μm in diameter, which are distributed throughout the nucleoplasm. These nuclear regions were found to be resistant to nuclease digestion and salt extraction. Using double-labeling experiments, it was shown that these nuclear regions differed from regions occupied by kinetochore antigens, sites of pre-mRNA processing factors, nucleoli, nuclear bodies, and chromosomes. Nuclear dots frequently appeared as pairs or doublets. The number of doublets observed increased when quiescent cells were serum-stimulated; however, the total number of dots did not increase in number. Nuclear dots were not observed in quiescent rat liver parenchymal cells, although dots were observed upon stimulation by chemical hepatectomy. Based on these data, it was suggested that nuclear dots may be related to the proliferative state of cells. However, the functional significance of these nuclear regions has not been determined.

In another study, Saunders et al (1991) used an antiserum, which recognized a set of autoantigens of 23–25 kd, to identify a subnuclear compartment within human cells, which they termed a PIKA (polymorphic interphase kakyosomal association). This nuclear domain was observed as a region of lower electron density than the surrounding nucleoplasm and was composed of fibers with an average diameter of 11–17 nm. PIKAs were occasionally found adjacent to nucleoli or heterochromatin and were sensitive to DNAase I digestion. The PIKAs ranged in size from a single large spherical region of 5 μm in diameter to dozens of smaller punctate foci. PIKAs were found to vary in size and number as cells progressed through the cell cycle. Nuclear regions occupied by the PIKA do not coincide with regions enriched in DNA replication factors, centromere antigens, RNA processing factors, or nuclear dot antigens described above.

At present little information is available with regard to the roles that the above domains play in the functioning of the cell nucleus. However, these nuclear regions provide the platform for potential future discoveries into the coordination of nuclear structure and function.

A MODEL FOR THE FUNCTIONAL ORGANIZATION OF THE NUCLEUS

On the basis of the data presented in the above review, a clearer understanding of the organization of the nucleus is beginning to emerge. Actively transcribing genes (RNA polymerase II) appear to be distributed at particular regions throughout the nucleoplasm that may be dependent upon transcriptional activity, cell cycle stage, and/or differentiation. Upon activation of a gene, splicing factors from a nearby interchromatin granule cluster (storage and/or assembly sites) may move to the site of the active gene and associate directly with the nascent transcripts that are visualized as perichromatin fibrils (Figure 4). Since the bulk of DNA is found adjacent to the interchromatin granule clusters, upon initiation of transcription, perichromatin fibrils appear as connections between larger speckles (interchromatin granule clusters) and may also localize in close proximity to the surface of the interchromatin granule clusters (Figure 4). The association of splicing factors with nascent RNA transcripts suggests that pre-mRNA splicing occurs in close proximity to the sites of transcription. Therefore, the organization of

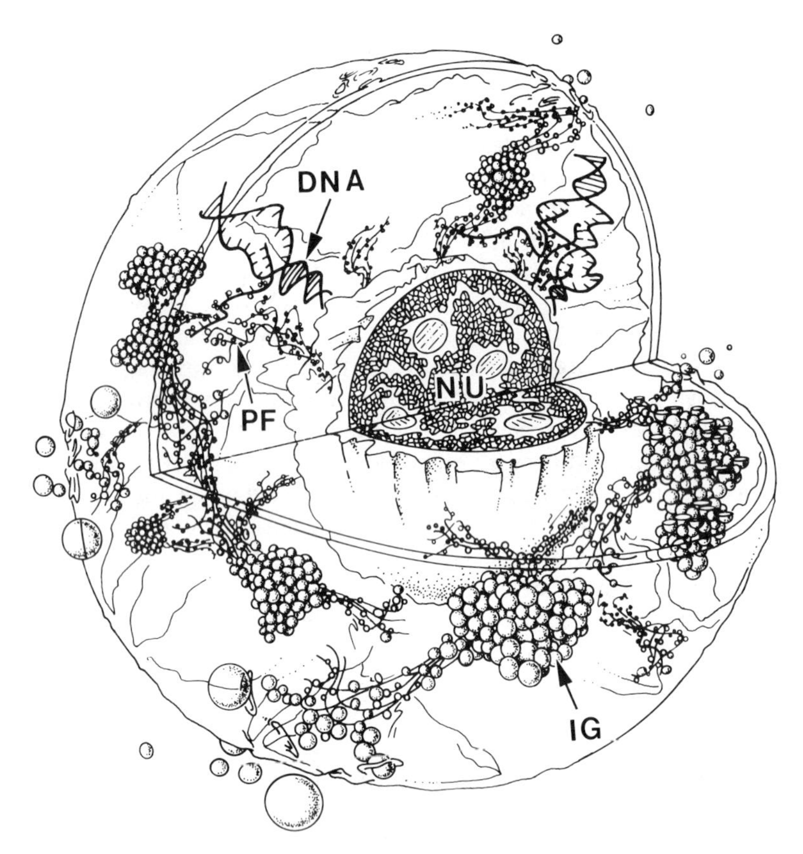

Figure 4 Model of the three-dimensional organization of the mammalian cell nucleus based on electron microscopic data. RNA polymerase I activity is restricted to the nucleolus (NU), which is attached to the nuclear envelope-lamina and contains fibrillar centers, dense fibrillar components, and granular components. Genes (DNA) transcribed by RNA polymerase II are distributed throughout the nucleoplasm. Upon activation of a gene (pol.II), transcripts (perichromatin fibrils) (PF) are produced. Splicing factors are recruited from storage and/or assembly sites, interchromatin granule clusters (IG), to the site of transcription where they associate with nascent transcripts so that pre-mRNA splicing can occur concomitantly with transcription. One IG cluster may service numerous genes that lie in its vicinity. The PFs in many cases appear to extend out from IG clusters and form connections between IG clusters and other nuclear regions. After the events of RNA processing are completed, the mature mRNA may leave the nucleus in a directed path.

the speckled pattern is a reflection of the transcriptional activity of the cell. Once spliced, RNA transcripts in some cases appear to exit the nucleus in a directed manner via tracks or paths that extend from the gene to the nuclear envelope, which suggests a highly organized and regulated means of transport of mRNA through the nucleoplasm to the nuclear pores. Upon their arrival at the NPC, mRNPs may dock by interacting with components of the NPC, such as the fibers extending from the fish-traps, and then be transported through the NPC into the cytoplasm.

While it is agreed that RNA polymerase I transcription occurs within the nucleolus, it is not clear whether transcription occurs in the fibrillar centers, the dense fibrillar component, or at the border where these two nucleolar regions meet. The granular region of the nucleolus contains pre-ribosomes as well as large and small ribosomal subunits.

No information is available with regard to potential nuclear domains involved in RNA polymerase III transcription.

In summary, while the nucleus contains domains associated with particular functions, there appears to be a degree of plasticity in this organization since in all likelihood the nuclear contents are dynamic. The specific orchestration of the dynamic movements of specific functional elements within the nucleus is sure to be realized in the forthcoming years.

ACKNOWLEDGMENTS

I thank Robert Goldman, Scott Henderson, Sui Huang, Raymond O'Keefe, Hans Ris, and Mona Spector for helpful comments on parts or all of the manuscript and Madeline Wisnewski for secretarial assistance. I thank Jim Duffy for drawing the nuclear model in Figure 4. Work from the author's laboratory was supported by grants from the National Institutes of Health (GM42694 and 5P30 CA45508) and the Council for Tobacco Research (3295).

Literature Cited

Aaronson, R, Blobel, G. 1975. Isolation of nuclear pore complexes in association with a lamina. *Proc. Natl. Acad. Sci. USA* 72:1007–11

Aebi, U, Cohn, J, Buhle, L, Gerace, L. 1986. The nuclear lamina is a meshwork of intermediate-type filaments. *Nature* 323:560–64

Agard, DA, Sedat, JW. 1983. Three-dimensional architecture of a polytene nucleus. *Nature* 302:676–81

Akey, CW. 1989. Interactions and structure of the nuclear pore complex revealed by cryo-electron microscopy. *J. Cell Biol.* 109:955–70

Akey, CW. 1990. Visualization of transport-related configurations of the nuclear pore transporter. *Biophys. J.* 58:341–55

Akey, CW, Goldfarb, DS. 1989. Protein import through the nuclear pore complex is a multistep process. *J. Cell Biol.* 109:971–82

Amero, SA, Raychaudhuri, G, Cass, CL, van Venrooij, WJ, Habets, WJ, et al. 1992. Independent deposition of heterogeneous nuclear ribonucleoproteins and small nuclear ribonucleoprotein particles at sites of transcription. *Proc. Natl. Acad. Sci. USA* 89:8409–13

Amrein, H, Gorman, M, Nothiger, R. 1988. The sex-determining gene *tra-2* of Drosophila encodes a putative RNA binding protein. *Cell* 55:1025–35

Andrade, LEC, Chan, EKL, Raska, I, Peebles, CL, Roos, G, Tan, EM. 1991. Human autoantibody to a novel protein of the nuclear coiled body: immunological characterization and cDNA cloning of p80-coilin. *J. Exp. Med.* 173:1407–19

Andrade, LEC, Tan, EM, Chan, EKL. 1993. Immunocytochemical analysis of the coiled body in the cell cycle and during cell proliferation. *Proc. Natl. Acad. Sci. USA* 90: 1947–51

Applebaum, J, Blobel, G, Georgatos, SD. 1990. In vivo phosphorylation of the lamin B receptor. *J. Biol. Chem.* 265:4181–84

Ascoli, CA, Maul, GG. 1991. Identification of a novel nuclear domain. *J. Cell Biol.* 112:785–95

Ashraf, M, Godward, MBE. 1980. The nucleolus in telophase, interphase and prophase. *J. Cell Sci.* 41:321–29

Avivi, L, Feldman, M. 1980. Arrangement of chromosomes in the interphase nucleus of plants. *Hum. Genet.* 55:281–95

Bachellerie, J-P, Puvion, E, Zalta, J-P. 1975. Ultrastructural organization and biochemical characterization of chromatin RNA protein complexes isolated from mammalian cell nuclei. *Eur. J. Biochem.* 58:327–37

Banfalvi, G, Tanke, H, Rapp, AK, Slats, J, van der Ploeg, M. 1990. Early replication signals in nuclei of Chinese hamster ovary cells. *Histochemistry* 94:435–40

Barr, ML, Bertram, EG. 1949. A morphological distinction between neurons of the male and female, and the behavior of the nuclear satellite during accelerated nucleoprotein synthesis. *Nature* 163:676–77

Bartholdi, MF. 1991. Nuclear distribution of centromeres during the cell cycle of human diploid fibroblasts. *J. Cell Sci.* 99:255–63

Berezney, R. 1984. Organization and functions of the nuclear matrix. In *Chromosomal Nonhistone Proteins,* ed. LS Hnilica, IV:119–80. Boca Raton, Fla.: CRC Press

Berezney, R, Basler, J, Buchholtz, LA, Smith, HC, Siegel, AJ. 1982. Nuclear matrix organization and DNA replication. In *The Nuclear Envelope and the Nuclear Matrix,* ed. GG Maul, 183–97. New York: Liss

Berman, SA, Bursztajn, S, Bowen, B, Gilbert, W. 1990. Localization of an acetylcholine receptor intron to the nuclear membrane. *Science* 247:212–14

Bernhard, W. 1969. A new staining procedure for electron microscopical cytology. *J. Ultrastruct. Res.* 27:250–65

Berrios, M, Fisher, PA. 1986. A myosin heavy chain-like polypeptide is associated with the nuclear envelope in higher eukaryotic cells. *J. Cell Biol.* 103:711–24

Berrios, M, Fisher, PA, Matz, EC. 1991. Localization of a myosin heavy chain-like polypeptide to *Drosophila* nuclear pore complexes. *Proc. Natl. Acad. Sci. USA* 88:219–23

Beyer, AL, Christensen, ME, Walker, BW, LeStourgeon, WM. 1977. Identification and characterization of the packaging proteins of core 40S hnRNP particles. *Cell* 11:127–38

Beyer, AL, Miller, OL, Jr, McKnight, SL. 1980. Ribonucleoprotein structure in nascent hnRNA is nonrandom and sequence dependent. *Cell* 20:75–84

Beyer, AL, Osheim, YM. 1988. Splice site selection, rate of splicing, and alternative splicing on nascent transcripts. *Genes Dev.* 2:754–65

Billia, F, De Boni, U. 1991. Localization of centromeric satellite and telomeric DNA sequences in dorsal root ganglion neurons, in vitro. *J. Cell Sci.* 100:219–26

Blobel, G. 1985. Gene gating: a hypothesis. *Proc. Natl. Acad. Sci. USA* 82:8527–29

Borden, J, Manuelidis, L. 1988. Movement of the X chromosome in epilepsy. *Science* 242:1687–91

Bostock, CJ, Prescott, DM. 1971. Buoyant density of DNA synthesized at different stages of the S phase of mouse L cells. *Exp. Cell Res.* 64:267–74

Bourgeois, CA, Laquerriere, F, Hemon, D, Hubert, J, Bouteille, M. 1985. New data on the in situ position of the inactive X chromosome in the interphase nucleus of human fibroblasts. *Hum. Genet.* 69:122–29

Bouteille, M, Kalifat, SR, Delarue, J. 1967. Ultrastructural variations of nuclear bodies in human diseases. *J. Ultrastruct. Res.* 19:474–86

Boveri, T. 1909. Die Blastomerenkerne von *Ascaris megalocephala* und die Theorie der Chromosomenindividualitat. *Arch. Exp. Zellforsch.* 3:181–268

Brasch, K, Harrington, S, Blake, H. 1989. Isolation and analysis of nuclear bodies from estrogen-stimulated chick liver. *Exp. Cell Res.* 182:425–35

Brasch, K, Ochs, RL. 1992. Nuclear bodies (NBs): a newly "rediscovered" organelle. *Exp. Cell Res.* 202:211–23

Bravo, R, Macdonald-Bravo, H. 1985. Changes in the distribution of cyclin (PCNA) but not its synthesis depend on DNA replication. *EMBO J.* 4:655–61

Bravo, R, Macdonald-Bravo, H. 1987. Existence of two populations of cyclin/proliferating cell nuclear antigen during the cell cycle: association with DNA replication sites. *J. Cell Biol.* 105:1549–54

Brenner, S, Pepper, D, Berns, MW, Tan, E, Brinkley, BR. 1981. Kinetochore structure, duplication, and distribution in mammalian cells: analysis by human autoantibodies from scleroderma patients. *J. Cell Biol.* 91:95–102

Brennwald, P, Porter, G, Wise, JA. 1988. U2 small nuclear RNA is remarkably conserved between *Schizosaccharomyces pombe* and mammals. *Mol. Cell. Biol.* 8:5575–80

Brinkley, BR, Ouspenski, I, Zinkowski, RP. 1992. Structure and molecular organization of the centromere-kinetochore complex. *Trends Cell Biol.* 2:15–21

Burke, B, Gerace, L. 1986. A cell free system

to study reassembly of the nuclear envelope at the end of mitosis. *Cell* 44:639–52

Busch, H. 1974–1984. *The Cell Nucleus*, Vols. 1–10. New York: Academic

Busch, H, Smetana, K. 1970. *The Nucleolus*. New York: Academic

Cairns, J. 1966. Autoradiography of HeLa cell DNA. *J. Mol. Biol.* 15:372–73

Cajal, S.R. 1903. Un sencillo metodo de coloracion selectiva del reticulo protoplasmico y sus efectos en los diversos organos nerviosos. *Trab. Lab. Invest. Biol.* 2:129–221

Calza, RE, Eckhardt, LA, DelGiudice, T, Schildkraut, CL. 1984. Changes in gene position are accompanied by a change in time of replication. *Cell* 36:689–96

Camargo, M, Cervenka, J. 1982. Patterns of DNA replication of human chromosomes. II. Replication map and replication model. *Am. J. Hum. Genet.* 34:757–80

Carmo-Fonseca, M, Pepperkok, R, Carvalho, MT, Lamond, AI. 1992. Transcription-dependent colocalization of the U1, U2, U4/U6, and U5 snRNPs in coiled bodies. *J. Cell Biol.* 117:1–14

Carmo-Fonseca, M, Pepperkok, R, Sproat, BS, Ansorge, W, Swanson, MS, Lamond, AI. 1991a. In vivo detection of snRNP-rich organelles in the nuclei of mammalian cells. *EMBO J.* 10:1863–73

Carmo-Fonseca, M, Tollervey, D, Barabino, SML, Merdes, A, Brunner, C, et al. 1991b. Mammalian nuclei contain foci which are highly enriched in components of the pre-mRNA splicing machinery. *EMBO J.* 10: 195–206

Carter, KC, Bowman, D, Carrington, W, Fogarty, K, McNeil, JA, et al. 1993. A three-dimensional view of precursor messenger RNA metabolism within the mammalian nucleus. *Science*. 259:1330–35

Carter, KC, Taneja, KL, Lawrence, JB. 1991. Discrete nuclear domains of poly(A) RNA and their relationship to the functional organization of the nucleus. *J. Cell Biol.* 115:1191–202

Celis, JE, Celis, A. 1985. Cell cycle-dependent variations in the distribution of the nuclear protein cyclin in cultured cells. *Proc. Natl. Acad. Sci. USA* 82: 3262–66

Chaly, N, Bladon, T, Setterfield, G, Little, JE, Kaplan, JG, Brown, DL. 1984. Changes in distribution of nuclear matrix antigens during the mitotic cell cycle. *J. Cell Biol.* 99:661–71

Chaly, N, Little, JE, Brown, DL. 1985. Localization of nuclear antigens during preparation of nuclear matrices in situ. *Can. J. Biochem. Cell Biol.* 63:644–53

Chaly, N, St. Aubin, G, Brown, DL. 1989. Ultrastructural localization of nuclear antigens during interphase in mouse 3T3 fibroblasts. *Biochem. Cell Biol.* 67:563–74

Chan, EKL, Sullivan KF, Tan, EM. 1989. Ribonucleoprotein SS-B/La belongs to a protein family with consensus sequences for RNA-binding. *Nucleic Acids Res.* 17:2233–44

Chang, T-H, Clark, MW, Lustig, AJ, Cusick, ME, Abelson, J. 1988. RNA11 protein is associated with the yeast spliceosome and is localized in the periphery of the cell nucleus. *Mol. Cell. Biol.* 8:2379–93

Choi, YD, Dreyfuss, G. 1984. Monoclonal antibody characterization of the C proteins of heterogeneous nuclear ribonucleoprotein complexes in vertebrate cells. *J. Cell Biol.* 99:1997–2004

Chung, SY, Wooley, J. 1986. Set of novel, conserved proteins fold premessenger RNA into ribonucleosomes. *Proteins* 1:195–210

Ciejek, EM, Nordstrom, JL, Tsai, M-J, O'Malley, BW. 1982. Ribonucleic acid precursors are associated with the chick oviduct nuclear matrix. *Biochemistry* 21:4945–53

Cohen, MM, Enis, P, Pfeifer, CG. 1972. An investigation of somatic pairing in the muntjak (*Muntiacus muntjak*). *Cytogenetics* 11: 145–52

Comings, DE. 1980. Arrangement of chromatin in the nucleus. *Hum. Genet.* 53:131–43

Comings, DE, Okada, TA. 1973. DNA replication and the nuclear membrane. *J. Mol. Biol.* 75:609–18

Compton, DA, Szilak, I, Cleveland, DW. 1992. Primary structure of NuMA, an intranuclear protein that defines a novel pathway for segregation of proteins at mitosis. *J. Cell Biol.* 116:1395–408

Cox, LS, Laskey, RA. 1991. DNA replication occurs at discrete sites in pseudonuclei assembled from purified DNA in vitro. *Cell* 66:271–75

Cremer, T, Cremer, C, Baumann, H, Luedtke, EK, Sperling, K, et al. 1982. Rabl's model of the interphase chromosome arrangement tested in Chinese hamster cells by premature chromosome condensation and laser-UV-microbeam experiments. *Hum. Genet.* 60: 46–56

Cremer, T, Tesin, D, Hopman, AH, Manuelidis, L. 1988. Rapid interphase and metaphase assessment of specific chromosomal changes in neuroectodermal tumor cells by in situ hybridization with chemically modified DNA probes. *Exp. Cell Res.* 176:199–220

Dabauvalle, MC, Benavente, R, Chaly, N. 1988. Monoclonal antibodies to a M_r 68,000 pore complex glycoprotein interferes with nuclear protein uptake in *Xenopus* oocytes *Chromosoma* 97:193–97

Dambaugh, T, Hennessy, K, Fennewald, S, Kieff, E. 1986. The virus genome and its expression in latent infection. In *The Epstein-Barr Virus: Recent Advances,* ed. MA Epstein, BG Achong, pp. 13–45. New York: Wiley & Sons

Daneholt, B. 1982. Structural and functional analysis of Balbiani ring genes in the salivary glands of *Chironomus tentans*. In *Insect Ultrastructure,* ed. R King, H Akai, 1:382–401. New York: Plenum

Davis, LI, Blobel, G. 1986. Identification and characterization of a nuclear pore complex protein. *Cell* 45:699–709

de The, G, Riviere, M, Bernhard, W. 1960. Examen au microscope electronique de la tumour VX2 du lapin domestique derivee du papillome de Shope. *Bull. Cancer* 47: 570–84

Derenzini, M, Thiry, M, Goessens, G. 1990. Ultrastructural cytochemistry of the mammalian cell nucleolus. *J. Histochem. Cytochem.* 38:1237–56

Dessev, G, Iovcheva-Dessev, C, Bischoff, JR, Beach, D, Goldman, R. 1991. A complex containing p34 cdc2 and cyclin B phosphorylates the nuclear lamin and disassembles nuclei of clam oocytes in vitro. *J. Cell Biol.* 112:523–33

Dijkwel, PA, Wenink, PW, Poddighe, PJ. 1986. Permanent attachment of replication origins to the nuclear matrix in BHK cells. *Nucleic Acids Res.* 14:3241–49

Dingwall, C, Laskey, RA. 1986. Protein import into the cell nucleus. *Annu. Rev. Cell. Biol.* 2:367–90

Dingwall, C, Laskey, R. 1992. The nuclear membrane. *Science* 258:942–47

Dooley, DC, Ozer HL. 1977. Replication kinetics of three DNA sequence families in synchronized mouse cells. *J. Cell Physiol.* 90:337–50

Dreyfuss, G. 1986. Structure and function of nuclear and cytoplasmic ribonucleoprotein particles. *Annu. Rev. Cell Biol.* 2:459–98

Dreyfuss, G, Choi, YD, Adam, SA. 1984. Characterization of hnRNA-protein complexes in vivo with monoclonal antibodies. *Mol. Cell. Biol.* 4:1104–14

Dreyfuss, G, Swanson, M, Piñol-Roma, S. 1988. Heterogeneous nuclear ribonucleoprotein particles and the pathway of mRNA formation. *Trends Biochem. Sci.* 13:86–91

Dwyer, N, Blobel, G. 1976. A modified procedure for the isolation of a pore complex-lamina fraction from rat liver nuclei. *J. Cell Biol.* 70:581–91

Elicieri, GL, Ryerse, JS. 1984. Detection of intranuclear clusters of Sm antigens with monoclonal anti-Sm antibodies by immunoelectron microscopy. *J. Cell. Physiol.* 121:449–51

Elliott, DJ, Bowman, DS, Abovich, N, Fay, FS, Rosbash, M. 1992. A yeast splicing factor is localized in discrete subnuclear domains. *EMBO J.* 11:3731–36

Fakan, S, Bernhard, W. 1971. Localization of rapidly and slowly labelled nuclear RNA as visualized by high resolution autoradiography. *Exp. Cell Res.* 67:129–41

Fakan, S, Hancock, R. 1974. Localization of newly-synthesized DNA in a mammalian cell as visualized by high resolution autoradiography. *Exp. Cell Res.* 83:95–102

Fakan, S, Leser, G, Martin, TE. 1984. Ultrastructural distribution of nuclear ribonucleoproteins as visualized by immunocytochemistry on thin sections. *J. Cell Biol.* 98:358–63

Fakan, S, Leser, G, Martin, TE. 1986. Immunoelectron microscope visualization of nuclear ribonucleoprotein antigens within spread transcription complexes. *J. Cell Biol.* 103:1153–57

Fakan, S, Nobis, P. 1978. Ultrastructural localization of transcription sites and of RNA distribution during the cell cycle of synchronized CHO cells. *Exp. Cell Res.* 113:327–37

Fakan, S, Puvion, E. 1980. The ultrastructural visualization of nucleolar and extranucleolar RNA synthesis and distribution. *Int. Rev. Cytol.* 65:255–99

Fakan, S, Puvion, E, Spohr, G. 1976. Localization and characterization of newly synthesized nuclear RNA in isolated rat hepatocytes. *Exp. Cell Res.* 99:155–64

Feldherr, C, Kallenbach, E, Schultz, N. 1984. Movement of a karyophilic protein through the nuclear pores of oocytes. *J. Cell. Biol.* 99:2216–22

Ferguson, M, Ward, DC. 1992. Cell cycle dependent chromosomal movement in premitotic human T-lymphocyte nuclei. *Chromosoma* 101:557–65

Fisher, DZ, Chaudhary, N, Blobel, G. 1986. cDNA sequencing of nuclear lamins A and C reveals primary and secondary structural homology to intermediate filament proteins. *Proc. Natl. Acad. Sci. USA* 83:6450–54

Foster, KA, Collins, JM. 1985. The interrelation between DNA synthesis rates and DNA polymerases bound to the nuclear matrix in synchronized HeLa cells. *J. Biol. Chem.* 260:4229–35

Fox, MH, Arndt-Jovin, DJ, Jovin, TM, Baumann, PH, Robert-Nicoud, M. 1991. Spatial and temporal distribution of DNA replication sites localized by immunofluorescence and confocal microscopy in mouse fibroblasts. *J. Cell Sci.* 99:247–53

Franke, WW. 1974. Structure, biochemistry and functions of the nuclear envelope. *Int. Rev. Cytol. Suppl.* 4:71–236

Franke, WW, Scheer, U. 1970. The ultrastructure of the nuclear envelope and the architecture of the nuclear periphery. *J. Ultrastruct. Res.* 30:288–316

Fu, XD, Maniatis, T. 1990. Factor required for mammalian spliceosome assembly is localized to discrete regions in the nucleus. *Nature* 343:437–41

Fu, XD, Maniatis, T. 1992. Isolation of a complementary DNA that encodes the mammalian splicing factor SC35. *Science* 256: 535–38

Fusconi, M, Cassani, F, Govoni, M, Caselli, A, Farabegoli, F, et al. 1991. Anti-nuclear antibodies of primary biliary cirrhosis recognize 78–92-kD and 96–100-kD proteins of nuclear bodies. *J. Immunol.* 83:291–97

Gall, JG. 1967. Octagonal nuclear pores. *J. Cell Biol.* 32:391–99

Gall, JG. 1991. Spliceosomes and snurposomes. *Science* 252:1499–500

Gall, JG. 1992. Organelle assembly and function in the amphibian germinal vesicle. In *Advances in Developmental Biochemistry*, ed. PM Wassarman, 1:1–29. Greenwich, Conn.: JAI Press

Gall, JG, Callan, HG. 1989. The sphere organelle contains small nuclear ribonucleoproteins. *Proc. Natl. Acad. Sci. USA* 86:6635–39

Ge, H, Zuo, P, Manley, JL. 1991. Primary structure of the human splicing factor ASF reveals similarities with Drosophila regulators. *Cell* 66:373–82

Gerace, L, Blobel, G. 1980. The nuclear envelope lamina is reversibly depolymerized during mitosis. *Cell* 19:277–88

Gerace, L, Blum, A, Blobel, G. 1978. Immunocytochemical localization of the major polypeptides of the nuclear pore complex-lamina fraction—interphase and mitotic distribution. *J. Cell Biol.* 79:546–66

Gerace, L, Burke, B. 1988. Functional organization of the nuclear envelope. *Annu. Rev. Cell Biol.* 4:335–74

Gerace, L, Comeau, C, Benson, M. 1984. Organization and modulation of nuclear lamina structure. *J. Cell Sci. Suppl.* 1:137–60

Gerace, L, Ottaviano, Y, Kondor-Koch, C. 1982. Identification of a major polypeptide of the nuclear pore complex. *J. Cell Biol.* 95:826–37

Ghetti, A, Piñol-Roma, S, Michael, WM, Morandi, C, Dreyfuss, G. 1992. hnRNP I, the polypyrimidine tract-binding protein: distinct nuclear localization and association with hnRNAs. *Nucleic Acids Res.* 20:3671–78

Gil, A, Sharp, PA, Jamison, SF, Garcia-Blanco, M. 1991. Characterization of cDNAs encoding the polypyrimidine tract-binding protein. *Genes Dev.* 5:1224–36

Goessens, G. 1979. Relations between fibrillar centres and nucleolus-associated chromatin in Ehrlich tumor cells. *Cell Biol. Int. Rep.* 3:337–43

Goessens, G. 1984. Nucleolar structure. *Int. Rev. Cytol.* 87:107–58

Goldberg, MW, Allen, TD. 1992. High resolution scanning electron microscopy of the nuclear envelope: demonstration of a new, regular, fibrous lattice attached to the baskets of the nucleoplasmic face of the nuclear pores. *J. Cell Biol.* 119:1429–40

Goldfarb, DS. 1989. Nuclear transport. *Curr. Opin. Cell Biol.* 1:441–46

Goldman, AE, Maul, G, Steinert, PM, Yang, HY, Goldman, RD. 1986. Keratin-like proteins that coisolate with intermediate filaments of BHK-21 cells are nuclear lamins. *Proc. Natl. Acad. Sci. USA* 83:3839–43

Goldman, AE, Moir, RD, Montag-Lowy, M, Stewart, M, Goldman, RD. 1992. Pathway of incorporation of microinjected lamin A into the nuclear envelope. *J. Cell Biol.* 119:725–35

Goldman, MA, Holmquist, GP, Gray, MC, Caston, LA, Nag, A. 1984. Replication timing of genes and middle repetative sequences. *Science* 224:686–92

Granboulan, N, Granboulan, P. 1965. Cytochimic ultrastructurale du nucleole. II. Etude des sites de synthese du RNA dans le nucleole et le noyau. *Exp. Cell Res.* 38:604–19

Gratzner, HG. 1982. Monoclonal antibody to 5-bromo and 5-iododeoxyuridine: A new reagent for detection of DNA replication. *Science* 218:474–75

Greber, UF, Senior, A, Gerace, L. 1990. A major glycoprotein of the nuclear pore complex is a membrane-spanning polypeptide with a large lumenal domain and a small cytoplasmic tail. *EMBO J.* 9:1495–502

Green, MR. 1991. Biochemical mechanisms of constitutive and regulated pre-mRNA splicing. *Annu. Rev. Cell Biol.* 7:559–99

Greider, CW, Blackburn, EH. 1987. The telomere terminal transferase of Tetrahymena is a ribonucleoprotein enzyme with two kinds of primer specificity. *Cell* 51: 887–98

Guo, W, Mulligan, GL, Wormsley, S, Helfman, DM. 1991. Alternative splicing of β-tropomyosin pre-mRNA: *cis*-acting elements and cellular factors that block the use of a skeletal muscle exon in nonmuscle cells. *Genes Dev.* 5:2096–107

Haaf, T, Schmid, M. 1989. Centromeric association and non-random distribution of centromeres in human tumour cells. *Hum. Genet.* 81:137–43

Haaf, T, Schmid, M. 1991. Chromosome topology in mammalian interphase nuclei. *Exp. Cell Res.* 192:325–32

Habets, WJ, Hoet, MH, De Jong, BAW, Van der Kemp, A, Van Venrooij, WJ. 1989.

Mapping of B cell epitopes on small nuclear ribonucleoproteins that react with human autoantibodies as well as with experimentally-induced mouse monoclonal antibodies. *J. Immunol.* 143:2560–66

Hadjiolov, A. 1985. The nucleolus and ribosome biogenesis. In *Cell Biology Monographs,* 12:1–263. Wien/New York: Springer Verlag

Hadlaczky, G, Went, M, Ringertz, NR. 1986. Direct evidence for the non-random localization of mammalian chromosomes in the interphase nucleus. *Exp. Cell Res.* 167:1–15

Hand, R. 1978. Eucaryotic DNA: organization of the genome for replication. *Cell* 15:317–25

Hardin, JH, Spicer, SS, Greene, WB. 1969. The paranucleolar structure, accessory body of Cajal, sex chromatin, and related structures in nuclei of rat trigeminal neurons: a cytochemical and ultrastructural study. *Anat. Rec.* 164:403–32

Hatanaka, M. 1990. Discovery of the nucleolar targeting signal. *BioEssays* 12:143–48

Hatton, KS, Dhar, V, Gahn, TA, Brown, EH, Mager, D, Schildkraut, CL. 1988. Temporal order of replication of multigene families reflects chromosomal location and transcriptional activity. *Cancer Cells* 6:335–40

Hay, ED, Revel, JP. 1966. The fine structure of the DNP component of the nucleus. An electron microscopic study utilizing autoradiography to localize DNA synthesis. *J. Cell Biol.* 16:29–51

He, D, Martin, T, Penman, S. 1991. Localization of heterogeneous nuclear ribonucleoprotein in the interphase nuclear matrix core filaments and on perichromosomal filaments at mitosis. *Proc. Natl. Acad. Sci. USA* 88:7469–73

Henderson, SC, Spector, DL. 1993. Non-random arrangement of DNA sequences in differentiating diploid human cells. *J. Cell Biol.* Submitted

Hernandez-Verdun, D. 1991. The nucleolus today. *J. Cell Sci.* 99:465–71

Heslop-Harrison, JS, Bennett, MD. 1990. Nuclear architecture in plants. *Trends Genet.* 6:401–5

Hilliker, AJ. 1985. Assaying chromosome arrangement in embryonic interphase nuclei of *Drosophila melanogaster* by radiation induced interchanges. *Genet. Res.* 47:13–18

Hilliker, AJ, Appels, R 1989. The arrangement of interphase chromosomes: structural and functional aspects. *Exp. Cell Res.* 185:297–318

Hinshaw, JE, Carragher, BO, Milligan, RA. 1992. Architecture and design of the nuclear pore complex. *Cell* 69:1133–41

Hochstrasser, M, Mathog, D, Gruenbaum, Y, Saumweber, J, Sedat, JW. 1986. Spatial organization of chromosomes in the salivary gland nuclei of *Drosophila melanogaster. J. Cell Biol.* 102:112–23

Hochstrasser, M, Sedat, JW. 1987. Three-dimensional organization of *Drosophila melanogaster* interphase nuclei. II. Chromosomal spatial organization and gene regulation. *J. Cell Biol.* 104:1455–70

Huang, S, Spector, DL. 1991. Nascent pre-mRNA transcripts are associated with nuclear regions enriched in splicing factors. *Genes Dev.* 5:2288–302

Huang, S, Spector, DL. 1992a. U1 and U2 small nuclear RNAs are present in nuclear speckles. *Proc. Natl. Acad. Sci. USA* 89: 305–8

Huang, S, Spector, DL. 1992b. Will the real splicing sites please light up? *Curr. Biol.* 2:188–90

Huberman, JA, Riggs, AD. 1968. On the mechanism of DNA replication in mammalian chromosomes. *J. Mol. Biol.* 32:327–41

Hutchison, N, Weintraub, H. 1985. Localization of DNAase I-sensitive sequences to specific regions of interphase nuclei. *Cell* 43:471–82

Izaurralde, E, Mattaj, IW. 1992. Transport of RNA between nucleus and cytoplasm. *Semin. Cell Biol.* 3:279–88

Jackson, DA. 1990. The organization of replication centres in higher eukaryotes. *BioEssays* 12:87–89

Jackson, DA, Cook, PR. 1986. Replication occurs at a nucleoskeleton. *EMBO J.* 5: 1403–10

Jarnik, M, Aebi, U. 1991. Toward a more complete 3-D structure of the nuclear pore complex. *J. Struct. Biol.* 107:291–308

Jiménez-García, LF, Spector, DL. 1993. In vivo evidence that transcription and splicing are coordinated by a recruiting mechanism. *Cell.* 73:47–59

Jones, RE, Okamura, CS, Martin, TE. 1980. Immunofluorescent localization of the proteins of nuclear ribonucleoprotein complexes. *J. Cell Biol.* 86:235–43

Jordan, EG. 1991. Interpreting nucleolar structure: where are the transcribing genes? *J. Cell Sci.* 98:437–42

Kass, S, Tyc, K, Steitz, JA, Sollner-Webb, B. 1990. The U3 small nucleolar ribonucleoprotein functions in the first step of preribosomal RNA processing. *Cell* 60: 897–908

Kedinger, C, Simard, R. 1974. The action of α-amanitin on RNA synthesis in Chinese hamster ovary cells. *J. Cell Biol.* 63:831–42

Kellum, R, Schedl, P. 1991. A position-effect assay for boundaries of higher order chromosomal domains. *Cell* 64:941–50

King, DS, Beggs, JD. 1990. Interactions of PRP2 protein with pre-mRNA splicing complexes in *Saccharomyces cerevisiae*. *Nucleic Acids Res*. 18:6559–64

Krainer, AR, Mayeda, A, Kozak, D, Binns, G. 1991. Functional expression of cloned human splicing factor SF2: homology to RNA-binding proteins, U1 70k, and Drosophila splicing regulators. *Cell* 66:383–94

Krystosek, A, Puck, TT. 1990. The spatial distribution of exposed nuclear DNA in normal, cancer, and reverse-transformed cells. *Proc. Natl. Acad. Sci. USA* 87:6560–64

Lafarga, M, Andres, MA, Berciano, MT, Maquiera, E. 1991. Organization of nucleoli and nuclear bodies in osmotically stimulated supraoptic neurons of the rat. *J. Comp. Neurol*. 308:329–39

Lamb, MM, Daneholt, B. 1979. Characterization of active transcription units in Balbiani rings of Chironomus tentans. *Cell* 17:835–48

Langer, PR, Waldrop, AA, Ward, DC. 1981. Enzymatic synthesis of biotin labelled polynucleotides: novel nucleic acid affinity probes. *Proc. Natl. Acad. Sci. USA* 78: 6633–37

Last, RL, Woolford, JL Jr. 1986. Identification and nuclear localization of yeast premessenger RNA processing components: RNA2 and RNA3 proteins. *J. Cell Biol*. 103:2103–12

Lawrence, JB, Singer, RH. 1991. Spatial organization of nucleic acid sequences within cells. *Semin. Cell Biol*. 2:83–101

Lawrence, JB, Singer, RH, Marselle, LM. 1989. Highly localized tracks of specific transcripts within interphase nuclei visualized by in situ hybridization. *Cell* 57:493–502

Lawrence, JB, Singer, RH, McNeil, JA. 1990. Interphase and metaphase resolution of different distances within the human dystrophin gene. *Science* 249:928–32

Lawrence, JB, Villnave, CA, Singer, RH. 1988. Sensitive high-resolution chromatin and chromosome mapping in situ: presence and orientation of two closely integrated copies of EBV in a lymphoma line. *Cell* 52:51–61

Lebel, S, Lampron, C, Royal, CA, Raymond, Y. 1987. Lamins A and C appear during retinoic acid-induced differentiation of mouse embryonal carcinoma cells. *J. Cell Biol*. 105:1099–104

Leno, GH, Laskey, RA. 1991. The nuclear membrane determines the timing of DNA replication in *Xenopus* egg extracts. *J. Cell Biol*. 112:557–66

Leonhardt, H, Page, AW, Weier, HU, Bestor, TH. 1992. A targeting sequence directs DNA methyltransferase to sites of DNA replication in mammalian nuclei. *Cell* 71: 865–73

Lerner, EA, Lerner, MR, Janeway, LA, Steitz, JA. 1981. Monoclonal antibodies to nucleic acid containing cellular constituents: probes for molecular biology and autoimmune disease. *Proc. Natl. Acad. Sci. USA* 78:2737–41

Leser, GP, Escara-Wilke, J, Martin, TE. 1984. Monoclonal antibodies to heterogeneous nuclear RNA-protein complexes. The core proteins comprise a conserved group of related polypeptides. *J. Biol. Chem*. 259:1827–33

Leser, GP, Fakan, S, Martin, TE. 1989. Ultrastructural distribution of ribonucleoprotein complexes during mitosis. snRNP antigens are contained in mitotic granule clusters. *Eur. J. Cell Biol*. 50:376–89

Lewin, B. 1980. Ribosomal gene clusters. In *Eucaryotic Chromosomes: Gene Expression*, pp. 865–906. New York: Wiley & Sons

Li, H, Bingham, PM. 1991. Arginine/serine-rich domains of the su(w^a) and tra RNA processing regulators target proteins to a subnuclear compartment implicated in splicing. *Cell* 67:335–42

Lichter, PL, Boyle, AL, Cremer, T, Ward, DC. 1991. Analysis of genes and chromosomes by nonisotopic in situ hybridization. *Genet. Analys. Tech. Appl*. 8:24–35

Lima-de-Faria, A, Jaworska H. 1968. Late DNA synthesis in heterochromatin. *Nature* 217:138–42

Lin, RJ, Lustig, AJ, Abelson, J. 1987. Splicing of yeast nuclear pre-mRNA in vitro requires a functional 40S spliceosome and several extrinsic factors. *Genes Dev*. 1:7–18

Lischwe, MA, Ochs, RL, Reddy, R, Cook, RG, Yeoman, LC, et al. 1985. Purification and partial characterization of a nucleolar scleroderma antigen (M_r=34,000; pI 8.5) rich in N^G,N^G-dimethylarginine. *J. Biol. Chem*. 260:14304–10

Lydersen, BK, Pettijohn, DE. 1980. Human specific nuclear protein that associates with the polar region of the mitotic apparatus: distribution in a human/hamster hybrid cell. *Cell* 22:489–99

Madsen, P, Celis, JE. 1985. S-phase patterns of cyclin (PCNA) antigen staining resemble topographical patterns of DNA synthesis. *FEBS Lett*. 193:5–11

Manuelidis, L. 1984. Different central nervous system cell types display distinct and non-random arrangements of satellite DNA sequences. *Proc. Natl. Acad. Sci. USA* 81: 3123–27

Manuelidis, L. 1985a. Indications of centromere movement during interphase and differentiation. *Ann. NY Acad. Sci.* 450:205–21

Manuelidis, L. 1985b. Individual interphase chromosome domains revealed by in situ hybridization. *Hum. Genet.* 71:288–93

Manuelidis, L. 1990. A view of interphase chromosomes. *Science* 250:1533–40

Manuelidis, L, Chen, TL. 1990. A unified model of eukaryotic chromosomes. *Cytometry* 11:8–25

Maquat, LE. 1991. Nuclear mRNA export. *Curr. Opin. Cell Biol.* 3:1004–12

Mariman, EC, van Eekelen, CAG, Reinders, RJ, Berns, AJM, van Venrooij, WJ. 1982. Adenoviral heterogeneous nuclear RNA is associated with the host nuclear matrix during splicing. *J. Mol. Biol.* 154:103–19

Marinozzi, V, Fiume, L. 1971. Effects of α-amanitin on mouse and rat liver cell nuclei. *Exp. Cell Res.* 67:311–22

Mathog, D, Hochstrasser, M, Gruenbaum, Y, Saumweber, H, Sedat, J. 1984. Characteristic folding pattern of polytene chromosomes in *Drosophila* salivary gland nuclei. *Nature* 308:414–21

Matunis, MJ, Michael, WM, Dreyfuss, G. 1992. Characterization and primary structure of the poly(C)-binding heterogeneous nuclear ribonucleoprotein complex K protein. *Mol. Cell. Biol.* 12:164–71

Maul, GG. 1977. The nuclear and cytoplasmic pore complex. Structure, dynamics, distribution and evolution. *Int. Rev. Cytol. Suppl.* 6:75–186

Mayeda, A, Zahler, AM, Krainer, AR, Roth, MB. 1992. Two members of a conserved family of nuclear phosphoproteins are involved in pre-mRNA splicing. *Proc. Natl. Acad. Sci. USA* 89:1301-1304

Mazzotti, G, Rizzoli, R, Galanzi, A, Papa, S, Vitale, M, et al. 1990. High-resolution detection of newly synthesized DNA by anti-bromodeoxyuridine antibodies identifies specific chromatin domains. *J. Histochem. Cytochem.* 38:13–22

McClintock, B. 1934. The relation of a particular chromosomal element to the development of the nucleoli in *Zea mays*. *Z. Zellforsch. Mikr. Anat.* 21:294–328

McKeon, FD, Kirschner, MW, Caput, D. 1986. Homologies in both primary and secondary structure between nuclear envelope and intermediate filament proteins. *Nature* 319:463–68

McKeon, FD, Tuffanelli, DL, Kobayashi, S, Kirschner, MW. 1984. The redistribution of a conserved nuclear envelope protein during the cell cycle suggests a pathway for chromosome condensation. *Cell* 36:83–92

McNeil, JA, Johnson, CV, Carter, KC, Singer, RH, Lawrence, J. B. 1991. Localizing DNA and RNA within nuclei and chromosomes by fluorescence in situ hybridization. *Genet. Analys. Tech. Appl.* 8: 41–58

Mehlin, H, Daneholt, B, Skoglund, U. 1992. Translocation of a specific premessenger ribonucleoprotein particle through the nuclear pore studied with electron microscope tomography. *Cell* 69:605–13

Mehlin, H, Skoglund, U, Daneholt, B. 1991. Transport of Balbiani ring granules through nuclear pores in *Chironomus tentans*. *Exp. Cell Res.* 193:72–77

Meier, UT, Blobel, G. 1992. Nopp140 shuttles on tracks between nucleolus and cytoplasm. *Cell* 70:127–38

Milligan, RA. 1986. A structural model for the nuclear pore complex. In *Nucleocytoplasmic Transport,* ed. R Peters, M Trendelenburg, pp. 113–22. London: Academic

Mills, AD, Blow, JJ, White, JG, Amos, WB, Wilcock, D, Laskey, RA. 1989. Replication occurs at descrete foci spaced throughout nuclei replicating in vitro. *J. Cell Sci.* 94:471–77

Mirre, C, Knibiehler B. 1982. A reevaluation of the relationships between the fibrillar centres and the nucleolus-organizing regions in reticulated nucleoli: ultrastructural organization, number, and distribution of the fibrillar centres in the nucleolus of the mouse Sertoli cell. *J. Cell Sci.* 55:261–76

Mirre, C, Stahl, A. 1981. Ultrastructural organization, sites of transcription and distribution of fibrillar centres in the nucleolus of the mouse oocyte. *J. Cell Sci.* 48:105–26

Miyawaki, H. 1974. Extranucleolar pyroninophilic substances in the liver cell nuclei of starve-refed mice as revealed by non-aqueous negative staining. *J. Ultrastruct. Res.* 47:255–71

Monneron, A, Bernhard W. 1969. Fine structural organization of the interphase nucleus in some mammalian cells. *J. Ultrastruct. Res.* 27:266–88

Moreno Diaz de la Espina, S, Sanchez Pina, A, Risueno, M. C. 1982. Localization of acid phosphatase activity, phosphate ions and inorganic cations in plant nuclear coiled bodies. *Cell Biol. Int. Rep.* 6:601–7

Moreno Diaz de la Espina, S, Sanchez Pina, A, Risueño, MC, Medina, FJ, Fernandez-Gomez, ME. 1980. The role of plant coiled bodies in the nuclear RNA metabolism. *Electron Micros.* 2:240–41

Mukherjee, AB, Parsa, NZ. 1990. Determination of sex chromosomal constitution and chromosomal origin of drumsticks, drumstick-like structures, and other nuclear bodies in human blood cells at interphase by fluorescence in situ hybridization. *Chromosoma* 99:432–35

Nakamura, H, Morita, T, Sato, C. 1986. Structural organizations of replicon domains during DNA synthetic phase in the mammalian nucleus. *Exp. Cell Res.* 165:291–97

Nakayasu, H, Berezney, R. 1989. Mapping replication sites in the eucaryotic cell nucleus. *J. Cell Biol.* 108:1–11

Newport, JW, Wilson, KL, Dunphy, WG. 1990. A lamin-independent pathway for nuclear envelope assembly. *J. Cell Biol.* 111:2247–59

Nickerson, JA, He, D, Fey, AG, Penman, S. 1990. The nuclear matrix. In *The Eukaryotic Nucleus, Molecular Biochemistry and Macromolecular Assemblies,* ed. PR Strauss, SH Wilson, 2:763–82. Caldwell: Telford

Nyman, U, Hallman, H, Hadlaczky, G, Pettersson, I, Sharp, G, Ringertz, NR. 1986. Intranuclear localization of snRNP antigens. *J. Cell Biol.* 102:137–44

O'Farrell, PH, Edgar, BA, Lakich, D, Lehner, CF. 1989. Directing cell division during development. *Science* 246:635–40

O'Keefe, RT, Henderson, SC, Spector, DL. 1992. Dynamic organization of DNA replication in mammalian cell nuclei: spatially- and temporally-defined replication of chromosome specific α-satellite DNA sequences. *J. Cell Biol.* 116:1095–110

Ochs, RL, Lischwe, MA, Spohn, WH, Busch, H. 1985. Fibrillarin: a new protein of the nucleolus identified by autoimmune sera. *Biol. Cell* 54:123–34

Ochs, RL, Press, RI. 1992. Centromere autoantigens are associated with the nucleolus. *Exp. Cell Res.* 200:339–50

Olins, AL, Olins, DE, Bazett-Jones, DP. 1992. Balbiani ring hnRNP substructure visualized by selective staining and electron spectroscopic imaging. *J. Cell Biol.* 117: 483–91

Olins, AL, Olins, DE, Franke, WW. 1980. Stereo-electron microscopy of nucleoli, Balbiani rings and endoplasmic reticulum in *Chironomus* salivary gland cells. *Eur. J. Cell Biol.* 22:714–23

Paddy, MR, Belmont, AS, Saumweber, H, Agard, DA, Sedat, JW. 1990. Interphase nuclear envelope lamins form a discontinuous network that interacts with only a fraction of the chromatin in the nuclear periphery. *Cell* 62:89–106

Paine, P, Moore, L, Horowitz, S. 1975. Nuclear envelope permeability. *Nature* 254: 109–14

Pardoll, DM, Vogelstein, B, Coffey, DS. 1980. A fixed site of DNA replication in eucaryotic cells. *Cell* 19:527–36

Pardue, ML, Gall, JG. 1970. Chromosomal localization of mouse satellite DNA. *Science* 168:1356–58

Patton, JG, Mayer, SA, Tempst, P, Nadal-Ginard, B. 1991. Characterization and molecular cloning of polypyrimidine tract-binding protein: a component of a complex necessary for pre-mRNA splicing. *Genes Dev.* 5:1237–51

Perraud, M, Gioud, M, Monier, JC. 1979. Structures intranucleaires reconnues par les autoanticorps anti-ribonucleoproteines: etude sur cellules de rein de singe en culture par les techniques d'immunofluorescence et d'immunomicroscopie electronique. *Ann. Immunol.* 130:635–47

Peter, M, Nakagawa, J, Doree, M, Labbe, JC, Nigg, EA. 1990. In vitro disassembly of the nuclear lamina and M phase-specific phosphorylation of lamins by cdc2 kinase. *Cell* 61:591–602

Petrov, P, Sekeris, CE. 1971. Early action of α-amanitin on extranucleolar ribonucleoproteins, as revealed by electron microscopic observation. *Exp. Cell Res.* 69: 393–401

Pinkel, D, Gray, JW, Trask, B, van den Engh, G, Fuscoe, J, et al. 1986. Cytogenetic analysis by in situ hybridization and fluorescently labeled nucleic acid probes. *Cold Spring Harbor Symp. Quant. Biol.* 51:151–57

Pinkel, D, Landegent, J, Collins, C, Fuscoe, J, Segraves, R, et al. 1988. Fluorescence in situ hybridization with human chromosome-specific libraries: detection of trisomy 21 and translocations of chromosome 4. *Proc. Natl. Acad. Sci. USA* 85:9138–42

Piñol-Roma, S, Choi, YD, Matunis, MJ, Dreyfuss, G. 1988. Immunopurification of heterogeneous nuclear ribonucleoprotein particles reveals an assortment of RNA-binding proteins. *Genes Dev.* 2:215–27

Piñol-Roma, S, Dreyfuss, G. 1991. Transcription-dependent and transcription-independent nuclear transport of hnRNP proteins. *Science* 253:312–14

Piñol-Roma, S, Dreyfuss, G. 1992. Shuttling of pre-mRNA binding proteins between nucleus and cytoplasm. *Nature* 355:730–32

Piñol-Roma, S, Swanson, MS, Gall, JG, Dreyfuss, G. 1989. A novel heterogeneous nuclear RNP protein with a unique distribution on nascent transcripts. *J. Cell Biol.* 109:2575–87

Popp, S, Scholl, HP, Loos, P, Jauch, A, Stelzer, E, et al. 1990. Distribution of chromosome 18 and X centric heterochromatin in the interphase nucleus of cultured human cells. *Exp. Cell Res.* 189:1–12

Potashkin, JA, Derby, RJ, Spector, DL. 1990. Differential distribution of factors involved in pre-mRNA processing in the yeast cell nucleus. *Mol. Cell. Biol.* 10:3524–34

Puvion, E, Bernhard, W. 1975. Ribonucleoprotein components in liver cell nuclei as visualized by cryoultramicrotomy. *J. Cell Biol.* 67:200–14

Puvion, E, Viron, A, Assens, C, Leduc, EH,

Jeanteur, P. 1984. Immunocytochemical identification of nuclear structures containing snRNPs in isolated rat liver cells. *J. Ultrastruct. Res.* 87:180–89

Puvion-Dutilleul, F, Bachellerie, JP, Puvion, E. 1991. Nucleolar organization of HeLa cells as studied by in situ hybridization. *Chromosoma* 100:395–409

Puvion-Dutilleul, F, Puvion, E. 1991. Sites of transcription of adenovirus type 5 genomes in relation to early viral DNA replication in infected HeLa cells. A high resolution in situ hybridization and autoradiographical study. *Biol. Cell* 71:135–47

Puvion-Dutilleul, F, Rousseu, R, Puvion, E. 1992. Distribution of viral RNA molecules during the adenovirus 5 infectious cycle in HeLa cells. *J. Struct. Biol.* 108:209–20

Rabl, C. 1885. Uber zellteilung. *Morphol. Jahrb.* 10:214–330

Rappold, GA, Cremer, T, Hager, HD, Davies, KE, Muller, CR, Yang, T. 1984. Sex chromosome positions in human interphase nuclei as studied by in situ hybridization with chromosome specific DNA probes. *Hum. Genet.* 67:317–25

Raska, I, Andrade, LEC, Ochs, RL, Chan, EKL, Chang, CM, et al. 1991. Immunological and ultrastructural studies of the nuclear coiled body with autoimmune antibodies. *Exp. Cell Res.* 195:27–37

Raska, I, Koberna, K, Jarnik, M, Petrasovicova, V, Raska, K, Bravo, R. 1989a. Ultrastructural immunolocalization of cyclin/PCNA in synchronized 3T3 cells. *Exp. Cell Res.* 184:81–89

Raska, I, Ochs, RL, Salamin-Michel, L. 1990. Immunocytochemistry of the cell nucleus. *Electron Micros. Rev.* 3:301–53

Raska, I, Reimer, G, Jarnik, M, Kostrouch, Z, Raska, K Jr. 1989b. Does the synthesis of ribosomal RNA take place within nucleolar fibrillar centers or dense fibrillar components? *Biol. Cell* 65:79–82

Rawlins, DJ, Highett, MI, Shaw, PJ. 1991. Localization of telomeres in plant interphase nuclei by in situ hybridization and 3D confocal microscopy. *Chromosoma* 100:424–31

Razin, SV. 1987. DNA interactions with the nuclear matrix and spatial organization of replication and transcription. *BioEssays* 6: 19–23

Reichelt, R, Holzenburg, A, Buhle, EL Jr., Jarnik, M, Engel, A, Aebi, U. 1990. Correlation between structure and mass distribution of the nuclear pore complex and of distinct pore complex components. *J. Cell Biol.* 110:883–94

Rendon, MC, Rodrigo, RM, Goenechea, LG, Garcia-Herdugo, G, Valdivia, MM, Moreno, FJ. 1992. Characterization and immunolocalization of a nucleolar antigen with anti-NOR serum in HeLa cells. *Exp. Cell Res.* 200:393–403

Reuter, R, Appel, B, Bringmann, P, Rinke, J, Luhrmann, R. 1984. 5′-terminal caps of snRNAs are reactive with antibodies specific for 2,2,7-trimethylguanosine in whole cells and nuclear matrices. *Exp. Cell Res.* 154:548–60

Ris, H. 1989. Three-dimensional imaging of cell ultrastructure with high resolution low voltage SEM. *Inst. Phys. Conf. Ser. No. 98,* Chpt. 16:657–62

Ris, H. 1990. Application of low voltage, high resolution SEM in the study of complex intracellular structures. *Proc. XIIth Int. Congr. Electron Micros.* pp. 18–19

Ris, H. 1991. The three-dimensional structure of the nuclear pore complex as seen by high voltage electron microscopy and high resolution low voltage scanning electron microscopy. *EMSA Bull.* 21:54–56

Rober, RA, Weber, K, Osborn, M. 1989. Differential timing of nuclear lamin A/C expression in the various organs of the mouse embryo and the young animal: a developmental study. *Development* 105: 365–78

Roth, MB, Murphy, C, Gall, JG. 1990. A monoclonal antibody that recognizes a phosphorylated epitope stains lampbrush chromosome loops and small granules in the amphibian germinal vesicle. *J. Cell Biol.* 111:2217–23

Ruby, SW, Abelson, J. 1991. Pre-mRNA splicing in yeast. *Trends Genet.* 7:79–85

Ruskin, B, Zamore, PD, Green, MR. 1988. A factor, U2AF, is required for U2 snRNP binding and splicing complex assembly. *Cell* 52:207–19

Sass, H, Pederson, T. 1984. Transcription-dependent localization of U1 and U2 small nuclear ribonucleoproteins at major sites of gene activity in polytene chromosomes. *J. Mol. Biol.* 180:911–26

Saunders, WS, Cooke, CA, Earnshaw, WC. 1991. Compartmentalization within the nucleus: discovery of a novel subnuclear region. *J. Cell Biol.* 115:919–31

Schatten, G, Maul, GG, Schatten, H, Chaly, N, Simerly, C, et al. 1985. Nuclear lamins and peripheral nuclear antigens during fertilization and embryogenesis in mice and sea urchins. *Proc. Natl. Acad. Sci. USA* 82:4727–31

Scheer, U, Benavente, R. 1990. Functional and dynamic aspects of the mammalian nucleolus. *BioEssays* 12:14–21

Scheer, U, Kartenbeck, J, Trendelenberg, MF, Stadler, J, Franke, WW. 1976. Experimental disintegration of the nuclear envelope: evidence for pore-connecting fibrils. *J. Cell Biol.* 69:1–18

Scheer, U, Messner, K, Hazan, R, Raska, I, Hansmann, P, et al. 1987. High sensitivity immunolocalization of double- and single-stranded DNA by a monoclonal antibody. *Eur. J. Cell Biol.* 43:358–71

Scheer, U, Raska, I. 1987. Immunocytochemical localization of RNA polymerase I in the fibrillar centers of nucleoli. In *Chromosomes Today,* ed. A Stahl, J Luciani, A Wagner-Capodano, 9:284–94. London: Allen & Unwin

Scheer, U, Rose, K. 1984. Localization of RNA polymerase I in interphase cells and mitotic chromosomes by light and electron microscopic immunocytochemistry. *Proc. Natl. Acad. Sci. USA* 81:1431–35

Schultz, MC. 1989. Ultrastructural study of the coiled body and a new inclusion, the "mykaryon," in the nucleus of the adult rat sertoli cell. *Anat. Rec.* 225:21–25

Sehgal, PB, Darnell, JE Jr, Tamm, I. 1976. The inhibition by DRB (5,6-dichloro-1-β-d-ribofuranosylbenzimidazole) of hnRNA and mRNA production in HeLa cells. *Cell* 9:473–80

Seite, R, Pebusque, MJ, Vio-Cigna, M. 1982. Argyrophilic proteins on coiled bodies in sympathetic neurons identified by Ag-NOR procedure. *Biol. Cell.* 46:97–100

Selig, S, Okumura, K, Ward, DC, Cedar, H. 1992. Delineation of DNA replication time zones by fluorescence in situ hybridization. *EMBO J.* 11:1217–25

Shelton, K, Higgins, L, Cochran, D, Ruffolo, D, Egle, P. 1980. Nuclear lamins of erythrocyte and liver. *J. Biol. Chem.* 255: 10978–83

Sinclair, GD, Brasch, K. 1978. The reversible action of α-amanitin on nuclear structure and molecular composition. *Exp. Cell Res.* 111:1–14

Skoglund, U, Andersson, K, Strandberg, B, Daneholt, B. 1986. Three-dimensional structure of a specific pre-messenger RNP particle established by electron microscope tomography. *Nature* 319:560–64

Sleeman, AM, Leno, GH, Mills, AD, Fairman, MP, Laskey, RA. 1992. Patterns of DNA replication in *Drosophila* polytene nuclei replicating in *Xenopus* egg and oocyte extracts. *J. Cell Sci.* 101:509–15

Smetana, K, Steele, WJ, Busch H. 1963. A nuclear ribonucleoprotein network. *Exp. Cell Res.* 31:198–201

Smith, HC, Harris, SG, Zillmann, M, Berget, SM. 1989. Evidence that a nuclear matrix protein participates in premessenger RNA splicing. *Exp. Cell Res.* 182:521–33

Smith, HC, Ochs, RL, Fernandez, EA, Spector, DL. 1986. Macromolecular domains containing nuclear matrix protein p107 and U-snRNP protein p28: further evidence for an in situ nuclear matrix. *Mol. Cell. Biochem.* 70:151–68

Smith, HC, Spector, DL, Woodcock, CLF, Ochs, RL, Bhorjee, J. 1985. Alterations in chromatin conformation are accompanied by reorganization of nonchromatin domains that contain U-snRNP protein p28 and nuclear protein p107. *J. Cell Biol.* 101:560–67

Spector, DL. 1984. Colocalization of U1 and U2 small nuclear RNPs by immunocytochemistry. *Biol. Cell.* 51:109–12

Spector, DL. 1990. Higher order nuclear organization: three-dimensional distribution of small nuclear ribonucleoprotein particles. *Proc. Natl. Acad. Sci. USA* 87:147–51

Spector, DL, Fu, XD, Maniatis, T. 1991. Associations between distinct pre-mRNA splicing components and the cell nucleus. *EMBO J.* 10:3467–81

Spector, DL, Lark, G, Huang, S. 1992. Differences in snRNP localization between transformed and nontransformed cells. *Mol. Biol. Cell* 3:555–69

Spector, DL, Schrier, WH, Busch, H. 1983. Immunoelectron microscopic localization of snRNPs. *Biol. Cell* 49:1–10

Spector, DL, Smith, HC. 1986. Redistribution of U-snRNPs during mitosis. *Exp. Cell Res.* 163:87–94

Spector, DL, Watt, RA, Sullivan, NF. 1987. The v- and c-*myc* oncogene proteins colocalize in situ with small nuclear ribonucleoprotein particles. *Oncogene* 1:5–12

Sperling, K, Ludtke, EK. 1981. Arrangement of prematurely condensed chromosomes in cultured cells and lymphocytes of the Indian muntjac. *Chromosoma* 83:541–53

Stahl, A, Wachtler, F, Hartung, M, Devictor, M, Schofer, C, et al. 1991. Nucleoli, nuclear chromosomes and ribosomal genes in the human spermatocyte. *Chromosoma* 101:231–44

Stefano, JE. 1984. Purified lupus antigen La recognizes an oligouridylate stretch common to the 3′ termini of RNA polymerase III transcripts. *Cell* 36:145–54

Stevens, BJ, Swift, H. 1966. RNA transport from nucleus to cytoplasm in *Chironomus* salivary glands. *J. Cell Biol.* 31:55–77

Stewart, C, Burke, B. 1987. Teratocarcinoma stem cells and early mouse embryos contain only a single major lamin polypeptide closely resembling lamin B. *Cell* 51:383–92

Stuurman, N, De Graff, A, Floore, A, Josso, A. Humbel, B, et al. 1992a. A monoclonal antibody recognizing nuclear matrix-associated nuclear bodies. *J. Cell Sci.* 101:773–84

Stuurman, N, de Jong, L, van Driel, R.

1992b. Nuclear frameworks: concepts and operational definitions. *Cell Biol. Int. Rep.* 16:837–52

Taylor, JH. 1960. Asynchronous duplication of chromosomes in cultured cells of Chinese hamster. *J. Biophys. Biochem. Cytol.* 7: 455–64

Ten Hagen, KG, Gilbert, DM, Willard, HF, Cohen, SN. 1990. Replication timing of DNA sequences associated with human centromeres and telomeres. *Mol. Cell. Biol.* 10:6348–55

Theissen, H, Etzerodt, M, Reuter, R, Schneider, C, Lottspeich, F, et al. 1986. Cloning of the human cDNA for the U1 RNA-associated 70K protein. *EMBO J* 5:3209–17

Thiry, M. 1992a. New data concerning the functional organization of the mammalian cell nucleolus: detection of RNA and rRNA by in situ molecular immunocytochemistry. *Nucleic Acids Res.* 20:6195–200

Thiry, M. 1992b. Ultrastructural detection of DNA within the nucleolus by sensitive molecular immunocytochemistry. *Exp. Cell Res.* 200:135–44

Thiry, M, Goessens, G. 1991. Distinguishing the sites of pre-rRNA synthesis and accumulation in Ehrlich tumor cell nucleoli. *J. Cell Sci.* 99:759–67

Thiry, M, Scheer, U, Goessens, G. 1988. Localization of DNA within Ehrlich tumour cell nucleoli by immunoelectron microscopy. *Biol. Cell* 63:27–34

Thiry, M, Scheer, U, Goessens, G. 1991. Localization of nucleolar chromatin by immunocytochemistry and in situ hybridization at the electron microscopic level. *Electron Micros. Rev.* 4:85–110

Thiry, M, Thiry-Blaise, L. 1989. In situ hybridization at the electron microscope level: an improved method for precise localization of ribosomal DNA and RNA. *Eur. J. Cell Biol.* 59:235–43

Tousson, A, Zeng, C, Brinkley, BR, Valdivia, MM. 1991. Centrophilin: a novel mitotic spindle protein involved in microtubule nucleation. *J. Cell Biol.* 112:427–40

Trask, B, van den Engh, G, Pinkel, D, Mullikin, J, Waldman, F, et al. 1988. Fluorescence in situ hybridization to interphase cell nuclei in suspension allows flow cytometric analysis of chromosome content and microscopic analysis of nuclear organization. *Hum. Genet.* 78:251–59

Tubo, RA, Berezney, R. 1987. Identification of 100 and 150 S DNA polymerase α-primase megacomplexes solubilized from the nuclear matrix of regenerating rat liver. *J. Biol. Chem.* 262:5857–65

Turner, BM, Franchi, L. 1987. Identification of protein antigens associated with the nuclear matrix and with clusters of interchromatin granules in both interphase and mitotic cells. *J. Cell Sci.* 87:269–82

Tyc, K, Steitz, JA. 1989. U3, U8 and U13 comprise a new class of mammalian snRNPs localized in the cell nucleolus. *EMBO J.* 8:3113–19

Unwin, PNT, Milligan, RA. 1982. A large particle associated with the perimeter of the nuclear pore complex. *J. Cell Biol.* 93:63–75

Vagner-Capodano, AM, Stahl, A. 1982. Numerical relationship between nucleolar organizer regions and fibrillar centres in porcine thyroid cells cultivated in vitro. *Biol. Cell* 45:111

van Dekken, H, van Rotterdam, A, Jonker, RR, van der Voort, HTM, Brakenhoff, GJ, Bauman, JGJ. 1990. Spatial topography of a pericentromeric region (1Q12) in hemopoietic cells studied by in situ hybridization and confocal microscopy. *Cytometry* 11: 570–78

van Dierendonck, JH, Keyzer, R, van de Velde, CJH, Cornelisse, CJ. 1989. Subdivision of S-Phase by analysis of nuclear 5-bromodeoxyuridine staining patterns. *Cytometry* 10:143–50

Vaughn, JP, Dijkwel, PA, Mullenders, LHF, Hamlin, JL. 1990. Replication forks are associated with the nuclear matrix. *Nucleic Acids Res.* 18:1965–69

Vazquez-Nin, GH, Echeverria, OM, Fakan, S, Leser, G, Martin, TE. 1990. Immunoelectron microscope localization of snRNPs in the polytene nucleus of salivary glands of *Chironomus thummi*. *Chromosoma* 99:44–51

Verheijen, R, Kuijpers, H, Vooijs, P, van Venrooij, W, Ramaekers, F. 1986. Distribution of the 70k U1 RNA-associated protein during interphase and mitosis. Correlation with other U RNP particles and proteins of the nuclear matrix. *J. Cell Sci.* 86:173–90

Vogelstein, B, Hunt, BF. 1982. A subset of small nuclear ribonucleoprotein particle antigens is a component of the nuclear matrix. *Biochem. Biophys. Res. Comm.* 105:1224–32

Wachtler, F, Mosgoller, W, Schwarzacher, HG. 1990. Electron microscopic in situ hybridization and autoradiography: localization and transcription of rDNA in human lymphocyte nucleoli. *Exp. Cell Res.* 187: 346–48

Wachtler, F, Schofer, C, Mosgoller, W, Weipoltshammer, K, Schwarzacher, HG, et al. 1992. Human ribosomal RNA gene repeats are localized in the dense fibrillar component of nucleoli: light and electron microscopic in situ hybridization in human sertoli cells. *Exp. Cell Res.* 198:135–43

Wang, J, Cao, LG, Wang, YL, Pederson, T. 1991. Localization of pre-messenger RNA at discrete nuclear sites. *Proc. Natl. Acad. Sci. USA* 88:7391–95

Ward, GK, McKenzie, R, Zannis-Hadjopoulos, M, Price, GB. 1990. The dynamic distribution and quantitation of DNA cruciforms in eukaryotic nuclei. *Exp. Cell Res.* 188:235–46

Ward, GK, Shihab-el-Deen, A, Zannis-Hadjopoulos, M, Price, G. B. 1991. DNA cruciforms and the nuclear supporting structure. *Exp. Cell Res.* 195:92–98

Weber, K, Geisler, N. 1985. Intermediate filaments: structural conservation and divergence. *Ann. NY Acad. Sci.* 455:126–43

Worman, HJ, Evans, CD, Blobel, G. 1990. The lamin B receptor of the nuclear envelope inner membrane: a polytopic protein with eight potential transmembrane domains. *J. Cell Biol.* 111:1535–42

Worman, HJ, Yuan, J, Blobel, G, Georgatos, SD. 1988. A lamin B receptor in the nuclear envelope. *Proc. Natl. Acad. Sci. USA* 85:8531–34

Wu, Z, Murphy, C, Callan, HG, Gall, JG. 1991. Small nuclear ribonucleoproteins and heterogeneous nuclear ribonucleoproteins in the amphibian germinal vesicle: loops, spheres, and snurposomes. *J. Cell Biol.* 113:465–83

Wurtz, T, Lonnroth, A, Ovchinnikov, L, Skoglund, U, Daneholt, B. 1990. Isolation and initial characterization of a specific premessenger ribonucleoprotein particle. *Proc. Natl. Acad. Sci. USA* 80:6436–40

Xing, Y, Johnson, CV, Dobner, PR, Lawrence, JB. 1993. Higher level organization of individual gene transcription and RNA splicing. *Science.* 259:1326–30

Xing, Y, Lawrence, JB. 1991. Preservation of specific RNA distribution within the chromatin-depleted nuclear substructure demonstrated by in situ hybridization coupled with biochemical fractionation. *J. Cell Biol.* 112:1055–63

Yang, CH, Lambie, EJ, Snyder, M. 1992. NuMA: an unusually long coiled-coil related protein in the mammalian nucleus. *J. Cell Biol.* 116:1303–17

Yuan, J, Simos, G, Blobel, G, Georgatos, SD. 1991. Binding of lamin A to polynucleosomes. *J. Biol. Chem.* 266:9211–15

Zahler, AM, Lane, WS, Stolk, JA, Roth, MB. 1992. SR proteins: a conserved family of pre-mRNA splicing factors. *Genes Dev.* 6:837–47

Zamore, PD, Green, MR. 1989. Identification, purification and biochemical characterization of U2 small nuclear ribonucleoprotein auxiliary factor. *Proc. Natl. Acad. Sci. USA,* 86:9243–47

Zamore, PD, Green, MR. 1991. Biochemical characterization of U2 snRNP auxiliary factor: an essential pre-mRNA splicing factor with a novel intranuclear distribution. *EMBO J.* 10:207–14

Zamore, PD, Patton, JG, Green, MR. 1992. Cloning and domain structure of the mammalian splicing factor U2AF. *Nature* 355: 609–14

Zamore, PD, Zapp, ML, Green, MR. 1990. RNA binding: βs and basics. *Nature* 348: 485–86

Zeitlin, S, Parent, A, Silverstein, S, Efstratiadis, A. 1987. Pre-mRNA splicing and the nuclear matrix. *Mol. Cell. Biol.* 7:111–20

Zeitlin, S, Wilson, RC, Efstratiadis, A. 1989. Autonomous splicing and complementation of in vivo-assembled spliceosomes. *J. Cell Biol.* 108:765–77

Zhang, M, Zamore, PD, Carmo-Fonseca, M, Lamond, AI, Green, MR. 1992. Cloning and intracellular localization of the U2 small nuclear ribonucleoprotein auxiliary factor small subunit. *Proc. Natl. Acad. Sci. USA* 89:8769–73

Zieve, G, Sauterer, RA. 1990. Cell biology of the snRNP particles. *Crit. Rev. Biochem. Mol. Biol.* 25:1–46

Zinkowski, RP, Meyne, J, Brinkley, BR. 1991. The centromere-kinetochore complex: a repeat subunit model. *J. Cell Biol.* 113: 1091–110

Annu. Rev. Cell Biol. 1993. 9:317–343

TUMOR NECROSIS FACTOR, OTHER CYTOKINES AND DISEASE

Kevin J. Tracey[*] *and Anthony Cerami*
Picower Institute for Medical Research, Manhasset, New York 11030

[*]Laboratory of Biomedical Science, Department of Surgery, North Shore University Hospital-Cornell University Medical College, Manhasset, New York 11030

KEY WORDS: shock, cachexia, sepsis

CONTENTS

INTRODUCTION

Disease morbidity and mortality is frequently caused by an overexpression of cytokines. Cytokines are a family of pleiotropic host-derived proteins,

0743–4634/93/1115–0317$05.00

or glycoproteins, produced by immunologically competent cells during the host response to infection, invasion, injury, or inflammation, that mediate metabolic and biochemical changes in responding cells when present at low concentrations. Early investigators presumed that cytokines acted in the "best interest" of the host to combat disease and promote healing. More recent studies have painted a slightly different picture, however, which suggests that cytokines are often directly pathogenic via mechanisms that may injure or kill the host. The general acceptance of this viewpoint paved the way for advances in the cellular biology of cytokines and for the development of novel experimental strategies for the therapy of disease.

An enormous amount of literature has been published in the field of cytokines and disease, and a comprehensive review of the subject is beyond the scope of this review. For clarity and simplicity, we focus on one cytokine, tumor necrosis factor-alpha (TNF), as representative of other cytokines, since it has been widely studied for its pathogenic role in disease. Since TNF's beneficial and injurious roles in disease are often intertwined with the effects of other cytokines, this discussion offers a general paradigm for considering cytokine biology in both the development of disease and the host defense response.

CYTOKINES AND EFFECTS

Cell Sources and Stimuli

The predominate manifestations of illness (e.g. fever, shock, and weight loss) are often not directly attributable to the invading pathogen or neoplastic cell, but rather to a complex, redundant, interactive, and occasionally deranged cascade of cytokine effects. (An abridged list of cytokines that have been implicated as pathogenic mediators of disease is given in Table 1.) Cytokines are produced in response to a wide range of tumor- and pathogen-derived molecules including endotoxin, enterotoxin, toxic shock syndrome toxin, and antigenic determinants present on tumors, viruses, parasites, and fungi. Assorted other stimuli include gamma irradiation, peroxidases, products of complement activation, free radicals, and cytokines themselves, which induce the biosynthesis of other cytokines in an autocrine manner.

Once induced, cytokines have been detected in virtually every body fluid and cavity including serum, cerebrospinal fluid, urine, joint exudates, peritoneal fluid, middle ear fluid, and alveolar fluid (Scuderi et al 1986; Marano et al 1990; Romero et al 1989; Peter et al 1991; Neale et al 1989). The time course of TNF, IL-1, and IL-6 appearance in serum after intravenous administration of a lethal dose of live *E. coli* or endotoxin (LPS)

Table 1 Abridged list of cytokines and some of their biological effects that have been implicated in disease

Cytokine	Effect(s) in disease
Tumor necrosis factor-alpha (TNF)	Fever, lethal shock and tissue injury, anorexia, catabolic state, increases expression of IL-1, IL-6, IFN-gamma
TNF-beta	Myelin injury; bone resorption
Interleukin-1 (IL-1)	Fever, transient hypotension, anorexia, increases expression of IL-2, IL-6
Interleukin-2 (IL-2)	Hypotension, anorexia
Interleukin-6 (IL-6)	Fever, transient hypotension, anorexia, catabolic state, increases expression of TNF and IL-1
Interleukin-10 (IL-10)	Decreases expression of TNF and IL-1
Interferon-gamma (IFN-gamma)	Increase toxicity of TNF, catabolic state
Macrophage inflammatory protein-1 (MIP-1)	Fever, recruitment of inflammatory cells
Transforming growth factor-beta	Decreases expression of TNF, decreases TNF-mediated macrophage production of superoxide

has been well characterized. Serum cytokine levels are not detectable prior to LPS, but within minutes after endotoxemia, TNF levels rise, peak at 90 min, then fall to basal levels by 4–6 hr; IL-1 has a broad peak at 3–4 hr; IFN-gamma and IL-6 levels continue to rise throughout an 8 hr period (Fong et al 1989b; Hesse et al 1988; Creasey et al 1991). Endotoxin is a potent inducer of TNF, IL-1, and IL-6 from isolated monocytes, but passive immunization with TNF mAbs during endotoxemia in vivo attenuates the appearance of IL-1 and IL-6 (Fong et al 1989b). This is one example of the interdependence of cytokine effects on cytokine biosynthesis; in this case when one pivotal cytokine (TNF) is removed from the system, the later systemic appearance of IL-1 and IL-6 is also prevented. The serum half-life of TNF and other cytokines in humans and laboratory animals is short (on the order of 6–20 min), and clearance of radio-labeled cytokines is predominately via the kidneys (Spriggs et al 1987; Beutler et al 1985a; Ulich et al 1987). The rapid decline of circulating TNF after peak levels has been attributed to binding by high affinity receptors, present in nearly all tissues, to the systemic release of TNF-binding proteins (TNF-BPs) that neutralize serum TNF (see below) and to renal clearance.

Biosynthesis

The principal cells producing TNF are macrophages/monocytes, but TNF is also made by other cells including, B cells, T cells, NK cells, Kupffer cells, glial cells, and adipoyctes. Human TNF is translated as a 233 amino acid precursor containing a hydrophobic sequence that is proteolytically cleaved to the mature 157 amino acid form (17 kd). The biologically active secreted form of TNF is a noncovalently bound trimer. A cell membrane-associated form of TNF (26 kd) possesses bioactivity and is proteolytically cleaved to release the active secreted moiety (17 kd) and a smaller membrane-associated fragment (14 kd) recognizable by antibodies against the TNF pro-hormone (Jue et al 1990; Perez et al 1990; Kriegler et al 1988). The proteolytic enzyme has not been isolated, so it is not known how cleavage activity is regulated. Since the prohormone sequence predicts a peptide of M_r 8.7 K, but the membrane fragment behaves as 14 K, the peptide may be acylated during cellular processing as occurs in IL-1 biosynthesis (Jue et al 1990). Subsequent studies confirmed that the prohormone is myristylated on lysines K19 and K20, which lie just downstream from the hydrophobic membrane-spanning segment of the propiece (Stevenson et al 1992). Presumably, myristic acylation participates in trafficking the prohormone to the plasma membrane.

Bacterial endotoxin (LPS) is perhaps the most widely studied and potent inducer of TNF and is frequently used as a stimulus in studies of the regulation of TNF biosynthesis. Regulation occurs at the levels of gene transcription and translation, so that exposure of monocytes to LPS leads to increased rates of transcription (3-fold), mRNA content (100-fold), and protein secretion (10,000-fold) (Beutler et al 1986b; Beutler et al 1986a; Sariban et al 1988, respectively). LPS-responsive elements in the 5′ -flanking sequence are highly conserved including, but not limited to TATA and GC boxes, and nuclear factor kappa-B elements (Jongeneel et al 1989; Kruys et al 1992). A highly conserved sequence present in the 3′ -untranslated region (3′ -UTR) of TNF and other cytokines affects post-transcriptional regulation of gene expression (Caput et al 1986). Studies using a construct of the 3′ -UTR and the TNF promoter attached to a CAT reporter revealed that the 3′ -UTR inhibits translation in the inactivated state, but increases the efficiency of translation in the activated state (Brown & Beutler 1990). The stability of TNF mRNA is regulated by major and minor degradative pathways: the major degradative process is inhibitable by actinomycin D and is not specific for the 3′ -untranslated region; the minor degradative process is accelerated by endotoxin, is not inhibited by actinomycin D, and is specific for the 3′ -untranslated region (Sariban et al 1988; Han et al 1991a). Stabilization of the TNF message has been observed after stimulation of macrophages with LPS and in T cells activated by anti-CD28 (Lindstein et al 1989).

Other factors influence the quantity of TNF produced in response to LPS. Some decrease (glucocorticoids, pentoxifylline, cyclosporine A, PGE2, PGI2, dibutyryl cAMP, dietary n-3 polyunsaturated fatty acid, 2-aminopyrine, inhibitors of protein kinase C), and others increase (LTB4, platelet-activating factor, interferon-gamma, bradykinin, phospholipase A2) LPS-induced TNF production. Interaction between these factors influences TNF biosynthesis in various disease states. For example, glucocorticoids, which possess potent anti-inflammatory properties, suppress TNF transcription and translation via interaction with sequences present in the TNF promoter and 3′ -untranslated regions (Han et al 1991a,b). By contrast, interferon-gamma, which is proinflammatory, has a synergistic effect on LPS-induced TNF biosynthesis and is capable of overriding the suppressive effects of glucocorticoids, which allows TNF biosynthesis to proceed unimpeded (Leudke & Cerami 1990). Thus the net production of TNF and other cytokines is inextricably linked to the balance of factors that either up- or down-regulate biosynthesis.

Receptors

Cytokines interact with target cells via specific high-affinity membrane receptors that function as transducing elements to mediate cellular responses. In most cases, maximal cellular biological responses occur when less than 10% of the total receptor number is occupied. Soluble fragments of the TNF receptor(s) were initially isolated by TNF-affinity purification of TNF-binding proteins (TNF-BP) present in the urine of patients with febrile illnesses, or the serum of patients with cancer (Seckinger et al 1990; Engelmann et al 1990; Gatanaga et al 1990). The sequences of these TNF-BPs were exploited in the subsequent isolation, identification, and cloning of two TNF transmembrane receptors (55 and 75 kd) (Smith et al 1990a,b; Gray et al 1990; Lewis et al 1991; Schall et al 1990; Loetscher et al 1990). Both TNF receptors are composed of a hydrophobic signal peptide, an extracellular domain that corresponds to the soluble TNF-BP, a transmembrane segment, and an intracellular domain. Surface expression of both types of receptor is up-regulated by interferon-gamma, IL-2, activators of protein kinase A, and concanavalin A (Thoma et al 1990; Tsujimoto et al 1986; Unglaub et al 1987). The DNA sequences of these receptors suggest that they are members of a larger receptor family that includes nerve growth factor, CD40, and the Fas gene product (Howard et al 1991).

Ultimately the cellular biological effects of TNF receptor-ligand interaction are dependent upon altered gene expression in responding cells. Studies of signal transduction have implicated altered expression of transcription fac-

Table 2 Abridged list of cellular responses to TNF

Myocyte	Reduced resting membrane potential, enhanced lactate production, net loss of glucagon and protein
Adipocyte	Suppression of LPL, enhanced release of FFA and glycerol
Hepatocyte	Enhanced lipogenesis, enhanced acute phase protein biosynthesis, increased glucagon-mediated amino acid uptake
Endothelial cell	Rearrangement of cytoskeleton, increased permability to albumin and water, enhanced expression of activation antigens, induction of surface procoagulant activity, IL-1 release
Pituicyte	Stimulation of ACTH and prolactin, inhibition of TSH and GH

tors, tyrosine kinases, serine kinases, phospholipase A2, GTP-binding proteins, and sphingomyelinase (Mathias et al 1993, 1991; Dressler et al 1992; Andrews et al 1990; Gray et al 1990; Hohmann et al 1990). It is not known precisely how individual cytokine receptor signaling cascades mediate intranuclear genetic regulatory events, but it is apparent that no single second messenger acts as a final common signal pathway mediating cellular effects. Rather, cellular responses to various cytokines are tissue-specific and dependent upon both the responding cell type and the metabolic state of the cell. Some of the biological effects attributable to TNF in a variety of cell types are summarized in Table 2.

REGULATORS AND EFFECTORS OF CYTOKINE EFFECTS

In either cell culture systems or in vivo experimental models, the net biological effect of TNF and other cytokines is ultimately determined by a complex interplay between (*a*) the biological effects of the cytokine on specific cells; (*b*) cytokine interaction with other factors or cytokines that enhance bioactivity; (*c*) cytokine interaction with other factors or cytokines that suppress bioactivity; (*d*) the location of cytokine production in specific tissues; (*e*) the development of tolerance or tachyphylaxis; and (*f*) the effects of secondary mediator cascades that amplify and propagate the response to a cytokine(s). Therefore, to understand cytokine biology in disease, one must consider how these interactions contribute to the observed biological response.

Factors that Enhance Effects of TNF

The disastrous consequences of inflammatory pathways activated during disease are often the result of synergistic toxicity between cytokines. For

Table 3 Factors or conditions that influence the biological response to TNF

	Decrease TNF effects	Increase TNF effects
Cytokines	Transforming growth factor-beta, interleukin-10, TNF-BPs, CNTF	Interleukin-1, interferon-gamma
Drugs/agents	Glucocorticoids, non-steroidal anti-inflammatory drugs, hydrazine, anti-TNF antibodies, TNF-BPs	D-galactosamine, leukotrienes, prostaglandins, platelet-activating factor, endotoxin/lipopolysaccharide
Disease states	Not determined	Chronic infection, cancer
Physiological states	Not determined	Hypophysectomy, adrenalectomy, hypoglycemia, dehydration

instance, small doses of TNF are minimally toxic, but lethal shock and tissue injury occur when these innocuous doses of TNF are co-administered with either IL-1, IFN-gamma, or LPS (Rothstein & Schreiber 1987; Waage & Espevik 1988; Talmadge et al 1989; Tredget et al 1989). The mechanisms underlying this increased sensitivity to TNF after exposure to other cytokines is unknown. It has been suggested that mediation of increased TNF receptor expression by IFN-γ is an obvious pathway for up-regulating the TNF response, but experimental studies have yet to identify a correlation between TNF receptor number and sensitivity to TNF (Patton et al 1986; Tsujimoto et al 1986). Moreover, since most cells express 1000–10,000 cytokine receptors, but maximal biological responses occur after binding to only 10% of receptors, one would not necessarily expect a correlation between receptor number and biological response. An abridged list of factors that increase sensitivity to TNF is given in Table 3. A review of this list reveals no obvious common factor that might contribute to enhance TNF responsiveness. Moreover, the broad spectrum of agents and physiological states that influence the response to TNF suggest that caution must be exercised when interpreting the role of an individual cytokine in the diseased host. It has been suggested by teleological reasoning that these additive and synergistic interactions between cytokines evolved to heighten immune responsiveness during invasion or infection. But synergistic toxicity during cytokine excess also contributes directly to the manifestations of disease.

Factors that Suppress Effects of TNF

Contrarily, a number of factors have been identified that directly or indirectly suppress the biological effects of TNF and other cytokines (Table 3).

Presumably these factors provide mechanisms that evolved to prevent excessive cytokine activity and offer yet another level of cytokine regulation that acts to protect the host against unwanted toxicity. Recent interest in these suppressive factors is kindled by the desire to utilize natural anticytokines as therapy for diseases of cytokine excess.

The TNF-BPs are generated in vivo by proteolytic cleavage of the receptor; the putative responsible enzyme has not been identified. Shedding is induced by a variety of agents including LPS, phorbol esters, GM-CSF, and TNF itself (Lantz et al 1990). Circulating levels of TNF-BP increase in the circulation of humans with sepsis, cancer, AIDS, other immunological illnesses, or after the injection of small doses of LPS (Aderka et al 1991; Digel et al 1992; Kalinkovich et al 1992; van Zee et al 1992). The biological effects of the TNF-BP are dependent upon the number released, since low concentrations augment the biological effects of TNF by stabilizing the bioactive trimeric form and slowing conversion to inactive monomers (Aderka et al 1992), but higher concentrations are capable of neutralizing TNF in models of inflammation and shock (van Zee et al 1992; Peppel et al 1991; Ashkenazi et al 1991).

IL-1 inhibitory activity present in the supernatants of monocytes cultured on adherent IgG and in the urine of patients with fever led to the identification of a 22-kd protein that binds to the IL-1 type 1 receptor, elicits no agonist activity, and blocks Il-1 binding (Arend et al 1990). Termed the IL-1 receptor antagonist (IL-1ra), this peptide is induced by LPS, TGF-beta, and other invasive stimuli, and is released systemically within minutes after LPS injection in humans (Fischer et al 1991). Inhibitory effects are observed when IL-1ra is present in upwards of 100- to 1000-fold molar excess, in animal models, where it confers modest protection against cytokine toxicity of endotoxic shock, cachexia, and inflammation (Alexander et al 1991; Arend et al 1990; Wakabayashi et al 1991; Gershenwald et al 1990; Henricson et al 1991). It is widely believed that the protective effects of the IL-1ra in these studies are the result of neutralizing the synergistic toxicity between IL-1 and TNF, since IL-1 alone possesses little toxicity even at very high doses (Waage & Espevik 1988). The existence of comparable receptor antagonists for TNF and other cytokines has been postulated, but to date none has been identified.

Classical endocrine hormones may counteract the biological actions of cytokines. Melanocyte-stimulating hormone (MSH), a pituitary-derived hormone, functions by unknown mechanisms to attenuate the development of inflammatory acute phase responses induced by the intravenous or intracerebroventricular administration of TNF or IL-1 (Robertson et al 1988). Glucocorticoid hormones increase survival of animals injected with otherwise lethal doses of IL-1 or TNF; it is not known whether these protective effects of glucocorticoids are the result of an interruption of cytokine cascades, of cellular protection against cytokine-mediated cytotoxicity, or by the induction of circulating inhibitors (Zuckerman et al 1989; Butler et al 1989). Insulin reverses the biological effects of chronically administered TNF including interstitial pneumonitis, loss of body weight, and capillary leakage syndrome (Fraker et al 1989). Taken together, these data suggest that the classical endocrine system participates in a bioregulatory feedback loop that may prevent unwanted toxicity from cytokine excess.

TGF-beta ablates the respiratory burst of macrophages induced by TNF (Tsunawaki et al 1988). Since LPS-induced macrophages release TGF-beta, it has been suggested that TGF-beta acts to down-regulate the effects of TNF (Tsunawaki et al 1988). Ciliary neurotrophic factor (CNTF) (374), a neural growth factor, has recently been added to the list of cytokines that protect against TNF-induced cytotoxicity (Louis et al 1993). During neural development, a number of oligodendrocytes in the brain undergo natural death, and this type of apoptosis can be initiated by TNF. CNTF, a normal product of astrocyte activation, attenuates TNF-induced cytotoxicity against oligodendrocytes, which suggests that interaction between TNF and CNTF may modulate the rate of oligodendrocyte death (Louis et al 1993).

Secondary Mediators

The effects of TNF, IL-1, and other cytokines in vivo are pleiotropic, but many of these biological responses are attributable to the activities of secondary mediators induced by the cytokines. A partial list of factors induced by TNF is given in Table 4. In some cases the most striking signs and symptoms of disease are mediated by the direct action of these secondary factors. For instance, administration of TNF causes fever mediated directly by increased hypothalamic prostaglandin biosynthesis (Dinarello et al 1986; Kettelhut & Goldberg 1988). This initial temperature elevation is followed by a second fever spike caused by IL-1 produced in response to the increased levels of circulating TNF (Dinarello et al 1986). At even higher doses of TNF that cause shock and tissue injury, inhibitors of nitric oxide or platelet-activating factor reduce the severity of hypotension and tissue necrosis (Sun & Hsueh 1988; Yue et al 1990; Kilbourn et al 1990). Thus it is sometimes difficult to assign a mediation role directly as either a primary cytokine effect or a secondary mediator effect in the biology of disease. For example, it is uncertain whether TNF-induced increases of adrenocorticotropin output from the pituitary are mediated at the level of the hypothalamus (via increased release of corticotropin-releasing factor) or by a direct stimulus to the pituicyte-ACTH biosynthesis (Feuerstein et al 1990; Michie et al 1988; Milenkovic et al 1989; Pang et al 1989; Darling et al 1989).

Table 4 Abridged list of factors induced by TNF that participate as secondary mediators in disease

Cytokines	IL-1, IL-6, IFN-γ
Growth factors	PDGF, EGF, NFG
Eicosanoids	Prostaglandins, leukotrienes
Classical hormone	Epinephrine, cortisol, glucagon
Other	Platelet-activating factor (PAF), nitric oxide, TNF-BPs, IL-1ra

Location of Production in Tissues

After the widespread acceptance that cytokines occupy an important role in causing disease, investigators wanted to be able to predict outcome based on measurement of serum cytokine levels. It has become apparent, however, that the net biological effects of cytokines may be primarily determined by the body compartment in which they are produced, and not the ambient serum level. This was dramatically illustrated by experiments done with a Chinese hamster ovary cell line that was genetically engineered to secrete h-TNF constitutively (Tracey et al 1990; Oliff et al 1987). When these cells were implanted in nude mice, they formed tumors that secreted TNF, and the net biological effects were dependent upon the site of the tumor and not the circulating TNF level (Tracey et al 1990). When implanted in skeletal muscle, TNF-secreting tumors caused elevated serum h-TNF levels and the development of cachexia over many weeks, but since only trace amounts of h-TNF cross the intact blood brain barrier (Beutler et al 1985a), little anorexia was observed (Tracey et al 1990). By contrast, when h-TNF-secreting tumors were implanted directly into the brain, serum h-TNF levels were elevated because h-TNF escaped through the disrupted blood brain barrier at the site of brain injection. But the higher levels of TNF in brain caused profound anorexia with death by starvation in 14 days (Tracey et al 1990). In addition, nude mice with elevated brain TNF levels grew hair; hair growth was not directly attributable to serum h-TNF since hair growth was not observed in animals with leg tumors and elevated TNF. Thus an isolated measurement of serum h-TNF was not predictive of the host's net biological response. Rather, the concentration of TNF in specific tissue compartments (e.g. CNS) determined the response.

Localized TNF production is also an important component of the host's armamentarium for clearance of viral or bacterial infection from tissues. For instance, including the gene for murine TNF in a recombinant Vaccinia virus construct led to more efficient clearance of virus and attenuated viral lethality as compared to controls injected intravenously with the virus devoid of the TNF gene (Sambhi et al 1991). In separate experiments, contact sensitivity to the hapten trinitrophenol (TNP) was suppressed by passive immunization against TNF, presumably by interference with triggering of the cytokine cascade (Bromberg et al 1992). Local production of TNF in the infected tissues of mice with cutaneous *Leishmaniasis* or *Legionella pneumophila* (Blanchard et al 1988) has been found to improve the clearance by the host of the invading pathogen (Titus et al 1989), and TNF produced in peripheral tissue may prevent the exponential growth of TNF-secreting experimental tumors in mice (Teng et al 1991). Moreover, the local production of intraperitoneal TNF has been implicated in the host defense against bacterial peritonitis induced by cecal ligation and puncture (Cross et al 1989; Sheppard et al 1989; Echtenacher et al 1990; Bagby et al 1991). These and other beneficial effects of locally produced TNF serve to defend the host as part of a complicated immune response that may not have a proportional correlation to the ambient serum level of TNF.

Development of Tolerance

Sublethal doses of endotoxin induce a state of tolerance or desensitization that confers protection against subsequent doses of LPS that would be lethal in a non-tolerized animal. For years the biological basis of LPS tolerance was completely unknown, but recent studies implicate both TNF and IL-1. These cytokines participate in the development of two types of LPS-induced tolerance: tolerance to further cytokine production, and cellular tolerance to cytokine action.

TOLERANCE TO CYTOKINE PRODUCTION When monocytes are exposed to LPS there is a rapid release of TNF, followed by a refractory period or LPS-tolerant state during which further exposure to LPS does not trigger additional TNF release (Mathison et al 1990; Takasuka et al 1991; Zuckerman & Evans 1992). The development of LPS tolerance to TNF production is differentially regulated from other cytokines, so that LPS-tolerized cells recover the ability to produce IL-1 before they are able to produce TNF (Zuckerman & Evans 1992). Transcription rates of TNF are decreased in tolerized cells, with a concomitant decrease in the expression of the nuclear transcription factor NF-kappa-B (Zuckerman & Evans 1992). Haas and colleagues found that LPS-induced desensitization is attenuated by cyclooxygenase inhibitors, which suggests that PGE-2, which inhibits TNF mRNA appearance, may participate in the development of tolerance; however, this observation has not been confirmed by others (Mathison et al 1990; Haas et al 1989). Salkowski & Vogel found that exposure to LPS led to enhanced expression of macrophage glucocorticoid receptors, which suggests that increased sensitivity to the inhibitory effects of glucocorticoids on macrophages may contribute to the refractory state of early endotoxin tolerance (Salkowski & Vogel 1992; Gelin et al 1991a). A similar form of tolerance develops in the serum compartment of endotoxemic animals and humans. Within minutes after endotoxemia, TNF appears as a burst in the serum that peaks at 90 min and then returns to undetectable levels within 3 hr. Animals rechallenged with endotoxin at 4 hr do not show another TNF burst (Zuckerman et al 1989), and macrophages harvested from mice previously treated with endotoxin produce less TNF than non-tolerized controls (Mathison et al 1990). Mathison and colleagues (1990) found that hyporesponsiveness to LPS in elicited rabbit peritoneal macrophages occurred after exposure to concentrations of LPS that were 1000-fold less than LPS concentrations required to induce TNF production. They further characterized this tolerant state by showing that during the development of tolerance, TNF mRNA half-life was unchanged and that the development of tolerance was unaffected by indomethacin (Mathison et al 1990). Moreover, tolerance to LPS-induced TNF production may be specific to the stimulus since exposure of LPS-tolerized cells to heat-killed Gram-positive organisms induced a normal pattern of TNF biosynthesis (Mathison et al 1990).

TOLERANCE TO CYTOKINE EFFECTS Repeated exposure to TNF and IL-1 induces a tolerant or hyporesponsive state in cell culture systems and in vivo. Early studies showed that chronic twice daily administration of low doses of h-TNF to rats was followed by the induction of tachyphylaxis manifested by a desensitization to the anorectic effects of TNF, so that increasingly higher doses had to be administered in order to achieve the same degree of anorexia or toxicity (Tracey et al 1988; Fraker et al 1988; Wallach et al 1988). Tolerance also developed when TNF was administered via continuous intravenous infusion (Hoshino et al 1991), but not in animals bearing a tumor that constitutively secreted TNF (Oliff et al 1987; Tracey et al 1990). Fraker and colleagues found that the serum half-life of exogenously administered TNF in tolerized animals was similar to the non-tolerized controls, which suggested that tolerance was not the result of increased rates of clearance or degradation (Fraker et al 1988). Mengozzi & Ghezzi found that TNF-tolerized animals failed to develop increased serum corticosteroid levels in response to the TNF, which suggests that glucocorticoids alone do not account for the development of tolerance (Mengozzi & Ghezzi, 1991). Grunfeld and co-workers found that tolerance was tissue-specific because tolerance to the hypertriglyceridemic effects of TNF mediated by increased hepatic lipid biosynthesis did not occur even after tolerance to the anorectic effects had developed (Grunfeld et al 1989). The molecular basis for TNF hyporesponsiveness remains unclear, but tolerized animals are also protected against experimental septicemia or endotoxemia (Doherty et al 1989; Sheppard et al 1989).

CYTOKINES AND DISEASE

The following is a brief review of the role of TNF and other cytokines in diseases selected to highlight a diversity of pathophysiological effects.

Shock and Tissue Injury

Shock is commonly classified by its association with an underlying cause: septic, hemorrhagic, neurogenic, or cardiogenic. Shock is defined as inadequate tissue perfusion, usually in the presence of low blood pressure which, if not reversed, causes widespread tissue injury, organ failure, and death. Septic shock syndrome develops in association with invasive infections and has a mortality rate ranging from 30 to 80%. Recent evidence indicates that TNF and the secondary cytokine cascade in large part determine the course and lethality of septic shock.

TNF fulfills Koch's postulates for having a causative role in the development of septic shock syndrome: (*a*) TNF is produced during septic shock syndrome; (*b*) TNF itself causes the development of the syndrome when given to uninfected animals; (*c*) neutralizing TNF in septic animals prevents the development of septic shock syndrome. Highly purified recombinant TNF causes shock and tissue injury when administered to every mammal studied thus far including mouse, rat, dog, pig, sheep, cow, monkey, baboon, and human (Tracey et al 1986, 1987a,b; Mathison et al 1988; Creasey et al 1991; Natanson et al 1989; Warner et al 1988; Stephens et al 1988; Remick et al 1987; Yue et al 1990). The TNF toxicity syndrome is nearly indistinguishable from the septic shock syndrome as manifested by hypotension, acidosis, oliguria, hemorrhagic necrosis in vital organs including kidney and lung, disseminated intravascular coagulopathy, and hormonal responses of a catabolic stress state (reviewed in Tracey 1991; Tracey & Lowry 1990). Humans given nonlethal doses of TNF develop fever, headache, anorexia, myalgia, hypotension, capillary leakage syndrome, and increased rates of lipolysis and skeletal muscle protein degradation (Warren et al 1987; Starnes et al 1988; Spriggs et al 1988; Bauer et al 1989). Death from TNF-induced shock and tissue injury occurs by respiratory arrest, while the heart continues to beat for several minutes, in a manner that is indistinguishable from death caused by endotoxin or a lethal injection of live bacteria (Tracey et al 1986, 1987b). Although IL-1 induces a transient decline of blood pressure at very high doses, TNF is the only cytokine or host-derived mediator identified thus far that is capable of triggering the entire spectrum of hemodynamic, metabolic, and pathological sequelae of septic shock.

Protection against endotoxic or bacteremic shock and tissue injury has been achieved by passive immunization with anti-TNF antibodies or TNF-BPs. Early studies used polyclonal TNF antibodies in models of endotoxic shock in mice and rabbits, and monoclonal TNF antibodies in a preclinical model of live *E. coli* bacteremia in baboons (Tracey et al 1987a; Beutler et al 1985b; Mathison et al 1988; Hinshaw et al 1989; Emerson et al 1992). These studies all indicate that when TNF is completely neutralized, the development of shock and tissue injury is prevented even when endotoxin or live bacteria persist in the circulation. More recent studies have employed TNF-BPs or TNF-BP fusion proteins to neutralize TNF (Peppel et al 1991; van Zee et al 1992). Neutralization of TNF is presumably protective to the host by (*a*) attenuating TNF cytotoxicity; (*b*) inhibiting the release of toxic secondary factors induced by TNF; and (*c*) preventing the release of other cytokines that synergistically increase the toxicity of TNF. Since preventing the TNF-induced cytokine cascade eliminates the lethal effects of the septic shock syndrome even in the presence of ongoing endotoxemia or bacteremia, then the development of shock and tissue injury can be dissociated from the invasive stimuli themselves and be directly attributed to the cascade of toxic events initiated by an acute overproduction of TNF. The efficacy of neutralizing TNF in septic humans is unknown, but early data suggest that it may be of benefit (Exley et al 1990). The final determination of efficacy will be made pending the completion of ongoing multicenter clinical trials in the United States and Europe.

Besides inhibiting TNF, other anti-cytokine strategies are being explored for potentially beneficial effects in septic shock syndrome including the IL-1 receptor antagonist (IL-1ra) and interferon-gamma antibodies (Alexander et al 1991; Grau et al 1989; Heinzel 1992; Doherty et al 1992; Silva & Cohen 1992). But neither of these cytokines triggers lethal septic shock and tissue injury when given alone, so any observed benefit of antagonist therapies may be the result of reducing synergistic toxicity with TNF. The efficacy or toxicity of these agents in human septic shock is unknown.

Cachexia

Cachexia is a state of weight loss, weakness, and anemia that complicates illnesses caused by infection, inflammation, cancer, or injury. In spite of reductions in food intake, the cachectic host continues to catabolize vital lipid and protein stores, which leads to a state of starvation that directly contributes to organ failure and suppressed immunological functions. The primary difference between unstressed starvation and cachexia is that in starvation there are normal protective mechanisms that lead to the conservation of whole-body protein with a preferential catabolism of lipid, but in cachexia, protein catabolism proceeds unimpeded (Brennan & Ekman 1984; Wilmore 1991; Tracey 1992). Serum factors were implicated as mediators of cachexia in studies of parabiotic union between normal and cachectic tumor-bearing animals by the observation that catabolic changes occurred in the non-tumor bearing animals (Norton et al 1985).

Cytokines were implicated as humoral mediators of cachexia by the observation that repeated administration of macrophage supernatants (now known to contain TNF, IL-1, IL-6, IFN-gamma, and other cytokines) induced anorexia and weight loss (Cerami et al 1985). The availability of recombinant cytokines and genetically engineered cell lines that constitutively secrete cytokines has enabled studies of the precise role of these individual factors. It is clear that chronic exposure of the host to TNF, IL-1, IL-6, or IFN-gamma are each capable of inducing a cachectic state characterized by anorexia and net catabolism of whole-body lipid and protein (Tracey et al 1988, 1990; Socher et al 1988; Fong et al 1989a; Strassmann et al 1992; Matthys et al 1991a,b; Hoshino et al 1991). From these studies of the metabolic sequelae of cytokine-induced cachexia has come a better understanding of the molecular basis of cachexia. Net protein losses induced by TNF appear to be mediated by a secondary factor or pathway in vivo since incubation of TNF with isolated skeletal muscle does not induce protein loss, but protein loss does occur when TNF is administered in vivo (Tracey et al 1988; Moldawer et al 1987). Skeletal muscle protein synthesis in vivo is decreased after chronic exposure to TNF or IL-1; further characterization of this response reveals both decreased expression of mRNA for myofibrillar proteins and increased rates of whole-body protein breakdown (Tracey et al 1988; Fong et al 1989a; Flores et al 1989). Glucocorticoids are synergistic with TNF in causing losses of muscle protein, but glucocorticoid antagonists do not completely prevent TNF-mediated protein loss (Angeras et al 1990). Peripheral clearance of lipids is inhibited by TNF-mediated suppression of lipoprotein lipase, and hepatic lipid production is increased (Grunfeld & Feingold 1991). TNF causes increases in hepatic gluconeogenesis, acute phase protein biosynthesis, and a net shunting of amino acid precursors from peripheral tissues to liver (Brenner et al 1990; Fong et al 1990; Fukushima et al 1992). The catabolic effects of continually infused TNF are dissociable from the development of anorexia since nitrogen excretion during TNF infusion persisted even when caloric needs were met by total parenteral nutrition (Matsui et al 1993). Finally, TNF is capable of causing anemia of chronic disease by depressing erythropoiesis and increasing rates of red blood cell degradation (Tracey et al 1988; Moldawer et al 1989; Johnson et al 1989).

Anti-cytokine strategies have been investigated for the ability to attenuate the development of cachexia in animal models of cancer or chronic inflammation (Sherry et al 1989; Gelin et al 1991a,b; Gershenwald et al 1990; Matthys et al 1991b). The ultimate goal of anti-cachexia therapy is to prevent or reverse muscle protein loss, but several problems have hampered efforts to develop inhibitors of human cachexia including difficulty in (*a*) identifying excessive cytokine production in individual patients; (*b*) devising a method for chronically administering the agent at doses high enough to completely neutralize the cytokines produced in tissue during a protracted course of cachectic illnesses (months); and (*c*) identifying the mechanism(s) that are causative of increased muscle protein loss.

Insulin Resistance

Insulin resistance is a metabolic state defined in vivo by suppressed peripheral glucose disposal rates and relatively increased hepatic glucose production rates. It develops in patients with non-insulin-dependent diabetes mellitus, obesity, cachexia, hypertension, cancer, and infection. A role for TNF in the development of insulin resistance is suggested by three lines of evidence: (*a*) chronic infusion of TNF impairs insulin action on peripheral glucose uptake and hepatic glucose output (Lang et al 1992); (*b*) the Glut4 glucose transporter is down-regulated by TNF in adipose and muscle (Stephens & Pekala 1991); and (*c*) TNF is overexpressed in adipocytes from insulin-resistant obese animals (Hotamisligil et al 1993). The pleiotropic actions of TNF may also contribute to insulin resistance by stimulating the mobilization of free fatty acids that in turn mediate insulin resistance (McGarry 1992).

Transplantation

TNF has been implicated in the mediation of allograft rejection and graft-vs-host disease (GVHD). Elevated levels of TNF occur during rejection of transplanted organs (Maury & Teppo 1987), and administration of anti-TNF antibodies improves cardiac allograft survival in rats (Bolling et al 1992). GVHD occurs as a reaction of transplanted foreign T cells to the immunocompromised host and occurs as a complication after bone marrow transplantation. TNF is produced in GVHD, and serum TNF levels correlate to the severity of the disease (Symington et al 1990). Antibodies to TNF reduce mortality and suppress the pathological manifestations of GVHD (Piguet et al 1987).

Cancer

A role for TNF as a broad-spectrum anti-neoplastic agent has not been defined and, in general, the results of therapeutic application of TNF in cancer have been disappointing. The major problem, which was not unexpected based on the studies of TNF in the development of shock and cachexia, has been that the toxicity of TNF is not limited to tumors or tumor endothelium, but affects normal host tissues and endothelium as well (North & Havell 1988; Havell et al 1988). Thus tumoricidal doses are toxic to the host. Several strategies have been proposed to avoid the limitations imposed by toxicity including (*a*) high dose administration intraarterially directly into the tumor compartment for liver or brain neoplasia (Ohkawa et al 1989); (*b*) identification of tumors that bear exquisite sensitivity to the cytotoxic rather than growth-promoting effects of TNF; (*c*) administration of nontoxic doses of TNF in combination with other agents that are tumoricidal or trigger apoptosis; and (*d*) developing methods that dissociate the toxicity of TNF from tumor cytotoxicity. These approaches have not been evaluated in prospective clinical trials.

HIV Infection

TNF and other cytokines have been implicated in AIDS for their potential roles in mediating cachexia, activation of latent infection, and protective antiviral effects. TNF is produced in patients with AIDS, and there is some suggestion that the amount of TNF produced correlates to the severity of the disease (Lahdevirta et al 1988). Monocytes isolated from patients with HIV infection have been found to make more (Roux Lombard et al 1989; Wright et al 1988) or less (Ammann et al 1987; Haas et al 1987) TNF when stimulated by LPS; the basis for these divergent observations is unknown. A great deal of interest has focused on the ability of TNF to activate the HIV long terminal repeat leading to increased HIV expression from a chronically infected T cell line that produces low levels of HIV (Folks et al 1987). Up-regulation of HIV expression may also occur through TNF activation of nuclear factor kappa-B (Griffin et al 1989) and provide a theoretical link connecting cytokine expression to the conversion from latent to active HIV infection (Folks et al 1989; Matsuyama et al 1989). Other research efforts have been directed toward harnessing the antiviral effects of TNF to treat HIV infection (Mestan et al 1986), but early enthusiasm has waned with the observation that administration of TNF was associated with rising levels of HIV p24 antigen and no evidence of clinical improvement (Aboulafia et al 1989).

Adult Respiratory Distress Syndrome and Reperfusion Injury

Adult respiratory distress syndrome (ARDS), a pulmonary disease characterized by hypoxia, increased capillary permeability, and pulmonary edema, is difficult to treat and frequently fatal. Early investigations focused on the roles of oxygen free radicals, eicosanoids, and proteases in the mediation of the ARDS, but recent evidence implicates TNF and other cytokines in triggering ARDS. TNF and IL-1 are produced locally in the lung after intratracheal instillation of LPS, and patients with ARDS have elevated levels of TNF in bronchoalveolar fluid (Ulich et al 1991). Pulmonary endothelial injury and edema occur in mammals given intravenous TNF (Tracey et al 1986; Stephens et al 1988), and in an experimental model of ARDS, isolated guinea pig lung perfused with TNF developed increased capillary permeability and increased lung water content (Stephens et al 1988). The cellular basis for these TNF-mediated responses involves endothelial cell toxicity, enhanced expression of adherence and activation antigens on the endothelial surface, margination of leukocytes, degranulation of polymorphonuclear leukocytes, superoxide radical formation by macrophages, and increased eicosanoid biosynthesis (Goldblum et al 1989; Zheng et al 1990; Redl et al 1990; Rubin et al 1990; Fong et al 1990; Solomkin et al 1985).

Reperfusion injury occurs in an organ when blood flow is restored after a period of transient ischemia. This type of tissue injury occurs in myocardial infarction and stroke precipitated by transient arterial occlusion followed by lysis of the clot with restoration of blood flow. During hypoxia, ATP is degraded to hypoxanthine, but with reperfusion it is degraded to uric acid and xanthine by xanthine oxidase. Superoxide radicals are produced during this step, and these radicals are lethal to the hypoxia-injured cells. TNF production is triggered by hypoxia and circulating levels are elevated in a model of transient hepatic ischemia (Colletti et al 1990). In this study, anti-TNF antibodies protected against the development of reperfusion injury in liver and lung, which suggests that TNF occupies a role as a mediator in the common final pathway of tissue necrosis (Colletti et al 1990). Although additional data are needed to characterize the biological role of TNF in human reperfusion injury, TNF levels are elevated in patients with myocardial infarction (Maury & Teppo 1989) and congestive heart failure (Levine et al 1990).

Diseases of the Central Nervous System

TNF and other cytokines have been implicated in the pathogenesis of multiple sclerosis, meningitis, brain tumors, and cerebral edema. Astrocytes and glial tumors secrete a variety of cytokines and, depending on the state of the cell, respond to TNF as a stimulus to growth, apoptosis, differentiation, or immunological activation. Recent evidence suggests that neurones also secrete TNF, where it may have a role in signaling brain centers that control autonomic function and appetite (Breder & Saper 1990). TNF cytotoxicity to oligodendrocyte appears to contribute to the development of pathological demyelination disease occurring in multiple sclerosis (Ruddle et al 1990; Selmaj & Raine 1988; Selmaj et al 1991). During meningitis, TNF and other cytokines accumulate in the cerebrospinal fluid, mediating hemorrhagic tissue necrosis, disruption of the blood-brain barrier, and irreversible tissue injury (Waage et al 1989; Leist et al 1988; Mustafa et al 1989; McCracken et al 1989; Saukkonen et al 1990). Brain tumors produce cytokine growth factors that act locally as autocrine stimuli and may disrupt the blood-brain barrier, thereby causing transcapillary fluid extravasation and cerebral edema (Kanno et al 1988; Lachman et al 1987; Barna et al 1990).

FUTURE OF CYTOKINES IN THE BIOLOGY OF DISEASE

Cellular biological studies have extended the breadth of understanding how cytokines may cause the most dramatic manifestations of disease. A new consensus has been reached with the general acceptance of the theory that the net effects of the host response to invasion, infection, or injury may ultimately depend upon a balance between the beneficial and injurious effects of cytokines. Therefore, research into novel therapies based on the molecular mechanisms of disease pathogenicity are focused not only on eliminating the pathogen, tumor, or external threat, but on modifying the immunological response as well. One may reasonably expect future therapeutic agents to be classified by their activities as enhancing the beneficial effects and/or neutralizing the injurious effects of cytokines. The rational development of these and other strategies will be facilitated by continued advances in the cellular biology of cytokines.

Literature Cited

Aboulafia, D, Miles, S, Saks, SR, Mitsuyasu, RT. 1989. Intravenous recombinant tumor necrosis factor in the treatment of AIDS-related Kaposi's sarcoma. *J. Acquired Immune Defic. Syndr.* 2:54–58

Aderka, D, Engelmann, H, Hornik, V. 1991. Increased serum levels of soluble receptors for tumor necrosis factor in cancer patients. *Cancer Res.* 51:5602–7

Aderka, D, Engelmann, H, Maor, Y, Brakebusch, C, Wallach, D. 1992. Stablization of the bioactivity of tumor necrosis factor by its soluble receptors. *J. Exp. Med.* 175:323–29

Alexander, HR, Doherty, GM, Buresh, CM, Venzon, DJ, Norton, JA. 1991. A recombinant human receptor antagonist to interleukin 1 improves survival after lethal endotoxemia in mice. *J. Exp. Med.* 173:1029–32

Ammann, AJ, Palladino, MA, Volberding, P, Abrams, D, Martin, NL, Conant, M. 1987. Tumor necrosis factors alpha and beta in acquired immunodeficiency syndrome (AIDS) and aids-related complex. *J. Clin. Immunol.* 7:481–85

Andrews, JS, Berger, AE, Ware, CF. 1990. Characterization of the receptor for tumor necrosis factor (TNF) and lymphotoxin (LT) on human T lymphocytes: TNF and LT differ in their receptor binding properties and the induction of MHC class I proteins on a human CD4+ T cell hybridoma. *J. Immunol.* 144:2582–91

Angeras, MH, Angeras, U, Zamir, O, Hasselgren, PO, Fischer, JE. 1990. Interaction between corticosterone and TNF stimulated protein breakdown in rat skeletal muscle, similar to sepsis. *Surgery* 108:460–66

Arend, WP, Welgus, HG, Thompson, RC, Eisenberg, SP. 1990. Biological properties of recombinant human monocyte-derived interleukin 1 receptor antagonist. *J. Clin. Invest.* 85:1694–97

Ashkenazi, A, Marsters, SA, Capon, DJ. 1991. Protection against endotoxic shock by a tumor necrosis factor receptor immunoadhesin. *Proc. Natl. Acad. Sci. USA* 88: 10535–39

Bagby, GJ, Plessala, KJ, Wilson, LA, Thompson, JJ, Nelson, S. 1991. Divergent efficacy of antibody to tumor necrosis factor-Alpha in intravascular and peritonitis models of sepsis. *J Infect. Dis.* 163:83–88

Barna, BP, Estes, ML, Jacobs, BS, Hudson, S, Ransohoff, RM. 1990. Human astrocytes proliferate in response to tumor necrosis factor alpha. *J. Neuroimmunol.* 30:239–43

Bauer, KA, ten Cate, H, Barzegar, S, Spriggs, DR, Sherman, ML, Rosenberg, RD. 1989. Tumor necrosis factor infusions have a procoagulant effect on the hemostatic mechanism of humans. *Blood* 74:165–72

Beutler, B, Krochin, N, Milsark, IW, Luedke, C, Cerami, A. 1986a. Control of cachectin (tumor necrosis factor) synthesis: mechanisms of endotoxin resistance. *Science* 232: 977–80

Beutler, B, Milsark, IW, Cerami, A. 1985a. Cachectin/tumor necrosis factor: production, distribution, and metabolic fate in vivo. *J. Immunol.* 135:3972–77

Beutler, B, Milsark, IW, Cerami, AC. 1985b. Passive immunization against cachectin/ tumor necrosis factor protects mice from lethal effect of endotoxin. *Science* 229:869–71

Beutler, B, Tkacenko, V, Milsark, I, Krochin, N, Cerami, A. 1986b. Effect of gamma interferon on cachectin expression by mononuclear phagocytes. Reversal of the lpsd (endotoxin resistance) phenotype. *J. Exp Med.* 164:1791–96

Blanchard, DK, Djeu, JY, Klein, TW, Friedman, H, Stewart, WE. 1988. Protective effects of tumor necrosis factor in experimental Legionella pneumophila infections of mice via activation of PMN function. *J. Leukoc. Biol.* 43:429–35

Bolling, SF, Kunkel, SL, Lin, H. 1992. Prolongation of cardiac allograft survival in rats by anti-TNF and cyclosporin combination therapy. *Transplantation* 53:283–86

Breder, CD, Saper, CB. 1990. TNF immunoreactive innervation in the mouse brain. *Soc. Neurosci. Abstr.* 144:1280

Brennan, MF, Ekman, L. 984. Metabolic consequences of nutritional support of the cancer patient. *Cancer* 54:2627–34

Brenner, DA, Buck, M, Feitelberg, SP, Chojkier, M. 1990. Tumor necrosis factor-alpha inhibits albumin gene expression in a murine model of cachexia. *J. Clin. Invest.* 85:248–55

Bromberg, JS, Chavin, KD, Kunkel, SL. 1992. Anti-TNF antibodies suppress cell-mediated immunity in vivo. *J. Immunol.* 148:3412–17

Brown, HT, Beutler, B. 1990. Endotoxin-responsive sequences control cachectin/tumor necrosis factor biosynthesis at the translational level. *J. Exp. Med.* 171:465–75

Butler, LD, Layman, NK, Riedl, PE, Cain, RL, Shellhaas, J, et al. 1989. Neuroendocrine regulation of in vivo cytokine production and effects: I. In vivo regulatory networks involving the neuroendocrine system, interleukin-1 and tumor necrosis factor-Alpha. *J. Neuroimmunol.* 24:143–53

Caput, D, Beutler, B, Hartog, K, Thayer, R, Brown Shimer, S, Cerami, A. 1986. Identification of a common nucleotide sequence in the 3′-untranslated region of mRNA molecules specifying inflammatory mediators. *Proc. Natl. Acad. Sci. USA* 83:1670–74

Cerami, A, Ikeda, Y, Le Trang, N, Hotez, PJ, Beutler, B. 1985. Weight loss associated with an endotoxin-induced mediator from peritoneal macrophages: the role of cachectin (tumor necrosis factor). *Immunol. Lett.* 11:173–77

Colletti, LM, Remick, DG, Burtch, GD, Kunkel, SL, Strieter, RM, Campbell, DA Jr. 1990. Role of tumor necrosis factor-alpha in the pathophysiologic alterations after hepatic ischemia/reperfusion injury in the rat. *J. Clin. Invest.* 85:1936–43

Creasey, AA, Stevens, P, Kenney, J, Allison, AC, Warren, K, et al. 1991. Endotoxin and cytokine profile in plasma of baboons challenged with lethal and sublethal *Escherichia coli*. *Circ. Shock* 33:84–91

Cross, AS, Sadoff, JC, Kelly, N, Bernton, E, Gemski, P. 1989. Pretreatment with recombinant murine tumor necrosis factor-alpha/cachectin and murine interleukin 1-alpha protects mice from lethal bacterial infection. *J. Exp. Med.* 169:2021–27

Darling, G, Goldstein, DS, Stull, R, Gorschboth, CM, Norton, JA. 1989. Tumor necrosis factor: Immune endocrine interaction. *Surgery* 106:1155–60

Digel, W, Porszolt, F, Schmid, M, Herrmann, F, Lesslauer, W, Brockhaus, M. 1992. High levels of circulating soluble receptors for tumor necrosis factor in hairy cell leukemia and type B chronic lymphocytic leukemia. *J. Clin. Invest.* 89:1690–93

Dinarello, CA, Cannon, JG, Wolff, SM, Bernheim, HA, Beutler, B, et al. 1986. Tumor necrosis factor (cachectin) is an endogenous pyrogen and induces production of interleukin 1. *J. Exp. Med.* 163:1433–50

Doherty, GM, Lange, JR, Langstein, HN, Alexander, HR, Buresh, CM. 1992. Evidence for IFN-gamma as a mediator of the lethality of endotoxin and tumor necrosis factor-α. *J. Immunol.* 149:1666

Doherty, PC, Allan, JE, Clark, IA. 1989. Tumor necrosis factor inhibits the development of viral meningitis or induces rapid death depending on the severity of inflammation at time of administration. *J. Immunol.* 142:3576–80

Dressler, KA, Mathias, S, Kolesnick, RN. 1992. Tumor necrosis factor-α activates the sphingomyelin signal transduction pathway in a cell-free system. *Science* 255: 1715–18

Echtenacher, B, Falk, W, Maennel, DN, Krammer, PH. 1990. Requirement of endogenous tumor necrosis factor/cachectin for recovery from experimental peritonitis. *J. Immunol.* 145:3762–66

Emerson, TE, Lindsey, DC, Jesmok, GJ, Duerr, ML, Fournel, MA. 1992. Efficacy of monoclonal antibody against TNF-α in an endotoxemic baboon model. *Circ. Shock* 38:75–84

Engelmann, H, Holtmann, H, Brakebusch, C, Avni, YS, Sarov, I, et al. 1990. Antibodies to a soluble form of a tumor necrosis factor (TNF) receptor have TNF-like activity. *J. Biol Chem.* 265:14497–504

Exley, AR, Cohen, J, Buurman, W, Owen, R, Hanson, G, et al. 1990. Monoclonal antibody to TNF in severe septic shock. *Lancet* 1:1275–77

Feuerstein, G, Hallenbeck, JM, Vanatta, B, Rabinovici, R, Perera, PY, Vogel, SN. 1990. Effect of gram-negative endotoxin on levels of serum corticosterone, TNF-alpha, circulating blood cells, and the survival of rats. *Circ. Shock* 30:265–78

Fischer, E, Marano, MA, Barber, AE, Hudson, A, Lee, K, et al. 1991. Comparison between effects if interleukin-1alpha administration and sublethal endotoxemia in primates. *Am. J. Physiol.* 261:R442–52

Flores, EA, Bistrian, BR, Pomposelli, JJ, Dinarello, CA, Blackburn, GL, Istfan, NW. 1989. Infusion of tumor necrosis factor/cachectin promotes muscle catabolism in the rat: A synergistic effect with interleukin 1. *J. Clin. Invest.* 83:1614–22

Folks, TM, Clouse, KA, Justement, J, Rabson, A, Duh, E, et al. 1989. Tumor necrosis factor alpha induces expression of human immunodeficiency virus in a chronically infected T-cell clone. *Proc. Natl. Acad. Sci. USA* 86:2365–68

Folks, TM, Justement, J, Kinter, A, Dinarello, CA, Fauci, AS. 1987. Cytokine-induced expression of HIV-1 in a chronically infected promonocyte cell line. *Science* 238: 800–2

Fong, Y, Moldawer, LL, Marano, MA, Wei, H, Barber, A, et al. 1989a. Cachectin/TNF or IL-1 alpha induces cachexia with redistribution of body proteins. *Am. J. Physiol.* 256:R659–65

Fong, Y, Tracey, KJ, Moldawer, LL, Hesse, DG, Manogue, KR, et al. 1989b. Antibodies to cachectin/TNF reduce interleukin-1-beta and interleukin-6 appearance during lethal bacteremia. *J. Exp. Med.* 170:1627–33

Fong, Y, Marano, MA, Moldawer, LL, Wei, H, Calvano, SE, et al. 1990. The acute splanchnic and peripheral tissue metabolic response to endotoxin in humans. *J. Clin. Invest.* 85:1896–904

Fraker, DL, Merino, MJ, Norton, JA. 1989. Reversal of the toxic effects of cachectin by concurrent insulin administration. *Am. J. Physiol.* 256:E725–31

Fraker, DL, Stovroff, MC, Merino, MJ, Norton, JA. 1988. Tolerance to tumor necrosis factor in rats and the relationship to endotoxin tolerance and toxicity. *J. Exp. Med.* 168:95–105

Fukushima, R, Saito, H, Taniwaka, K, Hiramatsu, T, Morioka, Y, et al. 1992. Different roles of IL-1 and TNF on hemodynamics and interorgan amino acid metabolism in awake dogs. *Am. J. Physiol.* 2625:E275–81

Gatanaga, T, Hwang, C, Kohr, W, Cappuccini, F, Lucci, JA, et al. 1990. Purification and characterization of an inhibitor (soluble tumor necrosis factor receptor) for tumor necrosis factor and lymphotoxin obtained from the serum ultrafiltrates of human cancer patients. *Proc. Natl. Acad. Sci. USA* 87:8781–84

Gelin, J, Andersson, C, Lundholm, K. 1991a. Effects of indomethacin, cytokines, and cyclosporin A on tumor growth and the subsequent development of cancer cachexia. *Cancer Res.* 51:880–85

Gelin, J, Moldawer, LL, Loennroth, C, Sherry, B, Chizzonite, R, Lundholm, K. 1991b. Role of endogenous tumor necrosis factor alpha and interleukin 1 for experimental tumor growth and the development of cancer cachexia. *Cancer Res.* 51:415–21

Gershenwald, JE, Fong, Y, Fahey, TJ III, Calvano, SE, Chizzonite, R, et al. 1990. Interleukin 1 receptor blockade attenuates the host inflammatory response. *Proc. Natl. Acad. Sci. USA* 87:4966–70

Goldblum, SE, Hennig, B, Jay, M, Yoneda, K, McClain, CJ. 1989. Tumor necrosis factor alpha-induced pulmonary vascular endothelial injury. *Infect. Immun.* 57:1218–26

Grau, GE, Heremans, H, Piguet, PF, Pointaire, P, Lambert, PH, et al. 1989. Monoclonal antibody against interferon gamma can prevent experimental cerebral malaria and its associated overproduction of tumor necrosis factor. *Proc. Natl. Acad. Sci. USA* 86:5572–74

Gray, PW, Barrett, K, Chantry, D, Turner, M, Feldman, M. 1990. Cloning of human tumor necrosis factor (TNF) receptor cDNA and expression of recombinant soluble TNF-binding protein. *Proc. Natl. Acad. Sci. USA* 87:7380–84

Griffin, GE, Leung, K, Folks, TM, Kunkel, S, Nabel, GJ. 1989. Activation of HIV gene expression during monocyte differentiation by induction of NF-kappa B. *Nature* 339: 70–73

Grunfeld, C, Feingold, KR. 1991. The metabolic effects of tumor necrosis factor and other cytokines. *Biotherapy* 3:143–58

Grunfeld, C, Wilking, H, Neese, R, Gavin, LA, Moser, AH, et al. 1989. Persistence of the hypertriglyceridemic effect of tumor necrosis factor despite development of tachyphylaxis to its anorectic/cachectic effects in rats. *Cancer Res*. 49:2554–60

Haas, JG, Riethmuller, G, Ziegler Heitbrock, HW. 1987. Monocyte phenotype and function in patients with the acquired immunodeficiency syndrome (AIDS) and AIDS-related disorders. *Scand. J. Immunol.* 26:371–79

Haas, JG, Thiel, C, Blomer, K, Weiss, EH, Riethmuller, G, Ziegler Heitbrock, H.W. 1989. Downregulation of tumor necrosis factor expression in the human Mono-Mac-6 cell line by lipopolysaccharide. *J. Leukoc. Biol.* 46:11–14

Han, J, Beutler, B, Huez, G. 1991. Complex regulation of TNF mRNA turnover in lipopolysaccharide-activated macrophages. *Biochim. Biophys. Acta* 1090:22–28

Han, J, Huez, G, Beutler, B. 1991. Interactive effects of the tumor necrosis factor promoter and 3′-untranslated regions. *J. Immunol.* 146:1843–48

Havell, EA, Fiers, W, North, RJ. 1988. The antitumor function of tumor necrosis factor (TNF): I. Therapeutic action of TNF against an established murine sarcoma is indirect, immunologically dependent, and limited by severe toxicity. *J. Exp. Med.* 167:1067–85

Heinzel, FP. 1992. The role of IFN-gamma in the pathology of experimental endotoxemia. *J. Immunol.* 145:2920

Henricson, BE, Neta, R, Vogel, SN. 1991. An interleukin-1 receptor antagonist blocks lipopolysaccharide-induced colony-stimulating factor production and early endotoxin tolerance. *Infect. Immunity* 59:1188–91

Hesse, DG, Tracey, KJ, Fong, Y, Manogue, KR, Palladino, MA Jr, et al. 1988. Cytokine appearance in human endotoxemia and primate bacteremia. *Surg. Gynecol. Obstet.* 166:147–53

Hinshaw, L, Olson, P, Kuo, G. 1989. Efficacy of post-treatment with anti-TNF monoclonal antibody in preventing the pathophysiology and lethality of sepsis in the baboon. *Circ. Shock* 27:362–69

Hohmann, H-P, Brockhaus, M, Baeuerle, PA, Remy, R, Kolbeck, R, Van Loon, APGM. 1990. Expression of the types A and B tumor necrosis factor (TNF) receptors is independently regulated, and both receptors mediate activation of the transcription factor NF-kappaB. TNFAlpha is not needed for induction of a biological effect via TNF receptors. *J. Biol Chem.* 265:22409–17

Hoshino, E, Pichard, C, Greenwood, CE, Kuo, GC, Cameron, RG, et al. 1991. Body composition and metabolic rate in rat during a continuous infusion of cachectin. *Am. J. Physiol* 260:E27–36

Hotamisligil, GS, Shargill, NS, Spiegelman, BM. 1993. Adipose expression of tumor necrosis factor-α direct role in obesity-linked insulin resistance. *Science* 259:87–91

Howard, ST, Chan, YS, Smith, GL. 1991. Vaccinia virus homologues of the Shope fibroma virus inverted terminal repeat proteins and a discontinuous ORF related to the tumor necrosis factor receptor family. *Virology* 180:633–47

Johnson, RA, Waddelow, TA, Caro, J, Oliff, A, Roodman, GD. 1989. Chronic exposure to tumor necrosis factor in vivo preferentially inhibits erythropoiesis in nude mice. *Blood* 74:130–38

Jongeneel, CV, Shakhov, AN, Nedospasov, SA, Cerottini, JC. 1989. Molecular control of tissue-specific expression at the mouse TNF locus. *Eur. J. Immunol.* 19:549–52

Jue, D-M, Sherry, B, Luedke, C, Manogue, KR, Cerami, A. 1990. Processing of newly synthesized cachectin/tumor necrosis factor in endotoxin-stimulated macrophages. *Biochemistry* 29:8371–77

Kalinkovich, A, Engelmann, H, Harpaz, N. 1992. Elevated serum levels of tumor necrosis factor receptors (zTNF-R) in patients with HIV infection. *Clin. Exp. Immunol.* 89:351–55

Kanno, H, Kuwabara, T, Yasumitsu, H, Umeda, M. 1988. Transforming growth factors in urine from patients with primary brain tumors. *J. Neurosurg.* 68:775–80

Kettelhut, IC, Goldberg, AL. 1988. Tumor necrosis factor can induce fever in rats without activating protein breakdown in muscle or lipolysis in adipose tissue. *J. Clin. Invest.* 81:1384–89

Kilbourn, RG, Gross, SS, Jubran, A, Adams, J, Griffith, OW, et al. 1990. N-methyl-L-arginine inhibits tumor necrosis factor-induced hypotension: Implications for the involvement of nitric oxide. *Proc. Natl. Acad. Sci. USA* 87:3629–32

Kriegler, M, Perez, C, DeFay, K, Albert, I, Lu, SD. 1988. A novel form of TNF/cachectin is a cell surface cytotoxic transmembrane protein: ramifications for the complex physiology of TNF. *Cell* 53: 45–53

Kruys, V, Kemmer, K, Shakhov, A, Jongeneel, V, Beutler, B. 1992. Constitutive activity of the tumor necrosis factor promoter is cancelled by the 3′ untranslated region in nonmacrophage cell lines; a trans-dominant factor overcomes this suppressive effect. *Proc. Natl. Acad. Sci. USA* 673:677

Lachman, LB, Brown, DC, Dinarello, CA. 1987. Growth-promoting effect of recombinant interleukin 1 and tumor necrosis factor for a human astrocytoma cell line. *J. Immunol.* 138:2913–16

Lahdevirta, J, Maury, CP, Teppo, AM, Repo, H. 1988. Elevated levels of circulating cachectin/tumor necrosis factor in patients with acquired immunodeficiency syndrome. *Am. J. Med.* 85:289–91

Lang, CH, Dobrescu, C, Bagby, GJ. 1992. Tumor necrosis factor impairs insulin action on peripheral glucose. *Endocrinology* 130: 43–52

Lantz, M, Malik, S, Slevin, ML, Olsson, I. 1990. Infusion of tumor necrosis factor (TNF) causes an increase in circulating TNF-Binding protein in humans. *Cytokines* 2:402–6

Leist, TP, Frei, K, Kam Hansen, S, Zinkernagel, RM, Fontana, A. 1988. Tumor necrosis factor alpha in cerebrospinal fluid during bacterial, but not viral, meningitis. Evaluation in murine model infections and in patients. *J. Exp. Med.* 167:1743–48

Leudke, CE, Cerami, A. 1990. Interferon-gamma overcomes glucocorticoid suppression of cachectin/tumor necrosis factor biosynthesis by murine macrophages. *J. Clin. Invest.* 86:1234–40

Levine, B, Kalman, J, Mayer, L, Fillit, HM, Packer, M. 1990. Elevated circulating levels of TNF in severe chronic heart failure. *N. Engl. J. Med.* 323:236–41

Lewis, M, Tartaglia, LA, Lee, A, Bennett, GL, Rice, GC, et al. 1991. Cloning and expression of cDNAs for two distinct murine tumor necrosis factor receptors demonstrate one receptor is species-specific. *Proc. Natl. Acad. Sci. USA* 88:2830–34

Lindstein, T, June, CH, Ledbetter, JA, Stella, G, Thompson, CB. 1989. Regulation of lymphokine messenger RNA stability by a surface-mediated T cell activation pathway. *Science* 244:339–43

Loetscher, H, Pan, Y-CE, Lahm, H-W, Gentz, R, Brockhaus, M, et al. 1990. Molecular cloning and expression of the human 55 kd tumor necrosis factor receptor. *Cell* 61:351–59

Louis, JC, Majal, E, Takayama, S, Varon, S. 1993. CNTF protection of oligodendrocytes against natural and TNF-induced death. *Science* 259:689–92

Marano, MA, Fong, Y, Moldawer, LL, Calvano, SE, Tracey, KJ, et al. 1990. Serum cachectin/TNF in critically ill patients with burns correlates with infection and mortality. *Surg. Gynecol. Obstet.* 170: 32–38

Mathias, S, Dressler, KA, Kokesnick, RN. 1991. Characterization of a ceramide-activated protein kinase: Stimulation by tumor necrosis factor-α. *Proc. Natl. Acad. Sci. USA* 88:10009–13

Mathias, S, Younes, A, Kan, CC, Orlow, I, Joseph, C, Kolesnick, RN. 1993. Activation of the sphingomyelin signaling pathway in intact EL4 cells and in a cell-free system by IL-1β. *Science* 259:519–22

Mathison, JC, Virca, GD, Wolfson, E, Tobias, PS, Glaser, K, Ulevitch, J. 1990. Adaptation to bacterial lipopolysaccharide controls lipopolysaccharide-induced tumor necrosis factor production in rabbit macrophages. *J. Clin. Invest.* 85:1108–18

Mathison, JC, Wolfson, E, Ulevitch, RJ. 1988. Participation of tumor necrosis factor in the mediation of gram-negative bacterial lipopolysaccharide-induced injury in rabbits. *J. Clin. Invest.* 81:1925–37

Matsui, J, Cameron, RG, Kurian, R, Kuo, GC, Jeejeebhoy, KN. 1993. Nutritional, hepatic, and metabolic effects of cachectin/TNF in rats receiving total parenteral nutrition. *Gastroenterology* 104:235–43

Matsuyama, T, Hamamoto, Y, Soma, G, Mizuno, D, Yamamoto, N, Kobayashi, N. 1989. Cytocidal effect of tumor necrosis factor on cells chronically infected with human immunodeficiency virus (HIV): enhancement of HIV replication. *J. Virol.* 63:2504–9

Matthys, P, Dukmans, R, Proost, P, Damme, JV, Heremans, H, et al. 1991a. Severe cachexia in mice inoculated with interferon-gamma producing tumor cells. *Int. J. Cancer* 49:77–82

Matthys, P, Heremans, H, Opdenakker, G, Billiau, A. 1991b. Anti-interferon-gamma antibody treatment, growth of Lewis lung tumours in mice and tumour-associated cachexia. *Eur. J. Cancer* 27:182–87

Maury, CP, Teppo, AM. 1987. Raised serum levels of cachectin/tumor necrosis factor alpha in renal allograft rejection. *J. Exp. Med.* 166:1132–37

Maury, CP, Teppo, AM. 1989. Circulating tumour necrosis factor-alpha (cachectin) in myocardial infarction. *J. Intern. Med.* 225: 333–36

McCracken, GH Jr, Mustafa, MM, Ramilo, O, Olsen, KD, Risser, RC. 1989. Cerebrospinal fluid interleukin 1-beta and tumor necrosis factor concentrations and outcome from neonatal gram-negative enteric bacillary meningitis. *Pediatr. Infect. Dis. J.* 8: 155–59

McGarry, JD. 1992. What if Minkowski had been ageusic? An alternative angle on diabetes. *Science* 258:766–70

Mengozzi, M, Ghezzi, P. 1991. Defective tolerance to the toxic and metabolic effects of interleukin-1. *Endocrinology* 128:1668–72

Mestan, J, Digel, W, Mittnacht, S, Hillen, H, Blohm, D, et al. 1986. Antiviral effects of recombinant tumour necrosis factor in vitro. *Nature* 323:816–19

Michie, HR, Manogue, KR, Spriggs, DR, Revhaug, A, O'Dwyer, S, et al. 1988. Detection of circulating tumor necrosis factor after endotoxin administration. *New Engl. J. Med.* 318:1481–86

Milenkovic, L, Rettori, V, Snyder, GD, Beutler, B, McCann, SM. 1989. Cachectin alters anterior pituitary hormone release by a direct action in vitro. *Proc. Natl. Acad. Sci. USA* 86:2418–22

Moldawer, LL, Marano, MA, Wei, H, Fong, Y, Silen, ML, et al., 1989. Cachectin/tumor necrosis factor-alpha alters red blood cell kinetics and induces anemia in vivo. *FASEB J.* 3:1637–43

Moldawer, LL, Svaninger, G, Gelin, J, Lundholm, KG. 1987. Interleukin 1 and tumor necrosis factor do not regulate protein balance in skeletal muscle. *Am. J. Physiol.* 253:C766–73

Mustafa, MM, Lebel, MH, Ramilo, O, Olsen, KD, Reisch, JS, et al. 1989. Correlation of interleukin-1 beta and cachectin concentrations in cerebrospinal fluid and outcome from bacterial meningitis. *J. Pediatr.* 115:208–13

Natanson, C, Eichenholz, PW, Danner, RL, Eichhacker, PQ, Hoffman, WD, et al. 1989. Endotoxin and tumor necrosis factor challenge in dogs simulate the cardiovascular profile of human septic shock. *J. Exp. Med.* 169:823–32

Neale, ML, Williams, BD, Matthews, N. 1989. Tumour necrosis factor activity in joint fluids from rheumatoid arthritis patients. *Br. J. Rheumatol.* 28:104–8

North, RJ, Havell, EA. 1988. The antitumor function of tumor necrosis factor (TNF): II. Analysis of the role of endogenous TNF in endotoxin-induced hemorrhagic necrosis and regression of an established sarcoma. *J. Exp. Med.* 167:1086–99

Norton, JA, Moley, JF, Green, MV, Carson, RE, Morrison, SD. 1985. Parabiotic transfer of cancer anorexia/cachexia in male rats. *Cancer Res.* 45:5547–52

Ohkawa, S, Wright, KC, Mahajan, H, Mavligit, GM, Wallace, S. 1989. Hepatic arterial infusion of human recombinant tumor necrosis factor-alpha. An experimental study in dogs. *Cancer* 63:2096–102

Oliff, A, Defeo-Jones D, , Boyer, M, Martinez, D, Kiefer, D, et al. 1987. Tumors secreting human TNF/cachectin induce cachexia in mice. *Cell* 50:555–63

Pang, XP, Pekary, AE, Mirell, C, Hershman, JM. 1989. Impairment of hypothalamic-pituitary-thyroid function in rats treated with human recombinant tumor necrosis factor-alpha (cachectin). *Endocrinology* 125:76–84

Patton, JS, Shepard, HM, Wilking, H, Lewis, G, Aggarwal, BB, et al. 1986. Interferons and tumor necrosis factors have similar catabolic effects on 3T3 L1 cells. *Proc. Natl. Acad. Sci. USA* 83:8313–17

Peppel, K, Crawford, D, Beutler, B. 1991. A tumor necrosis factor (TNF) Receptor-IgG heavy chain chimeric protein as a bivalent antagonist of TNF activity. *J. Exp. Med.* 174:1483–89

Perez, C, Albert, I, DeFay, K, Zachariades, N, Gooding, L, Kriegler, M. 1990. A nonsecretable cell surface mutant of tumor necrosis factor (TNF) kills by cell-to-cell contact. *Cell* 63:251–58

Peter, JB, Boctor, FN, Tourtellotte, WW. 1991. Serum and CSF levels of IL-2, sIL-2R, TNF-Alpha, and IL-1Beta in chronic progressive multiple sclerosis: Expected lack of clinical utility. *Neurology* 41:121–23

Piguet, PF, Grau, GE, Allet, B, Vassalli, P. 1987. Tumor necrosis factor/cachectin is an effector of skin and gut lesions of the acute phase of graft-vs-host disease. *J. Exp. Med.* 166:1280–89

Redl, H, Schlag, G, Lamche, H, Vogl, C, Paul, E, et al. 1990. TNF- and LPS-induced changes of lung vascular permeability: Studies in unanesthetised sheep. *Circ. Shock* 31:183–92

Remick, DG, Kunkel, RG, Larrick, JW, Kunkel, SL. 1987. Acute in vivo effects of human recombinant tumor necrosis factor. *Lab. Invest.* 56:583–90

Robertson, B, Dostal, K, Daynes, RA. 1988. Neuropeptide regulation of inflammatory and immunologic responses: The capacity of alpha-melanocyte-stimulating hormone to inhibit tumor necrosis factor and IL-1-inducible biologic responses. *J. Immunol.* 140: 4300–7

Romero, R, Manogue, KR, Mitchell, MD, Wu, YK, Oyarzun, E, et al. 1989. Infection and labor. IV. Cachectin-tumor necrosis factor in the amniotic fluid of women with intraamniotic infection and preterm labor. *Am. J. Obstet. Gynecol.* 161:336–41

Rothstein, JL, Schreiber, H. 1987. Relationship of tumour necrosis factor and endotoxin to macrophage cytotoxicity, haemorrhagic necrosis and lethal shock. *Ciba Found. Symp.* 131:124–39

Roux Lombard, P, Modoux, C, Cruchaud, A, Dayer, JM. 1989. Purified blood monocytes from HIV 1-infected patients produce high levels of TNF alpha and IL-1. *Clin. Immunol. Immunopathol.* 50:374–84

Rubin, DB, Wiener-Kronish, JP, Murray, JF, Green, DR, Turner, J, et al. 1990. Elevated von Willebrand factor antigen is an early

plasma predictor of acute lung injury in nonpulmonary sepsis syndrome. *J. Clin. Invest.* 86:474–80
Ruddle, NH, Bergman, CM, McGrath, KM, Lingenheld, EG, Grunnet, ML, et al. 1990. An antibody to lymphotoxin and tumor necrosis factor prevents transfer of experimental allergic encephalomyelitis. *J. Exp. Med.* 172:1193–200
Salkowski, CA, Vogel, SN. 1992. Lipoplysaccharide increases glucocorticoid receptor expression in murine macrophages. *J. Immunol.* 149:4041–47
Sambhi, SK, Kohonen-Corish, RJ, Ramshaw, IA. 1991. Local production of TNF encoded by recombinant Vaccinia virus is effective in controlling viral replication in vivo. *Proc. Natl. Acad. Sci. USA* 88:4025–29
Sariban, E, Imamura, K, Luebbers, R, Kufe, DW. 1988. Transcriptional and posttranscriptional regulation of tumor necrosis factor gene expression in human monocytes. *J. Clin. Invest.* 81:1506–10
Saukkonen, K, Sande, S, Cioffe, C, Wolpe, S, Sherry, B, et al. 1990. The role of cytokines in the generation of inflammation and tissue damage in experimental Gram-positive meningitis. *J. Exp. Med.* 171:439–48
Schall, TJ, Lewis, M, Koller, KJ, Lee, A, Rice, GC, et al. 1990. Molecular cloning and expression of a receptor for human tumor necrosis factor. *Cell* 61:361–70
Scuderi, P, Sterling, KE, Lam, KS, Finley, PR, Ryan, KJ, et al. 1986. Raised serum levels of tumour necrosis factor in parasitic infections. *Lancet* 2:1364–65
Seckinger, P, Zhang, J-H, Hauptmann, B, Dayer, J-M. 1990. Characterization of a tumor necrosis factor Alpha (TNF-Alpha) inhibitor: Evidence of immunological cross-reactivity with the TNF receptor. *Proc. Natl. Acad. Sci. USA* 87:5188–92
Selmaj, KW, Raine, CS. 1988. TNF mediates myelin and oligodendrocyte damage in vitro. *Ann. Neurol.* 29:339–46
Selmaj, K, Raine, CS, Cannella, B, Brosnan, CF. 1991. Identification of lymphotoxin and tumor necrosis factor in multiple sclerosis lesions. *J. Clin. Invest.* 87:949–54
Sheppard, BC, Fraker, DL, Norton, JA. 1989. Prevention and treatment of endotoxin and sepsis lethality with recombinant human tumor necrosis factor. *Surgery* 106:156–61
Sherry, BA, Gellin, J, Fong, Y, Marano, M, Wei, H, et al. 1989. Anticachectin/tumor necrosis factor-alpha antibodies attenuate development of cachexia in tumor models. *FASEB J.* 3:1956–62
Silva, AT, Cohen, J. 1992. Role of interferon-gamma in experimental Gram-negative sepsis. *J. Infect. Dis.* 166:331
Smith, CA, Davis, T, Anderson, D, Solam, L, Beckman, PM, et al. 1990a. A Receptor for TNF defines an unusual family of cellular and viral proteins. *Science* 248:1019–24
Smith, CA, Davis, T, Anderson, D, Solam, L, Beckmann, PM, et al. 1990b. A receptor for tumor necrosis factor defines an unusual family of cellular and viral proteins. *Science* 248:1019–23
Socher, SH, Friedman, A, Martinez, D. 1988. Recombinant human tumor necrosis factor induces acute reductions in food intake and body weight in mice. *J. Exp. Med.* 167:1957–62
Solomkin, JS, Cotta, LA, Satoh, PS, Hurst, JM, Nelson, RD. 1985. Complement activation and clearance in acute illness and injury: Evidence for C5a as a cell-directed mediator of the adult respiratory distress syndrome in man. *Surgery* 97:668–78
Spriggs, DR, Sherman, ML, Frei, E III, Kufe, DW. 1987. Clinical studies with tumour necrosis factor. In *Tumour Necrosis Factor and Related Cytotoxins (CIBA Found. Symp. 131)*, ed G Bock, J Marsh, pp. 206–27. Chichester: John Wiley & Sons
Spriggs, DR, Sherman, ML, Michie, HR, Arthur, KA, Imamura, K, et al. 1988. Recombinant human tumor necrosis factor administered as a 24-hour intravenous infusion. A phase I pharmacologic study. *J. Natl. Cancer Inst.* 80:1039–44
Starnes, HF Jr, Warren, RS, Jeevanandam, M, Gabrilove, JL, Larchian, W, et al. 1988. Tumor necrosis factor and the acute metabolic response to tissue injury in man. *J. Clin. Invest.* 82:1321–25
Stephens, JM, Pekala, PH. 1991. Transcriptional repression of the GLUT4 and C/EBP genes in 3T3-L1 adipocytes by TNF-α. *J. Biol. Chem.* 266:21839–45
Stephens, KE, Ishizaka, A, Larrick, JW, Raffin, TA. 1988. Tumor necrosis factor causes increased pulmonary permeability and edema. *Am. Rev. Resp. Dis.* 137:1364–70
Stevenson, FT, Burstein, SL, Locksley, RM, Lovett, DH. 1992. Myristyl acylation of the TNF α precursor on specific lysine residues. *J. Exp. Med.* 176:1053–62
Strassmann, G, Fong, M, Kenney, JS, Jacob, CO. 1992. Evidence for the involvement of interleukin 6 in experimental cancer cachexia. *J. Clin. Invest.* 89:1681–84
Sun, X-M, Hsueh, W. 1988. Bowel necrosis induced by tumor necrosis factor in rats is mediated by platelet-activating factor. *J. Clin. Invest.* 81:1328–31
Symington, FW, Pepe, MS, Chen, AB, Deliganis, A. 1990. Serum tumor necrosis factor alpha associated with acute graft-versus-host disease in humans. *Transplantation* 50:518–21

Takasuka, N, Tokunaga, T, Akagawa, S. 1991. Preexposure of macrophages to low doses of lipopolysaccharide inhibits the expression of tumor necrosis factor-α mRNA but not of IL-1β mRNA1. *J. Immunol.* 146:3824–30

Talmadge, JE, Bowersox, O, Tribble, H, Shepard, M, Liggitt, D. 1989. Therapeutic and toxic activity of tumor necrosis factor is synergistic with gamma interferon. *Pathol. Immunopathol. Res.* 8:21–34

Teng, MN, Park, BH, Koeppen, HKW, Tracey, KJ, Fendly, BM, Schreiber, H. 1991. Long-term inhibition of tumor growth by tumor necrosis factor in the absence of cachexia or T-cell immunity. *Proc. Natl. Acad. Sci. USA* 88:3535–39

Thoma, B, Grell, M, Pfizenmaier, K, Scheurich, P. 1990. Identification of a 60-kD tumor necrosis factor (TNF) receptor as the major signal transducing component in TNF responses. *J. Exp. Med.* 172:1019–23

Titus, RG, Sherry, B, Cerami, A. 1989. Tumor necrosis factor plays a protective role in experimental murine cutaneous Leishmaniasis. *J. Exp. Med.* 170:2097–104

Tracey, KJ. 1991. Tumor necrosis factor (cachectin) in the biology of septic shock syndrome. *Circ. Shock* 35:123–28

Tracey, KJ. 1992. TNF and other cytokines in the metabolism of septic shock and cachexia. *Clin. Nutr.* 11:1–11

Tracey, KJ, Beutler, B, Lowry, SF, Merryweather, J, Wolpe, S, et al. 1986. Shock and tissue injury induced by recombinant human cachectin. *Science* 234:470–74

Tracey, KJ, Fong, Y, Hesse, DG, Manogue, KR, Lee, AT, et al. 1987a. Anti-cachectin/TNF monoclonal antibodies prevent septic shock during lethal bacteraemia. *Nature* 330:662–64

Tracey, KJ, Lowry, SF. 1990. The role of cytokine mediators in septic shock. *Adv. Surg.* 23:21–56

Tracey, KJ, Lowry, SF, Fahey, TJ III, Albert, JD, Fong, Y, et al. 1987b. Cachectin/tumor necrosis factor induces lethal shock and stress hormone responses in the dog. *Surg. Gynecol. Obstet.* 164:415–22

Tracey, KJ, Morgello, S, Koplin, B, Fahey, TJ III, Fox, J, et al. 1990. Metabolic effects of cachectin/tumor necrosis factor are modified by site of production: Cachectin/tumor necrosis factor-secreting tumor in skeletal muscle induces chronic cachexia, while implantation in brain induces predominately acute anorexia. *J. Clin. Invest.* 86:2014–24

Tracey, KJ, Wei, H, Manogue, KR, Fong, Y, Hesse, DG, et al. 1988. Cachectin/tumor necrosis factor induces cachexia, anemia, and inflammation. *J. Exp. Med.* 167:1211–27

Tredget, EE, Yu, YM, Zhong, S, Burini, R, Okusawa, S, et al. 1989. Role of interleukin 1 and tumor necrosis factor on energy metabolism in rabbits. *Am. J. Physiol.* 255: E760–68

Tsujimoto, M, Yip, YK, Vilcek, J. 1986. Interferon-gamma enhances expression of cellular receptors for tumor necrosis factor. *J. Immunol.* 136:2441

Tsunawaki, S, Sporn, M, Ding, A, Nathan, CF. 1988. Deactivation of macrophages by transforming growth factor-beta. *Nature* 334:260–62

Ulich, TR, Del Castillo, J, Keys, M, Granger, GA, Ni, R-X. 1987. Kinetics and mechanisms of recombinant human interleukin 1 and tumor necrosis factor-alpha-induced changes in circulating numbers of neutrophils and lymphocytes. *J. Immunol.* 139: 3406–15

Ulich, TR, Yin, S, Guo, K, Del Castillo, J, Eisenberg, SP, Thompson, R.C. 1991. The intratracheal administration of endotoxin and cytokines III. The interleukin-1 (IL-1) receptor antagonist inhibits endotoxin- and IL-1-induced acute inflammation. *Am. J. Pathol.* 138:521–24

Unglaub, R, Maxeiner, B, Thoma, B, Pfizenmaier, K, Scheurich, P. 1987. Downregulation of tumor necrosis factor (TNF) sensitivity via modulation of TNF binding capacity by protein kinase C activators. *J. Exp. Med.* 166:1788–97

van Zee, KJ, Kohno, T, Fischer, E, Rock, CS, Moldawer, LL, Lowry, SF. 1992. Tumor necrosis factor soluble receptors circulate during experimental and clinical inflammation and can protect against excessive tumor necrosis factor-α in vitro and in vivo. *Proc. Natl. Acad. Sci. USA* 89:4845–49

Waage, A. Espevik, T. 1988. Interleukin 1 potentiates the lethal effect of tumor necrosis factor-alpha/cachectin in mice. *J. Exp. Med.* 167:1987–92

Waage, A, Halstensen, A, Shalaby, R, Brandtzaeg, P, Kierulf, P, Espevik, T. 1989. Local production of tumor necrosis factor-alpha, interleukin 1, and interleukin 6 in meningococcal meningitis. *J. Exp. Med.* 170:1859–967

Wakabayashi, G, Gelfand, JA, Burke, JF, Thompson, RC, Dinarello, CA. 1991. A specific receptor antagonist for interleukin 1 prevents *Escherichia coli*-induced shock in rabbits. *FASEB J.* 5:338–43

Wallach, D, Holtmann, H, Engelmann, H, Nophar, Y. 1988. Sensitization and desensitization to lethal effects of tumor necrosis factor and IL-1. *J. Immunol.* 140:2994–99

Warner, AE, DeCamp, MM Jr, Molina, RM, Brain, JD. 1988. Pulmonary removal of circulating endotoxin results in acute lung injury in sheep. *Lab. Invest.* 59:219–30

Warren, RS, Starnes, HF Jr, Gabrilove, JL, Oettgen, HF, Brennan, MF. 1987. The acute metabolic effects of tumor necrosis factor administration in humans. *Arch. Surg.* 122:1396–400

Wilmore, DW. 1991. Catabolic illness: Strategies for enhancing recovery. *N. Engl. J. Med.* 325:695–702

Wright, SC, Jewett, A, Mitsuyasu, R, Bonavida, B. 1988. Spontaneous cytotoxicity and tumor necrosis factor production by peripheral blood monocytes from AIDS patients. *J. Immunol.* 141:99–104

Yue, T-L, Farhat, M, Rabinovici, R, Perera, PY, Vogel, SN, Feuerstein, G. 1990. Protective effect of BN 50739:a new platelet-activating factor antagonist, in endotoxin-treated rabbits. *J. Pharmacol. Exp. Ther.* 254:976–81

Zheng, H, Crowley, JJ, Chan, JC, Hoffmann, H, Hatherill, JR, et al. 1990. Attenuation of tumor necrosis factor-induced endothelial cell cytotoxicity and neutrophil chemiluminescence. *Am. Rev. Respir. Dis.* 142:1073–78

Zuckerman, SH, Evans, GF. 1992. Endotoxin tolerance: in vivo regulation of tumor necrosis factor and interleukin-1 synthesis is at the transcriptional level. *Cell Immunol.* 140:513–19

Zuckerman, SH, Shellhaas, J, Butler, LD. 1989. Differential regulation of lipopolysaccharide-induced interleukin 1 and tumor necrosis factor synthesis: effects of endogenous and exogenous glucocorticoids and the role of the pituitary-adrenal axis. *J. Immunol.* 19:301–5

Annu. Rev. Cell Biol. 1993. 9:345–75

SIGNAL TRANSDUCTION IN GUARD CELLS

Sarah M. Assmann[1]
Harvard University, Department of Organismic and Evolutionary Biology, The Biological Laboratories, 16 Divinity Avenue, Cambridge, Massachusetts 02138

KEY WORDS: guard cell, stomata, H^+ ATPase, K^+ channel, anion channel, Ca^{2+} channel, phosphatidylinositol pathway, G protein, calmodulin, calcineurin

---

CONTENTS

INTRODUCTION

Vertebrates typically have specialized sensory organs. For example, in humans, photoreception by pigments in the eyes initiates transduction of visual cues. In plants, however, perhaps because there are fewer specialized organs, environmental sensing is often widely distributed throughout the plant body. Leaves, stems, flowers, and fruits can all contain photoreceptive chlorophylls and associated carotenoids that capture light energy and thus

[1]After 8/16/93, Biology Department, Pennsylvania State University, 208 Mueller Laboratory, University Park, Pennsylvania 16802

0743–4634/93/1115–0345$05.00

power photosynthesis, surely one of the most complex and important biological processes. Such organs also contain blue-light absorbing pigments (cryptochromes) and red/far-red absorbing pigments (phytochromes), which transduce light signals responsible for the dramatic alterations in morphology and physiology that occur in plants growing under sunny vs shaded conditions.

Similarly, a lack of extensive cellular specialization in plants may be compensated by the ability of plant cells to respond to many types of signals, thus exhibiting great complexity at the cellular level. Nowhere is the complex nature of signal transduction in plants more evident than in stomatal guard cells, which sense signals as diverse as light, humidity, and carbon dioxide concentration, as well as a variety of plant hormones. Guard cells are located in pairs in the epidermes of above-ground plant parts and define and control the apertures of microscopic pores called stomata (Figure 1). A small open space called the substomatal cavity is found interior to each stomatal complex and is bordered by the epidermis on one side and by mesophyll cells on the other sides. Gases diffuse through the stomatal pore down concentration gradients existing between the substomatal cavity and the atmosphere. Water evaporates from the aqueous medium surrounding plant cells and saturates

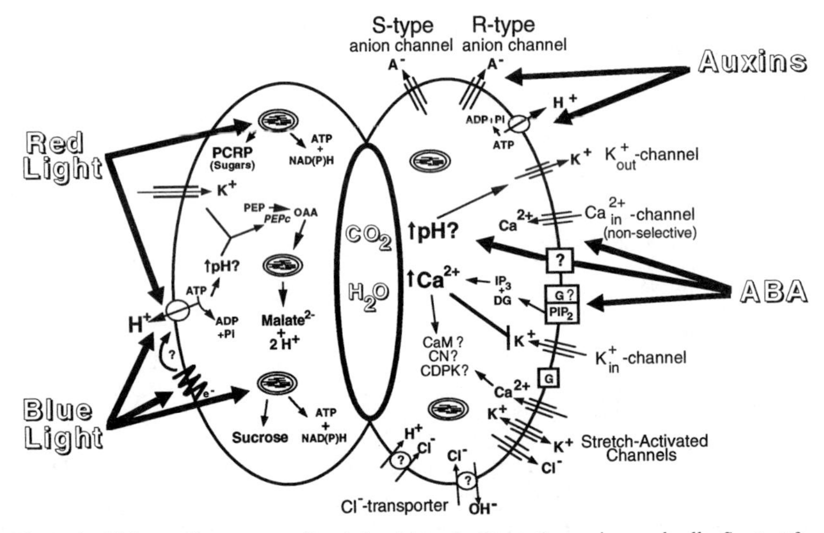

Figure 1 Light- and hormone-mediated signal transduction pathways in guard cells. See text for details. *Arrows* indicate activation, increase, or enhancement. *Blocked arrows* indicate inactivation, decrease, or inhibition. Abbreviations: PCRP = photosynthetic carbon reduction pathway; G = GTP-binding regulatory protein; PIP_2 = phosphatidylinositol 4,5-bisphosphate; IP_3 = inositol 1,4,5-trisphosphate; DG = diacylglycerol; CaM = calmodulin; CN = calcineurin; CDPK = calcium-dependent protein kinase.

air in the substomatal cavity, which loses water vapor through the stomatal pore to the drier external atmosphere. There is either net uptake of CO_2 and loss of O_2 by the plant, during photosynthesis, or the reverse when mitochondrial respiration predominates.

The role of the guard cells is to integrate environmental and endogenous signals into an appropriate stomatal aperture that neither starves the plant for CO_2 by restricting CO_2 influx nor desiccates the plant by allowing excessive water loss. Conditions that evoke a widening of the stomatal pore are typically those associated with an increased photosynthetic demand for CO_2, while conditions that stimulate stomatal closure are those associated with water stress or decreased photosynthetic demand. In intact systems, changes in pore aperture can typically be detected within a few minutes of stimulus application.

Signals are transduced by guard cells into alterations of ion channel and ion pump activity and into modulation of guard cell carbon metabolism. During stomatal opening, increases in intracellular osmotically active solutes drive water uptake down its free energy gradient and cause guard cell swelling. The source of the water and of the ions K^+ and Cl^-, which are taken up by the guard cells, is an extracellular fluid composed of water that contains dissolved gases and solutes including mM concentrations of inorganic nutrients (Atkinson et al 1989; Bowling 1987; Grignon & Sentenac 1991). This fluid is found in the interstices of the cell walls that surround plant cell membranes. These interstices are large enough that they are freely permeable to water and solutes, although some charge interactions do occur between ions and cell wall material (Grignon & Sentenac 1991). The lattice, which comprises the typical plant cell wall, is composed of carbohydrate polymers such as cellulose, small amounts of proteins and, in woody tissue, lignin, a polymer of aromatic alcohols. The guard cell wall is unusual in that its reinforcing cellulose microfibrils are radially distributed around the cell. Because of this arrangement, guard cell swelling and attendant build-up of turgor pressure result in a "bowing out" of the guard cells and an increase in pore aperture. A decrease in the concentrations of intracellular osmotica has the opposite effect, and water efflux results in guard cell deflation and stomatal closure.

In this review, I begin by describing current knowledge regarding guard cell mechanisms for ion transport and for production of organic osmotica. I then describe a widening sphere of intracellular, extracellular, and environmental signals that controls these processes. I focus on developments in our understanding of guard cell physiology gained over the past decade. Most of this knowledge has been obtained from studies on guard cells of fava bean (*Vicia faba*) and dayflower (*Commelina communis*). In these two species epidermes are easily removed. This allows the study of stomatal responses in isolation from non-epidermal influences and readily yields guard cell protoplasts upon enzymatic removal of the plant cell wall, a prerequisite for patch-clamp studies of ion fluxes.

Non-plant biologists reading this chapter will find some familiar players, such as G proteins, Ca^{2+}, and calmodulin, in an unfamiliar setting. One of the important questions in the field of plant signal transduction is the extent to which signal transduction mechanisms will be found to be similar between plants and animals. This question is of scientific interest in illuminating the evolutionary relatedness and antiquity of signal transduction pathways. It is also of practical interest in determining whether it is fruitful to investigate in plants, signal transduction paradigms first elucidated in animal systems, and vice-versa.

ION TRANSPORT MECHANISMS

Stomatal Opening

During stomatal opening, guard cells extrude H^+, take up K^+ and Cl^-, and produce malate^{2-}. Osmotica are stored to a large extent in the vacuole, thus uptake and efflux pathways across the tonoplast are integral to stomatal responses. However, less is known about these pathways (see Raschke et al 1988 for review), and attention is focused here on ion fluxes across the plasma membrane, at the interface between the guard cell and its environment.

Blatt and co-workers have proposed that K^+ uptake across the plasma membrane occurs via an energy-requiring carrier mechanism (Clint & Blatt 1989). Although such a mechanism may be important under some conditions, there is increasing evidence that ion uptake generally proceeds via a chemiosmotic mechanism (Pallas 1971; Zeiger et al 1978), whereby H^+ extrusion hyperpolarizes the membrane potential (V_m) to values on the order of -150 to -250 mV (Lohse & Hedrich 1992; Thiel et al 1992), thus creating an electrical gradient for passive K^+ uptake. These large negative potentials in combination with Cl^-concentrations that are generally greater inside the cell than outside dictate that Cl^- uptake cannot occur passively; the existence of either a Cl^-/OH^- antiport or a Cl^-/H^+ symport is presumed (Assmann & Zeiger 1987).

H^+ EXTRUSION Current data support the role of a plasma membrane P-type ATPase in H^+ extrusion. Signals promoting stomatal opening such as light (Assmann et al 1985; Serrano et al 1988), the phytohormone, auxin (Lohse & Hedrich 1992), and the fungal toxin, fusicoccin (Assmann & Schwartz 1992; Lohse & Hedrich 1992), activate a hyperpolarizing current (Figure

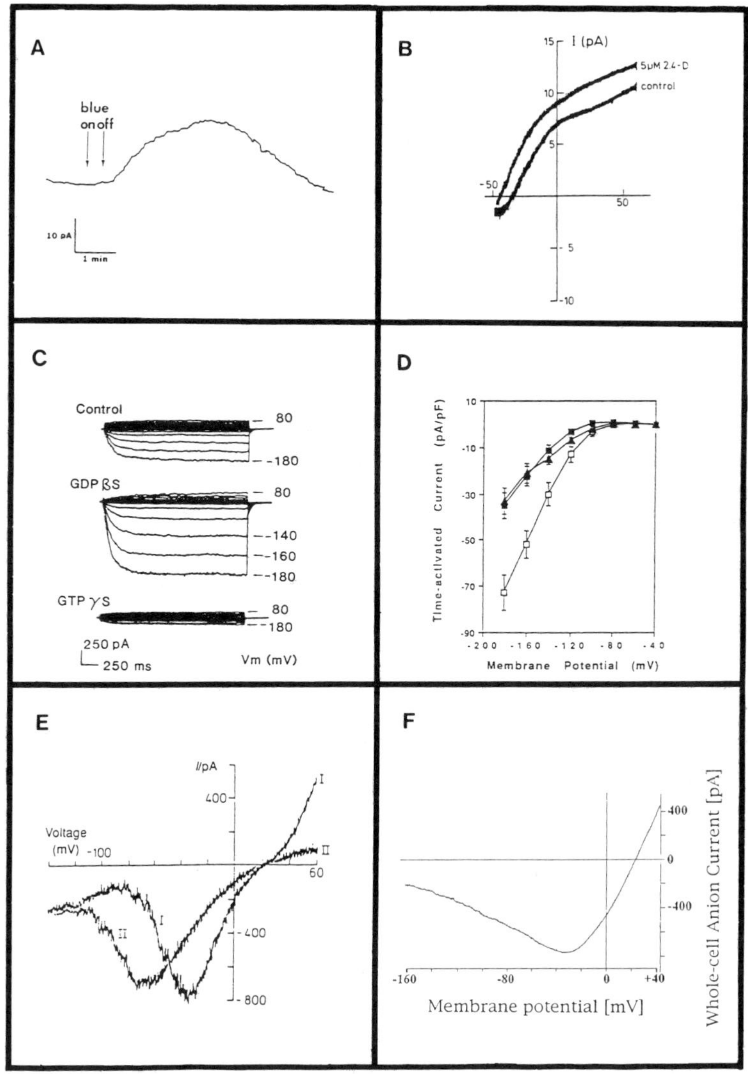

Figure 2 Patch-clamp measurements of ion transport processes at the *Vicia faba* guard cell membrane. *A*) Blue light-stimulated pump current from a guard cell in the slow whole-cell configuration (S.M. Assmann, unpublished observation). Solutions used were the K^+ glutamate solutions described in Assmann & Schwartz (1992). Membrane potential was held at −60 mV; *B*) Stimulation of pump current by the synthetic auxin 2,4-D (Lohse & Hedrich 1992); *C*) Enhancement of inward K^+ current by GDPβS, a G protein inactivator, and reduction by GTPγS, a G protein activator (Fairley-Grenot & Assmann 1991); *D*) Inhibition of inward K^+ current by elevation of Ca_i from 2 nM (*open squares*) to 200 nM (*closed squares*), or by introduction of constitutively active bovine calcineurin into the cytosol (*closed triangles*) (Luan et al 1993); *E*) voltage-dependence of R-type anion current in the absence (I) and presence (II) of 100 μM of the natural auxin IAA (Marten et al 1991); *F*) voltage-dependence of S-type anion current (J.I. Schroeder, unpublished observation). Conditions were essentially those described for Figure 3D of Schroeder & Keller (1992).

2a,b) in patch-clamped *V. faba* protoplasts that is dependent on cytosolic ATP and is curtailed by plasma membrane (P-type) H^+ ATPase inhibitors such as vanadate (Serrano et al 1988; Assmann & Schwartz 1992), which clearly points to the operation of a H^+ pump. H^+ accumulation by inside-out plasma membrane vesicles is similarly inhibited by vanadate (Becker et al 1993). These data oppose the alternative hypothesis that H^+ extrusion occurs via a plasma membrane redox chain (Gautier et al 1992; Raghavendra 1990). On a surface area basis, guard cell protoplasts show twofold higher levels of H^+ ATPase current than mesophyll protoplasts (Lohse & Hedrich 1992), which supports the importance of the plasma membrane H^+ ATPase in guard cell function. One question is whether guard cells possess a unique H^+ ATPase isoform, or perhaps several isoforms, each of which might be activated by a different opening signal. A family of H^+ ATPase genes with differential tissue expression has been identified in higher plants (Sussman & Harper 1989 and references therein), but little is known regarding the guard cell H^+ ATPase(s) beyond the observation that the plasma membrane of *V. faba* guard cells contains polypeptides of 100 and 92 kd that are recognized by antibody to a plasma membrane H^+ ATPase from corn coleoptile (Becker et al 1993).

INWARD K^+ CHANNELS Passive uptake of K^+ down an electrical gradient created by H^+ extrusion is mediated by K^+-selective ion channels (Schroeder et al 1984, 1987; Schroeder 1988; Schroeder & Fang, 1991). These channels are gated by voltage (Schroeder et al 1987) and K^+ (Schroeder & Fang 1991) such that they open at voltages that drive K^+ influx (Figure 2c,d). These inwardly rectifying or inward K^+ channels show significant selectivity for K^+ over Na^+ and other monovalent cations (Schroeder 1988; Schroeder et al 1984). The conductance and density of these channels are sufficient to account for the K^+ uptake observed during stomatal opening (Schroeder & Fang 1991), at least under typical (Bowling 1987) conditions of millimolar extracellular K^+ concentrations. Guard cells also possess stretch-activated (SA) K^+ channels (Cosgrove & Hedrich 1991) that could additionally contribute to K^+ uptake. It will be of interest to determine whether, as for the H^+ATPase, a family of genes encoding K^+ channel polypeptides (Schachtmann et al 1992) will be identified, with expression of particular members restricted to specific cell types or organs.

CARBON METABOLISM During stomatal opening, production of malate^{2-} and $2H^+$ replenishes H^+ lost via H^+ extrusion and provides a counterion for K^+ uptake, thus reducing energetically costly Cl^- uptake. Most of the necessary enzymes for malate^{2-} synthesis from starch have been identified in guard cells (Hedrich et al 1985; Robinson & Preiss 1987; Shimazaki et al 1989). Undetectable (Hedrich et al 1985) to low (Shimazaki et al 1989) levels of fructose 1,6 bisphosphatase are consistent with the recent observation that the six carbon compound glucose-6-phosphate, as well as typical three carbon compounds, are shuttled between the chloroplast and the cytoplasm in guard cells (Overlach et al 1993; Shimazaki et al 1989). Subsequent glycolysis of these starch breakdown products in the cytoplasm produces phosphoenolpyruvate (PEP) (Robinson & Preiss 1987), which is carboxylated to produce oxaloacetic acid (OAA), in a rate-limiting step catalyzed by cytosolic PEP carboxylase (Michalke & Schnabl 1990; Schnabl et al 1982; Schnabl & Kottmeier 1984). OAA reduction to malate^{2-} is catalyzed by cytosolic NAD-malate dehydrogenase, and/or chloroplastic NADP-malate dehydrogenase (Gotow et al 1985b).

Solute increases attributable to K^+ salts do not always account completely for the guard cell osmotic potential observed or predicted in open stomata (MacRobbie 1987 and references therein), which suggests a requirement for production of other solutes. Organic solutes such as sucrose might be imported, produced from starch breakdown, or synthesized from products of the photosynthetic carbon reduction pathway (PCRP) (also known as the Calvin-Benson cycle) (see section on Environmental Signals: Light).

Stomatal Closure

Stomatal closure does not simply require the inhibition or reversal of ion transport processes involved in opening, but rather invokes unique ion transport systems. Given the chemiosmotic theory that K^+ moves passively across the guard cell membrane, the K^+ loss associated with stomatal closure requires membrane depolarization.

H^+ ATPASE If the H^+ ATPase maintains some activity even when apertures are being held at steady state, then inhibition of this H^+ pump would result in membrane depolarization. However, the extent to which the pump contributes to the maintenance of apertures, as opposed to the attainment of such apertures, is a subject on which there is little information.

ANION CHANNELS Because the electrochemical gradient favors passive anion efflux, opening of plasma membrane anion channels is a mechanism for rapid membrane depolarization that could drive K^+ efflux. Several different types of anion channels have been described in guard cells of *V. faba* and cocklebur (*Xanthium strumarium*). The R-type or QUAC (rapid or quick) anion channel activates rapidly in response to depolarization and then inactivates; it also deactivates rapidly in response to hyperpolarization (Hedrich et al 1990; Linder & Raschke 1992; Keller et al 1989; Marten et al 1991; Schroeder & Keller 1992). R-type anion channels are maximally activated at voltages of −30 to −50 mV (Figure 2e). Thus, unless voltage sensitivity of these channels is shifted by closing signals to more negative voltages (see below), they would be expected to exhibit little activity under in vivo conditions. These channels are activated by either ATP or GTP (Hedrich et al 1990); this requirement may help to explain the observation (Karlsson & Schwartz 1988) that stomatal closure requires energy. R-type channels have the permeability sequence: $NO_3^- > Cl^- >$ malate^{2-} (Keller et al 1989); thus they are termed anion channels rather than Cl^- channels.

The second type of anion channel present in *V. faba* guard cells is the S-type (slow) (Schroeder & Hagiwara 1989; Schroeder & Keller 1992), which activates and deactivates slowly in response to voltage and is not voltage-inactivated. A channel with some similarities, dubbed SLAC (slowly-activating anion channel), has been described in *X. strumarium* (Linder & Raschke 1992). Like the R-type, S-type anion channels are not completely selective for Cl^- (Schroeder & Hagiwara 1989). Although S-type channel activity appears maximal around 0 mV (Schroeder & Keller 1992), the channels are active at potentials at least as negative as −160 mV (Schroeder & Keller 1992; Figure 2f). Because guard cells exhibiting pump activity can have membrane potentials negative of −160 mV (Lohse & Hedrich 1992; Thiel et al 1992), it is possible that the process of stomatal closure involves initial depolarization of the membrane by another mechanism (e.g. Ca^{2+} influx, see below), which then shifts membrane potential into the range where S-type and, perhaps eventually, R-type channels are activated.

The third type of anion channel described in guard cells is a SA anion channel, which exhibits decreased closed time as membrane stretch is increased (Cosgrove & Hedrich 1991). This channel is highly selective for Cl^- over K^+; its relative permeability to other anions has not been described. These channels are active at V_ms at least as negative as −100 mV (Cosgrove & Hedrich 1991). Given that these channels are activated by stretch, the question is whether stretch is a sufficient signal for channel activation, or whether channel opening (in vivo) additionally requires superimposition of a stomatal closing signal. Cosgrove & Hedrich (1991) suggest that because these channels show greater ion selectivity than typical SA channels of animal cells, they may not simply function as blow-off valves that prevent turgor pressure from exceeding a particular value, but have a regulated role in stomatal responses.

Ca^{2+} CHANNELS A concentration gradient for Ca^{2+} of several orders of magnitude exists across the cell membrane. Therefore, opening of Ca^{2+} channels will result in Ca^{2+} influx and membrane depolarization under physiological conditions. A SA channel that shows high selectivity for Ca^{2+}

over K^+ has been described (Cosgrove & Hedrich 1991). Opening of this channel is observed at V_ms as negative as −180 mV. The channel exhibits infrequent opening and its conductance is low in comparison to a stretch-insensitive channel with approximately equal permeability to Ca^{2+} and K^+ (Cosgrove & Hedrich 1991; Schroeder & Hagiwara 1990). Fairley-Grenot & Assmann (1992a) observed that the voltage-regulated inward K^+ channel described earlier also has limited Ca^{2+} permeability. Ca^{2+} permeability of this channel, which plays a major role in K^+ uptake and stomatal opening, might at first seem counterintuitive, but exactly because this channel is active at hyperpolarized V_ms, it will be available as a pathway for Ca^{2+} uptake upon reception of a closing signal by open stomata. We speculate that the relative Ca^{2+} permeability of this channel may be increased by signals that initiate stomatal closure.

OUTWARD K^+ CHANNELS A conduit for K^+ loss following depolarization is provided by outwardly rectifying K^+ channels (Blatt 1988; Hosoi et al 1988; Schroeder et al 1984, 1987), regulated by voltage and extracellular K^+ concentrations such that they open under conditions favoring K^+ efflux (Blatt 1988; Schroeder 1988; Schroeder et al 1987; Figure 2c). Guard cells also possess a SA K^+ channel with significant conductance (Cosgrove & Hedrich 1991), which may also contribute to K^+ efflux during stomatal closure.

CARBON METABOLISM Removal of organic osmotica during stomatal closure occurs via glycolysis and subsequent oxidative phosphorylation, by reincorporation into starch, and by loss to the apoplast. Based on enzyme activities, the glycolytic pathway is more active in malate^{2-} breakdown than is the gluconeogenic pathway in starch resynthesis (Hedrich et al 1985; Robinson & Preiss 1987; Schnabl et al 1982). Although loss of malate^{2-} represents a loss of energetically valuable fixed carbon, guard cells have been observed to lose as much as 60% of their malate^{2-} to the external medium upon stomatal closure (Van Kirk & Raschke 1978), presumably through malate^{2-}-permeable anion channels.

LOCAL AND SECOND MESSENGERS

Mechanisms involved in stomatal opening include a hyperpolarizing H^+ ATPase, K^+ channels mediating K^+ influx, a putative Cl^- carrier, and production of organic solutes. Stomatal closure requires membrane depolarization, which can be driven by H^+ pump cessation, anion efflux and/or Ca^{2+} entry, followed by loss of K^+ through outward K^+ channels, and regulated decreases in organic osmotica. Alterations in any of these processes can affect stomatal aperture; such alterations are triggered by alterations in the level or activity of second messengers such as Ca^{2+} and H^+ (pH).

Calcium

Ca_o Exogenous application of Ca^{2+} to the medium bathing isolated epidermes of *C. communis* inhibits opening of closed stomata and stimulates closure of open stomata, while application of the Ca^{2+} chelator, EGTA, has the opposite effects (see MacRobbie 1992; Mansfield et al 1990, for summaries and references). These effects may be mediated directly by extracellular Ca^{2+} concentrations (Ca_o): in patch-clamp experiments elevation of Ca_o from 1 to 10 mM partially blocks influx of K^+ through inward K^+ channels, apparently because Ca^{2+} permeates through the channel more slowly than K^+ (Fairley-Grenot & Assmann 1992a; see also Figure 5 of Marten et al 1991). This effect would tend to slow or reduce stomatal opening. However, the extent to which free Ca_o actually changes in vivo is difficult to assess because of the ability of the plant cell wall to bind Ca^{2+} (see Atkinson et al 1989 for discussion). The effects of exogenous Ca^{2+} application to epidermal peels may also be exerted via resultant alterations in intracellular free Ca^{2+} concentrations (Ca_i) (Gilroy et al 1991). Therefore, it is relevant to discuss what effects Ca_i has on the processes underlying stomatal aperture change, and how guard cells may regulate Ca_i in vivo.

Ca^{2+} EFFECTS ON ION TRANSPORT *Ca^{2+} effects on K^+ channels* In guard cells of *V. faba* and maize (*Zea mays*), elevation of Ca_i in whole-cell patch clamp experiments inhibits inward K^+ channels (Schroeder & Hagiwara 1989; Fairley-Grenot & Assmann 1992b). This effect would tend to prevent or delay K^+ uptake, consistent with Ca^{2+} inhibition of stomatal opening. Elevated Ca_i inhibits inward K^+ channels in isolated membrane patches (Schroeder & Hagiwara 1989), but whether this reflects direct interaction of cytosolic Ca^{2+} with the K^+ channel protein as well as Ca^{2+}-activation of regulatory mechanisms is not known. Recent experiments using immunosuppressants that interfere with Ca^{2+} signaling pathways provide evidence that one mechanism of the Ca^{2+} effect on K^+ channels involves a Ca^{2+}-activated phosphatase (Luan et al 1993). In animal systems, the immunosuppressants, cyclosporin A and FK506, bind to endogenous ligands, cyclophilins, and FKBPs. The resultant complexes specifically bind to and inactivate the Ca^{2+}/calmodulin-dependent phosphatase, calcineurin (Liu et al 1991; Schreiber et al 1991). When cyclosporin A or FK506 plus FKBP12 are administered to the guard cell cytosol in whole-cell patch-clamp experiments, the Ca^{2+} inactivation of inward K^+ channels is prevented. These results correlate well with observation of significant amounts of endogenous cyclophilins A and B in *V. faba* leaf tissue, but only low levels of the requisite FKBP, FKBP12 (Luan et al 1993). Further support for the involvement of a calcineurin-like phosphatase is provided by the observation (Luan et al 1993) that administration of constitutively active (Ca^{2+}/calmodulin-independent) bovine calcineurin to the guard cell cytosol results in K^+ channel inactivation even at low (2 nM) Ca_i (Figure 2d).

Guard cells contain relatively high concentrations of calmodulin (Ling & Assmann 1992), the other component necessary for calcineurin activation. Overlay assays with radioiodinated *V. faba* calmodulin did not identify any protein with the same molecular weight as animal calcineurin, perhaps reflecting either levels of the protein below the sensitivity of the assay, or a different molecular weight for the plant calcineurin-like protein. Several guard cell proteins did show Ca^{2+}-dependent labeling by ^{125}I-calmodulin; they included two protein bands, of M_r 88 and 39 K, unique to guard cells and thus candidates for guard cell-specific signal transduction pathways.

Ca^{2+} effects on anion channels Ca^{2+} can reduce stomatal apertures by the depolarizing effect its uptake has on V_m. However, elevated Ca_i also activates other depolarizing ion transport mechanisms. In patch-clamp experiments, elevated Ca_i activates S-type anion channels (Schroeder & Hagiwara 1989), with anion efflux, membrane depolarization and reduced stomatal apertures being the expected in vivo results. Elevated Ca_i also activates R-type anion channels (Hedrich et al 1990); this effect is observed in membrane patches (Keller et al 1989), which demonstrates that the response is membrane-delimited, i.e. mediated by components within or attached to the membrane.

CELLULAR MECHANISMS REGULATING Ca_i *Ca^{2+} ATPase and Ca^{2+} channels* A plasma membrane Ca^{2+}-extruding ATPase and mechanisms of intracellular Ca^{2+} sequestration (Askerlund & Evans 1992; Schumaker & Sze 1987) presumably lower Ca_i, but little information is available on how these processes function in guard cells. As described earlier, a variety of Ca^{2+}-permeable channels are present in the guard-cell plasma membrane, whose opening will increase Ca_i.

Phosphoinositide metabolism Release of Ca^{2+} from intracellular stores is an alternate mechanism for elevation of Ca_i and can be achieved in animal systems via activation of the phosphoinositide (PI) pathway. In the basic incarnation of the PI pathway (reviewed in Berridge 1987), excitation of a receptor results in activation of a receptor-coupled GTP-binding protein (G protein) or activation of kinase activity (Meldrum 1991). Consequent activation of the enzyme phospholipase C results in the production of diacylglycerol (DG) and inositol 1,4,5-trisphosphate (IP_3) from phosphatidylinositol 4,5-bisphosphate (PIP_2). IP_3 binds to receptors on the endoplasmic reticulum and triggers release of Ca^{2+} to the cytosol. DG increases the affinity of protein kinase C for Ca^{2+}, thereby lowering the effective Ca^{2+} concentration necessary for activation of this enzyme (Bell 1986), which phosphorylates protein targets.

Evidence for Ca^{2+} release mediated by the PI pathway in plants, in general, and guard cells, in particular, is accumulating. Information on the class of G proteins, Gq, that activates phospholipase C (Smrcka et al 1991) is lacking. However, binding of GTP by plant proteins, cross-reactivity of animal G protein antibodies with plant proteins, physiological effects of G protein regulators, and the cloning of an *A. thaliana* gene with significant homology to animal G protein alpha subunits all support a significant role for G proteins in plant signal transduction (see Ma et al 1990; Warpeha et al 1991; Fairley-Grenot & Assmann 1991; Li & Assmann 1993; and references therein). In *V. faba* guard cells, introduction of the G protein activator, GTPγS, to the guard cell cytosol results in the inactivation of inward K^+ currents, analogous to the effects of elevated Ca_i or of calcineurin, while GDPβS, which inactivates G proteins, enhances inward currents (Fairley-Grenot & Assmann 1991; Figure 2c). The efficacy of GTPγS is removed if it is introduced concomitantly with high concentrations of the Ca^{2+} chelator, 1,2-bis(2-aminophenoxy)ethane-N,N,N′,N′-tetraacetic acid (BAPTA) (Fairley-Grenot & Assmann 1991). These data are consistent with G protein activation of a pathway, such as the PI pathway, that elevates Ca_i, initiates a signal cascade that inactivates inward K^+ channels, and thus prevents stomatal opening. However, it should be noted that the patch-clamp data to date are equally consistent with a requirement for physiological Ca_i concentrations in order to observe the G protein effect, without invoking an elevation in Ca_i. Thus the GTPγS effect has not yet been directly linked with activation of the PI pathway.

In animal systems, some isoforms of phospholipase C are activated by G proteins, while others are activated by receptor-mediated phosphorylation (Meldrum et al 1991). Such detailed information is not available in plant systems, but it is known that plants possess both membrane-bound and cytosolic forms of phospholipase C (Coté & Crain 1993 and references therein). Constituent phospholipids of the PI pathway (PI, PIP, and PIP_2) are present in plant cells, as are the inositol phosphates IP, IP_2, and IP_3 (reviewed in Morse et al 1989; Coté & Crain 1993). Electrophysiology on intact *V. faba* guard cells shows that photorelease of caged IP_3 introduced to the cytosol via a microelectrode is associated with an inactivation of inward K^+ currents (Blatt et al 1990), similar to the inactivation of such currents by experimental elevation of Ca_i in patch-clamp experiments. In guard cells of *C. communis* microinjected with caged IP_3, photoactivated

IP_3 release does elevate Ca_i (as assayed with the fluorescent Ca^{2+} indicator, Fluo-3). Elevation of Ca_i to levels greater than 600 nM results in stomatal closure (Gilroy et al 1990). Guard cells are therefore competent to respond to an IP_3 signal via elevation of Ca_i. In other plant systems, and presumably in guard cells as well, the vacuole is a target for IP_3-mediated Ca^{2+} release (Schumaker & Sze 1987), and the chloroplasts and endoplasmic reticulum also represent organellar Ca^{2+} stores.

Ca^{2+} CAVEATS Despite the potency of exogenously applied Ca^{2+} in reducing stomatal apertures, there are also data suggesting that Ca^{2+} may play a role in stomatal opening. It is important to consider these results, since data that initially do not fit a paradigm are often those that are ultimately most important in providing new insights.

The idea that a class of second messengers can have both activating and inactivating effects on a process is already familiar from the G proteins, the classic example being the regulation of adenylate cyclase by both stimulatory and inhibitory G proteins (Gilman 1987). While the patch-clamp data of Fairley-Grenot & Assmann (1991) show GTPγS inhibition of inward K^+ currents in *V. faba,* consistent with inhibition of stomatal opening, Lee et al (1993) observed a small enhancement of stomatal opening after release of caged GTPγS microinjected into *C. communis* guard cells. Microinjected GDPβS had no effect on apertures. The patch-clamp data were obtained under conditions known to result in stomatal closure, whereas the microinjection experiments were performed under conditions designed to promote stomatal opening. Thus it is interesting to speculate that the different environmental conditions in the two types of experiments resulted either in proportionately altered activation of functionally distinct G proteins, or in redirection of the Ca^{2+} signal.

G protein-activated production of IP_3 from PIP_2 is accompanied by production of DG. Yet, whereas exogenous application of IP_3 results in Ca_i elevation and stomatal closure, exogenous application of DGs results in activation of the proton pump and stimulation of stomatal opening (Lee & Assmann 1991). The DG effect is produced only by DGs that are effective as signal transducing molecules in animal systems (Lee & Assmann 1991). One possibility is that exogenous DG may initiate a negative feedback pathway that inhibits phospholipase C activity and thus inhibits IP_3 production and Ca^{2+} release; this phenomenon is commonly observed in animal systems (Meldrum et al 1991 and references therein). Although IP_3 is not produced in vivo without concomitant DG production, the amounts of DG produced from PIP_2 breakdown are minimal compared to the amounts produced from hydrolysis of PIP, PI, or phosphatidylcholine (Nishizuka 1992); this may be especially true in plants, which have low amounts of PIP_2 (Morse et al 1989). Perhaps it is the balance of DG vs IP_3 that determines the balance between opening and closing processes. To resolve this issue, measurements of endogenous levels of IP_3, DG, and Ca_i following stimulation of guard cells with both opening and closing signals are required. Further research is also required on the question of whether plant cells possess C type protein kinases that are activated by DG and Ca^{2+} (Lawton et al 1989).

In the IP_3 experiments of Blatt et al (1990), IP_3 and Ca^{2+} release were also associated with a small decrease in outward K^+ current, an effect that would tend to oppose stomatal closure. In fact, an increase in Ca_i accompanied stomatal opening following application of fusicoccin or auxin to guard cells of the slipper orchid, *Paphiopedilum tonsum* (Irving et al 1992). Ca_i was assayed with acetomethoxy-esterfied forms of Fluo-3, and these results should be confirmed using dextran-linked indicator dyes, which cannot partition into plant organelles.

Further data implicating Ca^{2+} in stomatal opening were obtained by Shimazaki and colleagues (1992), who demonstrated that a variety of calmodulin antagonists with different modes of action inhibit H^+ extrusion by *V. faba* guard cell protoplasts and stomatal opening in *C. benghalensis* epidermal peels. Given the recent discovery that plants possess several different calmodulin genes and proteins (reviewed in Roberts & Harmon 1992), one could speculate that some calmodulin isoforms might be involved in stomatal opening while others might be involved in stomatal closure. Involvement in opening could include activation of the plasma membrane H^+ ATPase or processes of Ca_i regulation. For example, the Ca^{2+}-ATPase of the endoplasmic reticulum is a Ca^{2+}/calmodulin-activated enzyme in other plant systems (Askerlund & Evans 1992). In addition, PEP carboxylase, involved in $malate^{2-}$ production during stomatal opening, may be regulated by Ca^{2+}/calmodulin-stimulated phosphorylation (Echevarria et al 1988).

The story on Ca^{2+} and stomatal apertures is far from complete. While effects of application of Ca^{2+} or its chelators are often attributed to effects on free Ca^{2+} concentrations, the more important factors may not be effects on absolute Ca_i level, but rather disruptive effects of these compounds on spatial and temporal Ca^{2+} gradients within the guard cells (PK Hepler, personal communication). It is possible that Ca^{2+} both potentiates closure and stimulates opening, depending on the magnitude, location, and temporal pattern of Ca_i elevation.

pH

pH_o pH may act as an extracellular signal. Edwards et al (1988) speculate that a wave of extracellular pH (pH_o) change may mediate stomatal responses to stimuli that are transmitted along the leaf. Increasingly alkaline extracellular pH is associated with increasing water stress (Hartung et al 1988), which results in stomatal closure. Therefore, it is interesting that more alkaline pH_o decreases inward K^+ current magnitude in intact *V. faba* guard cells (Blatt 1992). Conversely, acidic pH_o activates inward K^+ channels (Blatt 1992); thus H^+ extrusion during stomatal opening, which can decrease extracellular pH from 7.2 to 5.1 (Edwards et al 1988), may additionally activate the pathway for K^+ influx.

In Blatt's experiments, acidic pH_o slightly decreased outward K^+ current. Patch-clamp experiments by Ilan et al (1993) reveal that pH_o decrease from 8.1 to 5.5 significantly reduces the magnitude of ouward K^+ currents, an effect that would tend to prevent or slow stomatal closure. Single channel data show that this effect can be attributed primarily to a decreased rate of channel activation under acidic pH_o caused by a change in the local electrical field sensed by the gating structures of the channel.

pH_i The extent to which cytosolic pH (pH_i) actually changes in guard cells is still not clear, although one report suggests that pH_i decreases by about 0.2 units during stomatal opening stimulated by fusicoccin or auxin (Irving et al 1992). Blatt has observed that outward K^+ current is blocked by acidic pH_i, imposed by loading with butyric acid (Blatt 1992) and attributes the decrease in outward K^+ current in the experiments with photoreleased IP_3, not to increased Ca_i, as reasoned above, but to an acidification of the cell cytosol upon photolysis of the cage (Blatt 1992; speech to Am. Soc. Plant Physiol.), thus providing an important cautionary note on the use of caged substances. He hypothesizes that stimuli that oppose stomatal closure may do so via cytosolic acidification, while closing stimuli may act via cytoplasmic alkalinization. However, it is also postulated that cytosolic alkalinization activates PEP carboxylase (see below), thus alkaline pH_i would stimulate $malate^{2-}$ production and drive stomatal opening, an effect opposite to that proposed by Blatt. Clearly, more data on pH_i measurements in responding guard cells are required.

ENDOGENOUS SIGNALS

Messengers that can be produced within the leaf or translocated from a distant site of synthesis include the plant hormones, as well as (other) xylem and/or phloem mobile substances. Here the focus is on the hormones abscisic acid (ABA) and auxin. Additional information is available in Assmann (1993), and the reader is also directed to the burgeoning literature on root-shoot communication (reviewed in Davies & Zhang 1991).

Abscisic Acid

ABA is produced in response to stress, particularly drought stress (Davies & Zhang 1991). ABA potently inhibits stomatal opening and a promotes stomatal closure. Because exogenous Ca^{2+} application has similar effects, an ABA-Ca^{2+} link has been eagerly sought (see summaries in MacRobbie 1991, 1992).

ABA AND Ca_i ABA-stimulation of Ca^{2+} uptake could cause membrane depolarization, required for stomatal closure and sufficient for inhibition of stomatal opening. To date, however, an obligate link between ABA application and Ca^{2+} uptake has not been demonstrated. Although Ca^{2+} channel blockers reduce the effectiveness of ABA in inhibiting stomatal opening in *C. communis* epidermal peels (DeSilva et al 1985), Fitzsimons & Weyers (1987) found that external Ca^{2+} was not required for ABA-induced volume regulation of guard cell protoplasts from this species. MacRobbie also observed no consistent effect of ABA application on $^{45}Ca^{2+}$ influx, with the reservation that small transients might have gone undetected (MacRobbie 1989). To date, the results of patch-clamp experiments show that in one third of the *V. faba* guard cells tested, ABA induces an inward current which, based on reversal potential and measurements of Ca_i with fura-2, can be attributed to Ca^{2+} uptake via a channel with permeability to both Ca^{2+} and K^+ (Schroeder & Hagiwara 1990). The absence of Ca_i elevation in two thirds of the population studied contrasts with uniform stomatal closure typically observed when epidermal peels of *C. communis* are incubated in ABA-containing solution (e.g. DeSilva et al 1985).

Under conditions where ABA does not stimulate Ca^{2+} uptake, it might still increase Ca_i by promoting Ca^{2+} release from intracellular stores. In a direct approach to the question of an ABA-PI pathway link, Y Lee and colleagues (personal communication) report a doubling in IP_3 levels 10 sec after application of ABA in the presence of Li^+ to isolated *V. faba* guard cells. MacRobbie and colleagues (MacRobbie 1992; Parmar & Brearley 1993) also present data indicative of ABA-stimulated turnover of inositol phospholipids and inositol phosphates. These important reports indicate that guard cells can utilize endogenous IP_3 in ABA-mediated signal transduction.

However, despite strong evidence that ABA can elevate Ca_i, Ca^{2+}-imaging studies, which should detect Ca_i elevated either by influx or by release from intracellular stores, show a variable ABA-Ca_i relationship. In four separate studies, an elevation of Ca_i following ABA application was observed in 80% (McAinsh et al 1990), 80% (McAinsh et al 1992), $\leq$ 25% (Gilroy et al 1991), and 70% (Irving et al 1992) of trials, whereas stomata uniformly closed. One caveat is that the absence of a definitive ABA-Ca_i relationship in these studies may merely reflect current technological limits on the spatial and temporal resolution of Ca_i imaging in plants.

ABA AND K^+ CHANNELS An alternate approach to the ABA-Ca_i question is to determine whether the effects of ABA application on ion transport mechanisms are the same as effects resulting from Ca^{2+} application. ABA inhibits inward K^+ currents in intact *V. faba* guard cells (Blatt 1990; Thiel et al 1992), an effect similar to that of elevated Ca_i, and one that could contribute to ABA inhibition of stomatal opening. Recent whole-cell patch-clamp studies by Lemtiri-chlieh, working in the laboratory of MacRobbie, show that ABA inhibition of inward K^+ currents in *V. faba* protoplasts is eliminated when high concentrations of the Ca^{2+} chelator, EGTA, are included in the patch pipette solution (Lemtiri-chlieh 1992). These results demonstrate a Ca_i requirement for the ABA effect on inward K^+ channels, but they are not quite proof that ABA elevates Ca_i, as opposed to merely requiring physiological Ca_i levels to be effective.

For outward K^+ channels, the scenario is different. Outward K^+ current is enhanced by ABA, consistent with ABA promotion of stomatal closure (Blatt 1990). However, Lemtiri-chlieh's patch-clamp experiments show that this ABA effect is not eliminated when Ca^{2+} is chelated (Lemtiri-chlieh 1992), consistent with Schroeder & Hagiwara's (1989) observation that increasing Ca_i from 0.1 to 1.5 μM did not alter outward K^+ currents in patch-clamp experiments. Blatt has described data (cited in Chasan & Schroeder 1992) suggesting that ABA elevates pH_i which, as described earlier, may enhance outward K^+ current by removing H^+ inhibition of the channels.

ABA AND ANION CHANNELS R-type anion channels are not directly regulated by ABA (Marten et al 1991), but to the extent that ABA elevates Ca_i, it will activate both R-type and S-type anion channels, which results in membrane depolarization and promotion of stomatal closure. In fact, membrane depolarization is an early electrical response following ABA application (Thiel et al 1992). However, because R-type anion channels do not appear to be active at large negative V_ms, they are not good candidates for initiating the depolarization in response to ABA. One candidate for the channel mediating the initial depolarization is the non-selective Ca^{2+} channel, which is not, however, ABA-activated in all trials (Schroeder & Hagiwara 1990). Other candidates are the inward K^+ channels, which have some Ca^{2+} permeability, and the SA Cl^- and Ca^{2+} channels. The SA Ca^{2+} channel and the S-type anion channels are active at voltages at least as negative as -160 mV (Cosgrove & Hedrich 1991; Schroeder & Keller 1992), and it would be interesting to investigate the ABA sensitivities of these channels. Finally, the possibility that ABA causes membrane depolarization by inhibiting H^+ ATPase activity has not been eliminated.

To summarize, there is good evidence that in some guard cells ABA stimulates Ca^{2+} influx via Ca^{2+} permeable channels. Perhaps ABA-stimulated Ca^{2+} influx potentiates further Ca^{2+} release via the PI pathway, as has been observed in animal systems (e.g. Iino & Endo 1992). Ca^{2+} influx and associated membrane depolarization activate anion channels that continue the depolarization. K^+ influx is curtailed by this membrane depolarization and/or by Ca_i inhibition of inward K^+ channels. Membrane depolarization provides a driving force for K^+ efflux. ABA, through a Ca_i-independent mechanism enhances this outward K^+ current. An alternative scenario, which does not invoke any Ca^{2+} involvement, would be that ABA stimulates a Ca^{2+}-independent anion channel (Blatt 1991) (the SA channel?); this results in membrane depolarization that prevents K^+ influx and promotes K^+ efflux through ABA-activated outward K^+ channels.

Only further research will allow us to distinguish between two possibilities. The first possibility is that guard cells possess both Ca^{2+}-dependent and Ca^{2+}-independent mechanisms of ABA-induced membrane depolarization. Such redundancy could reflect the vital role that ABA-induced stomatal closure plays in plant survival during drought stress. The second possibility is that in cases to date where ABA-stimulation of Ca^{2+} channel activity and/or of Ca_i increases have not been observed, this merely reflects current technological limitations and/or altered physiology of in vitro systems.

Auxins

Depending on the type of auxin and the auxin concentration, either stomatal opening or stomatal closure may be elicited (reviewed in Assmann 1993; Davies & Mansfield 1987). Indole-3-acetic acid (IAA) is a natural auxin and 1-naphtylacetic acid (1-NAA) and 2,4-dichlorophenoxyacetic acid (2,4-D) are synthetic auxins. While IAA concentrations from 10 μM to 1 mM stimulate stomatal opening (Levitt et al 1987), maximal stimulation of opening with 1-NAA occurs at 5 μM, and actual inhibition of opening occurs at 0.5 mM (Marten et al 1991). Preliminary data indicate that auxins do not affect the activity of either inward or outward K^+ channels in guard cells (Marten et al 1991). In other cell types, auxins stimulate the plasma membrane H^+ ATPase. Perhaps auxins that evoke stomatal opening are those that are more effective in stimulating the guard cell proton pump. To date, the only data available on auxin effects on the guard cell proton pump are those of Lohse & Hedrich (1992), which show that 5 μM 2,4-D stimulates pump current by a few pA (Figure 2b). Just as for the first measurement of a light-stimulated electrogenic current across the guard cell plasma membrane, this current is small, but experimental refinements will probably allow detection of larger effects (cf Figure 2a).

Closing effects stimulated by auxin may be associated with auxin regulation of the R-type anion channels. At 100 μM concentrations, the peak activation potential of these channels is shifted from −30 mV to −67 mV by 1-NAA, to −64 mV by 2,4-D, and to −51 mV by IAA (Marten et al 1991). If under in vivo conditions auxins function to shift these channels into a range where they are activated by the prevailing V_m, consequent anion efflux would drive stomatal closure. Thus I hypothesize that for a given auxin type and concentration, depending on whether the pump-activating or the anion channel-regulating properties of the auxin predominate, either stomatal opening or stomatal closure results.

In addition, in intact guard cells of *P. tonsum,* auxins have been reported to increase Ca_i and decrease pH_i (Irving et al 1992). Such alterations may have additional regulatory effects on channels in vivo that may not be observed in patch-clamp experiments, where these ions are often strongly buffered. In patch-clamp experiments, auxins modulate anion channel activity only when applied on the extracellular face of the plasma membrane, and they are effective in the isolated patch configuration, which implies a membrane-delimited signal transduction pathway (Marten et al 1991). It will be of interest to determine whether the R-type anion channel is a member of the recently described family of auxin-binding proteins, which has both intracellular and membrane-associated members (Barbier-Brygoo et al 1991; Hicks et al 1989; Inohara et al 1989). In keeping with this hypothesis, a 60-kd plant polypeptide that binds to IAA-23, also cross-reacts with polyclonal antibodies against kidney anion channels (Marten et al 1992).

ENVIRONMENTAL SIGNALS

Environmental signals with effects on stomatal aperture include water availability, light, CO_2, humidity, temperature, and pollutants. Although water stress undoubtedly also acts via mechanisms in addition to those of ABA and humidity, space limitations preclude further discussion of this complex topic (see Zeiger et al 1987a, for reviews). The following section discusses guard cell responses to light, CO_2, and humidity. Effects of temperature and pollutants have been reviewed elsewhere (Jarvis & Mansfield 1981; Zeiger 1983).

Light

PHOTORECEPTORS Red and blue light stimulate stomatal opening. Much evidence reviewed in detail elsewhere (Sharkey & Ogawa 1987; Zeiger 1983 1990) implicates chlorophyll as the pigment mediating red light-stimulated stomatal opening. Although guard cells contain the red/far-red absorbing pigment phytochrome (Saunders et al 1983), its role appears limited to involvement in the circadian rhythmicity of stomatal responses (Holmes & Klein 1985; Karlsson 1988; Deitzer & Frosch 1990).

Chlorophyll absorbs blue as well as red light, and thus mediates some of the stomatal opening stimulated by blue wavelengths. However, there is much evidence (see Zeiger 1983, 1990; Sharkey & Ogawa 1987; Zeiger et al 1987b for reviews of this literature) for the presence in guard cells of a blue light photoreceptor distinct from chlorophyll. The most definitive evidence is that the action spectrum for light-stimulated stomatal opening does not parallel the absorption efficiency of chlorophyll in the blue region of the spectrum (Hsiao et al 1973; Sharkey & Raschke 1981).

As with most blue light responses of higher plants, it is debated whether the stomatal blue light response is mediated by a flavin or a carotenoid. Plants treated with an inhibitor of carotenoid synthesis, norflurazon, still exhibit a stomatal response to blue light, which suggests that the response may be flavin-mediated (Karlsson et al 1983). However, Karlsson et al also isolated carotenoids from guard cell protoplasts and remarked that some carotenoids exhibited an absorption spectrum similar to the action spectrum for blue light-stimulated stomatal opening (Karlsson 1986; Karlsson et al 1992). Additionally, dithiothreitol (DTT) inhibits formation of the carotenoid zeaxanthin in guard cells; DTT also inhibits stomatal opening stimulated by blue light (Srivastava & Zeiger 1993). Srivastava & Zeiger (1992) report that blue light, but not red light, induces a phase of fast fluorescence quenching from isolated guard cell chloroplasts of *V. faba* and suggest that this may be a manifestation of the specific signal-transducing carotenoid.

A pulse of blue light given under anoxia does not immediately cause stomatal opening, but opening does occur once anoxic conditions are removed. This intriguing "memorization" of the blue light signal may assist in separating the initial phototransduction events from secondary responses associated with the mechanics of stomatal opening and aid in photoreceptor identification (Vavasseur et al 1990).

LIGHT AND ION TRANSPORT *H^+ extrusion* Both red and blue light stimulate an ATP-dependent outward current from guard cell protoplasts patch-clamped in the whole cell configuration (Assmann et al 1985; Serrano et al 1988). The currents are inhibited by H^+ ATPase inhibitors such as vanadate (Serrano et al 1988) and diethylstilbestrol (Assmann 1986); in addition, blue light-stimulated guard cell swelling, which is accompanied by net H^+ extrusion (Shimazaki et al 1986), is inhibited by vanadate (Amodeo et al 1992). These data indicate that light is activating a plasma membrane H^+ ATPase; whether the ATPase is identical for both red and blue light is not known. Although initial light-stimulated currents from guard cells were small, use of the slow whole-cell configuration (Lindau & Fernandez 1986), which better preserves cytosolic integrity, allows detection of blue light-stimulated currents of 10–12 pA (see Figure 2a and Schroeder 1988), large enough to hyperpolarize a membrane with a resistance of 10–20 GΩA by -100 to -240 mV.

In vivo, chloroplast electron transport and photophosphorylation

(Shimazaki & Zeiger 1985, and references therein), driven by both red and blue light, presumably provide ATP (Shimazaki et al 1989) to drive the H^+ pump. However, this is not the only role of light, since both blue and red light activate pump current even when excess ATP is provided to the guard cell cytosol in patch-clamp experiments (Assmann et al 1985; Serrano et al 1988). A few clues, but no definitive pictures, of intervening signal transduction processes are available. Based on the effects of DGs described earlier, DG might be a component of either the chlorophyll-mediated or the blue light-specific signal transduction chain. Red light-stimulated current is dependent on the provision of orthophosphate (P_i) in the patch pipette solution (Serrano et al 1988), which perhaps reflects pump activation in vivo via phosphorylation or via phosphorylated products, possibly including triose phosphates produced during photosynthetic carbon metabolism (see below).

A blue light-stimulated, vanadate-insensitive redox system has been reported in the plasma membrane of guard cells from several species, based on reduction of extracellular ferricyanide or tetrazolium (Gautier et al 1992; Raghavendra 1990; Vani & Raghavendra 1989, 1992). Pantoja & Willmer suggest that much of the apparent redox activity of *C. communis* protoplasts can be attributed to peroxidase activity (Pantoja & Willmer 1988, 1991), but Vani & Raghavendra contest this conclusion, at least in pea (*Pisum sativum*) (Vani & Raghavendra 1992). One possibility is that blue light-stimulated electron transport regulates the H^+ ATPase without being directly responsible for the majority of H^+ extrusion.

LIGHT AND MALATE^{2-} SYNTHESIS Guard cell malate^{2-} accumulation is greatly potentiated in the presence of light if KCl is present (Ogawa et al 1978). Poffenroth et al (1992) hypothesize that in the absence of K^+ and presence of light, malate^{2-} is still produced, but is metabolized in the mitochondria, thus providing energy for cytosolic sucrose synthesis.

Since PEP carboxylase in guard cells is not directly light regulated (Gotow et al 1985b; Michalke & Schnabl 1990), other mechanisms for the light stimulation of malate^{2-} accumulation have been hypothesized: (*a*) Schnabl & Kottmeier (1984) demonstrated that PEP carboxylase activity isolated from guard cell protoplasts of *V. faba* was stimulated by K^+. Light-stimulated H^+ ATPase activity drives K^+ uptake and may thereby activate the enzyme. (*b*) Light-stimulated H^+ extrusion may increase pH_i, activating PEP carboxylase, which has an alkaline pH optimum (Schnabl & Kottmeier 1984). (*c*) Hedrich et al (1985) report that the presence of (white) light significantly increases fructose 2,6 bisphosphate levels in guard cells. Increased concentrations of this regulatory metabolite would promote glycolysis and thus malate^{2-} production, as well as reduce gluconeogenesis

(Hedrich et al 1985). These mechanisms of light-stimulated $malate^{2-}$ formation are not mutually exclusive.

By unexplained mechanisms, blue light stimulates approximately eightfold higher rates of $malate^{2-}$ synthesis than does red light (Ogawa et al 1978). The chloroplastic NADP-malate dehydrogenase is activated by white light (Gotow et al 1985b), but the wavelength dependence of this phenomenon is not known. Alternatively, blue light could be more effective in activating the H^+ ATPase and thus initiating the first two control mechanisms described above. There is also a synergistic effect of red and blue light on $malate^{2-}$ formation (Ogawa et al 1978). One possibility is that the blue light photoreceptor, once activated, is further activated by red light (Iino et al 1985). Alternatively, the apparent differential effects of blue vs red light on PCRP activity (see below) might alter the availability of substrates for $malate^{2-}$ synthesis, although no definitive data are available.

LIGHT AND SUGAR PRODUCTION Light may also contribute to solute build-up via production of sugars from PCRP products. Ribulose-1,5-bisphosphate carboxylase/oxygenase (rubisco), the initial enzyme in the PCRP, and the most abundant protein on Earth, catalyses the formation of two molecules of the three carbon compound 3-phosphoglyceric acid (PGA) from CO_2, plus the five carbon compound ribulose 1,5-bisphosphate (RuBP). Rubisco has been localized to guard cells by immunological techniques and by the carboxyarabinitol bisphosphate assay (Reckmann et al 1990; Zemel & Gepstein 1985). Fluorescence of chlorophyll a in intact guard cells shows sensitivities to CO_2 and O_2 that would be predicted from the carboxylase/oxygenase properties of rubisco (Cardon & Berry 1992). In addition, Shimazaki et al report the presence of both rubisco and other PCRP enzymes in guard cells (Shimazaki 1989; Shimazaki et al 1989). Demonstration that isolated guard cell chloroplasts exhibit CO_2-dependent O_2 evolution in response to light (Wu & Assmann 1993) is also evidence for a functioning PCRP. Guard cell protoplasts fix ^{14}C into PGA, sugar phosphates, and triose phosphates, and contain RuBP, thus they contain expected metabolites of the PCRP (Gotow et al 1988).

Current debate revolves around the question of whether rubisco activity actually contributes significantly to the osmotic build-up associated with stomatal opening. Reckmann et al calculate from their data, and those of Gotow et al that for the rate of stomatal opening of *P. sativum* observed under 1,500 $\mu mol\ m^{-2}\ s^{-1}$ white light, rubisco would only contribute about 2% of the solute requirement for stomatal opening, whereas PEP carboxylase activity could contribute 100%. Outlaw (1989) also concludes that rubisco activity, if present, is insignificant.

Countering these conclusions, Zeiger's and Tallman's groups propose that signals, particularly light quality and K^+ availability, modulate PCRP activity in guard cells (Gotow et al 1985a, 1988; Shimazaki & Zeiger 1987; Poffenroth et al 1992; Tallman & Zeiger 1988). Poffenroth et al (1992) measured concentrations of soluble sugars in *V. faba* epidermal peels, prepared so that guard cells were the only intact cell type. A 2 hr illumination with red light in the presence of KCl resulted in sugar production that was completely inhibited by 3-(3,4-dichlorophenyl)-1,1-dimethylurea (DCMU), an inhibitor of noncyclic photosynthetic electron transport. In addition, under identical conditions, starch degradation was not detected by a staining technique (Tallman & Zeiger 1988). Within the limited sensitivity of this technique (Poffenroth et al 1992), these results suggest that the sugars produced were not derived from starch breakdown. Rates of solute production calculated from these data are six times higher than the rates calculated by Reckmann et al, and account for 12% of the necessary osmoticum if the osmotic requirements given by Reckmann et al are used. Under blue light minus KCl, DCMU-sensitive sugar production could account for about 10% of the requirements and DCMU-insensitive sugar production for another 10%. Under blue light plus KCl, none of the sugar production was DCMU-sensitive, although sugar production, presumably from DCMU-insensitive starch breakdown, occurred at a greater rate than in the absence of KCl. Poffenroth et al postulate that blue light is less effective in driving photosynthetic sugar production because P_i requirements of starch-degrading reactions activated specifically by blue light impose a P_i limitation on photosynthesis. This hypothesis would explain the observation that blue light, but not green light, suppresses fine structure in the fluorescence transient of chloroplasts of intact guard cells (Mawson & Zeiger 1991) that is associated, although not obligatorily (Reckmann et al 1990), with photosynthetic CO_2 reduction. The blue light suppression can be removed by provision of exogenous P_i (Mawson & Zeiger 1991).

Two conclusions can be drawn. First, percent osmotic contribution by the PCRP may be higher if different calculations for osmotic requirements are used (see Table III of Poffenroth et al 1992), but calculations still highlight the greater contributions of ion uptake and malate^{2-} synthesis during light-stimulated stomatal opening. Second, it appears that estimates of PCRP contributions are lower when rubisco activity, as opposed to sugar production, is measured. It is possible either that rubisco activity is higher in planta, or that DCMU-sensitivity is not an adequate criterion for involvement of the PCRP in sugar production.

Carbon Dioxide

Although exceptions occur under non-physiological conditions, elevated carbon dioxide typically stimulates stomatal closure, while lowered CO_2 concentrations have the opposite effect (see Mansfield et al 1990; Morison 1987 for review). Mott (1988) has elegantly shown in the intact leaf that it is the intercellular concentration of CO_2 (c_i), and not the external concentration, or the concentration within the pore, to which the guard cells are responding. Since c_i is influenced both by mesophyll photosynthesis and by ambient CO_2 concentrations (c_a), CO_2 can be thought of as both an endogenous and an exogenous signal. Recent global change predictions, as well as measurements indicating that c_a does change significantly throughout the day and in a gradient from soil to forest canopy (e.g. Bazzaz & Williams 1991) emphasize the importance of CO_2 as an exogenous signal.

The mechanism of the CO_2 response is unknown. Effects of physiologically relevant CO_2 concentrations on ion transport have simply not been studied. CO_2 stimulation of the PCRP might, at least under some conditions, divert NADPH away from malate^{2-} production and ATP away from pump activation. However, if as reasoned above, the PCRP provides alternate osmotica to drive stomatal opening, then mere detection of such diversion is not sufficient to explain why elevated CO_2 inhibits stomatal opening. In addition, CO_2 concentrations modulate stomatal apertures even in prolonged darkness where photosynthesis does not function.

Humidity

Reduction in ambient humidity causes stomatal closure. Of all the stomatal responses, the humidity response is the most mysterious. One challenge has been simply to define which humidity parameter guard cells actually sense. At a given temperature, a decrease in ambient relative humidity (RH = the actual amount of water vapor held by the surrounding atmosphere relative to the maximum amount it can hold) also increases vapor pressure difference (VPD). VPD is the difference in water vapor pressure between the intercellular air spaces (where air is assumed to be saturated with water vapor) and some reference point outside the leaf, usually the leaf surface.

Initially, by simple Ohm's law reasoning, an increase in driving force (VPD) results in increased transpirational water loss (E):

$$E_{total} = E_{stomata} + E_{cuticle} = VPD/R_{stomata} + VPD/R_{cuticle},$$

where R is the resistance to vapor flux. Ensuing stomatal closure response increases $R_{stomata}$, thus decreasing E_{total}. $R_{cuticle}$ is sufficiently high that the contribution of the second term to E_{total} can be ignored unless stomata are essentially closed. In some cases, stomatal closure increases $R_{stomata}$ so much that E_{total} actually decreases with increasing VPD, a phenomenon known as a "feedforward" response (Farquhar 1978) Thus the question is, are guard cells sensing RH, which is related to the chemical activity of water in the atmosphere, VPD, which is the gradient in vapor pressure from leaf to atmosphere, or E, which is the actual rate of water loss from the leaf? (In feedforward situations the E sensed would have to be $E_{cuticle}$ rather than $E_{stomata}$).

Ball et al (1987) modeled stomatal responses and showed an excellent empirical correlation between $1/R_{stomata}$ and RH at the leaf surface. However, experiments in which temperature was used to keep VPD constant at different RHs showed that VPD was the better predictor of stomatal response (Assmann & Grantz 1990; Aphalo & Jarvis 1991). These correlative studies are of interest, but fall short of defining causal relationships.

Most recently, Mott & Parkhurst (1991) studied stomatal responses in normal air and helox (79:21 He:O_2, with appropriate additions of H_2O and CO_2). In helox, molecular diffusion occurs 2.3 times faster than in air, thus E is 2.33 times greater for the same VPD as in normal air. Therefore, comparison of stomatal responses in air and helox allows separation of the VPD and E variables under non-feed-forward conditions. They found that stomata responded to E and not to VPD. In the case of feed-forward responses, where $E_{stomata}$ is not monotonic with $R_{stomata}$, they invoke the involvement of $E_{cuticle}$ as an effector of stomatal response.

Mott & Parkhurst's experiments seem to clearly define E as the important signal. They further conclude that stomatal responses to humidity simply involve decreased water availability leading to decreased guard cell turgor. However, this conclusion does not take into consideration the substantial body of literature (reviewed in Assmann & Gershenson 1991; Grantz 1990) that guard cells do respond metabolically to humidity, e.g. by losing K^+. Guard cells may be responding, at least in part, to regulatory metabolites such as ABA that are carried in the transpiration stream, as proposed by Grantz and others (Grantz & Schwartz 1988; Grantz 1990 and references therein).

CONCLUDING REMARKS

With the possible exception of auxins, reliable predictions can be made as to whether the signals discussed in this review will increase or diminish stomatal apertures. ABA promotes stomatal closure by directly or indirectly regulating K^+, anion, and Ca^{2+} channels. Auxins stimulate stomatal opening by activating the H^+ pump; they also shift the voltage sensitivity of R-type anion channels, although the impact of this effect on stomatal aperture is less clear. Red and blue light stimulate stomatal opening by activating a H^+ ATPase and by variably stimulating malate^{2-} synthesis, sucrose synthesis, and sugar production from the PCRP.

Yet despite this wealth of knowledge, the identities of all signal receptors in guard cells, except chlorophyll, remain unknown. Such identification will be one of the primary goals of future research. Receptor identification will also help to link signals with proposed second messenger pathways involving G proteins, calmodulin, kinases/phosphatases, Ca_i, and pH_i. Although current data connect Ca_i and pH_i most closely with the ABA response, future research may uncover primary involvement of these second messengers in transduction of other regulatory signals.

This review has dealt with each signal individually, but guard cells instantaneously and continuously integrate all of the signals described here, with the potential for interaction providing a new order of complexity. We have evidence of these interactions, but little information on underlying mechanisms. For example, why is it that red light enhances the extent of stomatal opening stimulated by blue light (Ogawa et al 1978; Karlsson 1988; Zeiger 1990)? Why is it that ABA and auxins alter CO_2 sensitivity of stomata (Davies & Mansfield 1987)?

Finally, although guard cells are uniquely situated at the interface between the plant interior and the aerial environment, they are not unique among plant cells in their capacity to sense hormones, light, and CO_2. Research on signaling systems of other cell types will help to determine which signal transduction pathways are guard cell-specific and which are variations on common themes in plant cell biology, or cell biology in general.

ACKNOWLEDGMENTS

I would like to thank the United States Department of Agriculture and the National Science Foundation for research support, and the members of my laboratory for their many helpful comments on this manuscript. I would also especially like to thank my fellow New England plant biologists at Harvard University, University of Massachusetts, and University of Connecticut, who have provided scientific inspiration and personal encouragement. The secretarial assistance of Ms. Lisa Ameden Claussen in the preparation of this manuscript is gratefully acknowledged.

Literature Cited

Amodeo, G, Srivastava, A, Zeiger, E. 1992. Vanadate inhibits blue light-stimulated swelling of *Vicia* guard cell protoplasts. *Plant Physiol.* 100:1567–70

Aphalo, PJ, Jarvis, PG. 1991. Do stomata respond to relative humidity? *Plant, Cell Environ.* 14:127–32

Askerlund, P, Evans, DE. 1992. Reconstitution and characterization of a calmodulin-stimulated Ca^{2+}-pumping ATPase purified from *Brassica oleracea* L. *Plant Physiol.* 100:1670–81

Assmann, SM. 1986. *Stomatal responses to light and carbon dioxide*. PhD thesis. Stanford Univ., Calif. 196 pp.

Assmann, SM. 1993. Hormonal regulation of ion channels and ion pumps: the guard cell system. In *Biochemistry and Molecular Biology of Plant Hormones*, ed. KR Libbenga, MA Hall. Amsterdam: Elsevier. In press

Assmann, SM, Gershenson, A. 1991. The kinetics of stomatal responses to VPD in *Vicia faba:* electrophysiological and water relations models. *Plant, Cell, Environ.* 14: 455–65

Assmann, SM, Grantz, DA. 1990. Stomatal response to humidity in sugarcane and soybean: effect of vapour pressure difference on the kinetics of the blue light response. *Plant, Cell Environ.* 13:163–69

Assmann, SM, Schwartz, A. 1992. Synergistic effect of light and fusicoccin on stomatal opening: epidermal peel and patch clamp experiments. *Plant Physiol.* 98:1349–55

Assmann, SM, Simoncini, L, Schroeder, JI. 1985. Blue light activates electrogenic ion pumping in guard cell protoplasts of *Vicia faba*. *Nature* 318:285–87

Assmann, SM, Zeiger, E. 1987. Guard cell bioenergetics. In *Stomatal Function,* ed. E Zeiger, GD Farquhar, IR Cowan, pp. 163–94. Stanford: Stanford Univ. Press

Atkinson, CJ, Mansfield, TA, Kean, AM, Davies, WJ. 1989. Control of stomatal aperture by calcium in isolated epidermal tissue and whole leaves of *Commelina communis* L. *New Phytol.* 111:9–17

Ball, JT, Woodrow, IE, Berry, JA. 1987. A model predicting stomatal conductance and its contribution to the control of photosynthesis under different environmental conditions. In *Progress in Photosynthesis Research,* ed. J Biggens, pp. 221–24. The Netherlands: Nijhoff

Barbier-Brygoo, H, Ephritikhine, G, Klämbt, D, Maurel, C, Palme, K, et al. 1991. Perception of the auxin signal at the plasma membrane of tobacco mesophyll protoplasts. *Plant J.* 1:83–93

Bazzaz, FA, Williams, WE. 1991. Atmospheric CO_2 concentrations within a mixed forest: Implications for seedling growth. *Ecology* 72:12–16

Becker, D, Zeilinger, C, Lohse, G, Depta, H, Hedrich, R. 1993. Identification and biochemical characterization of the plasma membrane H^+-ATPase in guard cells of *Vicia faba L. Planta.* 190:44–50

Bell, RM. 1986. Protein kinase C activation by diacylglycerol second messengers. *Cell* 45:631–32

Berridge, MJ. 1987. Inositol trisphosphate and diacylglycerol: two interacting second messengers. *Annu. Rev. Biochem.* 56:159–93

Blatt, MR. 1988. Potassium-dependent bipolar gating of K^+ channels in guard cells. *J. Membr. Biol.* 102:235–46

Blatt, MR. 1990. Potassium channel currents in intact stomatal guard cells: rapid enhancement by abscisic acid. *Planta* 180: 445–55

Blatt, MR. 1991. Ion channel gating in plants: Physiological implications and integration for stomatal function. *J. Membr. Biol.* 124: 95–112

Blatt, MR. 1992. K^+ channels of stomatal guard cells: Characteristics of the inward rectifier and its control by pH. *J. Gen. Physiol.* 99:615–44

Blatt, MR, Thiel, G, Trentham, DR. 1990. Reversible inactivation of K^+ channels of *Vicia* stomatal guard cells following the photolysis of caged inositol 1,4,5-trisphosphate. *Nature* 346:766–68

Bowling, DJF. 1987. Measurement of the apoplastic activity of K^+ and Cl^- in the leaf epidermis of *Commelina communis* in relation to stomatal activity. *J. Exp. Bot.* 38:1351–55

Cardon, ZG, Berry, J. 1992. Effects Of O_2 and CO_2 concentration on the steady-state fluorescence yield of single guard cell pairs in intact leaf discs of *Tradescantia albiflora. Plant Physiol.* 99:1238–44

Chasan, R, Schroeder, JI. 1992. Excitation in plant membrane biology. *Plant Cell* 4: 1180–88

Clint, GM, Blatt, MR. 1989. Mechanisms of fusicoccin action: evidence for concerted modulations of secondary K^+ transport in a higher plant cell. *Planta* 178:495–508

Cosgrove, DJ, Hedrich, R. 1991. Stretch-activated chloride, potassium, and calcium channels coexisting in plasma membranes of guard cells of *Vicia faba* L. *Planta* 186: 143–53

Coté, GC, Crain, RC. 1993. Biochemistry of phosphoinositides. *Annu. Rev. Plant Physiol. Plant Mol. Biol.* 44:333–56

Davies, WJ, Mansfield, TA. 1987 Auxins and stomata. See Zeiger et al 1987a, pp. 293–310

Davies, WJ, Zhang, J. 1991. Root signals and the regulation of growth and development of plants in drying soil. *Annu. Rev. Plant Physiol. Plant Mol. Biol.* 42:55–76

Deitzer, GF, Frosch, SH. 1990. Multiple action of far-red light in photoperiodic induction and circadian rhythmicity. *Photochem. Photobiol.* 52:173–79

DeSilva, DLR, Cox, RC, Hetherington, AM, Mansfield, TA. 1985. Suggested involvement of calcium and calmodulin in the responses of stomata to abscisic acid. *New Phytol.* 101:555–63

Echevarria, C, Vidal, J, Le Maréchal, P, Brulfert, J, Ranjeva, R, Gadal, P. 1988. The phosphorylation of sorghum leaf phosphoenolpyruvate carboxylase is a Ca^{2+}-calmodulin dependent process. *Biochem. Biophys. Res. Commun.* 155:835–40

Edwards, MC, Smith, GN, Bowling, DJF. 1988. Guard cells extrude protons prior to stomatal opening—A study using fluorescence microscopy and pH microelectrodes. *J. Exp. Bot.* 39:1541–47

Fairley-Grenot, K, Assmann, SM. 1991. Evidence for G-protein regulation of inward K^+ channel current in guard cells of fava bean. *Plant Cell* 3:1037–44

Fairley-Grenot, KA, Assmann, SM. 1992a. Permeation of Ca^{2+} through K^+ channels in the plasma membrane of *Vicia faba* guard cells. *J. Membr. Biol.* 128:103–13

Fairley-Grenot, KA, Assmann, SM. 1992b. Whole-cell K^+ current across the plasma membrane of guard cells from a grass: *Zea mays. Planta* 186:282–93

Farquhar, GD. 1978. Feedforward responses

of stomata to humidity. *Aust. J. Plant Physiol.* 5:787–800
Fitzsimons, PJ, Weyers, JDB. 1987. Responses of *Commelina communis* L. guard cell protoplasts to abscisic acid. *J. Exp. Bot.* 38:992–1001
Gautier, H, Vavasseur, A, Lascève, G, Boudet, AM. 1992. Redox processes in the blue light response of guard cell protoplasts of *Commelina communis* L. *Plant Physiol.* 98:34–38
Gilman, AG. 1987. G Proteins: Transducers of receptor- generated signals. *Annu. Rev. Biochem.* 56:615–49
Gilroy, S, Fricker, MD, Read, ND, Trewavas, AJ. 1991. Role of calcium in signal transduction of *Commelina* guard cells. *Plant Cell* 3:333–44
Gilroy, S, Read, ND, Trewavas, AJ. 1990. Elevation of cytoplasmic calcium by caged calcium or caged inositol trisphosphate initiates stomatal closure. *Nature* 346:769–71
Gotow, K, Sakaki, T, Kondo, N, Kabayashi, K, Syōno, K. 1985a. Light-induced alkalinization of the suspending medium of guard cell protoplasts from *Vicia faba* L. *Plant Physiol.* 79:825–28
Gotow, K, Tanaka, K, Kondo, N, Kobayashi, K, Syōno, K. 1985b. Light activation of NADP-malate dehydrogenase in guard cell protoplasts from *Vicia faba* L. *Plant Physiol.* 79:829–32
Gotow, K, Taylor, S, Zeiger, E. 1988. Photosynthetic carbon fixation in guard cell protoplasts of *Vicia faba* L. *Plant Physiol.* 86:700–5
Grantz, DA. 1990. Plant response to atmospheric humidity. *Plant, Cell Environ.* 13: 667–79
Grantz, DA, Schwartz, A. 1988. Guard cells of *Commelina communis* L. do not respond metabolically to osmotic stress in isolated epidermis: implications for stomatal responses to drought and humidity. *Planta* 174:166–73
Grignon, C, Sentenac, H. 1991. pH and ionic conditions in the apoplast. *Annu. Rev. Plant Physiol. Plant Mol. Biol.* 42:103–28
Hartung, W, Radin, JW, Hendrix, DL. 1988. Abscisic acid movement into the apoplastic solution of water-stressed cotton leaves. *Plant Physiol.* 86:908–13
Hedrich, R, Busch, H, Raschke, K. 1990. Ca^{2+} and nucleotide dependent regulation of voltage dependent anion channels in the plasma membrane of guard cells. *EMBO J.* 9:3889–92
Hedrich, R, Raschke, K, Stitt, M. 1985. A role for fructose 2,6-bisphosphate in regulating carbohydrate metabolism in guard cells. *Plant Physiol.* 79:977–82
Hicks, GR, Rayle, DL, Jones, AM, Lomax, TL. 1989. Specific photoaffinity labeling of two plasma membrane polypeptides with an azido auxin. *Proc. Natl. Acad. Sci. USA* 86:4948–52
Holmes, MG, Klein, WH. 1985. Evidence for phytochrome involvement in light-mediated stomatal movement in *Phaseolus vulgaris* L. *Planta* 166:348–53
Hosoi, S, Iino, M, Shimazaki, K-I. 1988. Outward- rectifying K^+ channels in stomatal guard cell protoplasts. *Plant Cell Physiol.* 29:907–11
Hsiao, TC, Allaway, WG, Evans, LT. 1973. Action spectra for guard cell Rb^+ uptake and stomatal opening in *Vicia faba. Plant Physiol.* 51:82–88
Iino, M, Endo, M. 1992. Calcium-dependent immediate feedback control of inositol 1,4,5-trisphosphate-induced Ca^{2+} release. *Nature* 360:76–78
Iino, M, Ogawa, T, Zeiger, E. 1985. Kinetic properties of the blue-light response of stomata. *Proc. Natl. Acad. Sci. USA* 82:8019–23
Ilan, N, Schwartz, A, Moran, N. 1993. External pH effects on depolarization-dependent K channels in guard cell protoplasts of *Vicia faba. J. Gen. Physiol.* Submitted
Inohara, N, Shimomura, S, Fukui, T, Futai, M. 1989. Auxin-binding protein located in the endoplasmic reticulum of maize shoots: Molecular cloning and complete primary structure. *Proc. Natl. Acad. Sci. USA* 86: 3564–68
Irving, HR, Gehring, CA, Parish, RW. 1992. Changes in cytosolic pH and calcium of guard cells precede stomatal movements. *Proc. Natl. Acad. Sci. USA* 89:1790–94
Jarvis, PJ, Mansfield, TA, eds. 1981 *Stomatal Physiology*. Cambridge: Cambridge Univ. Press. 295 pp.
Karlsson, PE. 1986. Blue light regulation of stomata in wheat seedlings. II. Action spectrum and search for action dichroism. *Physiol. Plant.* 66:207–10
Karlsson, PE. 1988. Phytochrome is not involved in the red-light-enhancement of the stomatal blue-light-response in wheat seedlings. *Physiol. Plant.* 74:544–48
Karlsson, PE, Bogomolni, RA, Zeiger, E. 1992. HPLC of pigments from guard cell protoplasts and mesophyll tissue of *Vicia faba* L. *Photochem. Photobiol.* 55:605–10
Karlsson, PE, Hoglund, H-O, Klockare, R. 1983. Blue light induces stomatal transpiration in wheat seedlings with chlorophyll deficiency caused by SAN 9789. *Physiol. Plant.* 57:417–21
Karlsson, PE, Schwartz, A. 1988. Characterization of the effects of metabolic inhibitors, ATPase inhibitors and a potassium-channel blocker on stomatal opening and closing in isolated epidermis of *Com-*

melina communis L. *Plant, Cell Environ.* 11:165–72

Keller, BU, Hedrich, R, Raschke, K. 1989. Voltage-dependent anion channels in the plasma membrane guard cells. *Nature* 341: 450–53

Lawton, MA, Yamamoto, RT, Hanks, SK, Lamb, CL. 1989. Molecular cloning of plant transcripts encoding protein kinase homologs. *Proc. Natl. Acad. Sci. USA* 86: 3140–44

Lee, H, Tucker, EB, Crain, RC, Lee, Y. 1993. Stomatal opening is induced in epidermal peels of *Commelina communis* by GTP analogues or pertussis toxin. *Plant Physiol* 102:95–100

Lee, Y, Assmann, SM. 1991. Diacylglycerols induce both ion pumping in patch-clamped guard cell protoplasts and opening of intact stomata. *Proc. Natl. Acad. Sci. USA* 88: 2127–31

Lemtiri-chlieh, F. 1992. Is extracellular calcium required for the action of abscisic acid on stomatal guard cells? *Ninth Int. Workshop Plant Membr. Biol. Monterey, Calif.*

Levitt, LK, Stein, DB, Rubenstein, B. 1987. Promotion of stomatal opening by indoleacetic acid and ethrel in epidermal strips of *Vicia faba* L. *Plant Physiol.* 85:318–21

Li, W, Assmann, SM. 1993. Characterization of a G-protein-regulated outward K^+ current in mesophyll cells of *Vicia faba* L. *Proc. Natl. Acad. Sci. USA* 90:262–66

Lindau, M, Fernandez, JM. 1986. IgE-mediated degranulation of mast cells does not require opening of ion channels. *Nature* 319:150–53

Linder, B, Raschke, K. 1992. A slow anion channel in guard cells, activating at large hyperpolarization, may be principal for stomatal closing. *FEBS Lett.* 313:27–30

Ling, V, Assmann, SM. 1992. Cellular distribution of calmodulin and calmodulin-binding proteins in *Vicia faba* L. *Plant Physiol.* 100:970–78

Liu, J, Farmer, JD Jr, Lane, WS, Friedman, J, Weissman, I, Schreiber, SL. 1991. Calcineurin is a common target of cyclophilin-cyclosporin A and FKBP-FK506 complexes. *Cell* 66:807–15

Lohse, G, Hedrich, R. 1992. Characterization of the plasma-membrane H^+-ATPase from *Vicia faba* guard cells. *Planta* 188:206–14

Luan, S, Li, W, Rusnak, F, Assmann, SM, Schreiber, SL. 1993. Immunosuppressants implicate protein phosphatase regulation of K^+ channels in guard cells. *Proc. Natl. Acad. Sci. USA*. 90:2202–6

Ma, H, Yanofsky, MF, Meyerowitz, EM. 1990. Molecular cloning and characterization of *GPA1*, a G protein α subunit gene from *Arabidopsis thaliana*. *Proc. Natl. Acad. Sci. USA* 87:3821–25

MacRobbie, EAC. 1987 Ionic relations of guard cells. See Zeiger et al 1987a, pp. 125–62

MacRobbie, EAC. 1989. Calcium influx at the plasmalemma of isolated guard cells of *Commelina communis:* Effects of abscisic acid. *Planta* 178:231–41

MacRobbie, EAC. 1991. Effect of ABA on ion transport and stomatal regulation. In *Abscisic Acid: Physiology and Biochemistry,* ed. WJ Davies, HG Jones, pp. 153–68. Oxford: BIOS Scientific

MacRobbie, EAC. 1992. Calcium and ABA-induced stomatal closure. *Philos. Trans. R. Soc. London Ser. B* 338:5–18

Mansfield, TA, Hetherington, AM, Atkinson, CJ. 1990. Some current aspects of stomatal physiology. *Annu. Rev. Plant Physiol. Plant Mol. Biol.* 41:55–75

Marten, I, Lohse, G, Hedrich, R. 1991. Plant growth hormones control voltage-dependent activity of anion channels in plasma membrane of guard cells. *Nature* 353:758–62

Marten, I, Zeilinger, C, Redhead, C, Landry, DW, Al-Awqati, Q, Hedrich, R. 1992. Identification and modulation of a voltage-dependent anion channel in the plasma membrane of guard cells by high-affinity ligands. *EMBO J.* 10:3569–75

Mawson, BT, Zeiger, E. 1991. Blue-light modulation of chlorophyll a fluorescence transients in guard cell chloroplasts. *Plant Physiol.* 96:753–60

McAinsh, MR, Brownlee, C, Hetherington, AM. 1990. Abscisic acid-induced elevation of guard cell cytosolic Ca^{2+} precedes stomatal closure. *Nature* 343:186–88

McAinsh, MR, Brownlee, C, Hetherington, AM. 1992. Visualizing changes in cytosolic-free Ca^{2+} during the response of stomatal guard cells to abscisic acid. *Plant Cell* 4:1113–22

Meldrum, E, Parker, PJ, Carozzi, A. 1991. The PtdIns-PLC superfamily and signal transduction. *Biochim. Biophys. Acta* 1092:49–71

Michalke, B, Schnabl, H. 1990. Modulation of the activity of phosphoenolypyruvate carboxylase during potassium-induced swelling of guard-cell protoplasts of *Vicia faba* L. after light and dark treatments. *Planta* 180: 188–93

Morison, JIL. 1987 Intracellular CO_2 concentration and stomatal response to CO_2. See Zeiger et al 1987a, pp. 229–51

Morse, MJ, Satter, RL, Crain, RC, Coté, GG. 1989. Signal transduction and phosphatidylinositol turnover in plants. *Physiol. Plant.* 76:118–21

Mott, KA. 1988. Do stomata respond to CO_2 concentrations other than intracellular? *Plant Physiol.* 86:200–3

Mott, KA, Parkhurst, DF. 1991. Stomatal

responses to humdidity in air and helox. *Plant, Cell Environ.* 14:509–19
Nishizuka, Y. 1992. Intracellular signaling by hydrolysis of phospholipids and activation of protein kinase C. *Science* 258:607–14
Ogawa, T, Ishikawa, H, Shimada, K, Shibata, K. 1978. Synergistic action of red and blue light and action spectra for malate formation in guard cells of *Vicia faba* L. *Planta* 142: 61–65
Outlaw, WH Jr. 1989. Critical examination of the quantitative evidence for and against photosynthetic CO_2 fixation by guard cells. *Physiol. Plant.* 77:275–81
Overlach, S, Diekmann, W, Raschke, K. 1993. Phosphate translocator of isolated guard-cell chloroplasts from *Pisum sativum* L. transports glucose-6-phosphate. *Plant Physiol.* 101:1201–4
Pallas, JE Jr. 1971. Induced guard cell activity of *Vicia faba* and apparent changes in neutral red uptake. *Bull. Ga. Acad. Sci.* 29: 209–28
Pantoja, O, Willmer, CM. 1988. Redox activity and peroxidase activity associated with the plasma membrane of guard-cell protoplasts. *Planta* 174:44–50
Pantoja, O, Willmer, CM. 1991. Ferricyanide reduction by guard cell protoplasts. *J. Exp. Bot.* 42:323–29
Parmar, PN, Brearley, CA. 1993. Identification of 3- and 4-phosphorylated phosphoinositides and inositol phosphates in stomatal guard cells. *Plant J.* In press
Poffenroth, M, Green, DB, Tallman, G. 1992. Sugar concentrations in guard cells of *Vicia faba* illuminated with red or blue light: Analysis by high performance liquid chromatography. *Plant Physiol.* 98:1460–71
Raghavendra, AS. 1990. Blue light effects on stomata are mediated by the guard cell plasma membrane redox system distinct from the proton translocating ATPase. *Plant, Cell Environ.* 13:105–10
Raschke, K, Hedrich, R, Reckmann, U, Schroeder, JI. 1988. Exploring biophysical and biochemical components of the osmotic motor that drives stomatal movements. *Bot. Acta* 101:283–94
Reckmann, U, Scheibe, R, Raschke, K. 1990. Rubisco activity in guard cells compared with the solute requirement for stomatal opening. *Plant Physiol.* 92:246–53
Roberts, DM, Harmon, AC. 1992. Calcium-modulated proteins: targets of intracellular calcium signals in higher plants. *Annu. Rev. Plant Physiol. Plant Mol. Biol.* 43:375–414
Robinson, NL, Preiss, J. 1987. Localization of carbohydrate metabolizing enzymes in guard cells of *Commelina communis*. *Plant Physiol.* 85:360–64
Saunders, MJ, Cordonnier, M-M, Palevitz, BA, Pratt, LH. 1983. Immunofluorescence visualization of phytochrome in *Pisum sativum* L. epicotyls using monoclonal antibodies. *Planta* 159:545–53
Schachtman, DP, Schroeder, JI, Lucas, WJ, Anderson, JA, Gaber, RF. 1992. Expression of inward-rectifying potassium channel by the *Arabidopsis KAT1* cDNA. *Science* 258:1654–58
Schnabl, H, Elbert, C, Krämer, G. 1982. The regulation of the starch-malate balances during volume changes of guard cell protoplasts. *J. Exp. Bot.* 33:996–1003
Schnabl, H, Kottmeier, C. 1984. Properties of phosphoenolpyruvate carboxylase in desalted extracts from isolated guard-cell protoplasts. *Planta* 162:220–25
Schreiber, SL, Liu, J, Albers, MW, Karmacharya, R, Koh, E, et al 1991. Immunophilin-ligand complexes as probes of intracellular signaling pathways. *Transplant. Proc.* 23:2839–44
Schroeder, JI. 1988. K^+ transport properties of K^+ channels in the plasma membrane of *Vicia faba* guard cells. *J. Gen. Physiol.* 92:667–83
Schroeder, JI, Fang, HH. 1991. Inward-rectifying K^+ channels in guard cells provide a mechanism for low-affinity K^+ uptake. *Proc. Natl. Acad. Sci. USA* 88:11583–87
Schroeder, JI, Hagiwara, S. 1989. Cytosolic calcium regulates ion channels in the plasma membrane of *Vicia faba* guard cells. *Nature* 338:427–30
Schroeder, JI, Hagiwara, S. 1990. Repetitive increases in cytosolic Ca^{2+} of guard cells by abscisic acid activation of nonselective Ca^{2+} permeable channels. *Proc. Natl. Acad. Sci. USA* 87:9305–09
Schroeder, JI, Hedrich, R, Fernandez, JM. 1984. Potassium selective single channels in guard cell protoplasts of *Vicia faba*. *Nature* 312:361–62
Schroeder, JI, Keller, BU. 1992. Two types of anion channel currents in guard cells with distinct voltage regulation. *Proc. Natl. Acad. Sci. USA* 89:5025–29
Schroeder, JI, Raschke, K, Neher, E. 1987. Voltage dependence of K^+ channels in guard cell protoplasts. *Proc. Natl. Acad. Sci. USA* 84:4108–12
Schumaker, KS, Sze, H. 1987. Inositol 1,4,5-trisphosphate releases Ca^{2+} from vacuolar membrane vesicles of oat roots. *J. Biol. Chem.* 262:3944–46
Serrano, EE, Zeiger, E, Hagiwara, S. 1988. Red light stimulates an electrogenic ion pump in *Vicia* guard cell protoplasts. *Proc. Natl. Acad. Sci. USA* 85:436–40
Sharkey, TD, Ogawa, T. 1987 Stomatal responses to light. See Zeiger et al 1987a, pp. 195–208
Sharkey, TD, Raschke, K. 1981. Effect of light quality on stomatal opening in leaves

of *Xanthium strumarium* L. *Plant Physiol.* 68:1170–74
Shimazaki, K-I, Iino, M, Zeiger, E. 1986. Blue light-dependent proton extrusion by guard cell protoplasts of *Vicia faba*. *Nature* 319:324–26
Shimazaki, K-I, Zeiger, E. 1985. Cyclic and noncyclic photophosphorylation in isolated guard cell chloroplasts from *Vicia faba* L. *Plant Physiol.* 78:211–14
Shimazaki, K-I. 1989. Ribulosebisphosphate carboxylase activity and photosynthetic O_2 evolution rate in *Vicia* guard-cell protoplasts. *Plant Physiol.* 91:459–63
Shimazaki, K-I, Kinoshita, T, Nishimura, M. 1992. Involvement of calmodulin and calmodulin-dependent myosin light chain kinase in blue light-dependent H^+ pumping by guard cell protoplasts from *Vicia faba* L. *Plant Physiol.* 99:1416–21
Shimazaki, K-I, Terada, J, Tanaka, K, Kondo, N. 1989. Calvin-Benson cycle enzymes in guard-cell protoplasts from *Vicia faba* L.: Implications for the greater utilization of phosphoglycerate/dihydroxyacetone phosphate shuttle between chloroplasts and the cytosol. *Plant Physiol.* 90:1057–64
Shimazaki, K-I, Zeiger, E. 1987. Red light-dependent CO_2 uptake and oxygen evolution in guard cell protoplasts of *Vicia faba* L.: Evidence for photosynthetic CO_2 fixation.*Plant Physiol.* 84:7–9
Smrcka, AV, Hepler, JR, Brown, KO, Sternweis, PC. 1991. Regulation of polyphosphoinositide-specific phospholipase C activity by purified G_q. *Science* 251:804–7
Srivastava, A, Zeiger, E. 1993. A role of zeaxanthin in blue light photoreception of guard cells. *Plant Physiol.* 102s:14
Srivastava, A, Zeiger, E. 1992. Fast fluorescence quenching from isolated guard cell chloroplasts of *Vicia faba* is induced by blue light and not by red light. *Plant Physiol.* 100:1562–66
Sussman, MR, Harper, JF. 1989. Molecular biology of the plasma membrane of higher plants. *Plant Cell* 1:953–60
Tallman, G, Zeiger, E. 1988. Light quality and osmoregulation in *Vicia* guard cells: Evidence for involvement of three metabolic pathways. *Plant Physiol.* 88:887–95
Thiel, G, MacRobbie, EAC, Blatt, MR. 1992. Membrane transport in stomatal guard cells: the importance of voltage control. *J. Membr. Biol.* 126:1–18
Vani, T, Raghavendra, AS. 1989. Tetrazolium reduction by guard cells in abaxial epidermis of *Vicia faba:* blue light stimulation of a plasmalemma redox system. *Plant Physiol.* 90:59–62
Vani, T, Raghavendra, AS. 1992. Plasma membrane redox system in guard cell protoplasts of pea (*Pisum sativum* L.). *J. Exp. Bot.* 43:291–97
Van Kirk, CA, Raschke, K. 1978. Release of malate from epidermal strips during stomatal closure. *Plant Physiol.* 61:474–75
Vavasseur, A, Lascève, G, Couchat, P. 1990. Different stomatal responses of maize leaves after blue or red illumination under anoxia. *Plant, Cell Environ.* 13:389–94
Warpeha, KMF, Hamm, HE, Rasenick, MM, Kaufman, LS. 1991. A blue-light-activated GTP-binding protein in the plasma membranes of etiolated peas. *Proc. Natl. Acad. Sci. USA* 88:8925–29
Wu, W, Assmann, SM. 1993. Photosynthesis by guard cell chloroplasts of *Vicia faba* L: Effects of factors associated with stomatal movements. *Plant Cell Physiol.* In press
Zeiger, E. 1983. The biology of stomatal guard cells. *Annu. Rev. Plant. Physiol.* 34:441–75
Zeiger, E. 1990. Light perception in guard cells. *Plant, Cell Environ.* 13:739–47
Zeiger, E, Bloom, AJ, Hepler, PK. 1978. Ion transport in stomatal guard cells: A chemiosmotic hypothesis. *What's New Plant Physiol.* 9:29–32
Zeiger, E, Farquhar, GD, Cowan, IR, eds. 1987a. *Stomatal Function*. Stanford: Stanford Univ. Press. 503 pp.
Zeiger, E, Iino, M, Shimazaki, K-I, Ogawa, T. 1987b. The blue-light response of stomata: Mechanism and function. See Zeiger et al 1987a, pp. 209–228
Zemel, E, Gepstein, S. 1985. Immunological evidence for the presence of ribulose bisphosphate carboxylase in guard cell chloroplasts. *Plant Physiol.* 78:586–90

Annu. Rev. Cell Biol. 1993. 9:377–410

STRUCTURE AND FUNCTION OF THE B CELL ANTIGEN RECEPTOR

Anthony L. DeFranco
Departments of Microbiology and Immunology, and of Biochemistry and Biophysics, and the George W. Hooper Foundation, University of California, San Francisco, California 94143-0552

KEY WORDS: receptor, signaling, endocytosis, tyrosine kinases

CONTENTS

INTRODUCTION

During the development of B lymphocytes, the immunoglobulin (Ig)[1] genes are rearranged in a sequential order that results in the formation of a single functional gene encoding Ig heavy chain and a single functional gene

[1]Abbreviations used: Ig, immunoglobulin; ER, endoplasmic reticulum; ARH1, antigen receptor homology 1; PI, phosphatidylinositol; PIP_2, phosphatidylinositol 4,5-bisphosphate; BiP, heavy chain binding protein; GAP, GTP-ase activating protein of $p21^{ras}$; ZAP70, ζ-associated protein, 70 kd; MHC, major histocompatability complex; CR2, complement receptor 2; PLC, phospholipase C; PKC protein kinase C; MAP kinase, mitogen-activated protein kinase; AP-1, activator protein 1; TRE, tetradecanoyl phorbol acetate-response element; SRF, serum response factor.

0743–4634/93/1115–0377$05.00

encoding Ig light chain. Each B cell then makes a unique Ig molecule that can be expressed either as a secreted antibody, or as a membrane Ig. The membrane Ig is a component of the antigen receptor of B cells, which functions in two distinct ways. Firstly, it serves as an efficient receptor that takes up the antigen by endocytosis, thereby allowing it to be processed and presented to helper T cells. Secondly, the antigen receptor transduces a signal that plays an important role in regulating B cell behavior.

The nature of the response of B cells to antigen depends upon the differentiation state of the B cell and on the nature of additional signals that are being delivered to the B cell by helper T cells. In immature B cells, the predominant result of antigen contact with the antigen receptor is inactivation of the cell, either through rapid programmed cell death (clonal deletion) or through entry into an unresponsive state (clonal anergy) (Goodnow 1992). Tests carried out with several different mice carrying Ig transgenes have shown that membrane forms of antigen induce B cell deletion, whereas soluble forms of antigen induce B cell anergy (Hartley et al 1991; Russell et al 1991). In contrast, antigen contact with mature B cells can lead to proliferation of the B cell, particularly in the presence of cytokines such as interleukin 4 (O'Garra et al 1988). Antibody production by antigen-stimulated mature B cells also requires differentiative signals provided by other cytokines.

In a distinct type of mature B cell, the germinal center B cell, antigen contact is necessary to prevent programmed cell death (Liu et al 1989). These cells undergo rapid mutagenesis of their immunoglobulin genes, and therefore the requirement of antigen contact for survival is probably a mechanism for purging cells with Ig gene mutations that weaken binding. Finally, the Ig heavy chain plays a critical role in B cell development (Kitamura et al 1991). Thus the antigen receptor of B lymphocytes plays a critical role in the regulation of B cell development, tolerance, activation, and survival.

STRUCTURE OF THE B CELL ANTIGEN RECEPTOR

Each of the five isotypes of immunoglobulins (IgM, IgD, IgG, IgA, and IgE) can exist as a membrane-bound form in which alternative splicing adds a hydrophobic transmembrane domain and a short cytoplasmic domain to the C-terminus of the classical immunoglobulin heavy chain (Wall & Kuehl 1983). In each case, the membrane Ig is thought to exist primarily in the canonical two heavy chain plus two light chain immunoglobulin structure. Higher order structures characteristic of secreted IgM and IgA are absent in the membrane forms. The antigen receptor of B lymphocytes is a multi-chain complex between membrane Ig and one or more heterodimers

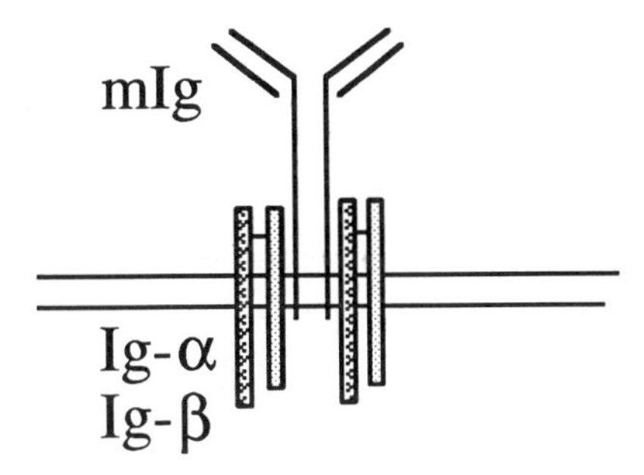

Figure 1 Structure of the B cell antigen receptor. The antigen receptor consists of membrane Ig (mIg), with its two transmembrane heavy chains and two light chains, bound in a complex with one or more disulfide-linked heterodimers of Ig-α and Ig-β. Two heterodimers are shown, although the stoichiometry has not been determined.

formed from two accessory proteins called Ig-α and Ig-β (Reth 1992). [A small fraction of the Ig-β molecules are present as slightly lower apparent molecular weight forms called Ig-γ (KS Campbell et al 1990)]. The exact composition of the antigen receptor complex is not known, although the presence of two identical heavy chains has led to the proposal that two Ig-α/Ig-β heterodimers are associated with one immunoglobulin unit (Figure 1).

The membrane forms of IgM and IgD, the forms present in the antigen receptors of primary B cells, have only three amino acid residues in their putative cytoplasmic domains (lysine-valine-lysine-COOH, in each case) (Reth 1992). Because the antigen receptor is capable of inducing vigorous signaling reactions, for a long time it was unclear how it interacted with cytoplasmic signaling components. Replacement of the transmembrane and cytoplasmic domains of μ_m, the heavy chain of membrane IgM, with the equivalent domains of other proteins results in chimeric molecules that fail to signal (Parikh et al 1991; Dubois et al 1992; JH Blum et al, submitted). In contrast, mutation of the cytoplasmic domain to other positively charged amino acids does not decrease signaling function (Shaw et al 1990) (JH Blum et al, submitted), which suggests that the particular transmembrane domain in μ_m is critical. Introduction of a stop codon at the beginning of the cytoplasmic domain abrogates signaling function (Shaw et al 1990) and also decreases cell surface expression (JH Blum, et al, submitted). This mutant molecule is processed to remove the transmembrane domain and to add a glycosyl-phosphatidylinositol tail (Mitchell et al 1991), which reinforces the notion that the transmembrane domain is functionally important to the signaling function of membrane Ig.

The transmembrane domains of the different membrane Ig isotypes are each more hydrophilic than the typical transmembrane domain. For example, 10 of the 25 or 26 amino acid residues of the membrane IgM transmembrane domain are serine, threonine, or tyrosine residues. It seems likely that these hydrophilic residues are involved in interactions with transmembrane regions of one or more other proteins (e.g. Ig-α and Ig-β). In addition, if the transmembrane regions of the various heavy chain isotypes are put into an α-helix, the most likely structure for a membrane crossing of this length, then a well-conserved patch of homology between each of the isoforms can be seen on one side of the helix (Reth 1992), which supports the view that this region is interacting with the transmembrane domain of another protein or proteins.

While the structures of the membrane Ig isotypes have been known for over ten years, only recently has the Ig-α/Ig-β heterodimer component of the antigen receptor been discovered. Reth and colleagues observed that introduction of a μ_m-expression vector into the J558 IgA-secreting B cell line failed to yield membrane IgM expression on the cell surface (Hombach et al 1988b). Typical membrane IgM molecules did assemble, but they were unable to exit the endoplasmic reticulum (ER). Using fluorescence-based cell sorting, they were able to isolate a variant cell line that did express membrane IgM on the cell surface. Immunoprecipitation of membrane IgM from the parental and variant cell line revealed co-precipitating proteins present in the variant cells, but not in the parental cells (Hombach et al 1988a). These proteins are now called Ig-α and Ig-β and are present in a disulfide linked heterodimer. Amino acid sequencing of purified Ig-α and Ig-β has shown that these proteins are the products of the *mb-1* and *B29* genes (Hombach et al 1990a; Campbell et al 1991a,b), each of which was previously isolated by subtractive hybridization approaches as mRNAs expressed in B cells and not in other cell types (Sakaguchi et al 1988; Hermanson et al 1988). Each of these genes encodes transmembrane polypeptides containing a hydrophobic signal sequence, a single extracellular Ig-like domain, a single transmembrane domain, and a moderate length cytoplasmic domain (61 and 48 amino acids, respectively). The cytoplasmic domains are noteworthy in that they contain a conserved sequence also found in polypeptides associated with the T cell antigen receptor and various Fc receptors (Reth 1989). This has been referred to as the antigen receptor homology 1 or ARH1 motif (Cambier 1992), or as the tyrosine-based activation motif (TAM) (Samelson et al 1992).

The J558 antibody-secreting cells fail to express *mb-1*, but transfection of the cells with a *mb-1* expression vector leads to cell surface expression of membrane IgM (Hombach et al 1990b). Moreover, transfection of Ig heavy and light chain *mb-1* and *B29* expression vectors into non-lymphoid cells results in membrane Ig cell surface expression (Venkitaraman et al

1991; Matsuuchi et al 1992). Each of the five isotypes of membrane Ig can pair with Ig-α/Ig-β to come to the cell surface (Venkitaraman et al 1991). In some cases, e.g. membrane IgD and membrane IgG2b, the heterodimer is not required for cell surface expression. In the case of membrane IgD, cell surface expression without the Ig-α/Ig-β heterodimer can occur in the form of a glycosyl-phosphatidylinositol-linked protein (Wienands & Reth 1992).

For membrane IgM, cell surface expression requires binding to Ig-α/Ig-β. This requirement can be abrogated by mutation of clusters of hydroxyl-containing amino acid residues in the transmembrane domain (Williams et al 1990). These observations suggest a model whereby the μ_m transmembrane domain binds to an ER resident protein in the absence of Ig-α/Ig-β. Binding of this protein would cause retention in the ER. Binding of Ig-α/Ig-β to membrane IgM would presumably prevent binding, or cause release of this retention protein and thereby allow transit to the cell surface (Figure 2).

The assembly of the antigen receptor appears to involve chaperones present in the ER. The best established of these is BiP (also called GRP78), a member of the HSP70 family. BiP was originally described as a protein that binds to free Ig heavy chains (Haas & Wabl 1983). The complex between Ig heavy chain and BiP is an intermediate in the assembly of the Ig molecule; BiP dissociates from heavy chain when it associates with light chain (Bole et al 1986; Hendershot 1990). BiP releases Ig heavy chains in vitro in an ATP-dependent manner (Munro & Pelham 1986), which suggests that an ATP-hydrolysis step involving BiP participates in Ig assembly. Recently, it became clear that unassembled Ig light chains also bind BiP.

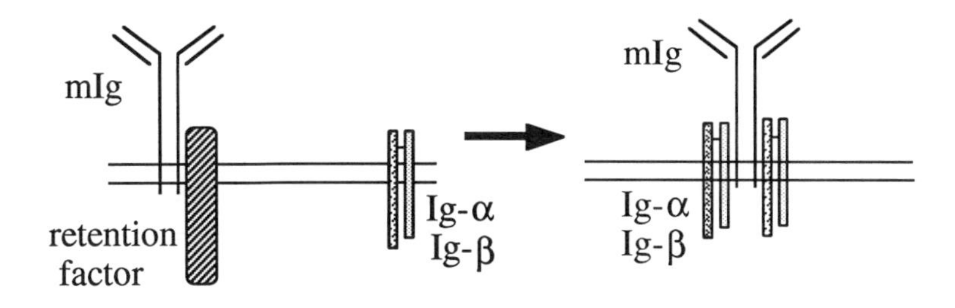

Figure 2 Model for assembly of the B cell antigen receptor. Membrane Ig (mIg) molecules are retained in the ER in the absence of the Ig-α/Ig-β heterodimer. This retention can be relieved by mutations to the transmembrane domain of the μ heavy chain, which suggests that an ER retention factor, possibly a molecular chaperone involved in multi-chain protein assembly, binds to the μ chain and holds it in the ER unless assembly with Ig-α and Ig-β competes for or displaces this retention factor.

For many light chains, this association is transient, and these free light chains are secreted in the absence of heavy chain. For some light chains, this interaction is stronger and appears to last until the light chains are degraded in the ER (Knittler et al 1992). A second ER resident protein, called IP90, has been found to bind to incompletely assembled T cell antigen receptor complexes and free MHC class I heavy chains (Hochstenbach et al 1992). IP90 also binds intracellularly to membrane Ig, although the nature of this interaction is not clear. IP90 could be responsible for ER retention of membrane IgM molecules that are not yet bound to the Ig-α/Ig-β heterodimer.

SIGNAL TRANSDUCTION FUNCTION OF THE ANTIGEN RECEPTOR

Signaling by the antigen receptor is thought to be induced by the aggregation of this receptor, which is induced by extracellular ligands that cause cross-linking of two or more receptor molecules in the membrane. The primary intracellular signaling event that occurs upon cross-linking of the antigen receptor is induced tyrosine phosphorylation of a series of proteins. These phosphorylations can be seen within 5 sec of stimulating the receptor with anti-Ig antibodies, which are anti-receptor antibodies used as a surrogate for antigen (Gold et al 1990; MA Campbell et al 1990; Lane et al 1990; Brunswick et al 1991). Among the targets of antigen receptor-triggered tyrosine phosphorylation are a number of proteins implicated in signal transduction by the tyrosine kinase growth factor receptors (see below).

The mechanism by which membrane Ig triggers tyrosine phosphorylation is only partially understood. A number of investigators have examined the association of various protein tyrosine kinases with the antigen receptor complex. Four *src*-family tyrosine kinases are associated with the antigen receptor complex, $p53/p56^{lyn}$, $p55^{blk}$, $p59^{fyn}$, and $p56^{lck}$ (Burkhardt et al 1991; Yamanashi et al 1991; Campbell & Sefton 1992; Leprince et al 1992; Li et al 1992). In addition, another intracellular tyrosine kinase, PTK72/*syk*, is associated with the antigen receptor (Hutchcroft et al 1992). In each of these cases, however, the fraction of tyrosine kinase molecules associated with the antigen receptor appears to be quite low (probably 1% or less). While the stoichiometry of the kinases to the receptor has not been measured precisely, it seems unlikely that the kinases are present in a large excess over the receptors. Thus either only a small minority of antigen receptors have a bound-tyrosine kinase, or the stability of these complexes is low and the amount of tyrosine kinase found associated with the antigen receptor is an underestimate of the true amount.

These observations argue against a model of signal initiation whereby two kinases are brought together and activate each other. In contrast, the EGF receptor and the PDGF receptor are induced to dimerize upon ligand binding, and once brought together in this way, the two cytoplasmic domains phosphorylate each other on tyrosine side chains. This in turn leads to the recruitment of key proteins to the autophosphorylation sites and the subsequent efficient phosphorylation of these substrates (Cantley et al 1991; Fantl et al 1992; Kazlauskas et al 1992). B cell antigen receptor signaling may be only partly analogous to this process. While cross-linking antigen receptors may not bring two pre-bound tyrosine kinases together, it may serve to bring together a kinase and one or more substrate sites on the receptor complex. Indeed, cross-linking membrane Ig does lead to strong tyrosine phosphorylation of Ig-α and less pronounced phosphorylation of Ig-β (Gold et al 1991).

As the cytoplasmic domain of μ_m is very short (3 amino acid residues) and can be mutated without loss of signaling function, it seems likely that the much larger cytoplasmic domains of Ig-α and Ig-β play important roles in signal initiation. This view has been borne out by studies in which chimeric proteins containing all or part of the intracellular domains of Ig-α and Ig-β have been created and transfected into B cell lines. Cross-linking of these chimeras induces tyrosine phosphorylation of a number of intracellular proteins (Kim et al 1993a; DA Law et al, submitted). Interestingly, the Ig-α and the Ig-β chimeras were able to induce tyrosine phosphorylation of the same set of targets, which suggests that these cytoplasmic domains are functionally redundant rather than being complementary (DA Law et al, submitted). This is in contrast to the growth factor receptors, where the distinct autophosphorylation sites serve as binding sites for different signaling components and thus are complementary in function to one another (Cantley et al 1991; Fantl et al 1992; Kazlauskas et al 1992).

Nonetheless, unique functions of Ig-α and Ig-β could exist for signaling events not yet examined. Indeed, analysis of proteins that bind in vitro to these cytoplasmic domains fused to glutathione S-transferase reveals that each fusion protein binds distinct sets of proteins (Clark et al 1992). Among the proteins that bind to the Ig-α chimera, but not the Ig-β chimera, are the *src*-family tyrosine kinases p53/p56^{lyn} and p59^{fyn}. This observation is somewhat surprising since Ig-β-containing chimeras appear to be as efficient as Ig-α-containing chimeras at activating tyrosine phosphorylation in vivo (DA Law et al, submitted). This observation suggests that the interactions of p53/p56^{lyn} and p59^{fyn} with the Ig-α and Ig-β cytoplasmic domains detected in vitro may not be critically important for function in vivo.

The cytoplasmic domains of Ig-α and Ig-β each contain a region of homology with polypeptides of the T cell antigen receptor, various Fc receptors, and several viral proteins (Figure 3). In the ζ component of the

T cell antigen receptor, there are three of these homology regions, called ARH1 motifs, and each is functional (Irving & Weiss 1991; Romeo & Seed 1991; Wegener et al 1992; Romeo et al 1992; Irving et al 1993). The other CD3 chains each have one ARH1 motif, and at least the one from CD3-ϵ is functional (Letourneur et al 1992). The mechanism by which ARH1 motifs function is discussed in detail below. Mutagenesis has shown that the two tyrosines that are part of this motif are essential to its function (Romeo et al 1992; Irving et al 1993). It is attractive to think that phosphorylation of one or both of these tyrosines may serve an important mechanistic role.

Interestingly, when either the Ig-α or the Ig-β cytoplasmic domain-containing chimeras is immunoprecipitated and tested for associated tyrosine kinase activity by incubation in vitro with γ^{32}P-ATP (in vitro kinase assay), it has low amounts of tyrosine kinases bound to it prior to cross-linking with antibodies directed to the extracellular portion of the chimera (DA Law

Antigen Receptor Homology 1 (ARH1) Motif

Ig-α	ENL	YEGL	NLDDCSM	YEDI
Ig-β	DHT	YEGL	NIDQTAT	YEDI
TCR-ζ1	NQL	YNEL	NLGRREE	YDVL
TCR-ζ2	EGV	YNAL	QKDKMAEAYSEI	
TCR-ζ3	DGL	YQGL	STATKDT	YDAL
CD3-ε	NPD	YEPI	RKGQRDL	YSGL
CD3-γ	EQL	YQPL	KDREYDQ	YSHL
CD3-δ	EQL	YQPL	RDREDTQ	YSRL
FcεRI-β	DRL	YEEL	NVYSPI	YSEL
FcεRI-γ	DAV	YTGL	NTRSQET	YETL
BLV-gp30	DSD	YQAL	LPSAPEI	YSHL
EBV-LMP2A	HSD	YQPL	GTQDQSL	YLGL
Consensus	D/E xx	Yxx L/I	xxxxxxx	Yxx L/I

Figure 3 The antigen receptor homology 1 (ARH1) motif. Comparison of sequences of various antigen receptor-associated polypeptide chains is shown, aligned to the two tyrosines of the motif. Also shown are ARH1 motifs found in membrane proteins of two viruses that induce B cell proliferation, bovine leukosis virus and Epstein-Barr virus. Other sequences are from the murine forms of the proteins. Sequences are from (Wegener et al 1992; Reth 1989; Longnecker et al 1991).

et al, submitted). Stimulation leads to dramatic increases in the amount of tyrosine kinase activity co-precipitating with either chimeric protein. Prominent bands labeled by this in vitro phosphorylation reaction include ones at 72 kd and several in the 55–60 kd range. Following labeling in the in vitro kinase assay, it is possible to disrupt the immune complex and see which of these bands precipitates with antibodies to candidate proteins. When this was done, labeled PTK72/*syk*, p53/p56^{lyn}, and p59^{fyn} (but not p55^{blk}) reprecipitated from these in vitro kinase reactions. When in vitro kinase phosphorylation and re-precipitation were done without prior cross-linking, some p59^{fyn} was detected, but no PTK72/*syk* or p53/p56^{lyn} was detected. The most straightforward interpretation of these results is that cross-linking of either chimera induced binding of these tyrosine kinases to the chimeric receptor. At this point, a large change in the activity of already bound protein kinases, or in their ability to become phosphorylated, cannot be ruled out as alternative explanations for these results.

Stimulation of the chimeric molecules also results in tyrosine phosphorylation of the Ig-α or Ig-β cytoplasmic domains. These observations suggest a model of signal initiation whereby the cross-linking of antigen receptor molecules brings together many ARH1 motifs and a small amount of pre-bound tyrosine kinases that act to phosphorylate the tyrosines in the motif (Figure 4). This phosphorylation would lead to binding of additional tyrosine kinase molecules which, in turn, would be a positive amplification loop since it would lead to further phosphorylation of ARH1 motifs in the complex. In agreement with this view, anti-phosphotyrosine immunofluorescence analysis of anti-IgM stimulated B cells revealed a concentration of tyrosine phosphorylation underneath the membrane Ig molecules that had been capped to one side of the cell (Takagi et al 1991). Although the mechanism of binding of PTK72/*syk* and p53/p56^{lyn} to the stimulated motifs is unknown, an attractive hypothesis is that the ARH1 motifs become phosphorylated on tyrosine residues and the kinase molecules bind via their SH2 domains (Songyang et al 1993).

PTK72 was first identified as a protein tyrosine kinase of 72 kd found in lymphoid tissues (Zioncheck et al 1986). The *syk* gene, cloned as a cDNA from pig spleen, appears to encode this kinase (Taniguchi et al 1991). PTK72/*syk* is highly homologous to a tyrosine kinase from T cells called ZAP70 (ζ-associated protein, 70 kd) (Chan et al 1992). ZAP70 was initially detected as a protein that binds to the ζ chain of the T cell antigen receptor complex following stimulation (Chan et al 1991). Thus cross-linking of the lymphocyte antigen receptors leads to tyrosine phosphorylation of the associated polypeptides and to the binding or PTK72/*syk* (in B cells) or ZAP70 (in T cells) to these chains.

Co-expression of ZAP70 and a *src*-family tyrosine kinase (either p56^{lck}

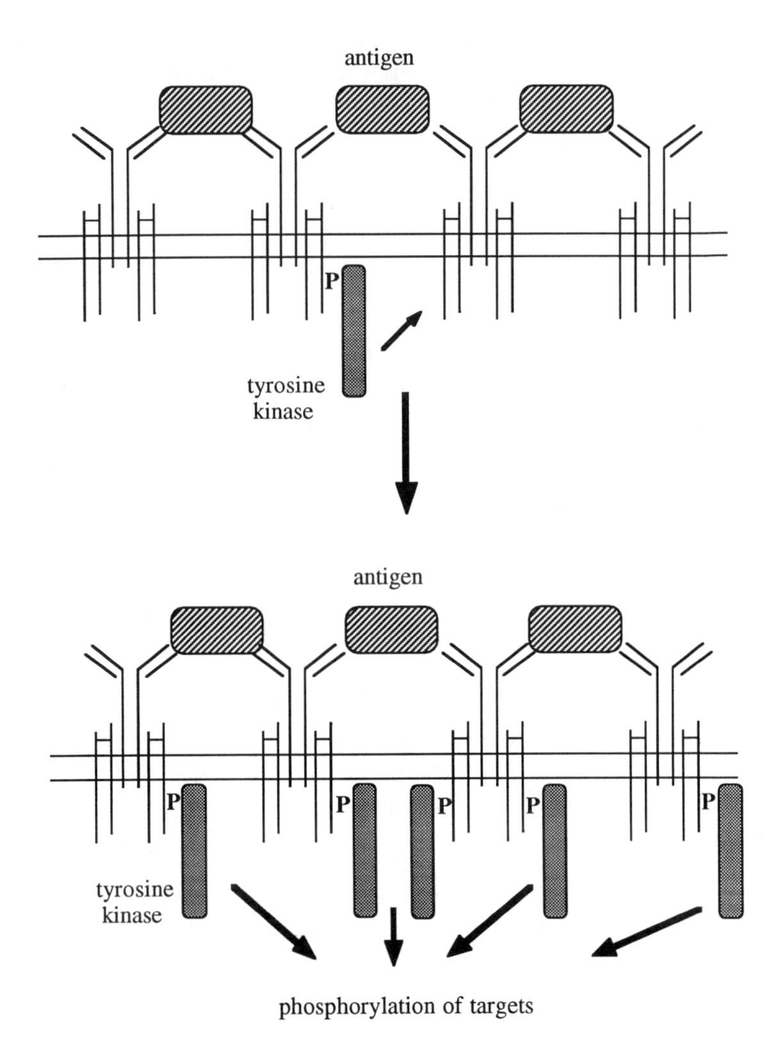

Figure 4 Model for antigen receptor signal initiation involving induced binding of a tyrosine kinase. According to this hypothesis, tyrosine kinases bind to the tyrosine phosphorylated forms of Ig-α/Ig-β heterodimers. Because of the equilibrium between tyrosine kinase action and tyrosine phosphatase action, a small number of heterodimers are tyrosine phosphorylated prior to stimulation. Multivalent antigen cross-links receptors, bringing the few bound tyrosine kinases adjacent to many cytoplasmic domains of Ig-α and Ig-β, which then serve as substrates for this tyrosine kinase. After phosphorylation of Ig-α and Ig-β, more tyrosine kinases bind to the cross-linked receptor, and this creates a positive feedback loop for phosphorylating Ig-α, Ig-β, and signaling targets.

or p59^{fyn}) in cos cells leads to dramatically activated tyrosine phosphorylation (Chan et al 1992). One possible explanation for this observation is that the *src*-family tyrosine kinase activates ZAP70 in some way. These observations suggest that PTK72/*syk* or ZAP70 becomes activated by binding to the phosphorylated, receptor-associated chains followed by phosphorylation by the same tyrosine kinase responsible for phosphorylating the receptor-associated chains (presumably, a *src*-family member) (Figure 5). Alternatively, the binding per se could activate these tyrosine kinases. Obviously, there are many possibilities for the mechanism of signal initiation, but the ARH1 motif seems to be intimately involved in this process, probably by serving as both a substrate for tyrosine phosphorylation and as a binding site for tyrosine kinases. Interestingly, two viruses that can infect B cells and induce their proliferation, bovine leukosis virus and Epstein-Barr virus, encode membrane proteins with ARH1 motifs that are also able to induce protein tyrosine phosphorylation as parts of chimeric transmembrane proteins (Kim et al 1993b).

The transmembrane tyrosine phosphatase CD45 seems to play an important role in antigen receptor signal initiation in lymphocytes. This has been demonstrated most clearly in T cells. Mutant T cells that lack CD45 also lack receptor-triggered tyrosine phosphorylation (Koretzky et al 1991) and

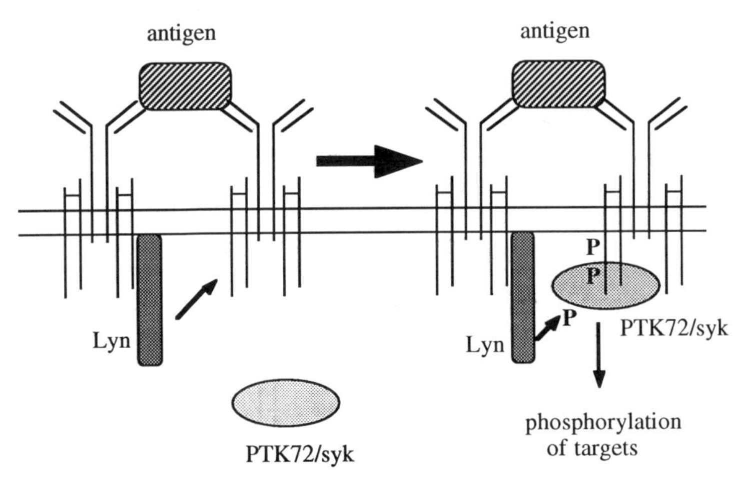

Figure 5 Model for the activation of PTK72/*syk*. The model illustrated in Figure 4 may be overly simplified in that it only includes one tyrosine kinase. An alternative possibility is that antigen receptor signaling involves an interplay between two types of tyrosine kinases, one of the *src*-family (e.g. either p53/p56^{lyn} or p59^{fyn}) and PTK72/*syk*. According to the model shown here, a *src*-family member acts first upon receptor stimulation to phosphorylate Ig-α or Ig-β. PTK72/*syk* then binds to the phosphorylated ARH1 motif, whereupon this tyrosine kinase is activated either by the binding event per se, or by its phosphorylation, possibly by the *src*-family tyrosine kinase.

interleukin 2 production and proliferation (Pingel et al 1989). Re-expression of CD45 restores signaling and biological responses (Koretzky et al 1992). The mechanism by which CD45 tyrosine phosphatase contributes to the activation of tyrosine kinases has not been established. One likely possibility is that CD45 acts to remove tyrosine phosphate from the negative regulatory site near the C-terminus of the *src*-family tyrosine kinases $p56^{lck}$ and $p59^{fyn}$. For example, in a CD45 negative mutant T cell line, Y505 of $p56^{lck}$ is hyperphosphorylated relative to the parental CD45-expressing cell line (Ostergaard et al 1989). Moreover, $p56^{lck}$ is required for T cell receptor signaling function (Straus & Weiss 1992), so dampened activity of $p56^{lck}$ would be expected to inhibit T cell receptor signaling. Thus regulation of CD45 activity could be a means of regulating T cell receptor signaling. Calcium elevation inhibits CD45 activity in T cells (Ostergaard & Trowbridge 1991). In addition, CD45 becomes tyrosine phosphorylated in T cells upon stimulation of the T cell antigen receptor, which suggests another regulatory event (Stover et al 1991).

In B cells, the role of CD45 in antigen receptor signaling is less well understood. In the μ_m and *mb-1*-transfected J558 cells expressing membrane IgM, anti-IgM induces tyrosine phosphorylation (Kim et al 1993b). In contrast to most B cell lines and splenic B cells, however, rapid elevation of intracellular calcium (a consequence of phosphoinositide breakdown) was not seen (Justement et al 1990). These cells are CD45 negative and transfection resulted in cells expressing CD45 that then shows a strong calcium response to anti-Ig stimulation (Justement et al 1991). Thus in one B cell line, CD45 seems to be required for some signaling reactions, but not for the tyrosine phosphorylation reactions thought to be the initial signaling event.

An important strategy employed by the tyrosine kinase growth factor receptors is that autophosphorylation sites on the receptor serve as binding sites for signaling proteins (Cantley et al 1991; Fantl et al 1992; Kazlauskas et al 1992). These signaling components bind to specific sites on the tyrosine phosphorylated receptor via their SH2 domains (Cantley et al 1991). One interesting possibility is that this function is assumed by other cell surface molecules of the B cell. Two candidates for such molecules are CD19 and CD22. Both of these molecules are associated with the antigen receptor complex in moderate amounts (Pesando et al 1989; Leprince et al 1993) and both become tyrosine phosphorylated upon stimulation of B cells via the antigen receptor (Schulte et al 1992; Leprince et al 1993; Tuveson et al 1993). The cytoplasmic tails of these molecules have numerous tyrosines embedded in potential SH2-binding motifs (van Noesel et al 1993; Stamenkovic et al 1989, Leprince et al 1993; Songyang et al 1993). Moreover, binding of phosphatidylinositol 3-kinase (a signaling target, see below) to the cytoplasmic tail of CD19 has been observed upon anti-Ig treatment of B cells (Tuveson et al 1993). Thus cell surface molecules such as CD19 and CD22 may participate in antigen receptor signal transduction by attracting targets to the activated antigen-receptor complex.

Another possible function for surface molecules such as CD19 or CD22 might be to function as co-receptors. In T cells, antigen is recognized by the T cell antigen receptor as an antigen-derived peptide bound to a major histocompatability complex (MHC) molecule. The CD4 (on helper T cells) or CD8 (on cytotoxic T cells) molecules serve as co-receptors by binding to the MHC molecule in a region distinct from the peptide binding site. Thus antigen recognition by the T cell brings together the T cell antigen receptor and CD4 or CD8 by virtue of their interaction with the same MHC molecule (Janeway 1989). The cytoplasmic tails of CD4 and CD8 bind efficiently to the *src*-family tyrosine kinase $p56^{lck}$. These observations suggest a model whereby the co-receptors serve to bring $p56^{lck}$ next to the ARH1 motifs of the T cell antigen receptor. The $p56^{lck}$ molecules would then phosphorylate the ARH1 motifs, and in this way initiate the tyrosine phosphorylation cascade (Figure 6).

By analogy, there may exist one or more potential co-receptors in B cells that bind *src*-family tyrosine kinases and can deliver them to the antigen receptor upon antigen binding. Such co-receptors might recognize comple-

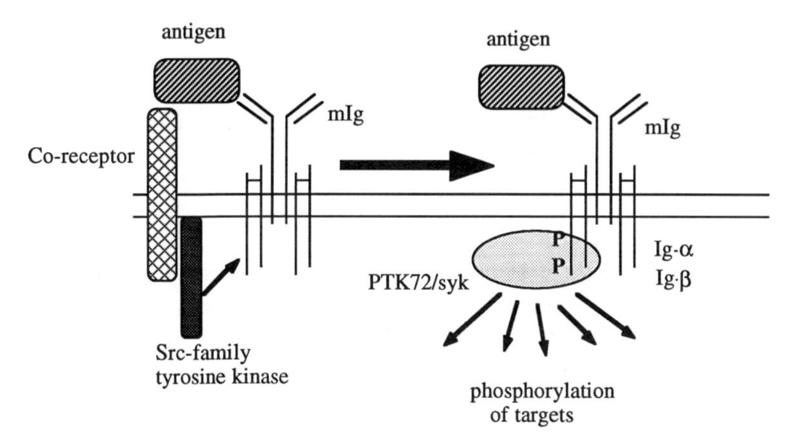

Figure 6 A co-receptor model for stimulation of B cells by small amounts of antigen. In T cells, CD4 and CD8 act as co-receptors for signaling by the T cell antigen receptor by efficiently delivering a tyrosine kinase ($p56^{lck}$) to the T cell receptor upon antigen binding. It is possible that B cells contain one or more potential co-receptors that can recognize antigens tagged in one way or another. For example, the presence of fragments of complement component C3b or of certain carbohydrate determinants unique to microorganisms could allow recognition by a co-receptor (CR2 in the former case, CD22 or some other molecule in the latter case). The co-receptor could then participate in signaling in response to those antigens.

ment components deposited upon antigens (van Noesel et al 1993) or carbohydrates unique to microorganisms, or some other feature of a subset of antigens. According to this hypothesis, a co-receptor would enhance the response of the B cell to antigen by efficiently delivering the initiating tyrosine kinase to the antigen-cross-linked receptors. This would trigger signaling at relatively low receptor occupancy (as is the case in T cells), whereas antigens that did not bind to co-receptors would need to be present at higher levels, or be more efficient at cross-linking the antigen receptor, in order to trigger a response of equal magnitude. In other words, putative co-receptors for B cells might reduce the antigen concentration/multivalency required for B cell activation. At this point there is no direct evidence for such a role for another B cell surface protein, but the complement receptor 2 (CR2)/CD19 complex is a good candidate. Co-cross-linking of the antigen receptor and CR2/CD19 with antibodies triggers synergistic signaling and B cell activation (Carter et al 1988; Carter & Fearon 1992). Moreover, injection of a soluble CR2/IgG chimeric protein inhibits antibody responses in vivo, possibly by preventing complement components from binding to CR2 on B cells (Hebell et al 1991). These results are consistent with a co-receptor model for antigen recognition by B cells (van Noesel et al 1993). It is also possible that putative B cell co-receptors act not by efficiently bringing a tyrosine kinase to the antigen-cross-linked antigen receptor, but by bringing their cytoplasmic domains to the receptor as substrates that can be tyrosine phosphorylated and then act to efficiently attract signaling targets, as described above.

TARGETS OF ANTIGEN RECEPTOR-STIMULATED TYROSINE PHOSPHORYLATION

The targets of antigen receptor-induced tyrosine phosphorylation include a number of proteins that have been implicated in tyrosine kinase growth factor receptor signaling (DeFranco 1992b). Among these are phospholipase Cγ1 and γ2, phosphatidylinositol 3-kinase, MAP kinase, $p95^{vav}$, rasGAP and two GAP-associated proteins, called p62 and p190. These targets are likely to mediate a number of distinct biological responses to antigen receptor stimulation of the B cell.

Stimulation of phospholipase C (PLC) γ1 and γ2 is probably responsible for the robust phosphatidylinositol 4,5-bisphosphate (PIP_2) hydrolysis that is seen upon antigen receptor stimulation (Bijsterbosch et al 1985). Inhibition of anti-Ig induced tyrosine phosphorylation with three different types of inhibitors, genestein, herbimycin A, and several tyrphostins, blocks anti-Ig induced PIP_2 breakdown (Lane et al 1991; Carter et al 1991; Padeh et al 1991). Moreover, PLC-γ1 and γ2 are among the most rapid targets of membrane Ig-induced tyrosine phosphorylation (Carter et al 1991; Hempel et al 1992; Roifman & Wang 1992; Kanner et al 1992; Coggeshall et al 1992). In murine B cells, PLC-γ2 seems to be more abundant than PLC-γ1 (Hempel et al 1992; Coggeshall et al 1992).

Tyrosine phosphorylation of PLC-γ1 and γ2 may not be the only events regulating PIP_2 breakdown in response to anti-Ig. In permeabilized B cells, GTP analogues stimulate anti-Ig induced tyrosine phosphorylation and GDP analogues inhibit it (Harnett & Klaus 1988). These results suggest that a GTP-binding protein may also participate in PLC activation in B cells. A precedent for combinatorial regulation of PLC-γ1 exists in hepatocytes, where EGF receptor appears to utilize G_i to activate PLC-γ1 (Yang et al 1991). Thus activation of PLC-γ1 or γ2 by the B cell antigen receptor could involve a G protein as well as tyrosine phosphorylation. If so, a G protein other than G_i would probably be involved, since pertussis toxin, which prevents G_i from interacting with receptors and blocks PLC activation by EGF in hepatocytes, does not block anti-Ig induced phosphoinositide breakdown (Harnett & Klaus 1988). Alternatively, it is possible that the B cell antigen receptor activates two isoforms of PLC, one via direct tyrosine phosphorylation and the other via a G protein. As the tyrosine kinase inhibitors readily block PIP_2 breakdown (Lane et al 1991; Padeh et al 1991; Carter et al 1991), this second pathway would also require tyrosine phosphorylation in some way.

Phosphoinositide breakdown leads to the generation of two second messengers, diacylglycerol, which activates protein kinase C isozymes, and inositol trisphosphate, which mediates elevation of intracellular free calcium (Berridge 1987). Calcium elevation was the first recognized signaling reaction observed for the B cell antigen receptor (Braun et al 1979; Pozzan et al 1982). Activation of protein kinase C has been suggested by the demonstration that anti-Ig induces translocation of PKC from the soluble fraction to the membrane-bound fraction, an event that generally accompanies PKC activation (Nel et al 1986; Chen et al 1986). Moreover, the pattern of proteins phosphorylated in response to anti-Ig is similar to the pattern seen upon treatment with PKC-activating phorbol diesters (Hornbeck & Paul 1986). Thus calcium elevation and PKC activation are both well-established as signaling reactions stimulated by cross-linking the antigen receptor.

Phosphoinositide breakdown is clearly an important signaling event in the B cell since many of the biological events induced by anti-Ig can be reproduced either partly or completely by using pharmacologic activators of PKC (phorbol esters) and ionophores to elevate cytosolic free calcium. For example, the combination of these pharmacologic agents induces resting splenic B cells to enter the G_1 phase of the cell cycle (Monroe & Kass 1985) and to proliferate upon addition of interleukin 4 (Paul et al 1986).

Under some circumstances, these agents can induce vigorous proliferation without addition of growth factors (Rothstein et al 1986). Correspondingly, anti-IgM antibodies induce proliferation at high doses by themselves or at low doses in combination with interleukin 4 (Howard et al 1982).

Most of the anti-Ig-induced tyrosine phosphorylated proteins are not seen when B cells are stimulated by pharmacologic agents that mimic the phosphoinositide breakdown pathway. The major exception is a prominent band at 42 kd. This protein is MAP kinase, a serine-threonine protein kinase that is activated by both tyrosine and threonine phosphorylations. Its function is not fully understood, although there is evidence that it can act on two receptor-regulated transcription factors, c-Jun and $p67^{TCF}$ (Pulverer et al 1991; Gille et al 1992). In B cells, as in fibroblasts, PKC appears to be capable of activating MAP kinase (Casillas et al 1991; Gold et al 1992). Interesting, PKC inhibitors block MAP kinase activation in response to phorbol esters, but only partially block its activation via antigen-receptor stimulation (Gold et al 1992). The most straightforward interpretation of this finding is that antigen receptor stimulation activates MAP kinase by two routes, only one of which is dependent upon PKC.

Many of the genes that are known to be rapidly induced in B cells by anti-Ig appear to be downstream of PIP_2 breakdown. For example, the early response genes *c-myc, c-fos, egr-1,* and *junB* are induced in response to anti-Ig (Kelly et al 1983; Monroe & Kass 1985; Seyfert et al 1989; Tilzey et al 1991). Induction of *c-fos* appears to require both PKC activation and elevation of intracellular free calcium, based on addition of the pharmacologic mimicking agents (Klemsz et al 1989; Mittelstadt & DeFranco 1993). In contrast, *egr-1* and *c-myc* induction appear to require only PKC activation (Seyfert et al 1990; Klemsz et al 1989). More recently, four additional early response genes from serum-stimulated fibroblasts have been found to be induced upon anti-IgM stimulation of B cells (Mittelstadt & DeFranco 1993). Two of these, *nur77* and *nup475,* are putative transcription factors; one, *3CH134*, appears to encode a protein phosphatase (Charles et al 1992); and a fourth (*pip92*) encodes a protein of unknown function. All four of these genes are induced by activating PIP_2 breakdown with an introduced M1 muscarinic acetylcholine receptor. In the case of *nur77, nup475,* and *pip92,* stimulating cells with phorbol diesters also gave full mRNA induction. In the case of *3CH134,* however, elevation of intracellular free calcium also played a role in the gene induction in splenic B cells. In 2PK3 cells, calcium elevation with a calcium ionophore gave full induction of *3CH134* (Mittelstadt & DeFranco 1993). Thus these four genes appear to be induced by the phosphoinositide pathway in B cells.

The mechanism by which PIP_2 breakdown leads to the induction of these early response genes in B cells is not known. The first family of transcription factors known to mediate phorbol diester-stimulated transcription is the Jun/Fos leucine zipper dimers that bind to AP-1 sites (also called TRE sites) (Ransone & Verma 1990). Anti-Ig treatment of B cells leads to an increase in the levels of proteins that can bind to an AP-1 site in vitro (Chiles et al 1991; Chiles & Rothstein 1992), but this increase occurs with delayed kinetics, which suggests that it involves production of c-Fos (etc) rather than the activation of pre-existing components such as c-Jun. Moreover, *jun-B* mRNA is also induced by anti-Ig in B cells (Tilzey et al 1991). Jun-B expression inhibits activation through a single AP-1 site, which suggests that it may be a transcriptional repressor (Chiu et al 1989). Nonetheless, the functional properties of Jun-B in the context of a complex enhancer are unknown, particularly as this could be influenced by the cellular context. In any case, the lack of AP-1 binding components evident in the period when early response gene transcription is occurring suggests that the role played by this family of transcription factors occurs later.

Another transcription factor family activated by phorbol diesters is NF-κB. In most cells, NF-κB is held in an inactive state in the cytoplasm by the inhibitor protein I-κB. Activation of NF-κB occurs by inactivation of I-κB and release of active NF-κB, which then goes to the nucleus, binds DNA, and activates transcription (Lenardo & Baltimore 1989). NF-κB is constitutively active in B cells and is thought to play a role in promoting transcription of the Ig κ locus because of its binding site in the κ intronic enhancer. In resting splenic B cells, however, some inactive NF-κB is also present, and this is activated upon addition of anti-Ig to B cells (Liu et al 1991; Rooney et al 1991). Thus NF-κB could play a role in early response gene induction. Indeed, *c-myc* induction by anti-Ig in the WEHI-231 B cell line has been shown to depend upon a DNA sequence that is a NF-κB binding site (Duyao et al 1990).

Serum response factor (SRF) is involved in phorbol diester and serum-induced early gene inductions in fibroblasts (Berk 1989). SRF is likely to play a role in induction of early response genes in B cells as well. The *egr-1* gene has several SRF binding sites (SREs) upstream of the promoter, and these are required for induction of this gene in response to anti-Ig (Seyfert et al 1990). Interestingly, there is evidence that MAP kinase can activate transcription through SRE sites by phosphorylating a protein called ternary complex factor ($p62^{TCF}$), which binds to SREs in vitro in the presence of SRF (Gille et al 1992).

Transcriptional activation occurring as a result of calcium elevation is even less well understood, although examples include *c-fos*, *3CH134*, and class II MHC molecules. One candidate here is the *ets-1* gene product. Anti-Ig treatment of B cells leads to rapid phosphorylation of Ets-1, and this can be reproduced by elevating calcium with a calcium ionophore, but not by treating cells with phorbol esters (Fisher et al 1991). In mitogen-stimulated thymocytes, calcium-induced phosphorylation of Ets-1 causes it to lose DNA binding ability (Pognonec et al 1989). If this is also the case in B cells, then calcium elevation would inactivate Ets-1, which would either shut off a gene whose activity depends on Ets-1, or induce a gene whose activity is repressed by Ets-1.

Another signaling pathway activated by antigen receptor signaling involves phosphorylation of inositol-containing phospholipids on the 3 position of the inositol ring. The enzyme that catalyzes this reaction is called phosphatidylinositol 3-kinase (PI 3-kinase), and in vitro it will act on PI, PI 4-P, and PI 4,5-P_2 to generate PI 3-P, PI 3,4-P_2, and PI 3,4,5-P_3 (Auger & Cantley 1991). Anti-Ig antibodies induce tyrosine phosphorylation of this enzyme on both the 85-kd regulatory subunit and on the 110-kd catalytic subunit (MR Gold et al, submitted; DA Law et al, submitted). Moreover, antigen receptor stimulation also leads to elevation of PI 3-P, PI 3,4-P_2, and PI 3,4,5-P_3, which indicates that this enzyme is activated in vivo (Tuveson et al 1993; MR Gold et al, submitted). Since these phospholipids accumulate during signaling, it may be that they are the second messengers of this signaling pathway.

The 3-phosphorylated versions of phosphatidylinositol are not good substrates for the phospholipase C pathway, so although this signaling route also uses phosphoinositides, it seems to be distinct from PIP_2 breakdown. Moreover, some receptors stimulate PLC and not PI 3-kinase or the converse, again suggesting that these are distinct signaling pathways (Cantley et al 1991). Recently it was discovered that a relative of PKC, called PKCζ, is activated by PI 3,4-P_2 and PI 3,4,5-P_3 rather than by diacylglycerol or phorbol esters (Nakanishi et al 1993). Thus receptor stimulation of PI 3-kinase leads to increased production of PI 3,4-P_2 and PI 3,4,5-P_3 and this may activate PKCζ. Downstream consequences of this signaling pathway are not well understood, but in fibroblasts, the ability of PDGF receptors to stimulate cell growth is abrogated by mutations of the receptor that selectively affect phosphorylation of PI 3-kinase (Fantl et al 1992). Similarly, point mutants of polyoma middle t antigen and *v-abl* that no longer activate PI 3-kinase also no longer transform fibroblasts (Cantley et al 1991). Thus PI 3-kinase seems to play an important role in mediating the ability of these agents to stimulate fibroblast growth.

Antigen receptor stimulation also leads to activation of $p21^{ras}$. This GTP-binding protein is active when GTP is bound and inactive when GDP is bound (Bourne et al 1991). Anti-Ig stimulation induces an increase in the bound GTP content of $p21^{ras}$ in B cells (AH Lazarus et al, submitted). Anti-Ig treatment of B cells also leads to capping of the antigen receptor molecules, and p21*ras* is found clustered underneath the caps (Graziadei et al 1990). The mechanism by which the antigen receptor regulates p21*ras* is unknown, but one possibility is that this regulation involves the GTPase-activating protein of p21*ras*, rasGAP. Anti-Ig treatment of B cells leads to increased tyrosine phosphorylation of rasGAP and also the formation of complexes between rasGAP and two other tyrosine phosphorylated proteins, p190 and p62 (Gold et al 1993). The tyrosine phosphorylation of p62 is strongly induced in response to membrane Ig cross-linking in most B cell lines. The p190 protein, on the other hand, is constitutively phosphorylated and associated with rasGAP in some B cell lines, and inducibly tyrosine phosphorylated in others (Gold et al 1993).

Similar or identical proteins also associate with rasGAP in growth factor-stimulated fibroblasts, and the genes encoding these proteins have been cloned (Settleman et al 1992b; Wong et al 1992). The function of p62 is unknown, although it has homology with an RNA-binding protein and has been shown to have RNA binding ability in vitro. The p190 protein has GAP activity toward a distinct group of *ras*-superfamily members, the *rac* and *rho* proteins (Settleman et al 1992a). These proteins have been implicated in the regulation of microfilament assembly and/or placement (Ridley & Hall 1992; Ridley et al 1992).

Regulation of microfilament assembly may be important for the capping phenomenon by which all of the stimulated antigen receptors are actively transported to one pole of the cell (Bourguignon et al 1984). One of the other targets of antigen receptor-induced tyrosine phosphorylation is p95*vav* (Bustelo & Barbacid 1992), which has homology to a yeast protein (the product of the CDC24 gene of *Saccharomyces cerevisiae)* that acts as a guanine nucleotide exchange factor for the yeast homologue of mammalian *rho* (Puil & Pawson 1992). This homology suggested that p95*vav* acts on the *rac* and *rho* proteins. Recently, however, it was found that p95*vav* is a guanine nucleotide exchange factor for p21*ras*. The activity of p95*vav* is rapidly stimulated by antigen receptor stimulation in T cells (Gulbins et al 1993), and antigen receptor-induced tyrosine phosphorylation appears to be responsible for this activation. Thus p95*vav* may mediate the action of the antigen receptor on p21*ras*.

In summary, stimulation of the antigen receptor leads to the activation of one or more protein tyrosine kinases, which selectively phosphorylate a number of potentially important targets. The exact processes within the cell that are acted upon by these targets are largely undefined at this time, although a number of transcriptional activation events have been shown to be downstream of PIP_2 breakdown. In addition, there are hints that some of the other signaling events may control cytoskeletal events.

ANTIGEN UPTAKE FUNCTION OF THE ANTIGEN RECEPTOR

Antibody responses to T cell-dependent antigens (such as most protein antigens) require interaction between the responding B cell and an antigen-specific helper T cell. After the T cell recognizes antigen presented by the B cell, it expresses a cell surface molecule, CD40 ligand, that provides a cell contact-dependent activation signal to the B cell (DeFranco 1992a). This molecule is defective in X-linked hyper-IgM syndrome, a genetic disease of humans in which IgG and IgA production is impaired (Aruffo et al 1993; Allen et al 1993; Korthauer et al 1993; DiSanto et al 1993; Fuleihan et al 1993). In addition, the stimulated helper T cell produces a battery of cytokines, some of which promote B cell proliferation and differentiation to the antibody-secreting state (Kishimoto & Hirano 1988; Vitetta et al 1989; Noelle & Snow 1991). Thus the process of antigen presentation to the helper T cell is central to many antibody responses.

Antigen presentation to helper T cells involves uptake of antigen, processing of antigen in endosomal or lysosomal compartments, binding of processed antigen-derived peptides to major histocompatability complex (MHC) class II molecules, and finally recognition of the peptide-MHC II complex by antigen-specific helper T cells (Lanzavecchia 1990; Myers 1991; Brodsky et al 1991). The mechanism of antigen uptake varies with the type of antigen-presenting cell. For example, macrophages take up antigen via phagocytosis, pinocytosis, or Fc receptor-mediated uptake. In the case of the antigen-specific B cell, this series of events is initiated by antigen binding to membrane Ig on the surface of the B cell (Figure 7).

Unlike signal transduction, antigen uptake via the antigen receptor does not require multivalent interaction with the antigen. For example, antigen presentation to rabbit IgG-specific T cells occurs with equal efficiency with bivalent $(Fab')_2$ fragments of rabbit anti-mouse IgM antibodies, or with monovalent Fab fragments (Tony et al 1985). Similarly, tetanus toxin-specific Epstein Barr virus-transformed B cell lines present monovalent tetanus toxin to antigen-specific T cells as well as they present polymerized forms of this antigen (Lanzavecchia 1985). Antigen presentation to helper T cells can also be achieved via pinocytosis of nonspecific rabbit IgG or tetanus toxin, but this requires 10^4-fold greater antigen concentrations than does membrane Ig-mediated uptake. Other molecules on the surface of the B cell can also facilitate antigen uptake, although usually less effectively. The ability of cell surface molecules to promote antigen uptake varies greatly, with some molecules such as CD45 and FcγRII (one of the Fc receptors for IgG) being particularly inefficient (Lanzavecchia 1990). Interestingly, FcγRII can mediate efficient uptake of antigen in antibody-antigen complexes

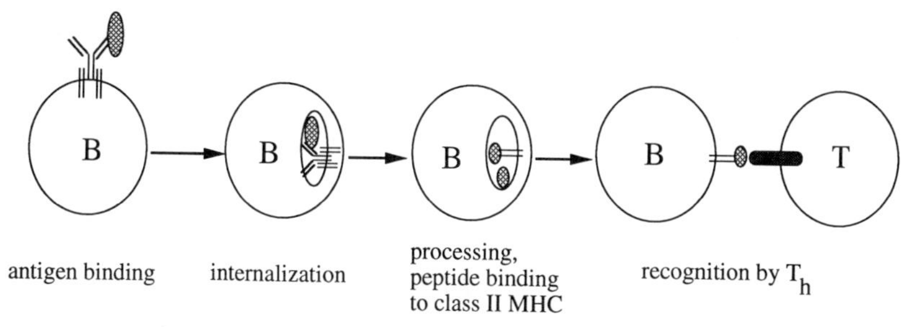

Figure 7 Antigen uptake and presentation by B cells. The B cell binds antigen via membrane Ig. Membrane Ig then mediates internalization of the antigen via its endocytic receptor function. The antigen and membrane Ig are then degraded by proteases in the late endosome or lysosome. Antigen-derived peptides can bind to class II MHC molecules and then go to the cell surface, where they serve as recognition structures that activate antigen-specific helper T cells. T cell activation leads to expression of CD40 ligand, which provides a cell contact-dependent activation signal to the B cell, and to synthesis and secretion of cytokines, which promote B cell proliferation and differentiation into antibody-secreting cells.

in macrophages, but not in B cells (Amigorena et al 1992). The B cell form of FcγRII contains 42 extra amino acids in its cytoplasmic tail because of alternative splicing, and these extra amino acids prevent antigen uptake, possibly by promoting interaction of the molecule with the cytoskeleton and thereby preventing endocytosis (Miettinen et al 1992).

Antigen uptake by membrane Ig appears to involve constitutive recycling of membrane Ig between the plasma membrane and early endosomes (Davidson et al 1990). In the Epstein-Barr virus-transformed tetanus toxin-specific B cell lines, nearly one half of the recycling membrane Ig is inside the cell at steady state, and the recycling time has been estimated at 8.5 min (Davidson et al 1990). Although it may be that the recycling is somewhat exaggerated in Epstein Barr virus-transformed cells, primary resting or antigen-stimulated B cells also appear to exhibit constitutive recycling of membrane Ig since they show efficient uptake of monovalent anti-Ig (Tony et al 1985). Without having any antigen bound, internalized membrane Ig efficiently returns to the cell surface. Interestingly, antigen-bound molecules are less likely to return to the cell surface in an intact form, and the larger the antigenic moiety, the more likely is intracellular degradation of the mIg-antigen complex (Davidson et al 1990). It is not known how this selective degradation is achieved.

Antigen presentation requires proteolytic processing of the antigen to peptides that can bind to the class II MHC molecule (Myers 1991; Brodsky et al 1991). As the low pH of the endosome is probably not sufficient to induce dissociation of antigen from a moderate- or high-affinity membrane

Ig molecule, it seems likely that degradation of the antigen is initiated in the antigen-membrane Ig complex. Indeed, there is considerable evidence to indicate that the specificity of the membrane Ig influences the nature of the proteolytic processing in the antigen, as would be expected by protection of the bound region of the molecule (Berzofsky 1983; Lanzavecchia 1990). In the process, the membrane Ig part of the complex is also degraded.

The location of antigen degradation and peptide association with class II MHC molecules is an issue of considerable current debate. Most evidence favors the late endosome as the most likely compartment where peptides are generated and combine with class II MHC molecules (Myers 1991; Brodsky et al 1991). It is possible that in some cases (such as phagocytosed antigen in the macrophage), antigenic peptides are created in the lysosome and flow back into the late endosome. Electron microscopy has demonstrated the co-localization in an endosomal compartment of anti-Ig bound to membrane Ig, MHC class II molecules, and cathepsin-D (Guagliardi et al 1990), which has been implicated in the processing of certain antigens (Myers 1991; Brodsky et al 1991). Although there may be some pickup of antigen by class II MHC molecules that have entered the endosome from the cell surface, most peptide is loaded onto newly synthesized class II MHC molecules (Myers 1991; Brodsky et al 1991; Davidson et al 1991).

During their biosynthesis, class II MHC molecules form a complex with a polypeptide called invariant chain. Current thinking is that invariant chain directs or participates in the routing of newly synthesized class II MHC molecules to the endosome and, moreover, that it prevents binding of peptides to the class II molecules in the endoplasmic reticulum or Golgi apparatus (Brodsky et al 1991). In the endosome, invariant chain is proteolytically cleaved and released, which then allows the class II MHC molecule to bind peptides.

The association and dissociation rates of peptides with purified class II MHC molecules is very slow (Buus et al 1986). While these rates are faster at the acidic pH characteristic of endosomes and lysosomes, the association is still considerably slower than would be expected from the kinetics of antigen processing and presentation seen with antigen presenting cells (DeNagel & Pierce 1992). An attractive hypothesis is that molecular chaperones accelerate the binding of peptides to MHC class II molecules. In support of this possibility, a peptide binding protein of 72–74 kd, called PBP72/74, has been isolated from antigen presenting cells (DeNagel et al 1991), and found to have amino acid sequence similarity to members of the HSP70 family of molecular chaperones (DeNagel & Pierce 1992). Moreover, this molecule is found in endosomes, as well as in the ER, Golgi, and on the cell surface (VanBuskirk et al 1991). This cellular localization is consistent with a role for PBP72/74 in facilitating peptide binding to class II MHC molecules.

Although membrane Ig signaling does not promote antigen uptake per se, it does significantly promote antigen presentation to helper T cells (Casten et al 1985). This observation probably reflects the summation of several different ways in which antigen receptor signaling promotes the antigen presenting capability of B cells. Firstly, the phosphoinositide signaling reactions stimulated by the antigen receptor induce increased expression of class II MHC molecules (Mond et al 1981). As newly synthesized class II MHC molecules preferentially pick up antigen-derived peptides, these new molecules would be primarily responsible for the binding of the antigenic peptides. In addition, membrane Ig signaling also enhances the affinity of the LFA-1 cell adhesion molecule on the B cell (Dang et al 1991), which can bind to ICAM-1 on the surface of the helper T cell. Stimulation of T cells through the T cell antigen receptor then increases the affinity of the T cell LFA-1 for ICAM-1 on the B cell (Dustin & Springer 1991). Indeed, analysis of conjugates formed between B cells and T cells reveals two types of interactions. In the absence of antigen presentation to the helper T cell, the interaction is weak and short-lived. However, if the T cell recognizes the correct peptide-MHC complexes on the B cell, the interaction is long-lived and strong, and exhibits extensive regions of membrane contact (Vitetta et al 1989; Kupfer & Singer 1989). Finally, antigen receptor signaling also induces expression of a cell surface molecule called B7 (Freedman et al 1987). B7 can provide a co-stimulatory signal to the helper T cell that considerably increases cytokine production (Liu & Linsley 1992). Thus although antigen uptake via membrane Ig does not involve receptor signaling, the overall process of antigen presentation is significantly enhanced by several signaling-mediated events.

ROLE OF THE B CELL ANTIGEN RECEPTOR IN B CELL DEVELOPMENT

B lymphocytes develop from hematopoietic stem cells in the bone marrow in an ordered process that is marked by sequential rearrangements of the immunoglobulin genes (Yancopoulos & Alt 1986; Rolink & Melchers 1991). The same recombinatorial machinery [called V(D)J recombinase] is used to rearrange Ig heavy and light chain loci and the T cell receptor loci (Schatz et al 1992). The action of the recombinase is believed to be controlled by regulation of its accessibility to the appropriate genetic loci. Once committed to the B cell lineage, pro-B cells preferentially rearrange their Ig heavy chain alleles until an in-frame Ig heavy chain gene is created. Once this occurs, the cell expresses the μ heavy chain protein intracellularly and is called a pre-B cell. Pre-B cells do not exhibit further rearrangement at the heavy chain alleles, but rather rearrange their light chain genes. One aspect of this process is that only one IgH allele and only one light chain gene are properly rearranged before the rearrangements cease. The use of only one of two Ig alleles is referred to as allelic exclusion.

This ordered process appears to involve regulatory events that sense when a particular immunoglobulin gene has been properly rearranged by the appearance of its product in the cell. Indeed, alterations of the Ig gene products disrupt B cell development. Thus B cell development is completely arrested in mice mutated by homologous recombination to prevent synthesis of the membrane form of μ heavy chain (Kitamura et al 1991). Rearrangements at the κ light chain genes in these mice are decreased by about 20-fold. These observations indicate that the membrane form of μ is needed to inform the cell that a successful IgH rearrangement has occurred and to promote movement to the next developmental stage. Similarly, some Abelson murine leukemia virus-transformed pro-B cell lines exhibit an increase in κ gene rearrangements upon introduction of a functional μ gene (Reth et al 1987; Iglesias et al 1991; Shapiro et al 1993). This increase is seen with genes that encode membrane μ, but not with genes that encode only secreted μ (Reth et al 1987).

When the μ_m mutation is expressed opposite a wild-type allele in a heterozygotic mouse, allelic exclusion of heavy chains is lost, which demonstrates that allelic exclusion is also caused by regulatory events directed by the membrane form of the μ heavy chain (Kitamura et al 1992b). Similarly, mice expressing functionally rearranged IgH transgenes have decreased rearrangement of endogenous heavy chain alleles, such that a majority of B cells express the transgene and no endogenously rearranged IgH gene (Storb 1987). This allelic exclusion is seen with a transgene that can make the membrane form of the μ chain, but not with a transgene that can only make the secretory form of μ chain (Nussenzweig et al 1987). Once again, the membrane form of the μ heavy chain is responsible for an important regulatory event of B cell development.

These observations indicate that expression of the μ_m heavy chain can redirect the V(D)J recombinase from the IgH locus to the IgL loci. This process involves two other polypeptides that form a complex with μ_m in the pre-B cell, the products of the λ5 and V_{preB1} genes (Venkitaraman 1992). These two genes are expressed exclusively in pre-B cells and have single Ig-like domains, the former resembling λ constant regions and the latter resembling variable regions. Unlike classical light chain genes, the λ5 and V_{preB1} genes do not undergo rearrangement. The similarity of the λ5 and V_{preB1} gene products to light chains has led to the hypothesis that they bind to μ_m chain in a way similar to authentic light chain, and hence to to calling them surrogate light chains, or alternative light chains (Venkitaraman 1992). Indeed, two small, appropriately sized polypeptides co-precipitate with μ chain in pre-B cells (Pillai et al 1987). The likely identity of these two polypeptides with the $\lambda 5$ and V_{preB1} gene products has been strongly supported by transfection experiments that demonstrate the ability of these proteins to pair with μ heavy chain (Tsubata & Reth 1990; Karasuyama et al 1990). Mice with the $\lambda 5$ gene disrupted by homologous recombination show a large decrease in the number of pre-B cells and in the rate of production of B cells (Kitamura et al 1992a). Unlike the μ_m knockout mice, which completely lack B cells, these mice do make small numbers of B cells. Presumably, the $\lambda 5$ gene product plays an important, although not absolute, role in the formation of a signaling complex in pre-B cells.

In the human genetic disease called X-linked agammaglobulinemia, B cell development is disrupted, with the result that there are very few mature B cells and very little antibody production. The gene defective in this disease, called *btk*, encodes a novel intracellular protein tyrosine kinase (Tsukada et al 1993; Vetrie et al 1993). It is possible that this tyrosine kinase mediates signaling by the $\mu/\lambda 5/V_{preB1}$ complex in pre-B cells. Alternatively, this tyrosine kinase could mediate some other signaling event (response to the growth factor interleukin 7, etc) required for growth, survival, or maturation of B cell precursors.

In addition to binding to the surrogate light chains, the μ chain in pre-B cells is complexed to the Ig-α and Ig-β heterodimer (Nakamura et al 1992). Most of the μ chain in pre-B cells is found intracellularly, probably in the ER, whereas only a small amount can be detected on the cell surface (Cherayil & Pillai 1991). The reason for this retention within the cell in not known, because one might expect a complex of μ, the surrogate light chains, Ig-α and Ig-β, to come to the cell surface (Tsubata & Reth 1990). An interesting question is where the μ_m molecules are located that promote B cell maturation and rearrangement of κ light chain genes. Do their signals arise from the cell surface where there are very few molecules, or from the ER as an intracellular form of signaling? Such intracellular signaling has been reported in circumstances where a cell expresses both a growth factor receptor and its ligand, which allows interaction to occur during their biosynthesis (Keating & Williams 1988; Bejcek et al 1989; Dunbar et al 1989).

Pre-B cell lines are capable of exhibiting an elevation of intracellular free calcium upon addition of anti-IgM, anti-Ig-α or anti-$\lambda 5$ antibodies (Takemori et al 1990; Nomura et al 1991; Misener et al 1991). Thus the complex of μ chain with surrogate light chains and Ig-α/Ig-β on the cell surface is capable of signaling in the pre-B cell. Interestingly, Tsubata et al (1992) have found that unlike wild-type μ chain, a truncated μ gene lacking the exons encoding the first two Ig-like domains (the V_H domain and the C_H1 domain) does not normally induce an increase in κ gene rearrangement. But an increase in κ gene rearrangement is induced in these cells upon addition of anti-μ antibody. This truncated μ heavy chain does not bind to the surrogate light chains, thus suggesting that these light chains bind to μ chain in a manner analogous to authentic light chain.

Taken together, the above observations suggest that the role of the surrogate light chains is to trigger the signaling function of μ and Ig-α/Ig-β, perhaps by binding to some ligand (exogenous or endogenous) or by cross-linking receptor molecules directly, and thereby triggering their signaling function. In any case, these experiments support a signaling model whereby the signaling function of μ chain and the Ig-α/Ig-β heterodimer are responsible for sending the signal that μ protein has been produced. This pre-B cell receptor may utilize a receptor signaling mechanism that shares features with antigen receptor signaling in mature B cells.

CONCLUSIONS

The antigen receptor of B lymphocytes is a multichain assembly of antigen-specific components (membrane Ig) and components involved in intracellular localization and signal transduction (Ig-α and Ig-β). This receptor functions both as a signaling receptor and as an endocytic receptor for uptake of antigen and presentation of processed antigen to helper T cells. In the latter role, the antigen receptor appears to recycle constitutively between the plasma membrane and endosomes. Complexes between the receptor and antigen are selectively degraded by an unknown mechanism. The signal transduction function of the receptor is mediated by a protein structure called the antigen receptor homology 1 (ARH1) motif, which is represented in the cytoplasmic domains of both Ig-α and Ig-β. Recent experiments suggest that this motif serves as a tyrosine phosphorylation site and then as a binding site for the intracellular tyrosine kinases $p53/56^{lyn}$, $p59^{fyn}$, and PTK72/*syk*. The exact roles of these different tyrosine kinases remain to be determined, but a number of targets of tyrosine phosphorylation have been identified as signaling components. These include PLC-γ, PI 3-kinase, rasGAP, two rasGAP-associated proteins, and the $p95^{vav}$ oncogene product. Thus downstream targets of B cell antigen receptor signaling resemble targets of the tyrosine kinase growth factor receptors.

The roles of these signaling components in mediating the biological responses of B cells to antigen contact remain to be determined. The ultimate result of antigen receptor signaling varies depending on the differentiation state of the B cell, but can include negative responses such as inactivation and apoptosis, or positive responses such as exit from the resting state, proliferation, and enhanced cell survival. Thus antigen receptor signaling plays a fundamental role in the regulation of B cell function. In addition, maturation of B cell precursors depends on expression of the transmembrane form of μ heavy chain, and this may reflect a signaling function of μ heavy chain in B cell development.

Literature Cited

Allen, RC, Armitage, RJ, Conley, ME, Rosenblatt, H, Jenkins, NA, et al. 1993. CD40 ligand gene defects responsible for X-linked hyper IgM-syndrome. *Science* 259:990–93

Amigorena, S, Bonnerot, C, Drake, JR, Choquet, D, Hunziker, W, et al. 1992. Cytoplasmic domain heterogeneity and functions of IgG Fc receptors in B lymphocytes. *Science* 256:1808–12

Aruffo, A, Farrington, M, Hollenbaugh, D, Li, X, Milatovich, A, et al. 1993. The CD40 ligand, gp39, is defective in activated T cells from patients with X-linked hyper-IgM syndrome. *Cell* 72:291–300

Auger, KR, Cantley, LC. 1991. Novel polyphosphoinositides in cell growth and activation. *Cancer Cells* 3:263–70

Bejcek, BE, Li, DY, Deuel, TF. 1989. Transformation by v-*sis* occurs by an internal autoactivation mechanism. *Science* 245: 1496–99

Berk, AJ. 1989. Regulation of eucaryotic transcription factors by post-translational modification. *Biochem. Biophys. Acta* 1009: 103–09

Berridge, M. 1987. Inositol trisphosphate and diacylglycerol: two interacting second messengers. *Annu. Rev. Biochem.* 56:159–93

Berzofsky, JA. 1983. T-B reciprocity. *Surv. Immunol. Res.* 2:223–29

Bijsterbosch, MK, Meade, CJ, Turner, GA, Klaus, GGB. 1985. B lymphocyte receptors and polyphosphoinositide degradation. *Cell* 41:999–1006

Bole, D, Hendershot, L, Kearny, J. 1986. Posttranslational association of immunoglobulin heavy chain binding protein with nascent heavy chains in nonsecreting and secreting hybridomas. *J. Cell Biol.* 102: 1558–66

Bourguignon, LYW, Bourguignon, GJ. 1984. Capping and the cytoskeleton. *Int. Rev. Cytol.* 87:195–224

Bourne, HR, Sanders, DA, McCormick, F. 1991. The GTPase superfamily: conserved structure and molecular mechanism. *Nature* 349:117–27

Braun, J, Sha'afi, R, Unanue, ER. 1979. Crosslinking by ligands to surface immunoglobulin triggers mobilization of intracellular $^{45}Ca^{2+}$ in B lymphocytes. *J. Cell. Biol.* 82:755–66

Brodsky, FM, Guagliardi, LE. 1991. The cell biology of antigen processing and presentation. *Annu. Rev. Immunol.* 9:707–44

Brunswick, M, Samelson, LE, Mond, JJ. 1991. Surface immunoglobulin crosslinking activates a tyrosine kinase pathway in B cells that is independent of protein kinase C. *Proc. Natl. Acad. Sci. USA* 88:1311–14

Burkhardt, AL, Brunswick, M, Bolen, JB, Mond, JJ. 1991. Anti-immunoglobulin stimulation of B lympocytes activates src-related protein-tyrosine kinases. *Proc. Natl. Acad. Sci. USA* 88:7410–14

Bustelo, XR, Barbacid, M. 1992. Tyrosine phosphorylation of the vav proto-oncogene product in activated B cells. *Science* 256: 1196–99

Buus, S, Sette, A, Colon, SM, Jenis, DM, Grey, HM. 1986. Isolation and characterization of antigen-Ia complexes involved in T cell recognition. *Cell* 47:1071–77

Cambier, JC. 1992. Signal transduction by T- and B-cell antigen receptors: converging structures and concepts. *Curr. Opin. Immunol.* 4:257–64

Campbell, KS, Cambier, JC. 1990. B lymphocyte antigen receptors (mIg) are non-covalently associated with a disulfide linked, inducibly phosphorylated glycoprotein complex. *EMBO J.* 9:441–48

Campbell, KS, Hager, EJ, Cambier, JC. 1991a. α-chains of IgM and IgD antigen receptor complexes are differentially N-glycosylated mb-1-related molecules. *J. Immunol.* 147:1575–80

Campbell, KS, Hager, EJ, Friedrich, RJ, Cambier, JC. 1991b. IgM antigen receptor complex contains phosphoprotein products of B29 and *mb-1* genes. *Proc. Natl. Acad. Sci. USA* 88:3982–86

Campbell, M-A, Sefton, BM. 1990. Protein tyrosine phosphorylation is induced in murine B lymphocytes in response to stimulation with anti-immunoglobulin. *EMBO J.* 9:2125–31

Campbell, M-A, Sefton, BM. 1992. Association between B-lymphocyte membrane immunoglobulin and multiple members of the Src family of protein tyrosine kinases. *Mol. Cell. Biol.* 12:2315–21

Cantley, LC, Auger, KR, Carpenter, C, Duckworth, B, Graziani, A, et al. 1991. Oncogenes and signal transduction. *Cell* 64: 281–302

Carter, RH, Fearon, DT. 1992. CD19: lowering the threshold for antigen receptor stimulation of B lymphocytes. *Science* 256: 105–07

Carter, RH, Park, DJ, Rhee, SG, Fearon, DT. 1991. Tyrosine phosphorylation of phospholipase C induced by membrane immunoglobulin in B lymphocytes. *Proc. Natl. Acad. Sci. USA* 88:2745–49

Carter, RH, Spycher, MO, Ng, YC, Hoffman, H, Fearon, DT. 1988. Synergistic interaction between complement receptor type 2 and membrane IgM on B lymphocytes. *J. Immunol.* 141:457–63

Casillas, A, Hanekom, C, Williams, K, Katz, R, Nel, AE. 1991. Stimulation of B-cells via the membrane immunoglobulin receptor or with phorbol myristate 13-acetate induces tyrosine phosphorylation and activation of a 42-kDa microtubule-associated protein-2 kinase. *J. Biol. Chem.* 266:19088–94

Casten, LA, Lakey, EK, Jelachich, ML, Margoliash, E, Pierce, SK. 1985. Anti-immunoglobulin augments the B-cell antigen-presentation function independently of receptor-antigen complex. *Proc. Natl. Acad. Sci. USA* 82:5890–94

Chan, AC, Irving, B, Fraser, JD, Weiss, A. 1991. The TCRζ chain associates with a tyrosine kinase and upon TCR stimulation associates with ZAP-70, a 70K M_r tyrosine phosphoprotein. *Proc. Natl. Acad. Sci. USA* 88:9166–70

Chan, AC, Iwashima, M, Turck, CW, Weiss, A. 1992. ZAP-70: A 70 kd protein-tyrosine kinase that associates with the TCRζ chain. *Cell* 71:649–62

Charles, CH, Abler, AS, Lau, LF. 1992. cDNA sequence of a growth-factor inducible immediate early gene and characterization of its encoded protein. *Oncogene* 7: 187–90

Chen, ZZ, Coggeshall, KM, Cambier, JC. 1986. Translocation of protein kinase C during membrane immunoglobulin-mediated transmembrane signaling in B lymphocytes. *J. Immunol.* 136:2300–04

Cherayil, BJ, Pillai, S. 1991. The ω/λ5 surrogate immunoglobulin light chain is expressed on the surface of transitional B lymphocytes in murine bone marrow. *J. Exp. Med.* 173:111–16

Chiles, T, Liu, J, Rothstein, T. 1991. Cross-linking of surface Ig receptors on murine B lymphocytes stimulates the expression of nuclear tetradecanol phorbol acetate-response element-binding proteins. *J. Immunol.* 146:1730–35

Chiles, TC, Rothstein, TL. 1992. Surface Ig receptor-induced nuclear AP-1-dependent expression in B lymphocytes. *J. Immunol.* 149:825–31

Chiu, R, Angel, P, Karin, M. 1989. Jun-B differs in its biological properties from, and is a negative regulator of, c-jun. *Cell* 59: 979–86

Clark, MR, Campbell, KS, Kazlauskas, A, Johnson, SA, Hertz, M, et al. 1992. The B cell antigen receptor complex: Association of Ig-α and Ig-β with distinct cytoplasmic effectors. *Science* 258:123–26

Coggeshall, KM, McHugh, JC, Altman, A. 1992. Predominant expression and activation-induced tyrosine phosphorylation of phospholipase C-γ2 in B lymphocytes. *Proc. Natl. Acad. Sci. USA* 90:5660–64

Dang, LH, Rock, KL. 1991. Stimulation of B lymphocytes through surface Ig receptors induces LFA-1 and ICAM-1-dependent adhesion. *J. Immunol.* 146:3273–79

Davidson, HW, Reid, PA, Lanzavecchia, A, Watts, C. 1991. Processed antigen binds to newly synthesized MHC class II molecules in antigen-specific B lymphocytes. *Cell* 67: 105–16

Davidson, HW, West, MA, Watts, C. 1990. Endocytosis, intracellular trafficking, and processing of membrane IgG and monovalent antigen/membrane IgG complexes in B lymphocytes. *J. Immunol.* 144:4101–09

DeFranco, AL. 1992a. Lymphocytes offer a helping hand. *Curr. Biol.* 2:477–79

DeFranco, AL. 1992b. Tyrosine phosphorylation and the mechanism of signal transduction by the B-lymphocyte antigen receptor. *Eur. J. Biochem.* 210:381–88

DeNagel, DC, Pierce, SK. 1991. Heat shock proteins implicated in antigen processing and presentation. *Semin. Immunol.* 3:65–71

DeNagel, DC, Pierce, SK. 1992. A case for chaperones in antigen processing. *Immunol. Today* 13:86–89

DiSanto, JP, Bonnefoy, JY, Gauchat, JF, Fischer, A, de Saint Basile, G. 1993. CD40 ligand mutations in X-linked immunodeficiency with hyper-IgM. *Nature* 361:541–43

Dubois, P, Stepinski, J, Urbain, J, Sibley, C. 1992. Role of the transmembrane and cytoplasmic domains of surface IgM in endocytosis and signal transduction. *Eur. J. Immunol.* 22:851–57

Dunbar, CE, Browder, TM, Abrams, JS, Nienhuis, AW. 1989. COOH-terminal-modified interleukin 3 is retained intracellularly and stimulates autocrine growth. *Science* 245:1493–96

Dustin, ML, Springer, TA. 1991. Role of

lymphocyte adhesion receptors in transient interactions and cell locomotion. *Annu. Rev. Immunol.* 9:27–66

Duyao, MP, Buckler, AJ, Sonenshein, G. 1990. Interaction of an NF-κB-like factor with a site upstream of the *c-myc* promoter. *Proc. Natl. Acad. Sci. USA* 87: 4727–31

Fantl, WJ, Escobedo, JA, Martin, GA, Turck, CW, del Rosario, M, et al. 1992. Distinct phosphotyrosines on a growth factor receptor bind to specific molecules that mediate different signaling pathways. *Cell* 69:413–23

Fisher, CL, Ghysdael, J, Cambier, JCC. 1991. Ligation of membrane Ig leads to calcium-mediated phosphorylation of the proto-oncogene product Ets-1. *J. Immunol.* 146: 1743–49

Freedman, AS, Freeman, G, Horowitz, JC, Daley, J, Nadler, LM. 1987. B7, a B cell-restricted antigen that identifies pre-activated B cells. *J. Immunol.* 139:3260–67

Fuleihan, R, Ramesh, N, Loh, R, Jabara, H, Rosen, FS, et al. 1993. Defective expression of the CD40 ligand in X chromosome-linked immunoglobulin deficiency with normal or elevated IgM. *Proc. Natl. Acad. Sci. USA* 90:2170–73

Gille, H, Sharrocks, AD, Shaw, PE. 1992. Phosphorylation of transcription factor $p62^{TCF}$ by MAP kinase stimulates ternary complex formation at *c-fos* promoter. *Nature* 358:414–17

Gold, MR, Crowley, MT, Martin, GA, McCormick, F, DeFranco, AL. 1993. Targets of B lymphocyte antigen receptor signal transduction include the $p21^{ras}$ GTPase-activating protein (GAP) and two GAP-associated proteins. *J. Immunol.* 150:377–86

Gold, MR, Law, DA, DeFranco, AL. 1990. Stimulation of protein tyrosine phosphorylation by the B-lymphocyte antigen receptor. *Nature* 345:810–13

Gold, MR, Matsuuchi, L, Kelly, RB, DeFranco, AL. 1991. Tyrosine phosphorylation of components of the B cell antigen receptors following receptor crosslinking. *Proc. Natl. Acad. Sci. USA* 88:3436–40

Gold, MR, Sanghera, JS, Stewart, J, Pelech, SL. 1992. Selective activation of p42 MAP kinase in murine B lymphoma cell lines by membrane immunoglobulin crosslinking. Evidence for protein kinase C-independent and -dependent mechanisms of activation. *Biochem. J.* 286:269–76

Goodnow, CC. 1992. Transgenic mice and analysis of B-cell tolerance. *Annu. Rev. Immunol.* 10:489–518

Graziadei, L, Riabowol, K, Bar-Sagi, D. 1990. Co-capping of ras proteins with surface immunoglobulins in B lymphocytes. *Nature* 347:396–400

Guagliardi, LE, Koppelman, B, Blum, JS, Marks, MS, Cresswell, P, Brodsky, FM. 1990. Co-localization of molecules involved in antigen processing and presentation in an early endocytic compartment. *Nature* 343: 133–39

Gulbins, E, Coggeshall, KM, Baier, G, Katzav, S, Burn, P, Altman, A. 1993. Tyrosine kinase-stimulated guanine nucleotide exchange activity of *vav* in T cell activation. *Science* 260:822–25

Haas, I, Wabl, M. 1983. Immunoglobulin heavy chain binding protein. *Nature* 306: 387–89

Harnett, MM, Klaus, GGB. 1988. G protein regulation of receptor signaling. *Immunol. Today* 9:315–20

Hartley, SB, Crosbie, J, Brink, R, Kantor, AB, Basten, A, Goodnow, CC. 1991. Elimination from peripheral lymphoid tissues of self-reactive B lymphocytes recognizing membrane-bound antigens. *Nature* 353: 765–69

Hebell, T, Ahearn, JM, Fearon, DT. 1991. Suppression of the immune response by a soluble complement receptor of B lymphocytes. *Science* 254:102–05

Hempel, WM, Schatzman, RC, DeFranco, AL. 1992. Tyrosine phosphorylation of phospholipase Cγ2 upon crosslinking of membrane Ig on murine B lymphocytes. *J. Immunol.* 148:3021–27

Hendershot, LM. 1990. Immunoglobulin heavy chain and binding protein complexes are dissociated in vivo by light chain addition. *J. Cell Biol.* 111:829–37

Hermanson, GG, Eisenberg, D, Kincade, PW, Wall, R. 1988. B29: A member of the immunoglobulin gene superfamily exclusively expressed on B-lineage cells. *Proc. Natl. Acad. Sci. USA* 85:6890–94

Hochstenbach, F, David, V, Watkins, S, Brenner, MB. 1992. Endoplasmic reticulum resident protein of 90 kilodaltons associates with the T- and B-cell antigen receptors and major histocompatability antigens during their assembly. *Proc. Natl. Acad. Sci. USA* 89:4734–38

Hombach, J, Leclercq, L, Radbruch, A, Rajewsky, K, Reth, M. 1988a. A novel 34-kd protein co-isolated with the IgM molecule in surface IgM-expressing cells. *EMBO J.* 7:3451–56

Hombach, J, Lottspeich, F, Reth, M. 1990a. Identification of the genes encoding IgM-α and Ig-β components of the IgM antigen receptor complex by amino-terminal sequencing. *Eur. J. Immunol.* 20:2795–99

Hombach, J, Sablitzky, F, Rajewsky, K, Reth, M. 1988b. Transfected plasmacytoma cells do not transport the membrane form of IgM to the cell surface. *J. Exp. Med.* 167:652–57

Hombach, J, Tsubata, T, Leclercq, L, Stappert, H, Reth, M. 1990b. Molecular components of the B-cell antigen receptor complex of the IgM class. *Nature* 343:760–62

Hornbeck, P, Paul, WE. 1986. Anti-immunoglobulin and phorbol esters induce phosphorylation of proteins associated with the plasma membrane and cytoskeleton in murine B lymphocytes. *J. Biol. Chem.* 261: 14817–24

Howard, M, Farrar, J, Hilfiker, M, Johnson, B, Takatsu, K, et al. 1982. Identification of a T cell-derived B cell growth factor distinct from interleukin 2. *J. Exp. Med.* 155:914–23

Hutchcroft, JE, Harrison, ML, Geahlen, RL. 1992. Association of the 72-kDa protein-tyrosine kinase PTK72 with the B cell antigen receptor. *J. Biol. Chem.* 267:8613–19

Iglesias, A, Kopf, M, Williams, GS, Bühler, B, Köhler, G. 1991. Molecular requirements for the μ-induced light chain gene rearrangement in pre-B cells. *EMBO J.* 10:2147–56

Irving, B, Chan, AC, Weiss, A. 1993. Functional characterization of a signal transducing motif present in the T cell receptor ζ chain. *J. Exp. Med.* 177:1093–1103

Irving, B, Weiss, A. 1991. The cytoplasmic domain of the T cell receptor ζ chain is sufficient to couple to receptor-associated signal transduction pathways. *Cell* 64:891–901

Janeway, CA Jr. 1989. The role of CD4 in T cell activation: accessory molecule or co-receptor? *Immunol. Today* 10:234–38

Justement, LB, Campbell, KS, Chien, NC, Cambier, JC. 1991. Regulation of B cell antigen receptor signal transduction and phosphorylation by CD45. *Science* 252: 1839–42

Justement, LB, Wienands, J, Hombach, J, Reth, M, Cambier, JC. 1990. Membrane IgM and IgD molecules fail to transduce Ca^{2+} mobilizing signals when expressed on differentiated B lineage cells. *J. Immunol.* 144:3272–80

Kanner, SB, Deans, JP, Ledbetter, JA. 1992. Regulation of CD3-induced phospholipase C-gamma1 (PLCγ1) tyrosine phosphorylation by CD4 and CD45 receptors. *Immunology* 75:441–47

Karasuyama, H, Kudo, A, Melchers, F. 1990. The proteins encoded by the VpreB and λ5 pre-B cell-specific genes can associate with each other and with μ heavy chain. *J. Exp. Med.* 172:969–72

Kazlauskas, A, Kashishian, A, Cooper, JA, Valius, M. 1992. GTPase-activating protein and phosphatidylinositol 3-kinase bind to a distinct region of the platelet-derived growth factor receptor β subunit. *Mol. Cell. Biol.* 12:2534–44

Keating, MT, Williams, LT. 1988. Autocrine stimulation of intracellular PDGF receptors in v-*sis* transformed cells. *Science* 239:914–16

Kelly, K, Cochran, BH, Stiles, CD, Leder, P. 1983. Cell-specific regulation of the c-myc gene by lymphocyte mitogens and platelet-derived growth receptor. *Cell* 35: 603–10

Kim, K-M, Alber, G, Weiser, P, Reth, M. 1993a. Differential signaling through the Ig-α and Ig-β components of the B cell antigen receptor. *Eur. J. Immunol.* 23:911–16

Kim, K-M, Alber, G, Weiser, P, Reth, M. 1993b. Signalling function of the B-cell antigen receptors. *Immunol. Rev.* 132:125–46

Kishimoto, T, Hirano, T. 1988. Molecular regulation of B lymphocyte response. *Annu. Rev. Immunol.* 6:485–512

Kitamura, D, Kudo, A, Schaal, S, Muller, W, Melchers, F, Rajewsky, K. 1992. A critical role of λ5 protein in B cell development. *Cell* 69:823–31

Kitamura, D, Rajewsky, K. 1992. Targeted disruption of μ membrane exon causes loss of heavy-chain allelic exclusion. *Nature* 356:154–56

Kitamura, D, Roes, J, Kuhn, R, Rajewsky, K. 1991. A B cell-deficient mouse by targeted disruption of the membrane exon of the immunoglobulin μ chain. *Nature* 350: 423–26

Klemsz, MJ, Justement, LB, Palmer, E, Cambier, JC. 1989. Induction of c-fos and c-myc expression during B cell activation by IL-4 and immunoglobulin binding ligands. *J. Immunol.* 143:1032–39

Knittler, MR, Haas, IG. 1992. Interaction of BiP with newly synthesized immunoglobulin light chain molecules: cycles of sequential binding and release. *EMBO J.* 11:1573–81

Koretzky, GA, Kohmetscher, MA, Kadleck, T, Weiss, A. 1992. Restoration of T cell receptor-mediated signal transduction by transfection of CD45 cDNA into a CD45-deficient variant of the Jurkat T cell line. *J. Immunol.* 149:1138–42

Koretzky, GA, Picus, J, Schultz, T, Weiss, A. 1991. The tyrosine phosphatase CD45 is essential for coupling T-cell antigen receptor and CD2 mediated activation of a protein tyrosine kinase and interleukin 2 production. *Proc. Natl. Acad. Sci. USA* 88:2037–41

Korthauer, U, Graf, D, Mages, HW, Briere, F, Padayachee, M, et al. 1993. Defective expression of T-cell CD40 ligand causes

X-linked immunodeficiency with hyper-IgM. *Nature* 361:539–41

Kupfer, A, Singer, SJ. 1989. Cell biology of cytotoxic and helper T cell functions: Immunofluorescence microscopic studies of single cells and cell couples. *Annu. Rev. Immunol.* 7:309–37

Lane, PJL, Ledbetter, JA, McConnell, FM, Draves, K, Deans, J, et al. 1991. The role of tyrosine phosphorylation in signal transduction through surface Ig in human B cells. *J. Immunol.* 146:715–22

Lane, PJL, McConnell, GM, Schieven, GL, Clark, EA, Ledbetter, JA. 1990. The role of class II molecules in human B cell activation: Association with phosphatidylinositol turnover, protein tyrosine phosphorylation and proliferation. *J. Immunol.* 144:3684–92

Lanzavecchia, A. 1985. Antigen-specific interaction between T and B cells. *Nature* 314:537–39

Lanzavecchia, A. 1990. Receptor-mediated antigen uptake and its effect on antigen presentation to class II restricted T lymphocytes. *Annu. Rev. Immunol.* 8:773–93

Lenardo, M, Baltimore, D. 1989. NF-κB: A pleiotropic mediator of inducible and tissue-specific gene control. *Cell* 58:227–29

Leprince, C, Draves, KE, Geahlen, RL, Ledbetter, JA, Clark, EA. 1993. CD22 associates with the human surface IgM-B cell antigen receptor complex. *Proc. Natl. Acad. Sci. USA* 90:3236–40

Leprince, C, Draves, KE, Ledbetter, JA, Torres, RM, Clark, EA. 1992. Characterization of the molecular components associated with surface immunoglobulin M in human B lymphocytes: presence of tyrosine and serine/threonine protein kinases. *Eur. J. Immunol.* 22:2093–99

Letourneur, F, Klausner, RD. 1992. Activation of T cells by a tyrosine kinase activation domain in the cytoplasmic tail of CD3ϵ. *Science* 255:79–82

Li, Z-H, Mahajan, S, Prendergast, MM, Fargnoli, J, Zhu, Z, et al. 1992. Cross-linking of surface immunoglobulin activates src-related tyrosine kinases in WEHI 231 cells. *Biochem. Biophys. Res. Comm.* 187:1536–44

Liu, J, Chiles, TC, Sen, R, Rothstein, TL. 1991. Inducible nuclear expression of NF-κB in primary B cells stimulated through the surface Ig receptor. *J. Immunol.* 146:1685–91

Liu, Y, Linsley, PS. 1992. Costimulation of T-cell growth. *Curr. Opin. Immunol.* 4:265–70

Liu, Y-J, Joshua, DE, Williams, GT, Smith, CA, Gordon, J, MacLennan, ICM. 1989. Mechanism of antigen-driven selection in germinal centres. *Nature* 342:929–31

Longnecker, R, Druker, B, Roberts, TM, Kieff, E. 1991. An Epstein-Barr virus protein associated with cell growth transformation interacts with a tyrosine kinase. *J. Virol.* 65:3681–92

Matsuuchi, L, Gold, MR, Travis, A, Grosschedl, R, DeFranco, AL, Kelly, RB. 1992. The membrane IgM-associated proteins MB-1 and Ig-β are sufficient to promote surface expression of a partially functional B-cell antigen receptor in a nonlymphoid cell line. *Proc. Natl. Acad. Sci. USA* 89:3404–8

Miettinen, HM, Matter, K, Hunziker, W, Rose, JK, Mellman, I. 1992. Fc receptor endocytosis is controlled by a cytoplasmic domain determinant that actively prevents coated pit localization. *J. Cell Biol.* 116:875–88

Misener, V, Downey, GP, Jongstra, J. 1991. The immunoglobulin light chain related protein λ5 is expressed on the surface of mouse pre-B cell lines and can function as a signal transducing molecule. *Int. Immunol.* 3:1–8

Mitchell, RN, Shaw, AC, Weaver, YK, Leder, P, Abbas, AK. 1991. Cytoplasmic tail deletion converts membrane immunoglobulin to a phosphatidylinositol-linked form lacking signaling and efficient antigen internalization functions. *J. Biol. Chem.* 266:8856–60

Mittelstadt, P, DeFranco, AL. 1993. Induction of early-response genes by cross-linking membrane immunoglobulin on B lymphocytes. *J. Immunol.* 150:4822–32

Mond, JJ, Seghal, E, Kung, J, Finkelman, FD. 1981. Increased expression of I-region associated antigen (Ia) on B cells after crosslinking of surface immunoglobulin. *J. Immunol.* 127:881–88

Monroe, JG, Kass, MJ. 1985. Molecular events in B cell activation I. Signals required to stimulate Go to G1 transition of resting B lymphocytes. *J. Immunol.* 135:1674–82

Munro, S, Pelham, HRB. 1986. An hsp70-like protein in the ER: Identity with the 78kd glucose-regulated protein and immunoglobulin heavy chain binding protein. *Cell* 46:291–300

Myers, CD. 1991. Role of B cell antigen processing and presentation in the humoral immune response. *FASEB J.* 5:2547–53

Nakamura, T, Kubagawa, H, Cooper, MD. 1992. Heterogeneity of immunoglobulin-associated molecules on human B cells identified by monoclonal antibodies. *Proc. Natl. Acad. Sci. USA* 89:8522–26

Nakanishi, H, Brewer, KA, Exton, JH. 1993. Activation of the ζ isozyme of protein kinase C by phosphatidylinositol 3,4,5-trisphosphate. *J. Biol. Chem.* 268:13–16

Nel, AE, Wooten, MW, Landreth, GE, Goldschmidt-Clermont, PJ, Stevenson, HC, et al. 1986. Translocation of phospholipid/Ca^{2+}-dependent protein kinase in B-lymphocytes activated by phorbol ester or cross-linkage of membrane immunoglobulin. *Biochem. J.* 233:145–49

Noelle, RJ, Snow, EC. 1991. T helper cell-dependent B cell activation. *FASEB J.* 5:2770–76

Nomura, J, Matsuro, T, Kubota, E, Kimoto, M, Sakaguchi, N. 1991. Signal transmission through the B cell-specific MB-1 molecule at the pre-B cell stage. *Int. Immunol.* 3:117–26

Nussenzweig, M, Shaw, A, Sinn, E, Danner, D, Holmes, K, et al. 1987. Allelic exclusion in transgenic mice that express the membrane form of immunoglobulin μ. *Science* 236:816–19

O'Garra, A, Umland, S, DeFrance, T, Christiansen, J. 1988. B cell factors are pleiotropic. *Immunol. Today* 9:45–54

Ostergaard, HL, Shackelford, DA, Hurley, TR, Johnson, P, Hyman, R, et al. 1989. Expression of CD45 alters phosphorylation of the *lck*-encoded tyrosine protein kinase in murine lymphoma T-cell lines. *Proc. Natl. Acad. Sci. USA* 86:8959–63

Ostergaard, HL, Trowbridge, IS. 1991. Negative regulation of CD45 protein tyrosine phosphatase activity by ionomycin in T cells. *Science* 253:1423–25

Padeh, S, Levitzki, A, Gazit, A, Mills, GB, Roifman, CM. 1991. Activation of phospholipase C in human B cells is dependent on tyrosine phosphorylation. *J. Clin. Invest.* 87:1114–18

Parikh, V, Nakai, C, Yokota, S, Bankert, R, Tucker, P. 1991. COOH terminus of membrane IgM is essential for an antigen-specific induction of some but not all early activation events in mature B cells. *J. Exp. Med.* 174:1103–09

Paul, WE, Mizuguchi, J, Brown, M, Nakanishi, K, Hornbeck, P, et al. 1986. Regulation of B-lymphocyte activation, proliferation, and immunoglobulin secretion. *Cell. Immunol.* 99:7–13

Pesando, J, Bouchard, L, McMaster, B. 1989. CD19 is functionally and physically associated with surface immunoglobulin. *J. Exp. Med.* 170:2159–64

Pillai, S, Baltimore, D. 1987. Formation of disulphide-linked μ2ω2 tetramers in pre-B cells by the 18K ω-immunoglobulin light chain. *Nature* 329:172–74

Pingel, JT, Thomas, ML. 1989. Evidence that the leukocyte-common antigen is required for antigen-induced T lymphocyte proliferation. *Cell* 58:1055–65

Pognonec, P, Boulukos, KE, Ghysdael, J. 1989. The c-ets-1 protein is chromatin associated and binds DNA in vitro. *Oncogene* 4:691–97

Pozzan, T, Arslan, P, Tsien, RY, Rink, TJ. 1982. Anti-immunoglobulin, cytoplasmic free calcium, and capping in B lymphocytes. *J. Cell Biol.* 94:335–40

Puil, L, Pawson, T. 1992. Vagaries of Vav. *Curr. Biol.* 2:275–77

Pulverer, BJ, Kyriakis, JM, Arruch, J, Nikolaki, E, Woodgett, JR. 1991. Phosphorylation of *c-jun* mediated by MAP kinases. *Nature* 353:670–74

Ransone, LJ, Verma, IM. 1990. Nuclear proto-oncogenes fos and jun. *Annu. Rev. Cell Biol.* 6:539–57

Reth, M. 1989. Antigen receptor tail clue. *Nature* 338:383–84

Reth, M. 1992. Antigen receptors on B lymphocytes. *Annu. Rev. Immunol.* 10:97–121

Reth, M, Petrac, E, Wiese, P, Lobel, L, Alt, FW. 1987. Activation of Vκ gene rearrangement in pre-B cells follows the expression of membrane-bound immunoglobulin heavy chains. *EMBO J.* 6:3299–305

Ridley, AJ, Hall, A. 1992. The small GTP-binding protein rho regulates the assembly of focal adhesions and actin stress fibers in response to growth factors. *Cell* 70:389–99

Ridley, AJ, Paterson, HF, Johnston, CL, Diekmann, D, Hall, A. 1992. The small GTP-binding protein rac regulates growth factor-induced membrane ruffling. *Cell* 70:401–10

Roifman, CM, Wang, G. 1992. Phospholipase C-γ1 and phospholipase C-γ2 are substrates of the B cell antigen receptor associated protein tyrosine kinase. *Biochem. Biophys. Res. Commun.* 183:411–16

Rolink, A, Melchers, F. 1991. Molecular and cellular origins of B lymphocyte diversity. *Cell* 66:1081–94

Romeo, C, Amiot, M, Seed, B. 1992. Sequence requirements for induction of cytolysis by the T cell antigen/Fc receptor ζ chain. *Cell* 68:889–97

Romeo, C, Seed, B. 1991. Cellular immunity to HIV activated by CD4 fused to T cell or Fc receptor polypeptides. *Cell* 64:1037–46

Rooney, JW, Dubois, PM, Sibley, CH. 1991. Cross-linking of surface IgM activates NF-κB in B lymphocytes. *Eur. J. Immunol.* 21:2993–98

Rothstein, TL, Baeker, TR, Miller, RA, Kolber, DL. 1986. Stimulation of murine B cells by the combination of calcium ionophore plus phorbol ester. *Cell. Immunol.* 102:364–73

Russell, DM, Dembic, Z, Morahan, G, Miller, JF, Burki, K, Nemazee, DA. 1991. Peripheral deletion of self-reactive B cells. *Nature* 354:308–11

Sakaguchi, N, Kashiwamura, S-I, Kimoto, M, Thalmann, P, Melchers, F. 1988. B lymphocyte lineage-restricted expression of mb-1, a gene with CD3-like structural properties. *EMBO J.* 7:3457–64

Samelson, LE, Klausner, RD. 1992. Tyrosine kinases and tyrosine-based activation motifs. *J. Biol. Chem.* 267:24913–16

Schatz, DG, Oettinger, MA, Schlissel, MS. 1992. V(D)J recombination:molecular biology and regulation. *Annu. Rev. Immunol.* 10:359–83

Schulte, RJ, Campbell, M-A, Fischer, WH, Sefton, BM. 1992. Tyrosine phosphorylation of CD22 during B cell activation. *Science* 258:1001–4

Settleman, J, Albright, CF, Foster, LC, Weinberg, RA. 1992a. Association between GTPase activators for Rho and Ras families. *Nature* 359:153–54

Settleman, J, Narasimhan, V, Foster, LC, Weinberg, RA. 1992b. Molecular cloning of cDNAs encoding the GAP-associated protein p190: Implications for a signaling pathway from Ras to the nucleus. *Cell* 69: 539–49

Seyfert, VL, McMahon, S, Glenn, W, Cao, X, Sukhatme, VP, Monroe, JG. 1990. Egr-1 expression in surface Ig-mediated B cell activation. Kinetics and association with protein kinase C activation. *J. Immunol.* 145:3647–53

Seyfert, VL, Sukhatme, VP, Monroe, JG. 1989. Differential expression of a zinc finger-encoding gene in response to positive versus negative signaling through receptor immunoglobulin in murine B lymphocytes. *Mol. Cell. Biol.* 9:2083–88

Shapiro, AM, Schlissel, MS, Baltimore, D, DeFranco, AL. 1993. Stimulation of κ light chain gene rearrangement by immunoglobulin μ heavy chain in a pre-B cell line. *Mol. Cell. Biol.* 13: In press

Shaw, AC, Mitchell, RN, Weaver, YK, Campos-Torres, J, Abbas, AK, Leder, P. 1990. Mutations of immunoglobulin transmembrane and cytoplasmic domains: Effects on intracellular signaling and antigen presentation. *Cell* 63:381–92

Songyang, Z, Shoelson, SE, Chaudhuri, M, Gish, G, Pawson, T, et al. 1993. SH2 domains recognize specific phosphoprotein sequences. *Cell* 72:767–78

Stamenkovic, I, Clark, EA, Seed, B. 1989. A B cell activation molecule related to the nerve growth factor receptor and induced by cytokines in carcinomas. *EMBO J.* 8: 1403–10

Storb, U. 1987. Transgenic mice with immunoglobulin genes. *Annu. Rev. Immunol.* 5:151–74

Stover, DR, Charbonneau, H, Tonks, NK, Walsh, KA. 1991. Protein-tyrosine-phosphatase CD45 is phosphorylated transiently on tyrosine upon activation of Jurkat T cells. *Proc. Natl. Acad. Sci. USA* 88:7704–07

Straus, D, Weiss, A. 1992. Genetic evidence for the involvement of the lck tyrosine kinase in signal transduction through the T cell antigen receptor. *Cell* 70:585–93

Takagi, S, Daibata, M, Last, TJ, Humphreys, RE, Parker, DC, Sairenji, T. 1991. Intracellular localization of tyrosine kinase substrates beneath crosslinked surface immunoglobulins in B cells. *J. Exp. Med.* 174:381–88

Takemori, T, Mizuguchi, J, Miyazoe, I, Nakanishi, M, Shigemoto, K, et al. 1990. Two types of μ chain complexes are expressed during differentiation from pre-B to mature B cells. *EMBO J.* 9:2493–500

Taniguchi, T, Kobayashi, T, Kondo, J, Takahashi, K, Nakamura, H, et al. 1991. Molecular cloning of a porcine gene *syk* that encodes a 72kDa protein-tyrosine kinase showing high susceptibility to proteolysis. *J. Biol. Chem.* 266:15790–96

Tilzey, JF, Chiles, TC, Rothstein, TL. 1991. *Jun-B* gene expression mediated by the surface immunoglobulin receptor of primary B lymphocytes. *Biochem. Biophys. Res. Commun.* 175:77–83

Tony, H-P, Phillips, NE, Parker, DC. 1985. Role of membrane immunoglobulin (Ig) crosslinking in membrane Ig-mediated, major histocompatability-restricted T cell-B cell cooperation. *J. Exp. Med.* 162:1695–708

Tsubata, T, Reth, M. 1990. The products of the pre-B cell specific genes (λ5 and V_{preB}) and the immunoglobulin μ chain form a complex that is transported onto the cell surface. *J. Exp. Med.* 172:973–76

Tsubata, T, Tsubata, R, Reth, M. 1992. Crosslinking of the cell surface immunoglobulin (μ-surrogate light chains complex) on Pre-B cells induces activation of V gene rearrangements at the immunoglobulin κ locus. *Int. Immunol.* 4: 637–41

Tsukada, S, Saffran, DC, Rawlings, DJ, Parolini, O, Allen, RC, et al. 1993. Deficient expression of a B cell cytoplasmic tyrosine kinase in human X-linked agammaglobulinemia. *Cell* 72:279–90

Tuveson, DA, Carter, RH, Soltoff, SP, Fearon, DT. 1993. CD19 of B cells as a surrogate kinase insert region to bind phosphatidylinositol 3-kinase. *Science* 260:986–9

VanBuskirk, AM, DeNagel, DC, Guagliardi,

LE, Brodsky, FM, Pierce, SK. 1991. Cellular and subcellular distribution of PBP72/74, a peptide-binding protein that plays a role in antigen processing. *J. Immunol.* 146:500–06

van Noesel, CJM, Lankester, AC, van Lier, RAW. 1993. Dual antigen recognition by B cells. *Immunol. Today* 14:8–11

Venkitaraman, A. 1992. Light chain surrogacy. *Curr. Biol.* 2:559–61

Venkitaraman, AR, Williams, GT, Dariavach, P, Neuberger, MS. 1991. The B cell antigen receptor of the five immunoglobulin classes. *Nature* 352:777–81

Vetrie, D, Vorechovsky, I, Sideras, P, Holland, J, Davies, A, et al. 1993. The gene involved in X-linked agammaglobulinemia is a member of the src family of protein-tyrosine kinases. *Nature* 361:226–33

Vitetta, ES, Fernandez-Botran, R, Myers, CD, Sanders, VM. 1989. Cellular interactions in the humoral immune response. *Adv. Immunol.* 45:1–105

Wall, R, Kuehl, M. 1983. Biosynthesis and regulation of immunoglobulins. *Annu. Rev. Immunol.* 1:393–422

Wegener, A-MK, Letourneur, F, Hoeveler, A, Brocker, T, Luton, F, Malissen, B. 1992. The T cell receptor/CD3 complex is composed of at least two autonomous transduction molecules. *Cell* 68:83–95

Wienands, J, Reth, M. 1992. Glycosyl-phosphatidylinositol linkage as a mechanism for cell-surface expression of immunoglobulin D. *Nature* 356:246–48

Williams, GT, Venkitaraman, AR, Gilmore, DJ, Neuberger, MS. 1990. The sequence of the μ transmembrane segment determines the tissue specificity of the transport of immunoglobulin M to the cell surface. *J. Exp. Med.* 171:947–52

Wong, G, Müller, O, Clark, R, Conroy, L, Moran, MF, et al. 1992. Molecular cloning and nucleic acid binding properties of the GAP-associated tyrosine phosphoprotein p62. *Cell* 69:551–58

Yamanashi, Y, Kakiuchi, T, Mizuguchi, J, Yamamoto, T, Toyoshima, K. 1991. Association of B cell antigen receptor with protein tyrosine kinase lyn. *Science* 251:192–94

Yancopoulos, GD, Alt, FW. 1986. Regulation of the assembly and expression of variable-region genes. *Annu. Rev. Immunol.* 4:339–68

Yang, L, Baffy, G, Rhee, SG, Manning, D, Hansen, CA, Williamson, JR. 1991. Pertussis toxin-sensitive Gi protein involvement in epidermal growth factor-induced activation of phospholipase C-γ in rat hepatocytes. *J. Biol. Chem.* 266:22451–58

Zioncheck, TF, Harrison, ML, Geahlen, R. 1986. Purification and characterization of a protein tyrosine kinase from bovine thymus. *J. Biol. Chem.* 261:15637–43

Annu. Rev. Cell Biol. 1993. 9:411–44

LIFE AT THE LEADING EDGE: The Formation of Cell Protrusions

John Condeelis

Department of Anatomy and Structural Biology, Albert Einstein College of Medicine, Bronx, New York 10461

KEY WORDS: actin, myosin, lamellipodia, motility, signal transduction

---

CONTENTS

INTRODUCTION

In this review we consider what is known about the mechanism of protrusion of the leading edge of motile cells. Protrusions at the leading edge have been given different names depending on their shape (Taylor & Condeelis 1979); for example, lobopodia are cylindrical, filopodia are needle-shaped, and pharopodia are flat veils. Pharopodia are also called lamellipodia. For the purposes of this review lobopodia and pharopodia are referred to as pseudopods since they appear to differ only in the degree of cytoplasmic

0743–4634/93/1115–0411$05.00

streaming accompanying their assembly, and both contain networks of actin filaments. Filopodia or filopods will refer only to needle-shaped structures containing a single bundle of unipolar actin filaments.

A number of studies with neutrophils (Marks & Maxfield 1990), *Dictyostelium* (Van Duijn & Van Haastert 1992), and keratocytes (Cooper & Schliwa 1986) demonstrate that extension of pseudopods and cell migration are separable, i.e. cells can migrate without changing shape, and pseudopod extension is not always followed by cell migration. In addition, pseudopods are three-dimensional structures that can be extended without the involvement of cell-substrate or cell-cell contacts. Therefore, in this review, the forces involved in pseudopod and filopod extension are considered without regard to mechanisms for cell migration, cell-substrate, and cell-cell contact.

In particular the following questions are discussed: What forces are most likely responsible for protrusive activity; Which of these are consistent with our current understanding of the mechanical and biochemical dynamics of the leading edge of the cell; What models have been proposed that integrate this information into testable hypotheses for protrusive activity?

Due to this narrow scope, many related issues are only touched on. Readers are referred to recent reviews for in-depth treatment of the following topics: actin binding proteins (Hartwig & Kwiatchowski 1991; Luna & Condeelis 1990); actin structure and function (Kabsch & Vandekerckhove 1992); actin-membrane interactions (Luna & Hitt 1992); mechanics of the leading edge (Oster & Perelson 1992; Peskin et al 1993); actin polymerization and cell movement (Cooper 1991); myosin (Cheney & Mooseker 1992); and amoeboid movement (Fukui 1993).

GENERATION OF FORCE FOR PROTRUSION

Three general mechanisms have been advanced for the generation of force for protrusive activity. The first proposes that the polymerization of actin is sufficient to generate a pushing force against the cell membrane to cause protrusion. The second proposes that the interaction between actin and myosin is required to generate force for extension of the processes. The third proposes that osmotic forces are involved in protrusive activity, in particular osmotic pressure generated in actin gels, or at the site of actin polymerization.

Forces Derived from Actin Polymerization

How much force is available as the result of actin polymerization, and is this sufficient to distort the cell membrane? The free energy of polymerization can be calculated from $\Delta G = RT \ln C1/Co$ (Hill & Kirschner 1982), where R is the gas constant = 8 • 3J/K/mol, T = temperature in degrees Kelvin, C1 = the monomer concentration, and Co = the critical concentration of 0.1μM. It has been estimated (Cooper 1991) that at ambient temperature in a typical 0.1μm diameter microvillus containing ten actin filaments, an equilibrium pressure[1] of 4×10^{-4} dyne/μm^2 or 4×10^{4} dyne/cm^2 is generated during the addition of a monomer to the barbed end of each filament in the presence 1μM monomer. A similar calculation done for polymerization of the barbed ends of filaments in the acrosomal process of Thyone sperm, where between 35 and 140 filaments pack the 0.1μm diameter process (Tilney et al 1973), predicts pressures between 14×10^{4} and 59×10^{4} dyne/cm^2. These values are large compared to a pressure of 400 dynes/cm^2 required to aspirate a hemispherical projection of 2μm in diameter from the granulocyte cell surface into a micropipette (Evans & Yeung 1989). Furthermore, the aspiration pressure is likely an overestimate of the stiffness of the cell membrane since the underlying cortical membrane skeleton must also be distorted in this experiment. Thus actin polymerization from structures normally associated with the cell cortex could generate force that is sufficient to protrude the cell surface.

The above thermodynamic argument that the free energy associated with actin polymerization adequately accounts for the force required to protrude the membrane provides no mechanism for how free energy is converted into mechanical pushing. However, a recently decribed model, the Brownian Ratchet Model (Peskin et al 1993), does propose such a mechanism. In this model, the ratchet is the mean excursion path of a thermally vibrating membrane. Each time the membrane is displaced a distance sufficient to permit addition of an actin monomer, the return of the membrane to its original position is prevented (Figure 1). Flicher spectra, as determined with healthy unfixed human erythrocytes, indicate that at room temperature the membrane oscillates over distances of about 0.1μm with a half-distance frequency of 5 Hz, which indicates that 0.1 μm× 5/sec = 0.5 μm/sec rates of membrane extension are possible if actin polymerizes against a thermally vibrating membrane (Fricke & Sackmann 1984), a rate actually faster than, but similar to, the 0.2 μm/sec observed for filopod extension in neuronal growth cones.

The conclusion that actin polymerization can generate sufficient force to protrude the cell membrane is supported by in vitro experiments. Actin monomers that have been encapsulated in lipid vesicles can be polymerized by increasing the internal ionic strength or by increasing temperature (Cortese

[1](pressure = (ΔG/distance extended by addition of one monomer per filament)÷area of microvillus tip = $(9.5\times10^{-13}\text{erg}/3\times10^{-7}\text{cm})\div7.9\times10^{-3}\ \mu m^2 = 4\times10^{-4}\text{dyne}/\mu m^2$)

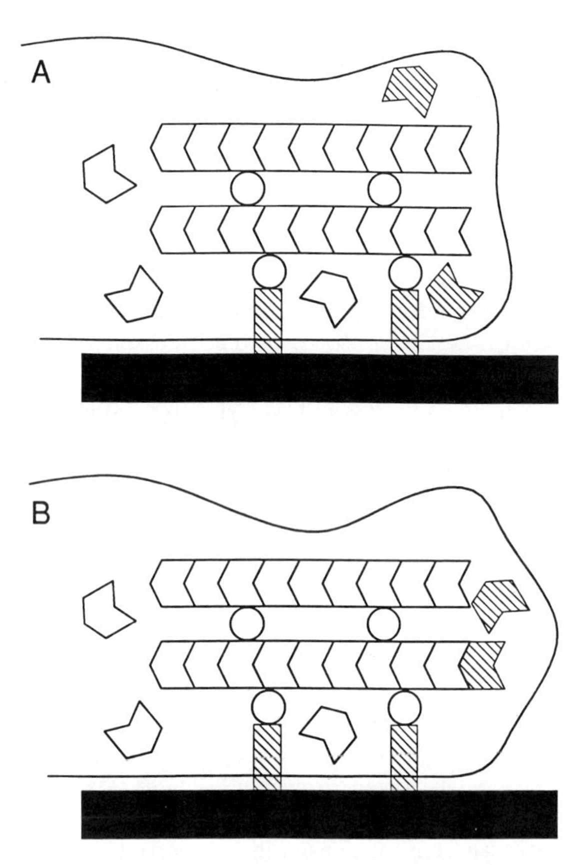

Figure 1 Protrusive force caused by actin polymerization. A cut-away of the leading edge is shown in profile. The solid bar represents the substrate. Actin filaments (*chevrons*) are attached via membrane proteins (*hatched bars*) through the membrane to the substrate and are cross-linked by actin-binding proteins (*circles*). Thermal vibration of the membrane permits monomer addition to the barbed ends of actin filaments, which prevents the membrane from rebounding. Filaments need not be in bundles since the polymerization and cross-linking of actin to form rigid filament networks could extend the membrane by the same mechanism. (*A*) before and (*B*) after monomer addition.

et al 1989; Miyata & Hotani 1992). Polymerization is correlated with a change in shape of the vesicles from spherical to asymmetric, and in some cases the vesicles form protrusions. In one study, the extent of change in shape is directly related to filament length as regulated by polymerization in the presence of various amounts of gelsolin (Cortese et al 1989).

Microvilli of membrane-intact brush borders that have been isolated from intestinal epithelial cells can be induced to elongate by the addition of monomeric actin (Mooseker et al 1982). The microvillar membrane is pushed in the direction of polymerization.

Unexpected support for an actin polymerization-based mechanism for protrusive activity comes from the behavior of infectious bacteria such as *Listeria* and *Shigella* in the cytoplasm of host cells, and from the behavior of polycationic beads on the surface of nerve growth cones. Phagocytosed bacteria that escape into cytoplasm recruit the host's actin to polymerize into a comet-like tail (Bernardini et al 1989; Tilney & Portnoy 1989). Polymerization of actin within this tail appears to propel the bacteria rapidly through the cytoplasm (Dabiri et al 1990; Sanger et al 1992). Photoactivation of labeled actin at a discrete site in the tail indicates that the tail is elongating adjacent to the bacterium at the rate of propulsion (Theriot et al 1992), consistent with propulsive force based on actin polymerization as proposed by Dabiri and co-workers (Dabiri et al 1990).

Interaction of polycationic microbeads with the dorsal surface of a neuronal growth cone usually causes the beads to move with the underlying centripetal flux of cytoplasmic actin at ~1μm/min as described below. However, some beads induce localized polymerization of actin that appears to propel the beads in random directions on the cell surface at velocities of 10μm/min or ~0.2 μm/sec (Forscher et al 1992).

Thus a strong case can be made for the direct involvement of actin polymerization in the generation of protrusive force based on theoretical considerations, in vivo models, and in vitro experiments.

Forces Derived From the Interaction of Actin and Myosin

It is generally agreed that there is sufficient force generated by the cross-bridge sliding interaction of actin and myosin to cause membrane protrusion (Oster & Perelson 1993). The issues are whether myosin is present in the right location at the right time in the cell to generate protrusive force, and whether actin filaments in that location have a polarity that can support protrusion by a sliding filament mechanism.

At least two types of myosin are found in most cell types. Myosin II is the conventional filament-forming myosin, while myosin I is a smaller protein with a single heavy chain that does not form filaments. Both will slide actin filaments in an ATP-dependent reaction and have similar amino terminal (active site) sequences (Korn & Hammer 1990).

Myosin II is localized in cultured cells in stress fibers and in general primarily behind the leading edge (Conrad et al 1989). Protrusive activity in cultured cells is usually confined to the leading edge, and the leading lamellipod is autonomous in its protruding and ruffling activities (Albrecht-Buehler 1980). In amoeboid cells, myosin II is localized primarily in lateral and tail portions of the cell, but little if any exists in the leading edge.

This has been well characterized in locomoting *Acanthamoeba* (Baines & Korn 1990), and in *Dictyostelium* during cell locomotion (Fukui et al 1989), chemotaxis (Rubino et al 1984), and ligand capping (Carboni & Condeelis 1985). Myosin II could participate in pseudopod extension even in this distant location by generating hydrostatic pressure through catalyzing tail contraction (Condeelis 1992). However, deletion of the single myosin II heavy chain gene by homologous recombination (DeLozanne & Spudich 1987) or inhibition of expression of the heavy chain by antisense interference (Knecht & Loomis 1987) results in *Dictyostelium* amoebae that are capable of pseudopod extension (Wessels et al 1988). Together, these results demonstrate that myosin II is not necessary for pseudopod extension.

Myosin I is localized in cultured cells at the cell periphery, in pseudopods and filopodia, and throughout the perinuclear cytoplasm (Wagner et al 1992; Conrad et al 1993). In amoeboid cells, myosin I has been localized with isoform-specific antibodies. In *Acanthamoeba,* myosin IA and IB show similar locations throughout the cytoplasm and in close apposition to the cell membrane (IA, Hagen et al 1986; IB, Miyata et al 1989). *Acanthamoeba* myosin IC is localized globally in close apposition to cell membrane and localized most intensely in association with the contractile vacuole (Baines & Korn 1990). In *Dictyostelium*, polyclonal antibodies prepared against *Dictyostelium* myosin I stain the leading edge and pseudopods as well as posterior cytoplasm (Fukui et al 1989). It is difficult to compare the results in *Dictyostelium* with those in *Acanthamoeba* since the isoform specificity of the *Dictyostelium* myosin I antibody is unknown. However, the pattern of localization is consistent with the binding of myosin I to cellular membranes and in vesicle transport as demonstrated by in vitro experiments (Cheney & Mooseker 1992). It is also consistent with a role for myosin I in the generation of force at the leading edge.

Several models have been proposed for the involvement of myosin I in protrusion. Many isoforms of myosin I contain two actin binding sites, an ATP-sensitive amino-terminal site common to all myosins, and a more C-terminal ATP-insensitive actin-binding site. These myosins I are ATP-sensitive actin cross-linking proteins that can slide one actin filament relative to another (Korn & Hammer 1990). Thus it has been proposed that myosin I-mediated sliding between parallel filaments could cause protrusion if all of the filaments have the same polarity with barbed ends adjacent to the membrane. No actin polymerization would be required for protrusive activity in this model (Egelhoff & Spudich 1991) (Figure 2).

In other models, a combination of actin polymerization and myosin I activity have been proposed. In these models, myosin I must be attached to a rigid membrane-associated helmet, and actin filaments must be parallel and unipolar with barbed ends adjacent to the membrane. If the membrane

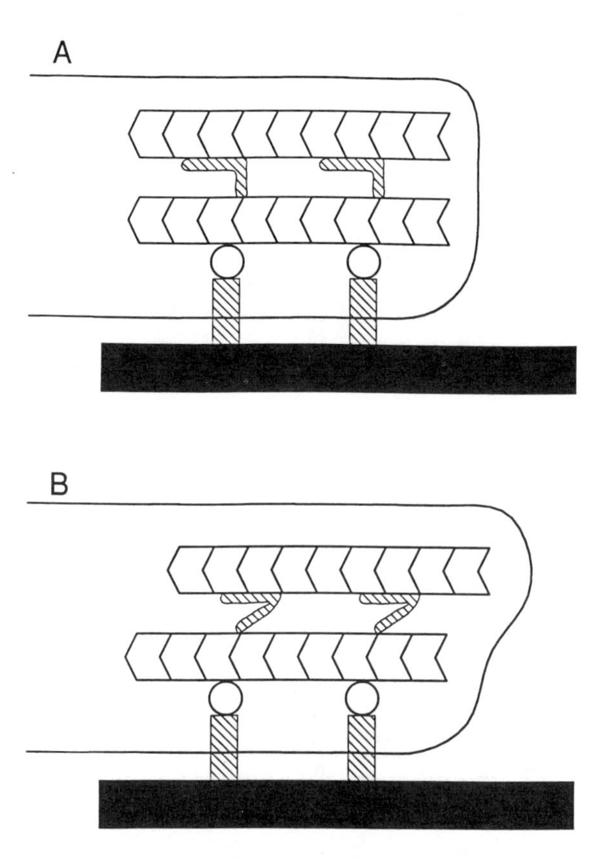

Figure 2 Protrusive force caused by the sliding of actin filaments by myosin I. Symbols are as defined in Figure 1. Myosin I is represented by the cross-hatched L-shaped structure cross-linking actin filaments. (*A*) before and (*B*) after myosin I-mediated filament sliding.

helmet forms a barrier to polymerization at the barbed ends of the filaments, actin polymerization and, therefore, protrusion will only occur when myosin I moves toward the barbed ends of the filaments pushing the rigid membrane helmet forward and away from the barbed ends (Mitchison & Kirschner 1988; Sheetz et al 1992) (Figure 3).

These models are consistent with (*a*) the presence of unipolar actin filament bundles with barbed ends adjacent to the membrane in filopodia; (*b*) the association of myosin I with filopodia and membranes (Hagen et al 1986; Miyata et al 1989; Wagner et al 1992; Cheney & Mooseker 1992; Conrad et al 1993); (*c*) the maximum rate of myosin I-mediated movement of 0.2 μm/sec measured in vitro (Korn & Hammer 1990), which is similar

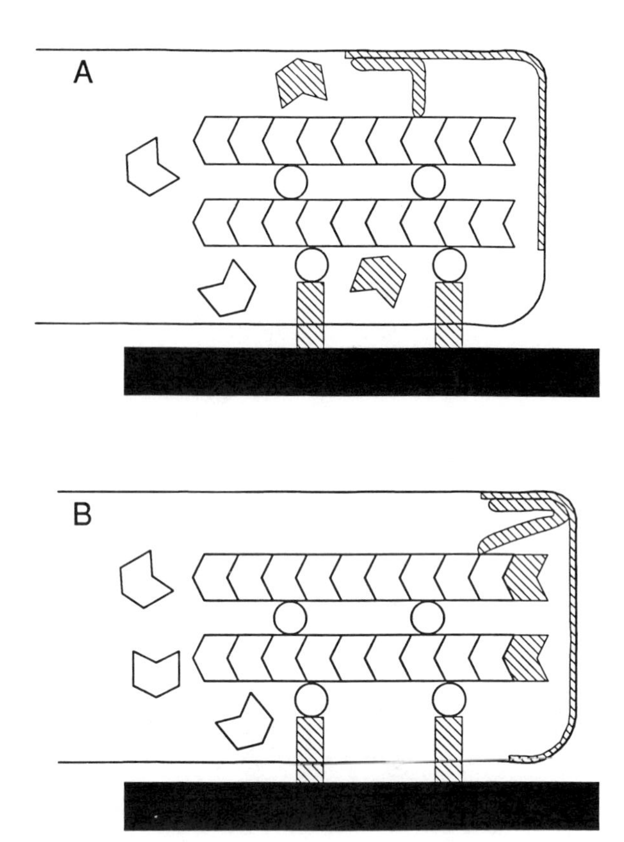

Figure 3 Protrusive force caused by a combination of the interaction between actin and myosin, and actin polymerization. Symbols as in Figures 1 and 2. The membrane-associated helmet is depicted as the hatched surface on the cytoplasmic surface of the membrane. Movement of myosin I, which links actin filaments to the membrane helmet, pushes the helmet away from barbed filament ends, thus allowing monomer addition. (*A*) before and (*B*) after myosin movement.

to the rate of filopod extension in neuronal growth cones; and (*d*) the forward transport of some surface particles toward the leading edge (Sheetz et al 1992). However, there is no evidence that a membrane presents a barrier to actin polymerization at ambient temperatures at either the barbed or pointed ends of filaments, or that a membrane-associated helmet exists with the above properties. In fact, as described above, polymerization of actin inside lipid vesicles causes protrusion of the lipid bilayer to form asymmetric shapes in the absence of myosin (Cortese et al 1989; Miyata & Hotani 1992). Furthermore, the elongation of microvilli in situ in isolated, but membrane intact, brush borders occurs because of actin polymerization upon the addition of exogenous monomeric actin, which indicates that the membrane at the tip of the microvillus does not represent a barrier to actin polymerization (Mooseker et al 1982).

The extent to which the above models for myosin I-mediated protrusion of filopods can be generalized to the protrusion of the leading lamella depends largely on the organization of filaments therein. Numerous studies have been made of the geometry of filaments in pseudopods by conventional thin sectioning of unextracted cells, negative staining of Triton-extracted cells, critical point drying of unextracted and extracted cells, and quick-freeze deep-etched extracted cells. There is general agreement that in preparations that are not collapsed by air drying, actin filaments exist as a dense three-dimensional network of filaments, the majority of which are oriented nearly orthogonal relative to each other. It should be noted that this geometry results in filaments that interact with the membrane in all possible ways; i.e. through barbed ends, pointed ends, and by lateral interactions (Hartwig & Yin 1988). This is a general picture that can be applied to macrophages (Hartwig & Shevlin 1986), platelets (Hartwig 1992), neutrophils (Cano et al 1992), fibroblasts (Small 1988), neuronal growth cones (Lewis & Bridgman 1992), and *Dictyostelium* amoebae (Rubino & Small 1987; Wolosewick & Condeelis 1986; Ogihara et al 1988). The orthogonal filament geometry is produced in vitro by well characterized proteins, ABP-120 in *Dictyostelium* (Wolosewick & Condeelis 1986), and ABP-280 in vertebrate nonmuscle cells (Niederman et al 1983), and these proteins are concentrated in networks with orthogonal filament geometries in situ (Ogihara et al 1988; Hartwig & Shevlin 1986). At present it is unclear how to apply models for filopod extension, which requires parallel unipolar geometries of actin filaments, to the extension of pseudopods containing actin filaments with random orientations. In particular, it is unclear how any myosin-like molecule could generate a force for protrusion on a scaffold of actin filaments of isotropic geometry and polarity.

The function of myosin I at the leading edge of the cell has been explored using *Dictyostelium* amoebae. Mutant cells whose myosin IB gene was disrupted by homologous recombination (Jung & Hammer 1990) appear capable of normal pseudopod extension and cell migration. However, quantitative video analysis demonstrates that mutant cells extend many more lateral pseudopods and do not exhibit optimal intracellular particle transport (Wessels et al 1991). Similar results have been obtained with myosin IA mutants (Titus et al 1993). This raises the possibility that other isoforms of myosin I such as IC or ID might compensate for deletion of either myosin IA or myosin IB alone. However, preliminary evidence obtained from double deletion mutants of myosins IA and IB, and of myosins IB and IC, indicate that mutant amoebae extend more pseudopods and have increased lateral turning rates consistent with extension of more lateral pseudopods (Novak & Titus 1992). These results suggest that myosin I is involved in the suppression of lateral pseudopods and particle transport. A role for myosin I in suppression of pseudopod extension is consistent with the ability of myosin I isoforms containing the ATP-insensitive actin-binding site to superprecipitate (collapse) networks of randomly oriented actin filaments in vitro (Fujisaki et al 1985).

The possibility that myosin I is involved in retraction of the cell edge is supported by careful studies on developing epithelial cells (Fath & Burgess 1993). Myosin I is associated with Golgi-derived vesicles in cells that are assembling brush borders. Upon interaction with preformed microvillar rootlets, which are composed of unipolarized actin filaments, the vesicles are proposed to translocate along the rootlet in the direction of the barbed filament ends. As the vesicles move apically on the rootlets, they fuse with the plasma membrane upon contact and link the membrane to the actin core bundles via myosin I. As a result of vesicle transport and fusion, the apical surface membrane retracts around the digits defined by the actin core bundles. In this view, microvilli are formed by retraction of membrane involving only the vesicle transport function of myosin I rather than myosin I-mediated extension of actin filaments.

The above considerations indicate that there is no compelling evidence either for or against the involvement of myosin I in the generation of force for protrusion of the cell surface. However, there is evidence for the involvement of myosin I in retraction forces at the cell surface and in intracellular vesicle transport. Further efforts to elucidate requirements for any type of myosin in the generation of force for protrusion will be complicated by the large number of unconventional myosins present in nonmuscle cells and their potential for functional overlap (Cheney & Mooseker 1992). Identification of the structures and organelles that bind to various types of myosin will be required to elucidate the functions of these individual myosins and to assemble a coherent understanding of why so many types of myosins co-exist in nonmuscle cells.

Forces Derived From Osmotic Pressure

The best characterized protrusive event is extension of the acrosomal process of Thyone sperm. Initially it was proposed that actin polymerization alone is sufficient to extend the process and, therefore, the rate of growth should be limited only by the diffusion of actin monomers from the periacrosomal cup to the tip of the growing acrosome where polymerization-competent barbed filament ends are located (Tilney & Inoue 1982). This diffusion-limited model predicts the observed relationship between protrusion rate (dL/dt) and the square root of time ($t^{1/2}$) where $L(t)$ is proportional to $t^{1/2}$.

However, a purely diffusion-limited elongation process predicts a d*L*/d*t* that is much slower than that observed (summarized in Oster & Perelson 1987). Therefore, Oster et al (1982) proposed a hydrostatic pressure model that requires both osmotic pressure and actin polymerization, in which osmotic pressure results in hydrostatic pressure to protrude the membrane and facilitate the diffusion of actin, and actin polymerization supports the membrane during extension. The force generated by swelling of the peri-acrosomal gel is large, calculated to be 5.3×10^5 dyne/cm^2 (Oster & Perelson 1987) and thus sufficient for membrane protrusion. This model correctly predicts the relationship between $L(t)$ and $t^{1/2}$ and the magnitude of d*L*/d*t*.

The osmotic model was tested by varying the osmolarity of the solution in which the acrosomal reaction occurred. d*L*/d*t* was found to depend on osmolarity as predicted, i.e. hypertonic solutions slowed the rate of extension, while hypotonic solutions increased the rate of extension. These results indicate that both actin polymerization and osmotic force contribute to protrusion of the acrosomal process (Tilney & Inoue 1985).

The osmotic model has been extended to explain the protrusion of pseudopods, structures in which actin filaments are arranged as networks, not bundles. Dramatic changes in volume of hydrated gels composed of cross-linked filament networks have been well characterized (Tanaka 1981; Li & Tanaka 1990). For actin gels, phase transitions resulting in large changes in volume could result from a number of physiologically relevant events that decrease the elastic modulus of the gel such as severing of filaments or breaking cross-links between filaments. As the elastic modulus is reduced, the gel will increase in volume because of the gel osmotic force until a new equilibrium volume is reached (Tanaka 1981; Oster 1984; Oster & Perelson 1985). This model predicts that $L(t)$ of the process will be proportional to $t^{1/2}$ in charge neutral gels, while in gels composed of charged polymers like actin, $L(t)$ will increase faster since they have an additional ion pressure term (Oster & Perelson 1993).

The involvement of gel osmotic pressure in the extension of pseudopods and filopods has been investigated. Addition of sorbitol to the culture medium quickly and reversibly inhibits the protrusion of lamellipodia by fibroblasts (Dipasquale 1975). In polymorphonuclear leukocytes, increases in solution osmolarity generated by a number of solutes inhibit cell locomotion (Bryant et al 1972) and pseudopod extension (Rabinovitch et al 1980). In similar experiments with primary cultures of chick ganglia, increases in solution osmolarity generated with a number of solutes inhibit lamellipod, but not filopod, extension from neuronal growth cones. In contrast, increases in osmolarity caused an increase in the rate and length of extension of filopods (Bray et al 1991). These results suggest that lamellipods and filopods are extended by different mechanisms. Lamellipod extension may have an osmotic component, while filopod extension may involve mechanisms that operate independently of osmotic force such as actin polymerization and myosin-mediated filament sliding. This is opposite from the result obtained with Thyone acrosomes, where extension is inhibited by increasing osmolarity. Thus it is dangerous to assume that extensions of the cell surface, which are supported by similar actin-containing structures, must have similar mechanisms of protrusion. Furthermore, the increase in rate and length of extension of filopods in hypertonic solutions is consistent with an actin polymerization-based mechanism for extension if the hypertonic load increases the actin concentration in the cell.

MECHANICAL DYNAMICS OF THE LEADING EDGE

The above considerations indicate that the force derived individually from either actin polymerization, the myosin cross-bridge cycle, or gel osmotic pressure is more than sufficient to account for the extension of pseudopods and filopods. In an effort to distinguish which of these mechanisms is operating in vivo, the results of detailed studies of the mechanical dynamics of the leading edge of motile cells are considered below.

Evidence for Actin Polymerization at Sites of Protrusion

The most compelling correlation between actin polymerization and protrusive activity comes from work on the Thyone acrosomal reaction and assembly of the mating tubule of *Chlamydomonas*. In both Thyone (Tilney & Inoue 1982) and *Chlamydomonas,* (Detmers et al 1983), actin polymerization from discrete nucleation sites in the cell cortex and cross-linking by actin-binding proteins generates a bundle of cross-linked actin filaments whose extension matches the growth of the cell process. Addition of cytochalasin, which primarily blocks addition of actin monomers to barbed ends of filaments, inhibits the extension of both processes, consistent with the conclusion that filaments are anchored near their pointed ends in the nucleation sites, while polymerizing at their barbed ends against the membrane at the tip of the process.

Pseudopod extension is correlated both spatially and temporally with the polymerization of actin in neutrophils (Coates et al 1992; Omann et al 1987), *Dictyostelium* amoebae (Hall et al 1988), platelets (Hartwig 1992), and malignant tumor cells (Jones et al 1991; Segall et al 1991). In these systems the correlations are striking since the onset of protrusive activity can be elicited synchronously by stimulatory ligands, which allows direct comparison of biochemical measurements of F-actin content with the location of newly assembled F-actin and cell behavior (Condeelis et al 1990). In general, actin polymerization is the earliest event following cell stimulation and always precedes pseudopod extension (Omann et al 1987; Hall et al 1988). Furthermore, this newly assembled actin is found in pseudopods that extend in response to stimulation (Coates et al 1992; Hall et al 1988). In tissue cells such as fibroblasts and neuronal growth cones, a spatial and temporal correlation between actin polymerization and protrusive activity has been established using microinjection of labeled actin into living cells (Okabe & Hirokawa 1991; Symons & Mitchison 1991), and by the behavior of certain polycationic beads on the cell surface (Forscher et al 1992). In macrophages, correlated video and electron microscopy has been used to document the rapid assembly of dense networks of actin filaments at sites of protrusion (Rinnerthaler et al 1991).

Movement of Cytoplasmic Mass and Actin-Containing Structures

Further attempts to observe the dynamics of actin filaments at the leading edge in vivo have led to different results and interpretations depending on the techniques and cell types employed (Table 1). FRAP analysis of rhodamine-labeled actin demonstrates a centripetal flux of actin away from the leading edge at rates of 0.8 μm/min in fibroblasts (Wang 1985) and 1.6 μm/min in neuronal growth cones (Okabe & Hirokawa 1991). However, DIC optics reveal a centripetal flux of cytoplasmic mass, usually visualized as variations in refractive index in the light microscope, away from the leading edge in neuronal growth cones at rates of 3–5 μm/min (Forscher & Smith 1988) and in fibroblasts at up to 12 μm/min (Fisher et al 1988). Experiments using caged resorufin-actin detect a filament population moving centripetally from the cell edge at rates of 2.5 μm/min in keratocytes (Theriot & Mitchison 1992) and 0.4–1 μm/min in fibroblasts (Theriot & Mitchison 1992). In the same fibroblast, the rate of centripetal flow of cytoplasmic mass is about three times greater than the centripetal flow of photoactivated resorufin-actin (Theriot & Mitchison 1992).

In general, these results indicate that there is no correlation between the rate of centripetal movement of labeled actin and the rate of centripetal flow of cytoplasmic mass. Furthermore, there is no correlation between the rate of centripetal movement of actin and the rate of protrusion in slowly moving fibroblasts and neuronal growth cones. However, in rapidly moving keratocytes, the rate of centripetal flow of actin away from the cell edge is approximately the same as the rate of cell locomotion over the substrate (Theriot & Mitchison 1992), which suggests that centripetal flow of actin may be a manifestation of the traction forces responsible for cell locomotion over the substrate (Table 1).

An exception to these general conclusions comes from work on immobilized *Aplysia* growth cones. Rapid centripetal flow of cytoplasmic mass

Table 1 Rates of protrusion and centripetal flow

Event	Rate (μm/min)	System	Reference
Protrusion relative to substrate	5	Primary cultures of chick and mouse fibroblasts	Abercrombie et al 1970; Felder & Elson 1990
	0.3	Lamellipod of neuronal growth cone	Okabe & Hirokawa 1991
		Wounded confluent fibroblasts	Theriot & Mitchison 1992
	0.4	IMR90	
	0.08	3T3	
	10–30	Keratocytes	Enteneuer & Schliwa 1984
	2.5	Keratocytes	Theriot & Mitchison 1991
	30	*Dictyostelium* amoebae	D. Wessels & D. Soll personal communication
Protrusion relative to cell center	2.5–6	Macrophages	Rinnerthaler et al 1991
	~0	Keratocytes	Enteneuer & Schliwa 1984
	12	Filopods of neuronal growth cones	Sheetz et al 1992
	500	Thryone acrosomal process	Tilney & Inoue 1982
Centripetal flow relative to leading edge			
Cytoplasmic mass	3–5	Neuronal growth cones	Forscher & Smith 1988
	12	Wounded confluent 3T3 fibroblasts	Fisher et al 1988
Fluorescently labeled actin after photobleaching	0.8	Nonconfluent IMR33 fibroblasts	Wang 1985
	1.6	Neuronal growth cone	Okabe & Hirokawa 1991
Fluorescently labeled actin after photoactivation		Wounded confluent fibroblasts	Theriot & Mitchison 1992
	1	IMR90	
	0.4	3T3	
	2.5	Keratocytes	Theriot & Mitchison 1993
Centripetal flow relative to the substrate			
Fluorescently labeled actin after photoactivation		Wounded Confluent fibroblasts	Theriot & Mitchison 1992
	0.66	IMR90	
	0.36	3T3	
	~0	Keratocytes	Theriot & Mitchison 1992

occurs at 3–5 μm/min. When cells are treated with high concentrations of cytochalasin there is a sudden detachment of actin-containing structures from the cell edge, and this mass of actin is transported centripetally at the same 3–5 μm/min rate as bulk cytoplasm (Forscher & Smith 1988). These results suggest a direct correlation between centripetal flow of cytoplasm and actin. However, this conclusion is not supported by FRAP experiments with fluorescently labeled actin in other neuronal growth cones where the two rates are different (Okabe & Hirokawa 1991).

Two possibilities are suggested by the variety of experiments described above. First, techniques employing photobleaching or photoactivation may detect only slowly moving populations of actin filaments such as those in the membrane skeleton of the dorsal and ventral plasma membrane, while rapidly moving populations of filaments involved in cytoplasmic flow escape detection. A population of filaments moving centripetally at rates corresponding to cytoplasmic flow might be associated with contraction waves or bulk retraction of the actin network (Conrad et al 1993), either of which could be involved in cell locomotion. Alternatively, the treatment of *Aplysia* growth cones with cytochalasin may cause a bulk contraction of the lamellar actomyosin at a rate that is coincidentally similar to that of centripetal cytoplasmic flow, but not mechanistically related to it.

Attempts to distinguish among the possible mechanisms for centripetal actin flow, such as treadmilling, local assembly, and disassembly of filaments, and contraction of an actin-containing network, by comparing the rates of actin flow in vivo to rate constants for actin polymerization and depolymerization in vitro, have been disappointing. In general, the in vitro rate constants for actin depolymerization are too slow to be consistent with a treadmilling model for centripetal flow of actin. Furthermore, the rate for actin polymerization measured in vitro near the critical concentration is at least an order of magnitude too slow to explain centripetal cytoplasmic flow (Wang 1985; Fisher et al 1988) (Table 1). However, in vivo, where macromolecular solute concentrations of 10% are likely, the rate constants for actin assembly and disassembly are likely to be much higher (Drenckhahn & Pollard 1986), and polymerization-competent actin may be released locally at concentrations much greater than the critical concentration (Table 2). In addition, the presence of proteins capable of severing and depolymerizing actin, such as gelsolin and the destrin/cofilin family, respectively, (Hartwig & Kwiatkowski 1991) may increase the apparent bulk off rate for actin depolymerization in vivo. Therefore, discarding any of the above mechanisms based on kinetic considerations is premature.

More compelling arguments against the treadmilling model for centripetal flow of actin are the isotropic geometry and random polarity of actin filaments in lamellipods (Small 1988; Hartwig & Shevlin 1986; Lewis & Bridgman 1992), the short half-life of actin filaments, and lack of sharp actin filament density gradients in lamellipods (Symons & Mitchison 1991; Theriot & Mitchison 1992). These considerations suggest that polymerization and depolymerization of actin filaments occurs everywhere within the volume

Table 2 Behavior of pseudopods predicted by different models

Model	Predicted rate (μm/min)	Predicted $L(t)$ dependence	Effect of hypertonicity on rate	Effect of permeabilization of cell
Actin polymerization	2–200[a]	t	Increase	None/inhibit[b]
Myosin I-based movements	1–14[c]	t	None	None
Gel osmotic swelling	6	t½	Decrease	None
Hydrostatic pressure	500[d]	t½	Decrease	Inhibit

[a] Lower rate is calculated for the barbed end from rate constants of Pollard (1986), 1.0 μM actin, 3 nm of length per actin monomer added. The higher rate is calculated at the barbed end for 10 μM actin from rate constants predicted in physiological macrosolute concentrations according to Drenckhahn & Pollard (1986). It is possible that 10 μM actin could be released suddenly and locally from sequestered pools of actin monomer. These are rates of linear polymerization. Rates for expansion of filament networks where most filaments are off axis would be much slower.

[b] None if actin monomers are not extracted. otherwise inhibition will occur.

[c] A broad range in rates of movement on actin filaments in vitro has been reported for myosins I (Korn & Hammer 1990). In general, myosin II-based rates of movement are much faster in the same assays (Warrick & Spudich 1987).

[d] This rate is based on the rate of extension of the Thyone acrosomal process where hydrostatic pressure may result from hydration of the periacrosomal compartment (Oster & Perelson 1987). Rates for other cell types would differ depending on how the hydrostatic pressure is generated. One possibility is an actomyosin-based contraction of the cell cortex as reviewed elsewhere (Condeelis 1992).

of the lamellipod and that filaments are relatively short in comparison to lamellipodial dimensions.

In summary, actin polymerization is correlated with protrusive activity in almost all cell types, while the significance of centripetal flow of cytoplasmic mass and actin, and their relationship to each other, remains unclear. Centripetal flow is likely to be related to the mechanism involved in locomoting the cell relative to its substrate, but not protrusive activity.

Protrusive Activity that Does Not Involve Actin Polymerization

There are three notable exceptions to the spatial and temporal relationship between actin polymerization and protrusive activity. The acrosomal reaction of *Mytilus* is one in which a unipolar bundle of preformed actin filaments is protruded in a myosin-independent mechanism proposed to involve a zippering interaction between the bundle and the cell membrane (Tilney et al 1987). Another is the acrosomal reaction in *Limulus,* where a conformational change in the preformed bundle of actin filaments, from a coiled to an extended orientation, is proposed to cause extension of a straight bundle that pushes against the cell membrane to protrude from the original cell surface (DeRosier et al 1982). However, the degree to which conformational changes in actin bundles and interactions between actin bundles and the cell membrane might contribute to protrusive force and motility in crawling cells in general has not been investigated.

Finally, the most unusual exception is pseudopod extension by spermatocytes of *Caenorhabditis* and *Ascaris*. Protrusive force in these cells does not involve either actin or myosin, but rather the polymerization of the major sperm proteins (MSPs) (King et al 1992) to form a network of 5–9 nm diameter filaments (Roberts & King 1991). The MSPs are presumably unique to these parasitic organisms.

Rates of Protrusion and Potential Models for Pseudopod Extension

Rates of protrusion of pseudopods and filopods from a variety of cell types are listed in Table 1. A striking feature of this data is the variation in rates of protrusion that may differ by several orders of magnitude. This makes it unlikely that any single mechanism can account for all of these events. Comparisons of the rates of protrusion predicted by different models of protrusion (Table 2) to actual rates of pseudopod extension (Table 1) might be considered useful in distinguishing mechanisms of protrusion. Unfortunately, this has not proven to be the case since most models predict a broad range of rates depending on the assumptions of each calculation. The dependence of the predicted rate of protrusion on time differs for some models, so this may be taken as a useful distinguishing feature of the different models. For example, careful measurements of the rate of lamellipod extension from fibroblasts indicates that the length of the process, $L(t)$, is directly proportional to time (Felder & Elson 1990). This appears to be inconsistent with pseudopod extension driven purely by gel osmotic pressure or by hydrostatic pressure, since in these two models $L(t)$ is predicted to depend on $t^{1/2}$ (Oster & Perelson 1985, 1987, respectively). However, during protrusion of the entire lamella, many heterogeneous events are occurring in different regions of the leading edge (Rinnerthaler et al 1991), which makes correlations between $L(t)$ and a single model impossible. Furthermore, the gel osmotic and hydrostatic models were derived to predict the behavior of a linear protrusion and do not apply in their simplest forms to three-dimensional protrusions such as the lamellipod (Oster & Perelson 1993). Generally, some measurements of extension of the leading edge may fail to distinguish protrusion relative to the substrate (locomotion of the cell body + protrusion) and protrusion relative to the cell body (protrusion). This obscures the identification of true protrusion in a background of cell locomotion (Table 1).

Evidence against a hydrostatic pressure model for pseudopod extension comes from experiments with permeabilized cells. Electropermeabilization of the cell membrane of *Dictyostelium* amoebae results in depolarization of the membrane potential and the passage of large molecules such as inositol. Under these conditions, permeabilized cells are capable of normal pseudopod extension and even respond to chemotactic stimulation by extending pseudopods (Van Duiyn & Van Haastert 1992), events that should not occur if hydrostatic pressure, which requires an impermeable membrane, is the driving force for pseudopod extension. Since gel osmotic pressure does not require an intact membrane, and actin polymerization is unaffected in similarly electroporated neutrophils (Downey et al 1990), these results do not exclude either actin polymerization or gel osmotic pressure as potential mechanisms for pseudopod extension.

BIOCHEMICAL DYNAMICS OF THE LEADING EDGE

Insights into mechanisms of protrusion have also been derived from biochemical analyses of changes in the actin cytoskeleton during pseudopod and filopod extension. In general, this work implicates localized actin polymerization, filament cross-linking and filament severing as major events during protrusive activity.

Regulation of Actin Polymerization

In order for polymerization to play a direct role in the generation of protrusive force, filament assembly must be controlled with spatial and temporal precision. This could be accomplished by regulation of the availability of polymerization-competent actin monomers in concert with proteins, such as actobindin (Lambooy & Korn 1988) that suppress the spontaneous nucleation of actin polymerization in order to prevent unwanted assembly. Another mechanism is regulation of assembly by limiting the availability of nucleation sites for actin polymerization. Nucleation sites could be formed de novo or by uncapping the barbed ends of pre-existing actin filaments. In situations where large amounts of F-actin are generated rapidly, both mechanisms might operate. An example of this coupling is the acrosome reaction in Thyone where polymerization-competent actin is suddenly released from a sequestered complex with profilin to polymerize from the barbed ends of filaments bundled into a preformed nucleation site, the actomere. How the nucleation activity of the actomere is regulated and how spontaneous nucleation of free actin monomer is suppressed are unclear but both monomer release/activation and nucleation are tightly coupled since actin polymerization only occurs in association with actomere-nucleated filaments.

The amount of polymerization-competent actin may be regulated by mechanisms that activate G-actin directly from an inactive form to one that is competent to polymerize. Phosphorylation of actin has been reported in *Physarum* (Furukashi et al 1992), *Amoeba proteus* (Sonobe et al 1986), and in *Dictostelium* (Schweiger et al 1992; Howard et al 1993). The residues on actin that are phosphorylated were shown to be serine, tryrosine, and threonine (Schweiger et al 1992; Howard et al 1993; Furukashi et al 1992). In some cases, this actin phosphorylation has been correletad with inhibition of polymerization in vitro (Sonobe et al 1986; Furukashi et al 1992) and shape changes in cells (Schweiger et al 1992).

In addition, there is evidence from chemical modification experiments that G-actin can exist in several different conformations, one of which may be required for polymerization (Shu et al 1992). However, there is no direct evidence that either phosphorylation or conformational changes associated with cation and nucleotide binding operate in vivo to regulate the amount of polymerization-competent actin.

A number of proteins such as profilin, destrin, depactin, actobindin, and cofilin are capable of sequestering actin in an unpolymerizable form (Hartwig & Kwiatkowski 1991). Most of these proteins are more likely to participate in depolymerization reactions since they are present in small amounts in cytoplasm. Profilin is relatively abundant and has been proposed to be the primary actin sequestering protein (Carlsson et al 1979). However, careful measurements indicate that profilin is present in amounts that are insufficient to account for the entire amount of unpolymerized actin in nonmuscle cells (Southwick & Young 1990; Goldschmidt-Clermont et al 1991b) and much of the profilin in vivo may be bound to phosphatidyl inositol 4, 5-bisphosphate (PIP_2) (Hartwig et al 1989; Goldschmidt-Clermont et al 1990). In fact, it is thought that profilin is involved in promoting actin polymerization rather than sequestering actin monomers since it catalytically enhances the exchange of bound nucleotide thereby increasing the amount of ATP-actin under physiological conditions (Goldschmidt-Clermont et al 1991b). ATP-actin elongates actin filaments faster than ADP-actin in vitro (Pollard 1986). Both possible roles for profilin function in regulation of actin assembly are supported by microinjection studies. Microinjection of excess profilin into living cells indicates that profilin can alternatively sequester actin monomers and enhance actin polymerization in vivo depending on the amount and form of profilin administered (Cao et al 1992).

Profilin may also be involved in regulating the accessibility of PIP_2 to hydrolysis by PLC-gamma$_1$. When profilin is bound to PIP_2, its interaction with actin is inhibited (Lassing & Lindberg 1985), and PIP_2 bound to profilin can not be hydrolyzed by PLC-gamma$_1$ (Goldschmidt-Clermont et al 1991a). However, phosphorylation of PLC-gamma$_1$ by tyrosine kinases such as EGF- or PDGF-receptor releases the profilin-mediated block to hydrolysis of PIP_2, which results in the production of diacylglycerol, IP_3 and, presumably, free profilin. Freed profilin could then enhance actin polymerization by promoting exchange of ADP for ATP on actin monomers, while diacylglycerol might promote the formation of membrane-associated nucleation sites for actin polymerization by an unknown mechanism (Schariff & Luna 1992). Together, these events could result in sudden localized bursts of actin polymerization.

The considerations discussed above suggest that there is an additional factor(s) responsible if large amounts of actin are sequestered in an unpolymerizable form in nonmuscle cells. The recent discovery of a 5-kd peptide that binds to monomeric actin and inhibits its polymerization may resolve this issue (Safer et al 1990). This peptide has been identified as Thymosin beta-4 (Tβ4), previously thought to be a thymic hormone (Safer et al 1991). Its wide spread occurrence, high concentration in cytoplasm (200–500 μM), and micromolar affinity for actin suggest that TB4 is responsible for sequestering the majority of unpolymerized actin in nonmuscle cells (Weber et al 1992; Cassimeris et al 1992; Yu et al 1993). Consistent with this conclusion is the observation that microinjection of excess TB4 into living cells results in a decrease in the content of filamentous actin throughout the cell (Sanders et al 1992).

TB4 blocks the dissociation of nucleotide bound to actin, while profilin catalytically promotes nucleotide exchange (Goldschmidt-Clermont et al 1992). Since both proteins bind to actin with moderate affinity, they might exchange rapidly among actin molecules in vivo thereby amplifying the effect of profilin in regulating the content of ATP-actin in cytoplasm. Otherwise the interaction between TB4 and the signal transduction pathways involved in actin polymerization may be passive since no modification in actin monomer sequestering activity has been detected following stimulation that leads to the bursts of actin polymerization (Cassimeris et al 1992). However, transient changes in TB4 activity have not been ruled out. Absence of changes in TB4 activity is consistent with a model in which actin polymerization is regulated by the availability of nucleation sites for actin polymerization such as the barbed ends of filaments, and emphasizes the importance of capping proteins in regulating actin polymerization in vivo.

Attempts to distinguish in vivo whether actin polymerization is regulated primarily by release and activation of sequestered actin monomers or by the availability of nucleation sites for actin polymerization have yielded mixed results. Microinjection of actin oligomers stabilized with phalloidin into Swiss 3T3 and NRK cells was done to determine if exogenous nucleation sites would influence actin polymerization in vivo. Microinjected nuclei induced neither an extensive polymerization nor change in distribution of actin filaments in vivo, which suggests that availability of nucleation sites is not the key step, but that the availability of polymerization-competent actin is of primary importance (Sanders & Wang 1990). However, in a closely related study, Handel et al (1990) microinjected glutaraldehyde cross-linked actin oligomers into PtK_2 cells and found a concentration-dependent disruption of endogenous actin filaments and assembly of filaments from the exogenous actin nuclei. These authors concluded that actin polymerization is regulated by the availability of nucleation sites in living cells. Such contradictory results might be explained by differences in the types of cells used, the larger number of nucleation sites injected by Handel et al, the ability of phalloidin-stabilized filaments to anneal, thereby decreasing the effective number of nucleation sites available, and differences in the ability of endogenous capping proteins to interact with phalloidin vs glutaraldehyde stabilized F-actin.

Other studies indicate that actin polymerization is regulated both spatially and temporally by the availability of actin nucleation sites. Studies on both neutrophils and *Dictyostelium* amoebae have led to similar conclusions. Stimulation of cells with chemoattractants generates a transient increase in actin nucleation activity seen in whole cell lysates. The new nucleation sites appear to be the barbed ends of actin filaments since nucleation is inhibited by low doses of cytochalasin, which is known to cap these ends. This stimulated nucleation activity is associated with the Triton-resistant actin cytoskeleton (Carson et al 1986; Hall et al 1989b), while an inhibitor of this stimulated nucleation activity is present in the Triton-extracted cytosol (Hall et al 1989b). Correlation between the location of newly assembled F-actin and pseudopod extension in synchronously responding *Dictyostelium* amoebae indicates that stimulated actin polymerization occurs in pseudopods, which suggests that the nucleation activity is present in pseudopods as well (Hall et al 1988). This conclusion is supported by fluorescence microscopy experiments on fibroblasts. Incubation of permeabilized cells with rhodamine-labeled actin results in cap Z-inhibitable polymerization of microinjected rhodamine-actin in the same locations as observed in vivo. This result suggests that the sites of polymerization are determined by the presence of free barbed ends of actin filaments rather than desequestration of actin and that these sites of polymerization are at the tips of pseudopods (Symons & Mitchison 1991).

Several capping proteins that bind to the barbed end of the actin filament, most notably gCap39, cap Z and gelsolin (Hartwig & Kwiatkowski 1991), might regulate the availability of the barbed ends, which nucleate actin polymerization. Many of these capping proteins bind to PIP and PIP_2, and this binding either inhibits their interaction with, or dissociates them from, the barbed end of the actin filament. Therefore, a synergistic interaction is possible between the PIP_2/profilin-mediated regulation of actin monomer polymerizability, and the PIP_2/capping protein-mediated increases in availability of free barbed ends. One of these proteins, gelsolin, has been analyzed in great detail and a hypothesis has been proposed for its function in vivo

(Stossel 1989). Gelsolin is a calcium-activated and polyphosphoinositide-inhibited, actin-severing and barbed end-capping protein. In this hypothesis, gelsolin, which is already capping the barbed ends of actin filaments in unstimulated cells, would be dissociated from the barbed ends during changes in PIP_2 levels that follow stimulation, provided that Ca^{2+} concentrations remain low. Alternatively, if the Ca^{2+} concentration were to rise to micromolar levels, free gelsolin would sever actin filaments and cap the newly created barbed ends, cap the barbed ends of pre-existing actin filaments, and/or bind to actin monomers to generate a heterotrimer with two actins that is capped at its barbed end. All of these events increase the number of capped barbed ends. To free these barbed ends, gelsolin would be dissociated by a subsequent rise in PIP_2 concentration and drop in Ca^{2+} levels. This hypothesis is consistent with the dissociation of pre-existing gelsolin-actin complexes during the initial burst of actin polymerization following chemotactic stimulation of human neutrophils (Howard et al 1990), and with the calcium-dependent actin nucleation activity in rabbit macrophages, which is stimulated by phorbol esters (Hartwig & Janmey 1989). The hypothesis is inconsistent with the lack of correlation between increases in bulk PIP and PIP_2 levels, actin polymerization, and the stability of gelsolin-actin complexes in A 431 cells following stimulation with epidermal growth factor (Dadabay et al 1991). The hypothesis is complicated by low levels of Ca^{2+} in resting and stimulated cells, usually less than 0.3 μM, and the lack of correlation between spatial (Marks & Maxfield 1990; Brundage et al 1991; Hahn et al 1992) and temporal (Schlatterer et al 1992; Jaconi et al 1988) changes observed in Ca^{2+} levels with that of actin polymerization and pseudopod extension. In general, increases in cytoplasmic calcium levels are not required for actin polymerization, but may cause depolymerization (Sha'afi et al 1986; Downey et al 1990), and pseudopod extension occurs in the absence of Ca^{2+} (Van Duijn & Van Haastert 1992). Gelsolin requires micromolar concentrations of Ca^{2+} for activation of its actin binding and severing activities. Lack of a calcium requirement for actin polymerization and pseudopod extension is also inconsistent with a direct role for gCap39 and other calcium-sensitive capping proteins in these events.

Direct tests of the gelsolin hypothesis in vivo by microinjection, underexpression and overexpression of gelsolin have resulted in conflicting results. Microinjection of cytoplasmic or plasma gelsolin from rabbits and humans appears to have no effect in fibroblasts and macrophages presumably because of in vivo Ca^{2+} levels, which are below those required to activate gelsolin in vitro, since microinjection of the N-terminal half of gelsolin, which is constitutively active in the absence of Ca^{2+}, caused disruption of the actin cytoskeleton (Cooper et al 1987). However, in a similar study, microinjection of human plasma gelsolin into fibroblasts and PtK_2 cells caused disruption of the actin cytoskeleton, but none of the other gelsolins used showed activity (Huckriede et al 1990). These results suggest that factors other than Ca^{2+}, such as PIP_2 and pH (Lamb et al 1992), regulate the actin binding and severing activity of gelsolin in vivo such that differences in the source and method of preparation of gelsolin may affect its final activity when microinjected into living cells.

Chemically mutagenized *Dictyostelium* cells that no longer express severin, a gelsolin-related protein, exhibit normal growth and morphogenesis and qualitatively normal cell motility (Andre et al 1989). However, increases in gelsolin levels in vivo achieved by transfection of fibroblasts do not alter their morphology, but do increase their rate of locomotion and this rate increase is proportional to the amount of gelsolin expressed over a narrow range (Cunningham et al 1991). A high level of overexpression actually disrupts the actin cytoskeleton.

In aggregate, these results indicate that gelsolin interacts with actin filaments in vivo and is capable of affecting the turnover of actin between its assembled and disassembled states. However, none of the results is of significant resolution to directly test the hypothesis that gelsolin regulates the number of free barbed ends of actin filaments in cells thereby regulating the timing and location of actin polymerization. In addition, these latter results with mutant cells indicate that the inability to detect a phenotype in cells that no longer express an actin-binding protein is not proof that a protein lacks a specific function in vivo. The possibility of compensation by related proteins (Witke et al 1992) and the use of inappropriate and/or insensitive assays for characterization of phenotypes make negative results with motility mutants particularly uninterpretable (e.g. compare the results of the different analyses in Jung & Hammer 1990 with Wessels et al 1991; Cox et al 1992 with Brink et al 1990).

In *Dictyostelium*, an inhibitor of stimulated actin nucleation activity is regulated by chemotactic stimulation in a manner reciprocal to the nucleation activity, which suggests that the inhibitor is a capping protein (Hall et al 1989b). Analysis of *Dictyostelium* signal transduction mutants indicates that the G protein regulating phospholipase C activity is required for stimulation of the nucleation activity (Hall et al 1989a). The inhibitory activity, aginactin, has been isolated by using its chemoattractant regulation to distinguish it from other capping activities. It is associated with a 70-kd protein that binds tightly to and caps the barbed end of the actin filament, but neither severs filaments nor nucleates actin polymerization. Ca^{2+} has no effect on the activities of the protein in vitro (Sauterer et al 1991). One

70-kd protein associated with aginactin has been cloned and sequenced, and sequence analysis shows that it is a member of the HSC-70 family of proteins (Eddy et al 1993). Since these proteins are ubiquitous and the cytosolic forms are abundant, HSC-70 may be involved in regulating the availability of barbed ends in a number of cell types.

Nucleation of actin filaments might also be regulated by the de novo formation of nucleation sites instead of uncapping pre-existing filament ends as discussed above. In *Dictyostelium*, the integral membrane protein, ponticulin, may associate with actin monomers to create oligomers of actin that can act as nuclei for polymerization when ponticulin becomes clustered (Schwartz & Luna 1988). Clustering of ponticulin may occur upon adhesion to the extracellular matrix or to other cells. A similar model has been proposed for the function of talin, a major protein of focal contacts and an early occupant of assembling focal contacts (DePasquale & Izzard 1991). Talin binds to actin monomers and nucleates polymerization presumably through the formation of actin oligomers when talin aggregates (Kaufmann et al 1992).

In summary, the location and timing of actin polymerization in cells is probably regulated by the concerted release of actin from sequestered and/or inactive forms and by the regulation of nucleation sites for actin polymerization. These reactions may be coordinated through the metabolism of PIP, usually under the control of receptors that regulate phospholipase activity. Additional signaling pathways are possible particularly for unstimulated pseudopod extension. Regulation of actin polymerization may also involve the small GTP-binding proteins (Ridley et al 1992), but not cGMP-regulated proteins (Segall 1992).

Filament Cross-linking and Severing

Cross-linking of actin filaments is an essential step if protrusive force is to be generated by either actin polymerization, myosin-mediated filament sliding, or osmotic gel force. Contrary to the model in Figure 2, which is oversimplified, individual actin filaments bend easily and have insufficient flexural rigidity to push against the plasma membrane to cause a protrusion (Gittes et al 1993). However, cross-linked actin filaments in the form of bundles and networks (gels) are extremely rigid and can form macroscopic gels in vitro. Numerous actin cross-linking proteins have been identified that are localized in the actin filament bundles and networks of filopods and pseudopods, respectively (Luna & Hitt 1992). Many of these actin cross-linking proteins probably localize in surface extensions after protrusion is complete to endow the actin cytoskeleton with special properties such as membrane binding, substrate adhesion, rigidity, etc. Others may associate with polymerizing filaments during the early stages of bundle and network formation, thereby contributing more directly to the initial forces of protrusion. In only a few cases (Condeelis et al 1990; Dharmawardhane et al 1989, 1991) have the kinetics of association of actin cross-linking proteins with the actin cytoskeleton been compared with the kinetics of pseudopod extension to distinguish proteins that are present in the pseudopod during its early stages of formation.

The orthogonal filament geometry seen in pseudopods requires the cross-linking activity of proteins like ABP-280 and ABP-120. These are related proteins that contain the same actin-binding site in the head and a similar cross beta sheet repeat in the tail (Noegel et al 1989; Bresnick et al 1990, 1991; Gorlin et al 1990). These proteins are concentrated in newly extending pseudopods (Stendahl et al 1980; Condeelis et al 1988), are concentrated in filament networks with orthogonal geometry in situ (Hartwig & Shevlin 1986; Ogihara et al 1988), and catalyze the formation of filament networks with the orthogonal geometry in vitro (Niederman et al 1983; Wolosewick & Condeelis 1986).

The function of ABP-280 has been investigated in human melanoma cells that spontaneously do not express ABP-280. These cells are unable to extend normal pseudopods, but exhibit heightened levels of blebbing and poor translocational locomotion. Since blebbing is the extension of an unstable fluid-filled hemispherical process that may involve hydrostatic pressure, blebbing is not considered to be normal pseudopod extension. Cells transfected with cDNA coding for ABP-280 under control of the human beta-actin promoter express ABP-280 near normal levels. Transfected cells are able to suppress blebbing, form normal pseudopods, and exhibit increased translocational locomotion, which demonstrates that the phenotype exhibited by the ABP-280-minus cells is due to ABP-280 (Cunningham et al 1992).

The function of ABP-120 in vivo has been analyzed in two studies. In the first, *Dictyostelium* cells lacking ABP-120 were prepared by exposure of normal cells to 1-methyl-3-nitro-1-nitrosoguanidine. ABP-120-minus cells, identified by colony blotting, were unable to incorporate actin into the cytoskeleton normally following stimulation with chemoattractant, although other cellular functions were reported to be grossly normal (Brink et al 1990).

In the second study, ABP-120-minus *Dictyostelium* cells were prepared by disrupting the native gene locus by homologous recombination with ABP-120 coding sequence. ABP-120-minus transformants were unable to cross-link F-actin into Triton-resistant cytoskeletons following stimulation and were unable to extend pseudopods or locomote normally. Instead of pseudopods, ABP-120-minus cells extended spikes and locomoted slowly (Cox et al 1992). Inability to extended pseudopods was correlated with the loss of normal filament networks with the orthogonal geometry from ABP-120-minus cells as compared to control cells (Cox et al 1993).

These studies indicate that the formation of actin gels containing dense filament networks, the result of the cross-linking activity of either ABP-280 or ABP-120, is required for normal pseudopod extension. This is consistent with the gel osmotic pressure mechanism for pseudopod extension in which formation of a gelled network of actin filaments increases in volume through uptake of water as the elastic modulus of the gel is reduced by severing of filaments and/or loss of cross-links between filaments (solation-expansion model of Oster 1984). However, the studies of ABP-280 and ABP-120 are also consistent with a model in which continuous polymerization and cross-linking of actin filaments in a rigid filament network could protrude the cell membrane as depicted in Figure 1.

Regulation of the cross-linking activities of ABP-280 and ABP-120 has not been well studied. However, severing proteins have been described that might regulate the elastic modulus of actin gels in vivo (Hartwig & Kwiatkowski 1991). As discussed above, there is evidence that gelsolin alters the assembly/disassembly dynamics of actin filaments in vivo. When gelsolin is overexpressed within a narrow range in 3T3 fibroblasts, the rate of cell migration increases (Cunningham et al 1991).

Filament severing accompanies the stimulation of pseudopod extension in platelets and neutrophils. Stimulation of platelets with thrombin or by spreading on glass causes the protrusion of lamellipods and filopods. This is accompanied by a large increase in the number of short actin filaments in networks in lamellipods and in bundles in filopods. Prevention of an increase in cytosolic Ca^{2+} during stimulation by preincubating cells with Quin-2 inhibits the formation of short filaments. Addition of Ca^{2+} to Quin-2 treated cells causes shortening of filaments and restores lamellipodia networks. Addition of A23187 and Ca^{2+} results in extensive severing of actin filaments. The severing activity may be caused by gelsolin, which is associated with these cortical actin filaments (Hartwig 1992).

In neutrophils, stimulation with chemoattractant causes pseudopod extension and a large increase in the number of short actin filaments within 90 sec. Filament length was inferred from depolymerization kinetics in cell lysates (Cano et al 1991). Although this approach is indirect, the simplest explanation of the data is the activation of severing proteins in the cortex of stimulated cells.

These results are consistent with the proposal that gelsolin and related severing proteins regulate the rate of swelling of actin gels under gel osmotic pressure by decreasing the elastic modulus of the gel when they sever actin filaments within (Oster & Perelson 1987; Condeelis 1992).

CONCLUDING REMARKS

The mechanical and biochemical events that I consider likely to explain the extension of pseudopods and filopods are summarized in Figure 4. This figure deals only with the earliest events in protrusion and proposes that there are three steps in the formation of initial protrusions.

1. The spatial and temporal regulation of protrusive activity is initially determined by the location and timing of appearance of the uncapped barbed ends of actin filaments, which nucleate the growth of actin

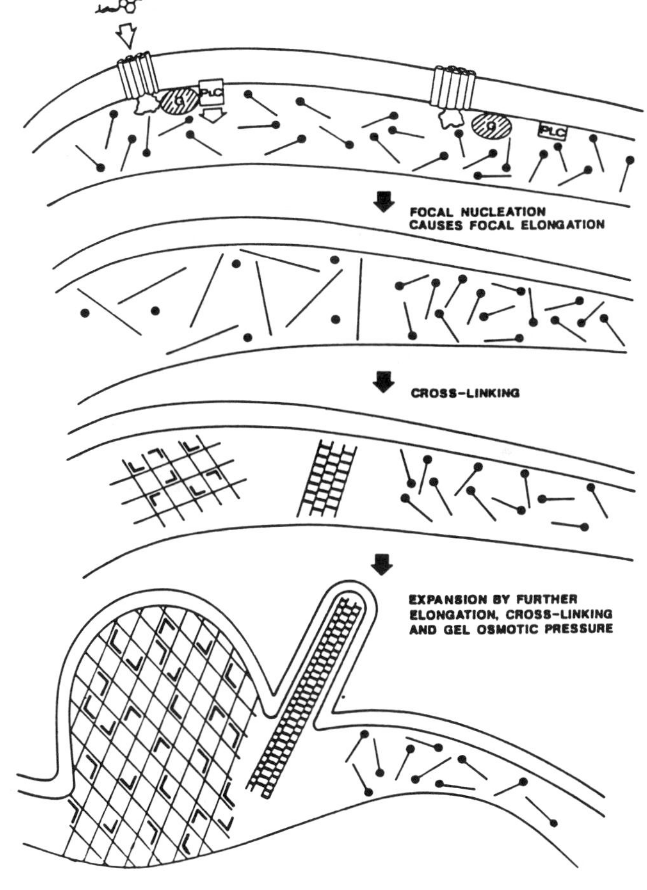

Figure 4 The cortical expansion model (Condeelis et al 1990; Jones et al 1991; Condeelis 1992). Reprinted from Jones et al (1991) with permission.

filaments. As discussed above, the appearance of free barbed ends probably results from uncapping of filament ends that are capped by proteins like gelsolin and aginactin and/or from de novo assembly of actin oligomers. Optimum filament elongation probably requires the release of actin monomers from the TB4 complex and profilin-catalyzed nucleotide exchange. Both nucleation and elongation would be regulated by the synergistic interplay between PIP metabolism, profilin, and capping proteins.

2. The cross-linking of the newly growing actin filaments, to form either filament networks with the space-filling orthogonal geometry, or bundles of filaments, is an essential intermediate step as emphasized by the disruption of normal pseudopod extension when key filament cross-linkers are absent.
3. Expansion of the volume of the structures containing cross-linked actin filaments generates the initial protrusions. Volume expansion by osmotic gel force probably contributes in protrusions containing filament networks. This requires the severing of some filaments within the network, which is consistent with the appearance of many short filaments in such structures soon after actin polymerization begins.

Further actin polymerization and filament cross-linking also contributes to volume expansion of the network as filaments continue to elongate against the membrane and within the gel itself. This would explain the retention of high filament densities in networks during pseudopod extension even though filament severing is occurring at this time. Actin polymerization probably dominates in the protrusion of filopods, which would explain the different behavior of pseudopods and filopods when under osmotic load.

As these protrusions mature, late events such as attachment of filaments through the membrane to substrate adhesion sites and actin depolymerization, when coupled to myosin-mediated retraction of the filament network or bundles, would locomote the cell forward. In the absence of stable substrate attachment, retraction and depolymerization would cause centripetal flow of these actin-containing structures.

The events described in Figure 4 may be regulated by several different signaling pathways. Stimulated pseudopod extension, such as that occurring during chemotaxis, involves crosstalk between specific receptors and key actin-binding proteins. Constitutive pseudopod extension, as seen during random cell locomotion, probably involves oscillations in the activity of these key actin-binding proteins that might be biased by regulatory proteins, such as the small G proteins, which are shared by a number of signaling pathways.

ACKNOWLEDGMENTS

The author is grateful to Drs. P. Conrad, J. Hartwig, G. Oster, D. Soll, T. Stossel, D. Wessels, and members of the Condeelis laboratory for editorial comments; to Drs. D. Wessels and D. Soll for providing unpublished data for Table 1; Dr. M. Weitzman for literature searchers; and Drs. P. Conrad, K. Fath, D. Burgess, Y. Fukui, E. Luna, and G. Oster for sharing unpublished manuscripts, and Drs. J. V. Small and F. Southwick for sending reprints. The author thanks the National Institutes of Health and the New York Lung Society for continued support.

Literature Cited

Abercrombie, M, Heaysman, J, Pegrum, S. 1970. The locomotion of fibroblasts in culture. I. movements of the leading edge. *Exp. Cell Res.* 59:393–98

Albrecht-Buehler, G. 1980. Autonomous movements of cytoplasmic fragments. *Proc. Natl. Acad. Sci. USA* 77:6639–43

Andre, E, Brink, M, Gerisch, G, Isenberg, G, Noegel, A, et al. 1989. A *Dictyostelium* mutant deficient in severin, an F-actin fragmenting protein, shows normal motility and chemotaxis. *J. Cell Biol.* 108:985–95

Baines, I, Korn, E. 1990. Localization of myosin IC and myosin II in *Acanthamoeba casellanii* by indirect immunofluorescence and immunogold electron microscopy. *J. Cell Biol.* 111:1895–904

Bernardini, M, Mounier, J, D'Hauteville, H, Coquis-Rondon, M, Sansonetti, P. 1989. Identification of ics A, a plasmid locus of *Shigella flexneri* that governs bacterial intra- and intercellular spread through interaction with F-actin. *Proc. Natl. Acad. Sci. USA* 86:3867–71

Bray, D, Money, N, Harold, F, Bamburg, J. 1991. Responses of growth cones to changes in osmolarity of the surrounding medium. *J. Cell Sci.* 98:507–15

Bresnick, A, Janmey, P, Condeelis, J. 1991. Evidence that 27-residue sequence is the actin-binding site of ABP-120. *J. Biol. Chem.* 266:12989–93

Bresnick, A, Warren, V, Condeelis, J. 1990. Identification of a short sequence essential for actin binding by *Dictyostelium* ABP-120. *J. Biol. Chem.* 265:9236–40

Brink, M, Gerisch, G, Isenberg, G, Noegel, A, Segall, J, et al. 1990. A *Dictyostelium* mutant lacking an F-actin cross-linking protein, the 120kD gelation factor. *J. Cell Biol.* 111:1477–89

Brundage, R, Fogarty, K, Tuft, R, Fay, F. 1991. Calcium gradients underlying polarization and chemotaxis of eosinophils. *Science* 254:703–6

Bryant, R, Sutcliffe, M, McGee, Z. 1972. Effect of osmolalities comparable to those of the renal medulla on function of human polymorphonuclear leucocytes. *J. Infect. Dis.* 126:1–10

Cano, M, Cassimeris, L, Fechheimer, M, Zigmond, S. 1992. Mechanisms responsible for F-actin stabilization after lysis of polymorphonuclear leukoctyes. *J. Cell Biol.* 116:1123–34

Cano, M, Lauffenburger, D, Zigmond, S. 1991. Kinetic analysis of F-actin depolymerization in polymorphonuclear leukocyte lysates indicates that chemoattractant stimulation increases actin filament number without altering the filament length distribution. *J. Cell Biol.* 115:677–87

Cao, L, Babcock, G, Rubenstein, P, Wang, Y. 1992. Effects of profilin and profilactin on actin structure and function in living cells. *J. Cell Biol.* 117:1023–29

Carboni, J, Condeelis, J. 1985. Ligand-induced changes in the location of actin, myosin, αactinin and 120K protein in amoebae of *Dictyostelium discoideum*. *J. Cell Biol.* 100:1884–93

Carlsson, L, Markey, F, Blikstad, I, Persson, T, Lindberg, U. 1979. Reorganization of actin in platelets stimulated by thrombin as measured by the DNase I inhibition assay. *Proc. Natl. Acad. Sci. USA* 76:6376–80

Carson, M, Weber, A, Zigmond, S. 1986. An actin nucleating activity in polymorphonuclear leukocytes is modulated by chemotactic peptides. *J. Cell Biol.* 103: 2707–14

Cassimeris, L, Safer, D, Nachmias, V, Zigmond, S. 1992. Thymosin beta -4 sequesters the majority of G-actin in resting human polymorphonuclear leukocytes. *J. Cell Biol* 119:1261–70

Cheney, R, Mooseker, M. 1992. Unconventional myosins. *Curr. Opin. Cell Biol.* 4:27–35

Coates, T, Watts, R, Hartman, R. Howard, T. 1992. Relationship of F-actin distribution to development of polar shape in human polymorphonuclear neutrophils. *J. Cell Biol.* 117:765–74

Condeelis, J. 1992. Are all pseudopods created equal? *Cell Motil. Cytoskel.* 22:1–6

Condeelis, J, Demma, B, Dharmawardhane, S, Eddy, R, Hall, A, et al. 1990. Mechanisms of amoeboid chemotaxis: An evaluation of the cortical expansion model. *Dev. Genet.* 11:333–40

Condeelis, J, Hall, A, Bresnick, A, Warren, V, Hock, R, et al. 1988. Actin polymerization and pseudopod extension during amoeboid chemotaxis. *Cell Motil. Cytoskel.* 10:77–90

Conrad, P, Giuliano, K, Fisher, G. Collins, K, Matsudaira, P, Taylor, D. 1993. Relative distribution of actin, myosin I and myosin II during the wound healing response of fibroblasts. *J. Cell Biol.* 120: 1381–91

Conrad, P, Nederlof, M, Herman, I, Taylor, D. 1989. Correlated distribution of actin, myosin, and microtubules at the leading edge of Swiss 3T3 fibroblasts. *Cell Motil. Cytoskeleton* 14:527–43

Cooper, J. 1991. The role of actin polymerization in cell motility. *Annu. Rev. Physiol.* 53:585–605

Cooper, J, Bryan, J, Schwab, W, Frieden, C, Loftus, D, Elson, E. 1987. Microinjection of gelsolin into living cells. *J. Cell Biol* 104:491–501

Cooper, M, Schliwa, M. 1986. Motility of cultured fish epidermal cells in the presence and absence of direct current electric fields. *J. Cell Biol.* 102:1384–99

Cortese, J, Schwab, B, Frieden, C, Elson, E. 1989. Actin polymerization induces a shape change in actin-containing vesicles. *Proc. Natl. Acad. Sci. USA* 86:5773–77

Cox, D, Condeelis, J, Hartwig, J. 1993. Genetic deletion of ABP-120 alters the three dimensional organization of actin filaments in *Dictyostelium* pseudopods. *Mol. Biol. Cell Abstr.* 4: Suppl.

Cox, D, Condeelis, J, Wessels, D, Soll, D, Kern, H, Knecht, D. 1992. Targeted disruption of the ABP-120 gene leads to cells with altered motility. *J. Cell Biol.* 116: 943–55

Cunningham, C, Gorlin, J, Kwiatkowski, D, Hartwig, J, Janmey, P, Stossel, T. 1992. Requirement for actin binding protein for cortical stability and efficient locomotion. *Science* 255:325–27

Cunningham, C, Stossel, T, Kwiatkowski, D. 1991. Enhanced motility in NIH 3T3 fibroblasts that overexpress gelsolin. *Science* 251:1233–36

Dabiri, G, Sanger, J, Portnoy, D, Southwick, F. 1990. *Listeria* monocytogenes moves rapidly through the host cell cytoplasm by inducing directional actin assembly. *Proc. Natl. Acad. Sci. USA* 87:6068–72

Dadabay, C, Patton, E, Cooper, J, Pike, L. 1991. Lack of correlation between changes in poly-phosphoinositide levels and actin/gelsolin complexes in A431 cells treated with epidermal growth factor. *J. Cell Biol.* 112:1151–56

DeLozanne, A, Spudich, J. 1987. Disruption of the *Dictyostelium* myosin heavy chain gene by homologous recombination. *Science* 236:1086–91

DePasquale, J, Izzard, C. 1991. Accumulation of talin in nodes at the edge of the lamellipodium and separate incorporation into adhesion plaques at focal contacts in fibroblasts. *J. Cell Biol.* 113:1351–59

DeRosier, D, Tilney, L, Bonder, E, Frankel, P. 1982. A change in twist of actin provides the force for the extension of the acrosomal process in *Limulus* sperm. *J. Cell Biol.* 93:324–37

Detmers, P, Goodenough, U, Condeelis, J. 1983. Elongation of the fertilization tubule in *Chlamydomonas*. *J. Cell Biol.* 97:522–32

Dharmawardhane, S, Demma, M, Yang, F, Condeelis, J. 1991. Compartmentalization and actin binding properties of ABP-50: The elongation factor-1 alpha of *Dictyostelium*. *Cell Motil. Cytoskel.* 20:279–88

Dharmawardhane, S, Warren, V, Hall, A, Condeelis, J. 1989. Changes in the association of actin-binding proteins with the actin cytoskeleton during chemotactic stimulation of *Dictyostelium discoideum*. *Cell Motil. Cytoskeleton* 13:57–63

Dipasquale, A. 1975. Locomotion of epithelial cells. Factors involved in extension of the leading edge. *Exp. Cell Res.* 95:425–39

Downey, G, Chan, C, Trudel, S, Grinstein, S. 1990. Actin assembly in electropermeabilized neutrophils: Role of intracellular calcuim. *J. Cell Biol.* 110:1975–82

Drenckhahn, D, Pollard, T. 1986. Elongation of actin filaments is a diffusion-limited reaction at the barbed end and is accelerated by inert macromolecules. *J. Biol. Chem.* 261:12754–58

Eddy, R, Sauterer, R, Condeelis, J. 1993. Aginactin, an agonist-regulated F-actin capping activity is associated with an HSP70 in *Dictyostelium*. *J. Biol. Chem.* In press

Egelhoff, T, Spudich, J. 1991. Molecular genetics of cell migration: *Dictyostelium* as a model system. *Trends Genet.* 7:161–66

Euteneuer, U, Schliwa, M. 1984. Persistent,

directional motility of cells and cytoplasmic fragments in the absence of microtubules. *Nature* 310:58–61

Evans, E, Young, A. 1989. Apparent viscosity and cortical tension of blood granulocytes determined by micropipet aspiration. *Biophys. J.* 56:151–60

Fath, C. Burgess, D. 1993. Golgi-derived vesicles from developing epithelial cells bind actin filaments and possess Myosin I as a cytoplasmically-oriented peripheral membrane protein. *J. Cell Biol.* 120:117–27

Felder, S, Elson, E. 1990. Mechanics of fibroblast locomotion: quantitative analysis of forces and motions at the leading lamellas of fibroblasts. *J. Cell Biol.* 111:2513–26

Fisher, G, Conrad, P, DeBiasio, R, Taylor, D. 1988. Centripetal transport of cytoplasm, actin and the cell surface in lamellipodia of fibroblasts. *Cell Motil. Cytoskeleton* 11: 235–47

Forscher, P, Lin, C, Thompson, C. 1992. Novel form of growth cone motility involving site-directed actin filament assembly. *Nature* 357:515–18

Forscher, P, Smith, S. 1988. Actions of cytochalasins on the organization of actin filaments and microtubules in a neuronal growth cone. *J. Cell Biol.* 107:1505–16

Fricke, K, Sackmann, E. 1984. Variation of frequency spectrum of erythrocyte flickering caused by aging, osmolarity, temperature and pathological changes. *Biochem. Biophys. Acta* 803:145–52

Fujisaki, H, Albanesi, J, Korn, E. 1985. Experimental evidence for the contractile activities of *Acanthamoeba* myosins IA and IB. *J. Biol. Chem.* 260:11183–89

Fukui, Y. 1993. Toward a new concept of cell motility: Cytoskeletal dynamics in amoeboid movement and cell division. *Int. Rev. Cytol.* 144:85–127

Fukui, Y, Lynch, T, Bryeska, H, Korn, E. 1989. Myosin I is located at the leading edges of locomoting *Dictyostelium* amoebae. *Nature* 341:328–31

Furukashi, K, Sadashi, H, Ando, S, Nishizawa, K, Inagaki, M. 1992. Phosphorylation by actin kinase of the pointed end domain on the actin molecule. *J. Biol. Chem.* 267:9326–30

Gittes, F, Mickey, B, Nettleton, J, Howard, J. 1993. Flexural rigidity of microtubules and actin filaments measured from thermal fluctuations in shape. *J. Cell Biol.* 120: 923–34

Goldschmidt-Clermont, P, Furman, M, Wochsstock, D, Safer, D, Nachmias, V, Pollard, T. 1992. The control of actin nucleotide exchange of Thymosin beta-4 and profilin. A potential regulatory mechanism for actin polymerization in cells. *Mol. Biol. Cell* 3:1015–24

Goldschmidt-Clermont, P, Kim, J, Machesky, L, Rhee, S, Pollard, T. 1991a. Regulation of phospholipase C-gamma 1 by profilin and tyrosine phosphorylation. *Science* 235: 1231–33

Goldschmidt-Clermont, P, Machesky, L, Baldassare, J, Pollard, T. 1990. The actin binding protein profilin binds to PIP_2 and inhibits its hydrolysis by phospholipase-C. *Science* 247:1575–78

Goldschmidt-Clermont, P, Machesky, L, Doberstein, S, Pollard, T. 1991b. Mechanism of the interaction of human platelet profilin with actin. *J. Cell Biol.* 113:1081–89

Gorlin, J, Yamin, R, Egan, S, Stewart, M, Stossel, T, et al. 1990. Human endothelial actin binding protein (ABP-280, non-muscle filamin): A molecular leaf spring. *J. Cell Biol.* 111:1089–105

Hagen, S, Kiehart, D, Kaiser, D, Pollard, T. 1986. Characterization of monoclonal antibodies to *Acanthamoeba* myosin I that cross-react with both myosin II and low molecular mass nuclear proteins. *J. Cell Biol.* 103:2121–28

Hahn, K, DeBiasio, R, Taylor, D. 1992. Patterns of elevated free calcium and calmodulin activation in living cells. *Nature* 359:736–38

Hall, A, Schlein, A, Condeelis, J. 1988. Relationship of pseudopod extension to chemotactic hormone-induced actin polymerization in amoeboid cells. *J. Cell Biochem.* 37:285–99

Hall, A, Warren, V, Condeelis, J. 1989a. Transduction of the chemotactic signal to the actin cytoskeleton of *Dictyostelium discoideum. Dev. Biol.* 136:517–25

Hall, A, Warren, V, Dharmawardhane, S, Condeelis, J. 1989b. Identification of actin nucleation activity and polymerization inhibitor in amoeboid cells: Their regulation by chemotactic stimulation. *J. Cell Biol.* 109:2207–13

Handel, S, Hendry, K, Sheterline, P. 1990. Microinjection of covalently crosslinked actin oligomers causes disruption of existing actin filament architecture in PtK_2 cells. *J. Cell Sci.* 97:325–33

Hartwig, J. 1992. Mechanisms of actin rearrangements mediating platelet activation. *J. Cell Biol.* 118:1421–42

Hartwig, J, Chambers, K, Hopcia, K, Kwiatkowski, D. 1989. Association of profilin with filament free regions of human leukocytes and platelet membranes and reversible membrane binding during platelet activation. *J. Cell Biol.* 109:1571–79

Hartwig, J, Janmey, P. 1989. Stimulation of a calcium-dependent actin nucleation activity by phorbol 12-myristate, 3-acetate in rabbit macrophage cytoskeletons. *Biochem. Biophys. Acta* 1010:64–71

Hartwig, J, Kwiatkowski, D. 1991. Actin binding proteins. *Curr. Opin. Cell Biol.* 3:87–97

Hartwig, J, Shevlin, P. 1986. The architecture of actin filaments and the ultrastructural location of actin-binding protein in the periphery of lung macrophages. *J. Cell Biol.* 103:1007–20

Hartwig, J, Yin, H. 1988. The organization and regulation of the macrophage actin skeleton. *Cell Motil. Cytoskeleton* 10: 117–25

Hill, T, Kirschner, M. 1982. Bioenergetics and kinetics of microtubule and actin filament assembly-disassembly. *Int. Rev. Cytol.* 78:1–123

Howard, P, Bartholomew, M, Sefton, B, Firtel, R. 1993. Tyrosine phosphorylation of actin in *Dictyostelium* associated with cell-shape changes. *Science* 259:241–44

Howard, T, Chaponnier, C, Yin, H, Stossel, T. 1990. Gelsolin-actin interaction in actin polymerization in human neutrophils. *J. Cell Biol.* 110:1983–92

Huckriede, A, Fuchtbauer, A, Hinssen, H, Chaponnier, C, Weeds, A, Jockusch, B. 1990. Differential effects of gelsolins on tissue culture cells. *Cell Motil. Cytoskeleton* 16:229–38

Jaconi, M, Rives, R, Schlegel, W, Wollheim, C, Pittet, D, Lew, P. 1988. Spontaneous and chemoattractant induced oscillations of cytosolic free calcium in single adherent human neutrophils. *J. Biol. Chem.* 263: 10557–60

Jones, J, Segall, J, Condeelis, J. 1991. Molecular analysis of amoeboid chemotaxis: parallel observations in amoeboid phagocytes and metastatic tumor cells. In *Cell Motility Factors,* ed. ID Goldberg, pp. 1–16. Basel: Birkhauser Verlag

Jung, G, Hammer, J. 1990. Generation and characterization of *Dictyostelium* cells deficient in a myosin I heavy chain isoform. *J. Cell Biol.* 110:1955–64

Kabsch, W, Vandekerekhove, J. 1992. Structure and function of actin. *Annu. Rev. Biophys. Biomol. Struct.* 21:49–76

Kaufmann, S, Kas, J, Goldmann, W, Sackmann, E, Isenberg, G. 1992. Talin anchors and nucleates actin filaments at lipid membranes. *FEBS Lett.* 314:203–5

King, K, Stewart, M, Roberts, T, Seavy, M. 1992. Structure and macromolecular assembly of two isoforms of the major sperm protein from the amoeboid sperm of the nematode, *Ascaris suum. J. Cell Sci.* 101: 847–57

Knecht, D, Loomis, W. 1987. Antisense RNA inactivation of myosin heavy chain gene expression in *Dictyostelium discoideum. Science* 236:1081–86

Korn, E, Hammer, J. 1990. Myosin I. *Curr. Opin. Cell Biol.* 2:57–61

Lamb, J, Allen, P, Tuan, B, Nakayama, T, Janmey, P. 1992. Low pH activates gelsolin in the absence of calcium. *Mol. Biol. Cell* 3:41a

Lambooy, P, Korn, E. 1988. Inhibition of an early stage of actin polymerization by actobindin. *J. Biol. Chem.* 263:12836–43

Lassing, I, Lindberg, U. 1985. Specific interaction between phosphatidylinositol 4,5 bisphosphate and profilactin. *Nature* 314:472–74

Lewis, A, Bridgman, P. 1992. Nerve growth cone lamellipodia contain two populations of actin filaments that differ in organization and polarity. *J. Cell Biol* 119:1219–43

Li, Y, Tanaka, T. 1990. Kinetics of swelling and shrinking gels. *J. Chem. Phys.* 92: 1365–71

Luna, E, Condeelis, J. 1990. Actin-associated proteins in *Dictyostelium discoideum. Dev. Genet.* 11:328–32

Luna, E, Hitt, A. 1992. Cytoskeleton-plasma membrane interactions. *Science* 258:955–64

Marks, P, Maxfield, F. 1990.. Transient increases in cytosolic free calcium appear to be required for the migration of adherent human neutrophils. *J. Cell Biol.* 110:43–52

Mitchison, T, Kirschner, M. 1988. Cytoskeletal dynamics and nerve growth. *Neuron* 1:761–71

Miyata, H, Bowers, B, Korn, E. 1989. Plasma membrane association of *Acanthamoeba* myosin I. *J. Cell Biol.* 109:1519–28

Miyata, H, Hotani, H. 1992. Morphological changes in liposomes caused by polymerization of encapsulated actin and spontaneous formation of actin bundles. *Proc. Natl. Acad. Sci. USA* 89:11547–51

Mooseker, M, Pollard, T, Whartan, K. 1982. Nucleated polymerization of actin from the membrane associated ends of microvillar filaments in the intestinal brush border. *J. Cell Biol.* 95:223–33

Niederman, R, Amrein, P, Hartwig, J. 1983. Three dimensional structure of actin filaments and of an actin gel made with actin binding protein. *J. Cell Biol.* 96:1400–13

Noegel, A, Rapp, S, Lottespeich, F, Schleicher, M, Stewart, M. 1989. The *Dictyostelium* gelation factor shares a putative actin-binding site with alpha-actinins and dystrophin and also has a rod domain containing six 100-residue motifs that appear to have cross-beta conformation. *J. Cell Biol.* 109:607–18

Novak, K, Titus, M. 1992. *Dictyostelium* cells devoid of two unconventional myosins show defects in growth. *Mol. Biol. Cell* 3:3a

Ogihara, S, Carboni, J, Condeelis, J. 1988. Electron microscopic localization of myosin II and ABP-120 in the cortical actin matrix of *Dictyostelium* amoebae using IgG-gold conjugates. *Dev. Genet.* 9:505–20

Okabe, S, Hirokawa, N. 1991. Actin dynamics in growth cones. *J. Neurosci.* 11:1918–29

Omann, G, Allen, R, Bokoch, G, Painter, R, Traynor, A, Sklar, L. 1987. Signal transduction and cytoskeletal activation in the neutrophil. *Physiol. Rev.* 67:285–322

Oster, G. 1984. On the crawling of cells. *J. Embryol. Exp. Morph.* 83:329–64

Oster, G, Perelson, A. 1985. Cell spreading and motility: a model lamellipod. *J. Math. Biol.* 21:383–88

Oster, G, Perelson, A. 1987. The physics of cell motility. *J. Cell Sci.* 8:35–54

Oster, G, Perelson, A. 1993. Cell protrusions. In *Lecture Notes In Biomathematics,* ed. S Levin, Vol. 100. New York/Berlin: Springer Verlag. In press

Oster, G, Perelson, A, Tilney, L. 1982. A mechanical model for elongation of the acrosomal process in Thyone sperm. *J. Math. Biol.* 15:259–65

Peskin, C, Odell, G, Oster, G. 1993. Cellular motions and thermal fluctuations: The Browinian ratchet. *Biophys. J.* In press

Pollard, T. 1986 Rate constants for the reactions of ATP and ADP-actin with the ends of actin filaments. *J. Cell Biol.* 103:2747–54

Rabinovitch, M, DeStefano, M, Dziezanowski, 1980. Neutrophil migration under agarose: Stimulation by lowered medium pH and osmolality. *J. Retic. Soc.* 27:189–200

Ridley, A, Paterson, H, Johnston, D, Diekmann, P, Hall, A. 1992. The small GTP binding protein rac regulates growth factor-induced membrane ruffling. *Cell* 70:401–10

Rinnerthaler, G, Herzog, M, Klappacher, M, Kunka, H, Small, J. 1991. Leading edge movement and ultrastructure in mouse macrophages. *J. Struct. Biol.* 106:1–16

Roberts, T, King, K. 1991. Centripetal flow and directed reassembly of the major sperm protein cytoskeleton in amoeboid sperm of the nematode *Ascaris suum. Cell Motil. Cytoskeleton* 20:228–41

Rubino, S, Fighetti, M, Unger, E, Cappauccinelli, P. 1984. Location of actin, myosin and microtubular structures during directed locomotion of *Dictyostelium* amoebae. *J. Cell Biol.* 98:282–90

Rubino, S, Small, J. 1987. The cytoskeleton of spreading *Dictyostelium* amoebae. *Protoplasma* 136:63–69

Safer, D, Elzinga, M, Nachmias, V. 1991. Thymosin beta-4 and Fx, an actin sequestering peptide, are indistinguishable. *J. Biol. Chem.* 266:4029–32

Safer, D, Golla, R, Nachmias, V. 1990. Isolation of a 5-kilodalton actin-sequestering peptide from human blood platelets. *Proc. Natl. Acad. Sci. USA* 87:2536–40

Sanders, M, Goldstein, A, Wang, Y. 1992. Thymosin beta 4 (Fx peptide) is a potent regulator of actin polymerization in living cells. *Proc. Natl. Acad. Sci. USA* 89:4678–82

Sanders, M, Wang, Y. 1990. Exogenous nucleation sites fail to induce detectable polymerization of actin in living cells. *J. Cell Biol.* 110:359–65

Sanger, J, Sanger, J, Southwick, F. 1992. Host cell actin assembly is necessary and likely to provide the propulsive force for intracellular movement of *Listeria* monocytogenes. *Infect. Immun.* 60:3609–19

Sauterer, R, Eddy, R, Hall, A, Condeelis, J. 1991. Purification and characterization of aginactin, a newly identified agonist-regulated actin capping protein from *Dictyostelium* amoebae. *J. Biol. Chem.* 266: 24533–39

Schariff, A, Luna, E. 1992. Diacylglycerol-stimulated formation of actin nucleation at plasma membranes. *Science* 256:245–47

Schlatterer, C, Knoll, G, Malchow, D. 1992. Intracellular calcium during chemotaxis of *Dictyostelium discoideum:* a new fura-2 derivative avoids sequestration of the indicator and allows long-term calcium measurements. *Eur. J. Cell Biol.* 58:172–81

Schwartz, M, Luna, E. 1988. How actin binds and assembles onto plasma membranes from *Dictyostelium discoideum. J. Cell Biol.* 107: 201–9

Schweiger, A, Mihalache, O, Ecke, M, Gerisch, G. 1992. Stage-specific tyrosine phosphorylation of actin in *Dictyostelium discoideum* cells. *J. Cell Sci.* 102:601–09

Segall, J. 1992. Behavioral responses of streamer F mutants of *Dictyostelium discoideum:* effects of cyclic GMP on cell motility. *J. Cell Sci.* 101:589–97

Segall, J, Jones, J, Condeelis, J. 1991. Analysis of responses of metastatic M27 tumor cells to the chemotactic peptide VGVAPG. *J. Cell Biol.* 115:346a

Sha'afi, R, Shefcyk, J, Yassin, R, Molski, T, Volpi, M, et al. 1986. Is a rise in intracellular concentration of free calcium necessary or sufficient for stimulated cytoskeletal-associated actin? *J. Cell Biol* 102:1459–63

Sheetz, M, Wayne, D. Pearlman, A. 1992. Extension of filopodia by motor-dependent actin assembly. *Cell Motil. Cytoskeleton* 22:160–69

Shu, W, Wang, D, Stracher, A. 1992. Chemical evidence for the existence of activated G-actin. *Biochem. J.* 283:567–73

Small, J. 1988. The actin cytoskeleton. *Electron Microsc. Rev.* 1:155–74

Sonobe, S, Tahahashi, S, Hatano, S, Kwada, K. 1986. Phosphorylation of amoeba G-actin and its effect on actin polymerization. *J. Biol. Chem.* 261:14837–43

Southwick, F, Young, C. 1990. The actin released from profilin-actin complexes is insufficient to account for the increase in F-actin in chemoattractant-stimulated polymorphonuclear leukocytes. *J. Cell Biol.* 110:1965–74

Stendahl, O, Hartwig, J, Brotschi, E, Stossel, T. 1980. Distribution of actin-binding protein and myosin in macrophages during spreading and phagocytosis. *J. Cell Biol.* 84:215–24

Stossel, T. 1989. From signal to pseudopod. *J. Biol. Chem.* 264:18261–64

Symons, M, Mitchison, T. 1991. Control of actin polymerization in live and permeabilized fibroblasts. *J. Cell Biol.* 114:503–13

Tanaka, T. 1981. Gels. *Sci. Am.* 244:124–38

Taylor, D, Condeelis, J. 1979. Cytoplasmic structure and contractility in amoeboid cells. *Int. Rev. Cytol.* 56:57–140

Theriot, J, Mitchison, T. 1992. Comparison of actin and cell surface dynamics in motile fibroblasts. *J. Cell Biol.* 118:367–77

Theriot, J, Mitchison, T, Tilney, L, Portnoy, D. 1992. The rate of actin-based motility of intracellular *Listeria* monocytogenes equals the rate of actin polymerization. *Nature* 357:257–60

Tilney, L, Fukui, Y, DeRosier, D. 1987. Movement of the actin filament bundle in Mytilus sperm: a new mechanism is proposed. *J. Cell Biol.* 104:981–93

Tilney, L, Hatano, S, Ishikawa, A, Mooseker, M. 1973. The polymerization of actin: its role in the generation of the acrosomal process of certain echinoderm sperm. *J. Cell Biol.* 59:109–26

Tilney, L, Inoue, S. 1982. Acrosomal reaction of Thyone sperm II. *J. Cell Biol.* 93:820–27

Tilney, L, Inoue, S. 1985. Acrosomal reaction of Thyone sperm. III The relationship between actin assembly and water influx during the extension of the acrosomal process. *J. Cell Biol.* 100:1273–83

Tilney, L, Portnoy, D. 1989. Actin filaments and the growth, movement and spread of the intracellular bacterial parasite, Listeria monocytogenes. *J. Cell Biol.* 109:1597–608

Titus, M, Wessels, D, Spudich, J, Soll, D. 1993. The unconventional myosin encoded by the myo A gene plays a role in *Dictyostelium* motility. *Mol. Biol. Cell* 4:233–46

Van Duijn, B, Van Haastert, P. 1992. Independent control of locomotion and orientation during *Dictyostelium discoideum* chemotaxis. *J. Cell Sci.* 102:763–68

Wagner, M, Barylko, B, Albanesi, J. 1992. Tissue distribution and subcellular localization of mammalian myosin I. *J. Cell Biol.* 119:163–70

Wang, Y. 1985. Exchange of actin subunits at the leading edge of living fibroblasts: Possible role of treadmilling. *J. Cell Biol.* 101:597–602

Warrick, H, Spudich, J. 1987. Myosin structure and function in cell motility. *Annu. Rev. Cell Biol.* 3:379–421

Weber, A, Nachmias, V, Pennise, C, Pring, M, Safer, D. 1992. Interaction of Thymosin beta-4 with muscle and platelet actin: implications for actin sequestration in resting platelets. *Biochemistry* 31:6179–85

Wessels, D, Murray, J, Jung, G, Hammer, J, Soll, D. 1991. Myosin IB null mutants of *Dictyostelium* exhibit abnormalities in motility. *Cell Motil. Cytoskeleton* 20:301–15

Wessels, D, Soll, D, Knecht, D, Loomis, W, Delozanne, A. Spudich, J. 1988. Cell motility and chemotaxis in *Dictyostelium* amoebae lacking myosin heavy chain. *Dev. Biol.* 128:164–77

Witke, W, Schleicher, M, Noegel, A. 1992. Redundancy in the microfilament system: abnormal development of Dictyostelium cells lacking two F-actin cross-linking proteins. *Cell* 68:53–62

Wolosewick, J, Condeelis, J. 1986. Fine structure of gels prepared from an actin binding protein and actin: Comparison to cytoplasmic extracts and cortical cytoplasm in amoeboid cells of *Dictyostelium discoideum. J. Cell. Biochem.* 30:227–43

Yu, F, Lin, S, Bogorad, M, Atkinson, M, Yin, H. 1993. Thymosin beta 10 and thymosin beta 4 are both actin monomer sequestering proteins. *J. Biol. Chem.* 268:502–9

Annu. Rev. Cell Biol. 1993. 9:445–478

PROTEIN IMPORT INTO PEROXISOMES AND BIOGENESIS OF THE ORGANELLE

Suresh Subramani

Department of Biology, University of California at San Diego, La Jolla, California 92093–0322

KEY WORDS: peroxisomal targeting signals, assembly of microbodies, peroxisomal disorders, in vitro import, PTS receptors, peroxisome proliferation

---

CONTENTS

0743–4634/93/1115–0445$05.00

INTRODUCTION

In eukaryotic cells, the trafficking of proteins to membrane-bound subcellular compartments, such as the endoplasmic reticulum (ER),[1] Golgi complex, lysosomes or vacuoles, peroxisomes, chloroplasts, mitochondria and the nucleus, is signal-dependent. These signals, endowed in proteins as part of their primary amino acid sequence or appended onto them as chemical modifications, are received and processed by receptors interacting with organelle-specific sorting machineries to achieve the selective targeting of proteins to defined subcellular locations.

The biogenesis of peroxisomes was last reviewed in this series by Lazarow & Fujiki (1985). Since that time tremendous advances have been made in the elucidation of the diverse biochemical reactions performed by peroxisomes (reviewed by Van den Bosch et al 1992), the targeting of proteins to the organelle (reviewed by Subramani 1992), the temporal events in the biogenesis of peroxisomes (Veenhuis & Goodman 1990), the development of in vivo (Walton et al 1992) and in vitro assays (Small et al 1987, 1988b; Imanaka et al 1987; Miyazawa et al 1989; Thieringer et al 1991; Wendland & Subramani 1993a) to monitor peroxisomal protein import, and in the discovery of human peroxisomal disorders in which an individual biochemical function or the biogenesis of the organelle is impaired (see Van den Bosch et al 1992). The isolation of peroxisome-deficient mammalian and yeast cells has been used to generate model systems that mimic the cellular phenotypes diagnostic of human peroxisomal disorders (Zoeller et al 1989; Tsukamoto et al 1990; Erdmann & Kunau 1992; McCollum et al 1993). Some of the molecular mutations responsible for these disorders are already known in human patients (Gartner et al 1992; Shimozawa et al 1992b). Finally, the conservation of peroxisomal targeting signals (PTSs) in evolutionarily diverse organisms (Gould et al 1988, 1989, 1990a,b; Keller et al 1991), the use of similar targeting signals for the sorting of proteins into the matrices of peroxisomes, glyoxysomes and glycosomes (Keller et al

[1]Abbreviations: CHO, Chinese hamster ovary; DHAP, dihydroxyacetonephosphate; DHFR, dihydrofolate reductase; ER, endoplasmic reticulum; HPA, hyperpipecolic acidemia; HSA, human serum albumin; IRD, infantile Refsum disease; NALD, neonatal adrenoleukodystrophy; NEM, N-ethylmaleimide; PC, phosphatidyl choline; PE, phosphatidyl ethanolamine; PMP, peroxisomal membrane protein; PPAR, peroxisome-proliferator-activated receptor; PTS, peroxisomal targeting signal; RAR, retinoic acid receptor; RXR, 9-cis retinoic acid receptor; TR, thyroid hormone receptor; VDR, vitamin D_3 receptor; VLCFA, very long chain fatty acid; ZS, Zellweger syndrome.

1991; Fung & Clayton 1991; Volokita 1991; Blattner et al 1992; Sommer et al 1992), and the conservation of many of the genes involved in peroxisome import and assembly are beginning to reveal unifying mechanisms for the biogenesis of microbodies. It is these advances that are the focus of this review.

FUNCTIONS OF PEROXISOMES

Peroxisomes are single-membrane-bound organelles that range in diameter from 0.1–1.0μM, are more dense (1.21–1.25 gm/cm^3) than mitochondria (1.18 gm/cm^3), and are found in all eukaryotes except archaezoa (Cavalier-Smith 1987). Although the organelle appears roughly circular in cross-section, electron micrographs of serial sections of rat sebaceous gland and liver provide evidence for a peroxisomal reticulum (Gorgas 1985; Yamamoto & Fahimi 1987). The organelle derives its name from the fact that many enzymes (especially oxidases) that generate toxic hydrogen peroxide, as well as catalase, the enzyme that decomposes H_2O_2 to water and oxygen, are residents of this compartment (DeDuve & Baudhuin 1966). However, the organelle is much more than the toxic-waste dump of the cell. It is now known to house close to 50 enzymes (Tolbert 1981), many of which participate in specific metabolic pathways. The biochemistry of peroxisomes was reviewed recently by Van den Bosch et al (1992) and is therefore mentioned here only briefly.

Peroxisomes contain the first two enzymes (alkyl dihydroxyacetone-phosphate synthase and dihydroxyacetonephosphate acyl transferase) involved in the synthesis of plasmalogens (1-alk-1′-enyl-acyl-sn-3-phosphoglycerolipid) (Hajra & Bishop 1982), which account for 19% of the total phospholipid content in adult human (Horrocks & Sharma 1982). The greatest abundance of these plasmalogens is in the brain and heart. In humans, the β-oxidation of fatty acids containing less than 18 carbon atoms occurs in mitochondria, but the β-oxidation of very long chain fatty acids (VLCFA) takes place in peroxisomes (Lazarow & De Duve 1976). In contrast all β-oxidation of fatty acids occurs solely in peroxisomes in fungi (Kunau et al 1988). Other peroxisomal enzymes participate in cholesterol and bile acid synthesis (reviewed by Krisans 1992), purine and amino acid catabolism (Takada & Noguchi 1986), glyoxylate utilization (Briedenbach & Beevers 1967), prostaglandin metabolism (Diczfalusy et al 1987; Schepers et al 1988), oxidation of L-pipecolic (Zaar et al 1986; Wanders et al 1988) and L-2-hydroxyphytanic acids (Draye et al 1987), and penicillin biosynthesis in *Penicillium chrysogenum* (Müller et al 1991).

Plant peroxisomes also play an important role in photorespiration. This pathway, catalyzing the recycling of phosphoglycolate, requires at least 13 enzymes distributed in three different organelles (chloroplasts, peroxisomes, and mitochondria) and about 20 passages of intermediates across organelle membranes (reviewed by Heupel & Heldt 1992). Phosphoglycolate hydrolysis in the stroma of chloroplasts produces glycolate, which is shuttled to peroxisomes for oxidation to glyoxylate and for transamination to glycine. This glycine enters the mitochondria where two molecules are oxidized to generate one molecule of serine, which is returned to the peroxisomes for transamination to hydroxypyruvate, followed by reduction to glycerate. The glycerate enters the chloroplast stroma where it is phosphorylated to 3-phosphoglycerate and is then used in the Calvin cycle (Heupel & Heldt 1992).

Peroxisomers and two other related organelles, the glyoxysomes of plants (Briedenbach & Beevers 1967) and the glycosomes of certain parasites (Opperdoes & Borst 1977) such as trypanosomes, belong to a more generic class of organelles known as microbodies, a term proposed by Rhodin to describe small bodies, distinct morphologically from mitochondria in nephron epithelia. It preceded the discovery of peroxisomes and originally did not have any biochemical connotation (Rhodin 1954). The biochemical thread that suggests a common evolutionary link for peroxisomes, glyoxysomes, and glycosomes is the presence of the enzymes involved in the β-oxidation of fatty acids (Lazarow & De Duve 1976; Opperdoes 1987; Cooper & Beevers 1969).

Glycosomes of certain *Trypanosomatides* (e.g. *T. brucei)* lack H_2O_2-producing oxidases and catalase, but those of related organisms, such as *Crithidia sp*. (Muse & Roberts 1973) and *Leptomonas samueli* (Souto-Padron & De Souza 1982), which belong to the same family, do have catalase. Glycosomes also contain the first two enzymes of the plasmalogen pathway (Opperdoes 1987). Thus while glycosomes and glyoxysomes grant residence to some of the same proteins as do peroxisomes, they are clearly specialized organelles. This is because the glyoxysomes also house the enzymes of the glyoxylate pathway (Briedenbach & Beevers 1967), whereas organisms that possess glycosomes harbor seven glycolytic enzymes in these organelles while also expressing (from related but distinct genes) cytosolic counterparts of the same enzymes (Opperdoes & Borst 1977).

As an added twist to this biochemical complexity, the number, volume, protein composition, and even the biochemical functions attributable to peroxisomes within a given cell can vary in response to the environmental milieu (Hartl & Just 1987; Angermüller & Fahimi 1988; Veenhuis et al 1989; Veenhuis & Goodman 1990, Goodman et al 1990). For example, the methylotrophic yeast *Pichia pastoris* has only one or a few small peroxisomes when grown on glucose (Gould et al 1992). Upon growth on methanol, there is a massive proliferation of several clustered peroxisomes accompanied by a substantial induction of several enzymes (alcohol oxidase, dihydroxyacetone synthase, and catalase) required for methanol utilization (Hazeu et al 1975; McCollum et al 1993). In contrast, growth on oleate as the sole carbon source results in several medium-sized peroxisomes (Gould et al 1992) that contain elevated levels of the β-oxidation enzyme (acyl CoA oxidase, the multifunctional β-oxidation enzyme and 3-ketoacyl CoA thiolase). Similar variations in protein composition have been documented in liver peroxisomes of rats treated with different hypolipidemic drugs (Angermüller & Fahimi 1988; Hartl & Just 1987; S. Krisans, personal communication). Thus while peroxisomes are certainly involved in the generation and degradation of H_2O_2 and also intimately linked to lipid metabolism, they are able to remain on call by proliferating as an adaptation to environmental conditions. Alternatively, they decrease in number when their functions can be placed on hold.

HUMAN PEROXISOMAL DISORDERS

During the last decade many recessive inherited diseases involving peroxisomes have been described (reviewed by Moser et al 1991; Van den Bosch et al 1992). The connection between the Zellweger syndrome, the prototypic peroxisomal disorder, and peroxisomes was first established by Goldfischer et al (1973). Underscoring the relevance of peroxisomes is the fact that human patients homozygous for mutations affecting genes responsible for these diseases usually succumb to the debilitating disorders. However, while the organelle is essential for survival at the organismal level, fibroblasts from these patients can be propagated readily in tissue culture, thereby providing an invaluable resource for genetic, cell biological, and biochemical studies.

The human peroxisomal disorders have been grouped into three broad categories (Van den Bosch et al 1992). The severest disorders (Group I) are those characterized by a generalized loss of peroxisomal functions. Multiple proteins that normally reside in peroxisomes are missing in peroxisomal fractions derived from cells of these patients (Wanders et al 1984, 1985, 1988; Tager et al 1985; Schram et al 1986; Santos et al 1988a; Mihalik et al 1989). Several of the matrix proteins are found instead in the cytosol (Wanders et al 1984, 1985; Schram et al 1986). The organelle itself appears to consist of membrane ghosts when viewed using indirect immunofluorescence (Santos et al 1988a, 1992; Yajima et al 1992). Peroxisomes from these cells migrate at aberrant densities upon density gradient centrifugation (Santos et al 1988b; Wiemer et al 1989; Gartner et al 1991), but contain several integral membrane proteins of molecular weight 22, 53, 69–70, and

140 K (Gartner et al 1991; Suzuki et al 1989; Santos et al 1988a; Wiemer et al 1989). Diseases in this class include classical Zellweger syndrome (ZS) (Opitz et al 1969; Zellweger 1987), neonatal adrenoleukodystrophy (NALD) (Kelley et al 1986), hyperpipecolic acidemia (HPA) (Gatfield et al 1968), and infantile Refsum disease (IRD) (Scotto et al 1982). These patients exhibit a variety of neurological, ocular, and liver abnormalities, display craniofacial dysmorphia, and die within the first decade after birth (Moser et al 1991; Van den Bosch et al 1992).

Complementation studies involving assessment of restoration of several peroxisomal functions have been undertaken using fibroblasts from patients with this generalized loss of peroxisomal function. The different criteria used include recovery of peroxisomes as judged by indirect immunofluorescencc with antibodies directed against peroxisomal matrix proteins (Brul et al 1988; Yajima et al 1992), restoration of β-oxidation activity (McGuiness et al 1990), reappearance of plasmalogen biosynthesis (Roscher et al 1989), and recovery of phytanic acid oxidation (Poll-Thé et al 1990). All the complementation tests proved consistent with each other. Based on these tests, eight different complementation groups were defined (Yajima et al 1992), and a ninth member was added subsequently to this list (Shimozawa et al 1992b). The total number of complementation groups has now increased to 11–13 (H. Moser, personal communication).

The second category (Group II) includes diseases exhibiting the loss of a smaller subset of peroxisomal functions. For example, in the Rhizomelic form of chondrodysplasia punctata (RCDP) (Spranger et al 1971), dihydroxyacetonephosphate (DHAP) acyl transferase, alkyl DHAP synthase, phytanic acid oxidase, the prethiolase protease and thiolase are absent from the peroxisomes (Heymans et al 1986; Hoefler et al 1988; Heikoop et al 1990). However, a patient with all the clinical symptoms of RCDP was found recently to lack only DHAP acyl transferase (Wanders et al 1992), which suggests that RCDP can also be caused by a single enzyme defect. If this observation holds, then RCDP patients missing this single enzyme would be more appropriately classified in the third group of peroxisomal disorders. In a second disease of this group, called Zellweger-like syndrome, DHAP acyl transferase and the β-oxidation enzymes were absent (Suzuki et al 1988).

The third group (Group III) comprises genetic diseases in which only a single peroxisomal enzyme is nonfunctional. These include acatalasemia (Eaton 1989), hyperoxaluria type I (Danpure et al 1989), X-linked adrenoleukodystropy (Moser et al 1984), and deficiencies in individual β-oxidation enzymes (Poll-Thé et al 1988; Watkins et al 1989; Schram et al 1987).

PEROXISOMES AS A MODEL FOR ORGANELLE ASSEMBLY AND DISASSEMBLY

While a great deal is known regarding the targeting of individual proteins to subcellular organelles, little is known about the mechanics of organelle assembly. Because most subcellular organelles are essential for normal vegetative growth, the steps in organelle assembly are being investigated by a combination of biochemical techniques and temperature-sensitive lethal mutations (Banta et al 1988; Pfanner & Neupert 1990; Douwe de Boer & Weisbeek 1991; Newport & Dunphy 1992; Bolotin-Fukuhara & Grivell 1992). In contrast to other subcellular organelles, peroxisomes are unique in that their functions are dispensable in certain nutritional environments. In normal cells, this means that they are maintained at a basal number of peroxisomes per cell. A variety of peroxisome assembly mutants of eukaryotic cells do maintain peroxisome remnants or ghosts, which are capable of division and segregation to daughter cells (Santos et al 1988b; McCollum et al 1993). Other peroxisome assembly mutants have no detectable peroxisomes (Erdmann & Kunau 1992). However, it is unclear whether these cells still retain peroxisome remnants that were not detected by the techniques used. Peroxisomes can also be made to proliferate (organelle assembly) or dissipate (organelle disassembly) in response to nutritional cues. Furthermore, the structural complexity of peroxisomes is probably simpler than that of other organelles. These features make it easy to obtain mutants that are impaired in the biogenesis of the organelle, while preserving the viability of the mutant cells.

Peroxisome Import and Assembly Mutants

HUMAN CELLS Because Zellweger syndrome fibroblasts lack many peroxisomal matrix proteins, they can be viewed as naturally-occurring peroxisome assembly and/or import mutants. All complementation groups belonging to Group I of the generalized peroxisomal disorders (see Human Peroxisomal Disorders) are deficient in the biosynthesis of plasmalogens (Roscher et al 1989), which were shown by Raetz and co-workers to protect cells from oxidative damage by scavenging singlet oxygen species (Zoeller et al 1988; Morand et al 1988). In addition they are deficient in β-oxidation of fatty acids, pipecolic and phytanic acid oxidation (Van den Bosch et al 1992), and cholesterol synthesis (Hodge et al 1991).

CHO CELLS Zoeller & Raetz (1986) used colony autoradiography as a screening assay to obtain Chinese hamster ovary (CHO) cell mutants defective in the peroxisomal plasmalogen biosynthetic enzyme, DHAP acyl transferase. These mutants were deficient in the assembly of functional peroxisomes and displayed reduced levels of peroxisomal DHAP acyl transferase, alkyl DHAP synthase, acyl CoA oxidase, particulate catalase, and plasmalogens. The mutants also mimicked the cellular phenotypes of Zellweger syndrome fibroblasts (Zoeller et al 1989). Somewhat surprisingly, however, all nine of the mutant lines isolated in this initial screen, as well as others obtained subsequently, belonged to a single complementation group (C. Raetz & R. A. Zoeller, personal communication). Tsukamoto et al (1990) used the same screen to isolate two CHO cell mutants (Z24 and Z65) deficient in peroxisome assembly. These fell into two complementation groups—one identical to that isolated by Raetz and co-workers and a new one (C. Raetz, personal communication). Subsequently a third complementation group of CHO mutants (ZP92) was found (Shimozawa et al 1992a). The Z24 and ZP92 lines correspond to the human complementation groups 1 and 4, respectively (Shimozawa et al 1992a), in the nomenclature of Yajima et al (1992), whereas the Z65 line is deficient in the PAF-1 gene (Tsukamoto et al 1991), which complements the ninth complementation group among the human disorders resulting in the generalized loss of peroxisome function (Shimozawa et al 1992b).

A direct selection for peroxisome biogenesis mutants devised by Morand et al (1990) is based on the principle that wild-type CHO cells incorporate the pyrene-labeled fatty acid analogue, 9-(1′-pyrene)nonanol (P9OH), into other lipids as the fatty acid alcohol and thereby become susceptible to the radical oxygen species generated by excitation of the pyrene with UV light. In contrast, plasmalogen-deficient mutants do not incorporate the pyrene-labeled analogue and are therefore resistant to UV light (Morand et al 1990; Zoeller & Raetz 1992).

All the peroxisome deficiency disorders in Groups I and II (ZS, NALD, IRD, HPA, and RCDP) have deficiencies in plasmalogen biosynthesis (Van den Bosch et al 1992). Cell lines from these patients and the CHO cell lines described above are sensitive to killing by exposure to 12-(1′-pyrene)dodecanoic acid (P12) followed by UV light because plasmalogens are required to protect cells from oxidative damage (Zoeller et al 1988; Tsukamoto et al 1991; Hoefler et al 1991). This property has been used to obtain genes that complement these peroxisomal disorders (Tsukamoto et al 1991).

SACCHAROMYCES CEREVISIAE The first peroxisome assembly (*pas*) mutants in yeast were isolated by Erdmann et al (1989). Because β-oxidation of fatty acids is exclusively peroxisomal in fungi (Kunau et al 1988), Erdmann et al (1989) obtained *o*leic-acid *n*on-*u*tilizers (*onu* mutants) and screened these by electron microscopy for the absence of peroxisomes and by crude biochemical fractionation for the absence of particulate catalase and thiolase. One in eight *onu* mutants turned out to be a *pas* mutant. Further analysis of these revealed 12 different complementation groups (Erdmann & Kunau 1992; Kunau & Hartig 1992). Van der Leij et al (1992) devised a positive enrichment based on the lethality of H_2O_2 produced in wild-type cells by the β-oxidation pathway. In the presence of 3-aminotriazole, a catalase inhibitor, the degradation of H_2O_2 in wild-type cells is prevented and accumulation of H_2O_2 causes cell death. Cells unable to accumulate H_2O_2, either because of a nonfunctional β-oxidation pathway or because of the inability to assemble functional peroxisomes, are able to survive and grow. Of those colonies that grew in the presence of 3-aminotriazole, 1% behaved like *pas* mutants. This led to the identification of three new complementation groups, bringing the total to 15 *pas* mutants in *S. cerevisiae* (Van der Leij 1992).

HANSENULA POLYMORPHA The methyltrophic yeast *H. polymorpha* has also been used to isolate peroxisome-deficient (*per*) mutants (Didion & Roggenkamp 1990; Cregg et al 1990; R. Rachubinski, personal communication). About 260 strains deficient in *m*ethanol *ut*ilization (Mut^-) (Cregg et al 1990) were screened using phase-contrast light microscopy to visualize peroxisomes and to look for large, cuboid, cytosolic crystalloids of alcohol oxidase. Strains lacking peroxisomes, or containing cytoplasmic alcohol oxidase, were then analyzed by transmission electron microscopy and by subcellular fractionation (Waterham et al 1992; Titorenko et al 1992; Veenhuis 1992). Of the 260 Mut^- mutants, 7.7% were *p*eroxisome-*im*port-deficient (Pim^- phenotype) and 22.3% were *per*oxisome-deficient (Per^- phenotype). Five complementation groups (*per*1-*per*5) of mutants with the Pim^- phenotype and 13 groups with the Per^- phenotype were defined. In addition, mutants affected in *p*eroxisome *s*ub*s*tructure (*pss*$^-$ mutants, five complementation groups) and conditional mutants (temperature-, cold- and pH-sensitive, six complementation groups) have been isolated (Veenhuis 1992). The distinction between these different phenotypes is blurred and complicated by the fact that different alleles within the *per*1 and *per*3 complementation groups display the Pim^- or Per^- phenotypes (Titorenko et al 1992). The true phenotype of mutations affecting genes in the various complementation groups will have to await the creation of null alleles using genes cloned by complementation of existing *per* mutants. It is also quite likely that there will be overlap between complementation groups exhibiting the Pim^-, Per^-, Pss^- and conditional phenotypes, so that the number of unique complementation groups affecting peroxisome biogenesis in *H.*

polymorpha is at least 20 (J. Cregg & R. Rachubinski, personal communication).

PICHIA PASTORIS In contrast to *S. cerevisiae* and *H. polymorpha,* which grow well on oleate and methanol, respectively, but not on both compounds, the methylotrophic yeast *P. pastoris* thrives on both. Utilization of each of these compounds as the sole carbon source requires distinct peroxisomal enzymes. To determine the phenotype of *pas* mutants in *P. pastoris,* J. Heyman (in my laboratory) isolated the homologue of the *PAS1* gene cloned from *S. cerevisiae* (Erdmann et al 1991), disrupted the gene, and found that the resulting cells, which grew on ethanol, glycerol, lactate or citrate but not on methanol or oleate, were peroxisome-deficient. Armed with this information, Gould et al (1992) and Liu et al (1992) simultaneously obtained *pas* and *per* mutants, respectively, in *P. pastoris*. The mutants from the two laboratories fall into 11 complementation groups.

Peroxisome assembly mutants are also being isolated in *Yarrowia lipolytica* (Nuttley et al 1993). The availability of these mutants affected in peroxisome import, assembly, and biogenesis in many different organisms has already led to the cloning of several new genes and promises to provide fundamental insights into the mechanisms of these processes.

PROTEIN IMPORT INTO PEROXISOMES

Unlike mitochondria and chloroplasts, peroxisomes are devoid of DNA. Consequently all the peroxisomal membrane and matrix proteins are encoded by nuclear genes. Following translation on free polysomes, both matrix and membrane proteins are targeted post-translationally to the organelle (Lazarow & Fujiki 1985). No evidence exists for glycosylation of peroxisomal proteins. The polypeptide composition of peroxisomes is distinct from that of the ER and mitochondria (Fujiki et al 1982). The lipids in the peroxisomal membranes have not been analyzed carefully or quantitatively. The phospholipid/protein ratio of rat liver peroxisomes is about 200 nmol/mg. The principal phospholipids are phosphatidyl choline (PC) and phosphatidyl ethanolamine (PE), as is the case in the ER and mitochondrial membranes. While mitochondria contain cardiolipin, peroxisomes do not (Fujiki et al 1982). The lipid composition of subcellular organelles of *S. cereviviae* has also been studied (Zinser et al 1991). It shows that of the total phospholipid in peroxisomes, 48.2% is PC, 22.9% is PE, 15.8% is phosphatidyl inositol, 4% is phosphatidyl serine, 7% is cardiolipin, and 1.6% is phosphatidic acid. The source of lipids and phospholipids required for peroxisome proliferation is not clear and is an area that deserves considerably more attention.

The glyoxysomes of cotton seed cotyledons contain as much as 36–62%

by weight of nonpolar lipid, nearly all of which is triacylglycerol or metabolites of triacylglycerol. Free fatty acids comprise the largest percent of these nonpolar lipids. PC and PE together constitute the majority (61% by weight) of the glyoxysomal membrane phospholipids. Interestingly, although PC and PE are normally synthesized from choline and ethanolamine in the ER, the nonpolar lipids and phospholipids required for glyoxysome proliferation are recruited directly from lipid bodies in which lipids are stored (Chapman & Trelease 1991).

Most microbody proteins are not proteolytically processed after import into the organelle. The exceptions are 3-ketoacyl CoA thiolase, acyl CoA oxidase, sterol carrier protein 2, and malate dehydrogenase. A surprising number of peroxisomal proteins are oligomeric and several have prosthetic groups such as the heme in catalase and FAD in acyl CoA oxidase and urate oxidase (Lazarow & Fujiki 1985). It is believed that monomeric subunits are translocated into the peroxisome matrix where they are then assembled with other subunits and cofactors (Lazarow & De Duve 1973; Bellion & Goodman 1987).

Targeting of Matrix Proteins

The discovery that firefly luciferase is targeted to peroxisomes of mammalian cells provided a model system for the elucidation of the peroxisomal targeting signal (Keller et al 1987). The first signal shown to be completely necessary and sufficient for peroxisomal targeting in mammalian cells was reported in 1987 by Gould et al (1987) at the C-terminus of firefly luciferase. In fact, it is just the last three amino acids, SKL (in the one letter code), of luciferase that function as a peroxisomal targeting signal (PTS) (Gould et al 1989). This sequence, called PTS1, was found to direct targeting to the peroxisomal matrix in mammalian (Gould et al 1989) and yeast cells (Distel et al 1992) and to glycosomes of trypanosomes (Fung & Clayton 1991; Blattner et al 1992; Sommer et al 1992). The ability of this peptide to function as a PTS has also been demonstrated in vitro using the import of rat acyl CoA oxidase (which also ends in SKL) into rat liver peroxisomes (Miyazawa et al 1989). Certain conservative substitutions in the SKL signal preserve the ability of the protein to be targeted into peroxisomes and glycosomes (Gould et al 1989; Swinkels et al 1992; Blattner et al 1992). In mammalian cells, S or A or C function in the first position, K or R or H in the second position, and L or M in the C-terminal location (Gould et al 1989; Swinkels et al 1992). These rules have also been confirmed for peroxisomal targeting to rat liver peroxisomes in vitro (Miura et al 1992). The ability of all 18 permutations of the consensus tripeptide PTS1 sequence to direct a cytosolic protein, chloramphenicol acetyltransferase, to mammalian peroxisomes was tested. A hierarchy of efficiencies was revealed. In general, S worked as efficiently as A, but better than C in position 1. K and R were equally efficient and better than H in position 2. Finally, L was more efficient than M in the last position (Swinkels et al 1992). While a peptide ending in the sequence SKL-COOH was able to compete for the import of proteins into peroxisomes in vivo (Walton et al 1992) or in vitro (Miura et al 1992; Wendland & Subramani 1993a), the same peptide with an amidated C-terminus failed to do so. This suggests that a free COOH group is required for recognition of PTS1 (Miura et al 1992).

Several other variants of PTS1 function in other organisms. In *H. polymorpha,* two peroxisomal enzymes, alcohol oxidase and dihydroxyacetone synthase, end in ARF and NKL tripeptides, respectively. Both tripeptides function as PTSs when fused onto the C-terminus of β-lactamase (Roggenkamp 1992; Hansen et al 1992). Rachubinski and co-workers showed that in *Candida tropicalis* the peroxisomal multifunctional β-oxidation enzyme ends in AKI (Nuttley et al 1988; Aitchison et al 1991b). This sequence is necessary for peroxisomal localization in *Candida sp.* (Aitchison et al 1991a). Mutagenesis of the DNA coding for this tripeptide reveals that A can be substituted by G in position 1 and Q can replace K at position 2 (Aitchison et al 1991a). Didion & Roggenkamp (1992) demonstrated the necessity of the C-terminal SKI for the targeting of *H. polymorpha* catalase to peroxisomes. Notably, neither SKI nor AKI functions as a PTS in mammalian cells (Gould et al 1989; Aitchison et al 1992).

In trypanosomes, the glycosomal phosphoglycerate kinase ends in the sequence NRWSSL (Osinga et al 1985; Swinkels et al 1988), which contains the glycosomal targeting signal (Blattner et al 1992). This sequence does not function in peroxisomal targeting in mammalian cells. The S in position 1 could be substituted by A or C (Blattner et al 1992), as was the case for peroxisomal targeting in mammalian cells (Gould et al 1989). A number of replacements (S, N, D, H, or Y) were functional at position 2 for glycosomal targeting (Blattner et al 1992). This stands in contrast to the strict requirement for a basic amino acid (K, R, or H) in this position for peroxisomal targeting in mammalian and yeast cells (Gould et al 1989; Distel et al 1992). Finally, L, M, or Y at the C-terminus directed glycosomal targeting (Blattner et al 1992). This parallels the ability of L or M (Y was not tested) to function in this position for peroxisomal targeting (Swinkels et al 1992). The different variations on the structure of the PTS1 tripeptide probably represent the evolutionary divergence in the use of this signal in different organisms, in much the same manner as the use of the KDEL, HDEL, or DDEL tetrapeptides for retention of proteins in the ER of different species (Munro & Pelham 1987; Lewis et al 1990). It will be interesting to see whether these variants are all recognized by the same class of receptors.

Conservation of the PTS1 Tripeptide Variants

Antibodies directed against the C-terminal SKL signal recognize the matrices of peroxisomes, glyoxysomes, and glycosomes. In each of these organelles many matrix proteins (15–40%) are recognized by the antibody (Gould et al 1990b; Keller et al 1991), presumably because they contain the SKL tripeptide at their carboxy terminus. Hydrogenosomes of *Trichomonas vaginalis* are not detected by these antibodies (Keller et al 1991). Anti-AKI or anti-SKL antibodies react with a number of peroxisomal proteins of *C. tropicalis, C. albicans, S. cerevisiae, Y. lipolytica, P. pastoris, Neurospora crassa,* and mammalian cells (Gould et al 1990b; Keller et al 1991; Aitchison et al 1992). This observation provides strong immunological evidence for the conservation of the PTS1 sequence in evolutionarily diverse organisms. Additionally, the utilization of the same type of signal for targeting of proteins to peroxisomes (Subramani 1992), glyoxysomes (Volokita 1991), and glycosomes (Fung & Clayton 1991; Blattner et al 1992; Sommer et al 1992) supports the common evolutionary origin of these three organelles and has led to the suggestion that the PTS1 sequence is more appropriately labeled as a microbody targeting sequence (Keller et al 1991).

The C-terminal location of the PTS1 sequence, which initially came as a surprise, has been confirmed directly for at least 12 proteins, thus proving beyond any doubt that this location was not just fortuitous in luciferase (Gould et al 1988; Aitchison et al 1991a; Volokita 1991; Fung & Clayton 1991; Roggenkamp 1992; Blattner et al 1992; Kragler et al 1993). The location also provides a satisfying explanation for the post-translational nature of import of most peroxisomal matrix proteins. A scan of computer data bases for the presence of PTS1 variants shows at least 40 microbody proteins in which this signal is conserved at the C-terminus (proteins cited in Subramani 1992; Aitchison et al 1992; Blattner et al 1992; Roggenkamp 1992; Kindl 1992; Legoux 1992; Steffan & McAllister-Henn 1992; De Hoop & Ab 1992). There also seems to be selection against the location of PTS1 peptides at the C-terminus of nonperoxisomal proteins (Subramani 1992).

The PTS1 peptide appears not to function at internal locations in proteins by several criteria. The addition of one or two amino acids to the C-terminus of luciferase caused the protein to remain cytoplasmic in mammalian cells (Gould et al 1989). Movement of this sequence to the amino-terminus or to several internal locations of mouse dihydrofolate reductase (DHFR) did not cause DHFR to localize in mammalian peroxisomes (J. Heyman et al, unpublished data). The ability of the SKL-COOH peptide, but not SKL-$CONH_2$, to compete for peroxisomal protein import in vitro argues that a C-terminal location of PTS1 is a prerequisite for function (Miura et al 1992). Finally, the cytosolic location of hundreds of proteins with internal PTS1-like sequences is a strong argument against the ability of this signal to function internally. However, a 12-amino acid sequence, CRYHLKPLQSKL-COOH, has been coupled, via the cysteinyl sulfhydryl, to the ϵ-NH_2 groups of lysines on human serum albumin (HSA-SKL conjugate) and used as a reporter for peroxisomal targeting in vivo (Walton et al 1992) and in vitro (Wendland & Subramani 1993a) with mammalian peroxisomes. Thus the signal does not need to be part of the primary sequence of a protein in order to function as a PTS.

Although a C-terminal location allows PTS1 to function, there is evidence for context dependence. Within cells, if a protein with a C-terminal PTS1-type signal also has an NH_2-terminal signal sequence that engages it in the cotranslational secretory pathway (e.g. α-amylase), then the cotranslational import must gain precedence, and the protein never finds the peroxisome. If a protein contains signals that could target it along two post-translational import pathways, then a simple kinetic competition probably explains how much of the protein finally resides in the two subcellular compartments. The enzyme alanine-glyoxylate aminotransferase is one such protein that is found both in mitochondria and peroxisomes in rodents (Oda et al 1990). Finally, certain luciferase mutants and protein fusions are not targeted to peroxisomes even though they contain the C-terminal SKL (Gould et al 1987). Similar context dependence of a nuclear targeting signal has been documented (Roberts et al 1987).

An Amino-Terminal PTS

Despite the fact that the PTS1 sequence is used by a vast majority of peroxisomal matrix proteins, it is not the only sequence that directs proteins to the organelle matrix. An 11-amino acid sequence at the amino-terminus of two peroxisomal 3-ketoacyl CoA thiolases also acts as a PTS (PTS2) and is both necessary and sufficient for peroxisomal targeting (Swinkels et al 1991; Osumi et al 1991a). Unlike most other peroxisomal proteins, these two thiolases contain 26 and 36-amino acid leaders, respectively, which are cleaved by a peroxisome-specific protease (J. Shackelford et al, unpublished data), after import of the proteins into the matrix. The PTS2 sequence, MHRLQVVLGHL, corresponds to the first 11 amino acids of the 26-amino acid leader and is found at an internal location (amino acids 11–21) in the 36-amino acid leader (Bodnar & Rachubinski 1990; Hijikata et al 1990). The PTS2 sequence is conserved completely between rat and human thiolases and resembles the sequences found at the amino-termini of yeast thiolases (Swinkels et al 1991). The glyoxysomal malate dehydrogenase from watermelon also has a weakly homologous prepiece that is cleaved (Gietl 1990) and may be involved in the transport of this protein into glyoxysomes.

Genetic evidence also points to the existence of two distinct signal-dependent pathways for the import of proteins into the peroxisomal matrix. The *pas8* mutant of *P. pastoris* (McCollum et al 1993) and the *pas10* mutant of *S. cerevisiae* (Van der Leij et al 1992) and certain complementation groups of Zellweger syndrome fibroblasts (Balfe et al 1990; Walton et al 1992) are defective in the transport of proteins containing PTS1, but not PTS2, into peroxisomes. Membrane proteins are imported faithfully in some, if not all, of these mutants (Santos et al 1988a,b). Conversely, the *pas7* mutant of *S. cerevisiae* is unable to import thiolase, which has the PTS2 sequence, but can import PTS1-containing proteins (Erdmann & Kunau 1992) and membrane proteins (Höhfeld et al 1992) into peroxisomes. It seems reasonable to assume that different receptors are involved in the recognition of PTS1 and PTS2. If this is so, the situation would be analogous to the involvement of MOM72 or MAS70 in the recognition of the mitochondrial targeting signal on the ADP/ATP carrier protein, and the specificity of MOM19 in the binding of the signal on the Fo/F_1 ATPase (see Pfanner & Neupert 1990).

Other Peroxisomal Matrix Targeting Signals

Small et al (1988b) and Kamiryo et al (1989) have described two internal PTSs in acyl CoA oxidase from *C. tropicalis* by monitoring peroxisomal import in vitro and in vivo, respectively. However, these sequences are embedded in fairly large peptide segments and have not been delimited further. Neither of these peptides appears to have sequences resembling PTS1 or PTS2. Catalase from *S. cerevisiae* has been reported to contain a C-terminal PTS (SSNSKF) bearing some resemblance to PTS1, as well as another internal PTS (Kragler et al 1993). It seems quite plausible that a few other PTSs remain to be discovered.

Evolutionary Use of Different PTSs by the Same Protein

It is interesting to note that acyl CoA oxidase from *C. tropicalis* appears to use a different PTS than the same protein from rat liver (Small et al 1988b; Gould et al 1988). Another example is the watermelon malate dehydrogenase, which probably has a PTS2 sequence (Gietl 1990), whereas the peroxisomal malate dehydrogenase from *S. cerevisiae* has a PTS1 sequence (SKL) at its C-terminus (Steffan & McAllister-Henn 1992). Two different PTSs have also been described in catalase from *S. cerevisiae* (Kragler et al 1993). Artificial constructs expressing bacterial chloramphenicol acetyltransferase fused to either PTS1 at its C-terminus, or to PTS2 at its N-terminus, have also been used to target the passenger protein to peroxisomes (Gould et al 1988; Swinkels et al 1991).

Targeting Signals For Peroxisomal Membrane Proteins

Several lines of evidence argue that peroxisomal membrane proteins (PMPs) must use signals other than PTS1 or PTS2. (*a*) The genes for five PMPs have been sequenced to date. These include PMP47 from *C. boidinii* (McCammon et al 1990), PMP22 from rat (Kaldi et al 1993), PMP70 from rat (Kamijo et al 1990), PAF1 from CHO (Tsukamoto et al 1991) and human cells (Shimozawa et al 1992b), and PAS3 from *S. cerevisiae* (Höhfeld et al 1991). None of these proteins has PTS1 or PTS2-like sequences at the C- or N-terminus. (*b*) The antibodies that recognize numerous peroxisomal proteins from different organisms decorate only the matrix and never the organelle membrane (Keller et al 1991). (*c*) Fibroblasts from Zellweger syndrome patients (Santos et al 1988a,b; Small et al 1988a), as well as several of the *pas* mutants of *P. pastoris* (McCollum et al 1993), contain peroxisome membrane ghosts that fail to import a majority of matrix polypeptides. (*d*) Finally, the C-terminal AKE and the internal SKL in PMP47 from *C. boidinii* were each mutated by K→A substitutions. Single or double mutations did not prevent PMP47 from assembling into peroxisomes (Goodman et al 1992). It is most likely that the transmembrane domains of these proteins will be necessary to anchor them in the membrane, but in addition, there must be a permanent or transient signal (e.g. chemical modification) that specifies association with only the peroxisomal membrane.

Receptors for PTSs

The recent cloning of the *PAS8* and *PAS10* genes from *P. pastoris* (McCollum et al 1993) and *S. cerevisiae* (H. Tabak, personal communication), respectively, reveals that they are homologous and that they encode members of the TPR (tetratricopeptide) family of proteins. These polypeptides contain a 34-amino acid repeat whose structure, but not entire sequence, is conserved among proteins involved in transcription, mitochondrial protein import, chromosome segregation, cell cycle control, and RNA processing (reviewed in Goebl & Yanagida 1991; Sikorski et al 1990).

The PAS8 protein of *P. pastoris* contains seven consecutive TPR domains in the C-terminal half of the protein, followed by a 52-amino acid tail. An important finding is that the PAS8 protein binds selectively to the SKL peptide directly, or after it is coupled to human serum albumin (HSA). No binding was detected to a control peptide lacking SKL or to HSA alone (McCollum et al 1993). These results suggest that PAS8 might be the much sought after SKL receptor protein. The tight association of PAS8 with the peroxisomal membrane, its localization to the cytosolic face of peroxisomes (E. Sock & S. Subramani, unpublished data), and the absence of an obvious transmembrane domain in the sequence of PAS8 suggest that it is a tightly-associated peripheral membrane protein that complexes with other peroxisomal membrane proteins without requiring its own PTS. The PAS8 protein is also inducible upon growth of *P. pastoris* on methanol or oleate (McCollum et al 1993). The localization of PAS8, its weak homology to MAS70 (a mitochondrial import receptor) (Hines 1992), and its inducibility upon peroxisome proliferation strongly support its proposed function as part of a complex receptor that must include specific peroxisomal membrane proteins (to which PAS8 probably binds).

Biochemical evidence also exists for the involvement of cytosolic SKL-binding factors in the import of proteins into peroxisomes of permeabilized CHO cells (Wendland & Subramani 1993a). The purification of such factors will reveal whether they are receptors or chaperonins involved in the import of matrix proteins into peroxisomes.

The PAS7 protein of *S. cerevisiae* is a candidate for the PTS2 receptor (Erdmann & Kunau 1992). However, no biochemical data exist at present to prove this point.

Two models have been proposed for the import of matrix proteins (McCollum et al 1993). They differ in whether one or two translocation machineries are involved in the import of PTS1- and PTS2-containing proteins. Although either of these models could be correct, the absence of additional pathway-specific (i.e. PTS1- or PTS2-specific) mutants in yeast is more consistent with a single shared translocation apparatus.

Peroxisomal Protein Import Deficiencies in Human Disease

A number of human peroxisomal disorders are now known to be caused by the inability to import peroxisomal proteins to the organelle or because of the mistargeting of a peroxisomal protein to the wrong subcellular compartment. Several peroxisomal matrix proteins involved in the β-oxidation of VLCFA were found to be missing in cells from patients suffering from the Group I disorders (ZS, NALD, HPA, and IRD) (Tager et al 1985). Several enzymes are synthesized but rapidly degraded in the cytosol (Schram et al 1986). However, catalase survives in the cytosol and is enzymatically active (Wanders et al 1984, 1985). Cells from at least five complementation groups (1 to 5) of patients contain peroxisomal membrane ghosts, as judged by indirect immunofluorescence with anti-PMP70 antibodies and by Western blots analyses of peroxisomal membrane constituents (Santos et al 1988a,b, 1992; Wiemer et al 1989; Suzuki et al 1989; Small et al 1988a). These observations suggested that fibroblasts from these patients might be deficient in protein import into peroxisomes. Direct proof of this point was provided using microinjection of peroxisomal proteins into fibroblasts of normal and

2 complementation groups (1 and 4) of ZS patients (Walton et al 1992). Two PTS1-containing proteins, luciferase and HSA-SKL, were transported into peroxisomes of normal human cells, but not into peroxisomes of these ZS cells. Wendland & Subramani (1993b) used permeabilized human cells to confirm these results for the same 2 complementation groups of ZS fibroblasts and to extend the same results to 4 other complementation groups of disorders in Group I (groups 2, 3, 6, and 8). However, at least one complementation group of ZS fibroblasts does import the thiolase precursor into membrane ghosts (Balfe et al 1990). Although this observation was also made in several other ZS fibroblasts (Gartner et al 1991), these lines were not placed into the existing complementation groups and are no longer available. Thus one can only state with certainty that at least one complementation group (and possibly more) among the Group I peroxisomal disorders displays a selective import deficiency for only PTS1-, but not for PTS2-containing proteins, or for membrane polypeptides. By analogy with the yeast mutants now available, it would not be surprising if several other complementation groups were deficient in the import of both PTS1- and PTS2-containing proteins.

Mutations Responsible for Some Human Peroxisomal Disorders

Some recent exciting advances have documented the molecular genetic defects responsible for two of the nine complementation groups of the generalized peroxisomal disorders. Two of 21 probands in complementation group 1 were defective in the PMP70 gene, which encodes a member of the multidrug-resistance family of ATPases (Gartner et al 1992). One patient had a splice-donor mutation, while the other had a missense mutation. Why no mutations were detected in 19 other probands in the same complementation group is a paradox (Gartner et al 1992). Another complementation group (called group F here to distinguish it from groups A-E in Yajima et al 1992) was deficient in the human PAF1 gene, which encodes a 35-kd PMP (Shimozawa et al 1992b). A premature termination codon was responsible for the mutation in this instance.

An unrelated lethal, autosomal recessive disorder, called primary hyperoxaluria I, is caused by the missorting of an essential peroxisomal protein, L-alanine-glyoxylate aminotransferase I to mitochondria (Danpure et al 1989; Takada et al 1990). The failure to transaminate glyoxylate leads to its conversion to oxalate whose low solubility results in hyperoxaluria (Williams & Wandzilak 1989). The targeting defect is the result of a polymorphism that reintroduces a mitochondrial targeting signal (this protein is both mitochondrial and peroxisomal in rodents in which there is only one gene, Oda et al 1990) and a second mutation that interferes with peroxisomal targeting (Purdue et al 1990; Takada et al 1990). These examples underscore the complex and lethal phenotypes that can result from failures in the fidelity of protein sorting to peroxisomes.

APPROACHES USED TO ELUCIDATE THE MECHANISM OF PROTEIN IMPORT INTO PEROXISOMES

In Vitro Systems

The import of ^{35}S-Met-labeled proteins translated in vitro into purified rat liver or yeast peroxisomes has been monitored by binding and protease-protection assays (Small et al 1987, 1988a; Imanaka et al 1987; Miyazawa et al 1989; Thieringer et al 1991; Miura et al 1992). The import was found to be signal-, time-, temperature- and ATP-dependent and did not require a membrane potential (Imanaka et al 1997; Small et al 1998b; Miyazawa et al 1999; Miura et al 1992). However, the in vitro systems have worked well and consistently in only a few laboratories with only a few proteins, acyl CoA oxidase (Imanaka et al 1987; Small et al 1988b; Miyazawa et al 1989), PMP22 (Fujiki et al 1989), and catalase A (Thieringer et al 1991). The principal problems appear to be the fragility of the peroxisomes, the dependence on protease-protection as the only hallmark of import, and the resistance of certain peroxisomal proteins to protease digestion. The cloning of the peroxisomal thiolase genes (Bodnar & Rachubinski 1990; Hijikata et al 1990) provided the hope that the removal of 26 or 36-amino acid leader peptides of the two peroxisomal thiolases would provide a size change in the protein diagnostic of import. Unfortunately, both thiolases and the protease that cleave the leader were found to leak from purified peroxisomes (J. Shackelford et al, unpublished data).

Microinjection

In order to overcome some of the inherent limitations of the in vitro systems, Walton et al (1992) resorted to microinjection of luciferase or HSA-SKL into mammalian cells and used indirect immunofluorescence to monitor the import of these proteins into peroxisomes. The import was found to occur in a time-, temperature-, and signal sequence-dependent manner. Furthermore, import was observed in normal human fibroblasts, but not in fibroblasts of two complementation groups of Zellweger syndrome cells. The peroxisomal localization of the imported proteins was demonstrated by colocalization with other peroxisomal matrix proteins (Walton et al 1992).

Semi-intact Cells

The success of the microinjection approach led Wendland & Subramani (1993a) to develop a permeabilized CHO cell system to investigate peroxisomal protein import. Similar work was done independently by Rapp et al (1993). CHO cells permeabilized with streptolysin O, a bacterial toxin, become permeable to reporters and antibodies while preserving the functional integrity of the peroxisomal membrane. Luciferase and HSA-SKL were transported into the organelles in a time-, temperature-, signal-, ATP-, and cytosol-dependent manner. Evidence was also presented for the existence of cytosolic SKL-binding factors that were necessary for the import of PTS1-containing proteins (Wendland & Subramani 1993a). The cytosol dependence of peroxisomal protein import had been missed in previous studies using in vitro systems, perhaps because proteins translated in reticulocyte lysates were used in these reactions. Wendland & Subramani (1993a) showed that reticulocyte lysate was an excellent source of cytosol. The fractionation of the cytosol and SKL-binding components should now allow the purification and characterization of proteins involved in facilitating peroxisomal protein import.

Mechanistic Aspects of Protein Import Into the Peroxisomal Matrix

The combined use of the approaches described above has begun to provide a picture of the steps and requirements of peroxisomal protein import. Most encouragingly, the conclusions have been confirmed using two or more of the techniques. Both PTS1-containing proteins (Miyazawa et al 1989; Walton et al 1992; Rapp et al 1993; Wendland & Subramani 1993a) and one PMP (Fujiiki et al 1989) are imported post-translationally into peroxisomes. The *C. tropicalis* acyl CoA oxidase, which contains a signal other than PTS1 or PTS2, is also imported (Small et al 1988b; Kamiryo et al 1989). However, no results have been reported as yet regarding the import of PTS2-containing proteins in these in vitro or semi-intact systems. The investigation of import using purified peroxisomes in vitro has revealed an ATP-independent binding step that occurs at 0–4°C with yeast and rat liver peroxisomes (Imanaka et al 1987; Small et al 1987). Experiments with the semi-intact CHO cells confirm this point, except that ATP-independent binding occurred at or above 20°C rather than at 0–4°C (Rapp et al 1993; Wendland & Subramani 1993a). The significance of the temperature difference is unclear. Because the binding step directly or indirectly measures the accumulation of the reporter protein on the surface of the peroxisome, it must reflect PTS-receptor interactions on the peroxisome membrane, or the interaction of the PTS-receptor complex with components of the translocation machinery.

The import is saturable and can be competed in the experiments involving microinjection (Walton et al 1992), semi-intact cells (Wendland & Subramani 1993a), or purified peroxisomes in vitro (Miura et al 1992). This implies that the receptor or the translocation machinery involved in the import of PTS1-containing proteins must be limiting.

The translocation step for transport of matrix proteins across the membrane requires ATP hydrolysis. Non-hydrolyzable ATP analogues or GTP will not substitute for ATP (Imanaka et al 1989; Rapp et al 1993; Wendland & Subramani 1993a). The ATP requirement suggests that the import of matrix proteins is an active process. It is unclear whether ATP hydrolysis is essential for the insertion of membrane proteins in the peroxisomal membrane (Fujiki et al 1989).

What are the important interactions between a PTS and its receptor? In mammalian cells, there is an obligatory requirement for a basic amino acid in the second position of the tripeptide and a free carboxyl group at the C-terminus. This suggests that ionic interactions are likely to be important for the PTS1-receptor interaction.

There is no N-ethylmaleimide (NEM) sensitivity of the cytosol, but the peroxisomes of semi-intact cells are NEM-sensitive (Wendland & Subramani 1993a). The identification of the relevant component of peroxisomes might help elucidate its role in unfolding or translocation.

The experiments with semi-intact human fibroblasts from normal humans or from patients with peroxisomal disorders (Group I) suggest that peroxisomes of all six complementation groups tested are defective in import. Cytosolic factors from these lines will substitute for the cytosol-dependence of peroxisomal protein import (Wendland & Subramani 1993b). Thus several peroxisome-associated components are likely to be involved, directly or indirectly, in the import of matrix proteins. This has already been presaged by the identification of mutations in the genes for 2 PMPs in human patients (see Mutations Responsible for Some Human Peroxisomal Disorders) (Gartner et al 1992; Shimozawa et al 1992b).

DISTINCT STEPS IN PEROXISOME BIOGENESIS

Peroxisome induction in all organisms appears to be an adaptive response to nutritional environments that require peroxisomal enzymes for growth and survival. The induction consists of two phases—proliferation of the organelle by budding from preexisting peroxisomes, and growth of the organelle by the import of matrix proteins.

In lower eukaryotes such as *S. cerevisiae*, growth on fatty acids (e.g. oleic acid) as the sole carbon source is sufficient to cause induction (Veenhuis et al 1987). In *H. polymorpha,* oleic acid does not induce peroxisomes, but other carbon and nitrogen substrates such as ethanol, primary amines, D-amino acids, and purines do (Veenhuis 1992). In the methylotrophic yeasts such as *Candida sp.*, *Hansenula sp.*, and *Pichia sp.*, growth on methanol as the sole carbon source also results in induction of peroxisomes. In these organisms, glucose repression prevents peroxisome induction (Trumbly 1992).

In mammalian cells, induction of peroxisomes and peroxisomal enzymes is caused by a large variety of hypolipidemic drugs (Reddy et al 1979, 1986; Hijikata et al 1990; Osumi et al 1991b; Tugwood et al 1992; B. Zhang et al 1992) and by retinoic acid (Hertz & Bar-Tana 1992). A family of proteins known as peroxisome-proliferator-activated-receptors (PPARs) are involved in mediating the response (Issemann & Green 1990; Green 1992; Dreyer et al 1992; Tugwood et al 1992). These peroxisome proliferators bind to PPARs, which are ligand-activated transcription factors of the steroid-hormone-receptor superfamily (Issemann & Green 1990; Dreyer et al 1992). Three different PPARs (α, β, and γ) have been cloned from *Xenopus laevis* (Dreyer et al 1992). PPARs and their closest relatives, receptors for thyroid hormone (TR), retinoic acid (RAR), and vitamin D_3 (VDR) (Issemann & Green 1990; Dreyer et al 1992), consist of a trans-activating domain, a DNA-binding region, and a ligand-binding domain. Hypolipidemic drugs, such as clofibrate or some cellular metabolite, activate PPAR by mimicking the action of an unknown endogenous cellular ligand. Following activation, the PPARs bind to peroxisome-proliferator-response-elements (PPREs) to activate transcription from nearby promoters (Dreyer et al 1992; Kliewer et al 1992b; Tugwood et al 1992; X. Zhang et al 1992).

Investigation of the sites of interaction of PPARs with DNA reveals that PPREs consist of two AGGTCA direct repeats (DR) separated by a spacer of one nucleotide (DR1) (Osumi et al 1991b; Tugwood et al 1992; X. Zhang et al 1992; Kliewer et al 1992b). The receptors for thyroid hormone, retinoic acid, and vitamin D_3 interact (Bugge et al 1992; X. Zhang et al 1992; Kliewer et al 1992a, Leid et al 1992) with members of the RXR family (α, β, and γ) (Mangelsdorf et al 1992; Leid et al 1992). The related PPARs also form heterodimers with RXRα (Kliewer et al 1992b; Marcus et al 1993), which binds to the ligand 9-cis retinoic acid (Heyman et al 1992). The PPAR-RXR heterodimer activates PPREs in response to clofibric acid or 9-cis retinoic acid. The presence of both ligands yields a synergistic induction of gene expression (Kliewer et al 1992b). Interestingly, although RXRα forms heterodimers with PPAR, VDR, TR, and RAR, the heterodimers interact the strongest with DR1, DR3, DR4, and DR5, where the number indicates the optimal length of the spacer between the direct repeats (Kliewer et al 1992b). There is also evidence that PPAR isoform-RXRα combinations may act positively or negatively to regulate gene expression. This makes it likely that combinatorial interactions between PPAR and RXR isoforms play an important role in peroxisome proliferation and lipid homeostasis (Marcus et al 1993).

The tissue and developmental regulation of members of the PPAR (Dreyer et al 1992) and RXR (Mangelsdorf et al 1992) families suggest that peroxisome induction is an important developmental response in mammalian development.

The steps in peroxisome proliferation have been better studied in yeasts than in mammals. These investigations have led to two models for the temporal order of events leading to peroxisome induction. In *H. polymorpha* shifted from glucose to methanol medium, individual peroxisomes grow to 80 times their original volume before they divide by fission into smaller organelles (Veenhuis et al 1979). Thus in the first model, based on events observed with *H. polymorpha,* growth of peroxisomes precedes proliferation (Veenhuis et al 1979).

In the alternative model, the proliferation of peroxisomes precedes matrix protein import and organelle growth (Veenhuis & Goodman 1990). An important study of the temporal events during peroxisome proliferation was undertaken with *C. boidinii* shifted from glucose to methanol medium. Electron microscopy was used to visualize the peroxisomes morphologically at various stages of induction, and Western blotting of representative membrane and matrix proteins was also performed. A number of specific stages in peroxisome induction were defined (Veenhuis & Goodman 1990).

Most glucose-grown *C. boidinii* cells (and other yeast cells) contain only one or a few peroxisomes. Within 0–1.5 hr after induction, peroxisomal membrane proteins (e.g. PMP47) are induced. This is accompanied by elongation and budding of one of the few small peroxisomes. Matrix proteins are not induced in this early proliferation step. Between 1.5–3 hr, the proliferation of the organelle begins, the number of peroxisomes increases, and low levels of matrix proteins are induced. Membrane proteins continue to be made (this step is termed late proliferation). In the next stage (3–10 hr), several matrix and membrane polypeptides are induced. Import of these into the proliferated organelles leads to peroxisome growth. Finally, the number of proliferated organelles per cell decreases as cells divide or are shifted to glucose. Several other yeasts, including *P. pastoris* (J. Heyman et al, unpublished data) appear to conform to this second model.

Although *H. polymorpha* and *C. boidinii* behave so differently, their behavior is not surprising because genetic and biochemical evidence suggest that microbody proliferation and growth are distinct steps that can not only be uncoupled, but also each can proceed in the absence of the other. For example, peroxisomes in the *pas5* (A. Spong & S. Subramani, submitted) and *pas8* mutants of *P. pastoris* (McCollum et al 1993), as well as those in ZS fibroblasts, are deficient in matrix protein import (and therefore growth), but are able to proliferate and segregate to daughter cells. Thus proliferation of peroxisomes does not require matrix protein import or growth.

Conversely, overexpression of alcohol oxidase in *S. cerevisiae* or *H. polymorpha* leads to peroxisome growth without proliferation (Distel et al 1988; Godecke et al 1989). One hypothesis that would explain both models is that peroxisome proliferation, and perhaps growth, requires the induction of specific proteins, and that it is the timing of the induction of these proteins that determines whether proliferation precedes growth, or vice versa.

GENERAL PROPERTIES OF PEROXISOME ASSEMBLY/IMPORT MUTANTS

A careful analysis of the phenotypes of peroxisome-deficient mutants isolated in *S. cerevisiae* (Erdmann & Kunau 1992; Van der Leij et al 1992), *H. polymorpha* (Cregg et al 1990; Veenhuis 1992), *P. pastoris* (Gould et al 1992; Liu et al 1992), and *Y. lipolytica* (Aitchison et al 1992) suggests that the mutants will fall into at least four classes. These are described below. For other views on the classification of such mutants, the reader is referred to Erdmann & Kunau (1992), Höhfeld et al (1992), and Veenhuis (1992).

Mutants Affected in Peroxisome Proliferation

These mutants would be impaired in the signal transduction pathway that receives and processes the peroxisome proliferation signal, or in transcription factors necessary for the induction of peroxisomal membrane proteins, or in the genes encoding membrane proteins required for organelle proliferation.

Example of mutants in this category would include the *pas1* mutants (see Table 1) of *S. cerevisiae* (Erdmann et al 1991) and perhaps *P. pastoris* (this mutant may be deficient in late proliferation or some aspects of peroxisome growth) (J. Heyman et al, unpublished data), the *pas3* mutant of *S. cerevisiae* (Höhfeld et al 1991), the *pas14* mutant of *S. cerevisiae* (Einerhand et al 1992), or yeast mutants deficient in ADR1, SNF1, or SNF4 (Simon et al 1991, 1992).

Mutants Affected in Peroxisome Growth

These mutants would be affected in the induction or import of matrix proteins, but would be capable of some membrane protein import, peroxisome proliferation, and segregation to daughter cells. The *pas7* mutant of *S. cerevisiae* (Erdmann & Kunau 1992), the *pas8* mutant of *P. pastoris* (McCollum et al 1993), the *pas5* mutant of *P. pastoris* (A. Spong & S. Subramani, submitted), the *pas10* mutant of *S. cerevisiae* (Van der Leij et

Table 1 Conservation of genes involved in peroxisome import and assembly

Gene[a]	Homologue	Remarks	Reference
S. cerevisiae PAS1	*P. pastoris PAS1*	Related to *SEC18*, NSF and other ATPases	Erdmann et al 1991; J. Heyman & S. Subramani, unpublished
S. cerevisiae PAS2	*P. pastoris PAS4*	Ubiquitin-conjugating enzyme	Wiebel & Kunau 1992; S. Gould, personal communication
S. cerevisiae PAS3	*P. pastoris PAS2*	Membrane protein	Hohfeld et al 1991; E. Wiemer & S. Subramani, unpublished data
S. cerevisiae PAS8	*P. pastoris PAS5* *Y. lipolytica PAY4*	Related to *PAS1*, *SEC18* NSF and other ATPases	S. Spong & S. Subramani, unpublished; R. Rachubinski, personal communication H. Tabak, personal communication
S. cerevisiae PAS10	*P. pastoris PAS8*	Binds SKL sequence	McCollum et al 1993; H. Tabak, personal communication
Human PAF1	Rat PAF1	35 kd PMP	Shimozawa et al 1992b; Tsukamoto et al 1991
Human PMP70	Rat PMP70	70 kd PMP	Kamijo et al 1990; Gartner et al 1992

[a] Other genes cloned from *S. cerevisiae* and their properties are listed in Kunau & Hartig 1992.

al 1992), and the *pay4* mutant of *Y. lipolytica* (R. Rachubinski, personal communication) would belong tentatively in this group.

Mutants Affected in Peroxisome Functions that Alter Assembly or Import Indirectly

Some of these mutants might cripple general peroxisome functions or the peroxisomal microenvironment (e.g. pH) so as to affect import of peroxisomal proteins without being directly involved in import itself.

Mutants Exhibiting Aberrant Peroxisome Disassembly

In all organisms in which peroxisomes are inducible, there is also a process of organelle disassembly that starts when the environmental milieu does not necessitate induced levels of peroxisomes for growth. Catabolite repression ensures that new peroxisomes are not induced and peroxisome disassembly shifts the cells to a state where peroxisomes are present at uninduced levels.

There is evidence in ZS fibroblasts (Heikoop et al 1992) and from *pas* mutants in *P. pastoris* (Gould et al 1992) that lysosomes or vacuoles, respectively, are involved in the degradation of peroxisomes or peroxisomal proteins (see Chiang & Schekman 1992). It is not known whether peroxisome degradation is a constitutive process, or whether it is triggered by some signal. Whatever the mechanism, an increase in the rate of degradation, relative to synthesis of peroxisomes, could also result in a *pas* phenotype. No mutants in this class are known yet, but they could easily have been missed.

SUMMARY AND FUTURE PROSPECTS

Remarkable progress has been made in the last five years in the areas of peroxisomal protein import, peroxisome biogenesis, and the genetic basis of human peroxisomal disorders. In the next five years, the yeast and mammalian cell mutants deficient in peroxisome assembly will yield a bonanza of new genes. The knowledge of the sequences of these genes from two or more organisms (see Table 1) will not only facilitate the cloning and characterization of homologues from other organisms, but will also contribute meaningfully in functional studies. The elucidation of the biochemical functions of the proteins encoded by these genes will require an array of molecular, cellular, and biochemical approaches. Of utmost importance for the field is a greater emphasis on in vitro assays that subdivide the import process into specific steps and attempt to reconstitute import into mutant peroxisomes via the addition of purified components.

It is also certain that as we endeavor to understand the molecular basis of human peroxisomal disorders, better diagnostic procedures, and eventually therapeutic intervention, will become feasible for these debilitating diseases. Much remains to be discovered regarding the principles of peroxisome assembly and disassembly and the physiological signals that set these processes in motion. The recruitment of lipids and phospholipids in peroxisome proliferation is also poorly understood. Finally investigation of the metabolic functions of peroxisomes and their role in development and lipid metabolism will open up new challenges.

Acknowledgments

I owe a great deal to the graduate students and post-doctoral fellows, past and present, for turning a serendipitous observation regarding the localization of luciferase in monkey cells into such a wealth of opportunities. Thanks are also due to special colleagues in the field—Stephen Gould, Gilbert Keller, Stuart Brody, Skai Krisans, Jim Feramisco, Paul Walton, Richard Rachubinski, Joel Goodman, Wolf Kunau, Ben Distel, Henk Tabak, Janardan Reddy, Jim Cregg, Hugo Moser, and others for sharing information, ideas, and reagents over the years. This work was supported by National Institutes of Health grant DK41737.

Literature Cited

Aitchison, JD, Murray, WW, Rachubinski, RA. 1991a. The carboxyl-terminal tripeptide Ala-Lys-Ile is essential for targeting *Candida tropicalis* trifunctional enzyme to yeast peroxisomes. *J. Biol. Chem.* 266: 23197–203

Aitchison, JD, Nuttley, WM, Szilard, RK, Brade, AM, Glover, JR, et al. 1992. Peroxisome biogenesis in yeast. *Mol. Microbiol.* 6:3455–60

Aitchison, JD, Sloots, JA, Nuttley, WM, Rachubinski, RA. 1991b. Sequence of the gene encoding *Candida tropicalis* peroxisomal trifunctional enzyme. *Gene* 105:135–36

Angermüller, S, Fahimi, HD. 1988. Heterogenous staining of D-amino acid oxidase in peroxisomes of rat liver and kidney. A light and electron microscopic study. *Histochemistry* 88:277–85

Balfe, A, Hoefler, G, Chen, WW, Watkins, PA. 1990. Aberrant subcellular localization of peroxisomal 3-ketoacyl-CoA thiolase in the Zellweger syndrome and rhizomelic chondrodysplasia punctata. *Pediatr. Res.* 27: 304–10

Banta, LM, Robinson, JS, Klionsky, DJ, Emr, SD. 1988. Organelle assembly in yeast: characterization of yeast mutants defective in vacuolar biogenesis and protein sorting. *J. Cell Biol.* 107:1369–83

Bellion, E, Goodman, JM. 1987. Proton ionophores prevent assembly of a peroxisomal protein. *Cell* 48:165–73

Blattner, J, Swinkels, B, Dorsam, H, Prospers, T, Subramani, S, et al. 1992. Glycosome assembly in trypanosomes: variations in the acceptable degeneracy of a COOH-terminal microbody targeting signal. *J. Cell Biol.* 119:1129–36

Bodnar, AG, Rachubinski, RA. 1990. Cloning and sequence determination of cDNA encoding a second rat liver peroxisomal 3-ketoacyl-CoA thiolase. *Gene* 91:193–99

Bolotin-Fukuhara, M, Grivell, LA. 1992. Genetic approaches to the study of mitochondrial biogenesis in yeast. *Antonie van Leeuwenhoek J. Microbiol. Serol.* 62:131–53

Briedenbach, RW, Beavers, H. 1967. Association of the glyoxylate cycle enzymes in a novel subcellular particle from castor bean endosperm. *Biochem. Biophys. Res. Commun.* 27:462–69

Brul, S, Westerveld, A, Strijland, A, Wanders, RJ, Schram, AW, et al. 1988. Genetic heterogeneity in the cerebrohepatorenal (Zellweger) syndrome and other inherited disorders with a generalized impairment of peroxisomal functions. A study using complementation analysis. *J. Clin. Invest.* 81: 1710–15

Bugge, TH, Pohl, J, Lonnoy, O, Stunnenberg, HG. 1992. RXR α, a promiscuous partner of retinoic acid and thyroid hormone receptors. *EMBO J.* 11:1409–18

Cavalier-Smith, T. 1987. The simultaneous symbiotic origin of mitochondria, chloroplasts and microbodies. *Ann. NY Acad. Sci.* 503:55–71

Chapman, KD, Trelease, RN. 1991. Acquisition of membrane lipids by differentiating glyoxysomes: role of lipid bodies. *J. Cell Biol.* 115:995–1007

Chiang, H, Schekman, R. 1992. Mechanism and regulation of import and degradation of cytosolic proteins in the lysosome/vacuole, In *Membrane Biogenesis and Protein Targeting,* ed. W Neupert, R Lill, pp. 149–64. Amsterdam: Elsevier

Cooper, TG, Beavers, H. 1969. β-oxidation in glyoxysomes from castor bean endosperm. *J. Biol. Chem.* 244:3514–20

Cregg, JM, van der Klei, IJ, Sulter, GJ, Veenhuis, M, Harder, W. 1990. Peroxisome-deficient mutants of *Hansenula polymorpha. Yeast* 6:87–97

Danpure, CJ, Cooper, PJ, Wise, PJ, Jennings, PR. 1989. An enzyme trafficking defect in two patients with primary hyperoxaluria type 1: Peroxisomal alanine/glyoxylate aminotransferase rerouted to mitochondria. *J. Cell Biol.* 108:1345–51

De Duve, C, Baudhuin, P. 1966. Peroxisomes (microbodies and related particles). *Physiol. Rev.* 46:323–57

De Hoop, MJ, Ab, G. 1992. Import of proteins into peroxisomes and other microbodies. *Biochem. J.* 286:657–69

Diczfalusy, U, Alexson, SEH, Pedersen, JI. 1987. Chain-shortening of prostaglandin F2α by rat liver peroxisomes. *Biochem. Biophys. Res. Commun.* 104:1206–13

Didion, T, Roggenkamp, R. 1990. Deficiency of peroxisome assembly in a mutant of the methylotrophic yeast *Hansenula polymorpha. Curr. Genet.* 17:113–17

Didion, T, Roggenkamp, R. 1992. Targeting signal of the peroxisomal catalase in the methylotrophic yeast *Hansenula polymorpha. FEBS Lett.* 303:113–16

Distel, B, Gould, SJ, Voorn-Brouwer, T, van der Berg, M, Tabak, HF, et al. 1992. The carboxyl-terminal tripeptide serine-lysine-leucine of firefly luciferase is necessary but not sufficient for peroxisomal import in yeast. *New Biol.* 4:157–65

Distel, B, Van der Ley, I, Veenhuis, M, Tabak, HF. 1988. Alcohol oxidase expressed under nonmethylotrophic conditions is imported, assembled, and enzymatically active in peroxisomes of *Hansenula polymorpha. J. Cell Biol.* 107:1669–75

Douwe de Boer, A, Weisbeek, PJ. 1991. Chloroplast protein topogenesis: import, sorting and assembly. *Biochim. Biophys. Acta* 1071:221–53

Draye, J, van Hoof, F, de Hoffman, E, Vamecq, J. 1987. Peroxisomal oxidation of L-2-hydroxyphytanic acid in rat kidney cortex. *Eur. J. Biochem.* 167:573–78

Dreyer, C, Krey, G, Keller, H, Givel, F, Helftenbein, G, et al. 1992. Control of the peroxisomal β-oxidation pathway by a novel family of nuclear hormone receptors. *Cell* 68:879–87

Eaton, JW. 1989. In *The Metabolic Basis of Inherited Disease,* ed. CR Scriver, AL Beaudet, WS Sly, D Valle, pp. 1151–61. New York: McGraw Hill 6th ed.

Einerhand, AW, Van der Leij, I, Kos, WT, Distel, B, Tabak, HF. 1992. Transcriptional regulation of genes encoding proteins involved in biogenesis of peroxisomes in *Saccharomyces cerevisiae. Cell. Biochem. Funct.* 10:185–91

Erdmann, R, Kunau, WH. 1992. A genetic approach to the biogenesis of peroxisomes in the yeast *Saccharomyces cerevisiae. Cell. Biochem. Funct.* 10:167–74

Erdmann, R, Veenhuis, M, Mertens, D, Kunau, W. 1989. Isolation of peroxisome-deficient mutants of *Saccharomyces cerevisiae. Proc. Natl. Acad. Sci. USA* 86: 5419–23

Erdmann, R, Wiebel, FF, Flessau, A, Rytka, J, Beyer, A, et al. 1991. PAS1, a yeast gene required for peroxisome biogenesis, encodes a member of a novel family of putative ATPases. *Cell* 64:499–510

Fujiki, Y, Fowler, S, Shio, H, Hubbard, AL, Lazarow, PB. 1982. Polypeptide and phospholipid composition of the membrane of rat liver peroxisomes: Comparison with endoplasmic reticulum and mitochondrial membranes. *J. Cell Biol.* 93:103–10

Fujiki, Y, Kasuya, I, Mori, H. 1989. Import of 22-kDa peroxisomal integral membrane protein into peroxisomes in vitro. *Agric. Biol. Chem.* 53:591–92

Fung, K, Clayton, C. 1991. Recognition of a peroxisomal tripeptide entry signal by the glycosomes of *Trypanosoma brucei. Mol. Biochem. Parasitol.* 45:261–64

Gartner, J, Chen, WW, Kelley, RI, Mihalik, SJ, Moser, HW. 1991. The 22-kD peroxisomal integral membrane protein in Zellweger syndrome—presence, abundance, and association with a peroxisomal thiolase precursor protein. *Pediatr. Res.* 29:141–46

Gartner, J, Moser, HW, Valle, D. 1992. Mutations in the 70K peroxisomal membrane protein gene in Zellweger syndrome. *Nature New Genetics* 1:16–22

Gatfield, PD, Taller, E, Hinton, GG, Wallace, AC, Abdelnour, GM, et al. 1968. Hyperpipecolacetemia: A new metabolic disorder associated with neuropathy and hepatomegaly. *Can. Med. Assoc. J.* 99: 1215–33

Gietl, C. 990. Glyoxysomal malate dehydrogenase from watermelon is synthesized with an amino-terminal transit peptide. *Proc. Natl. Acad. Sci. USA* 87:5773–77

Godecke, A, Veenhuis, M, Roggenkamp, R, Janowicz, ZA, Hollenberg, CP. 1989. Biosynthesis of the peroxisomal dihydroxyacetone synthase from *Hansenula polymorpha* in *Saccharomyces cerevisiae* induces growth but not proliferation of peroxisomes. *Curr. Genet.* 16:13–20

Goebl, M, Yanagida, M. 1991. The TPR snap helix: a novel protein repeat motif from mitosis to transcription. *Trends Biochem. Sci.* 16:173–77

Goldfischer, S, Moore, CL, Johnson, AB, Spiro, AJ, Valsamis, MP, et al. 1973. Peroxisomal and mitochondrial defects in the cerebro-hepato-renal syndrome. *Science* 182:62–64

Goodman, JM, Garrard, LJ, McCammon, MT. 1992. Structure and assembly of peroxisomal membrane proteins. See Chiang & Schekman 1992, pp. 221–29

Goodman, JM, Trapp, SB, Hwang, H, Veenhuis, M. 1990. Peroxisomes induced in *Candida boidinii* by methanol, oleic acid and D-alanine vary in metabolic function but share common integral membrane proteins. *J. Cell Sci.* 97:193–204

Gorgas, K. 1985. Serial section analysis of mouse hepatic peroxisomes. *Anat. Embryol.* 172:21–32

Gould, SJ, Keller, GA, Hosken, N, Wilkinson, J, Subramani, S. 1989. A conserved tripeptide sorts proteins to peroxisomes. *J. Cell Biol.* 108:1657–64

Gould, SJ, Keller, GA, Schneider, M, Howell, SH, Garrard, LJ, et al. 1990a. Peroxisomal protein import is conserved between yeast, plants, insects and mammals. *EMBO J.* 9:85–90

Gould, SJ, Keller, G, Subramani, S. 1987. Identification of a peroxisomal targeting signal at the carboxy terminus of firefly luciferase. *J. Cell Biol.* 105:2923–31

Gould, SJ, Keller, GA, Subramani, S. 1988. Identification of peroxisomal targeting signals located at the carboxy terminus of four peroxisomal proteins. *J. Cell Biol.* 107: 897–905

Gould, SJ, Krisans, S, Keller, GA, Subramani, S. 1990b. Antibodies directed against the peroxisomal targeting signal of firefly luciferase recognize multiple mam-

malian peroxisomal proteins. *J. Cell Biol.* 110:27–34
Gould, SJ, McCollum, D, Spong, AP, Heyman, JA, Subramani, S. 1992. Development of the yeast *Pichia pastoris* as a model organism for a genetic and molecular analysis of peroxisome assembly. *Yeast* 8:613–28
Green, S. 1992. Receptor-mediated mechanisms of peroxisome proliferators. *Biochem. Pharmacol.* 43:393–401
Hajra, AK, Bishop, JE. 1982. Glycerolipid biosynthesis in peroxisomes via the acyldihydroxyacetone pathway. *Ann. NY Acad. Sci.* 386:170–82
Hansen, H, Didion, T, Thiemann, A, Veenhuis, M, Roggenkamp, R. 1992. Targeting sequences of the two major peroxisomal proteins in the methylotrophic yeast *Hansenula polymorpha. Mol. Gen. Genet.* 235:269–78
Hartl, F, Just, WW. 1987. Integral membrane polypeptides of rat liver peroxisomes: topology and response to different metabolic states. *Arch. Biochem. Biophys.* 255:109–19
Hazeu, W, Batenburg van der Vegte, WH, Nieuwdorp, PJ. 1975. The fine structure of microbodies in the yeast *Pichia pastoris. Experientia* 31:926–27
Heikoop, JC, van den Berg, M, Strijland, A, Weijers, PJ, Just, WW, et al. 1992. Turnover of peroxisomal vesicles by autophagic proteolysis in cultured fibroblasts from Zellweger patients. *Eur. J. Cell Biol.* 57: 165–71
Heikoop, JC, van Roermund, CW, Just, WW, Ofman, R, Schutgens, RB, et al. 1990. Rhizomelic chondrodysplasia punctata. Deficiency of 3-oxoacyl-coenzyme A thiolase in peroxisomes and impaired processing of the enzyme. *J. Clin. Invest.* 86:126–30
Hertz, R, Bar-Tana, J. 1992. Induction of peroxisomal β-oxidation genes by retinoic acid in cultured rat hepatocytes. *Biochem. J.* 281:41–43
Heupel, R, Heldt, HW. 1992. Transfer of metabolites between chloroplasts, mitochondria and peroxisomes. In *Phylogenetic Changes in Peroxisomes of Algae-Phylogeny of Plant Peroxisomes,* ed. H Stabenau, pp. 13–26. Oldenberg, Germany: Univ. Oldenburg
Heyman, RA, Mangelsdorf, DJ, Dyck, JA, Stein, RB, Eichele, G, et al. 1992. 9-cis retinoic acid is a high affinity ligand for the retinoid X receptor. *Cell* 68:397–406
Heymans, HSA, Oorthuys, JWE, Nelck, G, Wanders, RJA, Dingemans, KP, et al. 1986. Peroxisomal abnormalities in rhizomelic chondrodysplasia punctata. *J. Inherit. Metab. Dis.* 9:328–31
Hijikata, M, Wen, JK, Osumi, T, Hashimoto, T. 1990. Rat peroxisomal 3-ketoacyl-CoA thiolase gene. Occurrence of two closely related but differentially regulated genes. *J. Biol. Chem.* 265:4600–6
Hines, V. 1992. The mitochondrial protein import machinery of *Saccharomyces cerevisiae.* See Chiang & Schekman 1992, pp. 241–52
Hodge, VJ, Gould, SJ, Subramani, S, Moser, HW, Krisans, SK. 1991. Normal cholesterol synthesis in human cells requires functional peroxisomes. *Biochem. Biophys. Res. Commun.* 181:537–41
Hoefler, G, Paschke, E, Hoefler, S, Moser, AB, Moser, HW. 1991. Photosensitized killing of cultured fibroblasts from patients with peroxisomal disorders due to pyrene fatty acid-mediated ultraviolet damage. *J. Clin. Invest.* 88:1873–79
Hoefler, S, Hoefler, G, Moser, AB, Watkins, PA, Chen, WW, et al. 1988. Prenatal diagnosis of rhizomelic chondrodysplasia punctata. *Prenat. Diagn.* 8:571–76
Höhfeld, J, Mertens, D, Wiebel, FF, Kunau, WH. 1992. Defining components required for peroxisome assembly. See Chiang & Schekman 1992, pp.185–207
Höhfeld, J, Veenhuis, M, Kunau, WH. 1991. PAS3, a *Saccharomyces cerevisiae* gene encoding a peroxisomal integral membrane protein essential for peroxisome biogenesis. *J. Cell Biol.* 114:1167–78
Horrocks, LA, Sharma, M. 1982. Plasmalogens and O-alkyl glycerolipids, In *Phospholipids,* ed. JN Hawthorne, GB Ansell, pp. 51–93. Amsterdam: Elsevier
Imanaka, T, Small, GM, Lazarow, PB. 1987. Translocation of acyl-CoA oxidase into peroxisomes requires ATP hydrolysis but not a membrane potential. *J. Cell Biol.* 105: 2915–22
Issemann, I, Green, S. 1990. Activation of a member of the steroid hormone receptor superfamily by peroxisome proliferators. *Nature* 347:645–50
Kamijo, K, Taketani, S, Yokota, S, Osumi, T, Hashimoto, T. 1990. The 70-kDa peroxisomal membrane protein is a member of the Mdr (P-glycoprotein)-related ATP-binding protein superfamily. *J. Biol. Chem.* 265:4534–40
Kamiryo, Y, Sakasegawa, Y, Tan, H. 1989. Expression and transport of *Candida tropicalis* peroxisomal acyl-coenzyme A oxidase in the yeast *Candida maltosa. Agric. Biol. Chem.* 53:179–86
Keller, G, Gould, SJ, Deluca, M, Subramani, S. 1987. Firefly luciferase is targeted to peroxisomes in mammalian cells. *Proc. Natl. Acad. Sci. USA* 84:3264–68
Keller, GA, Krisans, S, Gould, SJ, Sommer, JM, Wang, CC, et al. 1991. Evolutionary conservation of a microbody targeting signal that targets proteins to peroxisomes,

glyoxysomes, and glycosomes. *J. Cell Biol.* 114:893–904

Kelley, RI, Datta, NS, Dobyns, WB, Hajra, AK, Moser, AB, et al. 1986. Neonatal adrenoleukodystrophy: new cases, biochemical studies, and differentiation from Zellweger and related peroxisomal polydystrophy syndromes. *Am. J. Med. Genet.* 23:869–901

Kindl, H. 1992. Progress in studies on function and biosynthesis of peroxisomes. See Heupel & Heldt 1992, pp. 27–41

Kliewer, SA, Umesono, K, Mangelsdorf, DJ, Evans, RM. 1992a. Retinoid X receptor interacts with nuclear receptors in retinoic acid, thyroid hormone and vitamin D3 signalling. *Nature* 355:446–49

Kliewer, SA, Umesono, K, Noonan, DJ, Heyman, RA, Evans, RM. 1992b. Convergence of 9-cis retinoic acid and peroxisome proliferator signalling pathways through heterodimer formation of their receptors. *Nature* 358:771–74

Kragler, F, Langeder, A, Raupachova, J, Binder, M, Hartig, A. 1993. Two independent peroxisomal targeting signals in catalase A of *Saccharomyces cerevisiae*. *J. Cell Biol.* 120:665–73

Krisans, SK. 1992. The role of peroxisomes in cholesterol metabolism. *Am. J. Respir. Cell Mol. Biol.* 7:358–64

Kunau, WH, Buhne, S, de la Garza, M, Kionka, C, Mateblowski, M, et al. 1988. Comparative enzymology of β-oxidation. *Biochem. Soc. Trans.* 16:418–20

Kunau, WH, Hartig, A. 1992. Peroxisome biogenesis in *Saccharomyces cerevisiae*. *Antonie Van Leeuwenhoek J. Microbiol. Serol.* 62:63–78

Lazarow, PB, De Duve, C. 1973. The synthesis and turnover of rat liver peroxisomes. V. Intracellular pathway of catalase synthesis. *J. Cell Biol.* 59:507–24

Lazarow, PB, De Duve, C. 1976. A fatty acyl-CoA oxidizing system in rat liver peroxisomes: Enhancement by clofibrate, a hypolipidemic drug. *Proc. Natl. Acad. Sci. USA* 73:2043–46

Lazarow, PB, Fujiki, Y. 1985. Biogenesis of peroxisomes. *Annu. Rev. Cell Biol.* 1:489–530

Legoux, R, Delpech, B, Dumont, X, Guillemot, JC, Ramond, P, et al. 1992. Cloning and expression in *Escherichia coli* of the gene encoding *Aspergillus flavus* urate oxidase. *J. Biol. Chem.* 267:8565–70

Leid, M, Kastner, P, Lyons, R, Nakshatri, H, Saunders, M, et al. 1992. Purification, cloning, and RXR identity of the HeLa cell factor with which RAR or TR heterodimerizes to bind target sequences efficiently. *Cell* 68:377–95

Lewis, MJ, Sweet, DJ, Pelham, HR. 1990. The ERD2 gene determines the specificity of the luminal ER protein retention system. *Cell* 61:1359–63

Liu, H, Tan, X, Veenhuis, M, McCollum, D, Cregg, JM. 1992. An efficient screen for peroxisome-deficient mutants of *Pichia pastoris*. *J. Bacteriol.* 174:4943–51

Mangelsdorf, DJ, Borgmeyer, U, Heyman, RA, Zhou, JY, Ong, ES, et al. 1992. Characterization of three RXR genes that mediate the action of 9-cis retinoic acid. *Genes Dev.* 6:329–44

Marcus, S, Miyata, KS, Zhang, B, Subramani, S, Rachubinski, RA, et al. 1993. Diverse peroxisome proliferator activated receptors bind to the peroxisome proliferator responsive elements of the rat hydratase-dehydrogenase and fatty acyl-CoA oxidase genes but differentially induce expression. *Proc. Natl. Acad. Sci. USA* 90:5723–27

McCammon, MT, Dowds, CA, Orth, K, Moomaw, CR, Slaughter, CA, et al. 1990. Sorting of peroxisomal membrane protein PMP47 from *Candida boidinii* into peroxisomal membranes of *Saccharomyces cerevisiae*. *J. Biol. Chem.* 265:20098–105

McCollum, D, Monosov, E, Subramani, S. 1993. The *pas8* mutant of *Pichia pastoris* exhibits the peroxisomal protein import deficiencies of Zellweger syndrome cells - The PAS8 protein binds to the COOH-terminal tripeptide peroxisomal targeting signal, and is a member of the TPR protein family. *J. Cell Biol.* 121:761–74

McGuinness, MC, Moser, AB, Moser, HW, Watkins, PA. 1990. Peroxisomal disorders: complementation analysis using β-oxidation of very long chain fatty acids. *Biochem. Biophys. Res. Commun.* 172:364–69

Mihalik, SJ, Moser, HW, Watkins, PA, Danks, DM, Poulos, A, et al. 1989. Peroxisomal L-pipecolic acid oxidation is deficient in liver from Zellweger syndrome patients. *Pediatr. Res.* 25:548–52

Miura, S, Kasuya, AI, Mori, H, Miyazawa, S, Osumi, T, et al. 1992. Carboxyl-terminal consensus Ser-Lys-Leu-related tripeptide of peroxisomal proteins functions in vitro as a minimal peroxisome-targeting signal. *J. Biol. Chem.* 267:14405–11

Miyazawa, S, Osumi, T, Hashimoto, T, Ohno, K, Miura, S, et al. 1989. Peroxisome targeting signal of rat liver acyl-coenzyme A oxidase resides at the carboxy terminus. *Mol. Cell. Biol.* 9:83–91

Morand, OH, Allen, LA, Zoeller, RA, Raetz, CR. 1990. A rapid selection for animal cell mutants with defective peroxisomes. *Biochim. Biophys. Acta* 1034:132–41

Morand, OH, Zoeller, RA, Raetz, CR. 1988. Disappearance of plasmalogens from membranes of animal cells subjected to photosensitizcd oxidation. *J. Biol. Chem.* 263:11597–606

Moser, HW, Bergin, A, Comblath, D. 1991. Peroxisomal disorders. *Biochem. Cell Biol.* 69:463–74

Moser, HW, Moser, AE, Singh, I, O'Neill, BP. 1994. Adrenoleukodystrophy: Survey of 303 cases: Biochemistry, diagnosis and therapy. *Ann. Neurol.* 16:628–41

Müller, WH, van der Krift, TP, Krouwer, AJJ, Wosten, HAB, van der Voort, LHM, et al. 1991. Localization of the pathway of the penicillin biosynthesis in *Penicillium chrysogenum*. *EMBO J.* 10:489–96

Munro, S, Pelham, HRB. 1987. A C-terminal signal prevents secretion of luminal ER proteins. *Cell* 48:899–907

Muse, KE, Roberts, JF. 1973. Microbodies in *Crithidia fasciculata*. *Protoplasma* 78:343–48

Newport, J, Dunphy, W. 1992. Characterization of the membrane binding and fusion events during nuclear envelope assembly using purified components. *J. Cell Biol.* 116:295–306

Nuttley, WM, Aitchison, JD, Rachubinski, RA. 1988. cDNA cloning and primary structure determination of the peroxisomal trifunctional enzyme hydratase-dehydrogenase-epimerase from the yeast *Candida tropicalis* pK233. *Gene* 69:171–80

Nuttley, WM, Brade, AM, Gaillardin, C, Eitzen, GA, Glover, JR, et al. 1993. Rapid identification and characterization of peroxisomal assembly mutants in *Yarrowia lipolytica*. *Yeast*. 9:507–17

Oda, T, Funai, T, Ichiyama, A. 1990. Generation from a single gene of two mRNAs that encode the mitochondrial and peroxisomal serine:pyruvate aminotransferase of rat liver. *J. Biol. Chem.* 265:7513–19

Opitz, JM, Zurheim, BM, Vitale, L, Shahidi, NT, Howe, JJ, et al. 1969. The Zellweger syndrome (cerebro-hepato-renal syndrome), In *Malformation Syndromes, Part II*, ed. D Bergsma, 2:144. Baltimore: Williams & Wilkins

Opperdoes, FR. 1987. Compartmentation of carbohydrate metabolism in trypanosomes. *Annu. Rev. Microbiol.* 41:127–51

Opperdoes, FR, Borst, P. 1977. Localization of nine glycolytic enzymes in a microbody-like organelle in *Trypanosoma brucei*: the glycosome. *FEBS Lett.* 80:360–64

Osinga, KA, Swinkels, BW, Gibson, WC, Borst, P, Veeneman, GH, et al. 1985. Topogenesis of the microbody enzymes: a sequence comparison of the genes for the glycosomal (microbody) and cytosolic phosphoglycerate kinases of *Trypanosoma brucei*. *EMBO J.* 4:3811–17

Osumi, T, Tsukamoto, T, Hata, S, Yokota, S, Miura, S, et al. 1991a. Amino-terminal presequence of the precursor of peroxisomal 3-ketoacyl-CoA thiolase is a cleavable signal peptide for peroxisomal targeting. *Biochem. Biophys. Res. Commun.* 181:947–54

Osumi, T, Wen, JK, Hashimoto, T. 1991b. Two *cis*-acting regulatory sequences in the peroxisome proliferator-responsive enhancer region of rat acyl-CoA oxidase gene. *Biochem. Biophys. Res. Commun.* 175:866–71

Pfanner, N, Neupert, W. 1990. The mitochondrial protein import apparatus. *Annu. Rev. Biochem.* 59:331–53

Poll-Thé, BT, Roels, F, Ogier, H, Scotto, J, Vamecq, J, et al. 1988. A new peroxisomal disorder with enlarged peroxisomes and a specific deficiency of acyl-CoA oxidase (pseudo-neonatal adrenoleukodystrophy). *Am. J. Hum. Genet.* 42:422–34

Poll-Thé, BT, Skjeldal, OH, Stokke, O, Demaugre, F, Saudubray, JM. 1990. Complementation analysis of peroxisomal disorders and classical Refsum. *Prog. Clin. Biol. Res.* 321:537–43

Purdue, PE, Takada, Y, Danpure, CJ. 1990. Identification of mutations associated with peroxisome-to-mitochondrion mistargeting of alanine/glyoxylate aminotransferase in primary hyperoxaluria type 1. *J. Cell Biol.* *11* 1:2341–51

Rapp, S, Soto, U, Just, WW. 1993. Import of firefly luciferase into peroxisomes of permeabilized Chinese hamster ovary cells: A model system to study peroxisomal protein import in vitro. *Exp. Cell Res.* 205:59–65

Reddy, JK, Goel, SK, Nemali, MR, Camino, JJ, Laffler, TG, et al. 1986. Transcriptional regulation of peroxisomal fatty acyl-CoA oxidase and enoyl-CoA hydratase/3-hydroxyacyl-CoA dehydrogenase in rat liver by peroxisome proliferators. *Proc. Natl. Acad. Sci. USA* 83:1747–51

Reddy, JK, Rao, MS, Azarnoff, DL. 1979. Mitogenic and carcinogenic effects of a hypolipidemic peroxisome proliferator, 4[chloro-6-(2,3-xylidino)-2-pyrimidinyl-thio] acetic acid (Wy-14,643), in rat and mouse liver. *Cancer Res.* 39:152–61

Rhodin, J. 1954. *Correlation of ultrastructural organization and function in normal and experimentally changed proximal convoluted tubule cells of the mouse kidney*. PhD thesis. Karolinska Inst. Aktiebolaget Godvil, Stockholm

Roberts, BL, Richardson, WD, Smith, AE. 1987. The effect of protein context on nuclear location signal function. *Cell* 50:465–75

Roggenkamp, R. 1992. Targeting signals for protein import into peroxisomes. *Cell. Biochem. Funct.* 10:193–99

Roscher, AA, Hoefler, S, Hoefler, G, Paschke, E, Paltauf, F, et al. 1989. Genetic and phenotypic heterogeneity in disorders of peroxisome biogenesis—a complementation study involving cell lines from 19 patients. *Pediatr. Res.* 26:67–72

Santos, MJ, Hoefler, S, Moser, AB, Moser, HW, Lazarow, PB. 1992. Peroxisome assembly mutations in humans: structural heterogeneity in Zellweger syndrome. *J. Cell. Physiol.* 151:103–12

Santos, MJ, Imanaka, T, Shio, H, Lazarow, PB. 1988a. Peroxisomal integral membrane proteins in control and Zellweger fibroblasts. *J. Biol. Chem.* 263:10502–9

Santos, MJ, Imanaka, T, Shio, H, Small, GM, Lazarow, PB. 1988b. Peroxisomal membrane ghosts in Zellweger syndrome—aberrant organelle assembly. *Science* 239: 1536–38

Schepers, L, Casteels, M, Vamecq, J, Parmentier, G, van Velhoven, PP, et al. 1988. β-oxidation of the carboxyl side chain of prostaglandin E2 in rat liver peroxisomes and mitochondria. *J. Biol. Chem.* 263: 2724–31

Schram, AW, Goldfischer, S, van Roermund, CWT, Brouwer-Kelder, EM, Collins, J, et al. 1987. Human peroxisomal 3-oxoacyl-coenzyme A thiolase deficiency. *Proc. Natl. Acad. Sci. USA* 84:2494–96

Schram, AW, Strijland, A, Hashimoto, T, Wanders, RJA, Schutgens, RBH, et al. 1986. Biosynthesis and maturation of peroxisomal β-oxidation enzymes in fibroblasts in relation to the Zellweger syndrome and infantile Refsum disease. *Proc. Natl. Acad. Sci. USA* 83:6156–58

Scotto, JM, Hadchouel, M, Odievre, M, Laudat, MH, Saudabray, JM, et al. 1982. Infantile phytanic acid storage disease: three cases, including ultrastructural studies of the liver. *J. Inherit. Metab. Dis.* 5:83–90

Shimozawa, N, Tsukamoto, T, Suzuki, Y, Orii, T, Fujiki, Y. 1992a. Animal cell mutants represent two complementation groups of peroxisome-defective Zellweger syndrome. *J. Clin. Invest.* 90:1864–70

Shimozawa, N, Tsukamoto, T, Suzuki, Y, Orii, T, Shirayoshi, Y, et al. 1992b. A human gene responsible for Zellweger syndrome that affects peroxisome assembly. *Science* 255:1132–34

Sikorski, RS, Boguski, MS, Goebl, M, Hieter, P. 1990. A repeating amino acid motif in CDC23 defines a family of proteins and a new relationship among genes required for mitosis and RNA synthesis. *Cell* 60:307–17

Simon, M, Adam, G, Rapatz, W, Spevak, W, Ruis, H. 1991. The *Saccharomyces cerevisiae* ADR1 gene is a positive regulator of transcription of genes encoding peroxisomal proteins. *Mol. Cell. Biol.* 11:699–704

Simon, M, Binder, M, Adam, G, Hartig, A, Ruis, H. 1992. Control of peroxisome proliferation in *Saccharomyces cerevisiae* by ADR1, SNF1 (CAT1, CCR1) and SNF4 (CAT3). *Yeast* 8:303–9

Small, GM, Imanaka, T, Shio, H, Lazarow, PB. 1987. Efficient association of in vitro translation products with purified, stable *Candida tropicalis* peroxisomes. *Mol. Cell. Biol.* 7:1848–55

Small, GM, Santos, MJ, Imanaka, T, Poulos, A, Danks, DM, et al. 1988a. Peroxisomal integral membrane proteins in livers of patients with Zellweger syndrome, infantile Refsum's disease and X-linked adrenoleukodystrophy. *J. Inherit. Metab. Dis.* 11:358–71

Small, GM, Szabo, LJ, Lazarow, PB. 1988b. Acyl-CoA oxidase contains two targeting sequences each of which can mediate protein import into peroxisomes. *EMBO J.* 7:1167–73

Sommer, JM, Cheng, QL, Keller, GA, Wang, CC. 1992. In vivo import of firefly luciferase into the glycosomes of *Trypanosoma brucei* and mutational analysis of the C-terminal targeting signal. *Mol. Biol. Cell* 3: 749–59

Souto-Padron, T, De Souza, W. 1982. Fine structure and cytochemistry of peroxisomes (microbodies) in *Leptomonas samueli*. *Cell Tissue Res.* 222:153–58

Spranger, JW, Opitz, JM, Bidder, U. 1971. Heterogeneity of chondrodysplasia punctata. *Humangenetik* 11:190–212

Steffan, JS, McAllister-Henn, L. 1992. Isolation and characterization of the yeast gene encoding the MDH3 isozyme of malate dehydrogenase. *J. Biol. Chem.* 267:24708–15

Subramani, S. 1992. Targeting of proteins into the peroxisomal matrix. *J. Membr. Biol.* 125:99–106

Suzuki, Y, Shimozawa, N, Orii, T, Hashimoto, T. 1989. Major peroxisomal membrane polypeptides are synthesized in cultured skin fibroblasts from patients with Zellweger syndrome. *Pediatr. Res.* 26:150–53

Suzuki, Y, Shimozawa, N, Orii, T, Igarashi, N, Kono, N, et al. 1988. Zellweger-like syndrome with detectable hepatic peroxisomes: a variant form of peroxisomal disorder. *J. Pediatr.* 113:841–45

Swinkels, BW, Evers, R, Borst, P. 1988. The topogenic signal of the glycosomal (microbody) phosphoglycerate kinase of *Crithidia fasciculata* resides in a carboxy-terminal extension. *EMBO J.* 7:1159–65

Swinkels, BW, Gould, SJ, Bodnar, AG, Rachubinski, RA, Subramani, S. 1991. A novel, cleavable peroxisomal targeting signal at the amino-terminus of the rat 3-

ketoacyl-CoA thiolase. *EMBO J.* 10:3255–62
Swinkels, BW, Gould, SJ, Subramani, S. 1992. Targeting efficiencies of various permutations of the consensus C-terminal tripeptide peroxisomal targeting signal. *FEBS Lett.* 305:133–36
Tager, JM, Ten Harmsen van der Bock, WA, Wanders, RJA, Hashimoto, T, van den Bosch, H, et al. 1985. Peroxisomal β-oxidation enzyme proteins in the Zellweger syndrome. *Biochem. Biophys. Res. Commun.* 126:1269–75
Takada, Y, Kaneko, N, Esumi, H, Purdue, PE, Danpure, CJ. 1990. Human peroxisomal L-alanine: glyoxylate aminotransferase. Evolutionary loss of a mitochondrial targeting signal by point mutation of the initiation codon. *Biochem. J.* 208:517–20
Takada, Y, Noguchi, T. 1986. Ureidoglycollate lyase, a new metalloenzyme of peroxisomal urate degradation in marine fish liver. *Biochem. J.* 235:391–97
Thieringer, R, Shio, H, Han, YS, Cohen, G, Lazarow, PB. 1991. Peroxisomes in *Saccharomyces cerevisiae:* immunofluorescence analysis and import of catalase A into isolated peroxisomes. *Mol. Cell. Biol.* 11: 510–22
Titorenko, VI, Waterham, HR, Haima, P, Harder, W, Veenhuis, M. 1992. Peroxisome biogenesis in *Hansenula polymorpha:* different mutations in genes, essential for peroxisome biogenesis, cause different peroxisomal mutant phenotypes. *Fems. Microbiol. Lett.* 74:143–48
Tolbert, NE. 1981. Metabolic pathways in peroxisomes and glyoxysomes. *Annu. Rev. Biochem.* 50:133–57
Trumbly, RJ. 1992. Glucose repression in the yeast *Saccharomyces cerevisiae. Mol. Microbiol.* 6:15–21
Tsukamoto, T, Miura, S, Fujiki, Y. 1991. Restoration by a 35K membrane protein of peroxisome assembly in a peroxisome-deficient mammalian cell mutant. *Nature* 350: 77–81
Tsukamoto, T, Yokota, S, Fujiki, Y. 1990. Isolation and characterization of Chinese hamster ovary cell mutants defective in assembly of peroxisomes. *J. Cell Biol.* 110: 651–60
Tugwood, JD, Issemann, I, Anderson, RG, Bundell, KR, McPheat, WL, et al. 1992. The mouse peroxisome proliferator activated receptor recognizes a response element in the 5′ flanking sequence of the rat acyl CoA oxidase gene. *EMBO J.* 11:433–39
Van den Bosch, H, Schutgens, RBH, Wanders, RJA, Tager, JM. 1992. Biochemistry of peroxisomes. *Annu. Rev. Biochem.* 61: 157–97
Van der Leij, I, Van den Berg, M, Boot, R, Franse, M, Distel, B, et al. 1992. Isolation of peroxisome assembly mutants from *Saccharomyces cerevisiae* with different morphologies using a novel positive selection procedure. *J. Cell Biol.* 119:153–62
Veenhuis, M. 1992. Peroxisome biogenesis and function in *Hansenula polymorpha. Cell. Biochem. Funct.* 10:175–84
Veenhuis, M, Goodman, JM. 1990. Peroxisomal assembly: membrane proliferation precedes the induction of the abundant matrix proteins in the methylotrophic yeast *Candida boidinii. J. Cell Sci.* 96:583–90
Veenhuis, M, Keizer, I, Harder, W. 1979. Characterization of peroxisomes in glucose-grown *Hansenula polymorpha* and their levels after transfer of cells to methanol-containing media. *Arch. Microbiol.* 120: 167–75
Veenhuis, M, Mateblowski, M, Kunau, W, Harder, W. 1987. Proliferation of microbodies in *Saccharomyces cerevisiae. Yeast* 3:77–84
Veenhuis, M, Sulter, G, van der Klei, I, Harder, W. 1989. Evidence for functional heterogeneity among microbodies in yeasts. *Arch. Microbiol.* 151:105–10
Volokita, M. 1991. The carboxy-terminal end of glycolate oxidase directs a foreign protein into tobacco leaf peroxisomes. *Plant J.* 1:361–66
Walton, PA, Gould, SJ, Feramisco, JR, Subramani, S. 1992. Transport of microinjected proteins into peroxisomes of mammalian cells: inability of Zellweger cell lines to import proteins with the SKL tripeptide peroxisomal targeting signal. *Mol. Cell. Biol.* 12:531–41
Wanders, RJA, Kos, M, Roest, B, Meijer, AJ, Schrakamp, G, et al. 1984. Activity of peroxisomal enzymes and intracellular distribution of catalase in Zellweger syndrome. *Biochem. Biophys. Res. Commun.* 123: 1054–61
Wanders, RJA, Schumacher, H, Heikoop, J, Schutgens, RB, Tager, JM. 1992. Human dihydroxyacetonephosphate acyltransferase deficiency: a new peroxisomal disorder. *J. Inherit. Metab. Dis.* 15:389–91
Wanders, RJA, Schutgens, RBH, Tager, JM. 1985. Peroxisomal matrix enzymes in Zellweger syndrome: Activity and subcellular localization in liver. *J. Inherit. Metab. Dis.* 8:151–52
Wanders, RJA, van Roermund, CW, van Wijland, MJ, Schutgens, RB, van den Bosch, H, et al. 1988. Direct demonstration that the deficient oxidation of very long chain fatty acids in X-linked adrenoleukodystrophy is due to an impaired ability of peroxisomes to activate very long chain fatty acids. *Biochem. Biophys. Res. Commun.* 153:618–24

Waterham, HR, Keizer-Gunnink, I, Goodman, JM, Harder, W, Veenhuis, M. 1992. Development of multipurpose peroxisomes in *Candida boidinii* grown in oleic acid-methanol limited continuous cultures. *J. Bacteriol.* 174:4057–63

Watkins, PA, Chen, WW, Harris, CJ, Hoefler, G, Hoefler, S, et al. 1989. Peroxisomal bifunctional enzyme deficiency. *J. Clin. Invest.* 83:771–77

Wendland, M, Subramani, S. 1993a. Cytosol-dependent peroxisomal protein import in a permeabilized cell system. *J. Cell Biol.* 120: 675–85

Wendland, M, Subramani, S. 1993b. Presence of cytoplasmic factors functional in peroxisomal protein import implicates organelle-associated defects in several human peroxisomal disorders. *J. Clin. Invest.* In press

Wiebel, FF, Kunau, WH. 1992. The Pas2 protein essential for peroxisome biogenesis is related to ubiquitin-conjugating enzymes. *Nature* 359:73–76

Wiemer, EA, Brul, S, Just, WW, Van Driel, R, Brouwer-Kelder, E, et al. 1989. Presence of peroxisomal membrane proteins in liver and fibroblasts from patients with the Zellweger syndrome and related disorders: evidence for the existence of peroxisomal ghosts. *Eur. J. Cell Biol.* 50:407–17

Williams, HE, Wandzilak, TR. 1989. Oxalate synthesis, transport and the hyperoxaluric syndromes. *J. Urol.* 141:742–49

Yajima, S, Suzuki, Y, Shimozawa, N, Yamaguchi, S, Orii, T, et al. 1992. Complementation study of peroxisome-deficient disorders by immunofluorescence staining and characterization of fused cells. *Hum. Genet.* 88:491–99

Yamamoto, K, Fahimi, HD. 1987. Three-dimensional reconstruction of a peroxisomal reticulum in regenerating rat liver: Evidence of interconnections between heterogeneous segments. *J. Cell Biol.* 105:713–22

Zaar, K, Angermüller, S, Volkl, A, Fahimi, HD. 1986. Pipecolic acid is oxidized by renal and hepatic peroxisomes: implications for Zellweger's cerebro-hepato-renal syndrome (CHRS). *Exp. Cell Res.* 164:267–71

Zellweger, H. 1987. The cerebro-hepato-renal (Zellweger) syndrome and other peroxisomal disorders. *Dev. Med. Child. Neurol.* 29:821–29

Zhang, B, Marcus, SL, Sajjadi, FG, Alvares, K, Reddy, JK, et al. 1992. Identification of a peroxisome proliferator-responsive element upstream of the gene encoding rat peroxisomal enoyl-CoA hydratase/3-hydroxyacyl-CoA dehydrogenase. *Proc. Natl. Acad. Sci. USA* 89:7541–45

Zhang, XK, Hoffmann, B, Tran, PB, Graupner, G, Pfahl, M. 1992. Retinoid X receptor is an auxiliary protein for thyroid hormone and retinoic acid receptors. *Nature* 355: 441–46

Zinser, E, Sperka-Gottleib, CDM, Fasch, EV, Kohlwein, SD, Pallauf, F, Daum, G. 1991. Phospholipid synthesis and lipid composition of subcellular membranes in the unicellular eukaryote *Saccharomyces cerevisiae*. *J. Bacteriol.* 173:2026–34

Zoeller, RA, Allen, LA, Santos, MJ, Lazarow, PB, Hashimoto, T, et al. 1989. Chinese hamster ovary cell mutants defective in peroxisome biogenesis. Comparison to Zellweger syndrome. *J. Biol. Chem.* 264: 21872–78

Zoeller, RA, Morand, OH, Raetz, CR. 1988. A possible role for plasmalogens in protecting animal cells against photosensitized killing. *J. Biol. Chem.* 263:11590–96

Zoeller, RA, Raetz, CR. 1992. Strategies for isolating somatic cell mutants defective in lipid biosynthesis. *Meth. Enzymol.* 209:34–51

Zoeller, RA, Raetz, CR. H. 1986. Isolation of animal cell mutants deficient in plasmalogen biosynthesis and peroxisome assembly. *Proc. Natl. Acad. Sci. USA* 83: 5170–74

Annu. Rev. Cell Biol. 1993. 9:479–509

TRANSCRIPTIONAL REPRESSION IN EUKARYOTES

Brenda M. Herschbach
Department of Biochemistry and Biophysics, University of California, San Francisco, California 94143-0502

Alexander D. Johnson
Department of Biochemistry and Biophysics, and Department of Microbiology and Immunology, University of California, San Francisco, California 94143-0502

KEY WORDS: transcription initiation, negative regulation

CONTENTS

INTRODUCTION

Thirty years ago, Jacob, Monod, and their colleagues developed the idea of gene repressors and operators (Jacob & Monod 1961). So compelling were the arguments and so powerful the model that, at first, negative regulation was invoked to explain nearly all examples of genetic control in prokaryotes. Only gradually were examples of positive genetic control fully

0743–4634/93/1115–0479$05.00

acknowledged and incorporated into the general theory of prokaryotic gene regulation (Englesberg & Wilcox 1974).

In contrast, the studies of eukaryotic gene expression first emphasized positive control even though transcriptional repressors were among the first recognized eukaryotic gene regulatory proteins. Multicellular eukaryotic organisms employ hundreds of different cell types, each of which requires expression of a different collection of genes. The argument was presented that it would be much more efficient to turn the appropriate cell type-specific genes on in the proper cell type rather than to repress them in all other cell types (Alberts et al 1983). Therefore, it was proposed that positive control mechanisms should predominate in higher organisms. The discovery in 1981 of transcriptional enhancers (Banerji et al 1981)—DNA sequences that can activate transcription when positioned thousands of basepairs upstream of the transcription start site—supported this idea and also posed a fascinating series of mechanistic questions that attracted the attention of many molecular biologists. Finally, since most eukaryotic promoters require DNA-bound activator proteins to function in vivo, there was a natural reluctance to study repression, a process that disrupted a sequence of events that itself was only beginning to be understood.

It now appears that eukaryotic regulatory circuits may have evolved to maximize evolutionary flexibility rather that economy; negative regulatory mechanisms appear to be quite common in eukaryotes. In addition, it has now been shown that negatively acting DNA sequences—variously termed silencers, operators, extinguishers, etc—can, like enhancers, control transcription from a distance. Finally, the recent advances in our understanding of transcription initiation in eukaryotes provide an appropriate background for a review of negative control mechanisms. Due to the space limitations of this review, we have generally given only a single example of each type of negative control discussed. Where possible, we have chosen examples where the biology behind the regulatory circuit is understood and insights as to the molecular mechanism have been uncovered. In other cases, our choice of example was arbitrary, and we apologize for the inevitable omissions. We begin by emphasizing an important lesson from prokaryotic examples of negative control: for every step in transcription initiation, there probably exists a repressor that can block it.

NEGATIVE REGULATION IN PROKARYOTES

The initiation of transcription in bacteria involves a series of discreet, ordered steps (Figure 1; see Chamberlin 1974; McClure 1985; Krummel & Chamberlin 1989). In the first step of prokaryotic transcription initiation, the RNA polymerase holoenzyme (RNA polymerase core enzyme plus a sigma

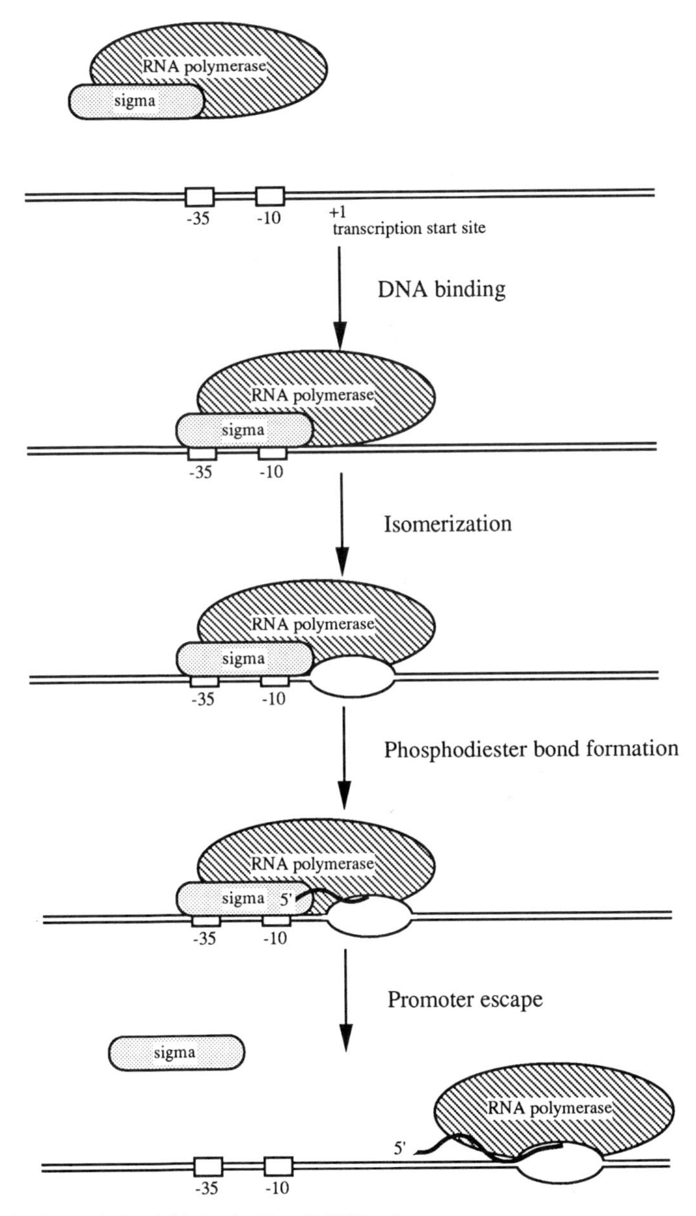

Figure 1 Transcription initiation by *E. coli* RNA polymerase.

factor that assists in promoter recognition) binds to the promoter. In the second step, the closed RNA polymerase-DNA complex isomerizes to an open form, a process that results in the unwinding of the DNA helix near the transcription start site. The first few phosphodiester bonds of the RNA transcript are formed in the third step of the reaction. The final step in prokaryotic transcription initiation is viewed as the escape of RNA polymerase from the promoter, with the concomitant release of the sigma factor.

Where along this pathway do known bacterial repressors act? The cI repressor of the coliphage lambda (bound at the $O_2 1$ and $O_2 2$ operators) blocks the initial binding of RNA polymerase to the promoter (Hawley et al 1985). The Arc repressor of bacteriophage P22 is thought to allow RNA polymerase binding, but to prevent the transition from the closed to the open complex (Vershon et al 1987). The *Escherichia coli* Gal repressor permits both RNA polymerase binding and isomerization of the RNA polymerase-promoter complex, but blocks the formation of the first phosphodiester bond (Choy & Adhya 1992). Finally, Lee & Goldfarb (1991) have argued that the *E. coli* Lac repressor prevents promoter escape at the lacUV5 promoter.

Thus studies of negative regulation in prokaryotic systems have identified transcriptional repressors that act at each step in the transcription initiation process. Whereas our understanding of repression mechanisms in eukaryotic transcription is not nearly as well developed, we might anticipate the same outcome. Below, we discuss the transcription initiation process at eukaryotic promoters and review some examples of negative control mechanisms, with an emphasis on the steps at which they may act to block transcription.

NEGATIVE REGULATION IN EUKARYOTES

Transcription initiation is more complex in eukaryotes than in prokaryotes. Unlike prokaryotic cells, which utilize a single RNA polymerase to synthesize all RNA molecules, eukaryotic cells contain three distinct RNA polymerases, each of which transcribes a different set of genes (reviewed in Sentenac 1985); RNA polymerase I transcribes a single gene encoding the large ribosomal RNA precursor; RNA polymerase II transcribes protein-coding genes; RNA polymerase III transcribes small genes encoding functional RNAs (tRNAs, 5SrRNA, U6 snRNA, etc). In this review, we limit our discussion to negative regulation of RNA polymerase II transcription.

RNA polymerase II consists of approximately 12 subunits, some of which are shared with RNA polymerases I and III (Sentenac 1985; Woychik et al 1990; Carles et al 1991). In contrast to bacterial RNA polymerases, eukaryotic RNA polymerase II cannot accurately initiate transcription on its own (Matsui et al 1980). Several additional proteins, termed general tran-

Table 1 The general transcription factors of eukaryotic RNA polymerase II

Factor	Native mass (kd)	Polypeptide composition (kd)	Function
TFIID	>700	TBP: 38 TAFs: 30–200	Binds to TATA in first step of pre-initiation complex assembly
TFIIB	33	33	Binds to TFIID-TATA complex
TFIIF	220	30 74	Recruits RNA polymerase IIa into the pre-initiation complex Role in transcriptional elongation
TFIIE	200	34 56	Binds to DBPolIIaF complex Kinase homology
TFIIH	230	90 62 43 41 35	Binds to DBPolIIaFE complex Associated CTD kinase activity 90-kd subunit identical to *ERCC-3* DNA repair helicase
TFIIJ	Unknown	Unknown	Binds to DBPolIIaFEH complex

scription factors (see Table 1), must first assemble at the promoter to allow transcription by RNA polymerase II (reviewed in Zawell & Reinberg 1993).

Figure 2 diagrams the initiation process at most eukaryotic promoters transcribed by RNA polymerase II. The first step involves binding of TFIID at the TATA element, typically located approximately 30 base pairs upstream of the transcription start site. In higher eukaryotes, TFIID consists of several polypeptides. The TATA binding protein (TBP) contacts DNA in the minor groove of the TATA element (DK Lee et al 1991; Starr & Hawley 1991; Nikolov et al 1992). The other components of TFIID, termed TBP associated factors (TAFs), somehow contribute specificity to the TFIID complex (for review, see Gill 1992; Sharp 1992; White & Jackson 1992; Rigby 1993) and may serve as the target for some transcriptional activator proteins (Hoey et al 1993).

Once TFIID has bound to the TATA element, the other general transcription factors assemble onto the complex in a prescribed order. TFIIB joins first. RNA polymerase II is then delivered to the complex in association with TFIIF. TFIIE, TFIIH, and TFIIJ follow, thus completing assembly of the pre-initiation complex. [Another factor, TFIIA, may play a role in stabilizing assembly of transcription complexes, perhaps by displacing an inhibitor associated with TFIID (for review, see Zawel & Reinberg 1993). TFIIA is not required for in vitro transcription systems that use recombinant TBP produced in bacteria.]

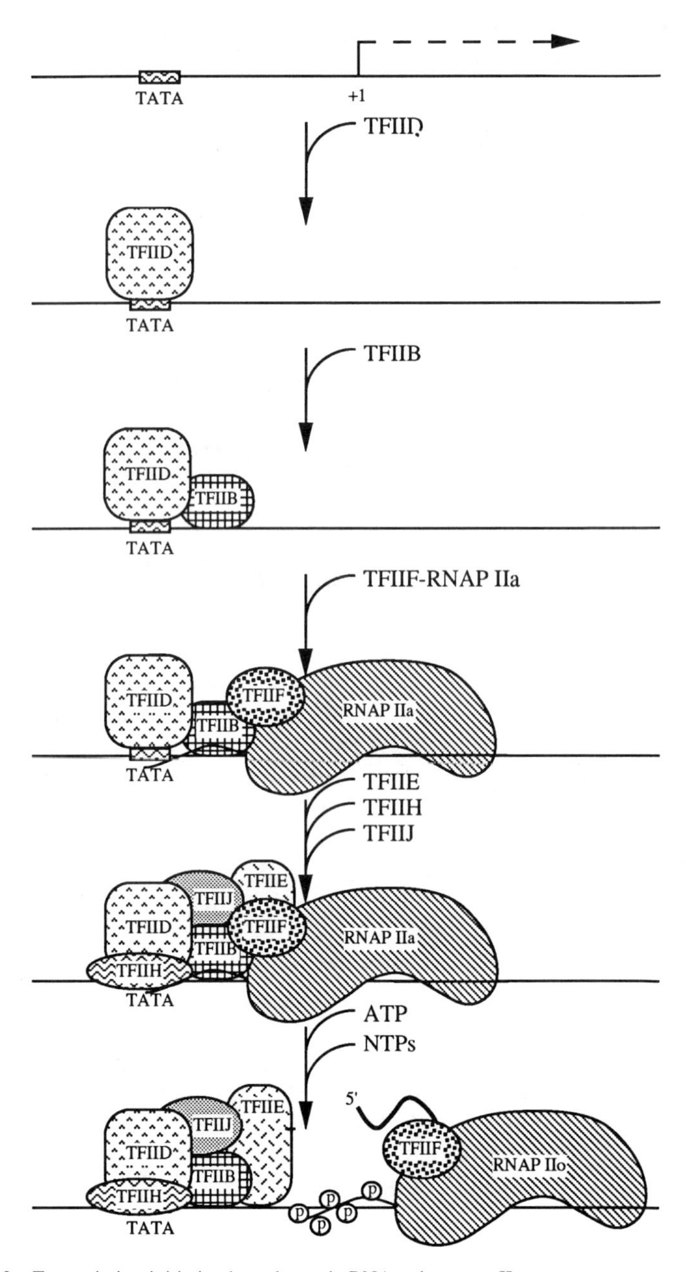

Figure 2 Transcription initiation by eukaryotic RNA polymerase II.

Assembly of this multi-component pre-initiation complex is analogous in certain ways to RNA polymerase binding at prokaryotic promoters. Initiation of transcription still requires unwinding of the DNA over the initiation site, formation of the first phosphodiester bond of the RNA transcript, and escape of RNA polymerase II from the promoter. The details of these later steps are not understood for eukaryotic systems; however, a reasonable model can be assembled from collected observations.

Recent work has demonstrated that the largest subunit of the general transcription factor TFIIH has helicase activity (Schaeffer et al 1993). This TFIIH helicase (also known as *ERCC-3*; Weeda et al 1990) is inhibited by the same concentration of the detergent Sarkosyl as blocks transcription in vitro (Schaeffer et al 1993 and references therein). Furthermore, both the transcription initiation reaction and the TFIIH-catalyzed DNA unwinding require hydrolysis of the β-γ bond of the ATP (Schaeffer et al 1993 and references therein). Thus it seems likely that the TFIIH helicase is responsible for unwinding the DNA helix over the start site during transcription initiation.

Promoter escape probably requires phosphorylation of RNA polymerase II. The C-terminal domain of the large subunit of RNA polymerase II contains, depending upon the species, between 26 and 52 copies of a heptapeptide repeat bearing the consensus sequence Tyr-Ser-Pro-Thr-Ser-Pro-Ser. Intact RNA polymerase II isolated from cells exists in two forms: IIo, in which the C-terminal tail is highly phosphorylated, and IIa, an unphosphorylated form. It is the unphosphorylated form (IIa) that associates with TFIIF and incorporates into assembling transcription complexes (Lu et al 1991; Chestnut et al 1992). However, elongating polymerases are highly phosphorylated in the C-terminal repeats, which suggests that phosphorylation occurs during the initiation process (Cadena & Dahmus 1987; Payne et al 1989; Laybourn & Dahmus 1990). Furthermore, the phosphorylated form of RNA polymerase II does not interact with TBP, although the non-phosphorylated form does (Usheva et al 1992). Thus it seems likely that phosphorylation of the C-terminal tail stimulates release of RNA polymerase II from the pre-initiation complex.

In addition to the helicase activity associated with its large subunit, the general transcription factor TFIIH contains a protein kinase activity associated with it 62-kd subunit that is capable of phosphorylating the C-terminal tail of RNA polymerase II (Feaver et al 1991; Lu et al 1992; Serizawa et al 1992). This activity is stimulated by TFIIE and by DNA containing a TATA box and a transcription start site. TFIIH can use either ATP or GTP as a phosphate donor. However, initiation of transcription by RNA polymerase II requires hydrolysis of ATP prior to formation of the first phosphodiester bond; GTP cannot be substituted for ATP. These observations indicate that phosphorylation of the C-terminal tail is not the step involving ATP hydrolysis in transcription initiation.

Once the pre-initiation complex has been formed, the DNA has been unwound, and RNA polymerase II has been phosphorylated, addition of nucleoside triphosphates (NTPs) allows elongation. Additional transcription factors have been described that affect elongation by promoting or hindering RNA polymerase II processivity (Rappaport et al 1987; Reinberg & Roeder 1987; Flores et al 1989; Price et al 1989; Bengal et al 1991).

Another difference between eukaryotic and prokaryotic transcription is that many, perhaps all, eukaryotic genes are expressed in vivo at very low levels (or not at all) unless stimulated by one or more transcriptional activators. Typically, such activators recognize specific DNA elements located upstream (sometimes several thousand base pairs) of the transcription start site. Activators stimulate either the rate of transcription complex assembly, or the fraction of functional complexes that assemble at a promoter in a given amount of time. Although the mechanisms of transcriptional activation are not understood in detail, at least some activators appear to act early in the pathway for assembly of the pre-initiation complex, perhaps stimulating DNA binding by TFIID, or helping to recruit TFIIB (Stringer et al 1990; Horikoshi et al 1991; WS Lee et al 1991; Lin & Green 1991; Ingles et al 1991; Lin et al 1991; Stringer et al 1991; Sundseth & Hansen 1992).

In principle, negative regulators of eukaryotic expression could inhibit transcription by interfering with any step in the transcription initiation pathway. From the breadth of repression mechanisms observed in prokaryotic systems, we anticipate the discovery of eukaryotic repressors working at most, if not all, of these steps. Some eukaryotic repressors might block transcriptional activation. Activator function could be affected at many levels, such as nuclear localization, DNA binding, or ability to stimulate transcription once bound to DNA. Other negative regulators might affect the general transcription machinery itself, thereby preventing formation of a functional pre-initiation complex (even in the presence of an activator). For example, repressors might occlude promoter DNA from the transcription apparatus. Alternatively, negative regulators could block association of one of the general transcription factors or of RNA polymerase II with the assembling pre-initiation complex. Some negative regulators might act late in the initiation pathway, perhaps interfering with the kinase activity of TFIIH and thus preventing the escape of RNA polymerase II from the promoter. Such late-acting repressors would be useful at genes requiring rapid induction in response to environmental stimuli.

We have arranged the following discussion of eukaryotic transcriptional repression into sections that correspond to the steps of the initiation process

Repressors that interfere with transcriptional activators

Interference with activator nuclear localization

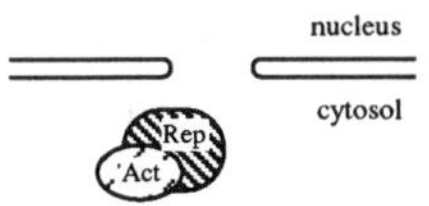

Interference with assembly of multisubunit activators

Interference with activator DNA binding

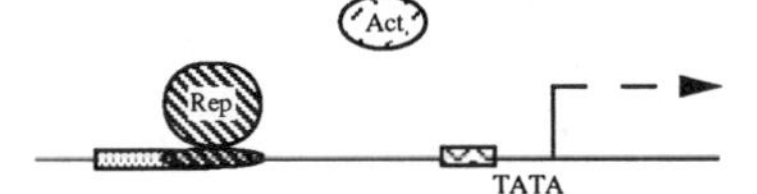

Interference with activity of DNA-bound activators

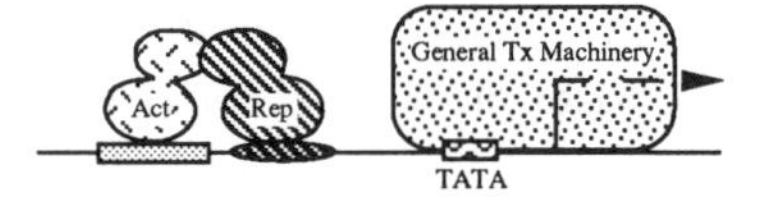

Repressors that interfere with the general transcription machinery

Interference with access of general transcription machinery to the DNA

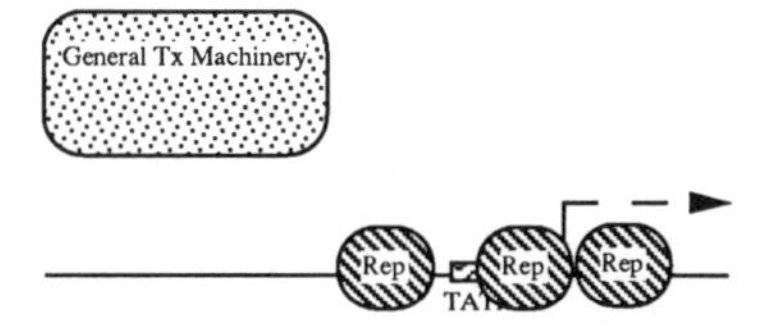

Interference with pre-initiation complex assembly

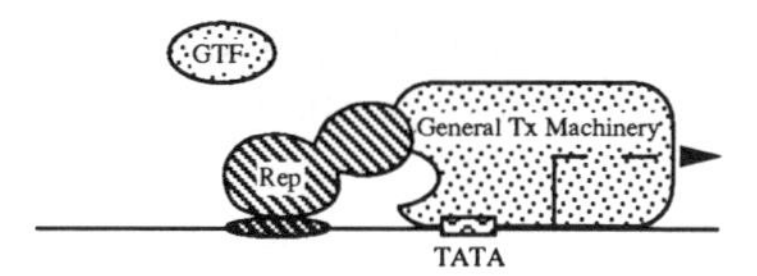

DNA Silencing

Figure 3 Mechanisms of transcriptional repression in eukaryotes. A schematic view of some of the molecular mechanisms described in the text.

at which negative regulators could, in theory, act (see Figure 3). For each step, we describe one or two examples that illustrate the principle. We have chosen, where possible, to describe examples in which the biological relevance of the regulatory circuit is understood. In those cases where clear-cut examples of a proposed repression mechanism are lacking, we have speculated as to the relevance of published instances of negative regulation. It is beyond the scope of this review to provide a catalogue of all known transcriptional repressors; again, we apologize for any blatant omissions.

Eukaryotic Repressors that Interfere with Transcriptional Activators

First, we discuss examples of repressors that interfere with the ability of transcriptional activators to stimulate transcription. Negative regulators use many different mechanisms to block activator function.

INTERFERENCE WITH ACTIVATOR NUCLEAR LOCALIZATION One of the earliest steps at which a repressor could interfere with the activity of a transcriptional activator is the transport of the activator from the cytoplasm into the nucleus. The IκB family of transcriptional inhibitors exemplifies this idea. IκBs block the nuclear import of members of the Rel family of transcriptional activators. The Rel family includes factors responsible for regulation of immune function and inflammation response genes in humans (NFκB; Ghosh et al 1990; Kieran ct al 1990; Nolan et al 1991), oncogenesis in chickens (v-*rel* and c-*rel*; Ballard et al 1990; Bull et al 1990), and determination of dorsoventral axis polarity in fruit flies (*dorsal;* Steward 1987). These proteins are related through an amino-terminal domain of approximately 300 amino acids, called the Rel homology domain. Their carboxy terminal domains are highly divergent.

The Rel homology domain contains sequences important for DNA binding, nuclear localization, and oligomerization. Rel domains interact with each other. NFκB, for example, is a heterodimer of two Rel proteins, p50 and p65. p50 can also homodimerize to activate transcription of a different set of genes as the positive regulatory factor KBF-1 (Kieran et al 1990). Different IκB proteins can interfere with the activities of different sets of Rel dimers (Zabel & Bauerle 1990; Davis et al 1991; Haskill et al 1991; Geisler et al 1992; Inoue et al 1992a; Kerr et al 1992; Kidd 1992; Tewari et al 1992).

How do IκB proteins prevent nuclear import of the Rel dimers? IκBs associate with a region of the Rel homology domain that contains the highly conserved Rel nuclear localization sequence (Beg et al 1992). Presumably interaction with an IκB masks the Rel nuclear localization signal and thus prevents nuclear import (Nolan et al 1991; Beg et al 1992; Inoue et al 1992b). Given the amino acid similarity among the Rel proteins, it might be anticipated that the IκB proteins would also be related to one another. Several IκB proteins have recently been cloned (human: *Mad-3,* Haskill et al 1991; bcl-3, Kerr et al 1992; mouse: IκBγ, Inoue et al 1992a; rat: RI/IF-1, Tewari et al 1992; chicken: pp40, Davis et al 1991; fruit fly: cactus, Geisler et al 1992; Kidd 1992). Each has five to eight copies of a 32 amino acid motif known as the ankyrin repeat. Ankyrin repeats are found in proteins with highly diverse functions that include putative integral membrane proteins, viral host-range factors, and multisubunit transcription factors (for review, see Bennett 1992; Blank et al 1992). The ankyrin motifs in IκB proteins mediate interaction with the Rel homology domain (Inoue et al 1992b). Presumably multiple interactions along the Rel/ankyrin interaction surface provide the specificity that dictates which IκB inhibits which type of Rel dimer.

In addition to masking Rel factor nuclear localization sequences, IκBs are capable of disrupting complexes of Rel factors bound to DNA (Zabel & Bauerle 1990). It is not clear that this activity is relevant in vivo, however, since IκBs are not known to enter the nucleus.

If IκB factors prevent transcriptional stimulation of genes controlled by Rel activators by sequestering Rel factors in the cytoplasm, how is this association reversed? The activity of IκB proteins is regulated by phosphorylation. Different IκBs seem to be inactivated by different treatments in vitro (Ghosh & Baltimore 1990; Link et al 1992). Presumably different phosphorylation cascades in the cell inactivate different IκBs. If only a subset of IκBs are inactivated in response to a particular physiological signal, only the appropriate Rel activators would be released to stimulate transcription.

INTERFERENCE WITH THE ASSEMBLY OF MULTISUBUNIT ACTIVATORS Many transcriptional activators consist of more than one polypeptide subunit. Some repressors work by competing for association with one of the activator subunits, thereby preventing the formation of a functional activator.

Members of the basic-region-helix-loop-helix (bHLH) family of transcriptional activators bind to DNA as dimeric (or higher-order oligomeric) complexes (Murre et al 1989b). The HLH domain, a conserved region of hydrophobic amino acids predicted to form two amphipathic helices separated by a loop, mediates oligomerization (Murre et al 1989ab; Voronova & Baltimore 1990); the basic region contributes to DNA sequence recognition (Lassar et al 1989; Davis et al 1990; Voronova & Baltimore 1990; Ferre-D'Amare et al 1993).

In the fruit fly *Drosophila melanogaster,* bHLH proteins play an essential role in the development of the peripheral nervous system. The body of the adult fly is lined with mechano- and chemo-sensory organs termed sensilla. The precursor cells that give rise to these sensilla develop during the late larval and early pupal stages from undifferentiated epithelial sheets that also give rise to ordinary epidermal cells (see Hartenstein & Posakony 1989). At some point, undifferentiated cells undergo the developmental decision to become either a sensillum precursor or an epidermal precursor. The *daughterless* (*da*) gene and the *achaete* (*ac* or T5), *scute* (*sc* or T4), and *asense* (*ase* or T8) genes of the *achaete-scute* complex (*AS-C*) are thought to encode subunits of transcriptional activators involved in the decision to develop into a sensillum precursor. Loss-of-function mutations in these genes result in the loss of sensory organs (García-Bellido & Santamaria 1978; Caudy et al 1988a, Dambly-Chaudiére et al 1988; Cline 1989; Romani et al 1987). Overexpression leads to the development of ectopic sensilla (García-Alonso & García-Bellido 1986; Campuzano et al 1986).

The *da* gene and the three *AS-C* genes each encode a bHLH protein (Villares & Cabrera 1987; Alonso & Cabrera 1988; Caudy et al 1988b; Murre et al 1989a). These proteins interact with each other in vitro and probably in vivo to form heterodimeric complexes that bind to DNA and presumably stimulate transcription of sensillum-specific genes (Dambly-Chaudiére et al 1988; Van Doren et al 1991).

The same genes that are activated in sensory organ precursors must be kept silent in epidermal precursors. Two genes, *extramachrochaetae* (*emc*) and *hairy* (*h*), are known to suppress sensory organ development. Thus *emc* and *h* are negative regulators of the genes activated by *da* and *AS-C*. Furthermore, the phenotype conferred by mutations in *emc* or *h* is sensitive to the wild-type dosage of *da* and *AS-C* (Moscoso del Prado & García-Bellido 1984). Such dosage-sensitive relationships suggest that the proteins encoded by these genes might physically interact, an intriguing possibility given their opposing developmental roles.

The amino acid sequence of the *emc* gene product suggests a model to explain these observations (Ellis et al 1990; Garrell & Modolell 1990). The *emc* protein contains an HLH dimerization motif, but lacks a nearby basic region necessary for DNA binding. This observation suggests that *emc* might be able to heterodimerize with other bHLH proteins (in this case with *da* and/or the *AS-C* proteins), thus creating complexes that cannot bind to DNA and therefore do not activate transcription (Van Doren et al 1991). By interacting with *da* or *AS-C* proteins, *emc* would prevent their association with each other and would, therefore, block formation of the transcriptional activators required for development of sensory organ precursors.

Members of other families of transcriptional regulatory proteins also form functional activators by complexing with themselves or with members of their protein family. By assembling transcriptional activators from multiple polypeptide subunits, each of which can be used in more than one activator, cells can elaborate complex transcriptional regulatory circuits with a limited number of proteins. This mechanism allows efficient generation of many different activators from a small number of cellular components and provides a convenient step at which repressors can regulate transcription (see, for example, Descombes & Schibler 1991; Nakabeppu & Nathans 1991; Ron & Habener 1992). Furthermore, a repressor that works by heterodimerization could in principle inactivate a whole family of transcriptional activators.

INTERFERENCE WITH ACTIVATOR DNA BINDING The *emc* protein described in the previous section represses transcription by interacting directly with individual subunits of the *da/AS-C* activators, thus forming complexes that are incapable of binding DNA. Other transcriptional repressors interfere at a later step, competing with a functional activator for access to the same DNA sequences.

For example, DNA-binding sites for the *Krüppel* (*Kr*) repressor protein of *Drosophila* often overlap binding sites for transcriptional activator proteins (Stanojevic et al 1989; Small et al 1991; Zuo et al 1991). A well studied example of this is found in the stripe 2 element of the *even-skipped* (*eve*) promoter. *eve* encodes a homeodomain protein that is first detected during embyronic nuclear cleavage cycle 12 when it is distributed uniformly in all nuclei. By cycle 14, *eve* has disappeared from both poles of the embryo. Twenty to thirty minutes later, *eve* expression is restricted to a series of seven transverse stripes along the length of the embryo. Each stripe is about 5–6 nuclei wide (Frasch & Levine 1987). Promoter fusion experiments have revealed that independent regulatory regions upstream of the *eve* promoter direct expression in individual stripes (Goto et al 1999; Harding et al 1999).

In addition to *Kr*-binding sites, the stripe 2 regulatory element of the *eve* promoter contains several binding sites for the gap genes *hunchback* (*hb*) and *giant* (*gt*) and for the maternal morphogen *bicoid* (*bcd*) (Stanojevic et al 1989; Small et al 1991). *bcd* and *hb* activate transcription of genes linked to the stripe 2 element. *gt* and *Kr* act as repressors (Frasch & Levine 1987; Small et al 1991). *gt,* probably in cooperation with other factors, determines the anterior boundary of stripe 2. *Kr* is responsible for shutting off *eve* expression at the posterior boundary. Virtually all of the *bcd* and *hb*-binding sites overlap with, or are closely linked to, a *Kr* or *gt* recognition sequence (Stanojevic et al 1989; Small et al 1991). Significantly, DNA-binding studies have demonstrated that *bcd* and *Kr* cannot co-occupy closely linked sites (Small et al 1991). Apparently *Kr* sets the posterior limit on *eve* expression by competing with activators for access to DNA. Because mutations in a single *bcd*-binding site can have dramatic effects on *eve* expression in stripe

2, it seems likely that *Kr* could effectively shut off the stripe 2 element by interfering with just one activator site (Small et al 1992).

One important aspect of this mode of repression is that repressor and activator sites must be tightly linked. For example, *Kr* bound to its sites in the stripe 2 element does not interfere with transcriptional activators bound in the stripe 3 element, about 1.5 kb away (Goto et al 1989; Harding et al 1989). This sort of short-range repression mechanism, which allows for complex promoters constructed of several autonomous modules, might also occur in mammalian promoters where regulatory regions often include interdigitated activation and repression elements (see, for example, Maniatis et al 1987).

INTERFERENCE WITH THE ACTIVITY OF DNA-BOUND ACTIVATORS Even after a transcriptional activator has been successfully imported into the nucleus, with its subunits properly assembled, and has bound to DNA, a repressor can still interfere with its ability to stimulate transcription.

In many promoters, repressor-binding sites are adjacent to, although not overlapping, binding sites for transcriptional activators. Repressors and activators can often occupy their sites simultaneously. In these cases, a repressor could block activator function by directly contacting the adjacent DNA-bound activator and masking the protein domain responsible for transcriptional stimulation.

Expression of the mammalian c-*myc* gene is turned off in terminally differentiated plasma cells. The identification of a derepressed c-*myc* allele in murine plasmacytomas indicates that inappropriate expression of c-*myc* probably contributes to tumor development (Kakkis et al 1988). A repressor, myc-PRF, binds to a DNA site upstream of the c-*myc* promoter and shuts off c-*myc* transcription (Kakkis & Calame 1987; Kakkis et al 1989). Notably, myc-PRF is absent in cell lines that represent early stages of B-cell development when c-*myc* is still expressed. This correlation suggests that myc-PRF is responsible for repression of c-*myc* in vivo in terminally differentiated plasma cells. The binding site for myc-PRF in the c-*myc* promoter is located immediately adjacent to a binding site for the widely expressed transcriptional activator myc-CF1 (Kakkis & Calame 1987; Riggs et al 1991). Studies of DNA-bound complexes indicate that myc-PRF and myc-CF1 can simultaneously occupy the same promoter; moreover, they physically interact (Kakkis et al 1989). It seems plausible that myc-PRF represses c-*myc* expression in terminally differentiated B cells by binding next to and, through direct physical interaction, masking the activation surface on myc-CF1.

Repressors need not bind adjacent DNA sequences in order to interact with and mask activating regions on transcriptional activators. Proteins bound at distant sites can interact by looping out the intervening DNA (Ptashne 1986; Choy & Adhya 1992; Schleif 1992). In fact, some repressors complex with DNA-bound activators, but do not themselves bind DNA at all. The Gal80 protein of the yeast *Saccharomyces cerevisiae* is perhaps the best-studied example of this. The Gal4 activator stimulates transcription of genes required for galactose metabolism in yeast (the *GAL* genes; reviewed in Johnston 1987). Gal4 recognizes a 17 bp DNA element found in several copies upstream of the *GAL* genes, from which it activates transcription when galactose is present in (and glucose is absent from) the growth medium. Gal80 interacts with the C-terminal region of Gal4, which also contains an acidic activation domain (SA Johnston et al 1987; Lu et al 1987; Ma & Ptashne 1987ab; Salmeron et al 1990). Presumably, association with Gal80 masks the nearby activation region of Gal4, thereby blocking transcriptional stimulation by Gal4.

Studies both in vivo and in vitro have revealed that Gal80 represses Gal4 activity even though Gal4 is bound to DNA (Giniger et al 1985; Lohr & Hopper 1985; Lu et al 1987). Furthermore, even during galactose induction, Gal80 remains associated with Gal4 (Chasman & Kornberg 1990; Leuther & Johnston 1992). It has been proposed that a conformational change, which may involve phosphorylation of Gal4 (Mylin et al 1989, 1990; Parthun & Jaehning 1992), allows exposure of the Gal4 activating region despite the continued association of Gal80. This seems a particularly efficient way for a repressor to respond to environmental signal; since the complex never dissociates, the repressor need not relocate the activator to reestablish repression.

Repressors that Interfere with the Assembly of the General Transcription Machinery

Above we have discussed examples of negative regulation by interference with activator proteins. While effective against individual activators, these repression mechanisms share a disadvantage: most eukaryotic genes respond to several different transcriptional activators. Full repression of such genes by activator interference would require a dedicated repressor for each different activator protein. A more efficient approach to repressing genes controlled by multiple activators would be to interfere directly with the assembly of the general transcription machinery.

INTERFERENCE WITH ACCESS OF THE GENERAL TRANSCRIPTION MACHINERY TO THE DNA Perhaps the first recognized transcriptional repressor in eukaryotes, the simian virus 40 T antigen (SV40 Tag), represses transcription by occluding promoter DNA from the general transcription machinery (Hansen et al 1981). Tag accumulates in the early stages of SV40 lytic growth. After reaching a threshold concentration, Tag both stimulates SV40 replication and represses transcription of the viral early genes (reviewed in Tjian 1981). Tag binds, probably as a tetramer, to three adjacent sites within the initiation region of the SV40 early promoter (Tjian 1978; Shalloway et al 1980; Hansen et al 1981). Although it is not clear exactly which components of the transcription machinery are excluded from the DNA when Tag is bound, the locations of the Tag-binding sites suggest that RNA polymerase II and possibly TFIID are likely to be affected.

Other negative regulators may also function by occluding promoter elements from components of the transcription machinery (see, for example, DeLuca & Schaeffer 1988; Roberts et al 1988; Shepard et al 1990; Ohkuma et al 1990; Kaufman & Rio 1991; Rijcke et al 1992).

The histone proteins can act as transcriptional repressors of eukaryotic genes, probably by preventing TFIID access to the DNA. The DNA of eukaryotic organisms is wrapped around octamers of histone proteins to form complexes termed nucleosomes. It has long been postulated that such packaging would interfere with the ability of DNA-binding proteins to recognize their sites. In particular, it has been proposed that packaged promoter DNA would be inaccessible to TFIID until the nucleosomes were removed. It has further been suggested that one role of transcriptional activator proteins might be to clear the promoter DNA of inhibitory nucleosomes, thereby allowing access of TFIID to the TATA. Consistent with this idea, Grunstein and colleagues observed that nucleosomes can be depleted from yeast cells in which histone genes have been put under control of a heterologous, experimentally regulatable promoter (Han & Grunstein 1988; Han et al 1988). Such nucleosome depletion induces transcription of many yeast genes (Han & Grunstein 1988; Han et al 1988; Durrin et al 1992). Furthermore, even genes whose upstream activator-binding sites had been removed are expressed when histones are depleted (Han & Grunstein 1999; Han et al 1998). These results suggest that, in the absence of histones, transcriptional activators are no longer required for expression of these yeast genes.

Studies of transcription in vitro also support the idea that nucleosomal structures might repress transcription by interfering with the assembly of the general transcription factors at the promoter (Knezetic & Luse 1986; Matsui 1987; Wasylyk & Chambon 1979, 1980). Incubation of TFIID with the template DNA prior to nucleosome assembly prevents the nucleosomal inhibition of transcription, which suggests that it is TFIID binding that is inhibited by the presence of histone complexes (Matsui 1987; Workman & Roeder 1987; Knezetic et al 1988).

Specifically positioned nucleosomes have been observed at some promoters and have been proposed to be involved in transcriptional regulation (see, for example, Almer et al 1986; Benezra et al 1986; Pérez-Ortín et al 1987; Matallana et al 1992). According to this idea, some gene repressor proteins might block transcription from a target promoter by directing the formation of a positioned nucleosome over the TATA box (Roth et al 1990, 1992; Shimizu et al 1991) (see below).

INTERFERENCE WITH PRE-INITIATION COMPLEX ASSEMBLY Even when promoter DNA is accessible to the general transcription machinery, negative regulators could effectively repress transcription by interfering with proper assembly of any one of the general transcription factors into the pre-initiation complex.

The *Drosophila* homeodomain protein *even-skipped* (*eve*) is an example of a eukaryotic transcriptional repressor that interferes with the assembly of a functional pre-initiation complex (Johnson & Krasnow 1992). *eve* is one of a large family of homeodomain proteins that control the early development of the *Drosophila* embryo (reviewed in Hayashi & Scott 1990).

Experiments carried out in vitro have revealed that *eve* represses transcription from promoters containing homeodomain-binding sites upstream of the TATA box (Biggin & Tjian 1989; Johnson & Krasnow 1992). Since transcription in these experiments apparently initiates without an activator protein, *eve* must be acting directly on components of the general transcription machinery. *eve* does not affect the kinetics of transcription initiation, but rather reduces the probability that a functional pre-initiation complex will assemble at the promoter (Johnson & Krasnow 1992). Additionally, pre-initiation complexes become resistant to *eve* repression early in their assembly pathway, which indicates that *eve* affects one of the first steps in the formation of pre-initiation complexes (Johnson & Krasnow 1992). Although the precise step has not yet been identified, DNA binding by TFIID or recruitment of TFIIB seem to be likely possibilities.

The multiple steps required to assemble the transcriptional machinery at a promoter provide many opportunities for negative regulation. Certainly, other examples of transcriptional repressors that interfere with the general transcription machinery will be forthcoming. One likely candidate is the *S. cerevisiae* repressor Ssn6/Tup1 (Keleher et al 1992). Ssn6/Tup1 is involved in transcriptional repression of several diverse sets of yeast genes including **a**-specific, haploid-specific, and glucose-repressible genes (Carlson et al 1984; Trumbly 1986; Mukai et al 1991; Keleher et al 1992). The Ssn6/Tup1 complex is believed to be recruited to the promoters it represses by interaction with other proteins that bind to DNA (Keleher et al 1992). Ssn6/Tup1 repression is equally effective against transcription catalyzed by RNA polymerases I and II, but not against transcription catalyzed by RNA polymerase III (Herschbach & Johnson 1993). This result suggests that the repressor interacts with some component common to the RNA polymerase I and II transcriptional machines. Since the activation systems used by these two RNA polymerases are not interchangeable (Butlin & Quincy 1991), it seems likely that Ssn6/Tup1 represses transcription not by blocking transcriptional activation, but rather by interfering with the activity of some component of the general transcriptional machinery that is similar for (or shared by) RNA polymerases I and II. Although the target of Ssn6/Tup1 repression has not been identified, recent work indicates that the pre-initiation complexes of the three eukaryotic RNA polymerases have more in common than was originally expected (Mann et al 1987; Woychik et al 1990; Carles et al 1991; Buratowski & Zhou 1992; Dequard-Chablat et al 1991; Colbert & Hahn 1992; Gill 1992; López-De-León 1992; Sharp 1992; White & Jackson 1992; Rigby 1993).

INTERFERENCE WITH LATE STEPS IN INITIATION Although negative regulators that block transcription initiation after assembly of the pre-initiation complex have not been identified, it is possible to predict several steps at which such regulation might occur. Repressors might interfere with unwinding of the DNA helix over the initiation site, or with the phosphorylation of the C-terminal domain of RNA polymerase II. Studies of the *hsp70* heat shock promoter in *Drosophila* have revealed that RNA polymerase II is bound at the promoter and has synthesized the first few phosphodiester bonds when the gene is transcriptionally inactive (Gilmour & Lis 1986; Rougive & Lis 1988). Might this polymerase be prevented from escaping the promoter by a late-acting repressor that blocks the transition from initiation to elongation? Mutation analysis of the *hsp70* promoter has indicated that sequences upstream of the *hsp70* TATA contribute to the formation of these "engaged" RNA polymerase complexes (Lee et al 1992). One possibility is that the GAGA factor, normally a transcriptional activator, represses transcription of this promoter by interacting so strongly with the general transcription machinery that it prevents promoter escape. Such a late-acting mechanism might make sense for promoters whose rapid induction is required for survival in stressful environmental conditions.

Some Eukaryotic Repressors Probably Interfere with More than One Step in the Transcription Initiation Reaction

It is important to point out the possibility that individual negative regulators might be able to repress transcription by more than one of the mechanisms outlined here (see, for example, Appel & Sakonju 1993). For example, as described above, the *Drosophila* protein *Kr* probably represses *eve* transcription at the posterior boundary of stripe 2 by competing with activators for access to DNA. However, there is also evidence that *Kr* can interfere with the activity of some activators even when their DNA-binding sites do not overlap (Licht et al 1990; Zuo et al 1991). In fact, in at least one case, *Kr* can prevent transcriptional stimulation by an activator without itself binding to DNA (Zuo et al 1991). Presumably, direct interaction between *Kr* and these activators masks the activation surface and thereby prevents transcriptional stimulation.

Similarly, the *eve* repressor described above blocks transcription from the *Ubx* promoter by interfering with assembly of the pre-initiation complex. However, *eve,* whose DNA-binding specificity overlaps that of other homeodomain proteins, can also repress transcription by competing with homeodomain activators for access to DNA sites (Han et al 1989). At some promoters, *eve* may simultaneously use both mechanisms, thus ensuring tight repression by blocking both activator binding and functional assembly of any pre-initiation complexes that may begin to form at the promoter despite the absence of an activator.

The Ssn6/Tup1 repressor may also use more than one mechanism to repress transcription. Simpson and colleagues have described a positioned nucleosome that forms adjacent to the DNA-binding site involved in Ssn6/Tup1 repression of the **a**-specific genes in yeast and have suggested that such a structure could contribute to transcriptional repression by Ssn6/Tup1 by obscuring neighboring DNA sequences important for expression of the downstream gene (Roth et al 1990, 1992; Shimizu et al 1991). Studies of a glucose-repressed gene also show a correlation between Ssn6/Tup1 repression and the presence of a positioned nucleosome in the initiation region (Pérez-Ortín et al 1987; Matallana et al 1992). Thus occlusion of promoter DNA by a positioned nucleosome might contribute to transcriptional repression by Ssn6/Tup1.

It seems likely that many negative regulators can interfere with transcription initiation by more than one mechanism and can thereby ensure highly efficient gene repression.

Position Effects and DNA Silencing

Thus far we have discussed mechanisms of negative regulation that shut off transcription at individual promoters. We now turn to larger scale repression mechanisms by which whole regions of DNA become refractory to transcription. This phenomenon is often referred to as transcriptional silencing or position effects, since it was first observed that gene expression can vary depending on chromosomal location. Silencing probably results from the folding of nucleosomal DNA into a form of especially compacted chromatin that obscures the DNA from the transcription machinery. In addition to being transcriptionally inert, silenced DNA often replicates very late, which suggests that both RNA and DNA polymerases have restricted access to silenced sequences.

Perhaps the most dramatic example of transcriptional silencing is found in female mammals, where one of the two X chromosomes is inactivated in every cell (Lyon 1961; reviewed in Grant & Chapman 1988; Rastan & Brown 1990). Female mammals have two X chromosomes, while males have one X and one Y. Presumably because a double dose of X information would be deleterious, female cells permanently silence one of the two X chromosomes, chosen at random. Once an X has been inactivated, this state is stably maintained and faithfully inherited in all subsequent cell divisions.

The molecular mechanism of X inactivation is not understood. Nor is it understood how the silenced state is faithfully inherited. Initiation of inactivation requires the presence of an X inactivation center (XIC in humans, Xic in mice) in *cis*. Recently, a gene that maps to the XIC/Xic has been cloned from humans (XIST) and mice (Xist) (Brown et al 1992; Brockdorff et al 1992) XIST/Xist is expressed only from the inactive X. Furthermore, the XIST/Xist RNA lacks any conserved open reading frame and is localized in the nucleus rather than with the cytoplasmic translational machinery. These observations suggest that the XIST/Xist gene product encodes a functional RNA, although the role this RNA molecule plays in X inactivation is not understood.

Maintenance of the silent state of the inactivated X chromosome involves methylation of cytosine (C) residues, primarily at CpG sites (Hockey et al 1989; Singer-Sam et al 1990; Norris et al 1991). Such methylation may help prevent re-activation of the inactivated chromosome by interfering, directly or indirectly, with DNA binding by transcriptional activators (Watt & Molloy 1988; Peifer et al 1990ab; Boyes & Bird 1991; Pfeifer & Riggs 1991). This two-tier system of X inactivation presumably ensures complete transcriptional silencing.

The use of compacted chromatin to turn off transcription of blocks of genes appears to be universal in eukaryotes. Both *Drosophila* and *S. cerevisiae* display transcriptional position effects, wherein the expression of a gene is affected by its chromosomal location. Examination of this phenomenon has revealed that some regions of fly and yeast chromosomes are refractory to transcription. As with the inactivated X, these transcriptionally silent regions replicate late and are packaged into complex chromatin structures (for reviews, see Henikoff 1990; Rivier & Rine 1992; Sandell & Zakian 1992).

Studies of X-ray-induced chromosomal translocations in *Drosophila* have revealed that the compacted chromatin structures associated with silenced sequences can spread along DNA. That is, genes that are normally expressed can be silenced if they are translocated near a region of compacted chromatin.

Moreover, the expression of such translocated genes is often variable; some cells express the gene, other cells do not, which indicates that the compacted chromatin structures have spread to different boundary points in different cells. This phenomenon, known as position effect variegation, is particularly striking when the translocated gene encodes an eye pigmentation protein. In such cases, the *Drosophila* eye contains clusters of pigmented and unpigmented cells. The existence of these clusters indicates that, once the extent of chromatin spreading has been set, it is stably inherited during subsequent cell divisions. Furthermore, there is a stochastic component to the decision; exactly which cells express the pigment gene and which do not varies from eye to eye. Although the molecular basis for position effect variegation is not yet understood, the proteins and DNA sequences involved in transcriptional silencing in *Drosophila* and yeast are beginning to be characterized (reviewed in Henikoff 1990; Rivier & Rine 1992; Sandell & Zakian 1992).

Genomic Imprinting

The chromatin-mediated repression mechanisms described above apparently allow cells to maintain developmental decisions by permanently inactivating regions of the genome. Other mechanisms for long-term gene inactivation probably exist. In particular, the phenomenon of genomic imprinting, wherein the expression of a mammalian gene depends on whether it was inherited from the mother of from the father, seems not to involve large-scale changes in chromatin structure (Sasaki et al 1992). Rather, methylation of CpG dinucleotides appears to be involved in the imprinting process (reviewed in Bird 1993). Because mammalian cells contain a maintenance methylase that acts only on hemimethylated CpG sequences, DNA methylation patterns can be faithfully inherited upon DNA replication (reviewed in Razin et al 1984). Furthermore, the expressed and unexpressed copies of an imprinted gene are often differentially methylated (Bartolomei et al 1991; Chaillet et al 1991; Sasaki et al 1992; Ferguson-Smith et al 1993).

The maternal and paternal copies of the mouse *Igf2r* gene, which encodes a receptor for insulin-like growth factor, are differentially expressed (only maternally-derived copy is active) and are also differentially methylated. Two clusters of CpG sites in the *Igf2r* gene, one covering the gene promoter and one located within a downstream intron, display different methylation patterns depending on their paternal origin (Stöger et al 1993). Only the paternal transcriptionally inactive copy of the promoter CpG sequence is methylated. For at least one other imprinted gene, the *H19* gene, methylation of CpG dinucleotides correlates with gene inactivity (Bartolomei et al 1991). However, both the *H19* CpG sequences and the *Igf2r* promoter CpG sites are unmethylated in the sperm and therefore cannot be the original imprinting signal (Ferguson-Smith et al 1993; Stöger et al 1993). Rather, this methylation is thought to be involved in maintenance of the imprinted state.

The intronic CpG sequence in the *Igf2r* gene, on the other hand, is methylated on the maternal, transcriptionally active copy of the gene (Stöger et al 1993). Furthermore, this methylation is observed in the oocyte and may actually serve as the original imprinting signal. The molecular mechanism by which one copy of an imprinted gene is transcriptionally repressed while the other copy is transcriptionally active is not understood.

Global Repression by Inactivating a Component of the General Transcription Machinery

When eukaryotic cells enter mitosis, transcription by all three RNA polymerases is shut down, presumably to allow easier separation of segregating chromosomes (Prescott & Bender 1962; Fink & Turnock 1977; LH Johnston et al 1987; White et al 1987). Recently it was shown that one of the general transcription factors of RNA polymerase III, TFIIIB, is inactivated during mitosis, probably by phosphorylation (Hartl et al 1993; J. Gottesfeld et al, personal communication). Since TFIIIB is required for initiation of RNA polymerase III transcription (Kassavetis et al 1990), inactivation of this factor is presumed to repress all RNA polymerase III transcription during the mitotic phase of the cell cycle. The TATA-binding protein and its RNA polymerase III-specific TAFs are essential components of TFIIIB. At least one component of TFIIIB that is phosphorylated in a mitotic extract is the same size as a previously identified RNA polymerase III-specific TAF (J. Gottesfeld et al, personal communication). Perhaps inactivation of TBP/TAF complexes by phosphorylation could serve as a global repression mechanism to repress transcription by all three nuclear RNA polymerases during mitosis.

CONCLUSIONS AND PERSPECTIVES

In this review, we have proposed that for each step in the pathway to transcription there exists a repressor that blocks it, and have described several examples of transcriptional repressors that are known to affect one or more steps in the transcription initiation pathway. The transcription initiation process can be considered as a linear series of equilibrium reactions. According to this view, negative regulators could reduce the overall level of transcription by shifting the position of any individual equilibrium, or offering alternate, non-productive reaction paths (see Figure 4). However, studies of transcription reactions in vitro have suggested that a linear reaction path may not be the most appropriate model for the transcription initiation reaction. Only a small fraction of available DNA templates are active in

ATP NTPs
I ⇌ II ⇌ III ⇌ ⇌ preIC → transcription

Modes of repression

A. Shifting an equilibrium

ATP NTPs
I ⇌ II ⇌ III ⇌ ⇌ preIC → transcription
Repressor

B. Offering an alternate, nonproductive reaction path

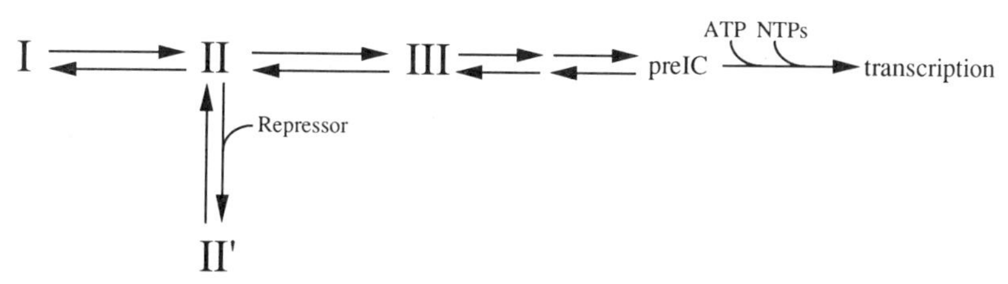

Figure 4 Transcription initiation as a linear series of equilibrium reactions. In this view, negative regulators could repress transcription either by shifting the position of any individual equilibrium, or by shunting assembling complexes off onto alternate, non-productive reaction paths.

typical in vitro transcription reactions, although most templates are bound in protein complexes (Hawley & Roeder 1987; Horikoshi et al 1988; Van Dyke et al 1988; Kadonaga 1990; Maldonado et al 1990). This observation suggests that the majority of assembling transcription complexes have branched off onto non-productive reaction pathways. Thus the transcription initiation reaction path may naturally contain branch points at which assembling transcription complexes partition between productive and non-productive forms (see Herschlag & Johnson 1993).

Whether negative regulators shift equilibria along a linear reaction pathway or influence partitioning ratios in a branched pathway, it is clear that there are many steps at which they can affect the initiation of transcription. In this review, we have described some of the known examples of molecular mechanisms used by transcriptional repressors. Further research will not only clarify the details of these mechanisms, but will also undoubtedly uncover new tactics used by repressors to block transcription initiation.

ACKNOWLEDGMENTS

The authors thank Burk Braun, Mike Chamberlin, Dan Herschlag, Kelly Komachi, Kevin Jarrell, Mark Ptashne, and Danny Reinberg for useful comments and suggestions.

Literature Cited

Alberts, B, Bray, D, Lewis, J, Raff, M, Roberts, K, Watson, JD, eds. 1983. The molecular organization of cells. Sect. 8, The cell nucleus. In *Molecular Biology of the Cell*. p. 440. New York: Garland

Almer, A, Rudolph, H, Hinnen, A, Horz, W. 1986. Removal of positioned nucleosomes from the yeast *PH05* promoter upon *PH05* induction releases additional upstream activating DNA elements. *EMBO J*. 5:2689–96

Alonso, MC, Cabrera, CV. 1988. The *achaete-scute* gene complex of *Drosophila melanogaster* comprises four homologous genes. *EMBO J*. 7:2585–91

Appel, B, Sakonju, S. 1993. Cell-type-specific mechanisms of transcriptional repression by the homeotic gene products UBX and ABD-A in *Drosophila* embryos. *EMBO J*. 12(3):1099–109

Ballard, DW, Walker, WH, Doerre, S, Sista, P, Molitor, JA, et al. 1990. The v-*rel* oncogene encodes a κB enhancer binding protein that inhibits NF-κB function. *Cell* 63:803–14

Banerji, J, Rusconi, S, Schaffner, W. 1981. Expression of a beta-globin gene is enhanced by remote SV40 DNA sequences. *Cell* 27:299–308

Bartolomei, MS, Zemel, S, Tilghman, SM. 1991. Parental imprinting of the mouse *H19* gene. *Nature* 351:153–55

Beg, AA, Ruben, SM, Scheinman, RI, Haskill, S, Rosen, CA, Baldwin, AS Jr. 1992. IκB interacts with the nuclear localization sequences of the subunits of NFκB: a mechanism for cytoplasmic retention. *Genes Dev*. 6:1899–913

Benezra, R, Cantor, CR, Axel, R. 1986. Nucleosomes are phased along the mouse β-major globin gene in erythroid and nonerythroid cells. *Cell* 44:697–704

Bengal, E, Flores, O, Krauskopf, A, Reinberg, D, Aloni, A. 1991. Role of the mammalian transcription factors IIF, IIS, and IIX during elongation by RNA polymerase II. *Mol. Cell. Biol*. 11:1195–1206

Bennett, V. 1992. Ankyrins: adaptors between diverse plasma-membrane proteins and the cytoplasm. *J. Biol. Chem*. 267:8703–6

Biggin, MD, Tjian, R. 1989. A purified Drosophila homeodomain protein represses transcription in vitro. *Cell* 58:433–40

Bird, AP. 1993. Genomic imprinting: imprinting on islands. *Curr. Biol*. 3:275–77

Blank, V, Kourilsky, P, Israel, A. 1992. NF-κB and related proteins: Rel/dorsal homologies meet ankyrin-like repeats. *Trends Biol. Sci*. 17:135–40

Boyes, J, Bird, A. 1991. DNA methylation inhibits transcription indirectly via a methyl-CpG binding protein. *Cell* 64:1123–34

Brockdorff, N, Ashworth, A, Kay, GF, McCabe, VM, Norris, DP, et al. 1992. The product of the mouse *XIST* gene is a 15 kb inactive X-specific transcript containing no conserved ORF and located in the nucleus. *Cell* 71:515–26

Brown, CJ, Hendrich, BD, Rupert, JL, Lafreniere, RG, Xing, Y, et al. 1992. The human *XIST* gene: analysis of a 17 kb inactive X-specific RNA that contains conserved repeats and is highly localized within the nucleus. *Cell* 71:527–42

Bull, P, Morley, KL, Hoekstra, F, Hunter, T, Verma, IM. 1990. The mouse c-*rel* protein has an N-terminal regulatory domain and a C-terminal transcriptional activation domain. *Mol. Cell. Biol*. 10:5473–85

Buratowski, S, Zhou, H. 1992. A suppressor of TBP mutations encodes an RNA polymerase III transcription factor with homology to TFIIB. *Cell* 71:221–30

Butlin, M, Quincey, R. 1991. The yeast rRNA gene enhancer does not function by recycling RNA polymerase I and cannot act as a UAS. *Curr. Genet*. 20:9–16

Cadena, DL, Dahmus, ME. 1987. Messenger RNA synthesis in mammalian cells is catalyzed by the phosphorylated form of RNA polymerase II. *J. Biol. Chem*. 262:12468–74

Campuzano, S, Balcells, L, Villares, R, Carramolino, L, García-Alonso, LA, Modolell, J. 1986. Excess function *hairy-wing* mutations caused by *gypsy* and *copia* insertions within structural genes of the *achaete-scute* complex of Drosophila. *Cell* 44: 303–12

Carles, C, Treich, I, Bouet, F, Riva, M, Sentenac, A. 1991. Two additional common subunits, ABC10α and ABC10β, are shared by yeast RNA polymerases. *J. Biol. Chem.* 266:24092–96

Carlson, M, Osmond, BC, Neigeborn, L, Botstein, D. 1984. A suppressor of SNF1 mutations causes constitutive high-level invertase synthesis in yeast. *Genetics* 107:19–32

Caudy, M, Grell, EH, Dambly-Chaudiére, C, Ghysen, A, Jan, LY, Jan, YN. 1988a. The maternal sex determination gene *daughterless* has zygotic activity necessary for the formation of peripheral neurons in *Drosophila*. *Genes. Dev.* 2:843–52

Caudy, M, Vässin, H, Brand, M, Tuma, R, Jan, LY, Jan, YN. 1988b. *Daughterless,* a Drosophila gene essential for both neurogenesis and sex determination, has sequence similarities to *myc* and the *achaete-schute* complex. *Cell* 55:1061–67

Chaillet, JR, Vogt, TF, Beier, DR, Leider, P. 1991. Parental-specific methylation of an imprinted transgene is established during gametogenesis and progressively changes during embryogenesis. *Cell* 66:77–83

Chamberlin, MJ. 1974. The selectivity of transcription. *Annu. Rev. Biochem.* 43:721–75

Chasman, DI, Kornberg, RD. 1990. Gal4 protein: purification, association with Gal80 protein, and conserved domain structure. *Mol. Cell. Biol.* 10:2916–23

Chesnut, JD, Stephens, JH, Dahmus, ME. 1992. The interaction of RNA polymerase II with the adenovirus-2 major late promoter is precluded by phosphorylation of the C-terminal domain of subunit IIa. *J. Biol. Chem.* 267:10500–6

Choy, HE, Adhya, S. 1992. Control of *gal* transcription through DNA looping: inhibition of the initial transcribing complex. *Proc. Natl. Acad. Sci. USA* 89:11264–8

Cline, TW. 1989. The affairs of *daughterless* and the promiscuity of developmental regulators. *Cell* 59:231–34

Colbert, T, Hahn, S. 1992. A yeast TFIIB-related factor involved in RNA polymerase III transcription. *Genes Dev.* 6:1940–49

Dambly-Chaudiére, C, Ghysen, A, Jan, LY, Jan, YN. 1988. The determination of sense organs in *Drosophila:* interaction of *scute* with *daughterless*. *Roux's Arch. Dev. Biol.* 197:419–23

Davis, N, Ghosh, S, Simmons, DL, Tempst, P, Liou, HC, et al. 1991. Rel associated pp40: an inhibitor of the Rel family of transcription factors. *Science* 253:1268–71

Davis, RL, Cheng, PF, Lassar, AB, Weintraub, H. 1990. The MyoD DNA binding domain contains a recognition code for muscle-specific gene activation. *Cell* 60: 733–46

DeLuca, NA, Schaffer, PA. 1988. Physical and functional domains of the herpes simplex virus transcriptional regulatory protein ICP4. *J. Virol.* 62:732–43

Dequard-Chablat, M, Riva, M, Carles, C, Sentenac, A. 1991. *RPC19,* the gene for a subunit common to yeast RNA polymerases A(I) and C(III). *J. Biol. Chem.* 266:15300–7

Descombes, P, Schibler, U. 1991. A liver-enriched transcriptional activator protein, LAP, and a transcriptional inhibitory protein, LIP, are translated from the same mRNA. *Cell* 67:569–79

Durrin, LK, Mann, RK, Grunstein, M. 1992. Nucleosome loss activates *CUP1* and *HIS3* promoters to fully induced levels in the yeast *Saccharomyces cerevisiae*. *Mol. Cell. Biol.* 12:1621–29

Ellis, HM, Spann, DR, Posakony, JW. 1990. Extramacrochaetae, a negative regulator of sensory organ development in Drosophila, defines a new class of helix-loop-helix proteins. *Cell* 61(1):27–38

Englesberg, E, Wilcox, G. 1974. Regulation: positive control. *Annu. Rev. Genet.* 8:219–42

Feaver, WJ, Gileadi, O, Li, Y, Kornberg, R. 1991. CTD kinase associated with yeast RNA polymerase II initiation factor b. *Cell* 67:1223–30

Ferguson-Smith, AC, Sasaki, H, Cattanach, BM, Surani, MA. 1993. Parental origin-specific epigenetic modification of the mouse *H19* gene. *Nature* 362:751–55

Ferre-D'Amare, AR, Prendergast, GC, Ziff, EB, Burley, SK. 1993. Recognition by Max of its cognate DNA through a dimeric b/HLH/z domain. *Nature* 363(6424):38–45

Fink, K, Turncock, G. 1977. Synthesis of transfer RNA during the synchronous nuclear division cycle in *Physarum polycephalum*. *Eur. J. Biochem.* 80:93–96

Flores, O, Ha, I, Reinberg, D. 1990. Factors involved in specific transcription by mammalian RNA polymerase II: purification and subunit composition of transcription factor IIF. *J. Biol. Chem.* 265:5629–34

Flores, O, Maldonado, E, Reinberg, D. 1989. Factors involved in specific transcription by mammalian RNA polymerase II: factors IIE and IIF independently interact with RNA polymerase II. *J. Biol. Chem.* 264:8913–21

Frasch, M, Levine, M. 1987. Complementary patterns of *even-skipped* and *fushi-taratzu* expression involve their differential regulation by a common set of segmentation genes in *Drosophila*. *Genes Dev.* 1:981–95

García-Alonso, LA, García-Bellido, A. 1986. Genetic analysis of the *hairy-wing* mutations. *Roux's Arch. Dev. Biol.* 197:328–38

García-Bellido, A, Santamaria, P. 1978. Developmental analysis of the *achaete-scute* system of *Drosophila melanogaster*. *Genetics* 88:469–86

Garrell, J, Modolell, J. 1990. The Drosophila extramacrochaetae locus, an antagonist of proneural genes that, like these genes, encodes a helix-loop-helix protein. *Cell* 61(1): 39–48

Geisler, R, Bergmann, A, Hiromi, Y, Nusslein-Volhard, C. 1992. *Cactus,* a gene involved in dorsoventral pattern formation of Drosophila, is related to the IκB gene family of vertebrates. *Cell* 71:613–21

Ghosh, S, Baltimore, D. 1990. Activation in vitro of NF-κB by phosphorylation of its inhibitor IκB. *Nature* 344:678–82

Ghosh, S, Gifford, AM, Rivere, LR, Tempst, P, Nolan, GP, Baltimore, D. 1990. Cloning of the p50 DNA binding subunit of NF-κB: homology to Rel and Dorsal. *Cell* 62:1017–29

Gill, G. 1992. Complexes with a common core. *Curr. Biol.* 2:565–67

Gilmour, DS, Lis, JT. 1986. RNA polymersae II interacts with the promoter region of the noninduced hsp70 gene in *Drosophila melanogaster* cells. *Mol. Cell. Biol.* 6: 3984–89

Giniger, E, Varnum, S, Ptashne, M. 1985. Specific DNA binding of Gal4, a positive regulatory protein of yeast. *Cell* 40:767–74

Goto, T, Macdonald, P, Maniatis, T. 1989. Early and late periodic patterns of *even-skipped* expression are controlled by distinct regulatory elements that respond to different spatial cues. *Cell* 57:413–22

Grant, S, Chapman, V. 1988. Mechanisms of X chromosome regulation. *Annu. Rev. Genet.* 22:199–233

Han, K, Levine, MS, Manley, JL. 1989. Synergistic activation and repression of transcription by Drosophila homeobox proteins. *Cell* 56:573–83

Han, M, Grunstein, M. 1988. Nucleosoms loss activates yeast downstream promoters in vivo. *Cell* 55:1137–45

Han, M, Kim, U-J, Kayne, P, Grunstein, M. 1988. Depletion of histone H4 and nucleosomes activates the *PH05* gene in *Saccharomyces cerevisiae*. *EMBO J.* 7: 2221–28

Hansen, U, Tenen, DG, Livingston, DM, Sharp, PA. 1981. T antigen repression of SV40 early transcription from two promoters. *Cell* 27:603–12

Harding, K, Hoey, T, Warrior, R, Levine, M. 1989. Auto-regulatory and gap response elements of the *even-skipped* promoter of *Drosophila*. *EMBO J.* 8:1205–12

Hartenstein, V, Posakony, JW. 1989. Development of adult sensilla on the wing and notum of *Drosophila melanogaster*. *Development* 107(2):389–405

Hartl, P, Gottesfeld, J, Forbes, DJ. 1993. Mitotic repression of transcription in vitro. *J. Cell. Biol.* 120:613–24

Haskill, S, Beg, AA, Tompkins, S, Morris, JS, Yurochko, AD, et al. 1991. Characterization of an immediate early gene induced in adherent monocytes that encodes IκB-like activity. *Cell* 65:1281–89

Hawley, DK, Johnson, AD, McClure, WR. 1985. Functional and physical characterization of transcription initiation complexes in the bacteriophage λ O_R region. *J. Biol. Chem.* 260:8618–26

Hawley, DK, Roeder, RG. 1987. Functional steps in transcription initiation and reinitiation from the major late promoter in a HeLa nuclear extract. *J. Biol. Chem.* 262:3452–61

Hayashi, S, Scott, MP. 1990. What determines the specificity of action of Drosophila homeodomain proteins? *Cell* 63:883–94

Henikoff, S. 1990. Position effect variegation after 60 years. *Trends Genet.* 6:422–26

Herschbach, BM, Johnson, AD. 1993. The yeast α2 protein can repress transcription by RNA polymerases I and II but not III. *Mol. Cell. Biol.* In press

Herschlag, D, Johnson, BF. 1993. Synergism in transcriptional activation: a kinetic view. *Genes Dev.* 7:173–79

Hockey, AJ, Adra, CN, McBurney, MW. 1989. Reactivation of *hprt* on the inactive X chromosome with DNA demethylating agents. *Somatic Cell Mol. Genet.* 14:421–34

Hoey, T, Weinzierl, RO, Gill, G, Chen, J-L, Dynlacht, BD, Tjian, R. 1993. Molecular cloning and functional analysis of Drosophila TAF110 reveal properties expected of coactivators. *Cell* 72:247–60

Horikoshi, M, Hai, T, Lin, Y-S, Green, MR, Roeder, RG. 1988. Transcription factor ATF interacts with the TATA factor to facilitate establishment of a preinitiation complex. *Cell* 54:1033–42

Horikoshi, N, Maguire, K, Kralli, A, Maldonado, E, Renberg, D, Weinmann, R. 1991. Direct interaction between adenovirus E1A protein and the TATA box binding transcription factor IIF. *Proc. Natl. Acad. Sci. USA* 88:5124–28

Ingles, CJ, Shales, M, Cress, WD, Triezenberg, S, Greenblatt, J. 1991. Reduced binding of TFIID to transcriptionally compromised mutants of VP16. *Nature* 351: 588–90

Inoue, J-I, Kerr, LD, Kakizuka, A, Verma, IM. 1992a. IκBY, a 70 kd protein identical to the C-terminal half of p110 NFκB: a new member of the IκB family. *Cell* 68:1109–20

Inoue, J-I, Kerr, LD, Rashid, D.. Davis. N, Bose, HR Jr, Verma, IM. 1992b. Direct association of pp40/IκBB with Rel/NF-κB transcription factors: role of ankyrin repeats in the inhibition of DNA binding activity. *Proc. Natl. Acad. Sci. USA* 89: 4333–37

Jacob, F, Monod, J. 1961 Genetic regulatory mechanisms in the synthesis of proteins. *J. Mol. Biol.* 3:318–56

Johnson, FB, Krasnow, MA. 1992. Differential regulation of transcription preinitiation complex assembly by activator and repressor homeo domain proteins. *Genes Dev.* 6:2177–89

Johnston, LH, White, JHM, Johnson, AL, Lucchini, G, Plevani, P. 1987. The yeast DNA polymerase I transcript is regulated both in the mitotic cell cycle and in meiosis and is also induced after DNA damage. *Nucleic Acids Res.* 15:5017–30

Johnston, M. 1987. A model fungal gene regulatory mechanism: the *GAL* genes of *Saccharomyces cerevisiae. Microbiol. Rev.* 51:458–76

Johnston, SA, Salmeron, SM, Dincher, SS. 1987. Interaction of positive and negative regulatory proteins in the galactose regulon of yeast. *Cell* 50:143–46

Kadonaga, JT. 1990. Assembly and dissasembly of the *Drosophila* RNA polymerase II complex during transcription. *J. Biol. Chem.* 265:2624–31

Kakkis, E, Calame, K. 1987. A plasmacytoma-specific factor binds the c-*myc* promoter region. *Proc. Natl. Acad. Sci. USA* 84:7031–35

Kakkis, E, Mercola, M, Calame, K. 1988. Strong transcriptional activation of translocated c-*myc* genes occurs without a strong nearby enhancer or promoter. *Nucleic Acids Res.* 16:77–96

Kakkis, E, Riggs, KJ, Gillespie W, Calame, K. 1989. A transcriptional repressor of c-*myc*. *Nature* 339:718–21

Kassevetis, GA, Braun, BR, Nguyen, LH, Geiduschek, EP. 1990. S. cerevisiae TFIIIB is the transcription initiation factor proper of RNA polymerase III, while TFIIIA and TFIIIC are assembly factors. *Cell* 60(2): 235–45

Kaufman, PD, Rio, DC. 1991. *Drosophila* P-element transposase is a transcriptional repressor in vitro. *Proc. Natl. Acad. Sci. USA* 88:2613–17

Keleher, CA, Redd, MJ, Schultz, J, Carlson, M, Johnson, AD. 1992. SSsn6-Tupl is a general repressor of transcription in yeast. *Cell* 68:709–19

Kerr, LD, Duckett, CS, Wamsley, P, Zhiang, Q, Chiao, P, et al. 1992. The proto-oncogene *BCL-3* encodes an IκB protein. *Genes Dev.* 6:2352–63

Kidd, S. 1992. Characterization of the Drosophila *cactus* locus and analysis of interactions between *cactus* and *dorsal* proteins. *Cell* 71:623–35

Kieran, M, Blank, V, Logeat, F, Vandekerckhove, J., Lottspeich, F, et al. 1990. The DNA binding subunit of NFκB is identical to factor KBF1 and homologous to the Rel oncogene product. *Cell* 62: 1007–18

Knezetic, JA, Jacob, GA, Luse, DS. 1988. Assembly of RNA polymerase II preinitiation complexes before assembly of nucleosomes allows efficient initiation of transcription on nucleosomal templates. *Mol. Cell. Biol.* 8:3114–21

Knezetic, JA, Luse, DS. 1986. The presence of nucleosomes on a DNA template prevents initiation by RNA polymerase II in vitro. *Cell* 45:95–104

Krummel, B, Chamberlin, MJ. 1989. RNA chain initiation by *Escherichia coli* RNA polymerase. Structural transitions of the enzyme in early ternary complexes. *Biochemistry* 28:7829–42

Lassar, AB, Buskin, JN, Lockshon, D, Davis, RL, Apone, S, et al. 1989. MyoD is a sequence-specific DNA binding protein requiring a region of *myc* homology to bind to the creatine kinase enhancer. *Cell* 58: 823–31

Laybourn, PJ, Dahmus, ME. 1990. Phosphorylation of RNA polymerase IIA occurs subsequent to interaction with the promoter and before the initiation of transcription. *J. Biol. Chem.* 265:13165–73

Lee, DK, Horikoshi, M, Roeder, RG. 1991. Interaction of TFIID in the minor groove of the TATA element. *Cell* 67:1241–50

Lee, H-S, Kraus, KW, Wolfner, MF, Lis, JT. 1992. DNA sequence requirements for generating paused polymerase at the start of *hsp70*. *Genes Dev.* 6:284–95

Lee, J, Goldfarb, A. 1991. lac repressor acts by modifying the initial transcribing complex so that it cannot leave the promoter. *Cell* 66:793–98

Lee, WS, Kao, CC, Bryant, GO, Liu, X, Berk, AJ. 1991. Adenovirus E1A activation domain binds the basic repeat in the TATA box transcription factor. *Cell* 67:365–76

Leuther, KK, Johnston, SA. 1992. Nondissociation of Gal4 and Gal80 gene regulatory proteins in vivo after galactose induction. *Science* 256:1333–35

Licht, JD, Grossel, MJ, Figge, J, Hansen, UM. 1990. *Drosophila* Krüppel protein is a transcriptional repressor. *Nature* 346:76–79

Lin, YS, Green, MR. 1991. Mechanism of action of an acidic transcriptional activator in vitro. *Cell* 64:971–81

Lin, YS, Ha, I, Maldonado, E, Reinberg, D, Green, MR. 1991. Binding of general transcription factor TFIIB to an acidic activating region. *Nature* 353:569–71

Link, E, Kerr, LD, Schreck, R, Zabel, U, Verma, I, Baeuerle, P. 1992. Purified IκB-6 is inactivated upon dephosphorylation. *J. Biol. Chem.* 267:239–46

Lohr, D, Hopper, JE. 1985. The relationship of regulatory proteins and DNAse I hypersensitive sites in the yeast *GAL1-10* genes. *Nucleic Acids Res.* 13:8409–23

López-De-León, A, Librizzi, M, Puglia, K, Willis, IM. 1992. *PCF4* encodes an RNA polymerase III transcription factor with homology to TFIIB. *Cell* 71:211–20

Lu, NF, Chasman, DI, Buchman, AR, Kornberg, RG. 1987. Interaction of Gal4 and Gal80 regulatory proteins in vitro. *Mol. Cell. Biol.* 7:3446–51

Lu, H, Flores, O, Weinmann, R, Reinberg, D. 1991. The nonphosphorylated form of RNA polymerase II preferentially associates with the preinitiation complex. *Proc. Natl. Acad. Sci. USA* 88:10004–8

Lu, H, Zawel, L, Fisher, L, Egly, J-M, Reinberg, D. 1992. Human general transcription factor IIH phosphorylates the C-terminal domain of RNA polymerase II. *Nature* 358:641–45

Lyon, MF. 1961. Gene action in the X-chromosome of the mouse (*Mus musculus L.*). *Nature* 190:372–73

Ma, J, Ptashne, M. 1987a. Deletion analysis of GALA defines two transcriptional activating segments. *Cell* 48:847–53

Ma, J, Ptashne, M. 1987b. The carboxy-terminal 30 amino acids of Gal4 are recognized by Gal80. *Cell* 50:137–42

Maldonado, E, Ha, I, Cortes, P, Weis, L, Reinberg, D. 1990. Factors involved in specific initiation by mammalian RNA polymerase II: role of transcription factors IIA, IID, and IIB during formation of a transcription-competent complex. *Mol. Cell. Biol.* 10:6335–47

Maniatis, T, Goodbourn, S, Fischer, JA. 1987. Regulation of inducible and tissue-specific gene expression. *Science* 236(4806):1237–45

Mann, C, Buhler, J-M, Treich, I, Sentenac, A. 1987. *RPC40,* a unique gene for a subunit shared between yeast RNA polymerases I and III. *Cell* 48:627–37

Matallana, E, Franco, L, Perez-Ortin, JE. 1992. Chromatin structure of the yeast *SUC2* promoter in regulatory mutants. *Mol. Gen. Genet.* 231:395–400

Matsui, T. 1987. Transcription of adenovirus 2 major late promoter and peptide IX genes under conditions of in vitro nucleosome assembly. *Mol. Cell. Biol.* 7:1401–8

Matsui, T, Segall, J, Weil, PA, Roeder, RG. 1980. Multiple factors required for accurate initiation of transcription by purified RNA polymerase II. *J. Biol. Chem.* 255:11992–96

McClure, WR. 1985. Mechanism and control of transcription initiation in prokaryotes. *Annu. Rev. Biochem.* 54:171–204

Moscoso del Prado, J, García-Bellido, A. 1984. Genetic regulation of the *achaete-scute* complex of *Drosophila. Roux's Arch. Dev. Biol.* 193:242–45

Mukai, Y, Harashima, S, Oshima, Y. 1991. Aar1/Tup1 protein with a similar structure to the β subunit of G proteins is required for α1-α2 and α2 repression in cell type control of *Saccharomyces cerevisiae. Mol. Cell. Biol.* 11:155–91

Murre, C, McCaw, PS, Baltimore, D. 1989a. A new DNA binding and dimerization motif in immunoglobulin enhancer binding, *daughterless, myoD,* and *myc* proteins. *Cell* 56:777–83

Murre, C, McCaw, PS, Vässin, H, Caudy, M, Jan, YN, et al. 1989b. Interactions between heterologous helix-loop-helix protein generate complexes that bind specifically to a common core sequence. *Cell* 58: 537–44

Mylin, LM, Bhat, JP, Hopper, JE. 1989. Regulated phosphorylation and dephosphorylation of Gal4, a transcriptional activator. *Genes Dev.* 3:1157–65

Mylin, LM, Johnston, M, Hopper, JE. 1990. Phosphorylated forms of Gal4 are correlated with ability to activate transcription. *Mol. Cell. Biol.* 10:4623–29

Nakabeppu, Y, Nathans, D. 1991. A naturally occurring truncated form of FosB that inhibits Fos/Jun transcriptional activity. *Cell* 64:751–59

Nikolov, DB, Hu, SH, Lin, J, Gasch, A, Hoffman, A, et al. 1992. Crystal structure of TFIID TATA-box binding protein. *Nature* 360:40–46

Nolan, GP, Ghosh, S, Liou, H-C, Tempst, P, Baltimore, D. 1991. DNA binding and IκB inhibition of the cloned p65 subunit of NFκB, a Rel-related polypeptide. *Cell* 64:961–69

Norris, DP, Brockdorff, N, Rastan, S. 1991. Methylation status of CpG-rich islands on active and inactive mouse X chromosomes. *Mamm. Genome* 1:78–83

Okhuma, Y, Horkoshi, M, Roeder, RG, Desplan, C. 1990. *Engrailed,* a homeodomain protein, can repress in vitro transcription by competition with the TATA box-binding protein transcription factor IID. *Proc. Natl. Acad. Sci. USA* 87:2289–93

Parthun, MA, Jaehning, JR. 1992. A transcriptionally active form of Gal4 is phosphorylated and associated with Gal80. *Mol. Cell. Biol.* 12:4981–87

Payne, JM, Laybourn, P-J, Dahmus, ME. 1989. The transition of RNA polymerase II from initiation to elongation is associated with phosphorylation of the carboxyl-terminal domain of subunit IIa. *J. Biol. Chem.* 264:19621–29

Pérez-Ortín, JE, Estruch, F, Matallana, E, Franco, L. 1987. An analysis of the chromatin structure of the yeast *SUC2* gene and of its changes upon derepression. Comparison between chromosomal and plasmid-inserted copies. *Nucleic Acids Res.* 15: 6937–56

Pfeifer, GP, Riggs, AD. 1991. Chromatin differences between active and inactive X chromosomes revealed by genomic footprinting of permeabilized cells using DNAse I and ligation-mediated PCR. *Genes Dev.* 5:1102–13

Pfeifer, GP, Steigerwald, SD, Hansen, RS, Gartler, SM, Riggs, AD. 1990a. Polymerase chain reaction-aided genomic sequencing of an X chromosome-linked CpG island: methylation patterns suggest clonal inheritance, CpG site autonomy, and an explanation of activity stability. *Proc. Natl. Acad. Sci. USA* 87:8252–56

Pfeifer, GP, Tanguay, RL, Steigerwald, SD, Riggs, AD. 1990b. In vivo footprint and methylation analysis by PCR-aided genomic sequencing: comparison of active and inactive X chromosomal DNA at the CpG island and promoter of human *PGK-1*. *Genes Dev.* 4:1277–87

Prescott, DM, Bender, MA. 1962. Synthesis of RNA and protein during mitosis in mammalian tissue culture cells. *Exp. Cell. Res.* 26:260–68

Price, DH, Sluder, AE, Greenleaf, AL. 1989. Dynamic interaction between a *Drosophila* transcription factor and RNA polymerase II. *Mol. Cell. Biol.* 9:1465–75

Ptashne, M. 1986. *A Genetic Switch.* Cambridge, Mass: Cell Press

Rappaport, J, Reinberg, D, Zandomeni, R, Weinmann, R. 1987. Purification and functional characterization of transcription factor IIS from calf thymus. *J. Biol. Chem.* 262: 5227–32

Rastan, S, Brown, SDM. 1990. The search for the mouse chromosome X-inactivation centre. *Genet. Res.* 56:99–106

Razin, A, Cedar, H, Riggs, AD, eds. 1984. DNA methylation. In *Biochemistry and Biological Significance.* New York: Springer-Verlag

Reinberg, D, Roeder, RG. 1987. Factors involved in specific transcription by RNA polymerase II. Transcription factor IIS stimulates elongation of RNA chains. *J. Biol. Chem.* 262:3331–37

Rigby, PWJ. 1993. Three in one and one in three: it all depends on TBP. *Cell* 72:7–10

Riggs, KJ, Merrell, KT, Gillespie, W, Calame, K. 1991. Common factor I is a transcriptional activator which binds in the c-*myc* promoter, the skeletal alpha-actin promoter, and the immunoglobulin heavy-chain enhancer. *Mol. Cell. Biol.* 11:1765–69

Rijcke, R, Seneca, S, Punyammalee, B, Glansdorff, N, Crabeel, M. 1992. Characterization of the DNA target site for the yeast ARGR regulatory complex, a sequence able to mediate repression or induction by arginine. *Mol. Cell. Biol.* 12:68–81

Rivier, DH, Rine, J. 1992, Silencing: the establishment and inheritance of stable, repressed transcription states. *Curr. Opin. Genet. Dev.* 2:286–92

Roberts, MS, Boundy, A, O'Hare, P. Pizzorno, MC, Ciufo, DM, Hayward, GS. 1988. Direct correlation between a negative autoregulatory response element at the cap site of the herpes simplex virus type IE175 ($\alpha 4$) promoter and a specific binding site for the IE175 (ICP4) protein. *J. Virol.* 62: 4307–20

Romani, S, Campuzano, S, Moldolell, J. 1987. The *achaete-scute* complex is expressed in neurogenic regions of *Drosophila* embryos. *EMBO J.* 6:2085–92

Ron, D, Habener, JF. 1992. CHOP, a novel developmentally regulated nuclear protein that dimerizes with transcription factors CIEBP and LAD and functions as a dominant negative inhibitor of gene transcription. *Genes Dev.* 6:439–53

Roth, SY, Dean, A, Simpson, RT. 1990. Yeast $\alpha 2$ repressor positions nucleosomes in TRP1/ARS1 chromatin. *Mol. Cell. Biol.* 10:2247–60

Roth, SY, Shimizu, M, Johnson, L, Grunstein, M, Simpson, RT. 1992. Stable nucleosome positioning and complete repression by the yeast $\alpha 2$ repressor are disrupted by amino-terminal mutations in histone H4. *Genes Dev.* 6:411–25

Rougive, AE, Lis, JT. 1988. The RNA polymerase II molecule at the 5′ end of the uninduced *hsp70* gene of Drosophila melanogaster is transcriptionally engaged. *Cell* 54:795–804

Salmeron, JM Jr, Leuther, KK, Johnston, SE. 1990. GAL4 mutations that separate the transcriptional activation and Gal80 interactive functions of the yeast Gal4 protein. *Genetics* 125:21–27

Sandell, LL, Bakian, VA. 1992. Telomeric position effect in yeast. *Trends Cell Biol.* 2:10–14

Sasaki, H, Jones, PA, Challiet, JR, Ferguson-Smith, AC, Barton, SC, et al 1992. Parental imprinting: potentially active chromatin of the repressed maternal allele of the mouse insulin-like growth factor (*Igf2*) gene. *Genes Dev.* 6:1843–56

Sawadogo, M, Roeder, RS. 1984. Energy

requirement for specific initiation by the human RNA polymerase II system. *J. Biol. Chem.* 259:5321–26

Schaeffer, L, Roy, R, Humbert, S, Moncollin, V, Vermeulen, W, et al. 1993. DNA repain helicase: a component of BTF2 (TFIIH) basic transcription factor. *Science* 260:58–63

Schleif, R. 1992. DNA looping. *Annu. Rev. Biochem.* 61:199–223

Sentenac, A. 1985. Eukaryotic RNA polymerases. *Crit. Rev. Biochem.* 18:31–91

Serizawa, H, Conaway, RC, Conaway, JW. 1992. A carboxyl-terminal-domain kinase associated with RNA polymerase II transcription factor δ from rat liver. *Proc. Natl. Acad. Sci. USA* 89:7476–80

Shalloway, D, Kleinburger, T, Livingston, DM. 1980. Mapping of SV40 replication origin binding sites for the SV40 T antigen by protection against exonuclease III digestion. *Cell* 20:411–22

Sharp, PA. 1992. TATA-binding protein is a classless factor. *Cell* 68:819–21

Shepard, AA, Imbalzano, AN, DeLuca, NA. 1989. Separation of primary structural components conferring autoregulation, transactivation, and DNA-binding properties to the herpes simplex virus transcriptional regulatory protein ICP4. *J. Virol.* 63:3714–28

Shimizu, M, Roth, SY, Szent-Yorgi, C, Simpson, RT. 1991. Nucleosomes are positioned with base pair precision adjacent to the α2 operator in *Saccharomyces cerevisiae*. *EMBO J.* 10:3033–41

Singer-Sam, J, Grant, M, LeBon, JM, Okuyama, K, Chapman, V, et al. 1990. Use of a *Hpaii*-polymerase chain reaction assay to study DNA methylation in the *Pgk-1* CpG island of mouse embryos at the time of X-chromosome inactivation. *Mol. Cell. Biol.* 10:4987–89

Small, S, Blair, A, Levine, M. 1992. Regulation of *even-skipped* stripe 2 in the Drosophila embryo. *EMBO J.* 11:4047–57

Small, S, Kraut, R, Hoey, T, Warrior, R, Levine, M. 1991. Transcriptional regulation of a pair-rule stripe in *Drosophila*. *Genes Dev.* 5:827–39

Sopta, M, Burton, Z, Greenblatt, J. 1989. Structure and associated DNA-helicase activity of a general transcription initiation factor that binds to RNA polymerase II. *Nature* 341:410–14

Stanojevic, D, Hoey, T, Levine, M. 1989. Sequence-specific DNA-binding activities of the gap proteins encoded by *hunchback* and *Krüppel* in *Drosophila*. *Nature* 341: 331–35

Starr, BD, Hawley, DK. 1991. TFIID binds in the minor groove of the TATA box. *Cell* 67:1231–40

Steward, R. 1987. *Dorsal,* an embryonic polarity gene in *Drosophila,* is homologous to the vertebrate proto-oncogene, c-*rel*. *Science* 238:692–94

Stöger, R, Kubicka, P, Liu, C-G, Kafri, T, Razin, A, et al. 1993. Maternal-specific methylation of the imprinted mouse *Igf2r* locus identifies the expressed locus as carrying the imprinting signal. *Cell* 73:61–71

Stringer, KF, Ingles, CJ, Greenblatt, J. 1990. Direct and selective binding of an acidic transcription activation domain to the TATA-box factor TFIID. *Nature* 345:783–86

Sundseth, R, Hansen, UM. 1992. Activation of RNA polymerase II transcription by the specific DNA-binding protein LSF. Increased rate of binding of the basal promoter factor TFIIB. *J. Biol. Chem.* 267: 7845–55

Tewari, M, Dobrzanski, P, Mohn, KL, Cressman, DE, Hsu, J-C, et al. 1992. Rapid induction in regenerating liver of RL/IF-I (an IκB that inhibits NF-κB, RelB-p50, and c-Rel-p50) and PHF, a novel κB site-binding complex. *Mol. Cell. Biol.* 12:2898–908

Tjian, R. 1978. The binding site on SV40 DNA of a T antigen-related protein. *Cell* 13:165–79

Tjian, R. 1981. T antigen binding and the control of SV40 gene expression. *Cell* 26:1–2

Trumbly, RJ. 1986. Isolation of *Saccharomyces cerevisiae* mutants constitutive for invertase synthesis. *J. Bacteriol.* 9:809–16

Usheva, A, Maldonado, E, Goldring, A, Lu, H, Houbavi, C, et al. 1992. Specific interaction between the nonphosphorylated form of RNA polymerase II and the TATA-binding protein. *Cell* 69:871–81

Van Doren, M, Ellis, HM, JW. Posakony. 1991. The *Drosophila extramachrochaetae* protein antagonizes sequence-specific DNA binding by *daughterless/achaete-scute* protein complexes. *Development* 113:245–55

Van Dyke, MW, Roeder, RG, Sawadogo, M. 1988. Physical analysis of transcription preinitiation complex assembly on a class II gene promoter. *Science* 241:1335–38

Vershon, AK, Liao, SM, McClure, WR, Sauer RT. 1987. Interaction of the bacteriophage P22 Arc repressor with operator DNA. *J. Mol. Biol.* 195:323–31

Villares, CR, Cabrera, CV. 1987. The *achaete-scute* gene complex of D. melanogaster: conserved domains in a subset of genes required for neurogenesis and their homology to *myc*. *Cell* 50:415–24

Voronova, A, Baltimore, D. 1990. Mutations that disrupt DNA binding and dimer formation in the E47 helix-loop-helix protein map to distinct domains. *Proc. Natl. Acad. Sci. USA* 87:4722–26

Wasylyk, B, Chambon, P. 1979. Transcription by eukaryotic RNA polymerases A and B of chromatin assembled in vitro. *Eur. J. Biochem.* 98:317–27

Wasylyk, B, Chambon, P. 1980. Studies on the mechanism of transcription of nucleosomal complexes. *Eur. J. Biochem.* 103:219–26

Watt, F, Molloy, PL. 1988. Cytosine methylation prevents binding to DNA of a HeLa cell transcription factor required for optimal expression of the adenovirus major late promoter. *Genes Dev.* 2:1136–43

Weeda, G. van Ham, RC, Masurel, R, Westerveld, A, Idijk, H, et al. 1990. Molecular cloning and biological characterization of the human excision repair gene *ERCC-3*. *Mol. Cell. Biol.* 10:2570–81

White, JHM, Green, SR, Barker, DG, Dumas, BL, Johnston, LH. 1987. The *CDC8* transcript is cell cycle regulated in yeast and is expressed coordinately with *CDC9* and *CDC21* at a point preceding histone transcription. *Exp. Cell. Res.* 171:223–31

White, RJ, Jackson, SP. 1992. The TATA-binding protein: a central role in transcription by RNA polymerases I, II, and III. *Trends Genet.* 8:284–88

Workman, JL, Roeder, RG. 1987. Binding of the transcription factor TFIID to the major late promoter during in vitro nucleosome assembly potentiates subsequent initiation by RNA polymerase II. *Cell* 55:613–22

Woychik, NA, Liao, S-M, Kolodziej, PA, Young, RA. 1990. Subunits shared by eukaryotic nuclear RNA polymerases. *Genes Dev.* 4:313–23

Zabel, U, Baeurle, PA. 1990. Purified human IκB can rapidly dissociate the complex of the NF-κB transcription factor with its cognate DNA. *Cell* 61:255–65

Zawel, L, Reinberg, D. 1993. Initiation of transcription by RNA polymerase II: a multi-step process. *Prog. Nucleic. Acids Res. Mol. Bio.* 44:68–108

Zuo, P, Stanojevic, D, Colgan, J, Han, K, Levine, M, Manley, JL. 1991. Activation and repression of transcription by the gap proteins *hunchback* and *Krüppel* in cultured *Drosophila* cells. *Genes Dev.* 5:254–64

Annu. Rev. Cell Biol. 1993. 9:511–540

MOLECULAR ASPECTS OF MESENCHYMAL-EPITHELIAL INTERACTIONS

C. Birchmeier
Max-Delbrück-Laboratorium in der Max-Planck-Gesellschaft, Carl-von-Linne-Weg 10, 5000 Köln 30, Germany

W. Birchmeier[1]
Institut für Zellbiologie (Tumorforschung), University of Essen Medical School, Virchowstrasse 173, 4300 Essen 1, Germany

KEY WORDS: organogenesis, extracellular matrix, receptor tyrosine kinases, transcription factors, cell adhesion molecules

---

CONTENTS

[1]After September 1: Max-Delbrück Centrum, Robert-Rössle Strasse 10, D–13122, Berlin, Germany

0743–4634/93/1115–0511$05.00

INTRODUCTION

Epithelia and mesenchymes are two distinct types of tissues found in virtually every organ: epithelia are composed of closely associated, largely immobile cells; in contrast, mesenchyme contains more mobile cells that form loosely associated agglomerations. During normal development, transitions of epithelia to mesenchyme can occur, and mesenchyme can differentiate into new epithelia. Such transitions are not confined to development. In particular, the loss of epithelial character in malignant carcinomas, which results in the appearance of invasive, motile cells, is of major importance in tumor progression (Birchmeier et al 1993).

In the early mammalian embryo, all cells are uniform and have the same developmental potential. At the blastula stage, only a single cell type resembling epithelia exists. During gastrulation, the three germinal cell layers, ectoderm, endoderm, and mesoderm, are formed. Whereas ectoderm and endoderm remain largely epithelial in character, the newly induced mesoderm is the first example in ontogenesis of conversion of epithelial cells to mesenchyme. Mesoderm formation has been studied extensively and the inductive signals have been elucidated in *Xenopus laevis*. In this review, we do not discuss this or other examples of the developmentally regulated conversion of epithelia into mesenchyme since the subject has been reviewed elsewhere (Gurdon 1987; Valles et al 1991; Green & Smith 1991; Moon & Christian 1992; Stern 1992; Jessell & Melton 1992).

During further embryogenesis, organs develop and often cells derived from more than one germinal layer contribute to their formation. In general, two major cell types are found in developing organs: mesenchyme, which derives from mesoderm or neural ectoderm, and epithelia, which can originate from all three germinal cell layers. An example of morphogenic interaction between epithelia and mesenchyme is found during development of the lung: outgrowth of epithelial cells from the endoderm into mesenchyme, the splanchnic mesoderm, generates the anlagen of the lung. All epithelia develop by further growth, branching and differentiating from the first endodermal buds. In parallel, the mesenchymal cell compartment also expands and differentiates (Spooner & Wessells 1970). This principle, i.e. epithelia that grow and branch in response to mesenchymal signals, is a theme found in the development of many other organs such as salivary gland, pancreas, prostate, pituitary, kidney, and breast (Grobstein 1953; Wessells & Cohen 1976; Lasnitzki & Mizuno 1980; Kusakabe et al 1985; Saxen 1987; Sakakura 1991). In this review we discuss the molecular nature of signals given by the mesenchyme and received by epithelia, which then respond by morphogenesis, differentiation, and growth during organ development.

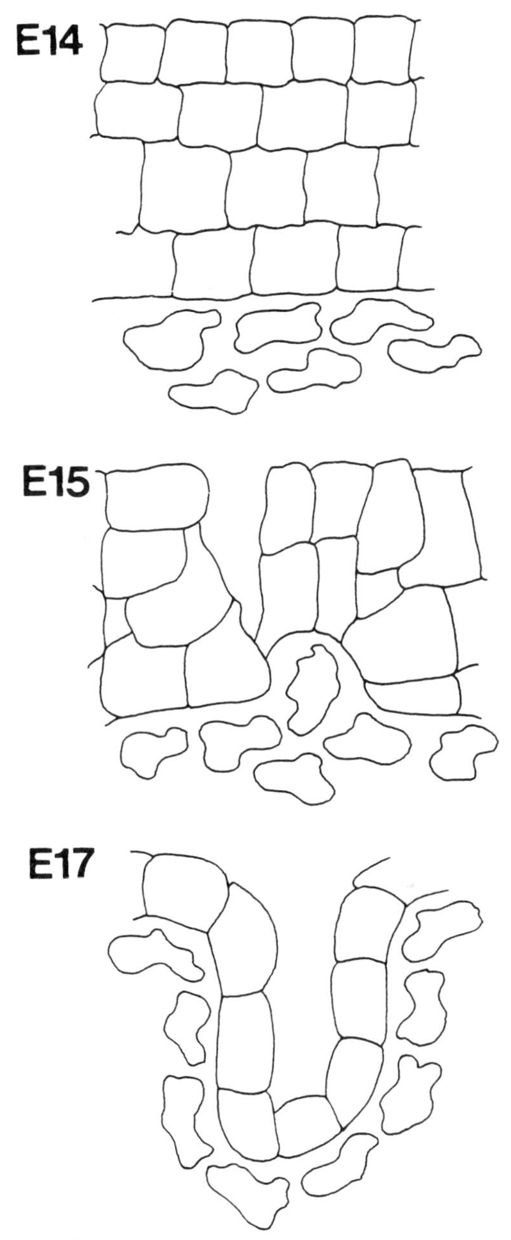

Figure 1 The appearance of the murine intestine at the indicated times in development is depicted schematically (modified after Mathan et al 1976). The cuboidal cells and the irregularily shaped cells represent epithelia and mesenchyme, respectively. During early stages of development, a multi-layered, poorly differentiated epithelium is observed; mesenchymal cells are located basally (E14). With the onset of terminal differentiation, the contacts between the epithelial cells are loosened and mesenchymal cells invade the epithelial cell layers (E15). The differentiated villi epithelium is single-layered (E16).

In kidney development, the epithelial ureter, a derivative of the Wolffian duct, invades the nephrogenic mesenchyme. In this case, both the epithelia and the mesenchyme derive from mesoderm. Then, the mesenchyme stimulates further growth and branching of the ureter epithelium, which differentiates into the collecting ducts. In contrast to the lung, not all epithelia of the kidney develop from these branching epithelia. Instead, inductive signals given by the tips of the growing ureter buds lead to the formation of epithelia from nephrogenic mesenchyme. The new epithelia differentiate into proximal and distal tubules as well as glomeruli (Saxen 1987; Ekblom 1992). This is an instance where conversion of mesenchyme to epithelia is observed.

The intestinal anlagen are formed early in development and consist of a poorly differentiated, stratified epithelium derived from the endoderm, which is surrounded by intestinal mesenchymal cells. During terminal differentiation, which occurs late in development, this multilayered epithelium is converted into the single-layered, columnar epithelium of the villi. The first signs of this differentiation process are the dissociation of the cell-cell contacts and the formation of secondary lumina in the multilayered epithelium (Figure 1). The mesenchymal cells then invade the loosened epithelium. Whereas degenerating superficial epithelial cells are exfoliated, the residual epithelial cells along the entire length of the newly formed villi proliferate (Mathan et al 1976), thereby providing a developmental model wherein one type of epithelium is converted into another type, a process guided by mesenchymal-epithelial interactions (Kedinger et al 1986). Mesenchymal-epithelial signaling also plays a role in the establishment of regional specificities in intestinal development (Yasugi 1993).

In these examples of organ development, the ordered growth, morphogenesis, and differentiation of epithelia and mesenchyme are essential and, in order to ensure this, communication between the two tissue types must occur. The first experimental evidence for such paracrine signal exchange came from organ culture and transplantation experiments.

Organ Culture and Transplantation Experiments

Organogenesis can be studied in vitro by the removal of the organ anlagen at the appropriate time in development and subsequent culture. For example, the murine kidney anlagen can be explanted on day 11 of embryogenesis, when the ureter has just invaded the nephrogenic mesenchyme, or on day 12, after the first branching of the ureter has occurred. During subsequent incubation in vitro, the ureter grows and branches, and the ureter buds induce the formation of new epithelia from the mesenchyme (Grobstein 1956; Saxen 1987). Alternatively, the ureter epithelium of the kidney anlagen can be dissociated manually, or by mild trypsinization from the surrounding mesenchyme and can be cultivated separately. The epithelia then flatten and spread, and no signs of branching morphogenesis are observed; after recombination with the mesenchymal cells, the epithelium grows, branches, and differentiates (Grobstein 1956; Saxen 1987). Heterologous epithelia and mesenchyme can be combined in such experiments, and it was thus demonstrated that the requirements of the epithelia from different organs vary. Since virtually every mesenchyme can support morphogenesis of epithelia from the pancreas primordia, the ureter epithelia of the kidney are absolutely dependent on kidney mesenchyme for growth and morphogenesis (Fell & Grobstein 1968; Saxen 1987). In certain cases, the mesenchymal factors not only permit growth and morphogenesis, but also give instructive signals that determine the further differentiation program of the epithelia. For example, the epithelial lung buds can be instructed to form not only bronchial epithelia, but also gastric glands (in the presence of stomach mesenchyme), villi epithelia (by intestinal mesenchyme), or hepatic cords (by liver mesenchyme) (Deuchar 1975). Mesenchymal contributions to epithelial differentiation are therefore complex, and it is obvious that many factors are involved.

Transplantation experiments show that the instructions sent by mesenchymal tissues can cross species barriers (e.g. Spemann's transplantation of presumptive oral ectoderm between frog and newt; Spemann & Schotte 1932). Extraordinarily, when epithelium from jaw-forming regions of chick embryos were combined with molar mesenchyme from mouse and grown at an ectopic site, teeth developed that were unlike mammalian teeth. Thus the chick epithelium has retained the potential to form teeth, a capability that it had not used for 100 million years (Kollar & Fisher 1980).

Nature of Paracrine Signals

It is evident from the previous considerations that paracrine factors originating from mesenchymal cells are responsible for the induction of growth, morphogenesis, and differentiation of the neighboring epithelial cells. Three modes of signal transduction have been suggested: (*a*) interactions mediated by direct cell-cell contact; (*b*) interactions mediated by the extracellular matrix; and (*c*) diffusion of soluble factors (Grobstein 1956; Saxen 1987; Barcellos-Hoff et al 1989; Hay 1990; Sakakura 1991; Trautman et al 1991; Montesano et al 1991b; Ekblom 1992; Hirai et al 1992). For example, direct cell-cell interactions between inducing mesenchymal and responding epithelial cells are observed during mammary gland development; interruptions of the basement membrane occur at the tips of the mammary sprouts on day 17–18 of embryonal development in rats, and cytoplasmic processes from both the mesenchymal and epithelial sides project through these discontinuities (Sakakura 1991). Functionally important direct cell-cell contacts appear to be responsible for the induction of new epithelia during kidney development (cf Saxen 1987, for a review). Transfilter experiments have shown, however, that direct cell-cell interactions are not essential for the mesenchymal induction of salivary gland morphogenesis (Takahashi & Nogawa 1991).

Signaling throught components of the extracellular matrix may occur during development of the tooth, the skin, and breast (Vainio et al 1992a; Streuli et al 1991; Hirai et al 1992). Epimorphin (Hirai et al 1992) is a particularly well characterized mesenchymal factor essential for morphogenesis of skin epithelia, which is a component of the cell surface and/or the extracellular matrix. The protein was identified by the use of monoclonal antibodies that inhibited the induction of hair follicles in embryonic skin epithelia by aggregated mesenchymal cells of the dermis. Epimorphin (M_r < 50.000) is strongly associated with collagen IV, and by this modification the apparent molecular weight of the protein raises to 150 K (Hirai et al 1993). Matrigel, the isolated extracellular matrix from Engelbreth-Holmes-Swarm sarcoma cells (EHS) (Kleinman et al 1986) contains epimorphin. The functional and molecular characteristics of epimorphin are discussed in detail below. Components of the extracellular matrix such as collagens and, in particular, the matrix of the EHS tumor, can support branching morphogenesis to quite a remarkable degree. For example, mouse mammary epithelia grown on collagen-coated tissue culture dishes form duct-like structures and respond to lactogenic hormones by the production of milk proteins. In collagen gels and EHS matrices, three-dimensional tubular and alveolar networks are formed (Sakakura et al 1991; Streuli et al 1991; Aggeler et al 1991; Kanazawa et al 1992). Sertoli cells plated in the EHS matrix form cord structures in which basal and luminal compartments are separated by tight junctions. Germ cells carried over during preparation move from basal regions to the luminal side of the cords while they differentiate (Hadley et al 1985).

However, detailed transfilter experiments have shown that, besides matrix components, morphogenesis requires soluble mesenchymal factors, for instance during salivary gland development (Takahashi & Nogawa 1991). Soluble factors have been postulated to participate in accurate histotypic morphogenesis of other organs as well (Grobstein 1956; Taub et al 1990; Montesano et al 1991b). Montesano and colleagues recently identified such a mesenchymal factor when they induced MDCK (kidney) epithelial cells to grow and branch within collagen matrices. The factor, produced by various fibroblast cell lines, was found to be identical to scatter factor/hepatocyte growth factor (SF/HGF; Montesano et al 1991a), a motility and growth factor for various epithelial cells (Stoker et al 1987; Weidner et al 1993). Its receptor has been identified as the product of the c-*met* protooncogene, a receptor-type tyrosine kinase (Bottaro et al 1991; Naldini et al 1991). The biology of SF/HGF and its receptor is discussed below in more detail. In addition, other tyrosine kinase receptors, originally identified because of their oncogenic potential, have recently been characterized in more detail and found to be expressed predominantly or exclusively in epithelia, e.g. keratinocyte growth factor receptor (Miki et al 1991), c-*ros* (E Sonnenberg et al 1991), c-*ret* (F Constantini et al personal communication) and c-*neu* (Press et al 1990). Direct and indirect evidence suggest that these receptors mediate paracrine signals given by mesenchyme and received by epithelia in embryogenesis. The c-*ros* receptor, whose ligand is as yet not identified, might even mediate a signal triggered by direct cell-cell contact since this has been demonstrated for the *Drosophila* homologue of c-*ros, sevenless* (Krämer et al 1991; see also below). These tyrosine kinase receptors and their ligands are discussed in detail below.

MESENCHYMAL AND EPITHELIAL CELLS

Structural Characteristics

Mesenchymal cells involved in mesenchymal-epithelial interactions are generally nonpolarized, loosely associated, and surrounded by extracellular matrix. Fibroblasts in culture can be considered representative of such mesenchymal cells: their adhesive contacts to the matrix are mediated primarily by cell-substrate adhesion molecules. For instance, fibroblasts express various integrin receptors of which some (e.g. $\alpha5\beta1$, a fibronectin receptor) are concentrated in the so-called focal contacts (Hynes 1992). Fibroblasts also produce various ligands for epithelial-specific receptor tyrosine kinases, e.g. SF/HGF, keratinocyte growth factor, and *neu* differentiation factor (see below).

Epithelial cells are morphologically entirely different; they form continuous cell layers, and they are generally polar and much less mobile (Simons & Fuller 1985; Rodriguez-Boulan & Nelson 1989). In single-layered epithelia (e.g. the mature intestine), apical and basolateral cell surfaces are separated by tight junctions. An example of a multilayered epithelium is the skin where basal cells (stem cells) are covered by layers of gradually differentiating cells. Epithelial cells are interconnected by adherens junctions and desmosomes, organelles responsible for strong intercellular adhesion (Tsukita et al 1992; Buxton & Magee 1992; Birchmeier et al 1993), and they form basement membranes at their basal site (Timpl 1989). Hemidesmosomes are responsible for cell contacts to the basement membrane (A Sonnenberg et al 1991).

CELL ADHESION MOLECULES IN THE BASEMENT MEMBRANE The basement membrane of epithelia represents a laminar strucure that can be divided, at the electron microscopy level, into a lamina lucida (close to the epithelial basal plasma membrane) and a lamina densa. A reticular lamina of various thickness connects the basement membrane with the mesenchyme. The characteristic constituents of the epithelial basement membrane are laminin, collagen IV, nidogen/entactin, and basement membrane proteoglycan (Timpl 1989). The mesenchymal reticular lamina contains collagen I and III as well as fibronectin. The epithelial receptors for basement membrane components are located in the basal plasma membrane, e.g. the integrins $\alpha 6\beta 1$ or $\alpha 6\beta 4$ (A Sonnenberg et al 1991), which are receptors for laminin, and the integrin $\alpha 1\beta 1$, which is a receptor for collagen (Hynes 1992). The hemidesmosomes are enriched for the integrin receptor $\alpha 6\beta 4$ (A Sonnenberg et al 1991).

CELL ADHESION MOLECULES IN THE INTERCELLULAR JUNCTIONS OF EPITHELIA
Adherens junctions are specialized regions of the epithelial plasma membranes where transmembrane cadherin molecules located on opposing cells contact each other (cf Tsukita et al 1992; Birchmeier et al 1993; for a recent review on desmosomes, see Buxton & Magee 1992). The cytoplasmic portion of E-cadherin is associated with a group of proteins, catenins, that make contact with the microfilament network through unknown linkage proteins (Ozawa et al 1989). Two catenins have recently been molecularly characterized: α-catenin shows structural similarity to fibroblast vinculin, which is also located in the focal contacts (Herrenknecht et al 1991; Nagafuchi et al 1991), and β-catenin is homologous to the *Drosophila* gene product *armadillo* (McCrea et al 1991; Butz et al 1992). γ-Catenin is possibly related or identical to the desmosomal protein plakoglobin (Knudsen & Wheelock 1992). Alterations in E-cadherin or the catenins can result in defects of the epithelial adherens junctions and have been studied recently by in vitro mutagenesis and by the analysis of non-adhesive carcinoma cells (Nagafuchi et al 1991; Ozawa et al 1990; Kintner 1992; Hirano et al 1992; Shimoyama et al 1992; Behrens et al 1991, 1989; Frixen et al 1991; Schipper et al 1991).

EXTRACELLULAR MATRIX AND CELL ADHESION MOLECULES IN MESENCHYMAL-EPITHELIAL INTERACTIONS

Epimorphin in Skin Development

A recent achievement in the molecular analysis of mesenchymal-epithelial interactions is the discovery of epimorphin (Hirai et al 1992). The starting point of this study was the observation that dermal mesenchymal cells of mouse skin have a strong potential to induce hair follicle formation in co-cultured epidermal epithelium (Jahoda et al 1984). The mesenchymal cells need to be in aggregates; cells cultured as monolayers have no inducing capacity. Hirai and colleagues immunized rats with dermal cells cultured as pellets and selected a monoclonal antibody (MC-1) that inhibits hair follicle formation. The antigen recognized by MC-1 is present in dermal cells cultured as pellets, but not on cells grown as monolayers. Antibody MC-1 recognizes a 150-kd protein (epimorphin) that is expressed in the mesenchymal compartment of skin, lung, intestine, and kidney, and is located on the cell surface and/or in the extracellular space. On day 13 of mouse development, highest expression is seen in mesenchymal cell condensates in front of the developing epithelia of the hair follicles.

With the help of the antibody MC-1, the epimorphin cDNA was isolated; consequent characterization of the cDNA predicted a protein with a M_r of 34 K after cleavage of the N-terminal signal sequence. The putative C-terminus contains a stretch of 23 hydrophobic amino acid residues, which suggests that epimorphin can be associated with the plasma membrane. A second, secreted isoform without the C-terminal sequences can be synthesized from an alternatively spliced mRNA. Expression of the cDNA in COS-1 cells results in the formation of recombinant epimorphin with an apparent M_r of 150 K. Remarkably, expression of the epimorphin cDNA in monolayer cultures of NIH3T3 cells enables them to induce hair follicle formation in co-cultured skin epidermis (Hirai et al 1992). Further analysis revealed that epimorphin is strongly (possibly covalently) associated with collagen IV; collagenase treatment reduced the M_r to < 50 K. The 150 K epimorphin is also present in the EHS matrix (Hirai et al 1993; cf also Pelham 1993; Spring et al 1993). Apparently, epimorphin is not one of the components involved in tissue-specific mesenchymal-epithelial interactions, since the component also induces lung bud branching and is expressed in many mesenchymes (Hirai et al 1992). These experiments are thus a convincing example of a molecular component of the cell surface and/or the extracellular matrix that is an important induction factor in epithelial morphogenesis.

Cell Adhesion Molecules

Cell adhesion molecules involved in epithelial morphogenesis have been extensively studied (cf Sakakura 1991; Ekblom 1992; Vainio & Thesleff 1992b; Birchmeier et al 1993). During conversion of mesenchyme to epithelia in kidney development, cells in the undifferentiated mesenchyme are surrounded by extracellular matrix containing types I and III collagen, as well as fibronectin, and they express the neural cell adhesion molecule (N-CAM) on their surfaces. With epithelial differentiation, collagen I and III, as well as N-CAM, are lost (Klein et al 1988a; Ekblom 1992). The cells then begin to express epithelium-specific cell adhesion molecules such as the laminin A chain, type IV collagen, basement membrane proteoglycan, α6β1 integrin, E-cadherin, and desmosomal proteins (Vestweber et al 1985; Klein et al 1988b; Garrod & Fleming 1990; Sorokin et al 1990). Interestingly, before morphological alterations are observed in the induced cells, the B1- and B2-chains, but not the A-chain of laminin, were found to be expressed. During condensation, the amount of A-chain increases 15-fold, and the presence of this chain apparently allows the formation of functionally active laminin, which is essential for epithelial morphogenesis. Thus antibodies that interfere with laminin function or binding to the α6β1 integrin were found to inhibit differentiation (Klein et al 1988b; Ekblom et al 1990; Sorokin et al 1990).

The cell adhesion molecule tenascin is a mesenchymal protein specifically expressed during mesenchymal-epithelial interactions, e.g. in the development of tooth (Chiquet-Ehrismann et al 1986), kidney (Aufderheide et al 1987), and gut (Aufderheide & Ekblom 1988). In the mammary gland, tenascin is expressed selectively in the dense mesenchyme surrounding the epithelial buds in the embryonic and juvenile stages, disappears from the mesenchyme in postnatal and adult stages, but reappears in the stroma of breast carcinomas (Mackie et al 1987; Kusakabe et al 1988; Chiquet-Ehrismann 1993). In vitro, embryonic mammary gland tissue induces tesnascin in surrounding feeder cells. When mammary carcinoma cells are injected into nude mice, tenascin is also induced in the neighboring connective tissue (Inaguma et al 1988). However, a functional role of tenascin during mesenchymal-epithelial interactions needs to be demonstrated and, furthermore, mice develop normally without tenascin following inactivation of the gene by homologous recombination (Saga et al 1992). A further cell adhesion molecule, syndecan, also undergoes remarkable changes of expression during mesenchymal-epithelial interactions in tooth, kidney, and mammary gland development (Vainio et al 1989a,b; Leppä et al 1992). Syndecan is expressed in ectodermal epithelia, but is lost when these cells differentiate into lens, nasal, and otic structures. During this period, syndecan transiently appears in the apposed condensing mesenchyme. When lost from mesenchyme, it reappears in the epithelial structures (Trautman et al 1991). It is apparent from these studies that the cell adhesion molecules mentioned are not involved in direct induction processes (e.g. as epimorphin); rather, their expression is modulated in response to these events. The cell adhesion molecules examined are not specific for the development of a particular organ; it is thus the proper combination and the defined time-course of expression that is characteristic for individual development processes.

TYROSINE KINASES AND THEIR LIGANDS: Role in Mesenchymal-Epithelial Interactions

Several tyrosine kinase receptors with exclusive or prevalent expression on epithelial cells have been characterized recently: c-*met*, c-*ros*, c-*ret*, c-*neu*, and the keratinocyte growth factor receptor (Kokai et al 1987; Press et al 1990; E Sonnenberg et al 1991, 1993a,b; F Constantini et al, personal communication). In addition, the corresponding ligands that are primarily synthesized by mesenchymal cells and act mainly or exclusively on epithelial cells were recently identified: SF/HGF, *neu* differentiation factor (NDF), and keratinocyte growth factor (KGF) (Stoker et al 1987; Miki et al 1991; Peles et al 1992; Weidner et al 1993). Since these receptors were first discovered because of their transforming potential (Cooper et al 1984; Schechter et al 1984; Takahashi et al 1985; Birchmeier et al 1986), they have commonly been associated with mediating mitogenic signals. However, it has recently become evident that they also give signals that direct differentiation, movement, or morphogenesis, i.e. these receptors are able to regulate decisive events in epithelial development. The evidence pointing toward a role of these ligand-receptor systems in mesenchymal-epithelial interactions is dicussed below.

The c-met *Receptor and Its Ligand, Scatter Factor/Hepatocyte Growth Factor*

The *met* gene was originally identified in a transfection/tumorigenicity assay (Blair et al 1982; Cooper et al 1984). This isolate, the product of a chromosomal rearrangement, encodes a truncated receptor tyrosine kinase with fused sequences derived from an additional gene, *tpr* (translocated promoter region) (Park et al 1986; Dean et al 1987). Similar translocations between c-*met* on chromosome 7 and *tpr* on chromosome 1 are frequently found in human gastric carcinomas (Soman et al 1991). The protooncogene encodes a transmembrane glycoprotein with a M_r of 190 K (Figure 2), which is posttranslationally cleaved into two subunits of 145 (β) and 50 K (α). The C-terminal part of the β chain contains the tyrosine kinase domain and is located intracellularly, whereas the N-terminal part of the β and the entire α chain are exposed on the cell surface (Gonzatti-Haces et al 1988; Giordano et al 1989).

The specific ligand for the c-*met* receptor has recently been identified: it is SF/HGF, a 90-kd secreted glycoprotein (Bottaro et al 1991; Naldini et al 1991). It consists of disulfide-linked heavy (H) and light (L) chains, which are generated by proteolytic cleavage from a single precursor molecule (Figure 2). The H-chain contains an N-terminal hairpin structure and four

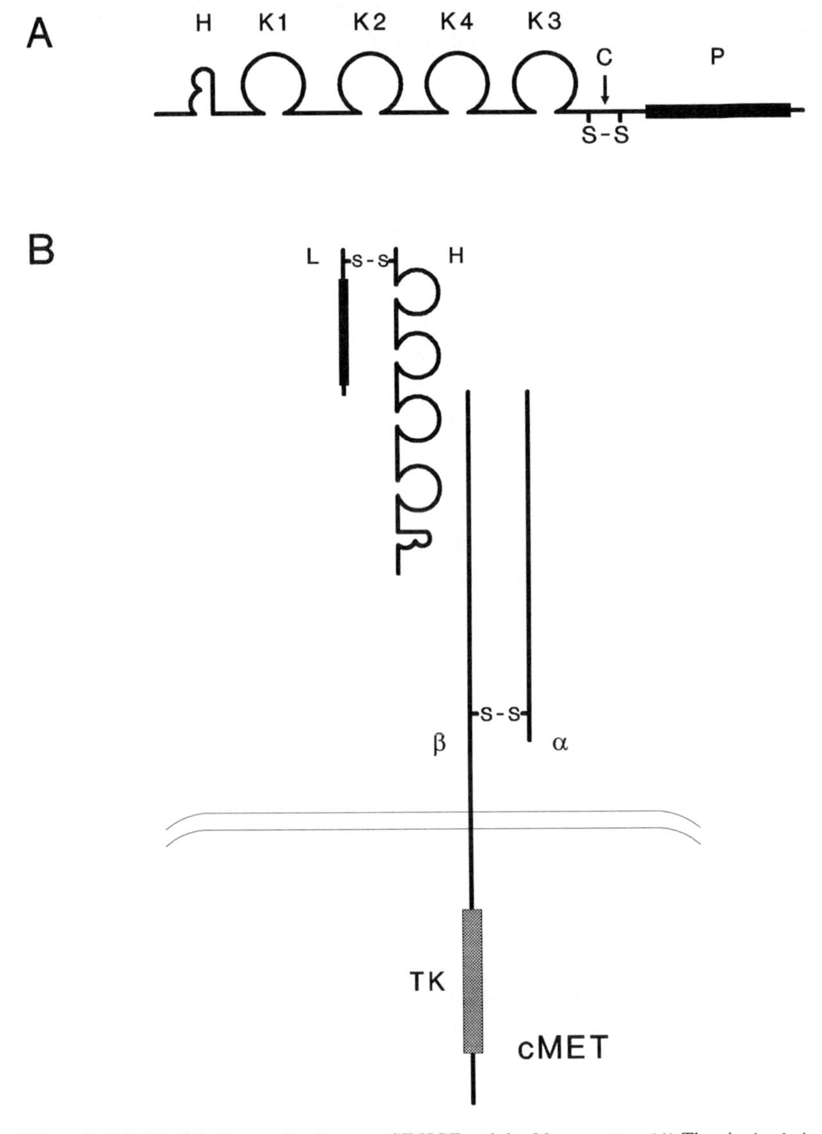

Figure 2 Model of the interaction between SF/HGF and the *Met* receptor. (*A*) The single chain, unprocessed SF/HGF (H, hairpin structure; K1–4, four kringle domains; P, protease homology domain; C, protease cleavage site between heavy and light chain) is activated into the heterodimeric molecule (*B*), which interacts with the *Met* receptor. Our data indicate that the N-terminal part of SF/HGF, consisting of a hairpin structure and two kringle domains, binds to the extracellular domain of the β-chain of *Met* (Hartmann et al 1992). SS, disulfide bridges between heavy (H) and light (L) chain of SF/HGF, and between α- and β-chain of *Met*; tyrosine kinase domain of *Met* is boxed (from Birchmeier et al 1993).

characteristic kringle domains; the L-chain has extensive homology to serine proteases. However, amino acids in the active site, which are essential for the catalytic activities of serine proteases, are absent in SF/HGF (Miyazawa et al 1989; Nakamura et al 1989; Weidner et al 1991). SF/HGF thus shows unique features; its structure is similar to proteases like plasminogen, but not to other known ligands for receptor-type tyrosine kinases. Two distinct activities, the ability to induce growth or movement of epithelial cells, have been used to independently isolate and molecularly characterize the factor (Gherardi et al 1989; Miyazawa et al 1989; Nakamura et al 1989; Zarnegar & Michalopoulos 1989; Weidner et al 1990, 1991; Rosen et al 1990; Rubin et al 1991; Hartmann et al 1992). SF/HGF can also increase invasiveness of epithelial cells and acts as a cytostatic factor on certain other cells (Weidner et al 1990; Higashio et al 1990; Shiota et al 1992).

The recent finding that SF/HGF is an inducer of epithelial tubulogenesis in vitro is an illustrative example of how factors involved in mesenchymal-epithelial interactions can be discovered. MDCK epithelial cells derive from the kidney and have functional properties similar to cells of distal tubules or collecting ducts. They grow slowly when cultured in collagen matrices and form cyst-like structures with smooth surfaces. Montesano et al (1991b) cultured these cells in collagen gels together with various fibroblasts, or in the presence of conditioned medium from fibroblasts. Under these conditions, the MDCK cells rapidly proliferated and formed complex networks of branching tubules. The single-layered epithelia of the tubules were well polarized; the cells expressed apical microvilli and junctional complexes (Figure 3). The paracrine fibroblast factor was then found to be identical to SF/HGF (Montesano et al 1991a). SF/HGF also stimulates tubulogenesis of liver and mammary gland epithelial cells. Outgrowth of tubules in the presence of SF/HGF can be prevented by the addition of a collagenase inhibitor, but not by a serine protease inhibitor (conversely, tubule formation in fibrin gels can be prevented by inhibitors of serine proteases, but not collagenases), which also suggests a role of proteases in epithelial morphogenesis under these conditions (Montesano et al 1991b; see below).

In situ hybridization analysis demonstrated that during mouse development c-*met* is expressed in many epithelia, whereas transcripts for the ligand are preferentially found in nearby mesenchymal cells (Sonnenberg et al 1993a,b). In particular, c-*met* is expressed in tubular epithelia, for example, in the branching epithelia of kidney, lung, salivary glands, and pancreas. In contrast, SF/HGF expression is found in the mesenchyme surrounding the tubular epithelia. Interestingly, in the developing kidney c-*met* transcripts are detected in the ureter as well as in the proximal and distal tubules, i.e. in all tubular epithelia, the ligand is expressed in undifferentiated mesenchyme (Figure 4a-c). Since SF/HGF can induce tubular structures in vitro

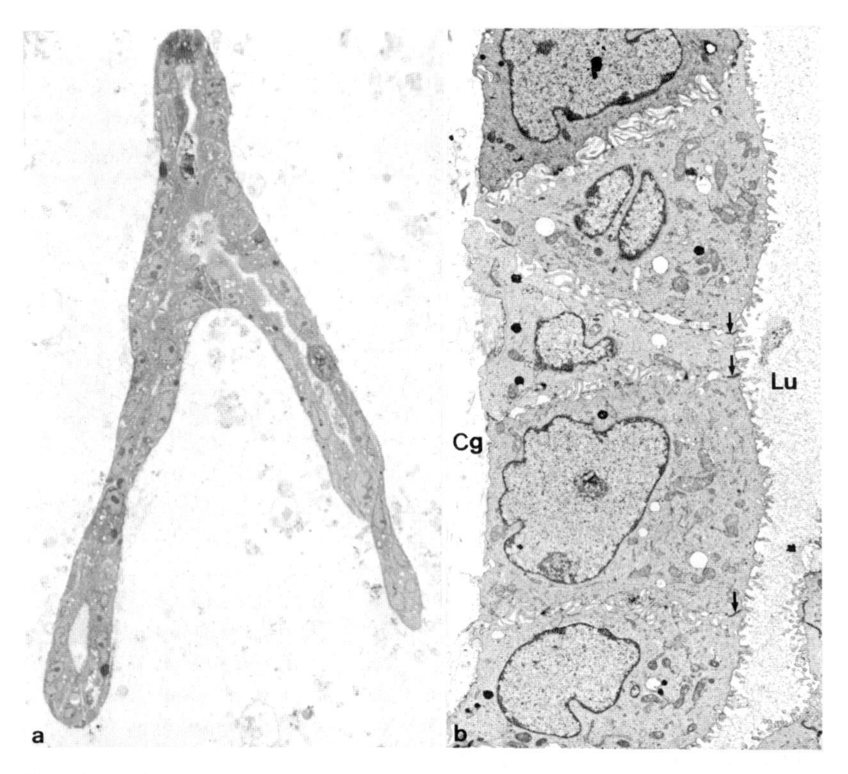

Figure 3 (*a*) Section of branching tubules formed by MDCK epithelial cells in collagen matrix co-cultured for 16 days with Swiss 3T3 fibroblasts (which produce SF/HGF). (*b*) The MDCK cells in these tubules are well polarized with a smooth basal surface in contact with the collagen matrix (Cg) and a microvilli-rich apical surface tracing the lumen (Lu) of the tubules. Note also the presence of apical junctional complexes (*arrows*). In the absence of exogenous or fibroblast-produced SF/HGF, no tubules are formed. Magnifications in (*a*) × 650, in (*b*) × 4400 (from Montesano et al 1991b, with permission).

(Montesano et al 1991b), the observed expression pattern is consistent with a function in epithelial tubulogenesis during development. Expression of the c-*met* receptor is also found in non-tubular epithelia; for example in the developing gut, the nasal cavities, or the teeth; SF/HGF expression is typically found in distinct mesenchymal cells in close vicinity (Sonnenberg et al 1993a,b). For example, during development of the intestine, the c-*met* receptor is expressed in the epithelia of all stages. During terminal differentiation, when the multilayered epithelium dissociates and mesenchyme invades, strong SF/HGF expression is found in distinct mesenchymal patches corresponding to the invading cells. Since a prominent activity of SF/HGF

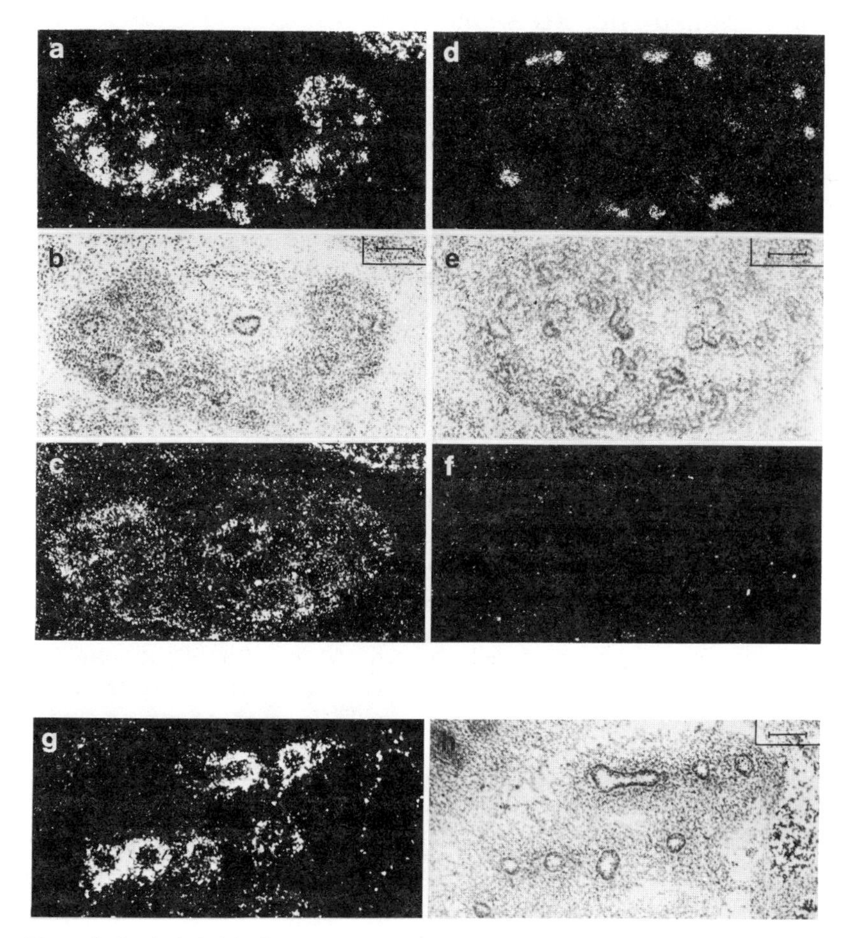

Figure 4 In situ hybridization analysis of tyrosine kinase receptors and ligands during murine organogenesis. Sections of embryos on day 13 showing the embryonal kidney (*a-f*) and on day 12 showing the embryonal lung (*g,h*) were hybridized to c-*met* (*a*); SF/HGF (*c*); c-*ros* (*d*); and *neu* differentiation factor (*g*) antisense probes. (*f*) is a control hybridized with a c-*ros* sense probe. The organs are shown in dark-field (*a,c,d,f,g*) and bright-field. Bars, 100 μm.

in vitro is the dissociation of adherent epithelial sheets (Stoker et al 1987), the observed loosening of the epithelial cells might be a direct consequence of SF/HGF production, which allows the underlying mesenchyme to penetrate. Thus SF/HGF and c-*met* might constitute a paracrine signaling system during organ development, a concept originally proposed by Stoker and his colleagues based on observations in cell culture (Stoker et al 1987).

*The c-*ros *and c-*ret *Receptors*

The first isolates of *ros* were transforming variants, which encode truncated tyrosine kinase receptors without *ros*-specific extracellular domains (Neckameyer & Wang 1985; Birchmeier et al 1986). The protein product of the *ros* protooncogene is an unusually large receptor-type tyrosine kinase, with a predicted M_r of 260,000, that shares extensive sequence and structural similarities with *sevenless* from *Drosophila* (Birchmeier et al 1990; Matsushime & Shibuya 1990; Chen et al 1991). These similarities indicate that *ros* and *sevenless* are homologous genes in vertebrates and invertebrates, respectively. However, the exact physiological roles of *ros* and *sevenless* have clearly diverged during evolution: whereas *sevenless* plays a role in differentiation and morphogenesis of the *Drosophila* eye (Basler & Hafen 1988), *ros* is implicated to function in the development of epithelia. During mouse embryogenesis, *ros* is expressed in a transient manner in epithelia of the urogenital tract, the intestine, and the lung (E Sonnenberg et al 1991). Most revealing in terms of a potential function is the pattern in the developing kidney, where *ros* transcripts are found only at the tips of the ureter located in the cortex (Figure 4d-f). Branching of the ureter is indeed only observed at these tips. The tips of the ureter are separated from the kidney capsule by 5–15 cell layers of uninduced mesenchyme, and it has been postulated that only these stem cells produce factors that stimulate growth and branching of the ureter (cf Ekblom 1992, for a review). However, c-*ros* is an orphan tyrosine kinase receptor; the specific ligand is as yet not known. Thc ligand for the *Drosophila* homologue *sevenless* is the protein product of *bride of sevenless* (*boss*), which is predicted to span the membrane seven times (Hart et al 1990). The *sevenless* tyrosine kinase and its membrane-bound ligand thus mediate a signal that is triggered by direct contact of neighboring cells (Krämer et al 1991). If the ligand of *ros* is also membrane-bound, it is expected to be produced in the mesenchyme, which is in direct contact with the *ros*-expressing epithelia.

Like *met* and *ros,* the gene for the *ret* receptor was originally identified because of its oncogenic potential. The transforming variant encodes a truncated tyrosine kinase receptor created by a DNA rearrangement (Takahashi et al 1985; Takahashi & Cooper 1987); chromosomal rearrangements involving the c-*ret* locus can be detected in human thyroid tumors (Wajjwalku et al 1992). The specific ligand for the receptor is as yet unknown. Expression of the *ret* protooncogene is regulated in the adult and embryonal organism, and specific transcripts can be detected in the embryonal, but not the adult, kidney. The expression pattern in the developing kidney is identical to the one observed for c-*ros,* i.e. specific transcripts are found at the tips of the branching ureter buds. They are thus confined to the same subset of ureter cells that receive mesenchymal signals for branching and give signals inducing the conversion of mesenchyme to epithelia (V Pachnis, A Schuchardt, F Constantini, personal communication). A targeted mutation was recently introduced into the murine c-*ret* locus by homologous recombination. Animals carrying two mutant alleles show severe hypoplasia or even aplasia of the kidneys (V Pachnis et al, personal communication). Therefore, the phenotypic consequences of mutant *ret,* together with the expression pattern of the intact gene, point toward a role of the receptor in mesenchymal-epithelial interactions during kidney development.

Other Receptor-type Tyrosine Kinases

THE *neu* RECEPTOR AND ITS LIGAND The *neu* gene was also isolated because of its oncogenic potential; it encodes a tyrosine kinase receptor with strong sequence similarity to the epidermal growth factor (EGF) receptor (Schechter et al 1984). The gene was isolated independently because it cross-hybridizes with EGF receptor sequences and was named EGF receptor 2 or *erbB2* (*erbB* is an oncogenic variant of the EGF receptor; Coussens et al 1985; Yamamoto et al 1986). The mechanism that activates the oncogenic potential of *neu* is unique because a point mutation in the sequence encoding the transmembrane domain can render the gene transforming (Bargman et al 1986). Amplification of the *neu* gene has been observed in carcinomas of different tissues, and this finding appears to correlate with a poor prognosis in patients with breast and ovarian cancers (Slamon et al 1987, 1989). After an elaborate search, the putative ligand of the *neu* receptor was recently identified (Peles et al 1992) and the corresponding cDNA sequenced (Wen et al 1992; Holmes et al 1992). This ligand is a soluble protein (M_r 43 K) that contains immunoglobulin and EGF sequence domains. A single internal hydrophobic sequence exists that might function as an internal signal peptide for translocation across the endoplasmic reticulum. Besides mitogenic activity, the factor can also induce differentiation of mammary carcinoma cells in culture. It was named *neu* differentiation factor (NDF) to emphasize this fact; an alternative name is heregulin (HER) (Wen et al 1992; Holmes et al 1992). The *neu* receptor is expressed predominantly in epithelia, both in the embryo and the adult, as well as on neuronal cells (Kokai et al 1987; Quirke et al 1989; Mori et al 1989; Press et al 1990). Our studies show that the ligand is expressed in mesenchyme surrounding embryonal epithelia, e.g. during lung development (Figure 4 g,h), and in neural cells (D Meyer & C Birchmeier, unpublished observations). In the embryo, the *neu* receptor and its ligand could therefore also mediate signals given by mesenchyme and received by epithelia.

KERATINOCYTE GROWTH FACTOR AND ITS RECEPTOR KGF was identified as a potent mitogenic factor for keratinocytes; it is a member of the fibroblast growth factor (FGF) family (Rubin et al 1989). However, the biological activities of KGF are distinct from other FGFs because it is not mitogenic for fibroblasts or endothelial cells, but is specific for epithelial cells. KGF is expressed in the mesenchymal compartment of embryonal and adult organs, as has been demonstrated by the analysis of cell lines derived from such sources. Targeting expression of KGF to keratinocytes elicits striking changes in epithelial differentiation in transgenic mice (Guo et al 1993). The specific receptor for KGF was identified using an elegant approach. Since NIH3T3 fibroblasts express KGF, but not the corresponding receptor, stable transfection with a keratinocyte cDNA expression library and consequent selection for foci were used to isolate a cDNA clone that encodes a high affinity KGF receptor. Subsequent molecular characterization demonstrated that the encoded protein is a member of the FGF tyrosine kinase receptor family (Miki et al 1991). This isolate corresponds to an alternatively spliced product of the human K-*sam* and the mouse *bek* gene, which were isolated from stomach carcinoma cells through gene amplification and from expression libraries with anti-phosphotyrosine antibodies, respectively (Kornbluth et al 1988; Hattori et al 1990). Besides KGF, this receptor isoform also binds acidic fibroblast growth factor (aFGF). Interestingly, aFGF induces motility and dissociation of rat bladder carcinoma cells in tissue culture (Jouanneau et al 1991).

The expression of the *bek* gene was analyzed during mouse embryogenesis. A probe was used that recognizes the mRNA for the KGF receptor as well as other alternatively spliced isoforms. A characteristic preference for expression in epithelia of ectodermal origin was observed, although some mesodermal cells also express the gene (Orr-Utreger et al 1991). The observed specificity of epithelial response to KGF in vitro, together with the finding that embryonal cells of mesenchymal origin produce the factor, make this receptor-ligand system a good candidate for mediating signal exchange between mesenchymal and epithelial cell compartments during embryogenesis.

OTHER COMPONENTS INVOLVED IN MESENCHYMAL-EPITHELIAL INTERACTIONS

Transforming Growth Factor beta (TGFβ)

In vitro, the members of the TGFβ family can exert mitogenic effects on various mesenchymal cells while they inhibit the proliferation of epithelial and endothelial cells and stimulate the production of extracellular matrix (cf Barnard et al 1990, for a review). The functional TGFβ receptors characterized so far are transmembrane serine/threonine-specific protein kinases (cf Massagué 1992). The various reports on possible roles of TGFβ in mesenchymal-epithelial interactions are somewhat diverse. During murine lung development, TGFβ1 was found to be expressed in bronchial mesoderm, TGFβ2 was expressed exclusively in epithelia, and TGFβ3 expression switched from mesenchyme to epithelia and then disappeared (Schmid et al 1991; see also Heine et al 1987; Pelton et al 1990). Other investigators report TGFβ expression exclusively in epithelia associated with active morphogenesis, e.g. in salivary glands and tooth buds (Millan et al 1991). In the mouse mammary gland, TGFβ1 and 3 were found to be expressed in the glands of young, mature, and pregnant, but not lactating, animals (Robinson et al 1991; Silberstein et al 1992). In principle, during mesenchymal-epithelial interaction the various TGFβ family members could give autocrine and paracrine signals leading to growth inhibition and differentiation of epithelial compartments on the one hand, and to the formation of extracellular matrix and the mitotic stimulation of mesenchymal cells on the other hand.

Proteases and Their Inhibitors

During branching morphogenesis, both the basement membrane and the matrix of the mesenchymal stroma must be degraded by secretory proteases in a directed manner to allow invasion of epithelial cells. Generally, branching morphogenesis can be inhibited in organ culture or in vitro by the addition of protease inhibitors (Nakanishi et al 1986; Montesano et al 1991b). The proteases can be produced by the branching epithelia, but examples of mesenchymally derived proteases are known as well. Recently, detailed expression analyses have been conducted on proteases during tumor progression. The mRNA for the 72-kd type IV collagenase is present in the stroma adjacent to the invasive nodules of skin carcinomas, but not in the carcinoma tissue itself (Pyke et al 1992). Stromelysin mRNA is specifically expressed in the stroma of breast cancer (Basset et al 1990). Similarly, urokinase-type plasmogen activator (u-PA) mRNA is expressed around the foci of colon adenocarcinomas, whereas the u-PA receptor mRNA is confined to the carcinoma tissue (Pyke et al 1991). Thus mesenchymally derived proteases bound to their receptors on epithelial cells could allow directed proteolysis, i.e. at the invasion fronts of carcinomas, and potentially also during epithelial morphogenesis in development. For instance, the 72-kd type IV collagenase is specifically expressed during tooth development (Reponen et al 1992; Sahlberg et al 1992). In the bell stage, the basement membrane is degraded, and this could allow signal exchange between odontoblasts and ameloblasts. Accordingly, expression of the collagenase preceded the disappearance of the basement membrane. Coordinated expression of matrix-degrading proteinases and their inhibitors also regulates the involution of the mammary epithelium (Talhouk et al 1992).

TRANSCRIPTION FACTORS AND MESENCHYMAL-EPITHELIAL INTERACTIONS

Inductive mesenchymal-epithelial interactions are the first steps in signaling cascades that induce differentiation events, which lead to changes in gene expression. The expression of several (putative) transcription factors seems to be restricted to or enhanced in cells that respond to such inductive signals during organogenesis.

N-*myc* is a member of a small gene family that includes c-*myc* and L-*myc* (Vennstrom et al 1982; Kohl et al 1983; Schwab et al 1983; Nau et al 1985). These genes encode nuclear proteins with site-specific DNA-binding activity (Blackwell et al 1990; Halazonetis & Kandil 1991; Kerkhoff et al 1991; Alex et al 1992; Ma et al 1992) enhanced by the formation of heterodimeric complexes with *max* or *myn* (Blackwood & Eisenman 1991; Prendergast et al 1991; Ma et al 1992). Whereas c-*myc* expression has been associated with cellular proliferation (Kelly et al 1983) and is found in a broad range of embryonal and adult cell types (Müller et al 1982; Pfeiffer-Ohlsson et al 1985; Zimmermann et al 1986; Schmid et al 1989; Semsei et al 1989), N-*myc* is expressed primarily in the developing embryo (Jakobowitz et al 1985; Mugrauer et al 1988; Downs et al 1989; Hirning et al 1991). By in situ hybridization, N-*myc* transcripts were localized to subsets of cells at early stages of differentiation; downregulation of the gene often correlates with terminal differentiation. In particular, N-*myc* is expressed in the primitive mesoderm, in developing neural tissues, and in many epithelia in organogenesis. During development of the lung, expression is found throughout the epithelium with the highest levels in the terminal regions of the branching tubules (Moens et al 1992). In the developing meso- and metanephric kidney, transcripts are confined to early stages of differentiating epithelia derived from mesenchyme (Mugrauer et al 1988).

By homologous recombination in embryonic stem cells, mutations have been generated in the murine N-*myc* gene: two null alleles (N-myc^{-}; Charron et al 1992; Stanton et al 1992) and a hypomorphic mutation (N-myc^{9a}; Moens et al 1992). N-myc^{-} and N-myc^{9a} homozygosity cause embryonic or postnatal lethality, respectively, and primarily affect organogenesis. The branching of the lung epithelia is inhibited (N-myc^{-}) or severely reduced (N-myc^{9a}) by the mutations. Whereas the N-myc^{9a} allele does not affect other organs, animals homozygous for N-myc^{-} show additional phenotypes in epithelial structures: the developing mesonephric tubules are reduced in numbers or fail to develop, and stomach and large intestine, as well as derivatives of the midgut, seem to be absent. Hypoplasia of neural structures is observed as well. Therefore, the absence or reduced activity of N-*myc* affects proliferation and differentiation of epithelia, which respond to inductive signals.

Pax2 and *Pax8* were identified because of their sequence similarities with other genes encoding proteins with paired box sequence (Dressler et al 1990; Plachov et al 1990). These genes were first discovered in *Drosophila* as genetic elements that determine segmentation (Bopp et al 1986) and were subsequently found in vertebrates as well. *Pax* proteins bind to DNA and can act as transcription factors (Treisman et al 1989, 1991; Chalepakis et al 1991), although biologically relevant target genes have not been identified. In the mouse, *Pax* genes are expressed in distinct spatio-temporal patterns suggesting that the encoded proteins act in developmentally important events (reviewed in Deutsch & Gruss 1991). In accordance, several well known mouse mutants and human congenital malformations are caused by altered *Pax* genes (reviewed in Gruss & Walther 1992). In particular, in the developing urogenital tract the *Pax2* and *Pax8* genes are expressed in distinct cells known to be involved in mesenchymal-epithelial interactions (Dressler et al 1990; Plachov et al 1990). A function of *Pax2* in development has been demonstrated directly by overexpression of the gene under the control of a constitutively active promoter in transgenic mice; primarily one organ, the kidney, is affected by this ectopic expression, which leads to disorganization of renal epithelial structures (Dressler et al 1993). In the mouse, the *Pax8* gene is located on chromosome 2 in close vicinity to a well characterized mutation, *Danforth's short tail* (*Sd*), which also affects kidney development. It has been speculated that *Sd* and *Pax2* are allelic (Gruss & Walther 1992). It is thus possible that both *Pax* genes are involved in the final steps of signal transduction mediated by mesenchymal-epithelial interactions.

Wilm's tumor is a rare embryonal malignancy of the kidney that is caused by malfunction of Wilm's tumor suppressor gene, WT1, located on chromosome 11p13. WT1 encodes a putative transcription factor with a zinc finger sequence motif (Call et al 1990; Gessler et al 1990). Wilm's tumors arise from mesenchyme induced to differentiate into epithelia and are observed to contain cells with variable epithelial or mesenchymal characteristics, apparently reflecting the differentiation state of the precursors. During normal kidney development, WT1 is expressed in the induced (condensed) mesenchyme and in the podocytes of the developing glomeruli (Pritchard-Jones et al 1990). The *Denys-Drash syndrome* is also associated with abnormal WT1 genes and is caused by mutant alleles that probably act in a dominant-negative fashion (Pelletier et al 1991). The syndrome caused by such mutations includes renal failure due to progressive glomerular sclerosis, pseudohermaphrodism and Wilm's tumor (Denys et al 1967; Drash et al 1970). Loss of WT1 function results in the loss of growth control and proper differentiation of the same epithelial cells that usually express the gene. Therefore, the WT1 gene product might have a crucial role in the initiation of the glomerular differentiation pathway.

SUMMARY AND FUTURE PERSPECTIVES

The biology of mesenchymal-epithelial interactions has been extensively studied, and it is well recognized that during development a variety of different factors participate in the determination of epithelial cell fate as well as the regulation of differentiation, morphogenesis, and growth. Organ culture experiments, where heterologous mesenchyme and epithelia have been combined, have demonstrated the existence of organ-specific factors or factor combinations. In addition, gradients of factors that determine differentiation and morphogenesis have to be postulated to explain regionalization of epithelial differentiation, for example in the development of the intestine.

We have concentrated in this review on a discussion of cell adhesion molecules and tyrosine kinase receptors. However, in many instances the results of the reported studies have remained essentially descriptive and now must be corroborated by functional experiments. The power of genetic approaches for the analysis of complex biological phenomena has been demonstrated in *Drosophila* and *C. elegans*. The recent technical advances that allow the production of transgenic mice carrying targeted mutations now allow the genetic analysis of the factors implicated in mesenchymal-epithelial interactions in mammals. Such genetic approachs should be complemented by organ culture experiments where interfering antibodies or recombinant antagonists can be used to test the functional importance of specific molecules. In addition, the search for new factors should not be neglected since the complexity of the biological phenomena indicates clearly that not all of the signaling molecules in mesenchymal-epithelial interactions are yet identified.

ACKNOWLEDGMENTS

We would like to thank Drs. R. Chiquet-Ehrismann, I. Thesleff, M. Bernfield, and F. Constantini for sharing unpublished data with us, and B. Lelekakis for excellent secretarial work. Our research is supported by the Bundesminister für Forschung und Technologie (BMFT) to C. and W. Birchmeier, the Deutsche Forschungsgemeinschaft (DFG) to W. Birchmeier, and the Deutsche Krebshilfe to W. Birchmeier.

Literature Cited

Aggeler, J, Ward, J, Blackie, LM, Barcellos-Hoff, MH, Streuli, CH, Bissell, M. 1991. Cytodifferentiation of mouse mammary epithelial cells cultured on a reconstituted basement membrane reveals striking similarities to development in vivo. *J. Cell Sci.* 99:407–17

Alex, R, Sözeri, O, Meyer, S, Dildrop, R. 1992. Determination of the DNA sequence recognized by the bHLH-zip domain of the N-Myc protein. *Nucleic Acids Res.* 20: 2257–63

Aufderheide, E, Chiquet-Ehrismann, R, Ekblom, P. 1987. Epithelial-mesenchymal interactions in the developing kidney lead to expression of tenascin in the mesenchyme. *J. Cell Biol.* 105:599–608

Aufderheide, E, Ekblom, P. 1988. Tenascin during gut development: Appearance in the mesenchyme, shift in molecular forms, and dependence on epithelial-mesenchymal interactions. *J. Cell Biol.* 107:2341–49

Barcellos-Hoff, MH, Aggeler, J, Ram, TG, Bissell, MJ. 1989. Functional differentiation and alveolar morphogenesis of primary mammary cultures on reconstituted basement membrane. *Development* 105:223–35

Bargman, CI, Hung, M-C, Weinberg, RA. 1986. The *neu* oncogene encodes an epidermal growth factor receptor-related protein. *Nature* 319:226–30

Barnard, JA, Lyons, RM, Moses, HL. 1990. The cell biology of transforming growth factor β. *Biochem. Biophys. Acta* 1032:79–87

Basler, K, Hafen, E. 1988. Control of photoreceptor cell fate by the *sevenless* protein requires a functional tyrosine kinase domain. *Cell* 54:299–311

Basset, P, Bellocq, JP, Wolf, C, Stoll, I, Hutin, P, et al. 1990. A novel metalloproteinase gene specifically expressed in stromal cells of breast carinomas. *Nature* 348:699–704

Behrens, J, Löwrick, O, Klein-Hitpass, L, Birchmeier, W. 1991. The E-cadherin promoter: Functional analysis of a GC-rich region and an epithelial cell-specific palindromic regulatory element. *Proc. Natl. Acad. Sci. USA* 88:11495–99

Behrens, J, Mareel, MM, Van Roy, F, Birchmeier, W. 1989. Dissecting tumor cell invasion: epithelial cells acquire invasive properties after the loss of uvomorulin-mediated cell-cell adhesion. *J. Cell Biol.* 108: 2435–47

Behrens, J, Vakaet, L, Friis, R, Winterhager, E, Van Roy, F, et al. 1993. Loss of epithelial differentiation and gain of invasiveness correlates with tyrosine phosphorylation of the E-cadherin/β-catenin complex in cells transformed with a temperature-sensitive v-SRC gene. *J. Cell Biol.* 120:757–66

Birchmeier, C, Birnbaum, D, Waitches, G, Fasano, O, Wigler, M. 1986. Characterization of an activated human *ros* gene. *Mol. Cell Biol.* 6:3109–16

Birchmeier, C, O'Neill, K, Riggs, M, Wigler, M. 1990. Characterization of ROS1 cDNA from a human glioblastoma cell line. *Proc. Natl. Acad. Sci. USA* 87:4799–803

Birchmeier, W, Weidner, KM, Hülsken, J, Behrens, J. 1993. Molecular mechanisms leading to cell junction (cadherin) deficiency in invasive carcinomas. *Sem. Cancer Biol.* 4: In press

Blackwell, TK, Kretzner, L, Blackwood, EM, Eisenman, RN, Weintraub, H. 1990. Sequence-specific DNA binding by the c-Myc protein. *Science* 250:1149–51

Blackwood, EM, Eisenman, RN. 1991. Max: A helix-loop-helix zipper protein that forms a sequence-specific DNA-binding complex with Myc. *Science* 251:1211–17

Blair, DG, Cooper, CS, Oskarsson, MK, Eader, LA, Vande Woude, GF. 1982. New method for detecting cellular transforming genes. *Science* 218:1122–25

Bopp, D, Burri, M, Baumgartner, S, Frigerio, G, Noll, M. 1986. Conservation of a large protein domain in the segmentation gene *paired* and in functionally related genes of Drosophila. *Cell* 47:1033–40

Bottaro, DP, Rubin, JS, Faletto, DL, Chan, AM-L, Kmiecik, TE, et al. 1991. Identification of the hepatocyte growth factor receptor as the c-*met* proto-oncogene product. *Science* 251:802–4

Butz, S, Stappert, J, Weissig, H, Kemler, R. 1992. Plakoglobin and βcatenin: Distinct but closely related. *Science* 257:1142–44

Buxton, RS, Magee, AI. 1992. Structure and interactions of desmosomal and other cadherins. *Sem. Cell Biol.* 3:157–67

Call, KM, Glaser, T, Ito, CY, Buckler, AJ, Pelletier, J, et al. 1990. Isolation and characterization of a zinc finger polypeptide gene at the human chromosome 11 Wilm's tumor locus. *Cell* 60:509–20

Chalepakis, G, Fritsch, R, Fickenscher, H, Deutsch, U, Goulding, M, Gruss, P. 1991. The molecular basis of the *undulated/Pax-1* mutation. *Cell* 66:873–84

Charron, J, Malynn, BA, Fisher, P, Stewart, V, Jeannotte, L, et al. 1992. Embryonic lethality in mice homozygous for a targeted disruption of the N-Myc gene. *Genes Dev.* 6:2248–57

Chen, J, Heller, D, Poon, B, Kang, L, Wang, L-H. 1991. The proto-oncogene c-*ros* codes for a transmembrane tyrosine protein kinase sharing sequence and structural homology with *sevenless* protein of *Drosophila melanogaster*. *Oncogene* 6:257–64

Chiquet-Ehrismann, R. 1993. Tenascin and other adhesion-modulating proteins in cancer. *Sem. Cancer Biol. 4: In press*

Chiquet-Ehrismann, R, Mackie, EJ, Pearson, CA, Sakakura, T. 1986. Tenascin: an extracellular matrix protein involved in tissue interactions during fetal development and oncogenes. *Cell* 47:131–39

Cooper, CS, Park, M, Blair, DG, Tainsky, MA, Huebner, K, et al. 1984. Molecular cloning of a new transforming gene from a chemically transformed human cell line. *Nature* 311:29–34

Coussens, L, Yang-Feng, TL, Liao, Y-C, Chen, E, Gray, A, et al. 1985. Tyrosine kinase receptor with extensive homology to EGF receptor shares chromosomal location with *neu* oncogene. *Science* 230:1132–39

Dean, M, Park, M, Vande Woude, GF. 1987. Characterization of the rearranged *tpr-met* oncogene breakpoint. *Mol. Cell Biol.* 7: 921–24

Denys, P, Malvaux, P, van den Berghe, H, Tanghe, W, Proesmans, W. 1967. De pseudohermaphrodisme masuclin, d'une tumeur de wilms, d'une nephropathie parenchymatuese et d'un mosaicisme XX/XY. *Arch. Fran. Ped.* 24:729–39

Deuchar, E. 1975. *Cellular Interactions in Animal Development*. London: Chapman & Hall. 262 pp.

Deutsch, U, Gruss, P. 1991. Murine paired domain proteins as regulatory factors of embryonic development. *Sem. Dev. Biol.* 2:413–24

Downs, KM, Martin, GR, Bishop, JM. 1989. Contrasting patterns of *myc* and N-*myc* expression during gastrulation of the mouse embryo. *Genes Dev.* 3:860–69

Drash, A, Sherman, F, Hartmann, WH, Blizzard, RM. 1970. A syndrome of pseudohermaphroditism, Wilm's tumor, hypertension, and degenerative renal disease. *J. Ped.* 76:585–93

Dressler, GR, Deutsch, U, Chowdhury, K, Nornes, HO, Gruss, P. 1990. *Pax2*, a new murine paired-box-containing gene and its expression in the developing excretory system. *Development* 109:787–95

Dressler, G, Wilkinson, FE, Rothenpieler, U, Patterson, L, William-Simons, L, Westphal, H. 1993. Deregulation of *Pax-2* expression in transgenic mice generates severe kidney abnormalities. *Nature* 362:65–67

Ekblom, M, Klein, G, Mugrauer, G, Fecker, L, Deutzmann, R, et al. 1990. Transient and locally restricted expression of laminin A chain mRNA by developing epithelial cells during kidney organogenesis. *Cell* 60: 337–46

Ekblom, P. 1992. In *The Kidney: Physiology and Pathophysiology*, ed. DW Seldin, G Giebisch, pp. 475–501. New York: Raven 2nd ed.

Fell, PE, Grobstein, C. 1968. The influence of extra-epithelial factors on the growth of embryonic mouse pancreatic epithelium. *Exp. Cell Res.* 53:301–4

Frixen, U, Behrens, J, Sachs, M, Eberle, G, Voss, B, et al. 1991. E-cadherin-mediated cell-cell adhesion prevents invasiveness of human carcinoma cell lines. *J. Cell Biol.* 117:173–85

Garrod, DR, Fleming, S. 1990. Early expression of desmosomal components during kidney tubule morphogenesis in human and murine embryos. *Development* 108:313–21

Gessler, M, Poustka, A, Cavenee, W, Neve, RL, Orkin, SH, Bruns, GAP. 1990. Homozygous deletion in Wilms tumours of a zinc-finger gene identified by chromosome jumping. *Nature* 343:774–78

Gherardi, E, Gray, JM, Stoker, M, Perryman, M, Furlong, R. 1989. Purification of scatter factor, a fibroblast-derived basic protein that modulates epithelial interactions and movement. *Proc. Natl. Acad. Sci. USA* 86:5844–48

Giordano, S, Ponzetto, C, Di Renzo, MF, Cooper, CS, Comoglio, PM. 1989. Tyrosine kinase receptor indistinguishable from the c-*met* protein. *Nature* 339:155–56

Gonzatti-Haces, M, Seth, A, Park, M, Copeland, T, Oroszlan, S, Vande Woude, GF. 1988. Characterization of the *TPR-MET* oncogene p65 and the *MET* protooncogene p140 protein-tyrosine kinases. *Proc. Natl. Acad. Sci. USA* 85:21–25

Green, JBA, Smith, JC. 1991. Growth factors as morphogens: do gradients and thresholds establish body plan? *Trends Genet.* 7:245–50

Grobstein, C. 1953. Morphogenetic interaction between embryonic mouse tissues separated by a membrane filter. *Nature* 172: 869–71

Grobstein, C. 1956. Inductive tissue interaction in development. *Adv. Cancer Res.* 4: 187–236

Gruss, P, Walther, C. 1992. Pax in development. *Cell* 69:719–22

Guo, L, Yu, Q-C, Fuchs, E. 1993. Targeting expression of keratinocyte growth factor to keratinocytes elicits striking changes in epithelial differentiation in transgenic mice. *EMBO J.* 12:973–86

Gurdon, JB. 1987. Embryonic induction—molecular prospects. *Development* 99:285–306

Hadley, MA, Byers, SW, Suarez-Quian, CA, Kleinman, HK, Dym, M. 1985. Extracellular matrix regulates sertoli cell differentiation testicular cord formation, and germ cell development in vitro. *J. Cell Biol.* 101:1511–22

Halazonetis, TD, Kandil, AN. 1991. Determination of the c-MYC DNA-binding site. *Proc. Natl. Acad. Sci. USA* 88:6162–66

Hart, AC, Krämer, H, Van Vactor, DL, Paidhungat, M, Zipursky, SL. 1990. Induction of cell fate in the *Drosophila* retina: the bride of sevenless protein is predicted to contain a large extracellular domain and seven transmembrane segments. *Genes Dev.* 4:1835–47

Hartmann, G, Naldini, L, Weidner, KM, Sachs, M, Vigna, E, et al. 1992. A functional domain in the heavy chain of scatter factor/hepatocyte growth factor binds the c-*Met* receptor and induces cell dissociation but not mitogenesis. *Proc. Natl. Acad. Sci. USA* 89:11574–78

Hattori, Y, Odagiri, H, Nakatani, H, Miyagawa, K, Naito, K, et al. 1990. K-*sam*, an amplified gene in stomach cancer, is a member of the heparin-binding growth factor receptor genes. *Proc. Natl. Acad. Sci. USA* 82:5983–87

Hay, ED. 1990. Role of cell matrix contacts in cell migration and epithelial-mesenchymal transformation. *Cell Diff. Dev.* 32:367–76

Heine, UI, Munoz, EF, Flanders, KC, Ellingsworth, LR, Lam, H-YP, et al. 1987. Role of transforming growth factor-β in the development of the mouse embryo. *J. Cell Biol.* 105:2861–76

Herrenknecht, K, Ozawa, M, Eckerskorn, C, Lottspeich, F, Lenter, M, Kemler, R. 1991. The uvomorulin-anchorage protein a catenin is a vinculin homologue. *Proc. Natl. Acad. Sci. USA* 88:9156–60

Higashio, K, Shima, N, Goto, M. Itagaki, Y, Nagao, M, Yasuda, H, Morinaga, T. 1990. Identity of a tumor cytotoxic factor from human fibroblasts and hepatocyte growth factor. *Biochem. Biophys. Res. Commun.* 170:397–404

Hirai, Y, Nakagawa, S, Takeichi, M. 1993. Reexamination of the properties of epimorphin and its possible roles. *Cell* 73:426–27

Hirai, Y, Takebe, K, Takashina, M, Kobayashi, S, Takeichi, M. 1992. Epimorphin: A mesenchymal protein essential for epithelial morphogenesis. *Cell* 69:471–81

Hirano, S, Kimoto, N, Shimoyama, Y, Hirohashi, S, Takeichi, M. 1992. Identification of a neural α-catenin as a key regulator of cadherin function and multicellular organization. *Cell* 70:293–301

Hirning, U, Schmid, P, Schulz, WA, Rettenberger, G, Hameister, H. 1991. A comparative analysis of N-*myc* and c-*myc* expression and cellular proliferation in mouse organogenesis. *Mech. Dev.* 33:119–26

Holmes, WE, Sliwkowski, MX, Akita, RW, Henzel, WJ, Lee, J, et al. 1992. Identification of heregulin, a specific activator of p 185^{erbB2}. *Science* 256:1205–10

Hynes, RO. 1992. Integrins: Versatility, modulation, and signaling in cell adhesion. *Cell* 69:11–25

Inaguma, Y, Kusakabe, M, Mackie, EJ, Pearson, CA, Chiquet-Ehrismann, R, Sakakura, T. 1988. Epithelial induction of stromal tenascin in the mouse mammary gland: from embryogenesis to carcinogenesis. *Dev. Biol.* 128:245–55

Jahoda, L, Horne, KA, Oliver, RF. 1984. Induction of hair growth by implantation of cultured dermal papilla cells. *Nature* 311: 560–62

Jakobovits, A, Schwab, M, Bishop, JM, Martin, GR. 1985. Expression of N-*myc* in teratocarcinoma stem cells and mouse embryos. *Nature* 318:188–91

Jessell, TM, Melton, DA. 1992. Diffusible factors in vertebrate embryonic induction. *Cell* 68:257–70

Jouanneau, J, Gavrilovic, J, Caruelle, D, Jaye, M, Moens, G, et al. 1991. Secreted or nonsecreted forms of acidic fibroblast growth factor produced by transfected epithelial cells influence cell morphology, motility, and invasive potential. *Proc. Natl. Acad. Sci. USA* 88:2893–97

Kanazawa, T, Hosick, HL. 1992. Transformed growth phenotype of mouse mammary epithelium in primary culture induced by specific fetal mesenchymes. *J. Cell Physiol.* 153:381–91

Kedinger, M, Simon-Assmann, PM, Lacroix, B, Marxer, A, Hauri, HP, Haffen, K. 1986. Fetal gut mesenchyme induces differentiation of cultured intestinal endodermal and crypt cells. *Dev. Biol.* 113:474–83

Kelly, K, Cochran, BH, Stiles, CD, Leder, P. 1983. Cell-specific regulation of the c-*myc* gene by lymphocyte mitogens and platelet-derived growth factor. *Cell* 35:603–10

Kerkhoff, E, Bister, K, Klempnauer, KH. 1991. Sequence-specific DNA binding by Myc proteins. *Proc. Natl. Acad. Sci. USA* 88:4323–27

Kintner, C. 1992. Regulation of embryonic cell adhesion by the cadherin cytoplasmic domain. *Cell* 69:225–36

Klein, G, Langegger, M, Goridis, C, Ekblom, P. 1988a. Neural cell adhesion molecules during embryonic induction and development of the kidney. *Development* 102:749–61

Klein, G, Langegger, M, Timpl, R, Ekblom, P. 1988b. Role of laminin a chain in the

development of epithelial cell polarity. *Cell* 55:331–41
Kleinman, HK, McGarvey, ML, Hassell, JR, Star, VL, Cannon, FB, et al. 1986. Basement membrane complexes with biological activity. *Biochemistry* 25:312–18
Knudsen, KA, Wheelock, MJ. 1992. Plakoglobin, or an 83-kD homologue distinct from β-catenin, interacts with E-cadherin and N-cadherin. J. *Cell Biol.* 118: 671–79
Kohl, NE, Kanda, N, Schreck, RR, Bruns, G, Latt, SA, et al. 1983. Transposition and amplification of oncogene-related sequences in human neuroblastomas. *Cell* 35:359–67
Kokai, Y, Cohen, JA, Drebin, JA, Greene, MI. 1987. Stage- and tissue-specific expression of the *neu* oncogene in rat development. *Proc. Natl. Acad. Sci. USA* 84:8498–501
Kollar, EJ, Fisher, C. 1980. Tooth induction in chick epithelium: Expression of quiescent genes for anamel synthesis. *Science* 207:993–95
Kornbluth, S, Paulson, KE, Hanafusa, H. 1988. Novel tyrosine kinase identified by phosphotyrosine antibody screening of cDNA libraries. *Mol. Cell Biol.* 8:5541–44
Krämer, H, Cagan, RL, Zipursky, SL. 1991. Interaction of *bride* of *sevenless* membrane-bound ligand and the *sevenless* tyrosine-kinase receptor. *Nature* 352:207–12
Kusakabe, M, Sakakura, T, Sano, M, Nishizuka, Y. 1985. A pituitary-salivary mixed gland induced by tissue recombination of embryonic pituitary epithelium and embryonic submandibular gland mesenchyme in mice. *Dev. Biol.* 110:382–91
Kusakabe, M, Yokoyama, M, Sakakura, T, Nomura, T, Hosick, HL, Nishizuka, Y. 1988. A novel methodology for analysis of cell distribution in chimeric mouse organs using a strain specific antibody. *J. Cell Biol.* 107:257–65
Lasnitzki, I, Mizuno, T. 1980. Prostatic induction: Interaction of epithelium and mesenchyme from normal wild-type mice and androgen-insensitive mice with testicular feminization. *J. Endocrinol.* 85:423–28
Leppä, S, Mali, M, Miettinen, HM, Jalkanen, M. 1992. Syndecan expression regulates cell morphology and growth of mouse mammary epithelial tumor cells. *Proc. Natl. Acad. Sci. USA* 89:932–36
Mackie, EJ, Chiquet-Ehrismann, R, Pearson, CA, Inaguma, Y, Taya, K, et al. 1987. Tenascin is a stromal marker for epithelial malignancy in the mammary gland. *Proc. Natl. Acad. Sci. USA* 84:4621–25
Massagué, J. 1992. Receptors for the TGF-β family. *Cell* 69:1067–70
Mathan, M, Moxey, PC, Trier, JS. 1976. Morphogenesis of fetal rat duodenal villi. *Am. J. Anat.* 146:73–92
Matsushime, H, Shibuya, M. 1990. Tissue-specific expression of *rat c-ros-1* gene and partial structural similarity of its predicted products with *sev* protein of *Drosophila melanogaster*. *J. Virol.* 64:2117–25
McCrea, PD, Turck, CW, Gumbiner, B. 1991. A homolog of the *armadillo* protein in *Drosophila* (plakoglobin associated with E-cadherin. *Science* 254:1359–61
Miki, T, Fleming, TP, Bottaro, DP, Rubin, JS, Ron, D, Aaronson, SA. 1991. Expression cDNA cloning of the KGF receptor by creation of a transforming autocrine loop. *Science* 251:72–75
Millan, FA, Denhez, F, Kondaiah, P, Akhurst, RJ. 1991. Embryonic gene expression patterns of TGF β1, β2 and β3 suggest different developmental functions in vivo. *Development* 111:131–44
Miyazawa, K, Tsubouchi, H, Naka, D, Takahashi, K, Okigaki, M, et al. 1989. Molecular cloning and sequence analysis of cDNA for human hepatocyte growth factor. *Biochem. Biophys. Res. Commun.* 163:967–73
Moens, CB, Auerbach, AB, Conlon, RA, Joyner, AL, Rossant, J. 1992. A targeted mutation reveals a role for N-*myc* in branching morphogenesis in the embryonic mouse lung. *Genes Dev.* 6:691–704
Montesano, R, Matsumoto, K, Nakamura, T, Orci, L. 1991a. Identification of a fibroblast-derived epithelial morphogen as hepatocyte growth factor. *Cell* 67:901–8
Montesano, R, Schaller, G, Orci, L. 1991b. Induction of epithelial tubular morphogenesis in vitro by fibroblast-derived soluble factors. *Cell* 66:697–711
Moon, RT, Christian, JL. 1992. Competence modifiers synergize with growth factors during mesoderm induction and patterning in Xenopus. *Cell* 71:709–12
Mori, S, Akiyama, T, Yamada, Y, Morishita, Y, Sugawara, I, et al. 1989. C-*erb*B-2 gene product, a membrane protein commonly expressed on human fetal epithelial cells. *Lab. Invest.* 61:93–97
Mugrauer, G, Alt, FW, Ekblom, P. 1988. N-*myc* proto-oncogene expression during organogenesis in the developing mouse as revealed by in situ hybridization. *J. Cell Biol.* 107:1325–35
Müller, R, Slamon, DJ, Tremblay, JM, Cline, MJ, Verma, IM. 1982. Differential expression of cellular oncogenes during pre- and postnatal development of the mouse. *Nature* 299:640–49
Nagafuchi, A, Takeichi, M, Tsukita, S. 1991. The 102 kd cadherin-associated protein:

Similarity to vinculin and posttranscriptional regulation of expression. *Cell* 65:849–57

Nakamura, T, Nishizawa, T, Hagiya, M, Seki, T, Shimonishi, M, Sugimura, A, Tashiro, K, Shimizu, S. 1989. Molecular cloning and expression of human hepatocyte growth factor. *Nature* 342:440–43

Nakanishi, Y, Sugiura, F, Kishi, J-I, Hayakawa, T. 1986. Collagenase inhibitor stimulates cleft formation during early morphogenesis of mouse salivary gland. *Dev. Biol.* 113:201–6

Naldini, L, Weidner, KM, Vigna, E, Gaudino, G, Bardelli, A, et al. 1991. Scatter factor and hepatocyte growth factor are indistinguishable ligands for the *MET* receptor. *EMBO J.* 10:2867–78

Nau, MM, Brooks, BJ, Battey, J, Sausville, E, Gazdar, AF, et al. 1985. L-*myc*, a new *myc*-related gene amplified and expressed in human small cell lung cancer. *Nature* 318:69–73

Neckameyer, WS, Wang, L. 1985. Nucleotide sequence of avian sarcoma virus UR2 and comparison of its transforming gene with other members of the tyrosine protein kinase oncogene family. *J. Virol.* 53:879–84

Orr-Urtreger, A, Givol, D, Yayon, A, Yarden, Y, Lonai, P. 1991. Developmental expression of two murine fibroblast growth factor receptors. *flg* and *bek*. *Development* 113:1419–34

Ozawa, M, Baribault, H, Kemler, R. 1989. The cytoplasmic domain of the cell adhesion molecule uvomorulin associates with three independent proteins structurally related in different species. *EMBO J.* 8:1711–17

Ozawa, M, Ringwald, M, Kemler, R. 1990. Uvomorulin-catenin complex formation is regulated by a specific domain in the cytoplasmic region of the cell adhesion molecule. *Proc. Natl. Acad. Sci. USA* 87:4246–50

Park, M, Dean, M, Cooper, CS, Schmidt, M, O'Brian, SJ, et al. 1986. Mechanism of *met* oncogene activation. *Cell* 45:895–904

Peles, E, Bacus, SS, Koski, RA, Lu, HS, Wen, D, et al. 1992. Isolation of the Neu/HER-2 stimulatory ligand: A 44 kd glycoprotein that induces differentiation of mammary tumor cells. *Cell* 69:205–16

Pelham, HRB. 1993. Is epimorphin involved in vesicular transport? *Cell* 73:425–26

Pelletier, J, Bruening, W, Kashtan, CE, Mauer, SM, Manivel, JC, et al. 1991. Germline mutations in the Wilms tumor suppressor gene are associated with abnormal urogenital development in Denys-Drash syndrome. *Cell* 67:437–47

Pelton, NW, Dickinson, ME, Moses, HL, Hogan, BLM. 1990. In situ hybridization analysis of TGFβ3 RNA expression during mouse development: comparative studies with TGFβ1 and β2. *Development* 110: 609–20

Pfeifer-Ohlsson, S, Rydnert, J, Goustin, AS, Larsson, E, Betsholtz, C, Ohlsson, R. 1985. Cell-type-specific pattern of *myc* protooncogene expression in developing human embryos. *Proc. Natl. Acad. Sci. USA* 82: 5050–54

Plachov, D, Chowdhury, K, Walther, C, Simon, D, Guenet, J-L, Gruss, P. 1990. *Pax8*, a murine paired box gene expressed in the developing excretory system and thyroid gland. *Development* 110:643–51

Prendergast, GC, Lawe, D, Ziff, EB. 1991. Association of myn, the murine homolog of max, with c-myc stimulates methylation-sensitive DNA binding and ras cotransformation. *Cell* 65:395–407

Press, MF, Cordon-Cardo, C, Slamon, DJ. 1990. Expression of the HER-2/*neu* protooncogene in normal human adult and fetal tissues. *Oncogene* 5:953–62

Pritchard-Jones, K, Fleming, S, Davidson, D, Bickmore, W, Porteous, D, et al. 1990. The candidate Wilm's tumour gene is involved in genitourinary development. *Nature* 346:194–97

Pyke, C, Kristensen, P, Ralfkiae, E, Grondahl-Hansen, J, Eriksen, J, et al. 1991. Urokinase-type plasminogen activator is expressed in stromal cells and its receptor in cancer cells at invasive foci in human colon adenocarcinomas. *Am. J. Pathol.* 138: 1059–67

Pyke, C, Ralfkiae, E, Huhtala, P, Hurskainen, T, Dano, K, Tryggvason, K. 1992. Localization of messenger RNA for M_r 72,000 and 92,000 type IV collagenases in human skin cancers by in situ hybridization. *Cancer Res.* 52:1336–41

Quirke, P, Pickles, A, Tuzi, NL, Mohamdee, O, Gullick, WJ. 1989. Pattern of expression of c-*erb*B-2 oncoprotein in human fetuses. *Br. J. Cancer* 60:64–69

Reponen, P, Sahlberg, C, Huhtala, P, Hurskainen, T, Thesleff, I, Tryggvason, K. 1992. Molecular cloning of murine 72-kDa Type IV collagenase and its expression during mouse development. *J. Biol. Chem.* 267:7856–62

Robinson, SD, Silberstein, GB, Roberts, AB, Flanders, KC, Daniel, CW. 1991. Regulated expression and growth inhibitory effects of transforming growth factor-β isoforms in mouse mammary gland development. *Development* 113:867–78

Rodriguez-Boulan, E, Nelson, WJ. 1989. Morphogenesis of the polarized epithelial cell phenotype. *Science* 245:718–25

Rosen, EM, Meromsky, L, Setter, E, Vinter,

DW, Goldberg, ID. 1990. Purified scatter factor stimulates epithelial and vascular endothelial cell migration. *Proc. Soc. Exp. Biol. Med.* 195:34–43

Rubin, JS, Chan, AM, Bottaro, DP, Burgess, WH, Taylor, WG, et al. 1991. A broad-spectrum human lung fibroblast-derived mitogen is a variant of hepatocyte growth factor. *Proc. Natl. Acad. Sci. USA* 88:415–19

Rubin, JS, Osada, H, Finch, PW, Taylor, WG, Rudikoff, S, Aaronson, SA. 1989. Purification and characterization of a newly identified growth factor specific for epithelial cells. *Proc. Natl. Acad. Sci. USA* 86: 802–6

Saga, Y, Takeshi, Y, Ikawa, Y, Sakakura, T, Aizawa, S. 1992. Mice develop normally without tenascin. *Genes Dev.* 6: 1821–31

Sahlberg, C, Reponen, P, Tryggvason, K, Thesleff, I. 1992. Association between the expression of murine 72 kDa type IV collagenase by odontoblasts and basement membrane degradation during mouse tooth development. *Arch. Oral Biol.* 37:1021–30

Sakakura, T. 1991. New aspects of stroma-parenchyma relations in mammary gland differentiation. *Int. Rev. Cytol.* 125:165–202

Saxen, L. 1987. *Organogenesis of the Kidney*. Cambridge: Cambridge Univ. Press. 173 pp.

Schechter, AL, Stern, DF, Vaidyanathan, L, Decker, SJ, Drebin, JA, et al. 1984. The *neu* oncogene, an *erb*-B-related gene encoding a 185,000-M, tumour antigen. *Nature* 312:513–16

Schipper, JH, Frixen, UH, Behrens, J, Unger, A, Jahnke, K, Birchmeier, W. 1992. E-cadherin expression in squamous cell carcinomas of head and neck; inverse correlation with tumor dedifferentiation and lymph node metastasis. *Cancer Res.* 51:6328–37

Schmid, P, Cox, D, Bilbe, G, Maier, R, McMaster, GK. 1991. Differential expression of TGF β1, β2 and β3 genes during mouse embryogenesis. *Development* 111: 117–30

Schmid, P, Schulz, WA, Hameister, H. 1989. Dynamic expression pattern of the *myc* protooncogene in midgestation mouse embryos. *Science* 243:226–29

Schwab, M, Alitalo, K, Klempnauer, K-H, Varmus, HE, Bishop, JM, et al. 1983. Amplified DNA with limited homology to *myc* cellular oncogene is shared by human neuroblastoma cell lines and a neuroblastoma tumour. *Nature* 305:245–48

Semsei, I, Ma, S, Cutler, RG. 1989. Tissue and age specific expression of the *myc* proto-oncogene family throughout the life span of the C57BL/6J mouse strain. *Oncogene* 4:465–70

Shimoyama, Y, Nagafuchi, A, Fujita, S, Gotoh, M, Takeichi, M, et al. 1992. Cadherin dysfunction in a human cancer cell line: possible involvement of loss of α-catenin expression in reduced cell-cell adhesiveness. *Cancer Res.* 52:1–5

Shiota, G, Rhoads, DB, Wang, TC, Nakamura, T, Schmidt, EV. 1992. Hepatocyte growth factor inhibits growth of hepatocellular carcinoma cells. *Proc. Natl. Acad. Sci. USA* 89:373–77

Silberstein, GB, Flanders, KC, Roberts, AB, Daniel, CW. 1992. Regulation of mammary morphogenesis: Evidence for extracellular matrix-mediated inhibition of ductal budding by transforming growth factor-β1. *Dev. Biol.* 152:354–62

Simons, K, Fuller, SD. 1985. Cell surface polarity in epithelia. *Annu. Rev. Cell Biol.* 1:243–88

Slamon, DJ, Clark, GM, Wong, SG, Levin, WJ, Ullrich, A, McGuire, WL. 1987. Human breast cancer: Correlation of relapse and survival with amplification of the HER-2/*neu* oncogene. *Science* 235:177–82

Slamon, DJ, Godolphin, W, Jones, LA, Holt, JA, Wong, SG, et al. 1989. Studies of the HER-2/*neu* proto-oncogene in human breast and ovarian cancer. *Science* 244:707–12

Soman, N, Correa, P, Ruiz, B, Wogan, G. 1991. The *TPR-MET* oncogenic rearrangement is present and expressed in human gastric carcinoma and precursor lesions. *Proc. Natl. Acad. Sci. USA* 88:4892–96

Sonnenberg, A, Calafat, J, Janssen, H, Daams, H, Van der Raaij-Helmer, LMH, et al. 1991. Integrin α6/β4 complex is located in desmosomes, suggesting a major role in epidermal cell-basement membrane adhesion. *J. Cell Biol.* 113:907–17

Sonnenberg, E, Gödecke, A, Walter, B, Bladt, F, Birchmeier, C. 1991. Transient and locally restricted expression of the *ros1* protooncogene during mouse development. *EMBO J.* 10:3693–702

Sonnenberg, E, Meyer, D, Weidner, KM, Birchmeier, C. 1993a. Scatter factor/hepatocyte growth factor and its receptor, the c-met tyrosine kinase, can mediate a signal exchange between mesenchyme and epithelia during mouse development. *J. Cell Biol.* In press

Sonnenberg, E, Weidner, K, Birchmeier, C. 1993b. In *Hepatocyte Growth Factor - Scatter Factor (HGF-SF) and the C-Met Receptor,* ed. ID Goldberg, EM Rosen, pp. 382–94. Basel: Birkhäuser Verlag

Sorokin, L, Sonnenberg, A, Aumailley, M, Timpl, R, Ekblom, P. 1990. Recognition of the laminin E8 cell-binding site by an integrin possessing the α_6 subunit is essential for epithelial polarization in developing kidney tubules. *J. Cell Biol.* 111:1265–73

Spemann, H, Schotte, O. 1932. Uber xenoplastische Transplantation als Mittel zur Analyse der embryonalen Induktion. *Naturwissenschaften* 20:31–37

Spooner, BS, Wessells, NK. 1970. Mammalian lung development: interactions in primordium formation and bronchial morphogenesis. *J. Exp. Zool.* 175:445–54

Spring, J, Kato, M, Bernfield, M. 1993. Epimorphin is related to a new class of neuronal and yeast vesicle targeting proteins. *Trends Biochem. Sci.* 18:124–25

Stanton, BR, Perkins, AS, Tassarollo, L, Sassoon, DA, Parada, LF. 1992. Loss of N-*myc* function results in embryonic lethality and failure of the epithelial component of the embryo to develop. *Genes Dev.* 6:2235–47

Stern, CD. 1992. Mesoderm induction and development of the embryonic axis in amniotes. *Trends Genet.* 8:158–63

Stoker, M, Gherardi, E, Perryman, M, Gray, J. 1987. Scatter factor is a fibroblast-derived modulator of epithelial cell mobility. *Nature* 327:239–42

Streuli, CH, Bailey, D, Bissell, MJ. 1991. Control of mammary epithelial differentiation: basement membrane induces tissue-specific gene expression in the absence of cell-cell interaction and morphological polarity. *J. Cell Biol.* 115:1383–95

Takahashi, M, Cooper, GM. 1987. *ret* Transforming gene encodes a fusion protein homologous to tyrosine kinases. *Mol. Cell Biol.* 7:1378–85

Takahashi, M, Ritz, J, Cooper, GM. 1985. Activation of a novel human transforming gene, *ret,* by DNA rearrangement. *Cell* 42: 581–88

Takahashi, Y, Nogawa, H. 1991. Branching morphogenesis of mouse salivary epithelium in basement membrane-like substratum separated from mesenchyme by the membrane filter. *Development* 111:327–35

Talhouk, RS, Bissell, MJ, Werb, Z. 1992. Coordinated expression of extracellular matrix-degrading proteinases and their inhibitors regulates mammary epithelial function during involution. *J. Cell Biol.* 118:1271–82

Taub, M, Wang, Y, Szczesny, TM, Kleinman, HK. 1990. Epidermal growth factor or transforming growth factor α is required for kidney tubulogenesis in matrigel cultures in serum-free medium. *Proc. Natl. Acad. Sci. USA* 87:4002–6

Timpl, R. 1989. Structure and biological activity of basement membrane proteins. *Eur. J. Biochem.* 180:487–502

Trautman, MS, Kimelman, J, Bernfield, M. 1991. Developmental expression of syndecan, an integral membrane proteoglycan, correlates with cell differentiation. *Development* 111:213–20

Treisman, J, Gönczy, P, Vashishta, M, Harris, E, Desplan, C. 1989. A single amino acid can determine the DNA binding specificity of homeodomain proteins. *Cell* 59: 553–62

Treisman, J, Harris, E, Desplan, C. 1991. The paired box encodes a second DNA-binding domain in the paired homeo domain protein. *Genes Dev.* 5:594–604

Tsukita, S, Tsukita, S, Nagafuchi, A, Yonemura, S. 1992. Molecular linkage between cadherins and actin filaments in cell-cell adherens junctions. *Curr. Opin. Cell Biol.* 4:834–39

Vainio, S, Jalkanen, M, Thesleff, I. 1989a. Syndecan and tenascin expression is induced by epithelial-mesenchymal interactions in embryonic tooth mesenchyme. *J. Cell Biol.* 108:1945–54

Vainio, S, Lehtonen, E, Jalkanen, M, Bernfield, M, Saxen, L. 1989b. Epithelial-mesenchymal interactions regulate the stage-specific expression of a cell surface proteoglycan, syndecan, in the developing kidney. *Dev. Biol.* 134:382–91

Vaino, S, Thesleff, I. 1992a. Sequential induction of syndecan, tenascin and cell proliferation associated with mesenchymal cell condensation during early tooth development. *Differentiation* 50:97–105

Vainio, S, Thesleff, I. 1992b. Coordinated induction of cell proliferation and syndecan expression in dental mesenchyme by epithelium: Evidence for diffusible signals. *Developmental Dynamics* 194:105–17

Valles, AM, Boyer, B, Thiery, JP. 1991. In *Cell Motility Factors,* ed. ID Goldberg, pp. 17–34. Basel: Birkhäuser Verlag

Vennstrom, B, Sheiness, D, Zabielski, J, Bishop, JM. 1982. Isolation and characterization of c-*myc,* a cellular homolog of the oncogene (v-*myc*) of avian myelocytomatosis virus strain 29. *J. Virol.* 42:773–79

Vestweber, D, Kemler, R, Ekblom, P. 1985. Cell adhesion molecule uvomorulin during kidney development *Dev. Biol.* 112:213–21

Wajjwalku, W, Nakamura, S, Hasegawa, Y, Miyazaki, K, Satoh, Y, et al. 1992. Low frequency of rearrangments of the RET and TRK protooncogenes in Japanese thyroid papillary carcinomas. *Jpn. J. Cancer Res.* 83:671–75

Weidner, KM, Arakaki, N, Hartmann, G, Vandekerckhove, J, Weingart, S, et al. 1991. Evidence for the identity of human scatter factor and human hepatocyte growth factor. *Proc. Natl. Acad. Sci. USA* 88: 7001–5

Weidner, KM, Behrens, J, Vandekerckhove, J, Birchmeier, W. 1990. Scatter factor: Molecular characteristics and effect on the invasiveness of epithelial cells. *J. Cell Biol.* 111:2097–108

Weidner, KM, Sachs, M, Birchmeier, W. 1993. The *Met* receptor tyrosine kinase transduces motility proliferation, and morphogenic signals of scatter factor/hepatocyte growth factor in epithelial cells. *J. Cell Biol.* 121:145–54

Wen, D, Peles, E, Cupples, R, Suggs, SV, Bacus, SS, et al. 1992. *Neu* differentiation factor: A transmembrane glycoprotein containing an EGF domain and an immunoglobulin homology unit. *Cell* 69: 559–72

Wessells, NK, Cohen, JH. 1967. Early pancreas organogenesis: morphogenesis, tissue interactions, and mass effects. *Dev. Biol.* 15:237–70

Yamamoto, T, Ikawa, S, Akiyama, T, Semba, K, Nomura, N, et al. 1986. Similarity of protein encoded by the human c-*erb*-B2 gene to epidermal growth factor receptor. *Nature* 319:230–34

Yasugi, S. 1993. Role of epithelial-mesenchymal interactions in differentiation of epithelium of vertebrate digestive organs. *Dev. Growth Diff.* 35:1–9

Zarnegar, R, Michalopoulos, G. 1989. Purification and biological characterization of human hepatopoietin A, a polypeptide growth factor for hepatocytes. *Cancer Res.* 49:3314–20

Zimmerman, K, Yancopoulos, GD, Collum, RG, Smith, RK, Kohl, NE, et al. 1986. Differential expression of *myc* family genes during murine development. *Nature* 319: 780–83

Annu. Rev. Cell Biol. 1993. 9:541–73

TUMOR CELL INTERACTIONS WITH THE EXTRACELLULAR MATRIX DURING INVASION AND METASTASIS

William G. Stetler-Stevenson, Sadie Aznavoorian, and Lance A. Liotta[1]

Laboratory of Pathology, Division of Cancer Biology, Diagnosis and Centers, National Cancer Institute, National Institutes of Health, Bethesda, Maryland 20892

KEY WORDS: cell adhesion, integrins, laminin receptor, cadherins, metalloproteinases, collagenases, gelatinases, tissue inhibitors of metalloproteinases, cell motility, autocrine motility

CONTENTS

INTRODUCTION

Malignancy is defined as neoplastic growth that tends to metastasize. Thus by definition metastatic ability is the correlate of malignant potential. The formation of metastatic foci is the most life-threatening aspect of malignant neoplasia. Occult metastatic tumor cells may persist in a dormant state for years after the resection or elimination of the primary tumor (Meltzer 1990; Zajicek 1987). They can then be activated by as yet unidentified stimuli and metastatic foci suddenly develop in an explosive fashion, which results in a rapid demise of the cancer patient. Most cancer deaths are due to the metastatic disease that remains resistant to conventional therapies. The primary aim of research into the mechanisms of tumor invasion and metastasis formation is to identify new strategies for more effective therapy against this most deadly aspect of human cancer.

The study of the genetic alterations associated with human tumor progression has yielded great insight into the mechanisms of oncogenesis (Bishop 1991; Fearon & Vogelstein 1990; Fidler & Radinsky 1990). Studies have clearly shown that the tumor development (tumorigenicity) and subsequent metastatic behavior (malignant potential) are under separate genetic control (Garbisa et al 1987; Muschel et al 1985). Genetic studies have demonstrated that establishment of a primary tumor focus is the result of multiple genetic alterations leading to uncontrolled tumor cell growth. These alterations include the loss or inactivation of anti-oncogenes as well as the activation of cellular proto-oncogenes. The genetic alterations associated with oncogenic transformation may occur in a random order and the resulting tumorigenicity is the net sum of these genetic changes. To date a single gene has not been identified that regulates the entire metastatic process. However, this is not surprising if we consider the multistep nature of metastasis formation, and that the genetic approach requires isolation and independent characterization of the multiple genetic alterations that may occur at each step.

SELECTION OF THE METASTATIC SUBPOPULATION: Clonal Dominance

Successful metastasis formation involves a series of linked sequential steps. This process requires a single tumor cell or groups of tumor cells to dissociate from the primary tumor, invade the surrounding extracellular matrix, including both basement membranes and interstitial compartments, enter the vascular or lymphatic space, escape immune surveillance and mechanical disruption, arrest at a distant site, escape from the vascular or lymphatic circulation, penetrate the target tissue and proliferate as a secondary colony that may itself subsequently metastasize. The fully competent metastatic tumor cell is capable of successfully negotiating all the steps in this sequence (Fidler & Radinsky 1990). Failure may occur at any step, which results in complete loss of metastatic behavior and elimination of the tumor cells. In fact, only a very small percentage of tumor cells that reach the circulation will survive and form metastases. Further complicating the study of metastatic phenotype development is the assumption that competency for each step of the cascade may be developed independently, in a random and reversible fashion. Although it is probable that genetic alterations may be associated with each individual step, the failure to clearly identify metastasis genes common in all tumors with this phenotype suggests that many of the alterations associated with metastatic cells may be quantitative in nature.

Fidler & Hart (1982) originally described tumor metastasis as a highly selective competition favoring the survival of a subpopulation of metastatic cells preexistent within the heterogeneous population of the primary tumor. This original view remains valid, and the metastatic subpopulation idea of Fidler is well-established by the consistent manner in which subclones of varying metastatic potential can be isolated from primary cultures of human tumors. Kerbel and colleagues (Kerbel 1990), using genetic markers for clonality, demonstrated that the metastatic subpopulation dominates the primary tumor mass early in its growth. More recently these investigators have shown that factors that behave as growth inhibitors for early stage benign tumors can switch function and act as mitogens for tumor cells of more advanced stages of disease progression, e. g. when tumor cells attain metastatic competence (Cornil et al 1991; Kerbel 1992). This effect confers a selective growth advantage to the small numbers of metastatically competent cells present within the primary tumor and accounts for the observed clonal dominance of these metastatic cells in the growing primary and at distant metastatic foci.

TUMOR-HOST INTERACTIONS MODULATE THE METASTATIC PROCESS

Orthotopic implantation of human colon carcinoma cells into nude mice demonstrates that the organ environment clearly influences the metastatic capacity of the injected tumor cells (Fidler 1991; Fidler et al 1990; Nakajima et al 1990; Stephenson et al 1992). Regardless of the malignant potential of the tumor cells in the original human patient, the transplanted human colon tumor cells did not metastasize unless they were injected into the cecum or spleen of the nude mouse. Heterotopic subcutaneous implantation resulted in a much lower frequency of metastasis formation. Recent experiments demonstrate that orthotopic implantation of human tumor cells from breast, stomach, pancreas, and prostate into nude mice favors recapitulation of the metastatic behavior and pattern of metastasis seen for these tumors in their human hosts (Fidler 1991; Fidler et al 1990; Furukawa et al 1993; Manzotti et al 1993; Nakajima et al 1990, Stephenson et al 1992). Further enhancement of this behavior was seen when fragments of tumor tissue were placed at the orthotopic site rather than a suspension of human tumor cells passaged in vitro (Furukawa et al 1993; Stephenson et al 1992). These findings suggest that in addition to tumor cell-host interactions the organization of tumor cells within the tumor itself and interaction with tumor-derived matrix may also influence metastatic behavior.

DEFINITION OF THE INVASIVE/METASTATIC PHENOTYPE: The Expanded Three-Step Hypothesis

Many of the steps in metastasis formation require specific interactions with the extracellular matrix, and the nature and degree of these matrix interactions will change from step to step during the metastatic process. Only the metastatically competent cells will successfully navigate each of these interactions. Escape of tumor cells from the primary tumor may require decreased adhesiveness to the tumor or stromal matrix. However, the arrest in the target organ that results in tumor-specific patterns of metastasis formation may be mediated by specific tumor-endothelial interactions and selective binding to specific matrix components (Pauli et al 1990; Zhu et al 1991). Tumor cells also may respond differently to various extracellular matrices and stromal cells that are encountered during metastasis formation. The concept of dynamic reciprocity, i.e. normal cells that produce extracellular matrix are also influenced by that matrix (Sage & Bornstein 1991), is also valid for tumor cells and the extracellular matrices that they encounter. However, the responses of tumor cells to various matrix components may also be anomalous when compared with that of normal cells, just as malignant tumor cell response to growth factors may be paradoxical when compared with that of benign or normal cells (Kerbel 1992).

Malignant tumor cell interactions with the extracellular matrix are different from those of normal or benign cells. The behavior of the malignant tumor cell is characterized by its tendency to cross tissue boundaries, intermix with cells of the various compartments, and metastasize to distant sites. It is now recognized that this invasive behavior is also shared by a number of normal cell types and occurs to a limited degree in other physiologic and pathologic conditions (Mareel et al 1990). For example, trophoblasts invade the endometrial stroma and blood vessels to establish contact with the maternal circulation during development of the hemocorial placenta.

Endothelial cells must invade basement membranes and interstitial stroma during angiogenesis. The striking similarity at the molecular level between these processes leads us to suggest that the ability of tumor cells to cross multiple tissue boundaries is the result of loss of control over the expression of the invasive phenotype observed in these normal cells (Liotta et al 1991). The invasive phenotype of tumor cells may be viewed as a quantitative escalation, or loss, of regulation over the normal invasive behavior of activated endothelial cells or trophoblasts, and acquisition of this phenotype is essential for successful completion of many steps in the metastatic cascade.

A successful strategy for studying the invasive phenotype has been to define and examine individual steps within this process. Historically, tumor cell interaction with the basement membrane is defined as the critical event of tumor invasion that signals the initiation of the metastatic cascade (Fidler & Radinsky 1990; Liotta et al 1991). The apparent emphasis on interaction of tumor cells with the basement membrane arises because of the presence of these connective tissue barriers at multiple key points in the metastatic cascade: escape from the primary tumor in epithelial malignancies, intravasation and extravasation during hematogenous dissemination, and perineural and muscular invasion. These basement membranes, composed of a dense meshwork of collagen-type IV, laminin, and heparan sulfate proteoglycans, do not normally contain pores that would allow passive tumor cell migration. Tumor cell traversal of basement membrane barriers is the result of acquisition of an invasive phenotype that can be separated into three steps: attachment, local proteolysis, and migration. However, it should be emphasized that these three steps describe tumor cell interaction with all extracellular matrices and not just basement membranes. The nature of the specific interaction (i.e. tumor cell types and type of matrix) may result in emphasis of some steps over others at particular points in the metastatic cascade. It is through the repetitive cycling of these three steps that tumor cells accomplish many of the processes necessary for successful metastatic behavior.

Metastasis is therefore a multistep process involving numerous tumor cell-host interactions. These interactions are defined by the invasive phenotype that is dominated by the ability of tumor cells to attach to the extracellular matrix, to degrade matrix components, and then migrate through these matrix defects. None of these functions is unique to tumor cell behavior. The difference between normal invasive processes and the pathologic nature of tumor cell invasion is therefore one of regulation. An understanding of the factors that control cellular processes essential to the invasive phenotype should allow identification of new therapeutic targets for prevention and treatment of metastasis formation.

CELL ADHESION DURING INVASION AND METASTASIS

A number of specific cell-surface-associated molecules that modulate cell-matrix and cell-cell interactions have been characterized. These include the integrins, a 67 kd laminin-binding protein, cadherins, Ig superfamily, and CD44. The role of these classes of molecules in tumor growth, invasion, and metastasis is under active investigation. As mentioned previously, tumor cells must show both decreased cell and matrix adhesive properties as well as enhancement of these functions at various stages of the metastatic process. Therefore, the apparent contribution of each class of cell adhesion molecules to the net cellular and matrix adhesiveness of tumor cells will be dependent on a variety of factors including the metastatic capacity of the tumor cell population under study and the model system used to study these cells.

Integrins are a family of cell-surface receptors that mediate cell adhesion. At least twenty different integrins have been characterized to date. They are formed by various noncovalent associations of 14 α and 8 β subunits to form heterodimers (Albelda 1993; Hynes 1992; Ruoslahti & Pierschbacher 1987). The integrins were originally identified as receptors for extracellular matrix proteins such as collagens, fibronectin, laminin, and vitronectin. Some integrins may also function as cell-cell adhesion molecules. Work has shown considerable redundancies within the integrin family in that most of the integrins bind to more than one ligand, and ligands can be recognized by more than one integrin.

The effects of integrin function on invasion and metastasis have been examined. A number of integrins bind via recognition of the RGD sequence common to a number of adhesive molecules including fibronectin, vitronectin, and other adhesion proteins (Albelda 1993; Ruoslahti & Pierschbacher 1987; Yamada 1991). The receptor functions associated with these integrins can be inhibited by synthetic RGD-containing peptides. These peptides have been used to disrupt integrin functions and successfully inhibit both in vitro and in vivo melanoma cell invasion (Gehlsen et al 1988; Humphries et al 1986, 1988; Saiki et al 1989). However, relatively high peptide concentrations (from 100–1000 $\mu g\ ml^{-1}$ for in vitro assays; 0.2–3 mg per injection with in vivo models) are required to achieve these effects. Recent work has shown a similar inhibitory activity of RGD-containing peptides, the so-called dysintegrins, obtained from snake venom.

Other evidence shows that expression of the vitronectin receptor ($a_v\beta_3$ integrin) is elevated in malignant melanoma cells (Albelda et al 1990; Gehlsen et al 1992). This receptor is also overexpressed in glioblastoma multiforme (Gladson & Cherech 1991). Most recently, experiments with the human melanoma cell line A375M have revealed that the enhanced ability of these cells to invade basement membranes in response to treatment with anti-$a_v\beta_3$ antibodies may be related in part to an increased expression of the matrix-degrading enzyme, 72-kd type IV collagenase (gelatinase A) (Seftor et al 1992). Signal transduction through this vitronectin receptor may modulate proteolytic enzyme production to enhance the invasive phenotype.

Cell surface receptors for laminin may mediate adhesion of tumor cells to the basement membrane prior to invasion (Rao et al 1982; Wewer et al 1986). Laminin is known to play a key role in cell attachment, cell spreading, mitogenesis, neurite outgrowth, morphogenesis, and cell movement. Many types of neoplastic cells contain cell-surface-binding sites for laminin with affinity constants in the nanomolar range. The isolated laminin receptor is a 67-kd protein that binds to the B chain (short arm) region of the laminin molecule (Wewer et al 1986, 1987). Breast carcinoma and colon carcinoma tissue contain a higher number of unoccupied receptors compared with benign lesions. The 67-kd laminin receptors of normal epithelium are polarized at the basal surface and occupied with laminin in the basement membrane. In contrast, the 67-kd laminin receptors on invading carcinoma cells are amplified and dispersed over the entire surface of the cell. Laminin adhesion can be shown experimentally to play a role in hematogenous metastasis (Barsky et al 1984; Liotta 1986). Pretreatment of tumor cells with very low concentrations of the receptor-binding fragment from the laminin molecule markedly inhibited or abolished the formation of lung metastasis from tumor cells injected intravenously. Recent studies using the MLuC5 monoclonal antibody, which is specific for the 67-kd laminin receptor, have shown that expression of this marker is associated with poor prognosis in human breast cancer patients (Martignone et al 1993). Multivariant analysis showed that the 67-kd laminin receptor is an independent prognostic factor, which indicates its predictive value in relation to overall survival.

Recent studies have indicated an inhibitory role for cell adhesion molecules (CAMs) of the cadherin family in the process of metastasis. Cadherins are calcium ion-dependent CAMs that mediate cell-cell binding (Takeichi 1990). Three subtypes (E-, N-, and P-cadherins) have been identified in mammals, and they are primarily distinguished by tissue distribution. E-cadherin, also termed uvomorulin, plays an anti-metastatic role in vitro with epithelial Madin-Darby canine kidney (MDCK) cells (Behrens et al 1989; Uleminckx et al 1991). When treated with monoclonal anti-E-cadherin antibodies, these cells acquired in vitro invasive capability (Takeichi 1990). Further studies manipulating the expression of E-cadherin have been informative (Uleminckx et al 1991). Transfection of plasmids containing the sense strand of E-cadherin messenger RNA in highly invasive clones resulted in overexpression of E-cadherin protein and the loss of invasive capacity. The partial down-regulation of E-cadherin by transfection of anti-sense mRNA in a noninvasive clone resulted in the acquisition of invasive behavior. These results indicate that enhancing the ability of tumor cells to bind to one another or to other host cells inhibits the ability of tumor cells to escape from their primary site to initiate invasion. Therefore, these studies provide evidence that E-cadherin can act as an invasion suppressor molecule.

Other cell adhesion molecules include the Ig superfamily members such as N-CAM, and VCAM-1. The Ig superfamily incorporates a wide variety of proteins that all share the immunoglobulin homology unit, which consists of 70 to 110 amino acids organized into 7–9 β-sheet structures (Albelda 1993). Family members include molecules involved in a variety of cell functions including cellular immunity and signal transduction, as well as cell adhesion (Hunkapiller & Hood 1989; Williams & Barclay 1988). This diversity makes generalizations about the role of Ig superfamily members in tumor cell invasion difficult. However, the role of VCAM-1 in facilitating the metastatic process appears straightforward. VCAM-1 (also known as INCM-110) was identified on endothelial cells as a cytokine-inducible, counter-receptor for the VLA-4 ($\alpha_4\beta_1$) integrin (Elices et al 1990; Osborn et al 1989; Rice & Bevilacqua 1989). VLA-4 is found primarily on white blood cells and functions in mediating leukocyte-endothelial cell attachment. Recently it was shown that malignant melanoma cells may also express VLA-4 (Albelda et al 1990; Osborn et al 1989). Thus VCAM-1 may serve as a tumor adhesion receptor, facilitating interaction of circulating melanoma cells with the endothelium in advance of tumor cell extravasation.

Enhancement of tumor cell attachment may also be facilitated by CD44. CD44 is a widely distributed integral membrane protein that appears to function as a receptor for hyaluronic acid (Aruffo et al 1990; Culty et al 1990; Gallatin et al 1991). It exists in a variety of M_r forms ranging from 85 to 160 K. A 90 K form found on leukocytes is thought to be important in lymphocyte homing. Larger isoforms are found on epithelial and mesenchymal cells. The epithelial form was more highly expressed in carcinoma, and overexpression was also seen in widely disseminated large-cell lymphomas (Horst et al 1990; Stamenkovic et al 1989). In another study, melanoma cells were sorted by fluorescence-activated cell sorting (FACS) analysis using CD44 antibodies (Birch et al 1991). Melanoma cells expressing higher CD44 levels resulted in significantly more lung colonies following intravenous injection into the nude mouse. These studies suggest that enhanced tumor cell binding to hyaluronic acid may effect tumor cell implantation and is important for attainment of full metastatic competency.

Altered tumor cell-matrix interactions may also contribute to the phenotypic changes associated with invasion. Kleinman and colleagues have shown that nontumorigenic NIH-3T3 cells will become tumorigenic, locally invasive, and form highly vascularized tumors following co-injection of the cells with reconstituted basement membrane, matrigel (Fridman et al 1992b). Investigators have also shown that a number of human tumor cell lines that are nontumorigenic in the nude mouse will become tumorigenic if co-injected with matrigel (Fridman et al 1991). The specific components of the basement membrane responsible for these effects have not been directly identified. However, a 19 residue synthetic peptide derived from the laminin A chain appears to enhance the metastatic phenotype of murine melanoma cells in vivo (Kanemoto et al 1991). The active sequence within this peptide has been identified as the hexapeptide sequence SIKVAV, which induces proteolytic activity and angiogenesis normally associated with the invasive phenotype (Sweeney et al 1991).

Recent work has shown that tumor cells and normal cells may be differentiated in vitro by the nature of their interactions with reconstituted basement membranes. Reconstituted basement membrane (matrigel) was used to culture normal and malignant breast cells and tissues (Petersen et al 1992). The normal cells were distinguished by their ability to re-express a differentiated phenotype as evidenced by formation of true acini within the matrix. Human breast cancer cell lines were easily distinguished by their lack of polarity and lack of a continuous basement membrane. As suggested, this system could be used to define and identify tumor supressor genes involved in preserving normal cell-matrix interaction and communication.

Thus the current evidence regarding the involvement of cell-matrix and cell-cell interactions in the process of metastasis has pointed to the complexity of the malignant phenotype. Studies of fibronectin, vitronectin, and laminin receptor functions indicate that inhibition of tumor cell adhesion results in less aggressive invasive behavior. Data suggest that both decreased tumor cell attachment through modulation of integrin receptor function and enhanced attachment via CD44 and VLA-4/VCAM systems are important in formation of tumor metastasis. Although these data appear to be somewhat contradictory, they most likely address the differences between growth control (tumorigenicity), which may be modulated by matrix receptor function, and the process of invasion, which is necessarily associated with the ability to attach to the extracellular matrix in order to arrest, extravasate, and migrate. The ability to invade may require an intermediate expression of adhesive capability, with too little resulting in the inability of a circulating tumor cell to arrest and invade a secondary site. At the other end of the spectrum, cells that are extremely adherent, either to the extracellular matrix or to each other, may be unable to begin the metastatic process and migrate from the primary tumor. The resolution to these conflicts lies in defining the specific control mechanisms through which cellular adhesion and motility are coupled to produce the invasive phenotype.

PROTEOYLSIS OF THE EXTRACELLULAR MATRIX DURING TUMOR INVASION

Although proteolysis and migration through tissue barriers are normal cell functions in specific physiologic circumstances, clearly a general aspect of malignant neoplasia includes a shift toward sustained invasive capacity. For invasion to take place, cyclic attachment to and subsequent release from matrix components must occur in a directed and controlled manner. This implies that proteolysis, although enhanced in tumor cells, is still tightly regulated in a temporal and spatial fashion with respect to cell attachment and migration. Proteolytic activity is the balance between the local concentration of activated enzymes and their endogenous inhibitors.

Positive correlation between tumor aggressiveness and protease levels has been documented for all four classes of proteases; thiol-, seryl- aspartyl- and metallo-proteases. All of the enzymes implicated in this association have been identified in normal cells. The association of these proteases with the invasive process is through inappropriate overexpression in the tumor tissue, either by the tumor cells, host cells intermixed or immediately adjacent to the invasion front, or both. In addition to protease enhancement, augmented heparanase activity has also been associated with malignant invasion (Nakajima et al 1990, 1987).

Over the last five years a significant body of evidence has accumulated that directly implicates members of the matrix metalloproteinase (MMP) family in tumor invasion and metastasis. This enzyme family currently includes eight members. The criteria used to define members of the MMP family include zinc metal-atom dependency, secretion as a zymogen, in vitro activation of the proenzyme by organomercurial reagents, auto-proteolytic removal of the N-terminus following activation, and inhibition by a specific class of biological inhibitors referred to as the tissue inhibitors of metalloproteinases (TIMPs). Profragment removal following organomercurial activation is the result of intramolecular endoproteolytic activity. Thus proenzyme activation must occur prior to removal of the profragment. Profragment loss is therefore indicative of, but not equivalent to, proenzyme activation.

Subgroups within the MMP family have traditionally been defined based on substrate specificity. This family of endopeptidolytic enzymes is now subdivided into three general classes: interstitial collagenases, stromelysins, and gelatinases (type IV collagenases). The members of the MMP gene family, their domain structure, substrate specificity, and activation mechanisms have been the subject of several excellent reviews to which the reader is referred (Matrisian 1992; Nagase et al 1991; Woessner 1991).

MATRIX METALLOPROTEINASES IN TUMOR INVASION

Evidence for the role of MMP enzymes in tumor invasion and metastasis comes from a variety of studies. These include many in vitro studies of murine and human tumor cell lines that transcribe, synthesize, and secrete MMP enzymes (Lyons et al 1991; Matrisian et al 1991; Sato et al 1992; Templeton et al 1990). In fact, several of the members of the MMP family were first identified, purified, and cloned from tumor cell lines. Certainly this is true of the type IV collagenases, now known as gelatinase A and gelatinase B, transin (the murine homologue of human stromelysin) as well as stromelysin-2 and matrilysin (formerly PUMP-1). These enzymes are also synthesized and secreted by normal cells under conditions that may be associated with physiologic tissue remodeling. The difference between enzyme production under physiologic and neoplastic conditions may be that in tumor cells the enzymes may be constitutively overexpressed or induced by autocrine growth factor stimulation. Tumor cells may also be unresponsive to signals from host cells and matrix that would down-regulate MMP expression of normal cells. While overexpression in tumor cells was an aid to their identification, isolation, and characterization, it may have led to the unfounded presumption that some MMPs are tumor cell-specific.

Studies have clearly demonstrated a positive correlation between MMP expression, invasive behavior, and metastatic potential in animal models (Bonfil et al 1989; Powell et al 1993; Sreenath et al 1992). Studies have also shown that growth factors can dramatically modulate MMP expression (Brown et al 1990; Matrisian 1992; Ogata et al 1992b, Weinberg et al 1990). For example, epidermal growth factor, transforming growth factor alpha, and platelet-derived growth factor have all been shown to upregulate transcription of MMPs, most notably interstitial collagenase and stromelysin. Autocrine growth factor stimulation, in addition to supporting autonomous growth necessary for clonal dominance (Kerbel 1992), may stimulate MMP production and thereby support the metastatic phenotype at several levels.

Extracellular matrix components, cell-matrix interactions, and the pericellular environment are also important determinants of MMP production. Stimulation of the $\alpha_v\beta_3$ integrin receptor (vitronectin receptor) enhanced gelatinase A production and stimulated melanoma cell invasion (Seftor et al 1992). Laminin A chain peptide fragments induced gelatinase A production, local invasiveness and augmented metastasis formation (Fridman et al 1991; Fridman et al 1992b; Sweeney et al 1991). Acidic culture environment and tumor necrosis have been shown to enhance production of MMP activities, most notably gelatinases A and B (Bonfil et al 1992; Kato et al 1992).

Host-tumor interactions may also greatly influence protease production. Indeed the MMP profile and metastatic competence of KM12SM human colorectal tumor cells is influenced by the site of implantation (Nakajima et al 1990). Subcutaneous implantation of tumor cells did not result in visceral metastases, whereas cecal implantation led to lymph node and hepatic metastases. The tumor explants from the cecum wall demonstrated an increase in the active forms of gelatinase A, as well as significant elevation of type IV collagenolytic and heparanase activities.

CORRELATION OF MATRIX METALLOPROTEINASE EXPRESSION WITH INVASIVE BEHAVIOR OF HUMAN TUMORS

Researchers have begun to examine MMPs in human tumor tissues and serum from cancer patients. These efforts have included immunoperoxidase staining (IPS) of tissue sections for localization of MMPs in human tumor tissues, Northern blot analysis of MMP transcripts in RNA samples extracted from human tumor samples, in situ hybridization (ISH) studies of MMP transcripts, and measurement of MMP levels in the body fluids of cancer patients.

Interstitial collagenase, which degrades the triple helical domains of the fibrillar collagens (types I, II, III, and X), is augmented in many human tumors. The level of proteolytic activity against soluble type I collagen demonstrated a statistically significant correlation with the degree of histologic differentiation in human colorectal tumors (van der Stappen et al 1990). IPS studies revealed enhanced staining for interstitial collagenase in the stromal cells and collagen fibers immediately adjacent to the malignant nests of colorectal tumor cells (Hewitt et al 1991). Little evidence for collagenase staining of normal, benign, or malignant epithelium was observed. Normal colorectal tissues and adenomas were negative for interstitial collagenase staining. Elevated interstitial collagenase transcripts have been observed in 40% of primary pulmonary malignancies, but not in samples from adjacent normal lung tissue (Urbanski et al 1992). ISH demonstrated elevated mRNA transcripts for interstitial collagenase in squamous cell carcinomas of the head and neck that localized to the stromal fibroblasts immediately adjacent to the malignant tumor masses (Gray et al 1992), thus confirming earlier reports (Muller et al 1991; Polette et al 1991).

Experimental results consistently localize interstitial collagenase production to the stromal fibroblasts immediately adjacent to the site of tumor invasion, which suggests that invasive tumor epithelial release is a stimulus for induction of fibroblast synthesis of this enzyme. A tumor cell-derived collagenase stimulatory factor has been characterized and partially sequenced (Nabeshima et al 1991). This sequence data shows no homology with known growth factors, motility factors, or collagenase stimulatory agents. Originally purified from a human lung carcinoma cell line, this stimulatory factor is released into tumor cell media and is also associated with the tumor cell membranes. This tumor cell-derived collagenase stimulatory agent also stimulates the synthesis and secretion of gelatinase A and stromelysin from human fibroblasts in culture (Karaoka et al 1993).

Members of the stromelysin group of MMP enzymes include stromelysins 1, 2, and 3, as well as matrilysin. Matrilysin lacks a C-terminal domain and is the smallest member of the MMP family. These enzymes have a fairly broad range of protease activity-degrading glycoproteins such as laminin and fibronectin, proteoglycans, and nonhelical domains of type IV collagen. Matrilysin can cleave urokinase to separate the catalytic and receptor-binding domains, which suggests that matrilysin activity may be important in regulating functional activity of this plasminogen activator (Marcotte et al 1992).

Expression of stromelysins 1 and 2 has been studied in carcinomas of the head and neck. In these studies, high levels of stromelysin mRNA transcripts were correlated with increased local invasiveness (Muller et al 1991). Transcripts were localized principally to the fibroblasts of the tumor stroma adjacent to areas of basement membrane disruption (Polette et al 1991). ISH studies have also shown that matrilysin, but not stromelysin 1 or stromelysin 2, are overexpressed in human gastric and colonic carcinomas (McDonnell 1991).

Matrilysin transcripts were also observed in 14 out of 18 RNA samples isolated from human prostate adenocarcinomas and in 3 out of 11 normal prostate biopsy samples (Pajouh et al 1991). ISH localization showed that matrilysin was expressed in the epithelial cells of the primary prostate adenocarcinoma and in some foci of epithelial dysplasia, but not in the stroma.

The most recently described member of the stromelysin subgroup, stromelysin 3, was discovered through its association with human breast cancer progression (Basset et al 1990). The endoproteolytic activity and substrate specificity of stromelysin 3 have yet to be defined. Like other stromelysins, stromelysin 3 expression was localized to the stromal cells surrounding invasive human breast carcinomas. Stromelysin 3 expression has also been studied in squamous cell carcinomas of the head and neck (Muller et al 1993), basal cell carcinomas (Wolf et al 1992), and primary pulmonary carcinomas (Urbanski et al 1992). In all of these studies there is a consistent association of stromelysin 3 expression with the stromal fibroblasts adjacent to the malignant epithelium. In the head and neck tumors, stromelysin 3 expression correlated with the degree of local invasiveness (Muller et al 1993). Although stromelysin 3 shows a consistent association with the invasive and malignant potential of human tumors, it is by no means specific for this process. Recent demonstration of stromelysin 3 in cutaneous wound healing (Wolf et al 1992) and apoptotic process of postlactating mammary gland involution (Lefebvre et al 1992) have led to the proposal that stromelysin 3 expression corresponds to a normal fibroblast response that is exacerbated in invasive carcinomas (Wolf et al 1992). This is consistent with our original proposal that tumor-associated MMP overexpression is a dysregulation of normal protease systems.

The third group of enzymes of the MMP gene family are the gelatinases A and B, formally referred to as type IV collagenases. These enzymes rapidly degrade denatured collagens (gelatin), as well as a number of native collagen types that contain helical disruptions. The 72-kd gelatinase A was first described through its ability to degrade pepsinized, triple-helical type IV collagen and its association with tumor cell invasion of the basement membrane. The two gelatinases arise from separate mRNA transcripts (Collier et al 1988; Wilhelm et al 1989) and are distinct from other members of the MMPs in that they possess a unique region immediately adjacent to the putative metal-binding domain that is homologous to the gelatin-binding domain of fibronectin and may function in substrate binding. Gelatinase A and B also differ from other members of the family by their ability to interact, as latent proenzymes, with the endogenous inhibitors of these enzymes, the TIMPs (Goldberg et al 1989; Stetler-Stevenson et al 1989; Wilhelm et al 1989). These proenzyme-inhibitor complexes are specific with progelatinase A binding TIMP-2 and progelatinase B binding TIMP-1.

Evidence for the expression of gelatinase A in human tumors is abundant. Most breast, colonic, and gastric adenocarcinomas are immunoreactive for 72-kd gelatinase A, whereas benign proliferative disorders of these tissues are negative (Campo et al 1992a,b; Clavel et al 1992; D'Errico et al 1991; Levy et al 1991; Monteagudo et al 1990; Ohori et al 1992; Pyke et al 1992). IPS for gelatinase A in malignant breast epithelium was reportedly more frequent (16 out of 22 cases) than either stromelysin 1 (12 cases) or interstitial collagenase (9 cases) (Clavel et al 1992). Strong immunostaining of the malignant epithelium both in invasive and preinvasive prostatic carcinoma has been reported (Boag & Young 1993). Recent studies of serous tumors of the ovary were unable to detect gelatinase A in benign cysts, yet invasive growths were positive (Campo et al 1992b). IPS studies of the 72-kd gelatinase A enzyme also has been found to correlate with tumor grade in neoplastic thyroid (Campo et al 1992a). However, gelatinase

A was also detected in benign disorders in which the tissue was undergoing remodeling and repair. Gelatinase A has also been demonstrated in the sclerosing and mucinous variants of brochioalveolar carcinoma, possibly contributing to the poorer prognoses of these subgroups (Ohori et al 1992). In summary, IPS studies have shown expression of gelatinase A in many types of human tumors, and this expression is usually limited to the malignant epithelial cells.

Elevated gelatinase mRNA transcripts have also been identified in RNA extracted from primary human pulmonary carcinomas (Urbanski et al 1992), colon carcinomas (Levy et al 1991), and primary breast cancers (Basset et al 1990). Gelatinase B was found in five of nine pulmonary tumors. Although gelatinase A transcripts were expressed in the majority of carcinomas, they were also occasionally present in normal uninvolved lung tissue. These observations are similar to those previously reported for the gelatinases when studied in primary breast cancers (Basset et al 1990).

Cellular localization of gelatinase production has been performed by ISH in a number of tumor systems. Signals for both gelatinases were distributed over the neoplastic epithelium as well as stromal elements of the primary pulmonary tumors (Urbanski et al 1992). ISH detected gelatinases A and B in infiltrating basal and squamous cell carcinomas (Pyke et al 1992) that localized to the stromal fibroblasts adjacent to the sites of tumor invasion. Gelatinase B has also been localized to eosinophils infiltrating the dermis in response to invasive basal cell carcinoma. ISH studies of human colon carcinomas demonstrate gelatinase A localized primarily to the stromal fibroblasts immediately adjacent to sites of invasion (Poulsom et al 1992). This stromal localization is distinct from the predominant IPS localization of gelatinase A in the malignant epithelium performed in the same report.

This discrepancy between the stromal localization of gelatinase A mRNA transcripts by ISH and the predominant IPS reactivity in the malignant epithelial cells has led some investigators to propose that the stromal fibroblasts at the invasive front are resposible for the bulk of gelatinase A expression associated with tumor invasion. It has also been suggested that this discrepancy is evidence for a cell-surface receptor on tumor cells for gelatinase A enzyme. Indeed, such a surface receptor for gelatinase A has been identified on human breast cancer cell lines MCF7 and MDA MB231, and preliminary characterization has been reported (Emonard et al 1992). However, alternative explanations for this discrepancy are equally likely. Given the diffuse IPS staining pattern for gelatinase A in many human tumors, it seems rather unlikely that stromal fibroblasts at the invasive front are responsible for producing enzyme that is detected at the center of the tumor mass. Thus it is unlikely that the stromal fibroblasts are the sole source for this enzyme. Another possible explanation for the discrepancy between ISH and IPS results is the relative sensitivity of these methods. Tumor cells may have low constitutive levels of gelatinse A transcriptional activity, whereas stromal fibroblasts may have a strong, highly induced, but short, temporal burst of gelatinase A transcriptional activity in response to stimuli at the invasion front. Fibroblasts that are reactive to the invasive stimulus may have high gelatinase A transcriptional activity, but low translational activity that results in low cytoplasmic stores of this enzyme. This question could be addressed by in vitro co-culture experiments and/or comparison of transcriptional and translational activity in tumor cells vs activated fibroblasts.

The correlative studies of MMP expression in human tumors suggest that these proteases are important in the biology of human tumors and may make a significant contribution to the invasive phenotype of such tumors. The picture obtained from these studies, however, is complex and still incomplete. Clearly no single member of the MMP family is responsible for the invasive phenotype of all human malignancies. Invasive tumors apparently utilize a number of different strategies for breakdown and crossing of matrix barriers as reflected in the heterogeneity of the MMP enzymes and other proteases expressed during this process.

Correlation of metalloproteinase expression with local recurrence, lymph node metastasis, distant metastasis, and/or patient survival allows assessment of the utility of these measurements as diagnostic or prognostic markers. Overexpression of gelatinase A has been correlated with local recurrence in human breast cancer (Daidone et al 1991) as well as with an increased frequency of lymph node metastasis in human gastric carcinoma (Otani 1990). Both gelatinase A and stromelysin expression have been correlated with lymph node metastasis and vascular invasion in human esophageal carcinoma (Shima et al 1992). IPS studies of gelatinase A in 46 patients with squamous carcinoma of the head and neck found high level expression in 77% of patients with lymph node metastases, but only 25% in patients without lymph node metastases (Kusukawa et al 1993). Expression of gelatinase A appears to be a useful marker for evaluating the malignant potential of individuals with squamous cancers of the oral cavity.

Gelatinase A and B enzymes are normal components of human plasma, and recent studies have begun to evaluate the diagnostic utility of measurement of these enzyme levels in plasma, serum, and other body fluids from cancer patients. Measurement of total serum gelatinase A levels using a substrate-capture assay correlated tumor burden in patients with primary pulmonary malignancies and may be a useful indicator of response to therapy (Garbisa et al 1992). Plasma levels of gelatinase A were not elevated in patients with breast or colon cancer (Zucker et al 1992), but gelatinase B levels were elevated in patients with cancer of the colon or breast, with no significant elevation in plasma levels of this enzyme in patients with lung tumors, genitourinary cancer, or leukemia/lymphomas (Zucker 1993). Significant elevations of the plasma gelatinase B levels were also observed during pregnancy. Concentrations of the gelatinases A and B in the pleural fluids of a variety of patients were independent of the serum concentration of these enzymes, but were not specific in discriminating nonmalignant from malignant pleural effusions (Hurewitz et al 1992). As a group, patients with transitional cell cancer of the bladder showed statistically significant elevation of gelatinase A and fragments of this enzyme in their urine (Margulies et al 1992).

PROENZYME ACTIVATION: An Important Control Point in Development of the Invasive Phenotype

All of the studies reviewed to this point have determined levels of total MMP enzyme protein or steady-state levels of MMP mRNA transcripts. However, it is known that all members of the MMP enzyme family are secreted in zymogen form and must be activated in the extracellular milieu prior to obtaining extracellular matrix degrading activity. This suggests that measurement of active enzyme species may be more informative with respect to the state of matrix turnover and possibly of more diagnostic or prognostic utility. Investigators have utilized quantitative gelatin zymography to assess the contribution of active gelatinase species to the invasive phenotype of human breast and non-small cell lung cancer (Brown et al 1993a,b; Davies et al 1993). These studies demonstrate that although many tumor tissues express both gelatinase A and B in zymogen form, only small amounts of the activated form of the 92-kd gelatinase B enzyme could be detected in some tumors. The fraction of total gelatinase A enzyme present as the 62-kd activated form of the enzyme was statistically elevated in malignant disease, and a higher proportion of this active enzyme species was detected in higher grade tumors (Davies et al 1993). Detection of activated forms of gelatinase A enzyme occurs more frequently in invasive human breast tumors than gelatinase B, which implies that activation of the gelatinase A is a feature of the invasive phenotype in breast cancer patients, whereas activation of gelatinase B is not. These studies emphasize that these enzymes are secreted as latent proenzymes and require activation prior to obtaining proteolytic activity. Overproduction of proenzyme species is necessary but not sufficient for the development of the invasive phenotype, and proenzyme activation is an important step in the acquisition of the invasive phenotype.

CELLULAR MECHANISM FOR ACTIVATION OF GELATINASE A (MMP-2)

Gelatinases A and B, like other members of the collagenase family, are secreted as latent proenzymes and must be activated extracellularly. But unlike other proenzymes in this family, progelatinase A and B may be complexed with endogenous inhibitors known as tissue inhibitors of metalloproteinases (TIMP-1 and TIMP-2). Progelatinase A selectively binds TIMP-2 and progelatinase B binds TIMP-1 preferentially (Goldberg et al 1989; Stetler-Stevenson et al 1989; Wilhelm et al 1989). The activation of these enzymes and/or enzyme-inhibitor complexes therefore constitutes an important, possibly unique, level of regulation for these enzymes. Recent in vitro studies of the mechanism of activation and domain structure of these enzymes has yielded new insights into the molecular basis of proenzyme latency and proenzyme-inhibitor interactions (Fridman et al 1992a; Howard & Banda 1991; Howard et al 1991; Kleiner et al 1992, 1993; Murphy et al 1992b).

Progelatinases A and B, like all matrix metalloproteinases with the exception of stromelysin-3, can be activated in vitro by a variety of agents including organomercurials, chaotropic agents, and other proteases. Proteases, specifically plasmin, have been shown to be responsible for the activation of both interstitial procollagenase and prostromelysin-1 in co-cultures of keratinocytes and dermal fibroblasts (He et al 1989; Murphy et al 1992a). The activation of the 92-kd gelatinase B has also been studied in vitro. Like interstitial collagenase and stromelysin-1, the 92-kd gelatinase can be activated by other proteinases such as trypsin and plasmin. Recent studies have shown that gelatinase B can be activated by stromelysin and that formation of TIMP-1/progelatinase B complex can modulate this activation (Goldberg et al 1992; Murphy et al 1992a; Ogata et al 1992a). Both plasmin and cathepsin B have been shown to activate prostromelysin, which suggests that either may initiate a cascade of MMP activation, similar to that of the blood clotting mechanism, as has been proposed by Matrisian (1992) and Murphy et al (1992a). However, plasmin and a variety of other proteases do not activate the 72-kd gelatinase A (Okada et al 1990).

Activation of gelatinase A or the gelatinase A/TIMP-2 complex does not appear to occur in the soluble phase as part of a proteolytic cascade similar to other members of the matrix metalloproteinase family (Brown et al 1990, 1993c; Murphy et al 1992b, Okada et al 1990; Ward et al 1991). Activation of this enzyme requires interaction with the cell surface, and this cellular activation mechanism requires the presence of an intact C-terminal domain on the gelatinase A (Murphy et al 1992b). Treatment of HT-1080 fibrosarcoma cells with phorbol ester or treatment of fibroblasts with concanavalin A induces processing of the 72-kd gelatinase A to a 62-kd activated form (Brown et al 1993c; Overall & Sodek 1990). The activity responsible for processing the latent complex to active enzyme was found to be confined to the cell monolayer and was not secreted into the soluble fraction. The existence of a plasmin-independent, cell-surface mechanism for the specific activation of a single collagenase family enzyme has important physiologic implications. Such a mechanism would give tight cellular control over matrix degradation by this enzyme and would limit proteolysis to the immediate vicinity of the cell surface. With respect to tumor invasion and metastasis, it might allow a tumor cell that expresses the activator, but not gelatinase A, to activate and utilize exogenous gelatinase A produced by stromal cells in response to the local presence of invading tumor cells. The identification of the molecular species responsible for the cellular activation of the 72-kd gelatinase A is likely to yield an important new series of target molecules in the development of treatments for invasive and degradative diseases.

TISSUE INHIBITORS OF METALLOPROTEINASES DEFINE THE FUNCTIONAL ROLE OF MMPS IN CELL INVASION

Activation of proenzyme is an important control point for development of the invasive phenotype, but activation alone may not be sufficient to obtain the invasive phenotype if this occurs in the presence of excess tissue inhibitors of matrix metalloproteinases (TIMPs), the endogenous and ubiquitous inhibitors of MMPs. Thus it is the balance of active enzyme and TIMP that will determine if local matrix degradation occurs. The numerous correlative studies outlined above suggest that overexpression of matrix metalloproteinase proenzymes and subsequent activation is the mechanism by which tumor cells achieve a balance in favor of proteolysis. Recent direct evidence, using exogenous TIMPs to alter this protease/inhibitor balance, has been obtained and establishes an effector role for MMP activities in tumor cell invasion.

TIMP-1, the first member of the TIMP family to be identified (Carmichael et al 1986; Cawston et al 1981; Welgus & Stricklin 1983) is a glycoprotein with an apparent molecular mass of 28.5 kd. TIMP-1 forms a complex of 1:1 stoichiometry with activated interstitial collagenase, activated stromelysin, and both activated and progelatinase B. The *Timp-1* gene has been mapped to the p11 region of the human X chromosome (Mahtani & Willard 1988; Willard et al 1989).

TIMP-2 is a 21 kd, non-glycosylated protein that shows 37% amino acid identity and 65.6% overall homology to TIMP-1, yet the proteins are immunologically distinct (Boone et al 1990; DeClerck et al 1989, 1991, 1992a; Goldberg et al 1989; Stetler-Stevenson et al 1990). The *Timp-2* gene has been localized to chromosome 11 in the mouse and human chromosome 17q25 (DeClerck et al 1992b, Stetler-Stevenson et al 1992). Northern blot analysis has revealed the existence of two *timp-2* transcripts of 3.5 and 1.0 kb. cDNA probes to *timp-2* are specific and do not detect the single 0.9 kb *timp-1* mRNA species. *timp-1* and *timp-2* expression are regulated independently. The pattern of expression and specific roles for these inhibitors in vivo remain areas of active investigation.

Native or recombinant TIMP-1 has been shown to inhibit in vitro invasion of human amniotic membranes (Mignatti & Robbins 1986; Schultz et al 1988) and in vivo metastasis in animal models (Alvarez et al 1990; Schultz et al 1988). Furthermore, transfection of antisense TIMP RNA into mouse 3T3 cells, which down-regulates TIMP-1 expression, enhances their ability to invade human amniotic membranes and to form metastatic tumors in athymic mice (Khokha et al 1989). TIMP-2 has been shown to successfully inhibit in vitro tumor cell invasion of extracellular matrices (Albini et al 1991; DeClerck et al 1991). Overexpression of TIMP-2 in invasive and metastatic *ras*-transformed rat embryo fibroblasts resulted in suppression of the ability of these cells to form lung colonies following intravenous injection in nude mice (DeClerck et al 1992a). Increased TIMP-2 levels significantly reduced the in vivo growth rate and invasive character of tumors following subcutaneous injection of these transfected cells. Recent studies have shown an inverse correlation between TIMP-2 expression and invasive potential of the HEp-3 human epidermoid carcinoma cell lines following extended in vitro culture and passage (Testa 1992). This suggests that in some tumors enhanced TIMP-2 expression may alter the balance of activated MMP and inhibitor, thus inhibiting matrix proteolysis.

Recent studies also suggest that TIMPs are capable of inhibiting angiogenesis, which has many functional aspects similar to the process of tumor cell invasion. Mignatti et al reported that TIMP-1 inhibited in vitro endothelial cell invasion of human amniotic membranes (Mignatti & Robbins 1986). Moses et al presented data showing that cartilage-derived inhibitor (CDI), a TIMP-related protein isolated from bovine articular cartilage, can block angiogenesis and also inhibit endothelial cell proliferation (Moses & Langer 1991; Moses et al 1990). Both TIMP-1 and TIMP-2 have been shown to inhibit chick yolk sac vessel morphogenesis in response to polyamines (Takigawa et al 1990). The balance of TIMPs and gelatinase activity appears to be an early and critical determinant of endothelial morphogenesis and tube formation during in vitro growth on matrigel matrix (Schnapper et al 1993). Collectively, these data support a role for collagenolytic activity in at least two functional processes contributing to metastasis. That is, collagenases are involved in tumor cell invasion as well as neovascularization, upon which solid tumor growth is dependent. These are important considerations in the design of proteolytic inhibitors for potential use as therapeutic agents.

THE INVASIVE PHENOTYPE IS A BALANCE OF ACTIVE PROTEASES AND THEIR INHIBITORS

The critical nature of the balance between active proteases and their inhibitors for the success of the invasive phenotype has been demonstrated in two experimental systems. The first system is an in vitro model of tumor cell invasion using amniotic membranes. Researchers have demonstrated a bimodal relationship between invasion of amniotic membranes and plasminogen activator activity (Tsuboi & Rifkin 1990). In these experiments it was shown that the in vitro invasive behavior of Bowes' melanoma cells, which produce large amounts of tissue plasminogen activator, and HT1080 fibrosarcoma cells, which produce large amounts of urokinase-type plasminogen activator, is not blocked but enhanced by the addition of plasmin inhibitors or anti-plasmin antibodies. Conversely, the invasive capacity of cells that produce low levels of these proteases was blocked by the addition of inhibitors. Successful cellular invasion requires a balance of proteases and protease inhibitors. Protease activity in excess of the optimal level may result in uncontrolled local matrix degradation and interrupt cell-matrix interactions necessary for invasion.

The second system is a model for angiogenesis. As mentioned previously, angiogenesis shares many functional similarities with tumor metastasis including the requirement for expression of the invasive phenotype and associated proteolytic activity. It has been shown that agents that induce angiogenesis, such as bFGF, will induce endothelial expression of both urokinase (uPA) and plasminogen-activator inhibitor-1 (PAI-1), with the balance slightly in favor of uPA (Pepper et al 1990). In these assays, a balance in favor of protease inhibition resulted in the formation of solid cords of endothelial cells rather than tubes. Montesano and co-workers (1990) studied the formation of angiomas (benign endothelial tumors) in fibrin gels. The angioma-forming cell lines produced hemangioma-like cystic structures in the fibrin gels. Addition of exogenous serine protease inhibitors resulted in the formation of endothelial cords instead of cystic structures. Thus protease/antiprotease balance can alter the morphology of the capillary tube with excessive proteolysis resulting in sac-like noninvasive structures.

THE ROLE OF TUMOR CELL MOTILITY IN INVASION AND METASTASIS

Following organogenesis and tissue differentiation, the migration of epithelial cells is a very rare event. This is in contrast with the active cell motility that is a necessary feature of invasive carcinoma cells. Active cell motility, coupled with matrix proteolysis, is required for the penetration of extracellular matrices during expression of the invasive phenotype. Invasive tumor cells also exhibit directional motility during intravasation and extravasation. The importance of tumor cell motility to invasion and metastasis has been appreciated for more than 40 years (Coman 1953; Enterline & Coman 1950). Tumor cell lines known to be more highly invasive and metastatic showed a higher degree of motility than their low metastatic counterparts, when parameters such as pseudopod extension, membrane ruffling, and vectorial translation were measured.

A variety of agents appear to stimulate motile responses in tumor cells in vitro including host-derived scatter factors (Rosen et al 1990; Weidner et al 1990), growth factors (Aznavoorian et al 1990; Jouanneau et al 1991; Kahan & Kramp 1987; Stracke et al 1988), components of the extracellular matrix (Lester et al 1989, 1991; McCarthy & Furcht 1984; McCarthy et al 1986; Taraboletti et al 1987; Yusa et al 1989), hyaluronan (Turley 1992; Turley et al 1991), and tumor-secreted factors (Atnip et al 1987; Liotta et al 1986; Ohnishi et al 1990; Seiki et al 1991; Siletti et al 1992; Strackc ct al 1992). Motility stimulated by each of these factors can be either random in nature (chemokinesis) or directed (chemotaxis). Chemotaxis is defined as directional migration of cells in response to concentration gradients of soluble factors. Tumor cells also migrate in a directional manner, in the absence of soluble attractant, towards substratum-bound, insoluble ECM proteins (haptotaxis) (Aznavoorian et al 1990; McCarthy et al 1985). These numerous stimuli could provide tumor cells with multiple opportunities for transiting different microenvironments during the metastatic process.

A family of tumor cell-derived motility-inducing cytokines has been discovered and termed autocrine motility factors (AMF). Studies with cultured human A2058 melanoma cells demonstrated that these cells produced an attractant material in serum-free media that was approximately 60 kd in molecular mass (Liotta et al 1986). This AMF stimulated both random (chemokinetic) and directed (chemotactic) motility in the same cells in which it was synthesized. The discovery of AMF led to the proposal that cells in the primary tumor presumably secrete AMF until the concentration rises enough to stimulate motility via receptors on the responding cells.

AMF stimulates chemotaxis in A2058 melanoma cells through a pertussis toxin- (PT) sensitive receptor (Stracke et al 1987), which results in directed pseudopodial protrusion (Guirguis et al 1987). However, a variety of agents that affect adenylate cyclase have no effect on AMF-stimulated motility, thus indicating that cAMP is not the necessary second messenger (Stracke et al 1987). As with phagocyte chemotaxis (Garcia-Castro et al 1983), methylation of phospholipids appears to be a component of the biochemical cascade of activation (Liotta et al 1986). AMF specifically induces enhanced motility in a variety of tumor cells, but fails to stimulate leukocyte migration (Liotta et al 1986). Other tumor-derived factors with a similar molecular mass have subsequently been reported and purified by several investigators (Atnip et al 1987; Ohnishi et al 1990; Schor et al 1988; Siletti et al 1992).

A potent new motility stimulator of approximately 120 kd was isolated from A2058 cell-conditioned medium, purified to homogeneity (Stracke et al 1992), and was termed autotaxin (ATX). ATX is a basic glycoprotein (pI ~7.7), active in the picomolar range, that stimulates both chemotactic and chemokinetic responses in A2058 cells. Identical to AMF, cells pre-treated with pertussis toxin lack a motile response to purified ATX. Direct sequencing of ATX peptides has shown no sequence homologies with known growth or motility factors and may represent a new member of the AMF family.

The use of ECM macromolecules as attractants in motility assays illustrates the fundamental differences between chemotaxis and haptotaxis with respect to post-receptor signal transduction events. A2058 melanoma cells are chemotactic and haptotactic on gradients of ECM proteins, such as laminin, fibronectin, type IV collagen (Aznavoorian et al 1990), and thrombospondin (Taraboletti et al 1987). When cells are pre-treated with pertussis toxin, the chemotactic response to laminin is diminished, and the response to type IV collagen is abolished (Aznavoorian et al 1990). In contrast, haptotaxis to these same proteins is completely insensitive to pertussis toxin. In the case of fibronectin, neither chemotaxis nor haptotaxis is affected by PT. One explanation of these data is that chemotaxis and haptotaxis to the same ECM protein are mediated by distinct cell-surface receptors that recognize different domains of the large multi-domain matrix proteins. Chemotaxis and haptotaxis receptors apparently generate motility signals through different transduction mechanisms. Studies with thrombospondin (Taraboletti et al 1987) identified distinct chemotaxis- and haptotaxis-promoting domains on this molecule. Differential inhibition of each type of migration with specific antibodies and/or cell-binding peptides prevented cell interaction with the relevant regions. Both haptotatic and chemotactic responses may be relevant to the metastatic phenotype. During the initial stages of metastasis, haptotactic migration over insoluble matrix proteins may be the more significant response. Later, partially degraded matrix proteins that result from proteolytic processing of the matrix could be involved in chemotactic responses that may then dominate the migratory phenotype.

Insulin-like growth factors and insulin stimulate a pertussis toxin-insensitive chemotactic response in A2058 cells (Stracke et al 1988). This response is strongest to IGF-I and appears to activate the cells through a type I IGF receptor (Stracke 1989). Both insulin and IGF-I have been implicated as necessary growth factors for culture of primary human melanoma cells (Rodeck et al 1987). This requirement suggests that IGF-I may serve as a kind of homing factor for tumor cells that have reached the vasculature. Secretion of IGF-1 could facilitate the extravasation of tumor cells into a secondary site that provides the necessary microenvironment for continued growth of the metastatic foci.

The detailed molecular mechanisms that stimulate cells to migrate directionally are incompletely understood. Morphological studies demonstrate that tumor cells, like leukocytes, exhibit amoeboid movement characterized by pseudopod extension (Hosaka et al 1979; Mohler et al 1987). This kind of motility requires the coordinated action of distinct steps including cellular protrusion at the leading edge, new adhesion formation, and release of old adhesions at the trailing edge, which result in the asymmetric morphology of motile cells observed on two-dimensional substrates. It is generally agreed that the machinery for cell locomotion in eukaryotic cells resides in the peripheral cytoplasm, or cell cortex, which consists of a network of polymerized, cross-linked actin filaments (Condeelis et al 1992; Cunningham 1992; Stossel 1990). For pseudopod protrusion and cell locomotion to occur, this network must be reversibly disassembled, or solated, to allow protrusion, then re-assembled to stabilize the resulting extension. The precise manner in which this is accomplished, as well as its linkage to a localized, receptor-mediated chemotactic stimulus, is still not fully understood, although studies with neutrophils and with the lower eukaryote *Dictyostelium discoideum* have provided some insights that may be applicable to tumor cells. Models proposed by Condeelis et al (1992) and Stossel (1990) envision the localized polymerization and cross-linking of actin filaments at the site of a chemotactic stimulus, which result in the directional protrusion of a pseudopod. Both models speculate that the signals generated by activated chemotactic receptors, and the resulting second messengers, regulate proteins that sever, cap, and cross-link actin filaments to generate pseudopods. The model based on studies with leukocytes links actin disassembly and re-assembly with the activation of the phosphatidylinositol (PI) cycle (Lester & McCarthy 1992). Thus in leukocytes, chemotactic receptor stimulation leads to activation of the PI cycle, actin polymerization, and motility; events that are mediated by a pertussis toxin-sensitive G protein (Goldman et al 1985; Spangrude et al 1985). In tumor cells, several lines of evidence also implicate the involvement of PT-sensitive G proteins in motility and invasion (Aznavoorian et al 1990; Lester et al 1989, 1991; Roos & Van De Pavert 1987; Stracke et al 1987, 1992). Indeed, using specific antibodies and oligonucleotide probes, G_ia_2 has been identified as the PT-sensitive G protein uniquely abundant in highly metastatic murine melanomas (Lester et al 1989, 1991), which leads to the speculation that this G protein contributes to the regulation of motility in highly invasive and metastatic tumor cells. However, as stated above, chemotactic migration of the A2058 human melanoma cell line to IGF-I and fibronectin (Stracke et al 1988) and haptotactic migration to laminin, fibronectin, and type IV collagen (Aznavoorian et al 1990) are insensitive to pertussis toxin treatment. Therefore, care must be taken not to implicate G_ia_2 as a general regulator of motility in all tumor cells. It appears, instead, that tumor cells are equipped to respond in a variety of ways to a diverse array of motility stimuli. This would obviously give tumor cells a great deal of flexibility as they encounter different microenvironments during invasion and metastasis, and would impart a selective advantage over cells that are less adaptable.

SUMMARY

Recent findings have produced great strides in developing an understanding of the molecular events involved in processes necessary for tumor cell invasion and subsequent metastasis formation. This information has been useful in developing new targets for therapeutic intervention such as disruption of tumor cell attachment by peptide analogues of cell adhesion molecules and the use of protease inhibitors to limit extracellular matrix proteolysis required for tumor cell invasion. Future efforts must focus on how the events of cell attachment, matrix proteolysis, and cell migration are controlled and integrated. This requires a better understanding of the transcriptional controls and cell signaling mechanisms that are involved in these events. Preliminary findings suggest that cell-matrix interactions influence gene expression and that the protease inhibitor balance can greatly influence cell-matrix interactions. Therefore it appears that all three steps in the invasive process are linked and interdependent. While this complicates the study of these processes, it is our belief that understanding this interdependence is critical for further development of metastasis research.

Literature Cited

Albelda, SM. 1993. Role of integrins and other cell adhesion molecules in tumor progression and metastasis. *Lab. Invest.* 68:4–17

Albelda, SM, Mette, SA, Elder, DE, Stewart, RM, Damjanovich, L, et al. 1990. Integrin distribution in malignant melanoma: association of the β-3-subunit with tumor progession. *Cancer Res.* 50:6757–64

Albini, A, Melchiori, A, Santi, L, Liotta, LA, Brown, PD, Stetler-Stevenson, WG. 1991. Tumor cell invasion inhibited by TIMP-2. *J. Natl. Cancer Inst.* 83:775–79

Alvarez, OA, Carmichael, DF, DeClerck, YA. 1990. Inhibition of collagenolytic activity and metastasis of tumor cells by a recombinant human tissue inhibitor of metalloproteinases. *J. Natl. Cancer Inst.* 82:589–95

Aruffo, A, Stamenkovic, E, Melnick, M, Underhill, CB, Seed, B. 1990. CD44 is the principal cell surface receptor for hyaluronate. *Cell* 61:1303–13

Atnip, KD, Carter, LM, Nicolson, GL, Dabbous, MK. 1987. Chemotactic response of rat mammary adenocarcinoma cell clones to tumor-derived cytokines. *Biochem. Biophys. Res. Commun.* 146:996–1002

Aznavoorian, S, Stracke, ML, Krutzsch, HC, Schiffman, E, Liotta, LA. 1990. Signal transduction for chemotaxis and haptotaxis by matrix macromolecules in tumor cells. *J. Cell Biol.* 110:1427–38

Basset, P, Bellocq, JP, Wolf, C, Stoll, I, Hutin, P, et al. 1990. A novel metalloproteinase gene specifically expressed in stromal cells of breast carcinoma. *Nature* 348:699–704

Behrens, J, Mareel, MM, Van Roy, FM, Birchmeier, W. 1989. Dissecting tumor cell invasion: Epithelial cells acquire invasive properties after the loss of uvomorulin-mediated cell-cell adhesion. *J. Cell. Biol.* 108: 2435–47

Birch, M, Mitchell, S, Hart, IR. 1991. Isolation and characterization of human melanoma cell variants expressing high and low levels of CD44. *Cancer Res.* 51:6660–67

Bishop, JM. 1991. Molecular themes in oncogenesis. *Cell* 64:235–48

Boag, AH, Young, ID. 1993. Immunohistochemical analysis of type IV collagenase expression in prostatic hyperplasia and adenocarcinoma. *Mod. Pathol.* 6:65–68

Bonfil, RD, Medina, PA, Gomez, DE, Farias, E, Lazarowski, A, et al. 1992. Expression of gelatinase/type IV collagenase in tumor necrosis correlates with cell detachment and tumor invasion. *Clin. Exp. Metastasis* 10: 211–20

Bonfil, RD, Reddel, RR, Ura, H, Reich, R, Fridman, R, et al. 1989. Invasive and metastatic potential of a v-Ha-*ras*-transformed human bronchial epithelial cell line. *J. Natl. Cancer Inst.* 81:587–94

Boone, TC, Johnson, MJ, DeClerck, YA, Langley, KE. 1990. cDNA cloning and expression of a metalloproteinase inhibitor related to tissue inhibitor of metalloproteinases. *Proc. Natl. Acad. Sci. USA* 87:2800–4

Brown, PD, Bloxidge, RE, Anderson, E, Howell, A. 1993a. Expression of activated gelatinase in human breast carcinoma. *Clin. Exp. Metastasis* 11:183–89

Brown, PD, Bloxidge, RE, Stuart, NSA, Gatter, KC, Carmichael, J. 1993b. Association between expression of activated 72-kilodalton gelatinase and tumor spread in non-small-cell lung carcinoma. *J. Natl. Cancer Inst.* 85:574–78

Brown, PD, Kleiner, DE, Unsworth, EJ, Stetler-Stevenson, WG. 1993c. Cellular activation of the 72 kDa type IV procollagenase/TIMP-2 complex. *Kidney Int.* 43:163–70

Brown, PD, Levy, AT, Margulies, IM, Liotta, LA, Stetler-Stevenson, WG. 1990. Independent expression and cellular processing of M_r 72,000 type IV collagenase and interstitial collagenase in human tumorigenic cell lines. *Cancer Res.* 50:6184–91

Campo, E, Merino, MJ, Liotta, L, Neumann, R, Stetler-Stevenson, WG. 1992a. Distribution of the 72-kd type IV collagenase in nonneoplastic and neoplastic thyroid tissue. *Hum. Pathol.* 23:1395–1401

Campo, E, Merino, MJ, Tavassoli, FA, Charonis, AS, Stetler-Stevenson, WG, Liotta, LA. 1992b. Evaluation of basement membrane components and the 72 kDa type IV collagenase in serous tumors of the ovary. *Am. J. Surg. Pathol.* 16:500–7

Carmichael, DF, Sommer, A, Thompson, RC, Anderson, DC, Smith, CG, et al. 1986. Primary structure and cDNA cloning of human fibroblast collagenase inhibitor. *Proc. Natl. Acad. Sci. USA* 83:2407–11

Cawston, TE, Galloway, WA, Mercer, E, Murphy, G, Reynolds, JJ. 1981. Purification of rabbit bone inhibitor of collagenase. *Biochem. J.* 195:159–65

Clavel, C, Polette, M, Doco, M, Binninger, I, Birembaut, P. 1992. Immunolocalization of matrix metallo-proteinases and their tissue inhibitor in human mammary pathology. *Bull. Cancer* 79:261–70

Collier, IE, Wilhelm, SM, Eisen, AZ, Marmer, BL, Grant, GA, et al. 1988. H-*ras* oncogene-transformed human bronchial epithelial cells (TBE-1) secrete a single metalloprotease capable of degrading basement membrane collagen. *J. Biol. Chem.* 263:6579–87

Coman, DR. 1953. Mechanisms responsible for the origin and distribution of blood-borne tumor metastases: a review. *Cancer Res.* 13:397–404

Condeelis, J, Jones, J, Segall, JE. 1992. Chemotaxis of metastatic tumor cells: Clues to mechanisms from the *Dictyostelium* paradigm. *Cancer Metastasis Rev.* 11:55–68

Cornil, I, Theodorescu, D, Man, S, Herlyn, M, Jambrosic, J, Kerbel, RS. 1991. Fibroblast cell interactions with human melanoma cells affect tumor cell growth as a function of tumor progression. *Proc. Natl. Acad. Sci. USA* 88:6028–32

Culty, M, Miyake, K, Kincade, PW, Silorski, E, Butcher, EC, Underhill, C. 1990. The hyaluronate receptor is a member of the CD44 (H-CAM) family of cell surface glycoproteins. *J. Cell. Biol.* 111:2765–74

Cunningham, CC. 1992. Actin structural proteins in cell motility. *Cancer Metastasis Rev.* 11:69–77

Daidone, MG, Silvestrini, R, D'Errico, A, Di Fronzo, G, Benini, E, et al. 1991. Laminin receptors, collagenase IV and prognosis in node-negative breast cancers. *Int. J. Cancer* 48:529–32

Davies, B, Miles, DW, Happerfield, LC, Naylor, MS, Bobrow, LG, et al. 1993. Activity of type IV collagenases in benign and malignant breast disease. *Br. J. Cancer*. In press

DeClerck, YA, Perez, N, Shimada, H, Boone, TC, Langley, KE, Taylor, SM. 1992a. Inhibition of invasion and metastasis in cells transfected with an inhibitor of metalloproteinases. *Cancer Res.* 52:701–8

DeClerck, Y, Szpirer, C, Aly, MS, Cassiman, J-J, Eeckhout, Y, Rousseau, G. 1992b. The gene for tissue inhibitor of metalloproteinases-2 is localized on human chromosome arm 17q25. *Genomics* 14:782–84

DeClerck, YA, Yean, T-D, Chan, D, Shimada, H, Langley, KE. 1991. Inhibition of tumor invasion of smooth muscle cell layers by recombinant human metalloproteinase inhibitor. *Cancer Res.* 51:2151–57

DeClerck, YA, Yean, T-D, Ratzkin, BJ, Lu, HS, Langley, KE. 1989. Purification and characterization of two related but distinct metalloproteinase inhibitors secreted by bovine aortic endothelial cells. *J. Biol. Chem.* 264:17445–53

D'Errico, A, Garbisa, S, Liotta, LA, Castronovo, V, Stetler-Stevenson, WG, Grigioni, WF. 1991. Augmentation of type IV collagenase, laminin receptor, and Ki67 proliferation antigen associated with human colon, gastric, and breast carcinoma progression. *Mod. Pathol.* 4:239–46

Elices, MJ, Osborn, L, Takada, Y, Crouse, C, Luhowsky, S, et al. 1990. VCAM-1 on activated endothelium interacts with the leukocyte intefrin VLA-4 at a site distinct from the VLA-4 fibronectin binding-site. *Cell* 60:577–84

Emonard, HP, Remacle, AG, Nöel, AC, Grimaud, JA, Stetler-Stevenson, WG, Foidart, JM. 1992. Tumor cell surface-associated binding site for the M_r 72,000 type IV collagenase. *Cancer Res.* 52:5845–48

Enterline, HT, Coman, DR. 1950. The amoeboid motility of human and animal neoplastic cells. *Cancer* 3:1033–38

Fearon, ER, Vogelstein, B. 1990. A genetic model for colorectal tumorigenesis. *Cell* 61: 759–67

Fidler, IJ. 1991. Orthotopic implantation of human colon carcinomas into nude mice provides a valuable model for the biology and therapy of metastasis. *Cancer Metastasis Rev.* 10:229–43

Fidler, IJ, Hart, IR. 1982. Biologic diversity in metastatic neoplasms-origins and implications. *Science* 217:998–1001

Fidler, IJ, Naito, S, Pathak, S. 1990. Orthotopic implantation is essential for the selection, growth and metastasis of human renal cell cancer in nude mice. *Cancer Metastasis Rev.* 9:149–65

Fidler, IJ, Radinsky, R. 1990. Genetic control of cancer metastasis. *J. Natl. Cancer Inst.* 82:166–68

Fridman, R, Fuerst, TR, Bird, RE, Hoyhtya, M, Oelkuct, M, et al. 1992a. Domain structure of human 72-kDa gelatinase/type IV collagenase. Characterization of proteolytic activity and identification of the tissue inhibitor of metalloproteinase-2 (TIMP-2) binding regions. *J. Biol. Chem.* 267:15398–15405

Fridman, R, Kibbey, MC, Royce, LS, Zain, M, Sweeney, TM, et al. 1991. Enhanced tumor growth of both primary and established human and murine tumor cells in athymic mice after coinjection with Matrigel. *J. Natl. Cancer Inst.* 83:769–74

Fridman, R, Sweeney, TM, Zain, M, Martin, GR, Kleinman, HK. 1992b. Malignant transformation of NIH-3T3 cells after subcutaneous co-injection with a reconstituted basement membrane (matrigel). *Int. J. Cancer* 51:740–44

Furukawa, T, Fu, X, Kubota, T, Watanabe, M, Kitajima, M, Hoffman, RM. 1993. Nude mouse metastatic models of human stomach cancer constructed using orthotopic implantation of histologically intact tissue. *Cancer Res.* 53:1204–8

Gallatin, WM, Rosenman, SJ, Ganji, A, St John, TP. 1991. Structure function relationships of the CD44 class of glycoproteins. *Cell Mol. Mech. Inf.* 2:131–50

Garbisa, S, Pozzatti, R, Muschel, R, Saffioti, V, Ballin, M, et al. 1987. Secretion of type IV collagenolytic protease and metastatic phenotype: induction by transfection with c-Ha-*ras* but not c-Ha-*ras* plus Ad2-E1a. *Cancer Res.* 47:1523–28

Garbisa, S, Scagliotti, G, Masiero, L, Di Francesco, C, Caenazzo, C, et al. 1992. Correlation of serum metalloproteinase levels with lung cancer metastasis and response to therapy. *Cancer Res.* 52:4548–49

Garcia-Castro, I, Mato, JM, Vasanthokumar, G, Wiesmann, WP, Schiffmann, E, Chiang, PK. 1983. Paradoxical effects of adenosine on neutrophil chemotaxis. *J. Biol. Chem.* 258: 4345–49

Gehlsen, KR, Argraves, WS, Piersbacher, MD, Ruoslahti, E. 1988. Inhibition of in vitro cell invasion by Arg-Gly-Asp-containing peptides. *J. Cell Biol.* 106: 925–30

Gehlsen, KR, Davis, GE, Sriramarao, P. 1992. Integrin expression in human melanoma cells with differing invasive and metastatic properties. *Clin. Exp. Metastasis* 10: 111–20

Gladson, CL, Cherech, DA. 1991. Glioblastoma expression of vitronectin and the alpha-v/β-3 integrin. *J. Clin. Invest.* 88: 1924–32

Goldberg, GI, Marmer, BL, Grant, GA, Eisen, AZ, Wilhelm, A, He, C. 1989. Human 72-kilodalton type IV collagenase forms a complex with a tissue inhibitor of metalloprotcinases designated TIMP-2. *Proc. Natl. Acad. Sci. USA* 86:8207–11

Goldberg, GI, Strongin, A, Collier, IE, Genrich, LT, Marmer, BL. 1992. Interaction of 92-kDa type IV collagenase with the tissue inhibitor of metalloproteinases prevents dimerization, complex formation with interstitial collagenase, and activation of the proenzyme with stromelysin. *J. Biol. Chem.* 267:4583–91

Goldman, DW, Chang, FH, Gifford, LA, Goetzl, EJ, Bourne, HR. 1985. Pertussis toxin inhibition of chemotactic factor-induced calcium mobilization and function in human polymorphonuclear leukocytes. *J. Exp. Med.* 162:145–56

Gray, ST, Wilkins, RJ, Yun, K. 1992. Interstitial collagenase gene expression in oral squamous cell carcinoma. *Am. J. Pathol.* 141:301–6

Guirguis, R, Margulies, I, Taraboletti, G, Schiffman, E, Liotta, LA. 1987. Cytokine-induced pseudopodial protrusion is coupled to tumor cell migration. *Nature* 329:261–63

He, C, Wilhelm, SM, Pentland, AP, Marmer, BL, Grant, GA, et al. 1989. Tissue cooperation in a proteolytic cascade activating human interstitial collagenase. *Proc. Natl. Acad. Sci. USA* 86:2632–36

Hewitt, RE, Leach, IH, Powe, DG, Clark, IM, Cawston, TE, Turner, DR. 1991. Distribution of collagenase and tissue inhibitor of metalloproteinases (TIMP) in colorectal tumours. *Int. J. Cancer* 49:666–72

Horst, E, Meijer, CJ, Radaszkiewicz, T, Osseloppele, GJ, Van Krieken, JH, Pals, ST. 1990. Adhesion molecules in the prognosis of diffuse large-cell lymphoma: expression of a lymphocyte homing receptor (CD44), LFA-1, (CD11a/18) and ICAM-1 (CD54). *Leukemia* 4:595–99

Hosaka, S, Suzuki, M, Sato, H. 1979. Leukocyte-like motility of cancer cells, with reference to the mechanism of extravasation. *Gann* 70:559–61

Howard, EW, Banda, MJ. 1991. Binding of tissue inhibitor of metalloproteinases 2 to two distinct sites on human 72-kDa gelatinase. Identification of a stabilization site. *J. Biol. Chem.* 266:17972–77

Howard, EW, Bullen, EC, Banda, MJ. 1991. Preferential inhibition of 72- and 92-kDa gelatinases by tissue inhibitor of metalloproteinases-2. *J. Biol. Chem.* 266:13070–75

Humphries, MJ, Olden, K, Yamada, KM. 1986. A synthetic peptide from fibronectin inhibits experimental metastasis of murine melanoma cells. *Science* 233:467–69

Humphries, MJ, Yamada, KM, Olden, K. 1988. Investigation of the biological effects of anti-cell adhesion synthetic peptides that inhibit experimental metastasis of B16-F10 murine melanoma cells. *J. Clin. Invest.* 81:782–90

Hunkapiller, TH, Hood, L. 1989. Diversity of the immunoglobulin superfamily. *Adv. Immunol.* 44:1–63

Hurewitz, AN, Zucker, S, Mancuso, P, Wu, CL, Dimassimo, B, et al. 1992. Human pleural effusions are rich in matrix metalloproteinases. *Chest* 102:1808–14

Hynes, RO. 1992. Integrins: Versatility, modulation, and signaling in cell adhesion. *Cell* 69:11–25

Jouanneau, J, Gavrilovic, J, Caruelle, D, Jaye, M, Moens, G, et al. 1991. Secreted or nonsecreted forms of acidic fibroblast growth factor produced by transfected epithelial cells influence cell morphology, motility and invasive potential. *Proc. Natl. Acad. Sci. USA* 88:2893–97

Kahan, DW, Kramp, DC. 1987. Nerve growth factor stimulation of mouse embryonal cell migration. *Cancer Res.* 47:6324–28

Kanemoto, T, Martin, GR, Hamilton, TC, Fridman, R. 1991. Effects of synthetic peptides and protease inhibitors on the interaction of a human ovarian cancer cell line (NIH:OVCAR-3) with a reconstituted basement membrane (Matrigel). *Invasion Metastasis* 11:84–92

Karaoka, H, DeCastro, R, Zucker, S, Biswas, C. 1993. The tumor cell-derived collagenase stimulatory factor, TCSF, increases expression of interstitial collagenase, stromelysin and 72 kDa gelatinase. *Cancer Res.* In press

Kato, Y, Nakayama, Y, Umeda, M, Miyazaki, K. 1992. Induction of 103-kDa gelatinase/type IV collagenase by acidic culture conditions in mouse metastatic melanoma cell lines. *J. Biol. Chem.* 267: 11424–30

Kerbel, R. 1990. Growth dominance of the metastatic cancer cell: cellular and molecular aspects. *Adv. Cancer Res.* 55:87–131

Kerbel, R. 1992. Expression of multicytokine resistance and multi-growth factor independence in advanced stage metastatic cancer. *Am. J. Pathol.* 141:519–24

Khokha, R, Waterhouse, P, Yagel, S, Lala, PK, Overall, CM, et al. 1989. Antisense RNA-induced reduction in metalloproteinase inhibitor causes mouse 3T3 cells to become tumorigenic. *Science* 243:947–50

Kleiner, DE Jr, Tuuttila, A, Tryggvason, K, Stetler-Stevenson, WG. 1993. Stability analysis of latent and active 72-kDa type IV collagenase: the role of tissue inhibitor of metalloproteinases-2 (TIMP-2). *Biochemistry* 32:1583–92

Kleiner, DE Jr, Unsworth, EJ, Krutzsch, HC, Stetler-Stevenson, WG. 1992. Higher-order complex formation between the 72-kilodalton type IV collagenase and tissue inhibitor of metalloproteinases-2. *Biochemistry* 31:1665–72

Kusukawa, J, Sasaguri, Y, Shima, I, Kameyama, T, Morimatsu, M. 1993. Expression of matrix metalloproteinase-2 related to lymph node metastasis of oral squamous cell carcinoma. A clinicopathologic study. *Am. J. Clin. Pathol.* 99:18–23

Lefebvre, O, Wolf, C, Limacher, J-M, Hutin, P, Wendling, C, et al. 1992. The breast cancer-associated stromelysin-3 gene is expressed during mouse mammary gland apoptosis. *J. Cell Biol.* 119:997–1002

Lester, BR, McCarthy, JB. 1992. Tumor cell adhesion to the extracellular matrix and signal transduction mechanisms implicated in tumor cell motility, invasion and metastasis. *Cancer Metastasis Rev.* 11:31–44

Lester, BR, McCarthy, JB, Sun, Z, Smith, RS, Furcht, KT, Spiegel, AM. 1989. G-protein involvement in matrix-mediated motility and invasion of high and low experimental metastatic B16 melanoma clones. *Cancer Res.* 49:5940–48

Lester, BR, Winstein, LS, McCarthy, JB, Sun, Z, Smith, RS, Furcht, LT. 1991. The role of G-protein in matrix-mediated motlilty of highly and poorly invasive melanoma cells. *Int. J. Cancer* 48:113–20

Levy, AT, Cioce, V, Sobel, ME, Garbisa, S, Grigioni, WF, et al. 1991. Increased expression of the M_r 72,000 type IV collagenase in human colonic adenocarcinoma. *Cancer Res.* 51:439–44

Liotta, LA. 1986. Tumor invasion and metastases-role of the extracellular matrix: Rhoads Memorial Award Lecture. *Cancer Res.* 46:1–7

Liotta, LA, Mandler, R, Murano, G, Katz, DA, Gordon, RK, et al. 1986. Tumor cell autocrine motility factor. *Proc. Natl. Acad. Sci. USA* 83:3302-6

Liotta, LA, Steeg, PS, Stetler-Stevenson, WG. 1991. Cancer metastasis and angiogenesis: an imbalance of positive and negative regulation. *Cell* 64:327–36

Lyons, JG, Birkedal-Hansen, B, Moore, WG, O'Grady, RL, Birkedal-Hansen, H. 1991. Characteristics of a 95-kDa matrix metalloproteinase produced by mammary carcinoma cells. *Biochemistry* 30:1449–56

Mahtani, MM, Willard, HF. 1988. A primary genetic map of the pericentormeric region of the human X chromosome. *Genomics* 2:294–301

Manzotti, C, Audisio, RA, Pratesi, G. 1993. Importance of orthotopic implantation for human tumors as model systems: relevance to metastasis and invasion. *Clin. Exp. Metastasis* 11:5–14

Marcotte, PA, Kozan, IM, Dorwin, SA, Ryan, JM. 1992. The matrix metalloproteinase pump-1 catalyzes formation of low molecular weight (Pro)urokinase in cultures of normal human kidney cells. *J. Biol. Chem.* 267:13803–6

Mareel, MM, Van Roy, FM, De Baetselier, P. 1990. The invasive phenotypes. *Cancer Metastasis Rev.* 9:45–62

Margulies, IMK, Hoyhtya, M, Evans, C, Stracke, ML, Liotta, LA, Stetler-Stevenson, WG. 1992. Urinary type IV collagenase: elevated levels are associated with bladder transitional cell carcinoma. *Cancer Epidemiol. Biomarkers Prev.* 1:467–74

Martignone, S, Menard, S, Bufalino, R, Cascinelli, N, Pellegrini, R, et al. 1993. Prognostic significance of the 67-kilodalton laminin receptor expression in human breast carcinomas. *J. Natl. Cancer Inst.* 85:398–402

Matrisian, LM. 1992. The matrix-degrading metalloproteinases. *BioEssays* 14:455–62

Matrisian, LM, McDonnell, S, Miller, DB, Navre, M, Seftor, EA, Hendrix, MJ. 1991. The role of the matrix metalloproteinase stromelysin in the progression of squamous cell carcinomas. *Am. J. Med. Sci.* 302:157–62

McCarthy, JB, Basara, ML, Palm, SL, Sas, DF, Furcht, LT. 1985. The role of cell adhesion proteins-laminin and fibronectin-in the movement of malignant and metastatic cells. *Cancer Metastasis Rev.* 4:125–52

McCarthy, JB, Furcht, LT. 1984. Laminin and fibronectin promote the haptotactic migration of B16 melanoma cells in vitro. *J. Cell Biol.* 98:1474–80

McCarthy, JB, Hager, SJ, Burcht, LR. 1986. Human fibronectin contains distinct adhesion- and motility-promoting domains for metastatic melanoma cells. *J. Cell Biol.* 102:179–88

McDonnell, S, Navre, M, Coffey, R Jr, Matrisian, LM. 1991. Expression and localization of the matrix metalloproteinase pump-1 (MMP-7) in human gastric and colon carcinomas. *Mol. Carcinog.* 4:527–33

Meltzer, A. 1990. Dormancy and breast cancer. *J. Surg. Oncol.* 43:181–88

Mignatti, P, Robbins, E, Rifkin, DB. 1986. Tumor invasion through the human amnion membrane: requirement for a proteinase cascade. *Cell* 47:487–98

Mohler, JL, Partin, AW, Coffey, DS. 1987. Prediction of metastatic potential by a new grading system of cell motility: validation in the Dunning R-3327 prostatic adenomcarcinoma model. *J. Urol.* 138:168–70

Monteagudo, C, Merino, MJ, San Juan, J, Liotta, LA, Stetler-Stevenson, WG. 1990. Immunohistochemical distribution of type IV collagenase in normal, benign, and malignant breast tissue. *Am. J. Pathol.* 136: 585–92

Montesano, R, Pepper, MS, Mohle-Steinlein, U, Risau, W, Wagner, WF, Orci, L. 1990. Increased proteolytic activity is responsible for the aberrant morphogenetic behavior of endothelial cells expressing the middle T oncogene. *Cell* 62:435–45

Moses, MA, Langer, R. 1991. A metalloproteinase inhibitor as an inhibitor of neovascularization. *J. Cell. Biochem.* 47: 230–35

Moses, MA, Sudhalter, J, Langer, R. 1990. Identification of an inhibitor of neovascularization from cartilage. *Science* 248:1408–10

Muller, D, Breathnach, R, Engelmann, A, Millon, R, Bronner, G, et al. 1991. Expression of collagenase-related metalloproteinase genes in human lung or head and neck tumours. *Int. J. Cancer* 48:550–56

Muller, D, Wolf, C, Abecassis, J, Millon, R, Engelmann, A, et al. 1993. Increased stromelysin 3 gene expression is associated with increased local invasiveness in head and neck squamous cell carcinomas. *Cancer Res.* 53:165–69

Murphy, G, Ward, R, Gavrilovic, J, Atkinson, S. 1992a. Physiological mechanisms for metalloproteinase activation. *Matrix* 1: 224–30

Murphy, G, Willenbrock, F, Ward, RV, Cockett, MI, Eaton, D, Docherty, AJ. 1992b. The C-terminal domain of 72 kDa gelatinase A is not required for catalysis, but is essential for membrane activation and modulates interactions with tissue inhibitors of metalloproteinases. *Biochem. J.* 283: 637–41

Muschel, RJ, Williams, JE, Lowy, DR, Liotta, LA. 1985. Harvey *ras* induction of metastatic potential depends upon oncogene activation and the type of recipient cell. *Am. J. Pathol.* 121:1–8

Nabeshima, K, Lane, WS, Biswas, C. 1991. Partial sequencing and characterization of the tumor cell-derived collagenase stimulatory factor. *Arch. Biochem. Biophys.* 285: 90–96

Nagase, H, Ogata, Y, Suzuki, K, Enghild, JJ, Salvesen, G. 1991. Substrate specificities and activation mechanisms of matrix metalloproteinases. *Biochem. Soc. Trans.* 19:715–18

Nakajima, M, Morikawa, K, Fabra, A, Bucana, CD, Fidler, IJ. 1990. Influence of organ environment on extracellular matrix degradative activity and metastasis of human colon carcinoma cells. *J. Natl. Cancer Inst.* 82:1890–98

Nakajima, M, Welch, DR, Belloni, PN, Nicholson, GL. 1987. Degradation of basement membrane type IV collagen and lung subendothelial matrix by rat mammary adenocarcinoma cell clones of differing metastatic potentials. *Cancer Res.* 47:4869–76

Ogata, Y, Enghild, JJ, Nagase, H. 1992a. Matrix metalloproteinase 3 (stromelysin) activates the precursor for the human matrix metalloproteinase 9. *J. Biol. Chem.* 267: 3581–84

Ogata, Y, Pratta, MA, Nagase, H, Arner, EC. 1992b. Matrix metalloproteinase 9 (92-kDa gelatinase/type IV collagenase) is induced in rabbit articular chondrocytes by cotreatment with interleukin 1 beta and a protein kinase C activator. *Exp. Cell Res.* 201:245–49

Ohnishi, R, Arita, N, Hayakawa, T, Izumoto, S, Taki, T, Yamamoto, H. 1990. Motility factor produced by malignant glioma cells: role in tumor invasion. *J. Neurosurg.* 73: 881–88

Ohori, NP, Yousem, SA, Griffin, J, Stanis, K, Stetler-Stevenson, WG, et al. 1992. Comparison of extracellular matrix antigens in subtypes of bronchioloalveolar carcinoma and conventional pulmonary adenocarcinoma. An immunohistochemical study. *Am. J. Surg. Pathol.* 16:675–86

Okada, Y, Morodomi, T, Enghild, JJ, Suzuki, K, Yasui, A, et al. 1990. Matrix metalloproteinase 2 from human rheumatoid synovial fibroblasts. *Eur. J. Biochem.* 194: 721–30

Osborn, L, Hession, C, Tizard, R, Vassallo, C, Luhowsky, S, et al. 1989. Direct expression cloning of vascular cell adhesion molecule 1, a cytokine-induced endothelial protein that binds to lymphocytes. *Cell* 59: 1203–11

Otani, Y. 1990. The collagenase activities, interstitial collagenase and type IV collagenase, in human stomach cancer: with special reference to local spreading and lymph node metastasis. *Keio J. Med.* 39: 159–67

Overall, CM, Sodek, J. 1990. Concanavalin A produces a matrix-degradative phenotype in human fibroblasts. Induction and endogenous activation of collagenase, 72-kDa gelatinase, and Pump-1 is accompanied by the suppression of the tissue inhibitor of matrix metalloproteinases. *J. Biol. Chem.* 265:21141–51

Pajouh, MS, Nagle, RB, Breathnach, R, Finch, JS, Brawer, MK, Bowden, GT. 1991. Expression of metalloproteinase genes in human prostate cancer. *J. Cancer Res. Clin. Oncol.* 117:144–50

Pauli, BU, Augustin-Voss, HG, El-Sabban, ME, Johnson, RC, Hammer, DA. 1990. Organ-preference of metastasis. The role of endothelial cell adhesion molecules. *Cancer Metastasis Rev.* 9:175–89

Pepper, MS, Belin, D, Montesano, R, Orci, L, Vassalli, J-D. 1990. Transforming growth factor-beta 1 modulates basic fibroblast growth factor-induced proteolytic and angiogenic properties of endothelial cells in vitro. *J. Cell Biol.* 111:743–55

Petersen, OE, Ronnov-Jessen, L, Howlett, AR, Bissell, MJ. 1992. Interaction with basement membrane serves to rapidly distinguish growth and differentiation pattern of normal and malignant human breast epithelial cells. *Proc. Natl. Acad. Sci. USA* 89:9064–68

Polette, M, Clavel, C, Muller, D, Abecassis, J, Binninger, I, Birembaut, P. 1991. Detection of mRNAs encoding collagenase I and stromelysin 2 in carcinomas of the head and neck by in situ hybridization. *Invasion Metastasis* 11:76–83

Poulsom, R, Pignatelli, M, Stetler-Stevenson, WG, Liotta, LA, Wright, PA, et al. 1992. Stromal expression of 72 kda type IV collagenase (MMP-2) and TIMP-2 mRNAs in colorectal neoplasia. *Am. J. Pathol.* 141: 389–96

Powell, WC, Knox, JD, Navre, M, Grogan, TM, Kittelson, J, et al. 1993. Expression of the metalloproteinase matrilysin in DU-145 cells increases their invasive potential in severe combined immunodeficient mice. *Cancer Res.* 53:417–22

Pyke, C, Ralfkiaer, E, Huhtala, P, Hurskainen, T, Danö, K, Tryggvason, K. 1992. Localization of messenger RNA for M_r 72,000 and 92,000 type IV collagenases in human skin cancers by in situ hybridization. *Cancer Res.* 52:1336–41

Rao, CN, Marguiles, IMK, Tralka, TS, Terranova, VP, Madri, JA, Liotta, LA. 1982. Isolation of a subunit of laminin and its role in molecular structure of tumor cell attachment. *J. Biol. Chem.* 257:9740–44

Rice, GE, Bevilavqua, MP. 1989. An inducible endothelial cell surface glycoprotein mediates melanoma adhesion. *Science* 246: 1303–6

Rodeck, U, Herlyn, M, Menssen, HD, Furlanetto, RW, Koprowski, H. 1987. Metastatic but not primary melanoma cell lines grow in vitro independently of exogenous growth factors. *Int. J. Cancer* 40:687–90

Roos, E, Van De Pavert, IB. 1987. Inhibition of lymphoma invasion and liver metastasis formation by pertussis toxin. *Cancer Res.* 47:5439–44

Rosen, RM, Meromsky, L, Setter, E, Vinter, DW, Goldberg, ID. 1990. Purified scatter factor stimulates epithelial and vascular endothelial cell migration. *Proc. Soc. Exp. Biol. Med.* 195:34–43

Ruoslahti, E, Pierschbacher, MD. 1987. New perspectives in cell adhesion: RGD and integrins. *Science* 238:491–97

Sage, EH, Bornstein, P. 1991. Extracellular proteins that modulate cell-matrix interactions. *J. Biol. Chem.* 266:14831–34

Saiki, I, Murata, J, Iida, J, Nishi, J, Sugimura, K, Azuma, I. 1989. The inhibition of murine lung metastasis by synthetic polypeptides [poly(arg-gly-asp) and poly(tyr-ile-gly-ser-arg)] with a core sequence of cell adhesion molecules. *Br. J. Cancer* 59:194–97

Sato, H, Kida, Y, Mai, M, Endo, Y, Sasaki, T, et al. 1992. Expression of genes encoding type IV collagen-degrading metalloproteinases and tissue inhibitors of metalloproteinases in various human tumor cells. *Oncogene* 7:77–83

Schnapper, HW, Grant, DS, Stetler-Stevenson, WG, Fridman, R, D'Orazi, GD, et al. 1993. Type IV collagenase(s) and TIMPs modulate endothelial cell morphogenesis in vitro. *J. Cell. Physiol.* In press

Schor, SL, Schor, AM, Grey, AM, Rushton, G. 1988. Fetal and cancer patient fibroblasts produce an autocrine migration-stimulating factor not made by normal adult cells. *J. Cell Sci.* 90:391–99

Schultz, RM, Silberman, S, Persky, B, Bajkowski, AS, Carmichael, DF. 1988. Inhibition by human recombinant tissue inhibitor of metalloproteinases of human amnion invasion and lung colonization by murine B16-F10 melanoma cells. *Cancer Res.* 48: 5539–45

Seftor, RE, Seftor, EA, Gehlsen, KR, Stetler-Stevenson, WG, Brown, PD, et al. 1992. Role of the alpha-v-beta-3 integrin in human melanoma cell invasion. *Proc. Natl. Acad. Sci. USA* 89:1557–61

Seiki, M, Sato, H, Liotta, LA, Schiffmann, E. 1991. Comparison of autocrine mechanisms promoting motility in two metastatic cell lines: human melanoma and *ras*-transfected NIH3T3 cells. *Int. J. Cancer* 49:717–20

Shima, I, Sasaguri, Y, Kusukawa, J, Yamana, H, Fujita, H, et al. 1992. Production of matrix metalloproteinase-2 and metalloproteinase-3 related to malignant behavior of esophageal carcinoma. A clinicopathologic study. *Cancer* 70:2747–53

Siletti, S, Watanabe, H, Hogan, B, Nabi, IR, Raz, A. 1992. Purification of B16-F10 melanoma autocrine motility factor and its receptor. *Cancer Res.* 51:3507-11

Spangrude, GJ, Sacchi, G, Hill, JR, Van Epps, DE, Daynes, RA. 1985. Inhibition of lymphocyte and neutrophil chemotaxis by pertussis toxin. *Cancer Res.* 135:4135–56

Sreenath, T, Matrisian, LM, Stetler-Stevenson, W, Gattoni-Celli, S, Pozzatti, RO. 1992. Expression of matrix metalloproteinase genes in transformed rat cell lines of high and low metastatic potential. *Cancer Res.* 52:4942–47

Stamenkovic, I, Amiot, M, Pesando, JM, Seed, B. 1989. A lymphocyte molecule implicated in lymph node homing is a member of the cartilage-like protein family. *Cell* 56:1057–62

Stephenson, RA, Dinney, CP, Gohji, K, Ordonez, NG, Killion, JJ, Fidler, IJ. 1992. Metastatic model for human prostate cancer using orthotopic implantation in nude mice. *J. Natl. Cancer Inst.* 84:951–57

Stetler-Stevenson, WG, Brown, PD, Onisto, M, Levy, AT, Liotta, LA. 1990. Tissue inhibitor of metalloproteinases-2 (TIMP-2) mRNA expression in tumor cell lines and human tumor tissues. *J. Biol. Chem.* 265:13933–38

Stetler-Stevenson, WG, Krutzsch, HC, Liotta, LA. 1989. Tissue inhibitor of metalloproteinase (TIMP-2). *J. Biol. Chem.* 264: 17374–78

Stetler-Stevenson, WG, Liotta, LA, Seldin, MF. 1992. Linkage analysis demonstrates that the Timp-2 locus is on mouse chromosome 11. *Genomics* 14:828–29

Stossel, TP. 1990. How cells crawl, with the discovery that the cellular motor contains muscle protein, we can begin to describe cell motility in molecular detail. *Am. Sci.* 78:408–23

Stracke, ML, Engel, JD, Wilson, LL, Rechler, MM, Liotta, LA, Schiffman, E. 1989. The type 1 insulin-like growth factor receptor is a motility receptor in human melanoma cells. *J. Biol. Chem.* 264:1544–49

Stracke, ML, Guirguis, R, Liotta, LA, Schiffman, E. 1987. Pertussis toxin inhibits stimulated motility independently of the adenylate cyclase pathway in human melanoma cells. *Biochem. Biophys.Res. Commun.* 146:339–45

Stracke, ML, Kohn, EC, Aznavoorian, S, Wilson, LL, Salomon, D, et al. 1988. Insulin-like growth factors stimulate chemotaxis in human melanoma cells. *Biochem. Biophys. Res. Commun.* 153:1076–83

Stracke, ML, Krutzsch, HC, Unsworth, EJ, Arestad, A, Cioce, V, et al. 1992. Identification, purification, and partial sequence analysis of autotaxin, a novel motility-stimulating protein. *J. Biol. Chem.* 267:2524–29

Sweeney, TM, Kibbey, MC, Zain, M, Fridman, R, Kleinman, HK. 1991. Basement membrane and the SIKVAV laminin-derived peptide promote tumor growth and metastases. *Cancer Metastasis Rev.* 10:245–54

Takeichi, M. 1990. Cadherins: A molecular family important in selective cell-cell adhesion. *Annu. Rev. Biochem.* 59:237–52

Takigawa, M, Nishida, Y, Suzuki, F, Kishi, J, Yamashita, K, Hayakawa, T. 1990. Induction of angiogenesis in chick yolk-sac membrane by polyamines and its inhibition by tissue inhibitors of metalloproteinases (TIMP and TIMP-2). *Biochem. Biophys. Res. Commun.* 171:1264–71

Taraboletti, G, Roberts, DD, Liotta, LA. 1987. Thrombospondin-induced tumor cell migration: haptotaxis and chemotaxis are mediated by different domains. *J. Cell Biol.* 105:2409–15

Templeton, NS, Brown, PD, Levy, AT, Margulies, IM, Liotta, LA, Stetler-Stevenson, WG. 1990. Cloning and characterization of human tumor cell interstitial collagenase. *Cancer Res.* 50:5431–37

Testa, J. 1992. Loss of the metastatic phenotype by a human epidermoid carcinoma cell line HEp-3, is accompanied by increased expression of tissue inhibitor of metalloproteinase-2. *Cancer Res.* 52:5597–5603

Tsuboi, R, Rifkin, DB. 1990. Bimodal relationship between invasion of the amniotic membrane and plasminogen activator activity. *Int. J. Cancer* 46:56–60

Turley, EA. 1992. Hyaluronan and cell locomotion. *Cancer Metastasis Rev.* 11:21–30

Turley, EA, Austen, L, Vandeligt, K, Clary, C. 1991. Hyaluronan and a cell-associated hyaluronan binding protein regulate the locomotion of *ras*-transformed cells. *J. Cell. Biol.* 112:1041–1047

Uleminckx, K, Vackat, L, Mareel, M, Fiers, W, Van Roy, FV. 1991. Genetic manipulation of E-cadherin expression by epithelial tumor cells reveals an invasion suppressor role. *Cell* 66:107–19

Urbanski, SJ, Edwards, DR, Maitland, A, Leco, KJ, Watson, A, Kossakowska, AE. 1992. Expression of metalloproteinases and their inhibitors in primary pulmonary carcinomas. *Br. J. Cancer* 66:1188–94

van der Stappen, JWJ, Hendriks, T, Wobbes, T. 1990. Correlation between collagenolytic activity and grade of histological differentiation in colorectal tumors. *Int. J. Cancer* 45:1071–1078

Ward, RV, Atkinson, SJ, Slocombe, PM, Docherty, AJP, Reynolds, JJ, Murphy, G. 1991. Tissue inhibitor of metalloproteinases-2 inhibits the activation of 72 kDa progelatinase by fibroblast membranes. *Biochim. Biophys. Acta* 1079:242–46

Weidner, KM, Behrens, J, Vandekerckhove, J, Birchmeier, W. 1990. Scatter factor: molecular characteristics and effect on the invasiveness of epithelial cells. *J. Cell Biol.* 111:2097–2108

Weinberg, WC, Brown, PD, Stetler-Stevenson, WG, Yuspa, SH. 1990. Growth factors specifically alter hair follicle cell proliferation and collagenolytic activity alone or in combination. *Differentiation* 45:168–78

Welgus, HG, Stricklin, GP. 1983. Human skin fibroblast collagenase inhibitor. Comparative studies in human connective tissues, serum, and amniotic fluid. *J. Biol. Chem.* 258:12259–64

Wewer, UM, Liotta, LA, Jaye, M, Ricca, GA, Drohan, WN, et al. 1986. Altered levels of laminin receptor mRNA in various human carcinoma cells that have different abilities to bind laminin. *Proc. Natl. Acad. Sci. USA* 83:7137–41

Wewer, UM, Taraboletti, G, Sobel, ME, Albrechtsen, R, Liotta, LA. 1987. Role of laminin receptor in tumor cell migration. *Cancer Res.* 47:5691–98

Wilhelm, SM, Collier, IE, Marmer, BL, Eisen, AZ, Grant, GA, Goldberg, GI. 1989. SV40-transformed human lung fibroblasts secrete a 92-kDa type IV collagenase which is identical to that secreted by normal human macrophages. *J. Biol. Chem.* 264: 17213–21

Willard, HF, Durfy, SJ, Mahtani, MM, Dorkins, H, Davies, KE, Williams, BRG. 1989. Regional localization of the TIMP gene on the human X chromosome. *Hum. Genet.* 81:234–38

Williams, AF, Barclay, AN. 1988. The immunoglobulin superfamily—domains for cell surface recognition. *Annu. Rev. Immunol.* 6:381–405

Woessner, JF Jr. 1991. Matrix metalloproteinases and their inhibitors in connective tissue remodeling. *FASEB J.* 5: 2145–54

Wolf, C, Chenard, M-P, de Grossouvre, PD, Bellocq, J-P, Chambon, P, Basset, P. 1992. Breast-cancer-associated stromelysin-3 gene is expressed in basal cell carcinoma and during cutaneous wound healing. *J. Invest. Dermatol.* 99:870–72

Yamada, KM. 1991. Adhesive recognition sequences. *J. Biol. Chem.* 266:12809–12

Yusa, R, Blood, CM, Zetter, BR. 1989. Tumor cell interactions with elastin: implication for pulmonary metastasis. *Am. Rev. Respir. Dis.* 140:1458–62

Zajicek, G. 1987. Long survival with micrometastases. At least 9% of breast cancer patients carry metastases for more than 10 years. *The Cancer J.* 1:414–15

Zhu, D, Cheng, C-F, Pauli, BU. 1991. Mediation of lung metastasis of murine melanomas by a lung-specific endothelial adhesion molecule. *Proc. Natl. Acad. Sci. USA* 88:9568–72

Zucker, S, Lysik, RM, Zarrabi, MH, Moll, U. 1993. M_r 92,000 type IV collagenase is increased in plasma of patients with colon cancer and breast cancer. *Cancer Res.* 53:140–46

Zucker, S, Lysik, RM, Zarrabi, MH, Stetler-Stevenson, WG, Liotta, LA, et al. 1992. Type IV collagenase/gelatinase (MMP-2) is not increased in plasma of patients with cancer. *Cancer Epidemiol. Biomarkers Prev.* 1:475–79

Annu Rev. Cell Biol. 1993. 9:575–99

THE ROLE OF GTP-BINDING PROTEINS IN TRANSPORT ALONG THE EXOCYTIC PATHWAY

Susan Ferro-Novick and Peter Novick
Department of Cell Biology, Yale University School of Medicine, 333 Cedar Street, New Haven, Connecticut 06510

KEY WORDS: vesicular transport, GTP-binding proteins

CONTENTS

INTRODUCTION

GTP-binding proteins have the ability to exist in either GDP- or GTP-bound forms, and thus are well suited to function as molecular switches (for a recent review of the general properties of GTP-binding proteins, see Bourne et al 1991). The nucleotide state of these proteins can be controlled by interaction with specific accessory proteins that alter the intrinsic rates of

0743–4634/93/1115–0575$05.00

GDP dissociation (GDP-dissociation inhibitor, GDI; GDP-dissociation stimulator, GDS) or GTP hydrolysis (GTPase-activating protein, GAP). In turn, the GTP-binding proteins control downstream effectors. This basic mechanism has been exploited in a wide range of cellular processes ranging from protein synthesis to the regulation of the cytoskeleton.

GTP-binding proteins have been implicated in the regulation of distinct events in transport on the exocytic, endocytic, and transcytotic pathways (for reviews, see Pfeffer 1992; Bomsel & Mostov 1992). However, this review focuses only on vesicular transport from the ER, through the Golgi apparatus, to the cell surface. Key questions in membrane traffic concern the molecular mechanism by which specific proteins are recruited from donor organelles and the cytoplasm to form carrier vesicles, and the mechanism by which these vesicles then migrate through the cytoplasm, recognize, and fuse with the appropriate target organelle. These events must be tightly regulated to maintain the unique composition and function of each organelle in the face of a constant flux of membrane traffic. GTP-binding proteins may be involved in these critical regulatory decisions.

GTP-binding proteins can be categorized by their subunit composition. Trimeric GTP-binding proteins (also known as classical G proteins) consist of an α subunit (~40 kd), which binds guanine nucleotide, as well as β and γ subunits. The monomeric GTP-binding proteins (also known as small GTP-binding proteins or *ras*-related GTP-binding proteins) are 20–25 kd. Both classes share certain conserved domains which, through the crystal structures of ras and elongation factor, are known to be involved in binding nucleotide. The first motif, $GX_4GK(S/T)$ interacts with the α and β phosphates. A tightly bound Mg^{2+} is coordinated to a conserved threonine in the second domain and to oxygens of the β and γ phosphates. The third motif, DX_2G, also interacts with the Mg^{2+} and directly with the γ phosphate. The fourth motif, (N/T)(K/Q)XD, interacts with the guanine ring. Because of this sequence and structural conservation, the study of GTP-binding proteins involved in vesicular transport has greatly benefited from the extensive biochemical analysis and mutagenesis studies done on the ras protein.

EARLY STAGES OF TRANSPORT

An Overview

The detailed mechanism by which proteins are transported from the endoplasmic reticulum (ER) to the Golgi complex and through the subcompartments of this organelle remains to be deciphered. However, through a combination of genetic and biochemical approaches, significant progress has been made in identifying and characterizing many of the players that participate in these events. A key advance that has catalyzed progress in this area is the establishment of transport assays that accurately reproduce vesicle fission and fusion between compartments. The pioneering work of Rothman and colleagues (Balch et al 1984) in the development of a mammalian in vitro intra-Golgi transport assay has served as a guide for the formulation of other assays that reconstitute vesicle-mediated transport at different stages of the secretory pathway.

In both yeast and higher eukaryotic cells, transport from the ER to the Golgi complex has been reconstituted in vitro using permeabilized cells (Beckers et al 1987; Simons & Virta 1987; Ruohola et al 1988; Baker et al 1988) or microsomes (Balch et al 1987; Ruohola et al 1988). Permeabilized cells contain intact organelles, but lack cytosol. In vitro transport is achieved when cytosol is added exogenously to these perforated cells in the presence of an energy source. In the yeast assay, functional transport vesicles exit from the cells and, in the presence of active Golgi membranes, bind to and fuse with this compartment to transfer their contents (Groesch et al 1990; Rexach & Schekman 1991). The advantage of such an assay is that the requirements for vesicle budding and vesicle consumption can be measured separately, thus making it possible to accurately determine the stage in transport in which a particular factor is required. In the mammalian intra-Golgi transport assay (Balch et al 1984), permeabilized cells are not employed. Instead, Golgi membranes that lack the *cis* Golgi marker N-acetylglucosaminyl (GlcNAc) transferase provide the donor compartment. Tritiated GlcNAc is incorporated into the vesicular stomatitis virus-viral encoded glycoprotein, VSV-G, when this protein is transferred from the earliest *cis* stack to the medial cisternae of wild-type Golgi membranes that contain a functional transferase.

The ability of the nonhydrolyzable analogue GTPγS to block both intra-Golgi membrane traffic and ER-to-Golgi transport in vitro complements genetic studies that have shown a requirement for proteins that bind GTP (Melançon et al 1987; Ruohola et al 1988; Baker et al 1988; Beckers et al 1989). Mutational analysis and suppression studies in *Saccharomyces cerevisiae* have led to the identification of a large collection of genes (called *SEC, BET, SAR, BOS, YPT, ARF, SLY,* and *USO*) whose products are required for transport from the ER to the Golgi complex (for review, see Pryer et al 1992). Of those genes that have been analyzed at a molecular and biochemical level some, such as *SEC23,* encode hydrophilic proteins, while others, such as *SEC12, BET1,* and *SEC22,* encode integral membrane proteins. For the purpose of this review, we only discuss those gene products that are GTP-binding proteins, the accessory proteins that interact with them, and the proteins that can compensate for the loss of a small GTP-binding protein. Mammalian homologues of these gene products (called *RAB* and *ARF*) and the in vivo and in vitro analysis of these proteins are also reviewed.

SAR1

SAR1 was identified by its ability to suppress the growth and secretion defect of the temperature-sensitive *sec12* mutant (Nakano & Muramatsu 1989). DNA sequence analysis has shown that *SAR1* is homologous to *ras* and *ras*-related proteins. While the overall homology of *SAR1* to these proteins is not high, there is significant homology in those regions that comprise the putative GTP-binding domain (Nakano & Muramatsu 1989). As predicted from the sequence, *E. coli*-expressed Sar1p (Sar1 protein) binds GTP (Oka et al 1991). A direct comparison of the 21 kd *SAR1* gene product to members of the *ARF* family has revealed that yeast *ARF* and *SAR1* are 34.5% identical over 168 residues. Both proteins are similar in that they contain an aspartic acid residue at the position corresponding to c-*ras*-H Gly-12 and, unlike members of the *YPT1* family, they are not prenylated at their COOH-termini. Instead, *ARF* has an amino-terminal motif (MGXXX[S/A]) that serves as a myristylation signal (Kahn & Gilman 1986). However, this signal is not present in Sar1p (Nakano & Muramatsu 1989).

In addition to suppressing *sec12*, overexpression of *SAR1* also suppresses *sec16* (Nakano & Muramatsu 1989), another temperature-sensitive yeast secretory mutant that fails to transport proteins from the ER to the Golgi complex (Novick et al 1980). The ability of *SAR1* to specifically suppress two different mutants that disrupt ER-to-Golgi transport suggests that its product may be required for membrane traffic at this stage of the pathway. To begin to address the cellular function of Sar1p, the consequences of depleting yeast of this protein were assessed. This study was performed by constructing a strain in which the sole copy of *SAR1* was placed under the control of the regulatable *GAL1* promoter. After several generations of growth in glucose-containing medium, these cells were depleted of Sar1p. In the absence of this small GTP-binding protein, the yeast pheromone pro-α-factor failed to be secreted, and the vacuolar protein carboxypeptidase Y (CPY) was blocked in transit to the vacuole. The form of the precursor that accumulated within the cell was consistent with a block in ER-to-Golgi transport (Nakano & Muramatsu 1989).

The role of Sar1p in ER-to-Golgi transport has been examined in vitro. In the first of these studies, overexpressed Sar1p was shown to suppress the temperature-sensitive transport defect of *sec12* mutant fractions (Oka et al 1991). *E. coli*-purified wild-type Sar1p also displayed the same transport activity. However, a mutant form of this protein that harbored a mutation in the third motif (DX_2G) of the GTP-binding domain ($Sar1p^{D73V}$) was defective in this suppression activity. Sar1p was also defective for suppression when it failed to contain nucleotide or was bound to GTPγS, which suggests that the hydrolysis of GTP is necessary for its function.

The overproduction of Sec12p in yeast leads to an increase in the membrane-bound pool of Sar1p, and this in turn results in an inhibition in the budding of transport vesicles from the ER membrane in vitro. The addition of Sar1p to the assay reverses this inhibition (d'Enfert et al 1991b; Barlowe et al 1993). Sec12p (Sec12 protein) is a 70-kd type II integral membrane glycoprotein. The N-terminal cytoplasmic domain of this membrane protein, but not the COOH-terminal domain, is required for its function (d'Enfert et al 1991a). Thus Sar1p may interact with the cytoplasmic domain of Sec12p to mediate the budding of transport vesicles from the ER membrane. In support of this hypothesis, it was recently shown that anti-Sar1p antibodies block vesicle formation (Barlowe et al 1993). Furthermore, the cytoplasmic domain of Sec12p stimulates nucleotide exhange on Sar1p (R. Schekman, personal communication).

SEC12 and *SAR1* genetically interact with an additional component of the secretory apparatus, *SEC23* (Kaiser & Schekman 1990). Sec23p is a cytosolic protein that associates with the periphery of membranes (Hicke & Schekman 1989). Two active forms of this protein have been purified; a monomer that is only present in cells that overproduce Sec23p, and a multimer. The multimeric form (260 kd) also contains a 105 kd polypeptide (105p) (Hicke et al 1992). Both the monomeric and multimeric forms of Sec23p have Sar1p GAP activity. Additionally, the *sec23* mutant is defective for this activity. This GAP is specific for Sar1p, since Sec23p does not hydrolyze GTP on Ypt1p or ARF (Yoshihisa et al 1993).

YPT1

YPT1 is an essential gene that was initially identified as the open reading frame adjacent to the actin gene (Gallwitz et al 1983). Interest in the analysis of this gene and its 23 kd protein product was prompted by its homology to the human *Ras* proto-oncogene. Although *YPT1* was shown to be essential for the vegetative growth of yeast cells, the study of conditional lethal mutants led to different hypotheses regarding the function of this protein.

A phenotypic analysis of *ypt1* mutants has shown that they accumulate membranes at their restrictive growth temperature. Additionally, these mutants display a defect in the transport of CPY to the vacuole and a partial block in the secretion of invertase. Although the *ypt1-1* mutant secretes an underglycosylated form of invertase (Segev et al 1988; Rossi et al 1991), an 80-kd core-glycosylated ER-form also accumulates internally (Segev et al 1988; Schmitt et al 1988; Rossi et al 1991). This form of invertase is found in other yeast secretory mutants that fail to transport proteins from the ER to the Golgi apparatus (Novick et al 1981). The temperature-sensitive growth defect of *ypt1* is rescued by the addition of calcium to the medium, which suggests that Ypt1p may regulate the level of intracellular calcium (Schmitt et al 1988). According to this hypothesis, the loss of Ypt1p function leads to an alteration in the intracellular concentration of calcium, which results in a block in vesicular transport.

In order to resolve the ambiguity concerning the function of Ypt1p, the role of this protein in vesicle-mediated transport was addressed in vitro in two different manners. First the ability of *ypt1* mutant fractions to support ER-to-Golgi transport in vitro was analyzed (Bacon et al 1989). Second, the efficacy of anti-Ypt1p antibody to block transport was also assessed (Baker et al 1990). Transport was monitored by following the processing of a precursor of the secreted pheromone α-factor. A 26-kd form of pro-α-factor, which resides within the lumen of the ER, is packaged into carrier vesicles that are released from permeabilized yeast cells. Upon fusing with the Golgi complex, pro-α-factor is converted to a 28-kd form that is subsequently processed to a high molecular weight species in a later subcompartment of this organelle. When two different alleles of the *ypt1* mutant were analyzed in vitro, they failed to support this reaction (Bacon et al 1989: Baker et al 1990). Although the mutant fractions efficiently packaged the 26-kd form of pro-α-factor into vesicles, only a portion was processed to the 28-kd species. The conversion of this later form of pro-α-factor to the high molecular weight species also failed to occur, and the addition of calcium to the assay mix did not rescue this block in transport (Bacon et al 1989). Therefore, these data indicate that Ypt1p is required for ER-to-Golgi and intra-Golgi membrane traffic and that it does not mediate vesicular transport by regulating the cellular level of calcium.

A subsequent study (Baker et al 1990) reported that affinity-purified anti-Ypt1p antibody, as well as Fab fragments, block transport in vitro. This inhibition was specific since it was competed by the addition of purified Ypt1p to the assay mix. A requirement for calcium in ER-to-Golgi transport was also demonstrated in vitro. However, Ypt1p and calcium were required at different steps in this reaction (Baker et al 1990). More recent studies indicate that Ypt1p functions after vesicle formation (Segev 1991) and may play a role in the targeting of vesicles to their acceptor Golgi complex (Rexach & Schekman 1991). Mutational analysis of the putative effector domain of Ypt1p suggests that the function of this small GTP-binding protein is regulated by a novel GAP that also recognizes Rab1 (Becker et al 1991).

Genes that Compensate for the Loss of YPT1

Proteins that physically interact with Ypt1p, or bypass the need of this small GTP-binding protein, may be encoded by genes that can compensate for the loss of *YPT1* function (Dascher at al 1991). *YPT1*-independent mutants were selected in a *GAL10-YPT1* fusion strain that fails to synthesize Ypt1p in the presence of growth medium that contains glucose. Four dominant mutants in one linkage group, named *SLY1,* were isolated (Dascher et al 1991). *SLY1* encodes a hydrophilic protein of 666 amino acids that is essential for ER-to-Golgi transport (Ossig et al 1991) and is homologous to Sec1, a protein required for post-Golgi secretion (Novick et al 1980; Aalto et al 1992).

In addition to *SLY1,* three multicopy suppressors (*SLY2, SLY12, SLY41*) that compensate for the loss of Ypt1p function were isolated (Dascher et al 1991). *SLY12,* but not *SLY2* or *SLY41,* is essential for the vegetative growth of yeast cells. *SLY2* was recently shown to be identical to *SEC22,* while *SLY12* is the same as *BET1* (Newman et al 1992a,b). Temperature-sensitive mutations in the *sec22* and *bet1* genes result in the proliferation of membranes at their restrictive growth temperature and a rapid block in protein transport between the ER and Golgi apparatus (Novick et al 1980; Newman & Ferro-Novick 1987; Newman et al 1990).

DNA sequence analysis has shown that *BET1* and *SEC22* encode small proteins that are structurally similar to synaptobrevin, an integral membrane protein that decorates the cytoplasmic surface of synaptic vesicles (Dascher et al 1991; Newman et al 1992a,b). Although Ypt1p and Sec22p are constituents of the ER-to-Golgi transport vesicles, Bet1p does not appear to colocalize to this comparment (Lian & Ferro-Novick 1993). Bet1p resides on the ER (Newman et al 1992b), but detectable quantities are not packaged into the transport vesicles as they bud from this membrane. These data suggest that Sec22p acts downstream of Ypt1p, while Bet1p may act on a parallel pathway. Alternatively, Bet1p may control the rate of vesicle budding or packaging of specific proteins, such as Sec22p, into the transport vesicles.

Rab1, Rab2

The first evidence suggesting that a Rab protein facilitates vesicular transport between the ER and Golgi complex in mammalian cells came from experiments in which synthetic peptide analogues were used (Plutner et al 1990). In this study, peptides that are designed to mimic various domains of Rab proteins were tested for their ability to inhibit membrane traffic between the ER and Golgi apparatus. In the assay that was used, transport of a cell surface glycoprotein, the G protein of VSV, was measured by monitoring the processing of the high mannose N-linked oligosaccharide ER Man_9 form to the Man_5 species that is trimmed in the *cis*-Golgi complex of semi-intact (permeabilized) cells (Beckers et al 1987). Peptides that correspond to the putative effector domain of the Rab protein family, but not those of Ras, Ral, or Rho, were found to block transport in vitro. Furthermore, peptides corresponding to domains of the Rab proteins that are analogous to nonessential regions of Ras also failed to disrupt transport, which indicates that the inhibition caused by the effector domain peptides was specific. Kinetic studies suggested that this block was in a late step in transport, possibly occurring at vesicle fusion.

To further explore the functional role of small GTP-binding proteins in ER-to-Golgi transport, antibodies to several specific Rab proteins were prepared. Of the Rab proteins identified, Rab1a and Rab1b are the closest in identity to *YPT1*. In fact, Rab1a complements the yeast *ypt1* mutant (Haubruck et al 1989). Fractionation experiments and indirect immunofluorescence studies have demonstrated that Rab1b can be found on several compartments. Rab1b resides on the ER membrane and Golgi apparatus, and it also co-localizes with Rab2 to a tubular-vesicular intermediate compartment that is believed to function in ER-to-Golgi transport. Additionally, anti-Rab1b antibody blocks transport at an early stage of membrane traffic, probably during the budding of transport vesicles from the ER membrane (Plutner et al 1991).

The in vivo role of Rab1 and Rab2 in ER-to-Golgi transport was addressed through the use of a transient expression system in which HeLa cells were cotransfected with recombinant vaccinia virus and mutant Rab DNA (Tisdale et al 1992). Ras contains four highly conserved domains that are essential for the binding and hydrolysis of GTP. Conserved amino acids in these regions were altered in several Rab proteins, and the consequences of these mutations were investigated. In particular, it was shown that mutations in Rab1a ($Rab1a_{121I}$) and Rab1b ($Rab1b_{121I}$) at position 121 (N→I in the NKXD guanine nucleotide-binding sequence) failed to bind GTP and were potent *trans*-dominant inhibitors of ER-to-Golgi transport. The equivalent mutation in Rab2 ($Rab2_{119I}$) behaved similarly. In contrast, mutations in the NKXD guanine nucleotide-binding sequence of Rab3a, Rab5, and Rab6, which are believed to function in other steps of vesicular transport, did not inhibit membrane traffic between the ER and Golgi complex.

PROTEIN PRENYLATION

Some GTP-binding proteins, such as Ypt1p, Sec4p, and their mammalian homologues are geranylgeranylated (Kinsella & Maltese 1991; Khosravi-Far et al 1991; Kohl et al 1991). Geranylgeranyl is a 20-carbon lipid that contains four isoprene moieties. When this lipid is attached to soluble proteins, it facilitates their membrane attachment. The transferase that modifies Ypt1p and Sec4p was originally identified as a gene whose product is required for the early steps of secretion in yeast (Newman & Ferro-Novick 1987; Rossi et al 1991). Using a variety of techniques, it was shown that the *bet2* mutant behaves as if it were deficient in both Ypt1p and Sec4p function (Rossi et al 1991). Since many small GTP-binding proteins bind to membranes to execute their function, it was reasoned that the activities of these proteins are reduced in *bet2* because the membrane attachment of Ypt1p and Sec4p is defective. When the cellular distribution of these proteins was analyzed, the soluble pool of both proteins was significantly larger in the mutant when compared to wild type.

The *BET2* gene was cloned and sequenced as a first step toward addressing the mechanism by which Bet2p mediates the membrane attachment of Ypt1p and Sec4p. This sequence analysis revealed that Bet2p shares 56.6% similarity and 34.1% identity with *RAM1* (*DPR1*), an essential component of the prenyltransferase that farnesylates (farnesyl is a 15-carbon lipid that contains three isoprene moieties) Ras (Table 1) (Powers et al 1986; Fujiyama et al 1987; Goodman et al 1988; Schafer et al 1990). Based on this finding and the phenotype of the *bet2* mutant, it was hypothesized that Bet2p is a component of a protein prenyltransferase that modifies both Ypt1p and Sec4p. More recent studies have shown that *bet2* mutant extracts are defective in the geranylgeranylation of Ypt1p and Sec4p (Kohl et al 1991; Y. Jiang, G. Rossi, S. Ferro-Novick, submitted).

Three different prenyltransferase activities have been characterized in eukaryotic cells: the farnesyltransferase and the type I and type II geranylgeranyltransferases (GGTase) (Table 1). The farnesyltransferase, which modifies mammalian Ras, has been purified from rat (Reiss et al 1990) and bovine brain cytosol (Moores et al 1991). It is composed of two nonidentical subunits: α (49 kd) and β (46 kd). The β subunit contains the substrate recognition site and is unstable in the absence of the α subunit (Chen et al 1991). The yeast homologue of the β subunit is the *RAM1* (*DPR1*) gene product (Powers et al 1986; Fujiyama et al 1987; Goodman et al 1988; Schafer et al 1990). Like the mammalian subunit, Ram1p recognizes the conserved carboxy terminal CAAX box (where CAAX is Cys-aliphatic-aliphatic, with a preference for methionine, phenylalanine, and serine at the X position—see Reiss et al 1991) of Ras by adding farnesyl to the cysteine residue. The CAAX box signals a series of modifications that include prenylation (farnesylation), proteolytic processing, and carboxymethylation. These events are required for the efficient membrane attachment of Ras (Hancock et al 1991).

Table 1 Prenyltransferases identified in yeast and mammals

		Mammals	Yeast
Farnesyltransferase		β_1	*DPRI(RAM1)*
(CAAX; X = methionine, phenylalanine, serine)		α	*RAM2*
Geranylgeranyltransferase (type I)		β_2	*CDC43*
(CAAX; X = leucine)		α	*RAM2*
Geranylgeranyltransferase (type II)	Component A	95 kd	?
(-CC-,-CXC-; Ypt1, Sec4, Rab1A, Rab3A)	Component B	38 kd	*BET2*
		60 kd	*MAD2*

The α subunit of the farnesyltransferase is shared with the GGTase-I (Seabra et al 1991), which recognizes CAAX (where X is leucine or isoleucine) consensus sequences (Reiss et al 1991; Moores et al 1991; Finegold et al 1991; Yokoyama et al 1991; Moomaw & Casey 1992). The *CDC43* gene product is believed to encode the β subunit of the yeast GGTase-I (Johnston et al 1991; Finegold et al 1991). A fourth gene product, *RAM2,* is 30% identical and 58% similar to the mammalian α subunit (Kohl et al 1991) and encodes the equivalent subunit of the yeast transferase (He et al 1991).

The GGTase-II in mammalian cells modifies Rab proteins that terminate in a CC or CXC motif (Seabra et al 1992b). This enzyme was recently purified (Seabra et al 1992a,b) and found to differ in subunit structure from the farnesyltransferase and the GGTase-I. The GGTase-II is composed of two components, A and B, that are required for activity. Component B contains two tightly bound polypeptides that are 38 and 60 kd. These proteins are probably analogous to the α and β subunits of the CAAX prenyltransferases (Seabra et al 1992b). Component A is 95 kd and has no apparent counterpart in the other transferases. The sequence of six peptides from component A were analyzed and found to be highly homologous to the choroideremia gene product (Seabra et al 1992b). Choroideremia is an X-linked recessive disease that leads to retinal degeneration. The choroideremia gene product also resembles the sequence of Rab3A GDI (GDP-dissociation inhibitor). GDI was originally identified as a protein that inhibits the dissociation of GDP from Rab3A (Araki et al 1990). These findings led to the proposal that component A binds to the Rab substrate, while component B transfers the geranylgeranyl group.

Recent characterization of the yeast GGTase-II has shown that *BET2* and *MAD2,* a gene that is required for the feedback control of mitosis (Li & Murray 1991), encode subunits of this enzyme. Several lines of evidence support this hypothesis. First, *MAD2* is homologous to *RAM2,* the α subunit of the farnesyltransferase (Boguski et al 1992). Second, mutations in either *BET2* or *MAD2* disrupt the membrane attachment of Ypt1p and Sec4p and block the transport of CPY to the vacuole (Rossi et al 1991; R. Li & A. Murray, personal communication). Third, co-expression of *BET2* and *MAD2* in *E. coli* leads to the formation of an enzyme complex that appears to bind geranylgeranyl pyrophosphate, but does not efficiently transfer geranylgeranyl onto Ypt1p (Y. Jiang, G. Rossi, S. Ferro-Novick, submitted). The efficient transfer of lipid onto Ypt1p requires the presence of a third component that is probably analogous to component A in mammalian cells (Y. Jiang G. Rossi, S. Ferro-Novick, submitted). These results are consistent with data that were obtained for the mammalian transferase in which both component A and component B were shown to be required for activity. Component B is composed of two tightly bound polypeptides that resemble the Bet2/Mad2 complex. A list of the prenyltransferases identified in both mammals and yeast is summarized in Table 1.

ARF

ADP ribosylation factor (ARF) was originally identified as a factor necessary for the ADP ribosylation of $G\alpha_s$ by cholera toxin in vitro (reviewed in Kahn 1990). This property was used as an assay during purification (Kahn & Gilman 1984). The ARF protein is a small (20 kd), monomeric, GTP-binding protein (Kahn & Gilman 1986). Sequence analysis established that ARF is related to *ras,* but it is the most divergent member of the *ras* superfamily (Price et al 1988; Sewell & Kahn 1988). While it shares the ability to bind guanine nucleotides with the other monomeric GTP-binding proteins, it does have a number of unique characteristics. Unlike the other members of the superfamily, ARF is not prenylated at its carboxy terminus, rather it is modified by myristylation of an amino terminal glycine (Kahn et al 1988). ARF interacts with phospholipid in a nucleotide-dependent fashion. The GTP bound form of ARF associates with liposomes, while the GDP bound form is a water-soluble monomer (Kahn 1991). Furthermore, GDP dissociation from ARF is stimulated by interaction with dimyristoyl phosphatidyl choline, which facilitates nucleotide exchange. The association of ARF with lipid vesicles probably involves not only the myristyl moiety, but also the amino terminal region (Kahn et al 1992). This region may form an amphipathic helix. The ability of ARF to associate with membranes in a nucleotide-dependent fashion may be critical in the cycle of ARF function as discussed below.

ARF is one of the most highly conserved eukaryotic proteins. The extent of identity between the yeast and bovine sequences is 74% (Sewell & Kahn 1988; Stearns et al 1990a), which suggests that it fufills an essential function that is conserved through evolution. The involvement of ARF in vesicular transport was first appreciated through genetic analysis of the two *ARF* genes in the yeast *Saccharomyces cerevisiae*. Deletion of both *ARF1* and *ARF2* is lethal, while deletion of only the more highly expressed *ARF1* results in a cold-sensitive phenotype (Stearns et al 1990a). The *arf1Δ* cells exhibit a partial defect in protein secretion (Stearns et al 1990b). The cell wall protein invertase is secreted, but there is also accumulation of an intracellular pool of invertase, which is indicative of a slowing of the secretory pathway. The secreted invertase lacks a substantial amount of the carbohydrate that is normally added in the Golgi, which suggests a partial loss of Golgi function. Transient depletion of all ARF protein leads to a block at the ER-to-Golgi stage of the secretory pathway. These results imply a role for ARF in both ER-to-Golgi transport and in the maintenance of Golgi function.

Immunocytochemical localization of ARF in NIH3T3 cells also supports a role in Golgi function (Stearns et al 1990b). ARF was found in association with the cytoplasmic face of *cis*-Golgi stacks and with Golgi-associated vesicles. Following cell lysis, ARF rapidly dissociates from Golgi membranes unless a nonhydrolyzable GTP analogue (such as GTPγS) is present (Donaldson et al 1991b). In the in vitro intra-Golgi transport assay, addition of GTPγS leads to the formation of vesicles bearing both ARF and an 800-kd complex of proteins known as the coatomer (Duden et al 1991; Serafini et al 1991; Waters et al 1991). While genetic depletion of ARF is not feasible in higher eukaryotic cells, a synthetic peptide derived from the amino terminus functions as a specific inhibitor of ARF function in vitro (Kahn et al 1992). This peptide interferes with the function of ARF as a cofactor in the ADP-ribosylation of $G\alpha_s$ by cholera toxin and also blocks intra-Golgi and ER-to-Golgi transport in vitro (Kahn et al 1992; Balch et al 1992). Formation of the coated vesicles, which are normally seen in response to the addition of GTPγS to a Golgi preparation, is blocked by the ARF peptide (Kahn et al 1992). In total, these data suggest that ARF facilitates the budding of vesicle from donor membranes in higher eukaryotic cells and is consistent with the studies of ARF depletion in yeast.

An important clue to the mechanism of ARF function was the observation that the association of ARF and coatomer with the Golgi apparatus could be disrupted by treatment of intact cells with the fungal metabolite brefeldin A (BFA) (Donaldson et al 1991a,b). Treatment of most higher eukaryotic cells with this drug results in dramatic changes in the morphology of the Golgi complex, a block in protein traffic from the ER to the Golgi, and an apparent collapse of the Golgi into the endoplasmic reticulum (Lippincott-Schwartz et al 1989, 1990). Dissociation of ARF and coatomer from the Golgi apparatus is seen after only brief treatment with BFA, well before the disappearance of the Golgi. GTPγS blocks the dissociation of ARF by BFA in permeabilized cells, but only if GTPγS is added prior to addition of BFA (Donaldson et al 1991a). BFA has recently been shown to inhibit a Golgi-associated ARF nucleotide exchange protein (Donaldson et al 1992b; Helms & Rothman 1992). In the absence of BFA, this activity catalyzes the dissociation of pre-bound GDP from ARF and thereby allows binding of GTP or GTPγS . Since the GTP-bound form of ARF binds membranes, nucleotide exchange may be coupled to membrane attachment. By analogy to the function of other small GTP-binding proteins, the GTP bound to ARF is probably subject to rapid hydrolysis as a result of the interaction with a GTPase-activating protein (GAP). Following GTP hydrolysis, ARF would assume its water-soluble GDP-bound conformation and dissociate from membranes. Under normal circumstances, these events cause ARF to undergo rapid cycles of nucleotide exchange, membrane attachment, GTP hydrolysis, and dissociation from membranes. BFA, by blocking nucleotide exchange, would prevent the membrane attachment of ARF. The non-hydrolyzable analogue, GTPγS, has the effect of locking ARF onto membranes in a form that is resistant to the effects of BFA. Since the ARF exchange protein has not been purified, it is not clear if BFA acts directly on the exchange protein, or indirectly, through intermediate components. It is also not clear if all of the effects of BFA are mediated through the ARF exchange protein, or if there are alternate targets as well. ARF function may be under regulation by one or more heterotrimeric GTP-binding proteins, as discussed in the following section.

Heterotrimeric GTP-Binding Proteins

Several lines of evidence have recently pointed to the involvement of heterotrimeric GTP-binding proteins in the regulation of vesicular transport along the exocytic pathway. In their well established roles in signal transduction, the binding of a ligand to a cell-surface receptor triggers dissociation of GDP from the α subunit and thereby allows binding of GTP (Gilman 1987). The α subunit, in its GTP-bound form, dissociates from the βγ dimer and, in most cases, acts to either stimulate or inhibit the generation of a second messenger by interaction with a downstream effector. In several well documented cases, it is the βγ dimer that acts as the regulator of a signal pathway. Hydrolysis of GTP by the α subunit allows rebinding to βγ and curtails the regulation of the signal pathway.

The first clue that a heterotrimeric G protein may regulate vesicular transport was the finding that both GTPγS and AlF_4^- block transport in the in vitro Golgi transport assay (Melançon et al 1987). While GTPγS could be acting through any one of the many monomeric GTP-binding proteins, or through a heterotrimeric G protein, the effects of AlF_4^- are thought to be limited to the heterotrimerics (Kahn 1991). In fact, the intra-Golgi transport assay is still sensitive to the effects of AlF_4^- if the cytosol is first depleted of ARF, although sensitivity to GTPγS is lost (Taylor et al 1992). AlF_4^- acts on the Gα subunit in its GDP-bound form by mimicking the γ phosphate of GTP and thus causing irreversible activation. These studies, therefore, suggest an inhibitory role for a heterotrimeric in transport. Subsequent studies indicated that two specific α subunits, $G\alpha_{i\text{-}2}$ and $G\alpha_{i\text{-}3}$ are associated with the Golgi apparatus (Ercolani et al 1990). Overexpression of $G\alpha_{i\text{-}3}$ leads to a slowing of the constitutive transport of proteoglycan through the Golgi (Stow et al 1991). Furthermore, inactivation of $G\alpha_i$ subunits by pertussis toxin-mediated ADP-ribosylation blocks the effects of $G\alpha_{i\text{-}3}$ overexpression and even leads to more rapid transport in normal cells. These results are consistent with an inhibitory role for $G\alpha_{i\text{-}3}$ protein in transport. Similar studies have implicated a heterotrimeric GTP-binding protein in the regulation of transport from the ER to the Golgi apparatus (Schwaninger et al 1992). Furthermore, in an assay that specifically measured the formation of vesicles from the *trans*-Golgi network, inhibition was seen by addition of GTPγS, AlF_4^-, or excess βγ (Tooze et al 1990; Barr et al 1991), which suggests that regulation by a G protein can occur at the budding stage of transport.

G proteins may exert their action on vesicular transport through the regulation of the binding of coatomer to membranes. Addition of AlF_4^- leads to the binding of coatomer to Golgi membranes in permeabilized cells or in extracts, and pretreatment with AlF_4^- blocks the release of coatomer by BFA (Donaldson et al 1991a,b). Mastoparan, a peptide that activates G proteins by catalyzing nucleotide exchange, also leads to coatomer binding to the Golgi (Ktistakis et al 1992). Pertussis toxin blocks the effects of mastoparan on coatomer binding, presumably by inactivation of a G protein.

Since ARF plays a key role in the attachment of coatomer to Golgi membranes (Donaldson et al 1992a), it is possible that a heterotrimeric G protein regulates coatomer binding by altering the cycle of ARF function. Incubation of Golgi membranes and cytosol in the presence of excess Gβγ prevents the GTPγS-stimulated binding of both coatomer and ARF (Donaldson et al 1991a). This suggests that the G protein acts upstream of ARF binding, which in turn is necessary for coatomer binding. A result in apparent conflict with this proposal is that AlF_4^- stimulates the binding of coatomer to Golgi membranes, but does not stimulate the redistribution of ARF (Donaldson et al 1991a). Nevertheless, the AlF_4^--induced binding of coatomer to Golgi membranes is blocked by an inhibitory peptide designed to mimic the amino terminus of ARF (Donaldson et al 1992a). This suggests an ARF requirement in the action of AlF_4^-. Furthermore, AlF_4^--stimulated coatomer binding requires addition of GTP. Since activation of a G protein by AlF_4^- does not, by itself, require GTP, this nucleotide requirement implies the involvement of another GTP-binding protein, possibly ARF, in the pathway leading to coatomer binding. Resolution of this pathway will require further biochemical dissection.

POST-GOLGI TRANSPORT

Sec4

Sec4 was the first specific GTP-binding protein to be implicated in vesicular transport (Salminen & Novick 1987). The gene was identified in a screen for temperature-sensitive yeast mutants defective in protein secretion (Novick et al 1980). The *sec4-8* mutant is blocked in transport from the Golgi apparatus to the cell surface, but is not defective in earlier transport events. *SEC4* encodes a protein that shares 32% sequence identity with *ras*. Higher levels of identity are seen with the yeast Ypt proteins and with the mammalian Rab proteins, thus defining a Sec4/Ypt/rab sub-family of the *ras* super-family. Like all members of the *ras* super-family, Sec4 binds both GDP and GTP with high affinity and only very slowly hydrolyzes bound GTP (Kabcenell et al 1990).

Sec4 was localized in yeast by a combination of subcellular fractionation and immunofluorescence (Goud et al 1988). Three pools were found; a small cytoplasmic pool, a pool associated with secretory vesicles, and a pool that resides on the plasma membrane. Pulse-chase studies (Goud et al 1988) suggest a cycle of localization in which the soluble pool binds to newly formed vesicles, the vesicles fuse with the plasma membrane, and Sec4 then recycles through the cytoplasm to bind to a new round of vesicles. Genetic evidence (Walworth et al 1989, 1992) supports the proposal (Bourne 1988; Novick et al 1988) that the cycle of localization is coupled to the cycle of GTP binding, hydrolysis, and nucleotide exchange (see Figure 1). A mutation that slows the hydrolysis of GTP by Sec4 leads to accumulation of the GTP-bound form and results in a loss, rather than a gain, of Sec4 function (Walworth et al 1992). This finding suggests that GTP hydrolysis is a key step in the Sec4 cycle and is not a mechanism of down-regulation, as it is in the case of *ras*.

Accessory proteins have been identified that appear to regulate each step of the proposed cycle. The very low intrinsic GTP hydrolysis rate of Sec4 is stimulated by interaction with a GTPase-activating protein (GAP), which is predominantly membrane bound (Walworth et al 1992). While this GAP has not been purified, competition studies suggest that the Sec4 GAP does not recognize Ypt1 (Walworth et al 1992). Interestingly, mammalian cells also possess distinct GAPs that act on Sec4 and Ypt1 (Jena et al 1992).

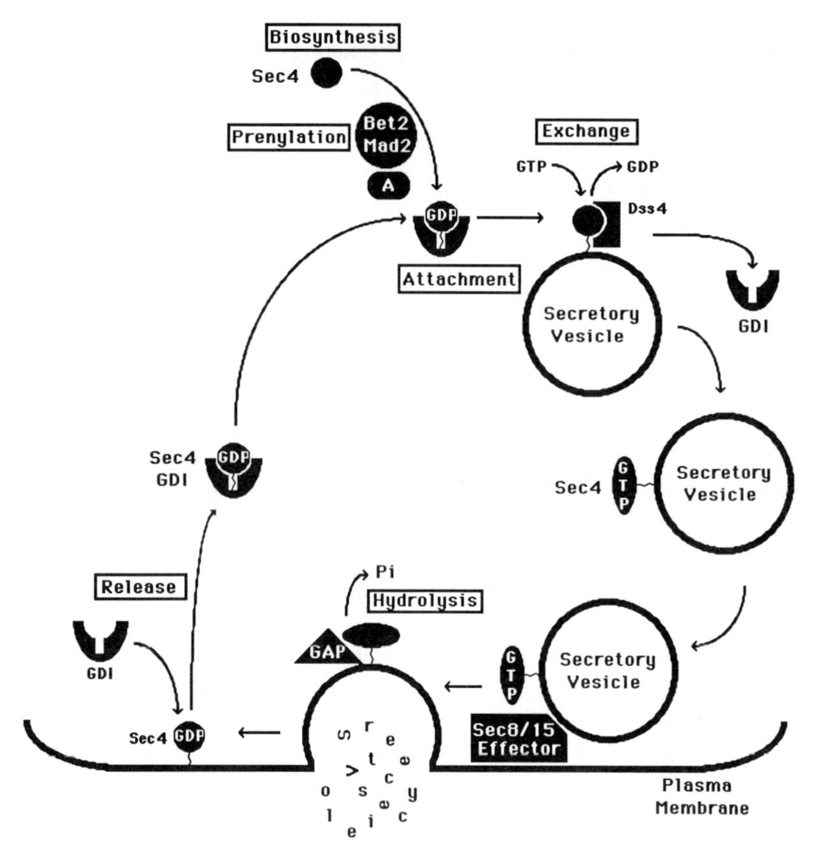

Figure 1 A model of Sec4 function. Following biosynthesis, Sec4 is modified by the addition of a geranylgeranyl moiety to one or both carboxy-terminal cysteines. This reaction is catalyzed by a geranylgeranyl transferase, which consists of the gene products of *BET2* and *MAD2* and a subunit analogous to component A of the mammalian enzyme. GDP-dissociation inhibitor (GDI) binds to the GDP-bound form of Sec4 and maintains the protein in a soluble complex. Dissociation of GDP from Sec4 is stimulated by interaction with Dss4. Binding of GTP causes dissociation of Sec4 from GDI, which in turn allows its attachment to the cytoplasmic surface of a secretory vesicle. Sec4, in its GTP-bound form, interacts with an effector complex on the cytoplasmic face of the plasma membrane, thus triggering events that lead to exocytosis. The effector may contain the products of the *SEC8* and *SEC15* genes. Interaction of Sec4 with its GTPase-activating protein (GAP) leads to hydrolysis of the bound GTP. Sec4 then dissociates from the plasma membrane by forming a soluble complex with GDI, which enables it to participate in another round of vesicular transport.

The intrinsic rate of GDP dissociation from Sec4 is quite high for a member of the *ras* superfamily. This observation prompted the suggestion that an inhibitor of GDP dissociation might exist to negatively regulate dissociation (Kabcenell et al 1990). A GDP-dissociation inhibitor (GDI) has been isolated from bovine brain (Sasaki et al 1990). While this protein was initially identified by its ability to slow the dissociation of GDP from the neuronal-specific protein, rab3A, it is active on a broad range of rab proteins, including Sec4 (Sasaki et al 1991). However, it is not active on members of the *ras* and Rho branches of the superfamily. In addition to slowing the dissociation of GDP, GDI also acts to solubilize Rab3A and other proteins of the Sec4/Ypt/Rab subfamily from membranes when they are in their GDP-bound form, but not in their GTP-bound form (Araki et al 1990; Regazzi et al 1992). The carboxy terminus of Rab3A forms a portion of the recognition site for GDI (Araki et al 1991). GDI may solubilize GDP bound Rab3A by masking the hydrophobic lipids that normally anchor the protein in the membrane. These two activities of GDI are consistent with the proposal that recycling of Sec4 from the plasma membrane to a new secretory vesicle occurs through the cytoplasm and is dependent upon prior hydrolysis of bound GTP. GDI has been well conserved through evolution. A *Drosophila* GDI (DGDI) has been purified, cloned, and sequenced (Zahner & Cheney 1993), and a yeast homologue has recently been cloned as well (M. Garrett et al, in preparation). They display respectively 68 and 50% amino acid sequence identity to the bovine GDI sequence (Matsui et al 1990).

A GDP-dissociation stimulator (GDS), which acts on Sec4, has been identified through suppressor analysis (Moya et al 1993). A dominant extragenic suppressor of the temperature-sensitive *sec4-8* mutant was selected at a semi-restrictive condition, and the suppressor gene, *DSS4-1,* was cloned. *DSS4-1* and the wild-type allele, *dss4,* encode 17-kd hydrophilic proteins that stimulate the rate of GDP dissociation from Sec4 (Moya et al 1993). While *dss4* is not an essential yeast gene, loss of *dss4* in a strain that is already partially defective in *SEC4* function dramatically reduces the growth rate. This suggests that Dss4 acts to facilitate Sec4 by catalyzing nucleotide exchange. One possibility is that by stimulating nucleotide exchange, Dss4 may release Sec4 from GDI and thereby allow reassociation of Sec4 with membranes (Figure 1). A mammalian homologue of Dss4, termed Mss4, was identified from a rat cDNA library by its ability to suppress *sec4-8*. The purified protein demonstrates biochemical properties similar to those of Dss4 (Burton et al 1993).

While the phenotype of *sec4-8* cells indicates a requirement for Sec4 in a step leading to exocytosis, the exact role is not yet clear. By analogy to the function of other GTP-binding proteins, the GTP-bound form may serve to activate a downstream effector pathway. A possible clue to the identity of some components of the effector pathway is the strong genetic interaction between *SEC4* and a number of the other *SEC* genes required at the post-Golgi stage of the yeast secretory pathway (Salminen & Novick 1987).

Two of these genes, *SEC8* and *SEC15* encode components of a 19.5*S* particle that peripherally associates with the plasma membrane (Bowser & Novick 1991; Bowser et al 1992). Indirect evidence suggests that Sec4 may interact with this complex.

Since different members of the Sec4/Ypt/Rab family appear to regulate different vesicular transport events, each protein could serve to direct a different class of carrier vesicles to their appropriate target organelle. By this hypothesis, every vesicle would be tagged with a specific protein. One aspect of this proposal was recently tested (Brennwald & Novick 1993). Various chimeras were generated that combine different domains of Sec4 and Ypt1. A single chimera was made that could efficiently complement a deletion of either *SEC4* or *YPT1* or even the simultaneous deletetion of both genes. This construct includes the loop 7 and hypervariable domains of Ypt1 (using the nomeclature based on the structure of *ras*) in a protein that otherwise consists of the Sec4 sequence. The tagging hypothesis, in its simplest form, would predict that cells carrying this bifunctional chimera would form ER-Golgi vesicles capable of binding and fusing with the plasma membrane. However, no evidence of such missorting was seen. This finding does not imply that Sec4 and the other members of the family are not involved in vesicle targeting, only that they do not, by themselves, specify a distinct target. In that light, it is interesting to note that several of the genes that can compensate for the loss of Ypt1 encode proteins that are associated with the ER-to-Golgi carrier vesicles.

Rab3

A large collection of Rab genes have been identified, and many have been localized to uniques stages of the exocytic or endocytic pathways (for a recent review see Pfeffer 1992). Rab3A (also called *smg* p25A) is a neuronal-specific protein that is found on the cytoplasmic surface of synaptic vesicles (Fischer von Mollard et al 1990; Mizoguchi et al 1990). Proteins related to Rab3A are found in other cell types that possess regulated secretory pathways (Darchen et al 1990; Baldini et al 1992). Rab3 may undergo a cycle of localization, analogous to that shown in Figure 1. Consistent with this hypothesis, a shift in distribution of Rab3 is observed following stimulation of synaptic vesicle discharge (Fischer von Mollard et al 1991). The cycle of Rab3 function is regulated by accessory proteins that control the binding and hydrolysis of nucleotides and its association with membranes. An exchange protein that stimulates the dissociation of GDP from Rab3A and a GAP that stimulates the hydrolysis of GTP bound to Rab3A have been biochemically detected, but not yet purified or cloned (Burnstein et al 1991; Burnstein & Macara 1992). A protein of 86 kd has been detected by its ability to be chemically cross-linked to Rab3A in its GTP-bound form, but not in its GDP-bound form (Shirataki et al 1992). Since this behavior might be predicted for a downstream effector, the protein has been named Rab3A target. GDI may function to release Rab3A from membranes after nucleotide hydrolysis and thus allow recycling, as discussed in the context of Sec4 function. The cycle of Rab3 function could be regulated by a second messenger pathway to trigger exocytosis of stored synaptic vesicles or secretory granules upon stimulation, but no such mechanism has been revealed to date.

GTP-Binding Proteins in Regulated Exocytosis

The earliest evidence for the involvement of GTP-binding proteins in vesicular transport came from the study of mammalian regulated exocytic systems. Discharge of stored granule contents in mast cells could be triggered by the addition of GTP analogues to the cytosol through a patch pipette (Fernandez et al 1984). This discharge was not blocked by treatment of the cells with an inhibitor of phospholipase C, a key component of the signal transduction pathway that normally triggers degranulation in mast cells. It was therefore reasoned that a hypothetical GTP-binding protein, termed G_E, was acting downstream of the signal transduction pathway to directly control the exocytic event. The stimulation of exocytosis by GTP analogues has also been observed in a variety of other cell types including neutrophils (Nusse & Lindau 1988), melanotrophs (Yamamoto et al 1987), eosinophils (Nusse et al 1990), HL-60 cells (Stutchfield & Cockroft 1987), parathyroid cells (Oetting et al 1986), and pancreatic acinar cells (Padfield et al 1992).

The identity of the postulated G_E is still unknown. However, recent results suggest that it may be related to the neuronal-specific GTP-binding protein, Rab3A. Addition of peptides related to the effector region of Rab3A triggers exocytosis when added to permeabilized pancreatic acinar cells (Padfield et al 1992), adrenal chromaffin cells (Senyshyn et al 1992), or when injected into mast cells through a patch pipette (Oberhauser et al 1992). It is interesting to note that while in vitro ER-to-Golgi transport is inhibited by these Rab3-related peptides (Plutner et al 1990), exocytosis is stimulated. This difference may reflect the fact that the in vitro ER-to-Golgi transport assay relies on repeated functional cycles of the rab effector protein. Exocytosis of stored granules, in contrast, may require only a single round of function of the G_E effector protein. Thus a reagent that traps the effector in an activated state may appear to stimulate exocytosis, yet inhibit ER-to-Golgi transport by allowing only one cycle of effector function.

In the nine years following the demonstration that a GTP analogue could trigger exocytosis in mast cells (Fernandez et al 1984), the involvement of GTP-binding proteins in vesicular transport has become one of the most exciting and fast-paced areas in cell biology. Through diverse approaches including genetic analysis, biochemical reconstitution, pharmacology, and physiology, GTP-binding proteins have been implicated at every stage of transport along the exocytic pathway; from protein translocation into the endoplasmic reticulum (ER), to exocytosis at the cell surface. Future studies must address the exact role of these proteins in vesicle budding, targeting, and fusion.

ACKNOWLEDGMENTS

This paper is dedicated to Rico Ferro and Robert Novick on the occasion of the seventieth anniversary of their birth.

Literature Cited

Aalto, MK, Keranen, S, Ronne, H. 1992. A family of proteins involved in intracellular transport. *Cell* 68:181–82

Araki, S, Kaibuchi, K, Sasaki, T, Hata, Y, Takai, Y. 1991. Role of the C-terminal region of *smg* p25A in its interaction with membranes and the GDP/GTP exchange protein. *Mol. Cell. Biol.* 11:1438–47

Araki, S, Kikuchi, A, Hata, Y, Isomura, M, Takai, Y. 1990. Regulation of reversible binding of *smg* p25A, a *ras* p21-like GTP-binding protein, to synaptic plasma membranes ands vesicles by its specific regulatory protein, GDP dissociation inhibitor. *J. Biol. Chem.* 265:13007–15

Bacon, RA, Salminen, A, Ruohola, H, Novick, P, Ferro-Novick, S. 1989. The GTP-binding protein Ypt1 is required for transport in vitro: the Golgi apparatus is defective in *ypt1* mutants. *J. Cell Biol.* 109:1015–22

Baker, D, Hicke, L, Rexach, M, Schleyer, M, Schekman, R. 1988. Reconstitution of *SEC* gene product-dependent intercompartmental protein transport. *Cell* 54:335–44

Baker, D, Wuestehube, L, Schekman, R, Botstein, D, Segev, N. 1990. GTP-binding protein Ypt1 protein and Ca^{2+} function independently in a cell-free protein transport reaction. *Proc. Natl. Acad. Sci. USA* 87: 355–59

Balch, WE, Dunphy, WG, Braell, WA, Rothman, JE. 1984. Reconstitution of the transport of protein between successive compartments of the Golgi measured by the coupled incorporation of N-acetylglucosamine. *Cell* 39:405–16

Balch, WE, Kahn, RA, Schwaninger, R. 1992. ADP-ribosylation factor is required for vesicular trafficking between the endoplasmic reticulum and the *cis*-Golgi compartment. *J. Biol. Chem.* 267:13053–61

Balch, WE, Wagner, KR, Keller, DS. 1987. Reconstitution of transport of vesicular stomatitis virus G protein from the endoplasmic reticulum to the Golgi complex using a cell-free system. *J. Cell Biol.* 104:749–60

Baldini, G, Hohl, T, Lin, HY, Lodish, HF. 1992. Cloning of a Rab3 isotype predominately expressed in adipocytes. *Proc. Natl. Acad. Sci. USA* 89:5049–52

Barlowe, C, d'Enfert, C, Schekman, R. 1993. Purification and characterization of SAR1p, a small GTP-binding protein required for transport vesicle formation from the endoplasmic reticulum. *J. Biol. Chem.* 268:873–79

Barr, FA, Leyte, A, Mollner, S, Pfeuffer, T, Tooze, SA, Huttner, WB. 1991. Trimeric G-proteins of the *trans*-Golgi network are involved in the formation of constitutive secretory vesicles and immature secretory granules. *FEBS Lett.* 293:239–43

Becker, J, Tan, TJ, Trepte, H-H, Gallwitz, D. 1991. Mutational analysis of the putative effector domain of the GTP-binding Ypt1 protein in yeast suggests specific regulation by a novel GAP activity. *EMBO J.* 10:785–92

Beckers, CJM, Balch, WE. 1989. Calcium and GTP: essential components in vesicular trafficking between the endoplasmic reticulum and Golgi apparatus. *J. Cell Biol.* 108: 1245–56

Beckers, CJM, Keller, DS, Balch, WE. 1987. Semi-intact cells permeable to macromolecules: use in reconstitution of protein transport from the endoplasmic reticulum to the Golgi complex. *Cell* 50:523–34

Boguski, MS, Murray, AW, Powers, S. 1992. Novel repetitive sequence motifs in the α and β subunits of prenyl-protein transferases and homology of the α subunit to the *MAD2* gene product of yeast. *New Biol.* 4:1–4

Bomsel, M, Mostov, K. 1992. Role of

heterotrimeric G proteins in membrane traffic. *Mol. Biol. Cell* 3:1317–28

Bourne, HR. 1988. Do GTPases direct membrane traffic in secretion? *Cell* 53:669–71

Bourne, HR, Sanders, DA, McCormick, F. 1991. The GTPase superfamily: conserved structure and molecular mechanism. *Nature* 349:117–27

Bowser, R, Müller, H, Govindan, B, Novick, P. 1992. Sec8p and Sec15p are components of a 19.5*S* particle that may function downstream of Sec4p to control exocytosis. *J. Cell. Biol.* 118:1041–56

Bowser, R, Novick, P. 1991. Sec15 protein, an essential component of the exocytotic apparatus, is associated with the plasma membrane and with a soluble 19.5*S* particle. *J. Cell Biol.* 112:1117–31

Brennwald, P, Novick, P. 1992. Interaction of 3 domains distinguishing the Ras-related GTP-binding proteins Ypt1 and Sec4. *Nature* 362:560–63

Burstein, ES, Linko-Stenz, K, Lu, Z, Macara, IG. 1991. Regulation of the GTPase activity of the *ras*-like protein p25*rab*3A: evidence for a *rab*3A specific GAP. *J. Biol. Chem.* 266:2689–92

Burnstein, ES, Macara, IG. 1992. Characterization of a guanine nucleotide-releasing factor and a GTPase-activating protein that are specific for the *ras*-related protein p25rab3A. *Proc. Natl. Acad. Sci. USA* 89: 1154–58

Burton, J, Roberts, D, Montaldi, M, Novick, P, De Camilli, PA. A mammalian guanine-nucleotide releasing protein enhances function of yeast secretory protein Sec4. *Nature* 361:464–67

Chen, WJ, Andres, DA, Goldstein, JL, Brown, MS. 1991. Cloning and expression of a cDNA encoding the α subunit of rat $p21^{ras}$ protein farnesyltransferase. *Proc. Natl. Acad. Sci. USA* 88:11368–72

Darchen, F, Zahraoui, A, Hammel, F, Monteils, M, Tavitian, A, Scherman, D. 1990. Association of the GTP-binding protein Rab3A with bovine adrenal chromaffin granules. *Proc. Natl. Acad. Sci. USA* 87: 5692–96

Dascher, C, Ossig, R, Gallwitz, D, Schmitt, HD. 1991. Identification and structure of four yeast genes (*SLY*) that are able to suppress the functional loss of *YPT1*, a member of the RAS superfamily. *Mol. Cell Biol.* 11:872–85

d'Enfert, C, Barlowe, C, Nishikawa, S, Nakano, A, Schekman, R. 1991a. Structural and functional dissection of a membrane glycoprotein required for vesicle budding from the endoplasmic reticulum. *Mol. Cell Biol.* 11:5727–34

d'Enfert, C, Wuestehube, L, Lila, T, Schekman, R. 1991b. Sec12p-dependent membrane binding of the small GTP-binding protein Sar1p promotes formation of transport vesicles from the ER. *J. Cell Biol.* 114:663–70

Donaldson, JG, Cassel, D, Kahn, RA, Klausner, RD. 1992a. ADP-ribosylation factor, a small GTP-binding protein, is required for binding of the coatomer protein β-COP to Golgi membranes. *Proc. Natl. Acad. Sci. USA* 89:6408–12

Donaldson, JG, Finazzi, D, Klausner, RD. 1992b. Brefeldin A inhibits Golgi membrane-catalysed exchange of guanine nucleotide onto ARF protein. *Nature* 360:350–52

Donaldson, JG, Kahn, RA, Lippincott-Schwartz, J, Klausner, RD. 1991a. Binding of ARF and β-COP to Golgi membranes: Possible regulation by a trimeric G protein. *Science* 254:1197–99

Duden, R, Griffiths, G, Frank, R, Argos, P, Kreis, TE. 1991. β-COP, a 110 kD protein associated with non-clathrin-coated vesicles and the Golgi complex, shows homology to β-adaptin. *Cell* 64:649–65

Donaldson, JG, Lippincott-Schwartz, J, Klausner, RD. 1991b. Guanine nucleotides modulate the effects of brefeldin A in semipermeable cells:regulation of the association of a 110-kD peripheral membrane protein with the Golgi apparatus. *J. Cell Biol.* 112: 579–88

Ercolani, L, Stow, JL, Boyle, JF, Holtzman, EJ, Lin, H, et al. 1990. Membrane localization of the pertussis toxin-sensitive G-protein subunits α_{i-2} and α_{i-3} and expression of metallothionein-α_{i-2} gene in LLC-PK1 cells. *Proc. Natl. Acad. Sci. USA* 87:4635–39

Fernandez, JM, Neher, E, Gomperts, BD. 1984. Capacitance measurements reveal stepwise fusion events in degranulating mast cells. *Nature* 321:453–55

Finegold, AA, Johnson, DI, Farnsworth, CC, Gelb, MH, Judd, SR, et al. 1991. Protein geranylgeranyltransferase of *Saccharomyces cerevisiae* is specific for Cys-Xaa-Xaa-Leu motif proteins and requires the *CDC43* gene product but not the *DPR1* gene product. *Proc. Natl. Acad. Sci. USA* 88:4448–52

Fischer von Mollard, G, Mignery, GA, Baumert, M, Perin, MS, Hanson, TJ, et al. 1990. rab3 is a small GTP-binding protein exclusively localized to synaptic vesicles. *Proc. Natl. Acad. Sci. USA* 87:1988–92

Fischer von Mollard, G, Südhof, TC, Jahn R. 1991. A small GTP-binding protein dissociates from synaptic vesicles during exocyosis. *Nature* 349:79–81

Fujiyama, A, Matsumoto, K, Tamanoi, F. 1987. A novel yeast mutant defective in the processing of ras proteins: assessment of the effect of the mutation on processing steps. *EMBO J.* 6:223–28

Gallwitz, D, Donrath, C, Sander, C. 1983. A yeast oncogene encoding a protein homologous to the human *c-has/bas* pro-oncogene product. *Nature* 306:704–9

Garrett, MD, Kabcenell, AK, Zahner, J, Sasaki, T, Takai, Y, et al. 1987. G proteins: transducers of receptor-generated signals. *Annu. Rev. Biochem.* 56:615–19

Goodman, LE, Perou, CM, Tamanoi, F. 1988. Structure and expression of yeast *DPR1*, a gene essential for the processing and intracellular localization of ras proteins. *Yeast* 4:271–81

Goud, B, Salminen, A, Walworth, NC, Novick, P. 1988. A GTP-binding protein required for secretion rapidly associates with secretory vesicles and the plasma membrane in yeast. *Cell* 53:753–68

Groesch, M, Ruohola, H, Bacon, R, Rossi, G, Ferro-Novick, S. 1990. Isolation of a functional vesicular intermediate that mediates ER to Golgi transport in yeast. *J. Cell Biol.* 111:45–53

Hancock, JF, Cadwallader, K, Marshall, CJ. 1991. Methylation and proteolysis are essential for efficient membrane binding of prenylated $p21^{K\text{-}ras(B)}$. *EMBO J.* 10:641–46

Haubruck, H, Prange, R, Vorgias, C, Gallwitz, D. 1989. The ras-related mouse ypt1 protein can functionally replace the *YPT1* gene product in yeast. *EMBO J.* 8: 1427–32

He, B, Chen, P, Chen, S-Y, Vancura, KL, Michaelis, S, Powers, S. 1991. *RAM2*, an essential gene of yeast, and *RAM1* encode the two polypeptide components of the farnesyltransferase that prenylates a-factor and Ras proteins. *Proc. Natl. Acad. Sci. USA* 88:11373–77

Helms, JB, Rothman, JE. 1992. Inhibition by brefelden A of a Golgi membrane enzyme that catalyses exchange of guanine nucleotide bound to ARF. *Nature* 360:352–54

Hicke, L, Schekman, R. 1989. Yeast Sec23p acts in the cytoplasm to promote protein transport from the ER to the Golgi complex in vivo and in vitro. *EMBO J.* 8:1677–84

Hicke, L, Yoshihisa, T, Schekman, R. 1992. Sec23p and a novel 105 kD protein function as a multimeric complex to promote vesicle budding and protein transport from the ER. *Mol. Cell. Biol.* 3:667–76

Jena, BP, Brennwald, P, Garrett, M, Novick, P, Jamieson, JD. 1992. Distinct GAPs in mammalian pancreas act on the yeast GTP-binding proteins Ypt1 and Sec4. *FEBS Lett.* 309:5–9

Johnston, DI, O'Brien, JM, Jacobs, CW. 1991. Isolation and sequence analysis of *CDC43*, a gene involved in the control of cell polarity in *Saccharomyces cerevisiae*. *Gene* 98:149–50

Kahn, RA. 1990. The ADP-ribosylation factor of adenylate cyclase: a 21 kDa GTP-binding protein, In *G-proteins*, ed. L Birnbaumer, R Iyengar, pp. 201–14. Orlando, Fla.: Academic

Kahn, RA. 1991. Fluoride is not an activator of the smaller (20–25 kDa) GTP-binding proteins. *J. Biol. Chem.* 266:15595–97

Kahn, RA, Gilman, AG. 1984. Purification of a protein cofactor required for ADP-ribosylation of the stimulatory regulatory component of adenylate cyclase by cholera toxin. *J. Biol. Chem.* 259:6228–34

Kahn, RA, Gilman, AG. 1986. The protein cofactor necessary for ADP ribosylation of Gs by cholera toxin is itself a GTP binding protein. *J. Biol. Chem.* 261:7906–11

Kahn, RA, Goddard, C, Newkirk, M. 1988. Chemical and immunological characterization of the 21 kDa ADP ribosylation factor of adenylate cyclase. *J. Biol. Chem.* 263: 8282–87

Kahn, RA, Randazzo, P, Serafini, T, Weiss, O, Rulka, C, et al. 1992. The amino terminus of ADP-ribosylation factor (ARF) is a critical determinant of ARF activities and is a potent and specific inhibitor of protein transport. *J. Biol. Chem.* 267:13039–46

Kabcenell, AK, Goud, B, Northup, JK, Novick, P. 1990. Binding and hydrolysis of guanine nucleotides by Sec4p, a yeast protein involved in the regulation of vesicular traffic. *J. Biol. Chem.* 265:9366–72

Kaiser, CA, Schekman, R. 1990. Distinct sets of *SEC* genes govern transport vesicle formation and fusion in the early secretory pathway. *Cell* 61:723–33

Khosravi-Far, R, Lutz, RJ, Cox, AD, Conroy, L, Bourne, JR, et al. 1991. Isoprenoid modification of *rab* proteins terminating in CC or CXC motifs. *Proc. Natl. Acad. Sci. USA* 88:6264–68

Kinsella, B, Maltese, WA. 1991. *rab* GTP-binding proteins implicated in vesicular transport are isoprenylated in vitro at cysteines within a novel carboxyl-terminal motif. *J. Biol. Chem.* 266:8540–44

Kohl, NE, Diehl, RE, Schaber, MD, Rands, E, Soderman, DD, et al. 1991. Structural homology among mammalian and *Saccharomyces cerevisiae* isoprenyl-protein transferases. *J. Biol. Chem.* 266:18884–88

Ktistakis, NT, Linder, ME, Roth, MG. 1992. Action of brefelden A blocked by activation of a pertussis-toxin-sensitive G protein. *Nature* 356:344–46

Li, R, Murray, AW. 1991. Feedback conrol of mitosis in budding yeast. *Cell* 66:519–31

Lian, JP, Ferro-Novick, S. 1993. Bos1p, an integral membrane protein of the endoplasmic reticulum to Golgi transport vesicles, is required for their fusion competence. *Cell* 73:735–45

Lippincott-Schwartz, J, Donaldson, JG,

Schweizer, A, Berger, EG, Hauri, H-P, et al. 1990. Microtubule-dependent retrograde transport of proteins into the ER in the presence of brefeldin A suggests an ER recycling pathway. *Cell* 60:821–36

Lippincott-Schwartz, J, Yuan, LC, Bonifacino, JS, Klausner, RD. 1989. Rapid redistribution of Golgi proteins into the ER in cells treated with brefeldin A: evidence for membrane cycling from Golgi to ER. *Cell* 56:801–13

Matsui, Y, Kikuchi, A, Araki, S, Hata, Y, Kondo, J, et al. 1990. Molecular cloning and characterization of a novel type of regulatory protein (GDI) for *smg* p25A, a *ras*-like GTP-binding protein. *Mol. Cell. Biol.* 10:4116–22

Melançon, P, Glick, BS, Malhotra, V, Weidman, PJ, Serafini, T, et al. 1987. Involvement of GTP-binding "G" proteins in transport through the Golgi stack. *Cell* 51: 1053–62

Mizoguchi, A, Kim, S, Ueda, T, Kikuchi, A, Yorifuji, H, et al. 1990. Localization and subcellular distribution of *smg* p25A, a *ras* p21-like GTP-binding protein, in rat brain. *J. Biol. Chem.* 265:11872–79

Moomaw, JF, Casey, PJ. 1992. Mammalian protein geranylgeranyltransferase. *J. Biol. Chem.* 267:17438–43

Moores, SL, Schaber, MD, Mosser, SD, Rands, E, O'Hara, MB, et al. 1991. Sequence dependence of protein isoprenylation. *J. Biol. Chem.* 266:14603–10

Moya, M, Roberts, D, Novick, P. 1993. *DSS4–1*, is a dominant suppressor of *sec4–8* that encodes a nucleotide exchange protein that aids Sec4p function. *Nature* 361:460–63

Nakano, A, Muramatsu, M. 1989. A novel GTP-binding protein, Sar1p, is involved in transport from the endoplasmic reticulum to the Golgi apparatus. *J. Cell Biol.* 109:2677–91

Newman, A, Ferro-Novick, S. 1987. Characterization of new mutants in the early part of the yeast secretory pathway isolated by a [^{3}H]mannose suicide selection. *J. Cell Biol.* 105:1587–94

Newman, AP, Graf, J, Mancini, P, Rossi, G, Lian, JP, Ferro-Novick, S. 1992a. *SEC22* and *SLY2* are identical. *Mol. Cell Biol.* 12:3663–64

Newman, AP, Groesch, M, Ferro-Novick, S. 1992b. Bos1p, a membrane protein required for ER to Golgi transport in yeast, co-purifies with carrier vesicles and with Bet1p and the ER membrane. *EMBO J.* 11:3609–17

Newman, AP, Shim, J, Ferro-Novick, S. 1990. *BET1, BOS1* and *SEC22* are members of a group of interacting genes required for transport from the ER to the Golgi complex. *Mol. Cell. Biol.* 10:3405–14

Novick, P, Ferro, S, Schekman, R. 1981. Order of events in the yeast secretory pathway. *Cell* 25:461–69

Novick, P, Field, C, Schekman, R. 1980. Identification of 23 complementation groups required for post-translational events in the yeast secretory pathway. *Cell* 21:205–15

Novick, P, Goud, B, Salminen, A, Walworth, NC, Nair, J, Potenza, M. 1988. Regulation of vesicular traffic by a GTP-binding protein on the cytoplasmic surface of secretory vesicles in yeast. *Cold Spring Harbor Symp. Quant. Biol.* 53:637–47

Nusse, O, Lindau, M. 1988. The dynamics of exocytosis in human neutrophils. *J. Cell Biol.* 107:2117–23

Nusse, O, Lindau, M, Cromwell, O, Kay, AB, Gomperts, BD. 1990. Intracellular application of guanosine-5′-O-(3-thiophosphate) induces exocytotic granule fusion in guinea pig eosinophils. *J. Exp. Med.* 171: 775–86

Oberhauser, AF, Monck, JR, Balch, WE, Fernandez, JM. 1992. Exocytotic fusion is activated by Rab3a peptides. *Nature* 360: 270–73

Oetting, M, LeBoff, M, Swiston, L, Preston, J, Brown, E. 1986. Guanine nucleotides are potent secretagogues in permeabilized parathyroid cells. *FEBS Lett.* 208:99–104

Oka, T, Nishikawa, S, Nakano, A. 1991. Reconstitution of GTP-binding Sar1 protein function in ER to Golgi transport. *J. Cell Biol.* 114:671–79

Ossig, R, Dascher, C, Trepte, H-H, Schmitt, HD, Gallwitz, D. 1991. The yeast *SLY* gene products, suppressors of defects in the essential GTP-binding Ypt1 protein, may act in endoplasmic reticulum-to-Golgi transport. *Mol. Cell. Biol.* 11:2980–93

Padfield, PJ, Balch, WE, Jamieson, JD. 1992. A synthetic peptide of the rab3a effector domain stimulates amylase release from permeabilized pancreatic acini. *Proc. Natl. Acad. Sci. USA* 89:1656–60

Pfeffer, SR. 1992. GTP-binding proteins in intracellular transport. *Trends Cell Biol.* 2: 41–46

Plutner, H, Cox, AD, Pind, S, Khosravi-Far, R, Bourne, JR, et al. 1991. Rab1b regulates vesicular transport between the endoplasmic reticulum and successive Golgi compartments. *J. Cell Biol.* 115:31–43

Plutner, H, Schwaninger, R, Pind, S, Balch, WE. 1990. Synthetic peptides of the Rab effector domain inhibit vesicular transport through the secretory pathway. *EMBO J.* 9:2375–83

Powers, S, Michaelis, S, Broek, D, Anna-A, SS, Field, J, et al. 1986. *RAM,* a gene of yeast required for a functional modification of *RAS* proteins and for production of mating pheromone a-factor. *Cell* 47:413–22

Price, SR, Nightingale, M, Tsai, SC, Williamson, KC, Adamik, R, et al. 1988. Guanine nucleotide-binding proteins that enhance choleragen ADP-ribosyltransferase activity: nucleotide and deduced amino acid sequence of an ADP-ribosylation factor cDNA. *Proc. Natl. Acad. Sci. USA* 85:5488–91

Pryer, NK, Wuestehube, LJ, Schekman, R. 1992. Vesicle-mediated protein sorting. *Annu. Rev. Biochem.* 61:471–516

Regazzi, R, Kikuchi, A, Takai, Y, Wolheim, CB. 1992. The small GTP-binding proteins in the cytosol of insulin-secreting cells are complexed to GDP dissociation inhibitor proteins. *J. Biol. Chem.* 267:17512–19

Reiss, Y, Goldstein, JL, Seabra, MC, Casey, PJ, Brown, MS. 1990. Inhibition of purified $p21^{ras}$ farnesyl: protein transferase by Cys-AAX tetrapeptides. *Cell* 62:81–88

Reiss, Y, Stradley, SJ, Gierasch, LM, Brown, MS. 1991. Sequence requirement for peptide recognition by rat brain $p21^{ras}$ protein farnesyltransferase. *Proc. Natl. Acad. Sci. USA* 88:732–36

Rexach, MF, Schekman, R. 1991. Distinct biochemical requirements for the budding, targeting, and fusion of ER-derived transport vesicles. *J. Cell Biol.* 114:219–29

Rossi, G, Jiang, Y, Newman, A, Ferro-Novick, S. 1991. Dependence of Ypt1 and Sec4 membrane attachment on Bet2. *Nature* 351:158–61

Ruohola, H, Kabcenell, AK, Ferro-Novick, S. 1988. Reconstitution of protein transport from the endoplasmic reticulum to the Golgi complex in yeast: the acceptor compartment is defective in the *sec23* mutant. *J. Cell Biol.* 107:1465–76

Salminen, A, Novick, P. 1987. A ras-like protein is required for a post-Golgi event in yeast secretion. *Cell* 49:527–38

Salminen, A, Novick, P. 1989. The Sec15 protein responds to the function of the GTP binding protein, Sec4, to control vesicular traffic. *J. Cell Biol.* 109:1023–36

Sasaki, T, Kaibuchi, K, Kabcenell, AK, Novick, P, Takai, Y. 1991. A mammalian inhibitory GDP/GTP exchange protein (GDI) for *smg* p25A is active on the yeast *SEC4* protein. *Mol. Cell. Biol.* 11:2909–12

Sasaki, T, Kikuchi, A, Araki, S, Hata, Y, Isomura, M, et al. 1990. Purification and characterization from bovine brain cytosol of a protein that inhibits dissociation of the GDP form and the subsequent binding of GTP to *smg* p25A, a *ras* p21-like GTP-binding protein. *J. Biol. Chem.* 265:2333–37

Schafer, WR, Trueblood, CE, Yang, C-C, Mayer, MP, Rosenberg, S, et al. 1990. Enzymatic coupling of cholesterol intermediates to a mating pheromone precursor and to the Ras protein. *Science* 249:1133–39

Schmitt, HD, Puzicha, M, Gallwitz, D. 1988. Study of a temperature-sensitive mutant of the *ras*-related *YPT1* gene product in yeast suggests a role in the regulation of intracellular calcium. *Cell* 53:635–47

Schwaninger, R, Plutner, H, Bokoch, GM, Balch, WE. 1992. Multiple GTP-binding proteins regulate vesicular transport from the ER to Golgi membranes. *J. Cell Biol.* 119:1077–96

Seabra, MC, Brown, MS, Slaughter, CA, Südhof, TC, Goldstein, JL. 1992a. Purification of component A of Rab geranylgeranyltransferase: possible identity with the choroideremia gene product. *Cell* 70: 1049–57

Seabra, MC, Goldstein, JL, Slaughter, CA, Südhof, T, Brown, MS. 1992b. Rab geranylgeranyltransferase. *J. Biol. Chem.* 267:14497–503

Seabra, MC, Reiss, Y, Casey, PJ, Brown, MS, Goldstein, JL. 1991. Protein farnesyltransferase and geranylgeranyltransferase share a common a subunit. *Cell* 65:429–34

Segev, N. 1991. Mediation of the attachment or fusion step in vesicular transport by the GTP-binding Ypt1 Protein. *Science* 252: 1553–56

Segev, N, Mulholland, J, Botstein, D. 1988. The yeast GTP-binding *YPT1* protein and a mammalian counterpart are associated with the secretion machinery. *Cell* 52:915–24

Senyshyn, J, Balch, WE, Holz, RW. 1992. Synthetic peptide of the effector-binding domain of rab enhance secretion from digitonin-permeabilized chromaffin cells. *FEBS Lett.* 309:41–46

Serafini, T, Stenbeck, G, Brecht, A. Lottspeich, F, Orci, L, et al. 1991. A coat subunit of Golgi-derived non-clathrin-coated vesicles with homology to the clathrin-coated vesicle protein β-adaptin. *Nature* 349:215–20

Sewell, JL, Kahn, RA. 1988. Sequences of the bovine and yeast ADP-ribosylation factor and comparison to other GTP-binding proteins. *Proc. Natl. Acad. Sci. USA* 85: 4620–24

Shirataki, H, Kaibuchi, K, Yamaguchi, T, Wada, K, Horiuchi, H, Takai, Y. 1992. A possible target protein for *smg-25A/rab3A* small GTP-binding protein. *J. Biol. Chem.* 267:10946–49

Simons, K, Virta, H. 1987. Perforated MDCK cells support intracellular transport. *EMBO J.* 6:2241–47

Stearns, T, Kahn, RA, Botstein, D, Hoyt, MA. 1990a. ADP ribosylation factor is an essential protein in *Saccharomyces cerevisiae* and is encoded by two genes. *Mol. Cell. Biol.* 10:6690–99

Stearns, T, Willingham, MC, Botstein, D, Kahn, RA. 1990b. ADP-ribosylation factor is functionally and physically associated with the Golgi complex. *Proc. Natl. Acad. Sci. USA* 87:1238–42

Stow, JL, de Almeida, JB, Narula, N, Holtzman, EJ, Ercolani, L, Ausiello, DA. 1991. A heterotrimeric G protein, Gα1–3, on Golgi membranes regulates the secretion of a heparan sulfate proteoglycan in LLC-PK1 epithelial cells. *J. Cell Biol.* 114:1113–24

Stutchfield, J, Cockcroft, S. 1988. Guanine nucleotides stimulate polyphosphoinositide phosphodiesterase and exocytotic secretion from HL-60 cells permeabilized with streptolysin O. *Biochem. J.* 250:375–82

Taylor, TC, Kahn, RA, Melançon, P. 1992. Two distinct members of the ADP-ribosylation factor family of GTP-binding proteins regulate cell-free intra-Golgi transport. *Cell* 70:69–79

Tisdale, EJ, Bourne, JR, Khosravi-Far, R, Der, CJ, Balch, WE. 1992. GTP-binding mutants of Rab1 and Rab2 are potent inhibitors of vesicular transport from the endoplasmic reticulum to the Golgi complex. *J. Cell Biol.* 119:749–61

Tooze, SA, Weiss, U, Huttner, WB. 1990. Requirement for GTP hydrolysis in the formation of secretory vesicles. *Nature* 347 :207–8

Walworth, NC, Brennwald, P, Kabcenell, AK, Novick, P. 1992. Hydrolysis of GTP by Sec4 protein plays an important role in vesicular transport and is stimulated by a GTPase activating protein in yeast. *Mol. Cell. Biol.* 12:2017–28

Walworth, NC, Goud, B, Kabcenell, AK, Novick, P. 1989. Mutational analysis of *SEC4* suggests a cyclical mechanism for the regulation of vesicular traffic. *EMBO J.* 8:1685–993

Water, MG, Serafini, T, Rothman, JE. 1991. "Coatomer:" a cytosolic protein complex containing subunits of non-clathrin-coated Golgi transport vesicles. *Nature* 349:248–51

Yamamoto, T, Furuki, Y, Kebabian, JW, Spatz, M. 1987. α-melanocyte-stimulating hormone release from permeabilized intermediate lobe cells of rat pituitary gland. *FEBS Lett.* 219:326–30

Yokoyama, K, Goodwin, GW, Ghomashchi, F, Glomset, JA, Gelb, MH. 1991. A protein geranylgeranyltransferase from bovine brain: implications for protein prenylation specificity. *Proc. Natl. Acad. Sci. USA* 88: 5302–6

Yoshihisa, T, Barlowe, C, Schekman, R. 1993. Requirement for a GTPase-activating protein in vesicle budding from the endoplasmic reticulum. *Science.* 259:1466–68

Zahner, JE, Cheney, CM. 1993. GDI, a regulatory protein for small GTP-binding proteins, is affected by the *Drosophila* mitotic mutant, *quartet*. *Mol. Cell. Biol.* 13:217–27

Annu. Rev. Cell Biol. 1993. 9:601–34

ROLE OF THE MAJOR HEAT SHOCK PROTEINS AS MOLECULAR CHAPERONES

C. Georgopoulos

Department of Medical Biochemistry, Centre Medical Universitaire, University of Geneva, 1 rue Michel-Servet, 1211 Geneva 4, Switzerland

W. J. Welch

Departments of Medicine and Physiology, University of California, Box 0854, San Francisco, California 94143

KEY WORDS: heat shock proteins, stress proteins, chaperone proteins, chaperone machines, protein folding, disaggregation

---

CONTENTS

0743–4634/93/1115–0601$05.00

INTRODUCTION

What began as a molecular curiosity in fruit flies over 30 years ago (Ritossa 1962), the so-called heat shock or stress response, now constitutes an active area of research in cell biology and biochemistry. The heat shock proteins (hsps), one of the most highly conserved group of proteins so far characterized, are being implicated as essential components in a number of diverse biological processes. Although referred to as heat shock proteins, most of these proteins in fact are expressed at rather significant levels in all cells maintained under normal growth conditions and are essential for cellular growth at all physiologically relevant temperatures. Much of the current interest in the hsps follows from recent studies demonstrating their role as molecular "chaperones", being intimately involved in various steps of protein maturation. Members of the Hsp70 (DnaK) and Hsp60 (GroEL) families, for example, participate in protein folding, protein translocation, and perhaps higher ordered protein assembly. Yet other members of the heat shock protein family, such as Hsp90, play important roles in the regulation of certain transcription factors and protein kinases. Here, we present our current understanding of the structure and function of those hsps that function as molecular chaperones and discuss their roles both within the normal cell, as well as in the cell experiencing metabolic stress. In addition, we discuss other possible roles of molecular chaperones as they relate to various macromolecular assembly/disassembly events that accompany a number of other important biologial processes in the cell. For more details of the biochemistry, the reader is referred to a recent review on protein folding (Jaenicke 1991) and recent reviews that cover in detail the functions of the various hsps (Hartl et al 1992; Gething & Sambrook 1992; Rothman 1989; Ellis & van der Vies 1991; Kelley & Georgopoulos 1992; Welch 1992).

Abnormally Folded Proteins Lead to Induction of the Heat Shock Response

While the history leading to the discovery of the heat shock response is interesting, we limit our discussion to the structure and function of those hsps that function as molecular chaperones (see Morimoto et al 1990 for a review of the biology of the heat shock response). Some discussion of the mechanisms by which the expression of the hsps is regulated in the cell, however, merits inclusion here since it helps to explain both their biochemical function as well as their collective role in providing for increased cellular protection.

Early work in *Drosophila melanogaster* demonstrated the rapid and selective increase in expression of a select group of proteins following exposure of fly larvae to temperatures above those optimal for normal growth and development (Tissieres et al 1974). These proteins were referred to as hsps because of the nature of the stimulus that caused their induction. Subsequent studies revealed that a large number of other metabolic insults could lead to very similar changes in gene expression. Exposure of *Drosophila* tissues to uncouplers of oxidative phosphorylation, inhibitors of electron transport, or agents known to inhibit the activities of different enzymes also resulted in an increased expression of one or more of the hsps (reviewed by Ashburner & Bonner 1979). Consequently, it was somewhat perplexing as to how so many different agents/treatments could lead to such similar changes in gene expression. The discovery that the exposure of cells to various amino acid analogues or the antibiotic puromycin also resulted in the increased expression of the hsps led to an hypothesis to explain how the heat shock response might be initiated (Kelley & Schlesinger 1978; Hightower 1980). Amino acid analogues were known to result in the production of proteins that exhibited extremely short half-lives, probably because of their inability to fold properly. The antibiotic puromycin was known to result in the premature release of nascent polypeptides from the translation machinery. These observations, along with the known adverse effects of heat on protein conformation (i.e. leading to protein aggregation), led to the suggestion that whenever abnormally folded proteins begin to accumulate, a heat shock-like response would be initiated. Perhaps the resultant increased expression and accumulation of the hsps somehow functioned to help the cell deal with its increased burden of abnormally folded proteins (Hightower 1980; Goff & Goldberg 1985; Pelham 1986).

Work over the next 10 years provided support for this so-called abnormal protein hypothesis put forward to explain the induction of the heat shock response. First, simply injecting a collection of denatured proteins into living cells was shown to result in an increased expression of the hsps (Ananthan et al 1986). Second, *S. cerevisiae* harboring mutations in one or more components involved in proteolytic pathways were found to increase their expression of the hsps (Finley et al 1984). Third, *Drosophila* cells expressing high levels of mutant actin exhibited an abnormally high constitutive expression of the hsps (Karlik et al 1984; Hiromi et al 1986). Finally, cells first incubated with agents known to stabilize protein conformation (e.g. glycerol or deuterium oxide) and then subjected to a heat shock treatment, expressed significantly lower levels of the hsps as compared to those cells heated in the absence of such agents (Edington et al 1989). Taken together, these observations are consistent with the idea that the accumulation of abnormally folded proteins within the cell leads to an increased expression of the hsps. Exactly how a cell is able to monitor or "sense" the accumulation of abnormally folded proteins and subsequently initiate those transcription/translation events that lead to high level expression of the hsps remains an important unanswered question (see below).

The Heat Shock Proteins

For largely historical reasons, the hsps from different organisms are referred to by a variety of different names. Unfortunately this often leads to some confusion when discussing the structure and function of a particular member of the heat shock protein family. Following the nomenclature first used in *Drosophila*, the various hsps in animal cells are referred to on the basis of their mode of induction, and apparent molecular mass on SDS [1] gels. Hence, their designation as heat shock protein 70 (Hsp70) and heat shock protein 90 (Hsp90), for example, refers to heat-inducible proteins of 70 or 90 kd, respectively. In eukaryotic cells, another class of stress-induced proteins, distinct from those induced by heat, were first observed to exhibit an increased expression in cells starved of glucose, and therefore were called the glucose-regulated proteins (grps) (Pouysségur et al 1977). Subsequent work revealed increased grps expression in cells under anoxic conditions, treatment with inhibitors of glycosylation, exposure to agents that perturb calcium homeostasis (e.g, calcium ionophores or EGTA), or exposure to different reducing agents (e.g. DTT, BME) (reviewed by Subjeck & Shyy 1986). The major grps, with apparent masses of 75, 78, and 94 kd, now are known to be related to the heat shock protein family and, like most of the other hsps, are also expressed in cells maintained under normal growth conditions.

In bacteria, the nomenclature used to refer to the hsps is based on earlier genetic studies, examining bacterial host functions that were essential for bacteriophage growth. Here, the genes coding for the constitutively expressed hsps were found to map within specific operons and shown to participate in various aspects of bacteriophage DNA replication and morphogenesis (reviewed by Friedman et al 1984; Georgopoulos et al 1990; Georgopoulos 1992). For example, the DnaK protein (prokaryotic homologue of Hsp70) is required for bacteriophage DNA replication, while the GroEL/GroES proteins (prokaryotic homologues of Hsp60/Hsp10) were shown to be necessary for the assembly of the bacteriophage head and tail components. Finally, in *S. cerevisiae*, the hsps are referred to by genetic nomenclature, or sometimes by their apparent size. In Table 1, we have summarized the nomenclature currently being used to refer to the two major stress protein families known to function as molecular chaperones in bacteria, *S. cerevisiae*, and higher eukaryotes. As is discussed below, the related hsps from bacteria to human have conserved their function.

[1]Abbreviations: hsps, heat shock proteins; DTT, dithiothreitol; βME, β-mercaptoethanol; grp, glucose-regulated protein; BPTI, bovine pancreatic trypsin inhibitor; TCP-1, tailless complex polypeptide-1; BiP, binding protein; HSF, heat shock transcription factor; SDS, sodium dodecyl sulfate; EGTA [ethylenebis (oxyethylenenitrilo)] tetraacetic acid; ER, endoplasmic reticulum.

Table 1 The Hsp70 (DnaK) and Hsp60 (GroEL) families

Family	Name	Locale	Size (kd)	Comments
Hsp70				
Prokaryotes	DnaK	Cytosol	70	Mutants grow poorly; constitutive/ inducible
Yeast	Ssa	Cytosol	70	Multiple members: essential family
	Ssb	Cytosol	70	Multiple members: mutants-cold sensitive
	Ssc	Mitochondria	70	Essential for viability
	Kar2	ER	70	Essential for viability
Higher eukaryotes	Hsc70	Cytosol/nucleus	70	Constitutive
	Hsp70	Cytosol/nucleus	70	Stress inducible
	Grp75	Mitochondria	75	Most homology to DnaK
	Hsp70	Chloroplasts	70	Constitutive
	Grp78(BiP)	ER	78	ADP-ribosylated
	?			additional members?
Hsp60				
Prokaryotes	GroEL	Cytosol	60	Essential for viability
Yeast	Mif4	Mitochondria	60	Essential for viability Constitutive/ inducible
	TCP-1α	Cytosol	60	Essential for viability
	TCP-1β	Cytosol	60	Essential for viability
	?			Additional members?
Higher eukaryotes	Hsp60	Mitochondria	60	Constitutive/inducible
	Rubisco binding protein	Chloroplasts	60	
	TCP-1	Cytosol	60	Additional members?
	?			Form present in ER?

The Hsp70 (DnaK) and Hsp60 (GroEL) proteins are the major chaperones of the Hsp70 (DnaK) and Hsp60 (GroE) chaperone machines respectively. Information on the additional members of each chaperone machine are given in the text.

The Problem of Protein Folding

The pioneering work of Anfinsen (1973) with the in vitro refolding of purified ribonuclease A left the long-lasting impression that the folding of a newly synthesized polypeptide was an intrinsic feature of its primary structure, independent of other factors (reviewed by Jaenicke 1991). Most of the in vitro protein refolding experiments are usually carried out by first denaturing a given purified polypeptide and then removing the denaturant. Under these conditions, most polypeptides quickly collapse into a compact structure, usually called "molten globule", thought to possess extensive secondary structure, but still exposing hydrophobic groups, which may lead to aggregation. The probability that a given unfolded polypeptide will fold properly increases at relatively low protein concentrations (which limit inter-polypeptide aggregation) and low temperature (which attenuates hydrophobic interactions). However, the relatively high protein concentrations in the cytosol [or the specialized organelles, i.e. endoplasmic reticulum (ER), mitochondria, chloroplasts] may subject the growing polypeptide chains, as they emerge from the ribosomes (or as they enter various cellular compartments), to premature interactions with other intra- or inter-polypeptide domains, thereby leading to misfolding and aggregation (reviewed by Jaenicke 1991).

To deal with these sorts of problems, a set of proteins, collectively called chaperones, has been identified and whose primary function is to ensure that polypeptides will fold or assemble properly in the cell. These chaperone proteins act primarily by binding to the reactive surfaces of polypeptides (such as the hydrophobic surfaces exposed by the molten globule intermediates of various proteins). In doing so, chaperones sequester these reactive sites from the rest of the reactive surfaces present in their vicinity, thus effectively preventing aggregation and favoring the proper folding pathway. The chaperone proteins act without covalently modifying their polypeptide substrates and without being part of the finished product (Ellis & van der Vies 1991). Because high temperatures tend to favor both protein unfolding on the one hand and hydrophobic interactions on the other, there is an extra need for chaperone protein function to prevent protein aggregation in vivo. This is most likely the reason why many protein chaperones are expressed at higher levels in cells after heat shock (reviewed by Ellis & van der Vies 1991; Jaenicke 1991; Kelley & Georgopoulos 1992; Welch 1991). Herein, we discuss in detail the properties of some of the stress proteins that function as molecular chaperones. While other members of the stress protein family (e.g. Hsp28, Hsp47) have been reported to exhibit chaperone-like properties (Jakob et al 1993; Nakai et al 1992), we limit our discussion to the Hsp70 (DnaK), Hsp60 (GroE) and Hsp90 families of stress proteins.

THE GroE (Hsp60) CHAPERONIN MACHINE

The original identification of the *groE* locus of *E. coli* was made in the early 1970s. Mutations in it blocked the morphogenesis of many bacteriophages. Subsequently the GroE locus was shown to comprise an operon of two genes, *groES* and *groEL* (S signifies the smaller and L the larger of the two gene products) (reviewed by Friedman et al 1984; Zeilstra-Ryalls et al 1991). The name *groE* was assigned because of the block in bacteriophage λ growth (*gro*), and because the first isolated λ compensatory mutations mapped in the λ*E* gene (hence, *groE*). In the case of λ, the *groE*-imposed block was traced to a failure in the assembly of the head-tail connector, an oligomer of 12 subunits that is essential for the correct assembly of the head. In the case of bacteriophage T4, the major capsid protein was found in the form of lumps associated with the cell membrane, whereas in the case of bacteriophage T5, the block was found at the level of tail assembly (reviewed by Zeilstra-Ryalls et al 1991). These observations gave the long-lasting impression that GroES/GroEL was uniquely involved in the oligomerization of proteins. Subsequent genetic and biochemical studies showed that the *groE* genes were necessary for host macromolecular synthesis as well (summarized by Zeilstra-Ryalls et al 1991). In particular, both *groE* genes were shown to be essential under all growth conditions tested (Fayet et al 1989). In addition, the genetic demonstration that overproduction of both GroE proteins suppressed the temperature-sensitive (Ts^-) phenotype of various missense mutations in either *E. coli* or *S. typhimurium* was indicative of a potential role in protein folding and stability (Van Dyk et al 1989; see below).

The GroEL (Hsp60) family of proteins was also identified in biochemical or genetic studies with eukaryotic organisms. (*a*) Studies on the biosynthesis of ribulose bisphosphate carboxylase (Rubisco) in chloroplasts culminated with the identification of the Rubisco-binding protein, a large oligomeric protein stably associated with the large subunits of Rubisco (reviewed by Ellis & van der Vies 1991). The Rubisco-binding protein was shown to be composed of two types of nuclear-coded subunits, α and β, each possessing a molecular mass of ~60 kd. The cloning and sequencing of the α-coding gene revealed an overall 46% identity at the amino acid sequence level with that of the GroEL protein of *E. coli* (Hemmingsen et al 1988), which demonstrates a dramatic conservation throughout evolution. Interestingly, the extent of similarity in amino acid sequence between the α and β subunits of the Rubisco subunits is at the same level as that exhibited by α or β towards GroEL (Martel et al 1990). (*b*) Studies with *Tetrahymena thermophila* led to the identification of a ~60 kd heat shock protein of mitochondrial origin that was structurally related to GroEL (McMullin & Hallberg 1988), and whose sequence was highly homologous to both GroEL and Rubisco-binding protein (Reading et al 1989). (*c*) The isolation of the *S. cerevisiae* mif4 mutation as a Ts^- lethal and its subsequent assignment to the GroEL family of proteins demonstrated that it is essential for yeast growth and led to studies demonstrating its key role in protein folding of all polypeptides imported into the mitochondria (Cheng et al 1989; reviewed by Hartl et al 1992). (*d*) The identification and characterization of an Hsp60 homologue in mitochondria of higher eukaryotes (Mizzen et al 1989).

The GroES (Hsp10) Protein

The native GroES protein structure is most likely a heptameric ring of 97 monomeric residues, possessing a donut-like shape, with a diameter of ~8 nm (Chandrasekhar et al 1986; Hemmingsen et al 1988). It does not possess any ATPase activity and is not known to bind on its own to any protein except GroEL and its eukaryotic homologues (Golubinoff et al 1989; see below). A GroES homologue protein recently has been isolated from bovine mitochondria on the basis of functionally interacting with *E. coli* GroEL protein (Lubben et al 1990; see below). Most likely this protein is identical to a mitochondrial heat shock protein, Hsp10, which was isolated from rats and also shown to functionally interact with *E. coli* GroEL (Hartman et al 1992). A larger GroES equivalent protein has also been isolated from pea and spinach chloroplasts and appears to have two distinct GroES domains arranged head-to-tail (Bertsch et al 1992).

The GroEL (Hsp60) Protein

The native GroEL (Hsp60) proteins are cylindrical structures, possessing an unusual sevenfold symmetrical axis, with a diameter of ~15 nm, and a longitudinal axis of ~15 nm (Hendrix 1979; Hohn et al 1979; Hutchinson et al 1989; Zwickl et al 1990; Saibil et al 1991; Langer et al 1992b; Ishii et al 1992; see Figure 1). The top view under the electron microscope is that of a star-like structure with a central hole (~6 nm in diameter). In a side view the protein is rectangular in appearance, with four equally spaced striations. Most likely this structure represents the tail-to-tail arrangement of two heptameric GroEL rings (see below).

The GroEL protein of *E. coli* was originally misidentified as *E. coli* RNA polymerase, but later was proven to be a polymerase contaminant, possessing a weak ATPase activity capable of hydrolyzing one ATP per 14-mer per sec (Ishihama et al 1976). Subsequently, the GroEL protein was purified by Hendrix (1979) and Hohn et al (1979) primarily because of its large size and the availability of overproducing strains.

The GroES/GroEL Proteins Physically Interact

The fact that mutations in either *groES* or *groEL* block bacteriophage λ head assembly at the same level demonstrates that the two gene products are needed in the same pathway, but not necessarily at the same step. Subsequent genetic and biochemical analyses amply demonstrated that the two proteins functionally and physically interact and constitute a chaperone

A

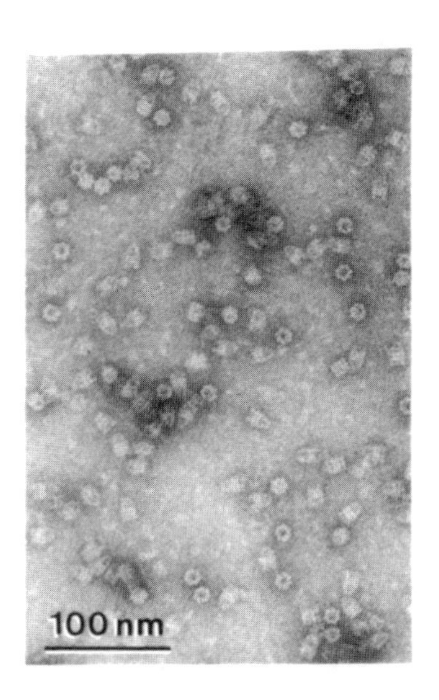

B

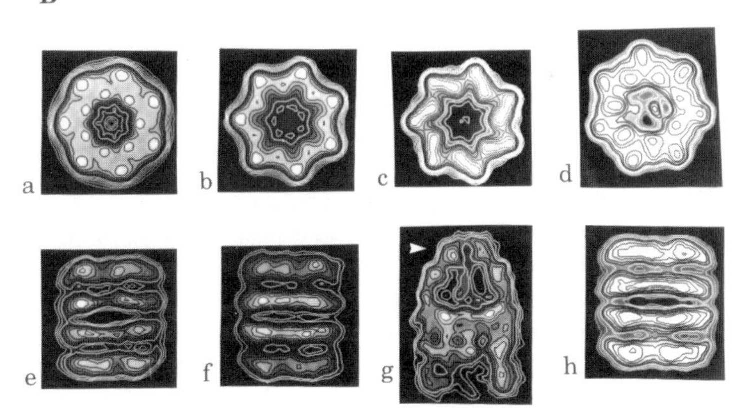

Figure 1 Electron microscopy of GroEL proteins. In panel A, an electron micrograph of negatively-stained GroEL/GroES complex is shown. Both end-on views and side-on views can be seen. In panel B, end-on (*a-d*) and side-on (*e-h*) views are shown, following averaging of electron microscopic images. (*a, e*) GroEL control; (*b, f*) GroEL in the presence of Mg-ADP, (*c, g*) GroEL-GroES, (*d, h*) GroEL with bound rhodanese. This panel has been constructed from Figures 4, 5, 6, and 9 of Langer et al (1992b) (consult this publication for actual details of representation and experimentation. These panels were provided by W. Baumeister).

machine (reviewed by Zeilstra-Ryalls et al 1991; Georgopoulos 1992; see below). The GroES/GroEL complex requires the presence of either ADP or ATP and is extremely stable (Chandrasekhar et al 1986; Viitanen et al 1990; Langer et al 1992b). Only one native molecule (i.e. heptamer) of GroES binds to one of the ends of the GroEL cylinder, thereby resulting in a bullet-shaped molecule (Saibil et al 1991; Ishii et al 1992; Langer et al 1992b; see Figure 1). Langer et al (1992b) provided additional evidence for this conclusion by examining the stoichiometry of the reaction and by demonstrating that GroES binding can only protect up to 50% of the proteolytic removal of a 50-amino acid long carboxyl-terminal fragment of GroEL. Although the GroEL site that binds GroES is not known, the GroES site that binds GroEL could be a recently identified ~20-amino acid long mobile loop (Landry et al 1993). This suggestion is based on three observations: (*a*) the GroES mobile loop is protected from proteolysis upon binding GroEL; (*b*) a synthetic 20-residue long mobile loop peptide binds GroEL, while a synthetic mutant peptide (based on the *groES*42 amino acid substitution) binds much more weakly; and (*c*) the sequencing of eight independently isolated *groES* mutations has shown that they are all located in this mobile loop. The fact that these mutations interfere with bacterial growth at elevated temperatures further attests to the importance of this amino acid segment in GroES function.

Substrate Recognition, Binding and Release

The Hsp60 (GroEL) family of proteins is known to bind promiscuously to many unfolded polypeptides, but not to their corresponding folded forms (Bochkareva et al 1988; Golubinoff et al 1989; Lecker et al 1989). This promiscuity of GroEL is exemplified by the studies of Viitanen et al (1992), which recently showed that GroEL will make stable quaternary complexes with over 50% of all *E. coli* proteins, provided the substrate proteins are first unfolded. Many laboratories have reached the conclusion that the GroEL family of proteins binds preferentially to the molten globule state of various polypeptides (Martin et al 1991; van der Vies et al 1992).

Recent data by Langer et al (1992b) and others (A. Horwich, personal communication) indicate that the polypeptide substrates most likely bind in the central hole of the GroEL structure, which is a good place to sequester them and minimize possible inadvertant inter-polypeptide aggregation. Most studies suggest that either one or two substrate molecules are bound by a GroEL 14-mer (Martin et al 1991; Bochkareva et al 1992; Zahn & Plückthun 1992). The GroEL cavity is sufficient to accommodate a protein, perhaps as large as the 75-kd alcohol oxidase (Langer et al 1992b). The role that GroES plays in the substrate-binding reaction is not precisely known, but it can vary from a stabilizing one, as in the case of pre-β-lactamase

(Bochkareva & Girshovich 1992), to a destabilizing one, as in the case of rhodanese (Bochkareva et al 1992).

Because both the strength and the quality of GroEL binding to the various polypeptide substrates will vary, and because of the known nucleotide-binding effects on GroEL conformation (Langer et al 1992b; Baneyx & Gatenby 1992; see Figure 1B), it is not surprising that the presence of either ADP or ATP will exert differential effects on the chaperone-substrate association. For example, although in most cases, ATP hydrolysis is needed to release the bound polypeptide, in the case of tryptophanase or dihydrofolate reductase, the addition of either ADP or nonhydrolyzable ATP analogues, respectively, results in their release from GroEL (Mizobata et al 1992; Viitanen et al 1991). Although the exact mechanism by which ATP hydrolysis results in the release of some polypeptide substrates is not known, it is likely due to the conformational change that GroEL undergoes (Langer et al 1992b; Baneyx & Gatenby 1992), which includes a lowering of the exposed GroEL hydrophobic surfaces, as judged by accessibility of the fluorescent probe ANS (Mendoza et al 1991). This probably minimizes GroEL-substrate interactions and may aid in substrate release. In general, when a polypeptide is released from GroEL, if it still exhibits surfaces that can be recognized by GroEL, it will rebind. Once the highly reactive surfaces have been sequestered (i.e. properly folded), the polypeptide will no longer bind to GroEL (Martin et al 1991).

In addition to ATP hydrolysis, most GroEL-bound polypeptides need GroES to be released (reviewed by Hartl et al 1992), although occasionally the GroEL-substrate interaction can be so strong that it cannot be disrupted (Frydman et al 1992). It is likely that GroES exerts its helpful role in this process by three different, not mutually exclusive mechanisms: (*a*) Saibil et al (1991) and Langer et al (1992b) showed that the binding of GroES to one of GroEL's ends causes a dramatic conformational change throughout the length of the GroEL structure. Such a conformational change may facilitate polypeptide folding in GroEL's cavity and/or facilitate polypeptide release; (*b*) GroES affects GroEL ATPase activity in two ways. On the one hand, it inhibits overall ATPase activity (Chandrasekhar et al 1986; Viitanen et al 1990), while on the other, it increases the cooperativity of ATP hydrolysis by GroEL subunits (Gray & Fersht 1991; Bochkareva et al 1992); such cooperative ATP hydrolysis may synchronize the disappearance of hydrophobic surfaces from GroEL, thus aiding the simultaneous release of the bound substrate; and (*c*) a direct interaction between GroES and the bound polypeptide substrate. Such a direct GroES/GroEL-bound substrate interaction leading to substrate displacement was implicit in the cogwheel model of GroES action (Georgopoulos & Ang, 1990). Recently, Bochkareva & Girshovich (1992), through crosslinking, demonstrated a physical interaction between GroES and the GroEL-bound pre-β-lactamase substrate. This interaction takes place on GroEL because it is only seen when all three proteins are present.

THE DnaK (Hsp70) CHAPERONE MACHINE

The DnaK chaperone machine (Figure 2) of *E. coli* consists of three members, DnaK, DnaJ, and GrpE. Its existence was originally inferred in various laboratories from genetic studies designed to detect host-bacteriophage interactions (reviewed by detail in Friedman et al 1984; Georgopoulos et al 1990). The original designation of the host-encoded genes was either *groP* (*gro* for blocking bacteriophage λ growth, and *P* because λ compensatory mutations mapped in the λ*P* gene), or *grp* (*groP*-like). The renaming to *dnaK* and *dnaJ* occurred later when it was found that mutations in these two complementation groups blocked *E. coli* DNA replication at 42°C (reviewed by Friedman et al 1984). Despite the fact that *grpE* mutations also block *E. coli* DNA replication at 42°C (Ang et al 1986), the name of this gene has not been changed. The exact role that DnaK, DnaJ, and GrpE play in λ DNA replication has been pinpointed to the disassembly of a λO-λP-DnaB helicase complex at the origin of λ replication, thus liberating the DnaB helicase whose unwinding activities allow DNA replication to commence (Alfano & McMacken 1989; Zylicz et al 1989); Dodson et al 1989; reviewed by Georgopoulos et al 1990; Ang et al 1991). This was one of the first clear demonstrations that the DnaK chaperone machine can act as a molecular crowbar or macromolecular detergent to disassemble protein complexes.

The various eukaryotic homologues of DnaK were discovered and, in some cases, rediscovered as performing various tasks. For example, an Hsp70 family member, now known to reside in the ER, was named Grp78 (glucose-regulated protein) because it accumulates following glucose starvation (Pouysségur et al 1977). The same protein was subsequently rediscovered as BiP, originally recognized as a protein that binds immunoglobulin heavy chains in the ER (Haas & Wabl 1983). Interestingly, its *S. cerevisiae* equivalent, Kar2, was first identified as being required for nuclear fusion during mating (Rose et al 1989a). The *S. cerevisiae* genome encodes at least seven additional Hsp70 family members (Craig 1990), one of which, Ssc1, resides in the mitochondria and has been shown to be essential for mitochondrial polypeptide import (Kang et al 1990). Similarly, various cytosolic members of Hsp70 were discovered using various biological assays, as in the case of bovine clathrin uncoating ATPase, which promotes the disassembly of clathrin lattices (Schlossman et al 1984; Chappell et al 1986), or in the case of Hsp70 members that maintain polypeptide precursors

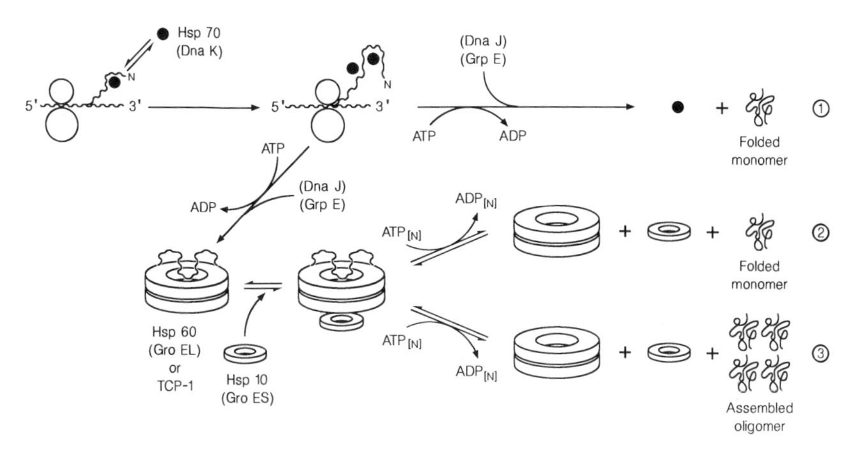

Figure 2 Current models by which molecular chaperones facilitate protein folding/assembly. During the course of protein synthesis, Hsp70 (DnaK) interacts with the nascent chain as it emerges from the ribosome. Such an interaction may prevent the premature folding of the nascent chain and/or prevent its inappropriate interaction with components of the translational machinery. For some proteins, folding of the completed polypeptide may commence upon its release from the Hsp70 chaperone and may not require any additional components. Release of the Hsp70 chaperone likely is mediated by additional factors such as DnaJ and GrpE (pathway 1). Alternatively, following their synthesis, some newly synthesized proteins, perhaps still bound to Hsp70 (DnaK), are transferred to the particular chaperonin (Hsp60, GroEL, TCP-1). Again, release and/or transfer of the nascent polypeptide from its Hsp70 chaperone likely requires the participation of additional components (DnaJ and GrpE). Binding of the unfolded polypeptide to the chaperonin triggers the recruitment of Hsp10 (GroES) to the opposite side of the chaperonin complex (it is possible that Hsp10/GroES may already be occupying this end of the chaperonin before substrate binding). Folding of the polypeptide, probably within the cavity of the chaperonin, may involve a series of binding and release events of the substrate, facilitated by many rounds of ATP hydrolysis, until folding is completed (pathway 2). Finally, it remains possible that the Hsp60 (GroEL), or TCP-1 chaperonins, preferentially participate in the folding and/or assembly of only those proteins that are part of homo- or hetero-oligomeric structures (pathway 3).

competent for translocation across the ER or mitochondrial membranes (Chirico et al 1988; Deshaies et al 1988; Zimmermann et al 1988). At least three Hsp70 homologues have been identified in the chloroplasts of plants (Marshall et al 1990; Amir-Shapira, et al 1990), and homologues exist in the chloroplasts of *Euglena* as well (Amir-Shapira et al 1990).

Various DnaJ eukaryotic homologues have also been discovered. For example, Blumberg & Silver (1991) found one family member in *S. cerevisiae*, Scj1, through its overexpression that resulted in missorting of a nuclear-targeted protein. Sec63 (an ER integral membrane protein), whose DnaJ homologous segment faces the ER lumen, was originally discovered to be essential for protein secretion in *S. cerevisiae* (Sadler et al 1989; Feldheim et al 1992). Other DnaJ homologues include Sis1, whose gene was discovered as a multicopy suppressor of the *sit4* mutation in *S. cerevisiae* (Luke et al 1991), and Mas5, whose gene was discovered as necessary for mitochondrial protein import in *S. cerevisiae* (Yaffe & Schatz 1984). The MAS5 gene turned out to be heat-inducible and identical to YDJ1 (Atencio & Yaffe 1992), which was identified because its product cross-reacted with antibodies prepared against a subfraction of the *S. cerevisiae* nuclear matrix (Caplan & Douglas 1991). The *S. cerevisiae* Zuotin homologue, yet another DnaJ homologue, was discovered as a putative Z-DNA-binding protein (Zhang et al 1992). One human DnaJ homologue was identified and shown to be a heat-inducible protein (e.g. Hsp40; Hattori et al 1991), while yet others were found accidently by screening cDNA expression libraries with antibodies thought to be specific for other unrelated proteins (HSJ1, HSJ2 in the brain, Cheetham et al 1992; HDJ1, Raabe & Manley 1991).

The *grpE* gene of *E. coli* was defined by a single mutation, *grpE*280, that blocked bacteriophage λ growth at all temperatures and bacterial growth at high temperature (reviewed by Friedman et al 1984; Ang et al 1986). No eukaryotic homologues have been discovered up to now.

The Weak ATPase of the DnaK (Hsp70) Proteins

The amino-terminal ~44 kd domain of this protein family has been highly conserved in evolution. It contains an ATPase active site and has been successfully crystallized (Flaherty et al 1990). The ATPase activity is extremely weak, 0.1 to 1.0 ATP molecules hydrolyzed per min per monomer, and is modulated by a variety of factors, e.g. (*a*) the presence of Ca^{+2} inhibits the ATPase activity of various members while simultaneously stimulating their weak autophosphorylation activity (Cegielska & Georgopoulos 1989; Kassenbrock et al 1988; Amir-Shapira, et al 1990; Mizzen et al 1991); (*b*) binding of unfolded protein substrates or small peptides results in a small (two to threefold) acceleration in the rate of ATP hydrolysis (Braell et al 1984; Flynn et al 1989; DeLuca-Flaherty et al 1990; Palleros et al 1991; Sadis & Hightower 1992); (*c*) the presence of wild-type DnaJ, but not mutant DnaJ259, protein leads to a modest one and one half to twofold increase in the ATPase activity of the DnaK protein, although no stable DnaJ/DnaK complex is observed. The molecular action of DnaJ has been pinpointed to a large acceleration in the rate of hydrolysis of DnaK-bound ATP (Liberek et al 1991a). This mode of DnaJ action is reminiscent of that of the GAP family of proteins in accelerating the rate of hydrolysis of GTP bound to members of the RAS family of proteins (Bourne et al 1990). In analogy with DnaJ, its Ydj1 *S. cerevisiae* homologue also stimulates Ssa1 (one of the cytolsolic Hsp70 proteins) ATPase activity tenfold (Cyr et al 1992); and (*d*) the presence of wild-type GrpE, but not mutant GrpE280, protein also leads to a modest one and one half to twofold increase in the ATPase activity of the DnaK protein. The molecular mechanism of action of GrpE has been pinpointed to a large acceleration in the rate of release of either ADP or ATP bound to DnaK (Liberek et al 1991a). This mode of GrpE action is reminiscent of that of the GNRP class of proteins that accelerates the release of GDP bound to members of the RAS family of proteins (Bourne et al 1990). In agreement with these observations, the simultaneous presence of both DnaJ and GrpE results in a dramatic, 50-fold acceleration in the rate of DnaK ATPase activity (Liberek et al 1991a). The purification of large quantities of the various Hsp70 (DnaK) members has been aided by their avid binding to ATP affinity columns (Welch & Feramisco 1985).

Chaperone Properties of the DnaK (Hsp70)

Over the last several years, considerable evidence has shown that the DnaK (Hsp70) class of proteins interacts with a variety of unfolded polypeptides in the cytosol, mitochondria, chloroplast, and ER (reviewed by Hartl et al 1992; Gething & Sambrook 1992; Welch 1992). The mammalian cytosolic Hsp70 members have been shown to transiently associate with most nascent, but not completed, polypeptide chains (Beckmann et al 1990). Similarly, Nelson et al (1992) showed that the Ssb1/Ssb2 pair of *S. cerevisiae* cytosolic Hsp70 proteins is associated with translating ribosomes. *S. cerevisiae* deleted for the SSB1/SSB2 genes grow slowly and contain a low number of translating ribosomes. One interpretation of these results is that the cytosolic forms of Hsp70 interact with the growing polypeptide chain as it emerges from the ribosome, thus facilitating its exit and preventing premature aggregation (see Figure 2; Beckmann et al 1990; Nelson et al 1992).

The *S. cerevisiae* Kar2, and Ssc1 homologues are needed for the actual transport process of polypeptides into their corresponding compartments (Kang et al 1990; Vogel et al 1990; Scherer et al 1990; Nguyen et al 1991). The Hsp70 proteins bind to the unfolded polypeptides as they enter their compartment, and probably through continuous cycles of binding and release, result in the efficient entry of the polypeptide. It is likely that the structure of a polypeptide as it exits from the ribosome is the same as that when it enters the various organelles, i.e. the peptide backbone is in a "stretched" conformation. Landry et al (1992) have demonstrated that when a specific 13-residue peptide binds to DnaK, it does so in a stretched conformation, whereas the same peptide when bound to GroEL assumes an α-helical conformation. This type of binding of DnaK to a peptide substrate is reminiscent of that proposed for the binding of peptides to human class I histocompatibility antigens (HLA) (Madden et al 1991).

There appears to be no strict specificity requirement in amino acid sequence for binding to members of the Hsp70 chaperone class as the studies of Flynn et al (1989, 1991) have revealed. Utilizing the stimulation of BiP ATPase as an assay, these workers first showed that a few randomly selected peptides were capable of binding, although their binding efficiency ranged at least 1,000-fold. In order to search for a pattern and size requirement, mixtures of almost random peptide libraries were constructed and tested. It was found that optimal stimulation of purified BiP ATPase activity occurred with peptides seven residues long. Subsequently, by using cycles of selective binding, it was shown that BiP's peptide-binding site selects for nonpolar, aliphatic residues, with an energy fingerprint strikingly similar to that found in the interior of a folded protein. This preference for hydrophobic regions was predicted by Pelham (1986) in his suggestions on the function of the Hsp70 family of proteins.

Properties of the DnaJ (Hsp40) Proteins

The DnaJ family of proteins comprises various members that share some features with the *E. coli* DnaJ dimeric archetype. These features include the most highly conserved 70-amino acid residue domain, present at the extreme amino-terminal portion of the *E. coli* DnaJ protein (reviewed by Bork et al 1992). Possession of this domain is the fingerprint of DnaJ family members, i.e. this is the only homology with DnaJ that the Sec63 protein of *S. cerevisiae* exhibits (Sadler et al 1989). The DnaJ chaperone has been shown to bind unfolded proteins such as rhodanese on its own, with a preference for the molten globule structure recognized by GroEL, and to protect them from aggregation (Langer et al 1992a). DnaJ can also bind to the unfolded form of the λP DNA replication protein (Hoffmann et al 1992).

The Ydj1 protein is farnesylated, and this farnesylation is needed for yeast growth at temperatures above 30°C (Caplan et al 1992b). Interestingly, in *S. cerevisiae* the presence of *E. coli* DnaJ protein or overproduced amounts of *S. cerevisiae* Sis1 protein will partially substitute for Ydj1 function by allowing growth at 30, but not 37°C (Caplan et al 1992a). Ydj1 stimulates Ssa1 (Hsp70) ATPase activity tenfold, thus resulting in the release of a Ssa1 (Hsp70)-bound polypeptide (Cyr et al, 1992). The fact that certain DnaJ homologues are anchored into various intracellular membranes may help to ensure the release of an Hsp70-bound substrate within the vicinity where membrane translocation commences.

The GrpE Protein

So far, GrpE homologues have only been detected in prokaryotic organisms. GrpE's significance as a bona fide member of the DnaK chaperone machine is exemplified by its ability to form a tight complex with DnaK, which is resistant to 2 M KCl, but is disrupted following ATP hydrolysis (reviewed by Georgopoulos et al 1990). The GrpE protein is able to compete with substrate binding to DnaK as exemplified in the case of λP protein (Liberek et al 1990; Osipiuk et al 1993) and unfolded BPTI and σ^{32} (Liberek et al 1991b, 1992). Although the presence of GrpE can enhance the release of an unfolded polypeptide substrate from its complex with DnaK/DnaJ (Langer et al 1992a; Hoffman et al 1992), other evidence suggests that GrpE may enable DnaK to bind with increased specificity to a potential substrate (Osipiuk et al 1993; Ziemienowicz et al 1993). Perhaps in the cytosol or specialized compartments, eukaryotic equivalents to GrpE do exist that similarly augment Hsp70's capacity and efficiency in performing its various tasks.

THE Hsp90 CHAPERONE MACHINE

Though expressed at relatively high levels in all cells maintained under normal growth conditions, the precise function of Hsp90 remains an enigma. Nevertheless, a substantial amount of phenomenology has been described for this protein, only some of which is discussed here. In *S. cerevisiae* and animal cells, there appear to be two highly related genes encoding Hsp90, with at least one gene being essential for growth at all temperatures (Rebbe et al 1987; Hickey et al 1986; Borkovich et al 1989). In bacteria the *htpG* gene product has been shown to be ~40% homologous to eukaryotic Hsp90, and is dispensible for bacterial growth, except at highly elevated temperatures (Bardwell & Craig 1988). In animal cells, Hsp90 is heavily phosphorylated, with the purified protein, at least *in vitro*, acting as a substrate for the so-called doubled-stranded DNA-activated kinase (Welch et al 1983; Walker et al 1985). While some investigators have reported that Hsp90 exhibits autophosphorylating activity, others have reported that the protein copurifies with, and can be phosphorylated by, casein kinase II (Csermely & Kahn 1991; Rose et al 1987; Dougherty et al 1987; Miyata & Yahara 1992). In its purified form, Hsp90 exists as a dimer with an unusually large Stokes radius, indicative of protein with an elongated or rod-like structure (Welch & Feramisco 1982).

Most germane to this discussion is the possible role of Hsp90 in controlling the activities of other macromolecules, most notably protein kinases and certain transcription factors. With respect to the former, early interest in Hsp90 followed from reports of its transient interaction with the virally-encoded tyrosine protein kinase, $pp60^{src}$, in cells infected with Rous sarcoma virus. Here Hsp90, along with another cellular protein of 50 kd, was observed to interact with newly synthesized $pp60^{src}$ within the cytoplasm of the infected cell. While present in this complex, $pp60^{src}$ function was suppressed as evidenced by a lack of tyrosine kinase activity. Once the complex of p50, Hsp90, and $pp60^{src}$ reached the plasma membrane, it dissociated, with pp60^{src} being deposited at the plasma membrane and exhibiting its full spectrum of tyrosine kinase activities. Subsequent studies have revealed that the tyrosine kinase proteins expressed by a number of other retroviruses all interact transiently with Hsp90 and the p50 proteins during the course of their maturation (reviewed by Brugge 1986).

Within the last few years, there have been a number of reports linking Hsp90 to the activities of other protein kinases. For example, as mentioned above, Hsp90 has been reported to copurify with and enhance the activities of casein kinase II. Other investigators have suggested that the heme-regulated protein kinase that phosphorylates the α subunit of the eukaryotic initiation factor EIF-2 also interacts with Hsp90 (Rose et al 1989b; Matts & Hurst 1989; Mendez et al 1992). Here, however, there is conflicting data on whether Hsp90 actually stimulates or inhibits the activity of this kinase in vivo. It is worth mentioning that in some cells after heat shock, an abrupt arrest of protein synthesis occurs, perhaps mediated by the increased phosphorylation of the EIF-2α subunit (Duncan & Hershey 1984). Whether Hsp90 somehow participates in this process is not known.

Much of the recent focus on Hsp90 concerns its interaction with, and apparent regulation of, various steroid hormone receptors. Different members of the steroid hormone family (e.g. progesterone, estrogen, testosterone) reside within the cytosol, and in the absence of their particular steroid hormone, exist in an inactive or nontransformed state. Once the appropriate steroid hormone diffuses into the cell, it binds to its receptor and results in the "transformation" of the receptor into an active transcription factor, capable of activating (or in some cases repressing) the expression of so-called steroid response genes (reviewed by Beato 1989; Pratt 1990). Work over the past few years has led to new insights into the mechanisms of steroid hormone receptor transformation and the apparent role of at least three different hsps in controlling this process in vivo. Briefly, while the inactive form of many steroid receptors appears to be a relatively large complex (8–10S), the active form of the receptor appears to exist as a homodimer (4S), with its steroid ligand tightly bound. Analysis of the purified 8–10S or inactive form of a number of different receptors has revealed the presence of the steroid-binding component, typically a 50–100 kd polypeptide, a number of polypeptides between 50–70 kd, and finally Hsp90 (Catelli et al 1985; Sanchez et al 1985). In addition, in the case of at least some steroid receptors (e.g. progesterone and glucocorticoid) two other hsps, Hsp70 and Hsp56/9 (Sanchez 1990), may also be present within the nontransformed steroid receptor complex (Kost et al 1989; Tai et al 1986).

While it is generally believed that in the absence of steroid hormone, Hsp90 serves to maintain the steroid receptor in an inactive or non-DNA-binding form, the exact mechanism by which this regulation occurs is not completely understood. For example, in the absence of steroid hormone, Hsp90 may interact directly with the DNA-binding domain, or perhaps a region within the receptor necessary for receptor dimerization, and thereby maintain the receptor in its inactive form. Whatever the case, following ligand binding, Hsp90 is released and the receptor, perhaps via the participation of Hsp70 and Hsp56/9, undergoes a conformational change leading to its dimerization and acquisition of a specific DNA-binding domain (Fawell et al 1990; DeMarzo et al 1991; Hutchison et al 1992). Hsp70's role in this process may be to mediate or stabilize those conformational changes that accompany receptor dimerization. Although the role of Hsp56/9 in the transformation process remains unclear, its gene has been recently cloned and shown to encode a member of a family of proteins referred to as peptidyl-proline isomerases (PPIase) or immunophilins (Lebeau et al 1992). Consistent with this, Hsp56/9 binds the immunosuppressant FK506 (Tai et al 1992; Yem et al 1992). Whether a proline isomerization event, either within the steroid receptor itself or perhaps within Hsp90, accompanies the transformation of the receptor into its biologically active form remains to be determined.

In addition to their apparent role in regulating steroid receptor activation, one or more of the hsps also may play a role in facilitating the synthesis and assembly of the non-transformed 8–10S steroid receptor complex. For example, in *S. cerevisiae* expressing 20-fold lower levels of Hsp90, the rate of synthesis of the transfected glucocorticoid receptor appears normal, yet its activation upon hormone addition is compromised (Picard et al 1990). Presumably under conditions where Hsp90 levels are limiting, proper assembly of the 8–10S form of the receptor is impaired. Indirect support for this idea comes from in vitro translation studies showing that the interaction of Hsp90 with the rat glucocorticoid receptor occurs either during or very shortly after the synthesis of the receptor (Dalmon et al 1989). These observations, along with the apparent interaction of cytosolic Hsp70 with nascent polypeptides makes it likely that the assembly of the inactive 8–10S steroid receptor complex occurs either during or shortly after the synthesis of the steroid-binding protein.

Recently, Wiech et al (1992) have shown that purified Hsp90 can prevent the aggregation of denatured polypeptides, thus resulting in increased yields of a folded enzyme. Since ATP has been shown to alter Hsp90's conformation (Csemerly et al 1993), it is possible that this conformational change in Hsp90, coupled with the actual process of substrate folding, weakens the strength of its interaction, thereby leading to substrate release. Surprisingly, Nadeau et al (1992) showed that Trypanosome Hsp90 is unique inasmuch as it possesses a substantial ATPase activity stimulated by peptides. This behavior is reminiscent of that of the Hsp70 family of proteins and raises the possibility that, at least in some cases, Hsp90 may play an important role in chaperoning protein folding in the cell.

COOPERATION BETWEEN THE CHAPERONE MACHINES

In a seminal paper, Langer et al (1992a) demonstrated the in vitro cooperation of the DnaK and GroE chaperone machines in chaperoning the ordered folding of some denatured proteins in vitro. The rationale for attempting these series of experiments was to mimic the series of events thought to take place during polypeptide import into mitochondria in vivo (Manning-Krieg et al 1991; Mizzen et al 1991; Kang et al 1990). Using an unfolded rhodanese polypeptide substrate, Langer et al (1992a) demonstrated a synergistic role of DnaK and DnaJ in binding to unfolded rhodanese and preventing its aggregation. The DnaJ/rhodanese/DnaK complex was found to be optimally stabilized in the presence of Mg-ATP, conditions known to favor ATP hydrolysis by DnaK. This was a surprising observation because normally ATP hydrolysis leads to the release of the DnaK-bound substrate (Liberek et al 1991b). A similar role of ATP in stabilizing a DnaJ/substrate/DnaK interaction has been seen with λP (Hoffmann et al 1992) and σ^{32} (Liberek & Georgopoulos 1993). The addition of the third DnaK chaperone machine member, GrpE, resulted in the release and refolding of a small percent of rhodanese. The role of GrpE in this reaction could be in its capacity to act as a DnaK substrate, thus resulting in the effective displacement and release of bound rhodanese. The further addition of the GroE chaperone machine (i.e. GroEL and GroES) resulted in the almost quantitative refolding of rhodanese.

The TCP family of proteins was originally identified by genetic means in mice. Mutations in TCP-1 were associated with mice exhibiting a tailless phenotype (hence the *t*ailless *c*omplex *p*olypeptide-1) (reviewed by Silver 1985). Although exhibiting weak sequence homology with members of the GroEL/Hsp60 family, native TCP-1 assumes the classical double-donut-like structure reminiscent of other members of the GroEL/Hsp60 family. Similar to the cooperation of the DnaK and GroE chaperone machines in facilitating protein folding in *E.coli*, the eukaryotic cytosolic Hsp70 and TCP-1 proteins likely function in a sequential fashion to facilitate protein folding or assembly (Trent et al 1991; Gao et al 1992; Yaffe et al 1992; Frydman et al 1992). Specifically, newly synthesized polypeptides made in the cytosol could be transferred from the Hsp70 chaperone machine to the TCP-1 chaperone machine. Since the ER does not contain either a GroEL or a TCP-1 family member, it is not clear what chaperone, if any, assists the Hsp70 chaperone machine (i.e. Grp78) there.

As mentioned above, the establishment of a cell-free reconstitution system for the assembly of the progesterone receptor complex has shown that in addition to Hsp90 and Hsp70, a 60 kd protein (distinct from the Hsp56/9 PPIase mentioned above) also associates (Smith et al 1992, 1993). The same workers used immunoaffinity purification of either Hsp70 or Hsp90 or the 60 kd protein to further show a functional interaction among the three proteins. It is possible that these three proteins constitute a super-chaperone machine, perhaps functioning together with the Hsp56/9 and 50 kd cohort proteins mentioned above, and shown to co-immunoprecipitate with anti-Hsp90 antibodies (Perdew & Whitelaw 1991; see below). The partial sequencing of the 60 kd protein showed a striking similarily to a 63 kd human protein, whose synthesis is sensitive to SV40 virus transformation (Honoré et al 1992) and which is related to a *S. cerevisiae* heat shock protein of unknown function (Nicolet & Craig 1989).

WHY ARE SOME MOLECULAR CHAPERONES UPREGULATED UNDER STRESS?

Knowing that many of the constitutively expressed hsps function as molecular chaperones within the normal, unstressed cell, investigators are beginning to address how and why these proteins are upregulated during stress. As was discussed earlier, many of the agents/treatments that lead to the induction of the stress response fall under the general category of protein denaturants, which probably lead to the intracellular accumulation of abnormally folded proteins. While some of these agents/treatments selectively affect the proper folding of only newly synthesized proteins (e.g. amino acid analogues, puromycin), other stresses (e.g. heat, heavy metals) may perturb the folding of both nascent as well as mature polypeptides. For example, proteins synthesized in the presence of amino acid analogues remain bound to their Hsp70 chaperone for a relatively long period of time (Beckmann et al 1992). Presumably because of the incorporation of the amino acid analogue, the newly synthesized protein is unable to properly fold and consequently remains as a continuous target for Hsp70.

Similar, if not identical observations have been reported for Grp78 (BiP) within the ER. Various secretory proteins that are underglycosylated, or contain mutations known to prevent their normal folding and maturation, also remain in a relatively stable complex with their Grp78 chaperone (Kozutsumi et al 1988; Kassenbrock et al 1988; Hurtley et al 1989). In the case of heat shock, both nascent as well as mature polypeptides begin to partially unfold and/or begin to aggregate as a consequence of thermal treatment. Such proteins then become targets for molecular chaperone interaction (Beckmann et al 1992). Recent evidence indicates that some proteins, either partially denatured or aggregated, can eventually be repaired or resurrected back into their native conformation via the action of Hsp70 (DnaK) or Hsp60 (GroE) chaperone machines (Skowyra et al 1990; Mendoza et al 1992; Martin et al 1992; Ziemienowicz et al 1993).

The fact that some hsps function as molecular chaperones helps explain both their collective role in providing cellular protection and their possible mode of regulation. For example, cells first subjected to a relatively mild, sublethal heat shock treatment, sufficient to upregulate the expression and accumulation of the hsps, appear better protected upon a subsequent and otherwise lethal heat shock exposure (Gerner & Scheider 1975). This phenomenon, referred to as acquired thermotolerance, clearly is dependent upon the increased expression of one or more of the hsps (Li & Werb 1982; Kusukawa & Yura 1988; reviewed in detail by Lindquist 1992). One suspects that the higher levels of the hsps provide the cell with an increased capability of dealing with the burden of increasing amounts of abnormally folded proteins after stress.

With regard to the regulation of their expression, a number of investigators have suggested that Hsp70 itself serves as a type of thermometer by which the cell senses a particular metabolic insult (reviewed by Craig & Gross 1991). For example, as high levels of unfolded proteins begin to accumulate in the cell under stress, they are recognized by and become bound to Hsp70. Consequently, as the levels of the free or available Hsp70 molecular chaperones are reduced, the eukaryotic heat shock transcription factor (HSF) is activated and the cell responds by increasing the expression of its hsps, many of which function as molecular chaperones. One currently favored model proposes that in an unstressed cell, HSF is maintained in an inactive state via its interaction with one of the hsps. After stress, and as the hsps are recruited to deal with the increased burden of abnormally folded proteins, HSF is liberated, undergoes oligomerization, and now is capable of binding to its target genes and activating their transcription. In support of this model, Abravaya et al (1992) and Baler et al (1992) have presented data suggesting a functional in vitro interaction between the mammalian HSF and cytosolic Hsp70 protein (reviewed by Morimoto 1993). The *S. cerevisiae* KAR2 gene, whose product is localized in the ER (i.e. Grp78), has been shown to be induced by heat, as well by the presence of misfolded polypeptides in the ER. In agreement with these observations, the promoter region of KAR2 has been shown to possess elements that respond to both stimuli (summarized by Kohno et al 1993).

In *E. coli* a great deal of recently accumulated genetic and biochemical evidence demonstrates that the DnaK chaperone machine is a key participant in the autoregulation of σ^{32}, which is responsible for activating the transcription of the hsp genes (reviewed by Gross et al 1990; Georgopoulos et al 1990; Craig & Gross 1991). Briefly, mutations in either *dnaK* or *dnaJ* or *grpE* often lead to an overexpression of the heat shock response, most likely by the stabilization of the otherwise extremely unstable σ^{32} factor (45–60 sec half life). The purified σ^{32} protein has been shown to form the following complexes with members of the DnaK chaperone machine: (*a*) a weak, ATP hydrolysis-sensitive complex with DnaK, (*b*) a strong, ATP-independent complex with DnaJ, and (*c*) a strong ATP hydrolysis-dependent complex with both DnaJ and DnaK (Liberek et al 1992; Gamer et al 1992; Liberek & Georgopoulos 1993). It has also been shown that the DnaK chaperone machine can effectively release and sequester σ^{32} from its complex with the RNA polymerase core (Liberek & Georgopoulos 1993). A likely scenario for the autoregulation of the *E. coli* heat shock response emerging from this molecular analysis is the following. In the absence of protein damage, the DnaK chaperone machine can sequester most available σ^{32}, and perhaps make it vulnerable to protease action. During or following protein damage, the DnaK chaperone machine is recruited elsewhere, and the liberated nascent σ^{32} protein can complex with RNA polymerase core, thus leading to increased heat shock gene expression, including DnaK, DnaJ, and GrpE. As the protein damage subsides, the elevated levels of the DnaK chaperone machine can resequester σ^{32}, thus leading to the attenuation of the heat shock response.

MOLECULAR CHAPERONES ARE INVOLVED IN OTHER BIOLOGICAL EVENTS DISTINCT FROM PROTEIN MATURATION

While clearly important factors in protein maturation, molecular chaperones also may participate in a number of other biological processes, in particular those that involve changes in macromolecular assembly/disassembly. As was discussed above, the bacterial form of Hsp70 (DnaK) was first identified on the basis of its role in facilitating bacteriophage λ DNA replication. The initiation of λ DNA replication is a complex process that first requires the orderly assembly of various proteins at the origin of replication. As mentioned before, the subsequent disassembly of this complex by the DnaK chaperone machine liberates the DnaB helicase and triggers the events that lead to DNA replication (reviewed by Georgopoulos et al 1990). Consequently, it seems plausible that both prokaryotic and eukaryotic DNA replication events may also require a similar participation of molecular chaperones in the initiation (and perhaps elongation) of DNA replication. Although little has been reported in this regard, a number of proteins expressed by different animal viruses and known to be important in either viral replication or transcription events have been shown to interact (or colocalize) with certain hsps, in particular the cytosolic and nuclear forms of Hsp70. Relevant examples include SV40 virus T antigen (Sawai & Butel 1989); polyomavirus middle T antigen (Walter et al 1987); the adenovirus EIA and EIB proteins (White et al 1988; W. J. Welch, unpublished), and the myelocytomatosis myc protein (Koskinen etal 1991). It is interesting that Zhang et al (1992) have shown that both small and large T antigens of an avian polyoma virus also possess the domain of DnaJ, common to all members of the DnaJ family, which therefore may function in recruiting the corresponding Hsp70 family member at the origin of replication.

The crowbar-like action of the DnaK protein, capable of disrupting some protein-protein interactions, has been observed in a number of other experimental systems. These include (*a*) the dissociation of bacteriophage P1-encoded RepA dimers to monomers, thus activating the DNA binding capacity of the protein (Wickner et al 1991); (*b*) the disaggregation of DnaA oligomers into monomers by DnaK and ATP (Hwang et al 1990); and (*c*) the disaggregation of heat-inactivated *E. coli* RNAP aggregates by DnaK and ATP (Skowyra et al 1990). The last disaggregation reaction cannot be carried out by the mutant DnaK756 protein and is vastly improved in the presence of both DnaJ and GrpE (Ziemienowicz et al 1993). A similar role of DnaK in preventing wholesale protein aggregation in vivo has been demonstrated by Gragerov et al (1992) and for certain proteins by Blum et al (1992).

Other biological properties associated with the DnaK (Hsp70) chaperone machine include (*a*) a key role in promoting thermotolerance in eukaryotic, as well as prokaryotic cells (critically reviewed by Lindquist 1992); (*b*) the activation of an otherwise inactive mutant DnaA46 form so that it can function in DNA replication (Hwang & Kaguni 1991); (*c*) the cytosolic forms of Hsp70 have been shown to be essential for the transport of certain nuclear-targeted polypeptides (Shi & Thomas 1992; Imamoto et al 1992). This function may be carried out by their ability to maintain polypeptides in a translocation competent form, perhaps by binding to the nuclear localization signal (NLS), or in its vicinity, thus helping its presentation to the nuclear import apparatus; (*d*) The polypeptide-binding protein Pbp72/74 Hsp70 mammalian homologue has been implicated as an important player in antigen presentation (VanBuskirk et al 1991); and (*e*) various studies in *E. coli* implicate all three members of the DnaK chaperone machine as important in the proteolysis of abnormal proteins (reviewed by Gross et al 1990; Sherman & Goldberg 1992). In the case of the proteolytic pathways, one favored interpretation is that the relatively stable binding of abnormally folded proteins to the DnaK (Hsp70) chaperone machine makes them better substrates for a variety of intracellular proteases, thus leading to their more efficient disposal.

The apparent role of a number of hsps in both the assembly and regulation of different steroid hormone receptors, an important class of transcription factors, raises the question as to whether other transcription factors are similarly dependent on molecular chaperones for their biological activities. For example, many transcription factors are known to be present in cells in an inactive state unable to bind to their target genes. Upon the appropriate biological stimulus, activation of the factor is observed. Modulation of transcription factor regulation occurs by a variety of mechanisms, some of which involve the release of negative regulatory components, and/or changes in phosphorylation (reviewed by Hunter & Karin 1992). Regardless of the mechanism employed, activation is usually accompanied by a conformational change of the transcription factor that results in its exposure of a specific DNA-binding domain. Moreover, for a large number of transcription factors, the activated form exists as a dimer or an even higher ordered structure. Consequently, one wonders whether molecular chaperones, either directly or indirectly, participate in the dynamic events that accompany transcription factor activation. In addition, once activated to bind its target DNA sequence, might molecular chaperones also participate in helping to position the particular transcription factor in its proper context with other more general transcription components and/or the subunits of the RNA polymerase II?

The recent work of Hupp et al (1992) and Shaknovich et al (1992) offers a fresh and potentially interesting biological insight into chaperone protein function. It appears that bacterially-produced p53 wild-type protein contains a cryptic DNA-binding site that can be activated by a variety of agents, including phosphorylation, binding of a specific monoclonal antibody, truncation of a short carboxy-terminal segment, or incubation with DnaK and ATP (reviewed by Hupp et al 1992). Thus although not detectable by physical means, a complex must form between wild-type p53 and DnaK, and the latter, utilizing the energy of ATP hydrolysis, can somehow guide p53 into assuming a different, DNA-binding conformation. Similarly, Shaknovich et al (1992) have shown that either full-length murine Hsp90, or even its carboxyl-terminal 196 amino acid portion, can somehow convert the MyoD1 transcription factor from an inactive to an active DNA-binding form. This action is surprising because the reaction does not have an energy requirement and because a small segment of Hsp90 is as functional as the full-length protein. These results suggest that chaperone proteins in general, and the Hsp70/Hsp90 members in particular, may peform hitherto unsuspected biological roles *in vivo* by modulating the extent of activation or deactivation of mature polypeptides rather than acting only on the folding or assembly of newly synthesized proteins.

Another area in biology where molecular chaperones may be involved is the structure and function of the cytoskeleton. The eukaryotic cytosolic TCP-1 chaperonin facilitates the synthesis of assembly-competent tubulin and actin (Yaffe et al 1992; Gao et al 1992). Consequently, one wonders whether molecular chaperones might also participate in the assembly/disassembly of actin microfilaments or tubulin microtubules in vivo. A number of observations indicate that this may indeed be the case. First, *S. cerevisiae* harboring mutations in TCP-1 appear unable to correctly assemble a mitotic spindle (Ursic & Culbertson 1991). Secondly, the cytosolic form of Hsp70 has been shown to copurify with microtubules subjected to repeated cycles of assembly/disassembly in vitro (Welch et al 1985). Finally, via indirect immunofluorescence studies, a portion of both Hsp70 and TCP-1 have been observed to be present within the centrosome either before or during mitosis (W. J. Welch, manuscript in preparation). Although the significance of this latter observation remains unclear, another recent study has shown that some cells, after a relatively severe heat shock treatment, exhibit multiple mitotic spindles and consequently give rise to multinucleated progeny. Interestingly, if the cells are first made thermotolerant (and therefore now contain relatively high levels of the various molecular chaperones), a similar severe heat shock treatment results in significantly fewer cells exhibiting such mitotic abnormalities (Vadair et al 1993). Thus it will be interesting to see whether the different molecular chaperones participate directly in facilitating the many dynamic events that are mediated by the various cytoskeletal systems.

Many other biological events involving dynamic changes in macromolecular assembly/disassembly events can easily be envisioned where molecular chaperones may have an important impact. Examples include the trafficking of proteins in and out of subcellular organelles, higher-ordered macromolecular assembly events such as those of ribonucleoprotein complexes (e.g. ribosomes, spliceosomes), antigen presentation, exo- and endocytosis, etc. Moreover, manipulation of molecular chaperone activities may eventually prove useful in our attempts to produce biologically active proteins via recombinant DNA technologies in bacteria, as well as facilitate their secretion out of the cell. Suffice it to say that discovery of the mechanisms by which macromolecules assume and maintain their final and biologically active structure in the cell will surely have an important impact on all biological processes.

ACKNOWLEDGMENTS

We thank many of our colleagues for supplying information in the form of reprints, preprints, and personal communication, and Professor W. Baumeister for supplying the prints that led to the formulation of Figure 1. Because of space limitation, we were unable to give credit to many contributions on various aspects of this exciting and fast moving field, and we sincerely apologize for this. We appreciate the help of Dr. D. B. Ang and Dr. W. L. Kelley in the preparation of the manuscript.

Literature Cited

Abravaya, K, Myers, MP, Murphy, SP, Morimoto, RI. 1992. The human heat shock protein hsp70 interacts with HSF, the transcription factor that regulates heat shock gene expression. *Genes Dev.* 6:1153–64

Alfano, C, McMacken, R. 1989. Heat shock protein-mediated disassembly of nucleoprotein structures is required for the initiation of bacteriophage λ DNA replication. *J. Biol. Chem.* 264:10709–18

Ananthan, J, Goldberg, AL, Voellmy, R. 1986. Abnormal proteins serve as eukaryotic stress signals and trigger the activation of heat shock genes. *Science* 232: 252–54

Amir-Shapira, D, Leustek, T, Dalie, B, Weissbach, H, Brot, N. 1990. Hsp70 proteins, similar to *Escherichia coli* DnaK, in chloroplasts and mitochondria of *Euglena gracilis*. *Proc. Natl. Acad. Sci. USA* 87: 1749–52

Anfinsen, CB. 1973. Principles that govern the folding of proteins chains. *Science* 181: 223–30

Ang, D, Chandrasekhar, GN, Zylicz, M, Georgopoulous, C. 1986. *Escherichia coli* grpE gene codes for heat shock protein B25.3, essential for both λ DNA replication at all temperatures and host growth at high temperature. *J. Bacteriol.* 167:25–9

Ang, D, Liberek, C, Skowyra, D, Zylicz, M, Georgopoulos, C. 1991. Function and regulation of the universally-conserved heat shock proteins. *J. Biol. Chem.* 266:24233–36

Ashburner, M, Bonner, JJ. 1979. The induction of gene activity in Drosophila by heat shock. *Cell* 17:241–54

Atencio, DP, Yaffe, MP. 1992. MAS5, a yeast homolog of DnaJ involved in mitochondrial protein import. *Mol. Cell Biol.* 12:283–91

Bardwell, JCA, Craig, EA. 1988. Ancient heat shock gene is dispensable. *J. Bacteriol.* 170:2977–83

Baler, R, Welch, WJ, Voellmy, R. 1992. Heat shock gene regulation by nascent polypeptides and denatured proteins: hsp70 as a potential autoregulatory factor. *J. Cell Biol.* 117:1151–59

Baneyx, F, Gatenby, AA. 1992. A mutation in GroEL interferes with protein folding by reducing the rate of discharge of sequestered polypeptides. *J. Biol. Chem.* 267:11637–44

Beato, M. 1989. Gene regulation by steroid hormones. *Cell* 56:335–44

Beckmann, RP, Lovett, M, Welch, WJ. 1992. Examining the function and regulation of hsp70 in cells subjected to metabolic stress. *J. Cell Biol.* 117:1137–50

Beckmann, RP, Mizzen, LA, Welch, WJ. 1990. Interaction of hsp70 with newly synthesized proteins: Implications for protein folding and assembly. *Science* 248:850–54

Bertsch, U, Soll, J, Seetharam, R, Viitanen, PV. 1992. Identification, characterization, and DNA sequence of a functional "double" groES-like chaperonin from chloroplasts of higher plants. *Proc. Natl. Acad. Sci. USA* 89:8696–700

Blum, P, Velligan, M, Lin, N, Matin, A. 1992. DnaK-mediated alterations in human growth hormone protein inclusion bodies. *Bio/Technol.* 10:301–4

Blumberg, H, Silver, PA. 1991. A homologue of the bacterial heat-shock gene DnaJ that alters protein sorting in yeast. *Nature* 349: 627–30

Bochkareva, ES, Girshovich, AS. 1992. A newly synthesized protein interacts with GroES on the surface of chaperonin GroEL. *J. Biol. Chem.* 267:25672–75

Bochkareva, ES, Lissin, NM, Flynn, GC, Rothman, JE, Girshovich, AS. 1992. Positive cooperativity in the functioning of molecular chaperone GroEL. *J. Biol. Chem.* 267:6796–800

Bochkareva, ES, Lissin, NM, Girshovich, AS. 1988. Transient association of newly synthesized unfolded proteins with the heat-shock GroEL protein. *Nature* 336:254–57

Bork, P, Sander, C, Valencia, A. 1992. A module of the DnaJ heat shock proteins found in malaria parasites. *Trends Biochem. Sci.* 17:129

Borkovich, KA, Furelly, FW, Finklestein, DB, Taulin, J, Lindquist, S. 1989. Hsp82 is an essential protein that is required in higher concentrations for growth of cells at higher temperature. *Mol. Cell Biol.* 9:3919–30

Bourne, HR, Sanders, DA, McCormick, F. 1990. The GTPase superfamily: a conserved switch for diverse cell functions. *Nature* 348:125–31

Braell, WA, Schlossman, DM, Schmid, SL, Rothman, JE. 1984. Dissociation of clathrin coats coupled to the hydrolysis of ATP: role of an uncoating ATPase. *J. Cell. Biol.* 99:734–41

Brugge, J. 1986. Interaction of the Rous sarcoma virus protein pp60^{src} with the cellular proteins pp50 and pp90. *Curr. Top. Microbiol. Immunol.* 123:1–22

Caplan, AJ, Douglas, MG. 1991. Characterization of YDJ1: a yeast homologue of the bacterial dnaJ protein. *J. Cell Biol.* 114: 609–21

Caplan, AJ, Cyr, DM, Douglas, MG. 1992a. YDJ1p facilitates polypeptide translocation across different intracellular membranes by a conserved mechanism. *Cell* 71:1143–55

Caplan, AJ, Tsai, J, Casey, PJ, Douglas, MG. 1992b. Farnesylation of YDJ1p is required for function at elevated growth temperatures in *Saccharomyces cerevisiae*. *J. Biol. Chem.* 267:18890–95

Catelli, MG, Binart, N, Jung-Testas, I, Renoir, J.-M, Baulieu, E.-E, Feramisco, JR, Welch, WJ. 1985. The common 90-kd protein component of non-transformed '8S' steroid receptors is a heat-shock protein. *EMBO J.* 4:3131–35

Cegielska, A, Georgopoulos, C. 1989. Genetic and biochemical characterization of the dnaK$^+$ and dnaK756 proteins of *Escherichia coli*. *J. Biol. Chem.* 264:21122–31

Chappell, TG, Welch, WJ, Schlossman, DM, Palter, KB, Schlesinger, MG, Rothman, JE. 1986. Uncoating ATPase is a member of the 70 kDa family of stress proteins. *Cell* 45:3–13

Chandrasekhar, GN, Tilly, K, Woolford, C, Hendrix, R, Georgopoulos, C. 1986. Purification and properties of the groES morphogenetic protein of *Escherichia coli*. *J. Biol. Chem.* 261:12414–19

Cheetham, ME, Brion, J.-P, Anderton, BH. 1992. Human homologues of the bacterial heat-shock protein dnaJ are preferentially expressed in neurons. *Biochem. J.* 284:469–76

Cheng, MY, Hartl, F.-U, Martin, J, Poloock, RA, et al. 1989. Mitochondrial heat shock protein hsp60 is essential for assembly of proteins imported into yeast mitochondria. *Nature* 337:620–25

Chirico, WJ, Waters, MG, Blobel, G. 1988. 70K heat shock related proteins stimulate protein translocation into microsomes. *Nature* 332:805–10

Craig, EA. 1990. Regulation and function of the HSP70 multigene family of *Saccharomyces cerevisiae*. In *Stress Proteins in Biology and Medicine*, ed. R Morimoto, A Tissieres, C Georgopoulos, pp. 301–21. Cold Spring Harbor Laboratory, NY.: Cold Spring Harbor Press

Craig, EA, Gross, CA. 1991. Is hsp70 the cellular thermometer? *Trends Biochem. Sci.* 16:135–39

Csermely, P, Kahn, CR. 1991. The 90 kDa heat shock protein (Hsp90) possesses an ATP binding site and autophosphorylating activity. *J. Biol. Chem.* 266:4943–50

Csermely, P, Kajtár, J, Hollósi, M, Jalsovszky, G, Holly, S, et al. 1993. ATP induces a conformational change of the 90-kDa heat shock protein (hsp90). *J. Biol. Chem.* 268:1901–7

Cyr, DM, Lu, X, Douglas, MG. 1992. Regulation of Hsp70 function by a eukaryotic DnaJ homolog. *J. Biol. Chem.* 267:20927–31

Dalmon, FC, Bresnick, EH, Patel, PD, Perdew, GH, et al. 1989. Direct evidence that the glucocorticoid receptor binds to Hsp90 at or near the termination of receptor translation in vitro. *J. Biol. Chem.* 264:19815–21

DeLuca-Flaherty, C, McKay, P, Parham, DB, Hill, BL. 1990. Uncoating protein (hsc70) binds a conformationally labile domain of clathrin light chain LCa to stimulate ATP hydrolysis. *Cell* 62:875–87

DeMarzo, AM, Beck, CA, Ornate, SA, Edwards, DP. 1991. Dimerization of the mammalian progesterone receptors occurs in the absence of DNA and is related to the release of the 90 kDa heat shock protein. *Proc. Natl. Acad. Sci. USA* 88:72–76

Deshaies, RJ, Koch, BD, Werner-Washburn, M, Craig, EA, et al. 1988. A subfamily of stress proteins facilitates translocation of secretory and mitochondrial precursor polypeptides. *Nature* 332:805–10

Dodson, M, McMacken, R, Echols, H. 1989. Specialized nucleoprotein structures at the origin of replication of bacteriophage λ. *J. Biol. Chem.* 264:10719–25

Dougherty, JJ, Rabideau, DA, Iannotti, AM, Sullivan, WP, Toft, DO. 1987. Identification of the 90 kDa substrate of rat liver type II casein kinase with the heat shock protein which binds steroid complexes. *Biochem. Biophys. Acta* 927:74–80

Duncan, K, Hershey, JWB. 1984. Heat shock induced translational attenuation in HeLa cells. *J. Biol. Chem.* 259:11882–89

Edington, BV, Whelan, SA, Hightower, LE. 1989. Inhibition of heat shock (stress) protein induction by deuterium oxide and glycerol: Additional support for the abnormal protein hypothesis of induction. *J. Cell Physiol.* 139:219–28

Ellis, RJ, van der Vies, SM. 1991. Molecular chaperones. *Annu. Rev. Biochem.* 60:321–47

Fawell, SE, Lees, JA, White, R, Parker, MG. 1990. Characterization and colocalization of steroid binding and dimerization activities in the mouse estrogen receptor. *Cell* 60: 953–66

Fayet, O, Ziegelhoffer, T, Georgopoulos, C. 1989. The groES and groEL heat shock genes of *Escherichia coli* are essential for bacterial growth at all temperatures. *J. Bacteriol.* 171:1379–85

Feldheim, D, Rothblatt, J, Schekman, R. 1992. Topology and functional domains of Sec63p, an endoplasmic reticulum membrane protein required for secretory protein translocation. *Mol. Cell. Biol.* 12:3288–96

Finley, D, Ciechanova, A, Varshavsky, A.

1984. Thermolability of ubiquitin activating enzyme from the mammalian cell cycle mutant ts85. *Cell* 37:43–55
Flaherty, KM, DeLuca, C, McKay, DB. 1990. Three-dimensional structure of the ATPase fragment of a 70 heat-shock cognate protein. *Nature* 346:623–28
Flynn, GC, Chappell, GT, Rothman, JE. 1989. Peptide binding and release by proteins implicated as catalysts of protein assembly. *Science* 245:385–90
Flynn, GC, Pohl, J, Flocco, MT, Rothman, JE. 1991. Peptide-binding specificity of the molecular chaperone BiP. *Nature* 353:726–30
Friedman, DI, Olson, ER, Tilly, K, Georgopoulos, C, Herskowitz, I, Banuett, F. 1984. Interactions of bacteriophage and host macromolecules in the growth of bacteriophage λ. *Microbiol. Rev.* 48:299–325
Fryfman, J, Nimmesgern, E, Erdjument-Bromage, H, Wall, J S, Tempst, P, Hartl, F-U 1992. Function in protein folding of TR:C, a cytosolic ring complex containing TP-1 and structurally related subunits. *EMBO J.* 11:4767-78
Gamer, J, Bujard, H, Bukau, B. 1992. Physical interaction between heat shock proteins DnaK, DnaJ, and GrpE and the bacterial heat shock transcription factor σ^{32}. *Cell* 69:833–42
Gao, Y, Thomas, JO, Chow, TL, Lee, GH, Cowan, NJ. 1992. A cytoplasmic chaperonin that catalyzes β-actin folding. *Cell* 69:1043–50
Georgopoulos, C. 1992. The emergence of the chaperone machines. *Trends Biochem. Sci.* 17:295–99
Georgopoulos, C, Ang, D. 1990. The *Escherichia coli* groE chaperonins. *Semin. Biol.* 1:19–25
Georgopoulos, C, Ang, D, Liberek, K, Zylicz, M. 1990. Properties of the *Escherichia coli* heat shock proteins and their role in bacteriophage lambda growth. See Craig 1990, pp. 191–221
Gerner, EW, Scheider, MJ. 1975. Induced thermal resistance in HeLa cells. *Nature* 256:500–2
Gething, M.-J, Sambrook, J. 1992. Protein folding in the cell. *Nature* 355:33–45
Goff, SA, Goldberg, AL. 1985. Production of abnormal protein in E. coli stimulates transcription of ion and other heat shock genes. *Cell* 41:587–95
Goloubinoff, P, Christeller, JT, Gatenby, AA, Lorimer, GH. 1989. Reconstitution of active dimeric ribulose biophosphate carboxylase from an unfolded state depends on two chaperonin proteins and Mg-ATP. *Nature* 342:884–89
Gragerov, A, Nudler, E, Komissarova, N, Gaitanaris, GA, Gottesman, ME, Nikiforov, V. 1992. Cooperation of GroEL/GroES and DnaK/DnaJ heat shock proteins in preventing protein misfolding in *Escherichia coli*. *Proc. Natl. Acad. Sci. USA* 89:10341–44
Gray, TE, Fersht, AR. 1991. Cooperativity in ATP hydrolysis by GroEL is increased by GroES. *FEBS Lett.* 292:254–58
Gross, CA, Straus, DB, Erickson, JW, Yura, T. 1990. The function and regulation of hsps in *Escherichia coli*. See Craig 1990, pp. 167–89
Haas, IG, Wabl, M. 1983. Immunoglobulin heavy chain binding protein. *Nature* 306: 387–89
Hartl, FU, Martin, J, Neupert, W. 1992. Protein folding in the cell: The role of molecular chaperones Hsp70 and Hsp60. *Annu. Rev. Biophys. Biomol. Struct.* 21: 293–322
Hartman, DJ, Hoogenraad, NJ, Condron, R, Hoj, PB. 1992. Identification of a mammalian 10-kDa heat shock protein, a mitochondrial chaperonin 10 homologue essential for assisted folding of trimeric ornithine transcarbamoylase in vitro. *Proc. Natl. Acad. Sci. USA* 89:3394–98
Hattori, H, Liu, Y.-C, Tohnai, I, Ueda, M, Kaneda, T, et al. 1992. Intracellular localization and partial amino acid sequence of a stress-inducible 40-kDa protein in HeLa cells. *Cell Struct. Funct.* 17:77–86
Hemmingsen, SM, Woolford, CA, van der Vies, SM, Tilly, K, Dennis, DJ, et al. 1988. Homologous plant and bacterial proteins chaperone oligomeric assembly. *Nature* 333:330–34
Hendrix, RW. 1979. Purification and properties of GroE, a host protein involved in bacteriophage assembly. *J. Mol. Biol.* 129: 375–92
Hickey, ES, Brandom, E, Potter, R, Stein, G, Stein, J, Weber, LA. 1986. Sequence and organization of genes encoding the human 27 kDa heat shock protein. *Nucleic Acids Res.* 14:4127–35
Hightower, LE. 1980. Cultured cells exposed to amino acid analogues or puromycin rapidly synthesize several polypeptides. *J. Cell Physiol.* 102:407–24
Hiromi, Y, Oxamoto, H, Gehring, WJ, HoHa, Y. 1986. Germline transformation with Drosophila mutant actin genes induces constitutive expression of heat shock genes. *Cell* 44:293–301
Hoffmann, HJ, Lyman, SK, Lu, C, Petit, M.-A, Echols, H. 1992. Activity of the Hsp70 chaperone complex-DnaK, DnaJ, and GrpE- in initiating phage λ replication by sequestering and releasing λP protein. *Proc. Natl. Acad. Sci. USA* 89:12108–11
Hohn, T, Hohn, B, Engel, A, Wurtz, M, Smith, PR. 1979. Isolation and character-

ization of the host protein GroE involved in bacteriophage lambda assembly. *J. Mol. Biol.* 129:359–73

Honoré, B, Leffers, H, Madsen, P, Rasmussen, HH, Vanderkerckhove, J, Celis, JE. 1992. Molecular cloning and expression of a transformation-sensitive human protein containing the TPR motif and sharing identity to the stress-inducible yeast protein STI1. *J. Biol. Chem.* 267:8485–91

Hunter, T, Karin, M. 1992. The regulation of transcription by phosphorylation. *Cell* 70:375–87

Hupp, TR, Meek, DW, Midgley, CA, Lane, DP. 1992. Regulation of the specific DNA binding function of p53. *Cell* 71: 875–86

Hurtley, SM, Bole, DG, Hoover-Lithy, H, Helenius, A, Copeland, C. 1989. Interactions of misfolded influenza virus hemagglutinin with binding protein (BiP). *J. Cell Biol.* 108:2117–26

Hutchinson, EG, Tichelaar, W, Hofhaus, G, Weiss, H, Leonard, KR. 1989. Identification and electron microscopic analysis of a chaperonin oligomer from *Neurospora crassa* mitochondria. *EMBO J.* 8:1485–90

Hutchison, KA, Czar, MJ, Pratt, WB. 1992. Evidence that the hormone binding domain of the mouse glucocorticoid receptor directly represses DNA binding activity in a major portion of receptors that are "misfolded" after removal of Hsp90. *J. Biol. Chem.* 267:3190–95

Hwang, DS, Crooke, E, Kornberg, A. 1990. Aggregated dnaA protein is dissociated and activated for DNA replication by phospholipase and dnaK protein. *J. Biol. Chem.* 265:19244–48

Hwang, DS, Kaguni, JM. 1991. DnaK protein stimulates a mutant form of dnaA protein in *Escherichia coli* DNA replication. *J. Biol. Chem.* 266:7537–41

Imamoto, N, Matsuoka, Y, Kurihara, T, Kohno, K, et al. 1992. Antibodies against 70-kD heat shock protein inhibit mediated nuclear import of karyophilic proteins. *J. Cell Biol.* 119:1047–61

Ishihama, A, Ikeuchi, T, Matsumoto, A, Yamamoto, S. 1976. A novel adenosine triphosphatase isolated from RNA polymerase of *Escherichia coli*. II. Enzymatic properties and molecular structure. *J. Biochem.* 79:927–36

Ishii, N, Taguchi, H, Sumi, M, Yoshida, M. 1992. Structure of holo-chaperonin studied with electron microscopy. *FEBS Lett.* 299:169–74

Jaenicke, R. 1991. Protein folding: local structures, domains, subunits, and assemblies. *Biochemistry* 30:3147–61

Jakob, U, Gaestel, M, Engel, K, Buchner, J. 1993. Small heat shock proteins are molecular chaperones. *J. Biol. Chem.* 268:1517–20

Karlik, CC, Coutu, MD, Fyrberg, EA. 1984. A nonsense mutation within the actin 88F gene disrupts myofibril formulation in Drosophila indirect flight muscles. *Cell* 38:711–19

Kang, P-J, Ostermann, J, Schilling, J, Neupert, W, Craig, EA, Pfanner, N. 1990. Requirement for hsp70 proteins in the mitochondrial matrix for translocation and folding of precursor proteins. *Nature* 348: 137–43

Kassenbrock, CK, Garcia, PD, Walter, P, Kelly, RB. 1988. Heavy-chain binding protein recognizes aberrant polypeptides translocated in vitro. *Nature* 333:90–93

Kelley, WL, Georgopoulos, C. 1992. Chaperones and protein folding. *Curr. Opin. Cell Biol.* 4:984–91

Kelley, PM, Schlesinger, MJ. 1978. The effect of amino acid analogs and heat shock on gene expression in chicken embryo fibroblasts. *Cell* 15:1277–86

Kohno, K, Normington, K, Sambrook, J, Gething, M-J, Mori, K. 1993. The promoter region of the yeast KAR2 (BiP) gene contains a regulatory domain that responds to the presence of unfolded proteins in the endoplasmic reticulum. *Mol. Cell. Biol.* 13: 877–90

Koskinen, PJ, Sistonen, L, Evan, G, Morimoto, R, Alitalo, K. 1991. Nuclear co-localization of cellular and viral myc proteins with hsp70 in myc-overexpressing cells. *J. Virol.* 65:842–48

Kost, SL, Smith, DF, Sullivan, WP, Welch, WJ, Toft, DO. 1989. Binding of heat shock proteins to the avian progesterone receptor. *Mol. Cell. Biol.* 9:3829–38

Kozutsumi, Y, Segal, M, Normington, K, Gething, MJ, Sambrook, J. 1988. The presence of malfolded proteins in the endoplasmic reticulum signals the induction of glucose regulated proteins. *Nature* 332:462–64

Kusukawa, N, Yura, T. 1988. Heat shock protein groE of *Escherichia coli*: key protective roles against thermal stress. *Genes Dev.* 2:874–82

Landry, SJ, Jordan, R, McMacken, R, Gierasch, L. 1992. Different conformations for the same polypeptide bound to chaperones DnaK and GroEL. *Nature* 355:455–57

Landry, SJ, Zeilstra-Ryalls, J, Fayet, O, Georgopoulos, C, Gierasch, L. 1993. Genetic and biophysical identification of a functionally important mobile domain of GroES. *Nature* 364:255–58

Langer, T, Lu, C, Echols, H, Flanagan, J, Hayer, MK, Hartl, F.-U. 1992a. Successive action of DnaK, DnaJ and GroEL along the pathway of chaperone-mediated protein folding. *Nature* 356:683–89

Langer, T, Pfeifer, G, Martin, J, Baumeister, W, Hartl, F.-U. 1992b. Chaperonin-mediated protein folding: GroES binds to one end of the GroEL cylinder, which accommodates the protein substrate within its central cavity. *EMBO J.* 11:4757–65

Lebeau, MC, Massol, N, Herrick, J, Faber, L, Renoir, JM, et al. 1992. P59, an Hsp90-binding protein. *J. Biol. Chem.* 267:4281–84

Lecker, S, Lill, R, Ziegelhoffer, T, Georgopoulos, C, Bassford Jr., PJ, Kumamoto, CA, Wickner, W. 1989. Three pure chaperone proteins of *Escherichia coli*, secB, trigger factor and groEL, form soluble complexes with precursor proteins in vitro. *EMBO J.* 8:2703–9

Li, CG, Werb, Z. 1982. Correlation between the synthesis of heat shock proteins and the development of thermotolerance in Chinese hamster fibroblasts. *Proc. Natl. Acad. Sci. USA* 79:3918–22

Liberek, K, Galitski, TP, Zylicz, M, Georgopoulos, C. 1992. The DnaK chaperone modulates *Escherichia coli's* heat shock response by binding to the σ^{32} transcription factor. *Proc. Natl. Acad. Sci. USA* 89:3516–20

Liberek, K, Georgopoulos, C. 1993. Autoregulation of the *Escherichia coli* heat shock response by DnaK, DnaJ and GrpE. *Proc. Natl. Acad. Sci. USA*. Submitted

Liberek, K, Marszalek, J, Ang, D, Georgopoulos, C, Zylicz, M. 1991a. The *Escherichia coli* DnaJ and GrpE heat shock proteins jointly stimulate DnaK's ATPase activity. *Proc. Natl. Acad. Sci. USA* 88: 2874–78

Liberek, K, Osipiuk, J, Zylicz, M, Ang, D, Skorko, J, Georgopoulos, C. 1990. Physical interactions among bacteriophage and *Escherichia coli* proteins required for initiation of λ DNA replication. *J. Biol. Chem.* 265: 3022–29

Liberek, K, Skowyra, D, Zylicz, M, Johnson, C, Georgopoulos, C. 1991b. The *Escherichia coli* DnaK chaperone protein, the Hsp70 eukaryotic equivalent, changes its conformation upon ATP hydrolysis, thus triggering its dissociation from a bound target protein. *J. Biol. Chem.* 266:14491–96

Lindquist, S. 1992. Heat-shock proteins and stress tolerance in microorganisms. *Curr. Opin. Gene Dev.* 2:748–55

Lubben, TH, Gatenby, AA, Donaldson, GK, Lorimer, GH, Viitanen, PV. 1990. Identification of a groES-like chaperonin in mitochondria that facilitates protein folding. *Proc. Natl. Acad. Sci. USA* 87:7683–87

Luke, MM, Sutton, A, Arndt, KT. 1991. Characterization of S1S1, a *Saccharomyces cerevisiae* homologue of bacterial dnaJ proteins. *J. Cell Biol.* 114:623–38

Madden, DR, Gorga, JC, Strominger, JL, Wiley, DC. 1991. The structure of HLA-B27 reveals nonamer "self-peptides" bound in an extended conformation. *Nature* 353: 321–25

Manning-Krieg, UC, Scherer, PE, Schatz, G. 1991. Sequential action of mitochondrial chaperones in protein import into the matrix. *EMBO J.* 11:3273–80

Marshall, JS, DeRocher, AE, Keegstra, K, Vierling, E. 1990. Identification of heat shock protein hsp70 homologues in chloroplasts. *Proc. Natl. Acad. Sci. USA* 87:374–78

Martel, R, Cloney, LP, Pelcher, LI, Hemmingsen, SM. 1990. Unique composition of plastid chaperonin-60: α and β polypeptide-encoding genes are highly divergent. *Gene* 94:181–87

Martin, J, Horwich, AC, Hartl, FU. 1992. Prevention of protein denaturation under heat stress by the chaperonin Hsp60. *Science* 258:995–98

Martin, J, Langer, T, Boteva, R, Schramel, A, Horwich, AL, et al. 1991. Chaperonin-mediated protein folding at the surface of groEL through a 'molten globule'-like intermediate. *Nature* 352:36–42

Matts, RL, Hurst, R. 1989. Evidence for the association of the heme-regulated eif-α kinase with the 90 kDa heat shock protein in rabbit reticulocyte lysate in situ. *J. Biol. Chem.* 264:15542–47

McMullin, TW, Hallberg, RL. 1988. A highly evolutionary conserved mitochondrial protein is structurally related to the protein encoded by the *Escherichia coli* groEL gene. *Mol. Cell. Biol.* 8:371–80

Mendez, R, Moreno, A, de Haro, C. 1992. Regulation of heme-controlled eukaryotic polypeptide chain initiation factor α-subunit kinase of reticulocyte lysates. *J. Biol. Chem.* 267:11500–7

Mendoza, JA, Lorimer, GA, Horwitz, PM. 1992. Chaperonin cpn60 from *Escherichia coli* protects the mitochondrial enzyme rhodanese against heat inactivation and supports folding of elevated temperatures. *J. Biol. Chem.* 267:17631–34

Mendoza, JA, Rogers, E, Lorimer, GH, Horowitz, PM. 1991. Chaperonins facilitate the in vitro folding of monomeric mitochondrial rhodanese. *J. Biol. Chem.* 266:13044–49

Miyata, Y, Yahara, I. 1992. The 90 kDa heat shock protein, Hsp90, binds and protects casein kinase II from self-aggregation and enhances its kinase activity. *J. Biol. Chem.* 267:7042–47

Mizobata, T, Akiyama, Y, Ito, K, Yumoto, N, Kawata, Y. 1992. Effects of the chaperonin GroE on the refolding of tryptophanase from *Escherichia coli*. *J. Biol. Chem.* 267:17773–79

Mizzen, LA, Chang, C, Garrels, J, Welch, WJ. 1989. Identification, characterization and purification of two mammalian stress proteins present within mitochondria: one related to hsp70 and the other to the bacterial GroEL protein. *J. Biol. Chem.* 264: 20664–75

Mizzen, LA, Kabiling, A, Welch, WJ. 1991. The two mitochondrial stress proteins, grp75 and hsp58, transiently interact with newly synthesized proteins. *Cell Reg.* 2: 165–79

Morimoto, RI. 1993. Cells in stress: Transcriptional activation of heat shock genes. *Science* 259:1409–10

Morimoto, R, Tissieres, A, Georgopoulos, C. 1990. The stress response, function of the proteins and perspectives. See Craig 1990, pp. 1–36

Nadeau, K, Sullivan, MA, Bradley, M, Engman, DM, et al. 1992. 83-kilodalton heat shock proteins of trypanosomes are potent peptide-stimulated ATPases. *Protein Sci.* 1:970–79

Nakai, A, Satoh, M, Hirayoshi, K, Nagata, K. 1992. Involvement of the stress protein hsp47 in procollagen processing in the endoplasmic reticulum. *J. Cell Biol.* 117:904–14

Nelson, RJ, Zeigelhoffer, T, Nicolet, C, Werner-Washburne, M, Craig, EA. 1992. The translation machinery and 70 kd heat shock protein cooperate in protein synthesis. *Cell* 71:97–105

Nguyen, TH, Law, DT, Williams, DB. 1991. Binding protein BiP is required for translocation of secretory proteins into the endoplasmic reticulum in *Saccharomyces cerevisiae*. *Proc. Natl. Acad. Sci. USA* 88: 1565–69

Nicolet, CM, Craig, EA. 1989. Isolation and characterization of STI1, stress-inducible gene from *Saccharomyces cerevisiae*. *Mol. Cell. Biol.* 9:3638–46

Osipiuk, J, Georgopoulos, C, Zylicz, M. 1993. Initiation of λ DNA replication: the *Escherichia coli* small heat-shock proteins, DnaJ and GrpE, increase DnaK's affinity for the λP protein. *J. Biol. Chem.* 268: 4821–27

Palleros, DR, Welch, WJ, Fink, AL. 1991. Interaction of Hsp70 with unfolded proteins: Effects of temperature and nucleotides on the kinetics of binding. *Proc. Natl. Acad. Sci. USA* 88:5719–23

Pelham, HR. 1986. Speculations on the functions of the major heat shock and glucose-regulated stress proteins. *Cell* 46:959–61

Perdew, GH, Whitelaw, ML. 1991. Evidence that the 90-kDa heat shock protein (hsp90) exists in cytosol in heteromeric complexes containing hsp70 and three other proteins with M_r of 63,000, 56,000, and 50,000. *J. Biol. Chem.* 250:6708–13

Picard, D, Khursheed, B, Garabedian, MJ, Fortin, MG, et al. 1990. Reduced levels of Hsp90 compromise steroid receptor action in vivo. *Nature* 348:166–68

Pouysségur, J, Shiu, RPC, Pastan, I. 1977. Induction of two transformation sensitive membrane polypeptides in normal fibroblasts by a block in glycoprotein synthesis or glucose deprivation. *Cell* 11:941–47

Pratt, WB. 1990. At the cutting edge: Interaction of Hsp90 with steroid receptors: Organizing some diverse observations and presenting the newest concepts. *Mol. Cell. Endocrinol.* 74:C69–76

Raabe, T, Manley, JL. 1991. A human homologue of the *Escherichia coli* DnaJ heat-shock protein. *Nucleic Acids Res.* 19:6645

Reading, DS, Hallberg, RL, Myers, AM. 1989. Characterization of the yeast HSP60 gene coding for a mitochondrial assembly factor. *Nature* 337:655–59

Rebbe, NF, Ware, J, Bertring, RM, Modrich, P, Stafford, DW. 1987. Nucleotide sequence of a cDNA for a member of the human 90 kDa heat shock protein. *Gene* 53:235–42

Ritossa, FM. 1962. A new puffing pattern induced by a temperature shock and DNP in *Drosophila*. *Experientia* 18:571–73

Rose, MD, Misra, LM, Vogel, JP. 1989a. KAR2, a karyogamy gene, is the yeast homolog of the mammalian BIP/GRP78 gene. *Cell* 57:1211–21

Rose, DW, Welch, WJ, Kramer, G, Hardesty, B. 1989b. Possible involvement of the 90-kDa heat shock protein in the regulation of protein synthesis. *J. Biol Chem* 264:6239–44

Rose, DW, Wettenhall, RE, Kudlicki, W, Kramer, G, Hardesty, B. 1987. The 90-kilodalton peptide of the heme-regulated eif-α kinase has sequence similarity with the 90-kilodalton heat shock protein. *Biochemistry* 26:6583–87

Rothman, JE. 1989. Polypeptide chain binding proteins: Catalysts of protein folding and related processes in cells. *Cell* 59:591–601

Sadis, S, Hightower, LE. 1992. Unfolded proteins stimulate molecular chaperone Hsc70 ATPase by accelerating ADP/ATP exchange. *Biochemistry* 31:9406–12

Sadler, I, Chiang, A, Kurihara, T, Rothblatt, JJ, Way, JP, Silver, P. 1989. A yeast gene important for protein assembly into the endoplasmic reticulum and the nucleus has homology to dnaJ, an *Escherichia coli* heat shock protein. *J. Cell Biol.* 109:2665–75

Saibil, H, Dong, Z, Wood, S, auf der Mauer, A. 1991. Binding of chaperones. *Nature* 353:25–26
Sanchez, ER. 1990. Hsp56: A novel heat shock protein associated with untransformed steroid receptor complexes. *J. Biol. Chem.* 265:22067–70
Sanchez, ER, Toft, DO, Schlesinger, MJ, Pratt, WB. 1985. Evidence that the 90-kDa phosphoprotein associated with the untransformed L-cell glucocorticoid receptor is a murine heat shock protein. *J. Biol. Chem.* 260:12398–401
Sawai, ET, Butel, JS. 1989. Association of a cellular heat shock protein with simian virus 40 large T-antigen in transformed cells. *J. Virol.* 63:3961–73
Scherer, PE, Krieg, UC, Hwang, ST, Vestweber, D, Schatz, G. 1990. A precursor protein partly translocated into yeast mitochondria is bound to a 70 kd mitochondrial stress protein. *EMBO J.* 8:4315–22
Schlossman, DM, Schmid, SL, Braell, WA, Rothman, JE. 1984. An enzyme that removes clathrin coats: purification of an uncoating ATPase. *J. Cell Biol.* 99: 723–33
Shaknovich, R, Shue, G, Kohtz, DS. 1992. Conformational activation of a basic helix-loop-helix protein (MyoD1) by the C-terminal region of murine HSP90 (HSP84). *Mol. Cell. Biol.* 12:509–68
Sherman, MY, Goldberg, AL. 1992. Involvement of the chaperonin dnaK in the rapid degradation of a mutant protein in *Escherichia coli*. *EMBO J.* 11:71–77
Shi, Y, Thomas, JO. 1992. The transport of proteins into the nucleus requires the 70-kilodalton heat shock protein or its cytosolic cognate. *Mol. Cell. Biol.* 12:2186–92
Silver, LM. 1985. Mouse *t* haplotypes. *Annu. Rev. Genet.* 19:179–208
Skowyra, D, Georgopoulos, C, Zylicz, M. 1990. The E. coli DnaK gene product, the Hsp70 homolog, can reactivate heat-inactivated RNA polymerase in an ATP-dependent manner. *Cell* 62:939–44
Smith, DF, Stensgard, BA, Welch, WJ, Toft, DO. 1992. Assembly of progesterone receptor with hsp's and receptor activation are ATP mediated events. *J. Biol. Chem.* 267:1350–56
Smith, DF, Sullivan, WP, Marion, TN, Zaitsu, K, et al. 1993. Identification of a 60-kDa stress-related protein, p60, which interacts with hsp90 and hsp70. *Mol. Cell. Biol.* 13:869–76
Subjeck, JR, Shyy, TT. 1986. Stress protein systems of mammalian cells. *Am. J. Physiol.* 250:C1–17
Tai, P.-K, Albers, MW, Chang, H, Faber, LE, Schreiber, SL. 1992. Association of a 59-kilodalton immunophilin with the glucocorticoid receptor complex. *Science* 256: 1315–18
Tai, PK, Maeda, Y, Nakao, K, Wakim, NG, et al. 1986. A 59-kilodalton protein associated with progesterone, estrogen, androgen, and glucorticoid receptors. *Biochemistry* 25:5269–75
Tissieres, A, Mitchel, HK, Tracy, UM. 1974. Protein synthesis in salivary glands of *Drosophila melanogaster:* relation to chromosome puffs. *J. Mol. Biol.* 84:389–98
Trent, JD, Nimmesgern, E, Wall, JS, Hartl, F-U, Horwich, AL. 1991. A molecular chaperone from a thermophilic archaebacterium is related to the eukaryotic protein t-complex polypeptide-1. *Nature* 354: 490–93
Ursic, D, Culbertson, MR. 1991. The Yeast homolog to mouse Tcp-1 affects microtubule mediated processes. *Mol. Cell Biol.* 11:2629–40
Vadair, CA, Doxey, SJ, Dewey, W. 1993. Heat shock alters centromere organization leading to mitotic disfunction and cell death. *J. Cell. Physiol.* 154:443–55
VanBuskirk, AM, De Nagel, DC, Gugliardi, LE, Brodsky, FM, Pierce, SK. 1991. Cellular and subcellular distribution of PBP72/74, a peptide-binding protein that plays a role in antigen processing. *J. Immunol.* 146:500–6
van der Vies, SM, Viitanen, PV, Gatenby, AA, Lorimer, GH, Jaenicke, R. 1992. Conformational states of ribulose bisphosphate carboxylase and their interaction with chaperonin 60. *Biochemistry* 31:3635–44
Van Dyk, TK, Gatenby, AA, LaRossa, RA. 1989. Demonstration by genetic suppression of interaction of GroE products with many proteins. *Nature* 342:451–53
Viitanen, PV, Donaldson, GK, Lorimer, GH, Lubben, TH, Gatenby, AA. 1991. Complex interactions between the chaperonin 60 molecular chaperone and dihydrofolate reductase. *Biochemistry* 30:9716–23
Viitanen, PV, Gatenby, AA, Lorimer, GH. 1992. Purified chaperonin 60 (GroEL) interacts with the nonnative states of a multitude of *Escherichia coli* proteins. *Protein Sci.* 1:363–69
Viitanen, PV, Lubben, TH, Reed, J, Goloubinoff, P, O'Keefe, DP, Lorimer, GH. 1990. Chaperonin-facilitated refolding of ribulosebisphosphate carboxylase and ATP hydrolysis by chaperonin 60 (groEL) are K^+-dependent. *Biochemistry* 29:5665–71
Volgel, JP, Misra, LM, Rose, MD. 1990. Loss of BiP/GRP78 function blocks translocation of secretory proteins in yeast. *J. Cell Biol.* 110:1885–95
Walker, AI, Hunt, T, Jackson, RJ, Anderson, CW. 1985. Double-stranded DNA induces the phosphorylation of several proteins including the 90,000 molecular weight heat

shock protein in animal cell extracts. *EMBO J.* 4:139–45

Walter, G, Carbone, A, Welch, WJ. 1987. Medium tumor antigen of polyomavirus transformation-defective mutant NG59 is associated with 73-kilodalton heat shock protein. *J. Virol.* 61:405–10

Welch, WJ. 1991. The role of heat shock proteins as molecular chaperones. *Curr. Opin. Cell Biol.* 3:1033–38

Welch, WJ. 1992. Mammalian stress response: cell physiology, structure/function of stress proteins and implications for medicine and disease. *Physiol. Rev.* 72:1063–81

Welch, WJ, Feramisco, JR. 1982. Purification of the major mammalian stress proteins. *J. Biol. Chem.* 257:14949–59

Welch, WJ, Feramisco, JK. 1985. Rapid purification of mammalian 70,000 dalton stress proteins: affinity of the proteins for nucleotides. *Mol. Cell. Biol.* 5:1229–37

Welch, WJ, Feramisco, JR, Blose, SH. 1985. The mammalian stress response and the cytoskeleton: alterations in intermediate filaments. *Int. Conf. Intermediate Filaments. Ann. NY Acad. Sci.* 455:57–67

Welch, WJ, Garrels, JI, Thomas, GP, Lin, JJ, Feramisco, JR. 1983. Biochemical characterization of the mammalian stress proteins and identification of two stress pro- teins as glucose- and Ca^{2+} ionophore-regulated proteins. *J. Biol. Chem.* 258: 7102–11

White, E, Spector, D, Welch, WJ. 1988. Differential localization of the adenovirus E1A proteins and association of E1A with the 70-kilodalton cellular heat shock protein in infected cells. *J. Virol.* 168:1475–80

Wickner, S, Hoskins, J, McKenney, K. 1991. Monomerization of RepA dimers by heat shock proteins activates binding to DNA replication origin. *Proc. Natl. Acad. Sci. USA* 88:7903–7

Wiech, H, Buchner, J, Zimmermann, R, Jakob, U. 1992. Hsp90 chaperones protein folding in vitro. *Nature* 358:169–70

Yaffe, MB, Farr, GW, Miklos, D, Horwich, AL. et al. 1992. TCP1 complex is a molecular chaperone in tubulin biogenesis. *Nature* 358:245–48

Yaffe, MP, Schatz, G. 1984. Two nuclear mutations that block mitochondrial protein import in yeast. *Proc. Natl. Acad. Sci. USA* 81:4819–23

Yem, AW, Tomasselli, AG, Heinrikson, RL, Zurcher-Neely, H, Ruff, VA, et al. 1992. The hsp56 component of steroid receptor complexes binds to immobilized FK506 and shows homology to FKBP-12 and FKBP-13. *J. Biol. Chem.* 267:2868–71

Zahn, R, Plückthun, A. 1992. GroE prevents the accumulation of early folding intermediates of pre-β-lactamase without changing the folding pathway. *Biochemistry* 31: 3249–55

Zeilstra-Ryalls, J, Fayet, O, Georgopoulos, C. 1991. The universally-conserved GroE chaperonins. *Annu. Rev. Microbiol.* 45: 301–25

Zhang, S, Lockshin, C, Herbert, A, Winter, E, Rich, A. 1992. Zuotin, a putative Z-DNA binding protein in *Saccharomyces cerevisiae*. *EMBO J.* 11:3787–96

Ziemienowicz, A, Skowyra, D, Zeilstra-Ryalls, J, Fayet, O, Georgopoulos, C, Zylicz, M. 1993. Either of the *Escherichia coli* GroEL/GroES and DnaK/DnaJ/GrpE chaperone machines can reactivate heat-treated RNA polymerase: different mechanisms for the same activity. *J. Biol. Chem.* In press

Zimmermann, RM, Sagstetter, MJ, Lewis, J, Pelham, HRB. 1988. Seventy-kilodalton heat shock proteins and an additional component from reticulocyte lysate stimulate import of M13 procoat protein into microsomes. *EMBO J.* 7:2875–80

Zwickl, P, Pfeifer, G, Lottspeich, F, Kopp, F, et al. 1990. Electron microscopy and image analysis reveal common principles of organization in two large protein complexes: groEL-type proteins and proteasomes. *J. Struct. Biol.* 103:197–203

Zylicz, M, Ang, D, Liberek, K, Georgopoulos, C. 1989. Initiation of λ DNA replication with purified host- and bacteriophage-encoded proteins: The role of the dnaK, dnaJ and grpE heat shock proteins. *EMBO J.* 8:1601–8

SUBJECT INDEX

U

V

W

X

Y

Z

CUMULATIVE INDEXES

CONTRIBUTING AUTHORS, VOLUMES 5–9

CHAPTER TITLES, VOLUMES 5–9

From:

Name ______________________________

Address ____________________________

__________________ Zip Code _________

Place Stamp Here

ANNUAL REVIEWS INC.
4139 EL CAMINO WAY
P. O. BOX 10139
PALO ALTO CA 94303-0897